Global Change – The IGBP Series

Springer-Verlag Berlin Heidelberg GmbH

W. Steffen · A. Sanderson · P. D. Tyson · J. Jäger · P. A. Matson · B. Moore III
F. Oldfield · K. Richardson · H. J. Schellnhuber · B. L. Turner II · R. J. Wasson

Global Change and the Earth System

A Planet Under Pressure

With 258 Figures

Springer

Authors

Steffen, Will

IGBP Secretariat,
Royal Swedish Academy of Sciences, Stockholm

Sanderson, Angelina

IGBP Secretariat,
Royal Swedish Academy of Sciences, Stockholm

Tyson, Peter

Climatology Research Group,
University of the Witwatersrand, Johannesburg, South Africa

Jäger, Jill

Coordinator,
Initiative on Science and Technology for Sustainability

Matson, Pamela

School of Earth Sciences
Stanford University, CA, USA

Moore III, Berrien

Institute for the Study of Earth, Oceans and Space (EOS),
University of New Hampshire, Durham, NH, USA

Oldfield, Frank

Department of Geography,
University of Liverpool, UK

Richardson, Katherine

Department of Marine Ecology,
Århus University, Denmark

Schellnhuber, H. John

Tyndall Centre for Climate Change Research,
University of East Anglia, Norwich, UK,
and
Potsdam Institute for Climate Impact Research, Germany

Turner, B. L. II

Graduate School of Geography
& George Perkins Marsh Institute,
Clark University, MA, USA

Wasson, Robert J.

Centre for Resource and Environmental Studies
The Australian National University, Canberra

Library of Congress Control Number: 2005927608

Additional material to this book can be downloaded from http://extras.springer.com

ISBN 978-3-662-30813-4 ISBN 978-3-540-26607-5 (eBook)

DOI 10.1007/978-3-540-26607-5

First ed. 2004, 2nd printing 2005
3-540-40800-2 1st printing 2004

Springer is a part of Springer Science+Business Media
springeronline.com

Originally published by Springer Berlin Heidelberg New York in 2005
Softcover reprint of the hardcover 1st edition 2005

Cover design: Erich Kirchner, Heidelberg
Typesetting: Büro Stasch · Bayreuth (stasch@stasch.com)
Production: Luisa Tonarelli

Printed on acid-free paper 32/2132/LT – 5 4 3 2 1 0

Preface

The relationship of humans with the Earth's environment has changed throughout the evolution of *Homo sapiens* and the development of societies. For virtually all of human existence on the planet, interaction with the environment has taken place at the local, or at most the regional, scale, except perhaps for one example in which regional-scale human activities were repeated to create global consequences in concert with climate change – the Holocene megafauna extinction. Apart from this possible example, the environment at the scale of the Earth as a whole – the passing of the seasons, the vagaries of weather and climate, the ebbing and flowing of river systems and glaciers, the rich diversity of life in all its forms – has been a framework within which humans have been able to evolve and develop social structures, subject only to the great forces of nature and the occasional perturbations of extraterrestrial origin. The Earth's environment has been a bountiful source of resources as well as a remarkably accommodating life support system that has allowed human civilisations to develop and flourish.

This book focuses on the profound transformation of Earth's environment that is now apparent, a transformation owing not to the great forces of nature or to extraterrestrial sources but to the numbers and activities of people – the phenomenon of *global change*. Begun centuries ago, this transformation has undergone a profound acceleration during the second half of the twentieth century. During the last 100 years the population of humans soared from little more than one to six billion and economic activity increased nearly 10-fold between 1950 and 2000. The world's population is more tightly connected than ever before via globalisation of economies and information flows. Half of Earth's land surface has been domesticated for direct human use and nearly all of it is managed by humans in one way or another. Most of the world's fisheries are fully or over-exploited and little pristine coastline exists outside of the high latitudes. The composition of the atmosphere – greenhouse gases, reactive gases, aerosol particles – is now significantly different from what it was a century ago. The Earth's biota is now experiencing the sixth great extinction event, but the first caused by another species: *Homo sapiens*. The evidence that these changes are affecting the basic functioning of the Earth System, particularly the climate, grows stronger every year. Evidence from several millenia shows that the magnitude and rates of human-driven changes to the global environment are in many cases unprecedented. There is no previous analogue for the current operation of the Earth System.

This book sets out what is known about global change and the nature of the Earth System. It addresses a number of important but difficult questions. How did the Earth System operate in the absence of significant human influence? How can human-driven effects be discerned from those due to natural variability? What are the implications of global change for human well-being? How robust is the Earth System in the face of these new internal forces of change? Can human activities trigger abrupt and potentially irreversible changes to which adaptation would be impossible? How serious is this inadvertant human experiment with its own life support system? By raising and attempting to address these questions in this volume, the authors hope to give some direction to the future of Earth System science and to challenge the global change research community to find answers to these questions.

Such an undertaking as this volume could not have been possible without the active involvement of a large number of people. The book's production has truly been a community effort. The project began as a synthesis of a decade of research undertaken under the auspices of the International Geosphere-Biosphere Programme (IGBP) but quickly grew to encompass contributions from the global change research community more generally, particularly IGBP's partner international programmes: DIVERSITAS, an international programme of biodiversity science; the International Human Dimensions Programme on Global Environmental Change (IHDP); and the World Climate Research Programme (WCRP). The acknowledgements section at the end of this volume is thus unusually long. The authors hope that the many contributions to the book have been properly acknowledged; any inadvertant oversights are the responsibility of the authors and are regretted.

Finally, this volume stands as one contribution of the many required to build the knowledge base to support the long-term, sustainable existence of the human enterprise on planet Earth. It argues that a truly global system of science is needed for coping with the challenges that lie ahead.

Will Steffen, Angelina Sanderson, Peter Tyson, Jill Jäger, Pamela Matson, Berrien Moore III, Frank Oldfield, Katherine Richardson, H. John Schellnhuber, B. L. Turner II, Robert Wasson

Stockholm, August 2003

Contents

Contributors to Expert Boxes

Alverson, Keith

PAGES (Past Global Changes) International Project Office,
Bern, Switzerland

Anisimov, Oleg

State Hydrological Institute,
St. Petersburg, Russia

Betts, Richard

Hadley Centre for Climate Prediction and Research,
UK Meteorological Office

Blunier, T.

Climate & Environmental Physics, Physics Institute,
University of Bern, Switzerland

Brasseur, Guy P.

Max Planck Institute for Meteorology,
Hamburg, Germany

Clark, William C.

Belfer Center for Science and International Affairs,
Harvard University, USA

Claussen, Martin

Potsdam Institute for Climate Impact Research,
Germany

Cole, Julia

Department of Geosciences,
University of Arizona, Tucson, USA

Crossland, Chris

Land-Ocean Interactions in the Coastal Zone (LOICZ)
International Project Office,
Texel, The Netherlands

Crutzen, Paul

Max Planck Institute for Chemistry,
Mainz, Germany

DeFries, R.

Department of Geography and Earth System Science
Interdisciplinary Center, University of Maryland, USA

Ehleringer, Jim

Department of Biology,
University of Utah, USA

Farquhar, Graham D.

Cooperative Research Centre for Greenhouse Accounting,
Research School of Biological Sciences,
The Australian National University, Canberra

Folke, Carl

Center for Transdisciplinary Environmental Studies (CTM) and
Department of Systems Ecology, Stockholm University, Sweden

Fu, Congbin

Institute of Atmospheric Physics,
Chinese Academy of Sciences, Beijing

Galloway, James

Department of Environmental Sciences,
University of Virginia, USA

Goodwin, Ian

The University of Newcastle,
Australia

Hanson, Roger B.

Joint Global Ocean Flux Study (JGOFS)
International Project Office, Bergen, Norway

Harvey, Nick

The University of Adelaide,
Australia

Heintzenberg, Jost

Leibniz-Institute for Tropospheric Research,
Leipzig, Germany

Held, H.

Potsdam Institute for Climate Impact Research,
Germany

Jansson, Åsa

The Centre for Research on Natural Resources and the
Environment, Stockholm University, Sweden

Jayaraman, A.

Physical Research Laboratory,
Ahmedabad, India

Kates, Robert W.

Independent Scholar, USA

Kaye, Jack A.

Research Division, Office of Earth Science,
NASA Headquarters, Washington, DC, USA

Lavorel, Sandra

Laboratoire d'Ecologie Alpine, CNRS,
Université Joseph Fourier, Grenoble, France

Leemans, Rik

Office for Environmental Assessment,
Bilthoven, The Netherlands

Leichenko, Robin

Department of Geography,
Rutgers University, USA

Lenton, Tim

Centre for Ecology and Hydrology,
Edinburgh, UK

Lohmann, Ulrike

Department of Physics and Atmospheric Science,
Dalhousie University, Canada

Messerli, Bruno

Institute of Geography,
University of Bern, Switzerland

Meybeck, Michel

Laboratoire de Géologie Appliquée,
Université de Paris, France

Mitra, A. P.

National Physical Laboratory,
New Delhi, India

Nobre, Carlos A.

Center for Weather Forecasting and Climate Research - CPTEC,
Cachoeira Paulista, SP, Brazil

Norby, Richard J.

Oak Ridge National Laboratory, USA

O'Brien, Karen

Center for International Climate and Environment Research,
Oslo, Norway

Pataki, Diane

Department of Biology,
University of Utah, USA

Petschel-Held, Gerhard

Potsdam Institute for Climate Impact Research,
Germany

Rahmstorf, Stefan

Potsdam Institute for Climate Impact Research,
Germany

Raynaud, D.

Laboratoire de Glaciologie et Géophysique de l'Environnement,
Centre National de la Recherche Scientifique and Université
Joseph Fourier, Saint-Martin-d'Hères, France

Richey, Jeffrey E.

School of Oceanography,
University of Washington, Seattle, USA

Roderick, Michael L.

Cooperative Research Centre for Greenhouse Accounting,
Research School of Biological Sciences,
The Australian National University, Canberra

Scholes, R. J.

Council for Scientific and Industrial Research,
South Africa

Serneels, Suzanne

University of Louvain, Belgium;
currently at the
International Livestock Research Institute, Kenya

Stocker, Thomas F.

Climate and Environmental Physics,
University of Bern, Switzerland

Vörösmarty, Charles

Complex Systems Research Center,
University of New Hampshire, USA

Watson, Andrew J.

School of Environmental Sciences,
University of East Anglia, Norwich, UK

Chapter 1

An Integrated Earth System

The interactions between environmental change and human societies have a long and complex history spanning many millennia. They vary greatly through time and from place to place. Despite this spatial and temporal variability, a global perspective has begun to emerge in recent years and to form the framework for a growing body of research within the environmental sciences. Crucial to the emergence of this perspective has been the dawning awareness of two aspects of Earth System functioning. First, that the Earth itself is a single system within which the biosphere is an active, essential component. Secondly, that human activities are now so pervasive and profound in their consequences that they affect the Earth at a global scale in complex, interactive and apparently accelerating ways; humans now have the capacity to alter the Earth System in ways that threaten the very processes and components, both biotic and abiotic, upon which the human species depends. This book describes what is known about the Earth System and the nature of the human-driven changes impacting it. It also considers the responses of the System, the consequences of these responses for the stability of the System and for human well-being, and some of the ways forward towards an Earth System science that can contribute to the goal of global sustainability.

1.1 A Research Agenda to Meet the Challenge of the Future

The urge to deepen the scientific understanding of the environment derives from a number of motivations whose relative importance has changed through time. Among the more readily discernible of these has been the wish to reveal more clearly the nature of what has been perceived to be divine order, the compulsion to satisfy simple curiosity and the need to solve, or at least respond to, problems that have already arisen from human interactions with the environment. Over the last two decades a new imperative has come to dominate environmental concerns with the growing awareness that human activities have an increasing influence on Earth System functioning, upon which human welfare and the future of human societies depend.

The most familiar and dramatic illustration of this awareness is the impact humans have had on the atmospheric concentration of greenhouse gases (Fig. 1.1). The heat-absorbing property of these gases and the empirical demonstration that their changing atmospheric concentrations been closely linked to climate change over the last 400 000 years combine to pose urgent questions about the future. A wide range of additional human impacts on the Earth System have had, and continue to have, dramatic and far-reaching cumulative effects. The ever-growing combined effects and implications of human activities bring into focus the nature of the problems currently facing environmental and other scientists.

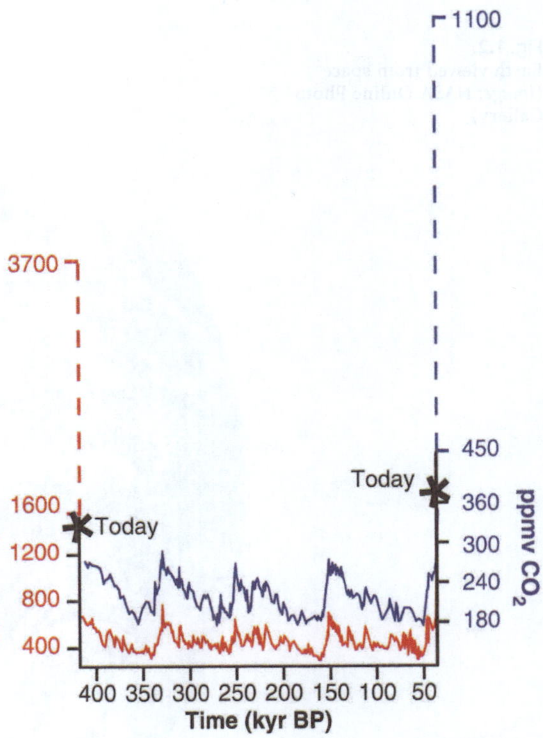

Fig. 1.1. Measurements of the atmospheric concentrations of the greenhouse gases CO_2 and CH_4 over the last four glacial-interglacial cycles from the Vostok ice core record, combined with current measurements and projections of future CO_2 and CH_4 levels based on IPCC 2000 scenarios (Petit et al. 1999; IPCC 2001). *Dashed lines* along the *y*-axis indicate the IPCC range of projections for CO_2 and CH_4 concentrations in 2100

The science required to address this future-oriented research agenda must transcend disciplinary boundaries, for it is concerned with issues that lie beyond any single field of study. All complex systems defy purely mechanistic analysis. The systems-level approach required to achieve an understanding of those aspects of Earth System functioning upon which human survival, and life in general, depend must encompass complex interactions, synergies between system components, non-linear responses and multiple feedbacks. It must also embrace both biophysical and anthropogenic drivers of change, not as separate influences but as closely interwoven and interactive processes.

Classical analytical science in which individual variables are isolated and their separate effects determined individually cannot cope with the challenges posed by Earth System science. This is often most clearly seen where responses to environmental problems have been designed to address specific, narrowly defined problems within a framework that fails to consider the full range of consequences inherent in a complex, interactive system. More generally, the emergent behaviour that often results from interactions among components of the system cannot be understood by studying the components of the system in isolation. The identification of cause-effect relationships is still useful, but they are embedded in complex systems in which synergies, interactions and non-linearities defy the classic, analytical approach.

Systems thinking and its application to the environment are not new. Ecosystem concepts go back to the 1930s (Tansley 1935), and systems formulations gained prominence in some fields (e.g., fluvial geomorphology) three decades ago (Chorley and Kennedy 1971). More general systems thinking was well expounded by the 1970s (Laszlo 1972), and the idea of global biogeochemistry, a major theme in this volume, was also discussed and described in broad outline some 50 years ago (Hutchinson 1944, 1954). However, each of these examples considers only a part of Earth System functioning. What is really new about perceptions of the Earth System over the last 10–15 years is the development of a perspective that embraces the Earth System *as a whole*. Several developments have led to this fundamental and accelerating change in scientific perception. These are that:

- the view of Earth from a spaceship, a blue-green sphere floating in blackness, triggers emotional *feelings* of a home teeming with life set in a lifeless void, as well as more analytical *perceptions* of a materially limited and self-contained entity (Fig. 1.2);

Fig. 1.2.
Earth viewed from space (*image:* NASA Online Photo Gallery)

- global observation systems allow scientists to apply *concepts* that were only previously applicable at subsystem, regional or local scale to the Earth as a whole. The Earth itself *is* a system;
- global databases address global scale phenomena with consistently acquired *data* that have the potential for harmonisation and comparison at a global scale;
- dramatic advances in the power to infer properties indicative of Earth processes in the past set contemporary observational snapshots in a *time continuum*; and
- enhanced computing power makes possible not only essential data assimilation, but increasingly sophisticated *models* that improve understanding of functional interactions and system sensitivities.

Science is at the threshold of a potentially profound shift in the perception of the human-environment relationship, operating across humanity as a whole and at the scale of the Earth as a single system.

1.2 The Earth as a System

The notion that global-scale cycles operate as systems is, again, not new. The fact that the hydrological and the carbon cycles, for example, each operate as planetary systems has been known for well over a century. However, the fact that these planetary cycles themselves are closely interlinked and the suggestion that life itself is an active and necessary player in planetary dynamics (e.g., the Gaia hypothesis; Lovelock 1979) are much more recent.

The extent to which, over the geologically recent past, the Earth behaves as a single, interlinked, self-regulating system was put into sharp focus in 1999 with the publication of the 420 000-year record from the Vostok ice core (Petit et al. 1999) (Figs. 1.1 and 1.3). These data,

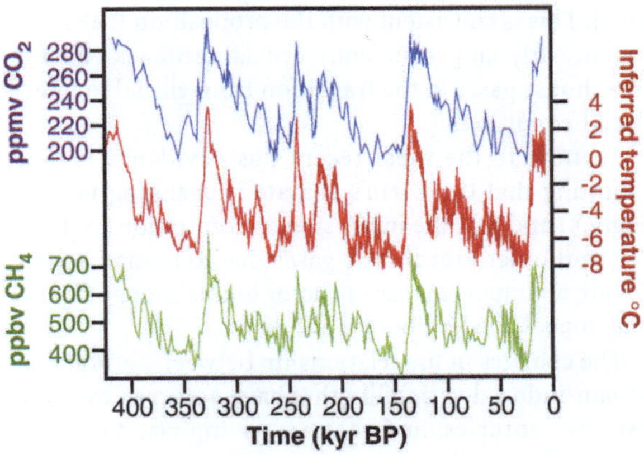

Fig. 1.3. The 420 000-year Vostok (Antarctica) ice core record, showing the regular pattern of atmospheric CO_2 and CH_4 concentration and inferred temperature through four glacial-interglacial cycles (adapted from Petit et al. 1999)

arguably among the most important produced by the global change scientific community in the twentieth century, provide a powerful temporal context and dramatic visual evidence for an integrated planetary environmental system.

The Vostok ice core data give a wealth of insights into the Earth System, and they will be used at several points throughout this volume to examine aspects of system dynamics in more detail. For now, three striking characteristics are immediately apparent. Together, they demonstrate beyond any doubt that the Earth is a system, with properties and behaviour that are characteristic of the system as a whole. In particular:

- The evidence for climate variability, as represented by a proxy for local temperature ($\delta^{18}O$) and the record of changes in the global carbon cycle, as represented by the atmospheric concentration of the trace gases carbon dioxide (CO_2) and methane (CH_4) trapped in air bubbles in the ice, show largely parallel temporal variations throughout (Fig. 1.3). In fact, the record from Vostok confirms that there has been a close coupling between the climate proxies and both trace gas and aerosol (dust and sulphate) concentrations, all of which are linked in part to biological processes.
- The main maxima and minima of temperature and trace gas concentrations, which mark the alternation between glacial and interglacial conditions, follow a regular, cyclic beat through time, each cycle spanning approximately 100 000 years. The smooth changes in the eccentricity of the Earth's orbit that are believed to be the primary forcing mechanism for this dominant periodicity are too slight and too smooth to generate the changes recorded without strong modulation by internal feedbacks. This is especially so when the abrupt shifts to interglacial conditions at the end of each glacial period are considered. This highly non-linear response of the Earth System to external forcing must involve interactions among biological, chemical and physical components.
- The range over which isotopically inferred temperature and trace gas concentrations vary is limited. Throughout all four cycles, each interglacial gives rise to similar peak values; each glacial culminates in comparable minima. This points to a high degree of self-regulation within the Earth System over the whole of the time interval recorded in the Vostok ice core.

As Chap. 2 will show, this systemic behaviour of Earth's environment is due to a combination of external forcing – primarily variations in solar radiation levels near the Earth's surface – and a large and complex array of feedbacks and forcings within Earth's environment itself. In fact, it is undoubtedly the internal dynamics that keep the planet habitable for life. Without the thin layer of ozone in the upper atmosphere, for example, much more harmful ultraviolet radiation would penetrate to the Earth's surface; and without the thin layer of heat-absorbing trace gases in the lower atmosphere, the planet's mean surface temperature would be about 30 °C lower than it is now.

Over the past few decades, evidence has been mounting that planetary-scale changes are occurring rapidly in response to the forcings and feedbacks that characterise the internal dynamics of the Earth System. As indicated in Fig. 1.1, key indicators, such as the concentrations of trace gases in the atmosphere, are changing dramatically, and in many cases the linkages of these changes to human activities are strong. It is clear that the Earth System is being subjected to an ever-increasing diversity of new planetary-scale forces that originate in human activities. Primarily, it is these activities that give rise to the phenomenon of *global change* that is considered in this volume.

1.3 The Nature of Global Change

Global change is more than climate change and its full extent and complexity has been realised only very recently. The origins of the concept are largely derived from the careful and consistent measurement of atmospheric CO_2 concentration at the Mauna Loa Observatory in Hawaii (Keeling et al. 1995; Keeling and Whorf 2000). These observations first demonstrated beyond doubt that human activities can have direct global-scale consequences for the environment (Fig. 1.4). The Mauna Loa record shows a steady increase in CO_2 concentration in the atmosphere over the past three decades and forms the link between the natural variability provided by the Vostok ice core data and the projections of future values in Fig. 1.1. Coupled with other lines of evidence, such as the isotopic composition of the carbon in atmospheric CO_2 and the latitudinal distribution of CO_2 (Prentice et al. 2001), the record points to fossil fuel combustion as the main source of the additional CO_2. Startling though it is, the Mauna Loa record on its own gives

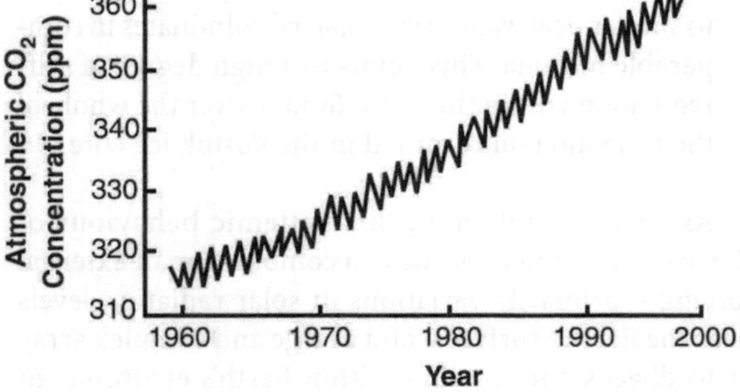

Fig. 1.4. Increase in atmospheric CO_2 concentration over the last 40 years as measured at the Mauna Loa Observatory, Hawaii (adapted from Keeling and Whorf 2000)

little hint of its implications for the functioning of the Earth System. Two other pieces of evidence are required to understand this.

As Fig. 1.1 shows, the Vostok ice core record of atmospheric CO_2 concentration over the past 420 000 years shows a normal operating range of about 180 to 280 ppmV (parts per million by volume). The addition of the Mauna Loa record to bring the Vostok data up to the present shows immediately that the scale of change in the composition of Earth's atmosphere is of the same order of magnitude as the full range of natural variability. The long palaeo-record places modern instrumental observations into a time continuum and thereby provides a much clearer perspective on the magnitude of recent human perturbations to the Earth System.

Secondly, the detailed Vostok ice core record, as shown more fully in Fig. 1.3, also points to a relationship between the atmospheric concentrations of CO_2 and CH_4 and the climate, at least as represented by temperatures in the neighbouring part of Antarctica. However, many records from both marine and continental sediments (Alverson et al. 2000, 2003) confirm that the sequence of major climate changes – the oscillations between glacial and interglacial conditions – on the Vostok time-scale was broadly synchronous over the whole globe. Moreover, recent data from ice core records with higher temporal resolution (Alverson et al. 2003) reinforce the view that there is a close coupling between atmospheric greenhouse gas concentrations and climate change even during the intervals of rapid warming at each glacial termination. Although the phasing and likely lags are still not precisely determined for each glacial-interglacial transition, careful analysis shows that increases in the atmospheric concentrations of CO_2 and CH_4 closely followed solar radiation changes associated with changes in the eccentricity of the Earth's orbit that appear to have initiated the transition to interglacial conditions, but preceded by several thousand years the onset of intense deglaciation (Petit et al. 1999; Pépin et al. 2002). This is consistent with the proposition that there was an early and recurrently crucial feedback role for greenhouse gases in the transition from glacial to interglacial conditions.

Turning to the most recent past, evidence is now mounting that the Earth's climate is changing rapidly (Fig. 1.5) and that the increasing concentrations of CO_2, CH_4 and other greenhouse gases due to human activities are an important causal factor in this change (Mann et al. 1999; Crowley 2000; IPCC 2001).

The changes in the relationship between natural and human-induced variability that have occurred over the last few centuries, and that are giving rise to global change, are complex and profound. They are almost certainly unprecedented in the history of the Earth. The expansion of humankind, both in numbers and per capita exploitation of the Earth's resources, has been re-

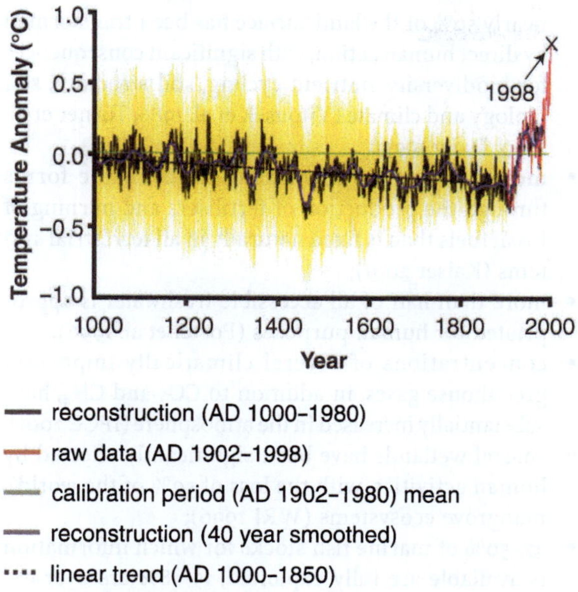

Fig. 1.5. Mean annual surface temperature over the northern hemisphere for the last thousand years (Mann et al. 1999)

markable. During the past three centuries human population increased 10-fold to 6 000 million. Concomitant with this population increase, the rate of consumption has risen even more sharply. Just as rapid and profound are other changes sweeping across human societies, many through the process often termed *globalisation* (Fig. 1.6). During the last 50 years alone, the increase in virtually every sphere of human activity has been considerable, such that:

- the world's population, currently 6 100 million, has doubled since 1960, tripled since 1930 and is projected to rise to over 9 000 million by 2050 (UNFPA 2001);
- since 1950 the global economy has increased by more than a factor of 15; real world Gross Domestic Product (GDP) grew from USD 2 trillion in 1965 to USD 28 trillion in 1995 (UNDP 2000; UNEP 2000);
- economic inequality is increasing. The richest nations have 15% of the global population but generate 50% of world GDP; between 1960 and 1994 the ratio of income of the richest 20% to the poorest 20% increased from 30:1 to 78:1 (World Bank 2002; UNDP 2000);
- world petroleum consumption has increased by a factor of 3.5 since 1960; global use of fuel wood has doubled in the last 50 years (EIA 2002; UNFPA 2001);
- transport accounts for 25% of world energy use, with motor vehicles accounting for 80% of the use. The number of motor vehicles has increased from 40 million in the late 1940s to 676.2 million in 1996 (IRF 1997);
- global communication has exploded with the development of the internet. In just six years, from 1993 to 1999, the number of people globally connected to the internet increased from three million to over 200 million (Internet Economy Indicators 2002);

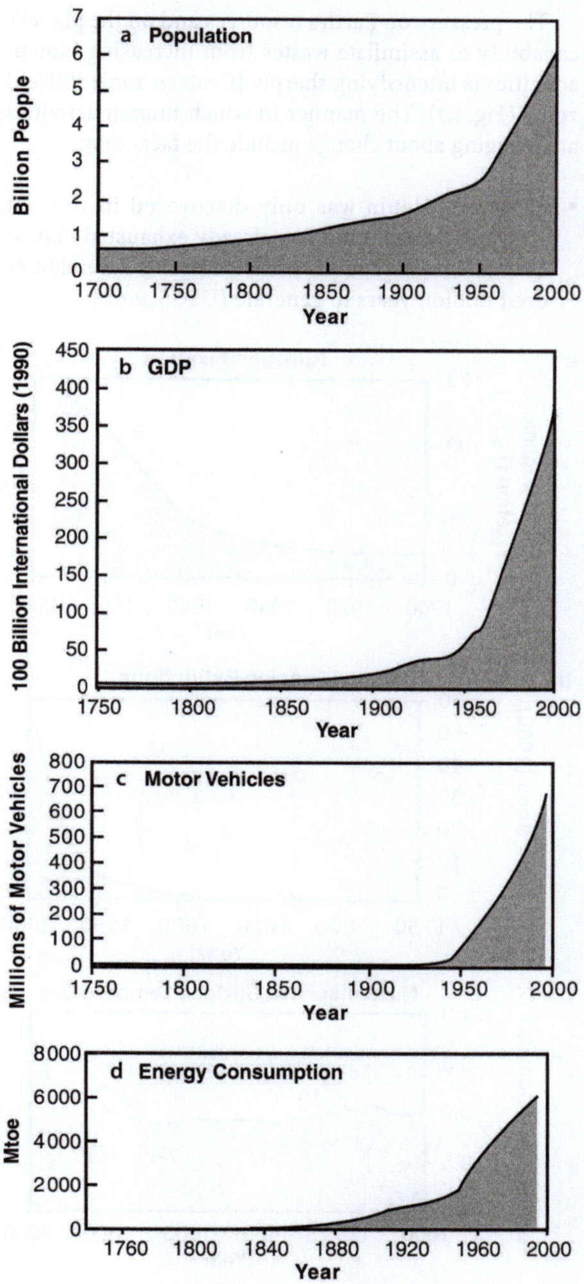

Fig. 1.6. Rate of increase in many spheres of human activity for the last 300 years: **a** population (US Bureau of the Census 2000); **b** world economy (Nordhaus 1997); **c** motor vehicles (UNEP 2000); and **d** energy consumption (Klein Goldewijk and Battjes 1997)

- urbanisation increased 10-fold in the twentieth century. From 1950 to 2000 the percentage of the world's population living in urban areas increased from 30% to 47% and is projected to increase to 56% by 2020. The number of megacities (10 million or more inhabitants) increased from five in 1975 to 19 in 2000. (FAOSTAT 2002; UNFPA 2001); and
- the interconnectedness of the cultures of the world is increasing rapidly with the increase in communication, travel and the globalisation of economies.

The pressure on Earth's resources and on the planet's capability to assimilate wastes from increasing human activities is intensifying sharply (Crutzen 2002; McNeill 2001) (Fig. 1.7). The manner in which human activities are bringing about change include the facts that:

- while petroleum was only discovered in the last 150 years, humankind has already exhausted almost 40% of known oil reserves that took over several hundred million years to generate (USGS 2000);

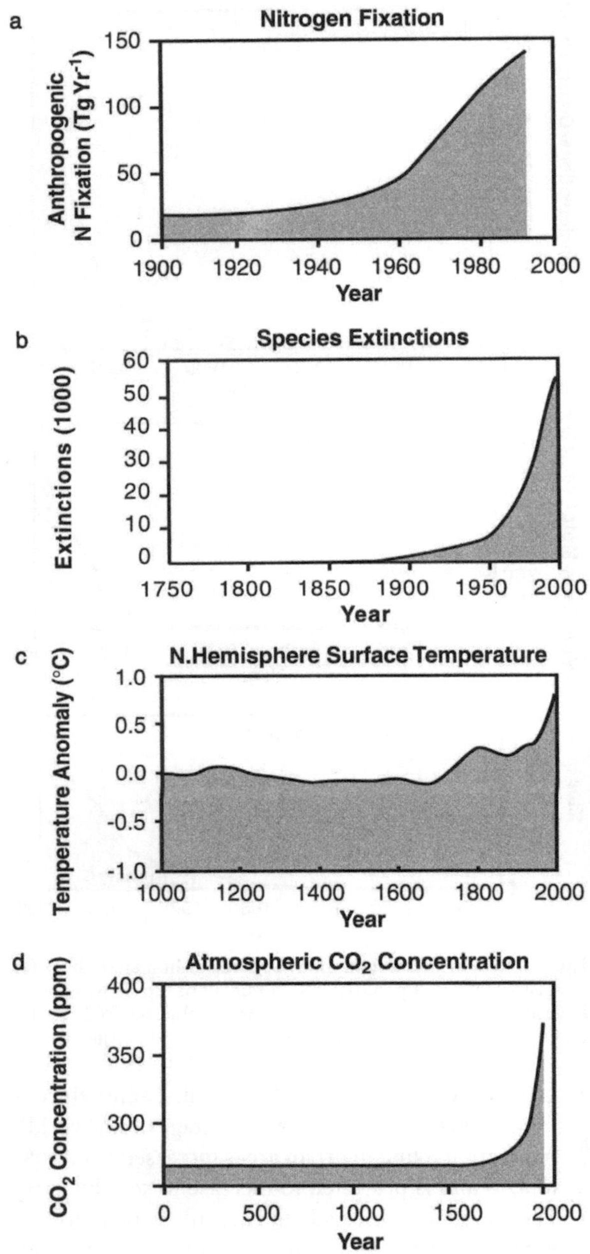

Fig. 1.7. Responses of the Earth System to increasing pressure from human activities: **a** nitrogen fixation (Vitousek 1994); **b** species extinctions (Smith 2002); **c** northern hemisphere surface temperature (Mann et al. 1999); and **d** atmospheric CO_2 concentration (adapted from Keeling and Whorf 2000)

- nearly 50% of the land surface has been transformed by direct human action, with significant consequences for biodiversity, nutrient cycling, soil structure, soil biology and climate (Vitousek et al. 1986; Turner et al. 1990; Daily 1995);
- more nitrogen is now fixed into available forms through the production of fertilisers and burning of fossil fuels than is fixed naturally in all terrestrial systems (Kaiser 2001);
- more than half of all accessible freshwater is appropriated for human purposes (Postel et al. 1996);
- concentrations of several climatically important greenhouse gases, in addition to CO_2 and CH_4, have substantially increased in the atmosphere (IPCC 2001);
- coastal wetlands have been significantly affected by human activities, with the loss of 50% of the world's mangrove ecosystems (WRI 1996);
- 47–50% of marine fish stocks for which information is available are fully exploited, 15–18% are over-exploited and 9–10% have been depleted or are recovering from depletion (FAO 2000); and
- extinction rates are increasing sharply in marine and terrestrial ecosystems around the world, with the present sixth great extinction event in the Earth's history being the first one caused by the activities of a biological species, in this case *Homo sapiens* (Lawton and May 1995; Pimm et al. 1995).

The extent to which human activities are influencing or even dominating many aspects of Earth's environment and its functioning has led to suggestions that a new geological era, the *Anthropocene* (Crutzen and Stoermer 2001), has begun.

Listing the broad suite of biophysical and socio-economic changes that is taking place fails to capture the complexity and connectivity of global change since the many linkages and interactions among the individual changes are not included. For example, most of the rapid urbanisation is occurring in the coastal zone, leading both to direct conversion of natural coastal ecosystems to urban areas and increasing demand for marine resources, which in turn leads to further conversion of natural coastal ecosystems. These changes are compounded by land-use changes upstream, altering the mix and amount of suspended and dissolved material entering the coastal zone. Both within-coastal zone ecosystem modification/conversion and land-use change upstream are impacting on biological diversity. The potential synergies between biophysical and socio-economic trends becomes startlingly apparent when the effects of land-use change upstream combine with probable sea-level rise and changes in the frequency of storm events to increase the vulnerability of coastal settlements and infrastructure. Similar webs of connectivity can be found for other clusters of global changes, leading to *syndromes* of change characteristic of particular regions (Petschel-Held et al. 1999).

Furthermore, many changes do not occur in a linear fashion, but rather, thresholds are passed and rapid, non-linear changes ensue. For example, the initial pockets of deforestation in a large tropical forest may have little or no impact on the number of species of animals and birds. However, as the forest becomes increasingly fragmented, the rate of species loss can increase sharply (Hobbs 1989; Noss 1996). Similar effects have been shown to occur in the responses of insects to changing climate. In Sweden the tick species which carries encephalitis normally requires two seasons to complete its life cycle. However, with a warming climate in the high latitudes, at a certain point in the warming trend the tick can complete its life cycle in a single season, causing an apparently unpredicted outbreak of tick populations (Lindgren 2000). The pattern of little or no change until a critical threshold followed by a sharp response is common in Earth System dynamics, and may be the rule rather than the exception, especially as global change intensifies. In addition, many effects of global change are cumulative; for example, slow addition over time to the atmosphere of seemingly innocent chemicals in the form of chlorofluorocarbons led eventually to a rapid and significant disruption of the atmospheric chemical system in the stratosphere with the production of the so-called ozone hole.

Finally, global change is being played out in contrasting ways at different locations, each with its own set of characteristics being impacted by a location-specific mix of interacting changes. The global environment and the phenomenon of global change are both heterogeneous, and the variety of human-environment relationships is vast. The potential for an almost infinite range of interactions is large. To cope with or adapt to global change requires analyses that couple the particular characteristics of a location or region with the nature of the systemic, globally connected changes to Earth's environment as they interact with other factors affecting the location or region. Any attempt at achieving sustainability at any scale requires that this is done and that human societies learn to live with global change.

1.4 Objectives and Structure of the Book

The intent of this book is to explore what has been learned about the fundamental workings of the Earth System (see Box 1.1 for a definition of the Earth System),

Box 1.1. The Earth System

Frank Oldfield · Will Steffen

In the context of global change, the *Earth System* has come to mean the suite of interacting physical, chemical, and biological global-scale cycles (often called biogeochemical cycles) and energy fluxes which provide the conditions necessary for life on the planet. More specifically, this definition of the Earth System has the following features:

- It deals with a materially closed system that has a primary external energy source, the sun.
- The major dynamic components of the Earth System are a suite of interlinked physical, chemical and biological processes that cycle (transport and transform) materials and energy in complex dynamic ways within the System. The forcings and feedbacks *within* the System are at least as important to the functioning of the System as are the external drivers.
- Biological/ecological processes are an integral part of the functioning of the Earth System, and not just the recipients of changes in the dynamics of a physico-chemical system. Living organisms are active participants, not simply passive respondents.
- Human beings, their societies and their activities are an integral component of the Earth System, and are not an outside force perturbing an otherwise natural system. There are many modes of natural variability and instabilities within the System as well as anthropogenically driven changes. By definition, both types of variability are part of the dynamics of the Earth System. They are often impossible to separate completely and they interact in complex and sometimes mutually reinforcing ways.
- Time scales considered in Earth System science vary according to the questions being asked. Many global environmental change issues consider time scales of decades to a century or two. However, a basic understanding of Earth System dynamics demands consideration of much longer time scales in order to capture longer-term variability of the System, to understand the fundamental dynamics of the System, and to place into context the current suite of rapid global-scale changes occurring within the System. Thus palaeo-environmental and prognostic modelling approaches are both central to Earth System science.

The term *climate system* is also used in connection with global change, and is encompassed within the Earth System. Climate usually refers to the aggregation of all components of weather – precipitation, temperature, cloudiness, for example – averaged over a long period of time, usually decades, centuries, or longer. The processes which contribute to climate comprise the climate system, and they are closely connected to biogeochemical cycles. However, there are some important differences between climate change and global change:

- Many important features of biogeochemical cycles can have significant impacts on Earth System functioning without any direct change in the climate system. Examples include the direct effects of changing atmospheric CO_2 concentration on carbonate chemistry and hence on calcification rates in the ocean and also the sharp depletion of stratospheric ozone from the injection of chlorofluorocarbons in the atmosphere.
- Many interactions between biology and chemistry can have profound impacts on ecological systems, and hence feedbacks to Earth System functioning, without any change in the climate system. Examples include the impact of nitrogen deposition on the biological diversity of terrestrial ecosystems and the effect of non-climate driven changes in terrestrial and marine biospheric emission of trace gases and hence to the chemistry of the atmosphere.
- Human societies and their activities are usually not considered to be a direct part of the climate system, although their activities certainly impact on important processes in the climate system (e.g., greenhouse gas emissions).

and the responses of that system to myriad anthropogenic changes as they interact with each other and with the patterns and processes of natural variability. In doing this, global change research over the past decade has been used, but not exclusively limited to this period. The results of research carried out in the International Geosphere-Biosphere Programme (IGBP, see Appendix), together with significant contributions from the IGBP's partner global change programmes (the World Climate Research Programme (WCRP), the International Human Dimensions Programme on Global Environmental Change (IHDP) and DIVERSITAS, an international programme of biodiversity science) are included and acknowledged. In addition, the book draws on the work of those not directly linked to any of the international programmes *per se*, but working within the numerous national and regional global change programmes around the world.

The theme of this book is the nature of Earth as a system, the evolving role of anthropogenic activities as an ever-increasing planetary-scale force in the System, and the consequences of rapid change for the future of the Earth's environment and for the well-being of human societies. As much of the research on global change has, until very recently, focused on components of the Earth System rather than on the System as a whole, this book continually attempts to balance reductionist, analytical approaches to the science with integrative, systems-level approaches. However, the book also acknowledges the fact that complete knowledge of the complex consequences of anthropogenic change has thus far proved to be elusive. In addition, the book attempts to put some solid scientific analyses into integrative systems approaches, which may otherwise lapse into vague generalities and lose the essential virtues of deductive science and especially hypothesis testing. This dual approach gives an indication of the type of science required for Earth System analysis in future.

The book takes the reader on a journey through time, beginning with a consideration of natural changes occurring over a half-million year time-frame, including glacial-interglacial cycle dynamics. It then passes through to the contemporary period of accelerating human dominance of many planetary processes. Following on from this come more speculative approaches to analysing and simulating the dynamics of the Earth System into the future. Figure 1.8 gives a visual representation of the structure of the book. The structural outline of the work may be summarised as follows:

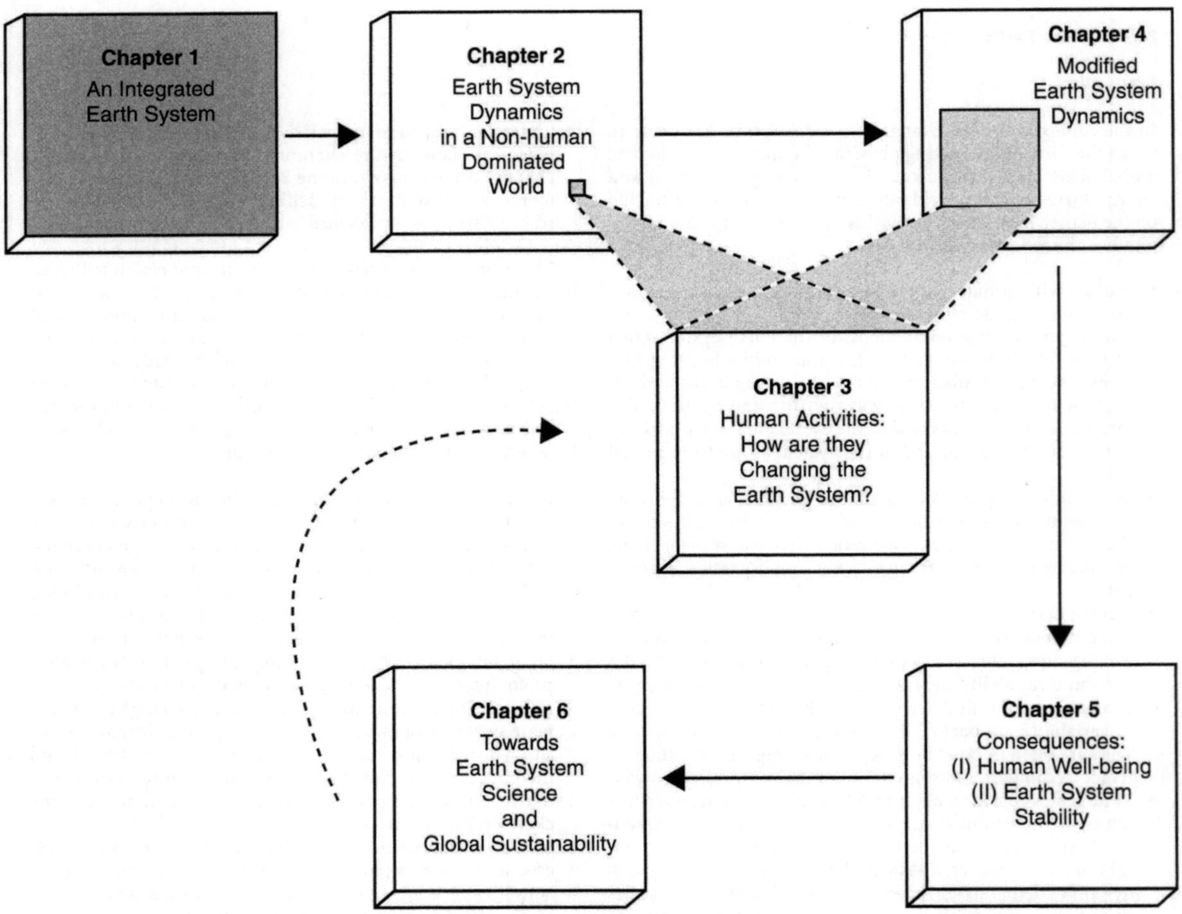

Fig. 1.8. Structural outline of the book

- Chapter 2 begins by examining Earth System dynamics as they functioned prior to significant human influence. In it the question of how the Earth's environment operated in the period before human numbers and activities became a planetary-scale force is considered. In addition, thought is given to how those natural processes least impacted by human activities operate at present.
- Chapter 3 focuses only on the human component of the Earth System and examines the nature of human alterations of the Earth System, the types and rates of anthropogenic changes, their interrelationships and teleconnections, and the direct and underlying human driving forces for these changes. It shows how human society has grown from insignificance to a force equal to or exceeding many of the great forces of nature.
- The much-expanded human enterprise is placed back into the Earth System in Chap. 4, which examines the responses of the Earth System to this new force. It deals mainly with how the major human driving forces associated with the burning of fossil fuels, on the one hand, and how deforestation and subsequent intensification of agriculture, on the other, affect the Earth System and how the Earth System responds. A central theme is how initial responses reverberate through the Earth System, sometimes damping the initial forcing and sometimes amplifying it. A related theme is how the complex webs of internal forcings and feedbacks that characterise the functioning of the Earth System are affected by the expanding human enterprise.
- The consequences of accelerating global change for human well-being and the stability of the Earth System are explored in Chap. 5. The main question addressed is how global change is likely to affect the goal of feeding, housing, clothing, educating and employing the expanding human population of the twenty-first century without compromising the sustainability of the Earth's life support system.
- The final chapter considers a new scientific approach aimed at a fully developed Earth System science. It points the way towards reducing the dichotomy between isolated empiricism and overarching theory; to linking skills and disciplines both biophysical and socio-cultural in exciting new combinations; and to treating biospheric functioning in complex, interactive, quantitative and trajectory terms. It seeks to identify key Earth System switches and triggers, to redefine creatively but rigorously the interaction between deductive and inductive scientific modes, as well as between model-based and empirically based research. It explores the ways in which this new approach can advance understanding of the dynamics of the planet, operationalise the growing knowledge base and use the resulting wisdom to help human societies develop in ways increasingly compatible with the natural dynamics of the Earth System.

References

Alverson K, Oldfield F, Bradley R (2000) Past global changes and their significance for the future. Quarternary Sci Rev 19:1–5

Alverson K, Bradley R, Pedersen T (2003) Paleoclimate, global change and the future. IGBP Global Change Series. Springer-Verlag, Berlin Heidelberg New York

Chorley R, Kennedy BA (1971) Physical geography: A systems approach. Prentice Hall, New Jersey

Crowley TJ (2000) Causes of climate change over the past 1 000 years. Science 289:270–277

Crutzen PJ (2002) Geology of mankind – the Anthropocene. Nature 415:23

Crutzen PJ, Stoermer E (2001) The "Anthropocene". International Geosphere Biosphere Programme, Global Change NewsLetter 41:12–13

Daily GC (1995) Restoring value to the world's degraded lands. Science 269:350–354

EIA (2002) Historical international data online database. Energy Information Administration. *http://www.eia.doe.gov/neic/historic/hinternational.htm*, 10 April 2002

FAO (2000) State of world fisheries and aquaculture 2000. Wijkstrom U, Gumy A, Grainger R (eds) Food and Agriculture Organization of the United Nations, Rome. *http://www.fao.org/DOCREP/003/X8002E00.htm*, 10 April 2002

FAOSTAT (2002) Statistical databases online. Food and Agriculture Organization of the United Nations, Rome. *http://www.apps.fao.org*, 10 April 2002

Hobbs RJ (1989) The nature and effects of disturbance relative to invasions. In: Drake JA, Mooney HA, di Castri F, Groves RH, Kruger FJ, Rejmanek M, Williamson M (eds) Biological invasions: A global perspective. John Wiley & Sons, Chichester, UK, pp 389–401

Hutchinson GE (1944) Nitrogen and the biogeochemistry of the atmosphere. Am Sci 32:284–306

Hutchinson GE (1954) The biogeochemistry of the terrestrial atmosphere. In: Kuiper GP (ed) The Earth as a planet. University of Chicago Press, Chicago, pp 371–433

Internet Economy Indicators (2002) Facts and figures. *http://www.internetindicators.com/factfigure.html*, 10 April 2002

IPCC (2001) Climate change 2001: The scientific basis. Contribution of Working Group I to the Third Assessment Report of the Intergovernmental Panel on Climate Change. Houghton JT, Ding Y, Griggs DJ, Noguer M, van der Linden PJ, Dai X, Maskell K, Johnson CA (eds) Intergovernmental Panel on Climate Change. Cambridge University Press, Cambridge New York

IRF (1997) World road statistics, 1997 edition. International Road Federation, Geneva Washington DC

Kaiser J (2001) The other global pollutant: Nitrogen proves tough to curb. Science 294:1268–1269

Keeling CD, Whorf TP (2000) Atmospheric CO_2 records from sites in the SIO air sampling network. In: US Department of Energy (ed) Trends: A compendium of data on global change. Carbon Dioxide Information Analysis Center, Oak Ridge National Laboratory, US Department of Energy, Oak Ridge, TN, USA

Keeling CD, Whorf TP, Wahlen M, van der Plicht J (1995) Interannual extremes in the rate of rise of atmospheric carbon dioxide since 1980. Nature 375:666–670

Klein Goldewijk CGM, Battjes JJ (1997) One hundred year database for integrated environmental assessments. National Institute for Public Health and the Environment (RIVM), Bilthoven, Netherlands

Laszlo E (1972) The systems view of the world. Basil Blackwell, Oxford

Lawton JH, May RM (eds) (1995) Extinction rates. Oxford University Press, Oxford

Lindgren E (2000) The new environmental context for disease transmission, with case studies on climate change and tick-borne encephalitis. PhD Thesis. Natural Resources Management, Department of Systems Ecology, Stockholm University, Sweden

Lovelock JE (1979) Gaia: A new look at life on Earth. Oxford University Press, Oxford

Mann ME, Bradley RS, Hughes MK (1999) Northern hemisphere temperatures during the past millennium: inferences, uncertainties, and limitations. Geophys Res Lett 26:759–762
McNeill JR (2001) Something new under the sun. WW Norton, New York, London
Nordhaus WD (1997) Do real wage and output series capture reality? The history of lighting suggests not. In: Bresnahan T, Gordon R (eds) The economics of new goods. University of Chicago Press, Chicago
Noss RF (1996) Conservation of biodiversity at the landscape scale. In: Szaro RC, Johnston DW (eds) Biodiversity in managed landscapes: Theory and practice. Oxford University Press, New York, pp 574–589
Pépin L, Raynaud D, Barnola J-M, Loutre MF (2002) Hemispheric roles of climate forcings during glacial-interglacial transitions as deduced from the Vostok record and LLN-2D model experiments. J Geophys Res 106:31885–31892
Petit JR, Jouzel J, Raynaud D, Barkov NI, Barnola J-M, Basile I, Bender M, Chappellaz J, Davis M, Delaygue G, Delmotte M, Kotlyakov VM, Legrand M, Lipenkov VY, Lorius C, Pépin L, Ritz C, Saltzman E, Stievenard M (1999) Climate and atmospheric history of the past 420 000 years from the Vostok ice core, Antarctica. Nature 399:429–436
Petschel-Held G, Block A, Cassel-Gintz M, Kropp J, Lüdecke MKB, Moldenhauer O, Reusswig F, Schellnhuber HJ (1999) Syndromes of global change – a qualitative modelling approach to assist global environmental management. Environ Model Assess 4:295–314
Pimm SL, Russell GJ, Gittleman JL, Brooks TM (1995) The future of biodiversity. Science 269:347–350
Postel S, Daily GC, Ehrlich P (1996) Human appropriation of renewable fresh water. Science 271:785–788
Prentice IC, Farquhar GD, Fasham MJR, Goulden ML, Heimann M, Jaramillo VJ, Khesghi HS, Le Quéré C, Scholes RJ, Wallace DWR (2001) The carbon cycle and atmospheric carbon dioxide. In: Houghton JT, Ding Y, Griggs DJ, Noguer M, van der Linden PJ, Dai X, Maskell K, Johnson CA (eds) Climate change 2001: The scientific basis. Contribution of Working Group I to the Third Assessment Report of the Intergovernmental Panel on Climate Change. Cambridge University Press, Cambridge New York
Smith N (2002) Species extinction. *http://www.whole-systems.org/extinctions.html*, 24 Jan 2003
Tansley AG (1935) The use and abuse of vegetational concepts and forms. Ecology 16:284–307
Turner II BL, Clark WC, Kates RW, Richards JF, Mathews JT, Meyer WB (1990) The Earth as transformed by human action: Global and regional changes in the biosphere over the past 300 years. Cambridge University Press, Cambridge
UNDP (2000) Economic reforms, globalization, poverty and the environment. Reed D, Rosa H (eds) United Nations Development Programme
UNEP (2000) Global environmental outlook 2000. Clarke R (ed) United Nations Environment Programme
UNFPA (2001) The state of world population 2001. Marshall A (ed) United Nations Population Fund
US Bureau of the Census (2000) International database. *http://www.census.gov/ipc/www/worldpop.htm*, data updated 10 May 2000
USGS (2000) U.S. Geological Survey world petroleum assessment 2000. U.S. Geological Survey World Energy Assessment Team, United States Geological Survey. *http://greenwood.cr.usgs.gov/energy/WorldEnergy/DDS-60/*, 2 April 2002
Vitousek PM (1994) Beyond global warming: Ecology and global change. Ecology 75:1861–1876
Vitousek PM, Ehrlich PR, Ehrlich AH, Matson PA (1986) Human appropriation of the products of photosynthesis. Bioscience 36:368–373
World Bank (2002) Economic growth research. *http://www.worldbank.org/research/growth*, 10 April 2002
WRI (1996) World resources 1996–1997. World Resources Institute. Oxford University Press, New York

Chapter 2

Planetary Machinery: The Dynamics of the Earth System Prior to Significant Human Influence

The properties and processes of the non-human dominated Earth System vary across a wide range of space and time scales. Nevertheless, the Earth System has functioned within domains characterised by well-defined limits and periodic patterns. Interconnections among physical, chemical and biological processes and between land, ocean and atmosphere, across both space and time, are ubiquitous and critical for the functioning of the System. Forcings and feedbacks are difficult to distinguish as one becomes the other in the cyclical dynamics of the System. Rapid, abrupt changes can occur as the Earth System reorganises into a new state.

2.1 The Natural Dynamics of the Earth System

This chapter focuses on the nature of the Earth System before human activities became an important feature of the system, that is, prior to the beginning of the Industrial Revolution (Fig. 2.1). Much of the Earth System research that has been carried out in recent decades has sought to understand the backdrop of biogeochemical cycling and climate processes against which all current and future anthropogenic changes must be evaluated. Central to this analysis is the evaluation of the response of biological systems to physical and chemical changes in the Earth System and the role of biological systems in biogeochemical and biophysical processes within the Earth System. Global change research over the past decade has demonstrated that the biosphere is an active and important contributor to the functioning of the Earth System. The persistent notion that the Earth System, especially the climate, is driven largely by the physics of coupled ocean-atmospheric dynamics, with very little role for biology, has been effectively dispelled by the discovery of many concrete mechanisms by which both the marine and terrestrial biospheres interact with physical and chemical processes, and indeed, even help to control some critical Earth System functions.

In this chapter attention is drawn to the ways in which the biogeochemical and biological systems interact with the hydrological cycle and climate over space and time. The examination of biogeochemical systems assumes familiarity with the global cycles of carbon, nitrogen, phosphorus and sulphur. These are not dealt with in detail here. Complete descriptions of these cycles and their most critical processes are available in a number of recent texts (e.g., Schlesinger 1999; Jacobson et al. 2000), in review articles (e.g., Schimel et al. 1995; Vitousek et al. 1997; Smil 2001), and in syntheses in the IGBP series of books (Tyson et al. 2002; Brasseur et al. 2003; Fasham 2003; Kabat et al. 2004; Pedersen et al. 2003). Here, rather, the focus is on examples of some of the most exciting and critical recent findings that give insight into how Earth works as an integrated whole.

Studying the dynamics of the Earth System in a non-human dominated state is not straightforward. As will be made clear later in Chap. 3, many critical processes in Earth System functioning are now significantly influenced by human actions, and observing them directly now could give an incorrect or misleading picture of how the Earth functioned prior to being significantly influenced by human activities. Consequently, this chapter relies heavily on two sources of information: (*i*) palaeo-records, which provide data from time periods where human activities had no or only minimal impact on global-scale processes and which provide the opportunity to evaluate the natural dynamics and variability of the System, and (*ii*) contemporary research in those areas of Earth System functioning where human influences are still thought to be relatively small. The timeframe considered here is the last one million years of the evolution of the Earth System, as this is the period against which the very recent era of significant human influence is most appropriately considered. At a few points in the chapter events earlier than one million years ago are mentioned, but only to help place more recent events in context.

In addition, this chapter introduces a style of analysis that is used throughout this volume. Initially, attention is focused on individual aspects of the Earth System. The functioning of these is described in some detail leaving aside, for the moment, the systems-level perspective. In essence, the reader is asked to take a magnifying lens and examine various segments of the Earth System individually. Then, at the end of each chapter, the reader is asked to put the magnifying glass down and look at the entire object – the Earth System – as a whole, in order to see how the individual segments examined contribute to the functioning of the System.

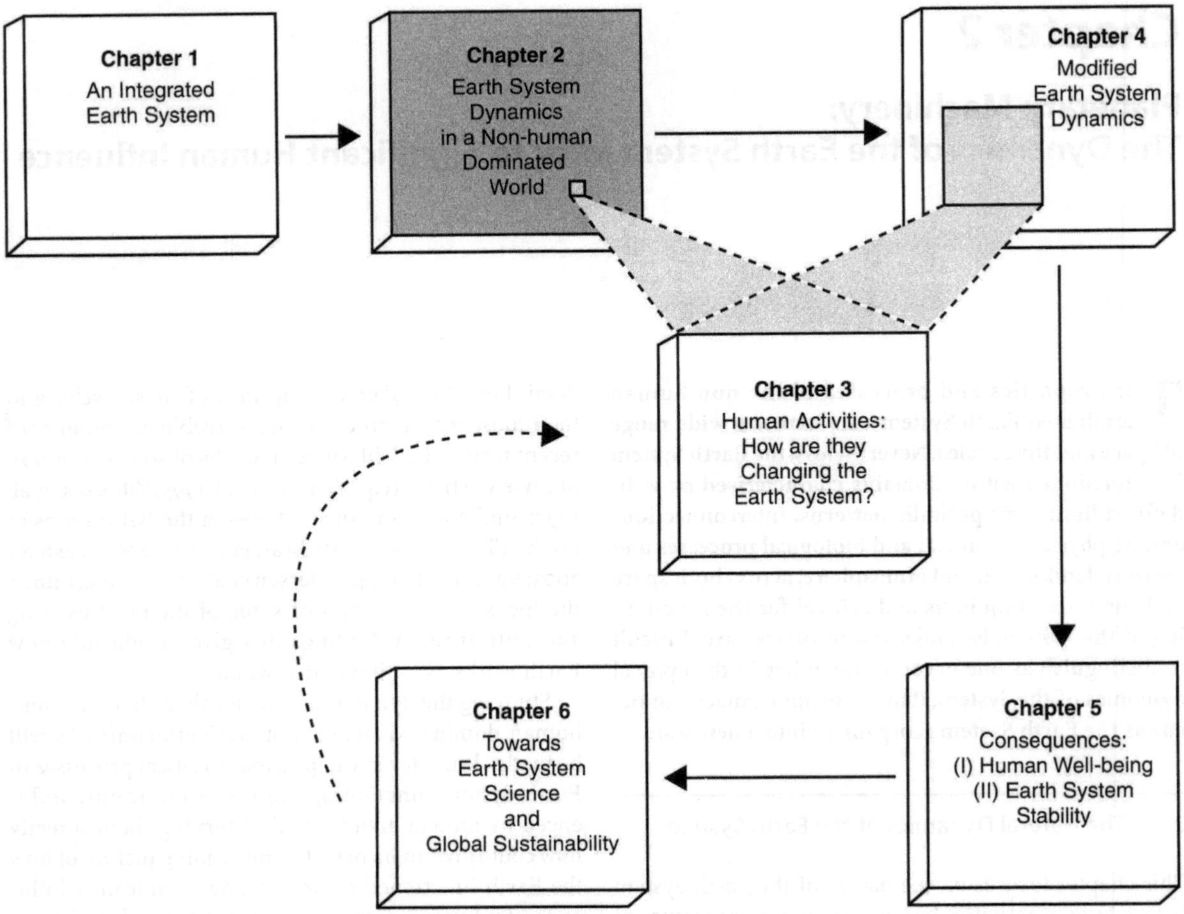

Fig. 2.1. Chapter 2 focus: Earth System functioning prior to significant human influence

2.2 New Insights in Temporal Variability of the Earth System

Recent research has illustrated the considerable natural variability and temporal dynamics of the Earth System. It has made clear that temporal change is a reality of the Earth System, and that static equilibria are unlikely to be a part of the System on almost any time scale and certainly not over the last 400 000 years. In the following sections, some of the insights gained by the great variety of palaeo-records that have been collected and evaluated over the past decade are considered.

2.2.1 The Long-Term Envelope of Natural Self-Regulation

The geological record shows that the functioning of the Earth System has varied continuously on all time-scales (Fig. 2.2). Changes in the Earth's orbit lead to changes in the latitudinal distribution of incoming solar radiation. The orbital parameters responsible for these changes have known periodicities, ~19 000 years and ~23 000 years for precession, ~40 000 years for obliquity and ~100 000 years for eccentricity. The changes in climate to which these orbital parameters give rise are far reaching and their periodicities can be detected in the record of climate variability contained in many environmental archives. As was pointed out in Chap. 1, climate changes are not linearly related to the external forcing. Interactions between biological, chemical and physical components of the Earth System, in response to subtle shifts in external forcing, have often given rise to non-linear changes that are abrupt and out of all proportion to the changes in incoming solar radiation. Nor are the climate changes necessarily synchronous with the orbital changes to which they ultimately respond, for the combination of internal thresholds, differing response times of Earth System components and strong, non-linear interactions can lead to lags between the onset of external changes and the responses of the Earth System.

Not all the past changes in climate that are detectable in the palaeo-record can be ascribed to orbitally driven changes. Many of the changes in temperature and

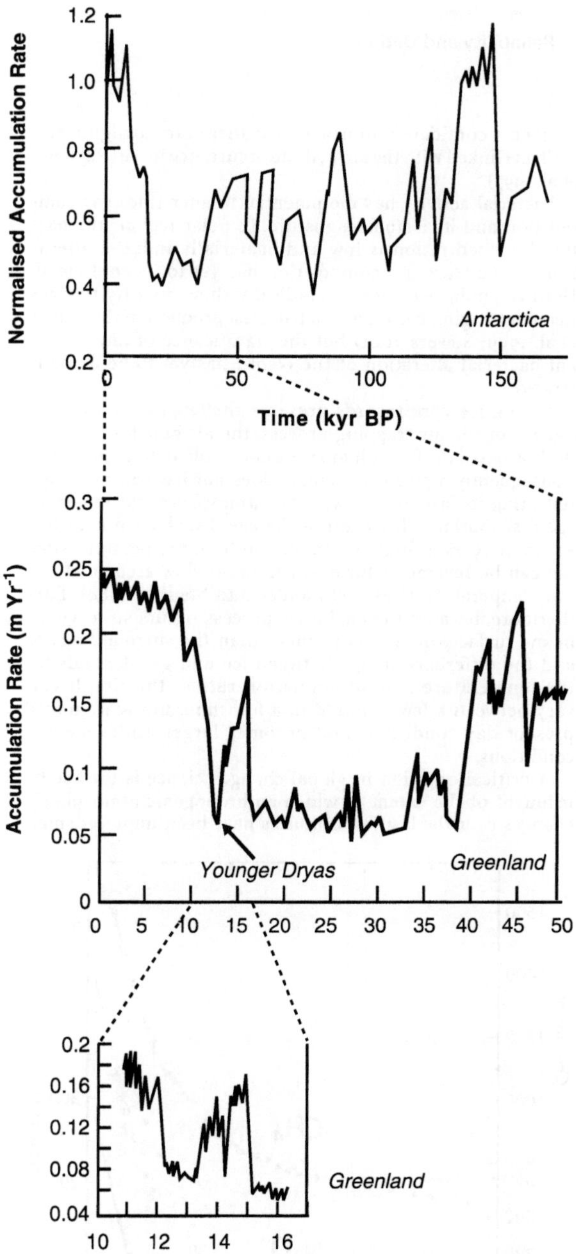

Fig. 2.2. Natural variability in Earth System functioning, from top to bottom: historic measurements of glacial ice accumulation rates, with insets focusing on the Younger Dryas cold interval and the very rapid termination of the last glaciation (adapted from Jacobson et al. 2000)

precipitation that are consistently recorded in environmental archives occur on shorter timescales, ranging from annual to millennial. Volcanicity influences climate on the shortest, mainly sub-decadal, timescales, but many other factors interact to drive the variability that has characterised the Earth's climate on sub-orbital forcing timescales, including short term changes in solar activity, ocean circulation and the terrestrial biosphere. The important point here is that regardless of the extent to which human activities may lead to changes in the Earth System, the functioning of the System will continue to vary in the future. Moreover, future changes, like those in the past, are likely to be characterised by highly non-linear responses to forcing, irrespective of the extent to which this is natural or anthropogenic. Documenting and understanding past variability, therefore, has a vital role to play in understanding present and predicting future change in Earth System functioning. This section focuses on the last 400 000 to 500 000 years, as this time period allows appraisal of four full glacial-interglacial cycles; these dynamics have been the normal operating mode of the Earth System throughout the later stages of human evolution.

The four-cycle Vostok ice core record from Antarctica, spanning the last 420 000 years (Fig. 1.3; Petit et al. 1999), provides the single most compelling template for multi-millennial scale natural variability in the geologically recent past. Of the many noteworthy features of the Vostok record, two provide an essential point of departure for any attempt to place contemporary trends into the longer-term context of natural variability. First, the atmospheric trace gas and isotopically inferred temperature records show that, over the whole of the last four glacial-interglacial cycles, values have oscillated within recurrently similar extreme values. Secondly, the changes in temperature and trace gas concentrations show a high degree of coherence throughout the sequence. Taken together, these observations demonstrate that the vast transformations involved in swings from interglacial to full glacial conditions have all taken place within a strongly self-limiting system; moreover, they show that the self-regulation of the system involves complex interactions in which atmospheric greenhouse gases play a significant part as climatic amplifiers and through essential feedback mechanisms. It is in the context of demonstrable self-regulation, involving maximum global atmospheric concentrations over almost half a million years of about 280 ppmv for CO_2 and 750 ppbv for methane, that any evaluation of contemporary values rising beyond 360 and 1 700, respectively, must be placed. These contemporary values are already significantly in excess of the peaks recorded in the Vostok series even when they are smoothed to replicate the processes affecting the trace gas record in the Vostok ice (Raynaud et al. 2003) (Box 2.1).

2.2.2 Millennial-Scale Oscillations and Abrupt Changes

While the Vostok record provides an essential long-term perspective, it fails to resolve the changes that have taken place on shorter time-scales. For these, archives that provide information with greater temporal resolution are

Box 2.1. The Ice Record of Atmospheric Greenhouse Trace Gases: Reliability and Dating

D. Raynaud · T. Blunier

Air trapped in glaciers and ice sheets is unique as it provides a clear record of the changes in greenhouse trace gas (CO_2, CH_4, N_2O) levels during the past. Not all glaciers and ice sheets provide an equally reliable record of greenhouse gas concentrations. Where melting occurs, gas content and gas composition may be altered by chemical reactions taking place in aquatic systems or by physical gas exchange between the gaseous and the aquatic sections.

Under dry and cold conditions of polar areas, essentially in Antarctica and Greenland, the top layer of the ice sheets (the first 50 to 130 m) results from the compaction of the surface snow through sintering. In this porous firn layer air readily enters the open spaces and records essentially the same composition as the atmosphere. Below this zone the air in the firn is static and mixes only by molecular diffusion. An equilibrium between molecular diffusion and gravitational settling is reached for each gas component (Craig et al. 1988; Schwander 1989). As a consequence, the air, just before being trapped at the base of the firn column, dates back several decades due to the slow diffusion through the firn pores and has a composition that departs slightly from the atmosphere because of the gravitational fractionation. The magnitude of this fractionation is well known, allowing an accurately corrected ice core record to be constructed.

Physi- or chemi-sorption of gases (especially CO_2) at the surface of the snow and firn grains, and subsequent expulsion of the attached gas molecules after recrystallisation of the grains in the ventilated top layer of firn may induce uncertainty into ice records of trace gases. Selective fractionation of the atmospheric ratios may also occur during the last stage of the closing of channels leading to the bubble close-off (Bender 2002). However, this effect does not significantly affect the concentrations of major trace gases. Despite these possibilities, no significant modifications of greenhouse trace gas concentrations during the trapping phase have been identified. This is demonstrated by the consistency between trace gas concentrations measured in air from the firn and the gas record directly measured in the atmosphere (Battle et al. 1996; Etheridge et al. 1996) and the generally good match of atmospheric and ice core data (Fig. 2.3).

Slow chemical reactions may alter CO_2 concentrations after the gas has been trapped in ice. This has been observed in Greenland where high concentrations of impurities in the ice may lead to significant *in situ* CO_2 production via acid-carbonate interactions and oxidation of organic material (Anklin et al. 1997; Delmas 1993; Haan and Raynaud 1998). Antarctic records provide the most reliable data of changes in global atmospheric CO_2 (Raynaud et al. 1993). Carbon dioxide measurements made several years apart on the same core show no significant changes. Antarctic results are consistent between sites to within a few parts per million by volume despite the coring sites having different ice accumulation rates, temperatures and concentrations of impurities.

At depth, air bubbles progressively disappear and air hydrates form as the ice molecular structure encapsulates the air molecules in the so-called brittle zone (in which recovered ice commonly contains a high density of cracks and fractures). Results obtained from the brittle zone of a single ice core have to be considered with caution as the presence of air hydrates in the deeper parts of long ice cores may induce artifacts in the record. However, multi-site data may be used with more confidence as the brittle zone usually does not have the same age in cores from different sites. Providing that appropriate extraction procedures are used, the agreement between records from different sites or measurements performed on the same core several years after the initial measurements confidently indicates that there are no significant artifacts linked with the air-hydrate occurrence in ice (Raynaud et al. 1993).

Bacterial activity has the potential to alter trace gas composition and its isotopic signature. In polar regions the bacterial concentration is low and bacterially induced alteration of the trace gas composition has yet to be confirmed. However, polar ice cores sporadically show high N_2O values that may originate from *in situ* bacterial production (Flückiger et al. 1999; Sowers 2001) but the significance of any resultant bacterial alteration of the record has yet to be demonstrated.

Dating ice-core records presents challenges. As a consequence of the air trapping process, the air bubbles found at the bottom of the firn column were closed off at different times. Consequently a given ice sample does not have an exact age reflecting its last contact with the atmosphere, but rather an age distribution. The width of the age distribution is as low as seven years at high accumulation/high temperature sites but can be several centuries for Antarctic low accumulation/low temperature sites (Schwander and Stauffer 1984). Furthermore, because the enclosure process occurs 50 to 130 m below surface, the gas is younger than the surrounding ice and the difference in age between ice and gas depends on the temperature and accumulation rate at the site. It can vary between a few hundred to a few thousand years under present day conditions and becomes larger under ice age conditions.

A critical question in global change science is the establishment of the extent to which anthropogenic atmospheric changes over the last few centuries have been unprecedented

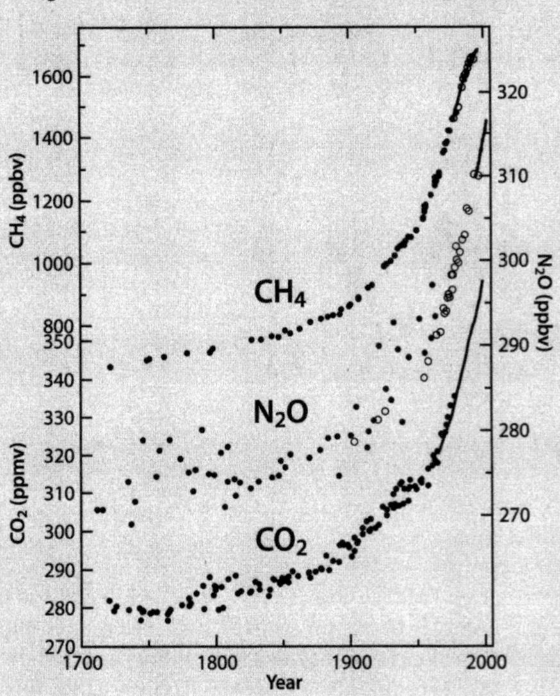

Fig. 2.3. Greenhouse trace gas records since AD 1700 (adapted from Raynaud et al. 2003). The ice core measurements overlap with direct atmospheric measurements in the contemporary period (*solid lines*)

◄

over the last 400 000 years. The record of atmospheric gas composition is smoothed during the air trapping process. Since in the case of the Vostok record this smoothing averages over centuries, it may be questioned whether an anthropogenic signal can be observed in the Vostok data. A simulation to obtain a smoothed Vostok-like record of the present anthropogenic CO_2 increase (Raynaud et al. 2003) shows that such an increase would be imprinted in the record through a CO_2 peak reaching concentrations higher than 315 ppmv with a very slow return toward the pre-industrial level. Such a CO_2 signal is not visible in the Vostok record (Fig. 2.4), although the time resolution of the record does not exclude a pulse-like atmospheric CO_2 signal of a few decades duration with concentrations as high as today. However, this would require both a large carbon release within a few decades (of the order of 200 Gt C) and an equally large and rapid uptake. Such an oscillation is not compatible with the present understanding of the global carbon cycle.

An important question is whether the time resolution of the ice record is precise enough to resolve the leads or lags between greenhouse gases and climate signals. The main uncertainties in this respect arise in connection with the age differences between enclosed air and the surrounding ice. In regard to the onsets of the last two glacial-interglacial transitions, the best-resolved records indicate an uncertainty of about hundred to a few hundred years and an Antarctic warming leading the CO_2 increase by a few hundred years (Monnin et al. 2001; Caillon et al. 2003). In the case of rapid signals like those of the Dansgaard/Oeschger events (cf. Sect. 2.2.2), the uncertainty in age difference is reduced to 10–20 years (Leuenberger et al. 1999; Severinghaus et al. 1998).

Despite the uncertainties and caveats discussed above, ice-core data provide an excellent record of past changes; the record can be used with confidence and provides the most direct and accurate evidence for past atmospheric change yet obtained. The atmospheric greenhouse gas record measured in ice cores is a smoothed temporal record whose accuracy is currently of the order of ±5 ppmv for CO_2 and ±10 ppbv for CH_4.

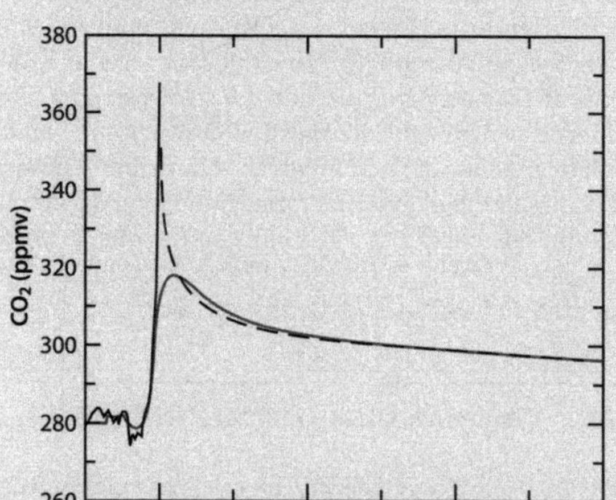

Fig. 2.4. Simulation of the minimum smoothed Vostok-like CO_2 record (in *grey*) that would result from the hypothetical incorporation of the present anthropogenic CO_2 increase (in *black*) in an ice core record (Raynaud et al. 2003)

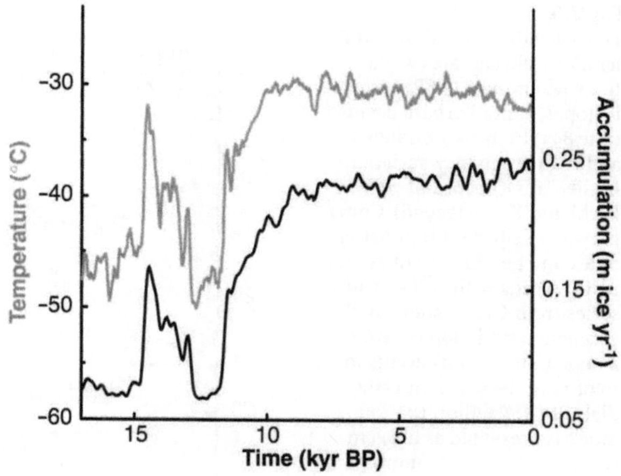

Fig. 2.5. Ice accumulation and oxygen isotope (interpreted as temperature) records from the GISP2 (central Greenland) ice core for the period between the present and 18 000 years ago (Alley et al. 1993; Alverson and Oldfield 2000), showing the abrupt climate shift at the termination of the Younger Dryas cold event some 11 600 years ago. In this record the event is manifested as a warming estimated to be as much as 15 °C, accompanied by a doubling in annual precipitation volume, that occurred in less than a decade

required. The ice cores from central Greenland provide a detailed, well-substantiated and repeatable record spanning the last glacial cycle of around 100 000 years. The rapid oscillations in inferred temperature (Dansgaard/Oeschger Cycles, see Fig. 2.5) recorded during the glacial part of the record often show over half the amplitude of the full glacial/interglacial cycle (Grootes et al. 1993). Apparently synchronous palaeo-oceanographic changes have been widely detected in the northern hemisphere (e.g., Behl and Kennett 1996; McManus et al. 1999; van Kreveld et al. 2000; see Fig. 2.6) and there are an increasing number of continental archives demonstrating parallel patterns of variability on land (e.g., Chen et al. 1997; Allen et al. 2000; Wang et al. 2001; Fang et al. 1999).

In Antarctica relatively high resolution records covering significant parts of the same time interval are, or will soon become, available from the Byrd Station, Taylor Dome, Law Dome (Indermuhle et al. 1999; Raynaud et al. 2000) (Fig. 2.7) and the first stages of analysis of the Dome C (Concordia) core (Flückiger et al. 2002). Now that data from the two hemispheres have been synchronised using common variations in the methane records (Blunier et al. 1998; Raynaud et al. 2000, 2003), several new insights have emerged. Although both sets of polar records show strong evidence for millennial-scale oscillations during the last glacial period in almost every parameter measured, the record from the opposite poles is often out of phase and, for some of the major oscillations, it is in antiphase. Moreover, during the glacial termination, the main warming trend in Antarctica precedes any rapid temperature increase in the northern hemisphere.

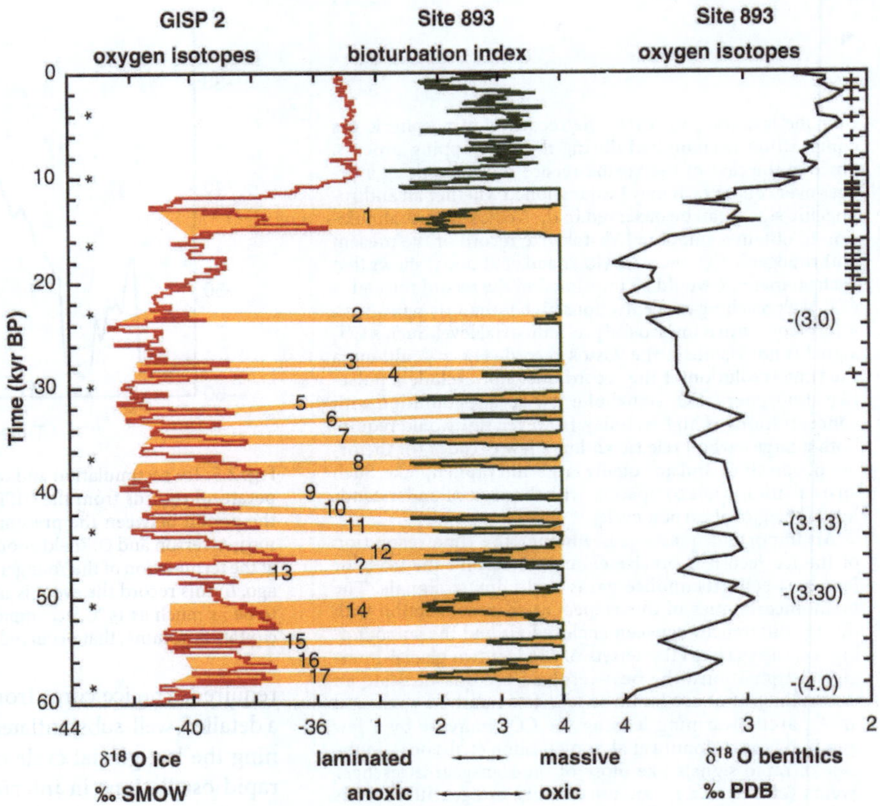

Fig. 2.6. Low latitude expression of millennial scale climate oscillations taken from GISP2 oxygen isotopes, Santa Barbara basin (Site 893) bioturbation index and high-frequency variations in the $CaCO_3$ record of 70KL (Behl and Kennett 1996). Comparison of site 893 bioturbation index and benthic foraminiferal $\delta^{18}O$ records with $\delta^{18}O_{ice}$ time series from GISP2, showing the excellent correlation of site 893 anoxia (lamination) events to 16 of 17 of the warm interstadials of GISP2. Bioturbation index is presented as a 49 cm (ca. 300–400 years) running average to dampen high-frequency variation and to match the resolution of the GISP2 record. Chronologies for GISP2 and site 893 were independently derived. Radiocarbon age control points (+) and SPECMAP data used for the site 893 age model are shown to the right. Numbers in () refer to standard data of the SPECMAP stratigraphy. The base of each core interval in Hole 893A is indicated by *arrows* to the left

These observations further refine the perspective on past global changes, for they show that:

- major switches in the earth's climate system occurred on much shorter time scales than the glacial/interglacial cycles;
- the recorded changes were often rapid and of high amplitude;
- the changes demonstrate widespread spatial coherence, but, when characterised with sufficient temporal resolution, they are not globally synchronous; and
- complex inter-hemispheric leads and lags occur that require feedback mechanisms for amplifying and propagating changes.

In terms of present day and future implications, these observations are especially important, for they raise the possibility of anthropogenically induced global changes triggering positive feedbacks capable of provoking sudden, dramatic switches in climate comparable to those that have occurred in the past.

Aspects of the broad pattern of change presented above call for closer attention, specifically the rapidity with which the Earth's climate system may undergo major rearrangement. The best-documented period of rapid major change falls at the end of the glacial period, some 11 600 years ago. At mid to high latitudes in the northern hemisphere, the rapid warming at the opening of the Holocene is the final, decisive step in a sequence of oscillations. It represents the culmination of a suite of changes that appear to begin with the first clearly detectable warming trend in the record from Antarctica some 6 000 years earlier. Many lines of evidence point to a rapid warming at the opening of the Holocene, with records from central Greenland indicating that dramatic changes occurred within only a few decades at most (Alley et al. 1993; Alley 2000; Fig. 2.8). Studies of the opening of the Holocene from a wider range of sites may well provide the strongest empirical evidence available for the speed with which the Earth's climate system can respond given a sufficiently powerful combination of external forcing and internal amplification. They may also shed light on the rate at which ecological systems can respond to such rapid changes (Amman and Oldfield 2000; Birks et al. 2000).

2.2.3 Climate Variability in Interglacial Periods

Most of the high variability in inferred temperature outlined above is characteristic of glacial intervals and terminations, but the Earth is currently in a major interglacial that has lasted for over eleven millennia. It is, therefore, not surprising that an increasing amount of attention has been devoted to characterising climate during the last, Eemian interglacial (Marine Isotope Stage 5e). Depending somewhat on definition and on the chronology adopted, this interglacial spanned

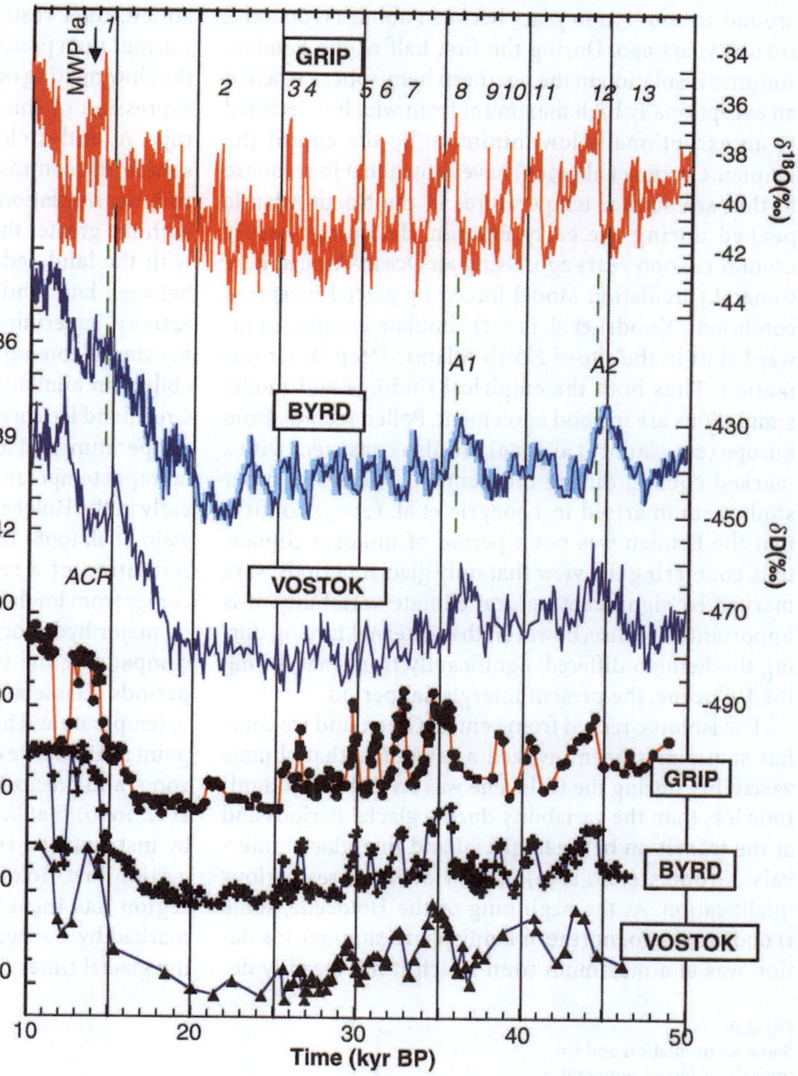

Fig. 2.7.
Inter-hemispheric phasing of the Antarctic Cold Reversal and the Younger Dryas and the timing of the Antarctic Cold Reversal and the atmospheric CO_2 increase with respect to the Younger Dryas event, as deduced from the CH_4 synchronisation, the isotopic records of the GRIP, Byrd and Vostok ice cores, together with the CO_2 record from Byrd (Blunier et al. 1998)

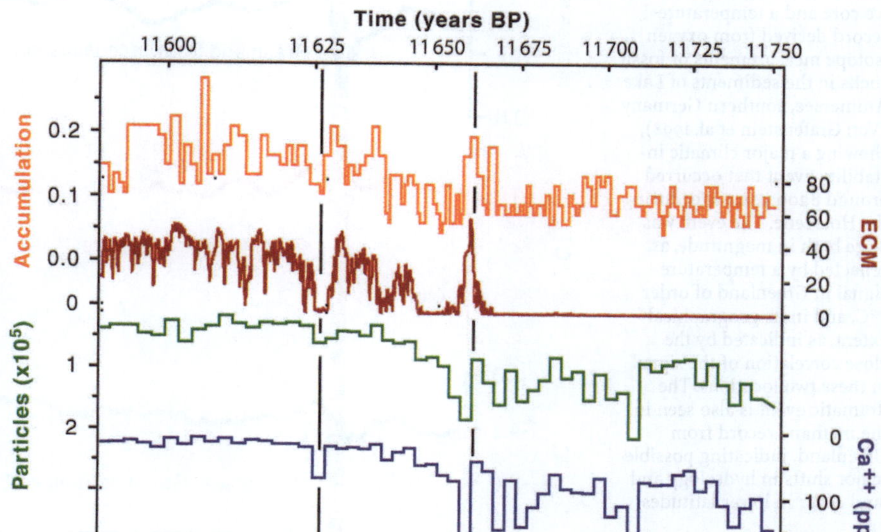

Fig. 2.8.
The end of the Younger Dryas in the GISP2 ice core from central Greenland (Alley 2000). Shown are selected curves illustrating the possible flickering behaviour near the transition: snow accumulation in m ice yr^{-1} (*orange*); electrical conductivity, ECM current in microamps (*red*); insoluble-particulate concentration in number ml^{-1} (*green*); soluble-calcium concentration in ppb (*blue*). The main step at the end of the Younger Dryas falls between 1677 and 1678 m depth, between the two probable flickers indicated by the *vertical lines* at 1676.9 and 1678.2 m depth

around 10 000–15 000 years between about 135 000 and 110 000 years ago. During the first half of the Eemian, summer insolation in the northern hemisphere reached an exceptionally high maximum from which it declined to an exceptionally low minimum by the end of the Eemian. Cortijo et al. (1999) have shown that in response to this, sea surface temperatures in the North Atlantic peaked during the early Eemian, declining steeply around 120 000 years ago. Using an Ocean-Atmosphere General Circulation Model forced by glacial inception conditions, Khodri et al. (2001) simulate a rapid southward shift in the site of North Atlantic Deep Water formation. Thus both the empirical findings and model simulations are in good agreement. Pollen records from Europe (e.g., Guiot et al. 1989) are also consistent with a marked cooling during the Eemian. These and other studies summarised in Labeyrie et al. (2003) confirm that the Eemian was not a period of uniform climate, thus countering the view that only glacial periods were marked by significant natural climate variability. It is important to realise, however, that external forcing during the Eemian differed significantly from that during the Holocene, the present interglacial period.

The isotopic record from central Greenland ice cores has sometimes been invoked as evidence that climate variability during the Holocene was an order of magnitude less than the variability during glacial periods and at the transition between glacial and interglacial intervals (Grootes et al. 1993). This view requires serious qualification. At the beginning of the Holocene, some 11 600 years ago, northern hemisphere summer insolation was at a maximum from which it has steadily declined. As a result, the early- to mid-Holocene was warmer than present in northern high latitudes, though the slow melting of the Laurentide Ice Sheet delayed the expression of this in parts of North America and Europe. At mid- to lower latitudes in the northern hemisphere, the contrast between high northern hemisphere summer insolation and low winter insolation, combined with the greater thermal inertia of the ocean compared with the land, led to increased temperature contrasts between land and sea and hence stronger monsoonal activity. Superimposed on these changes largely linked to external forcing is abundant evidence for climate variability on annual to millennial timescales. The central Greenland ice core record itself shows a rapid change in temperature and ice accumulation, correlated to a similar rapid temperature change in Germany and elsewhere, early in the Holocene some 8 200 years ago (Von Grafenstein et al. 1998) (Fig. 2.9). However, the most dramatic refutation of a relatively constant Holocene climate comes from low latitudes, where the Holocene is marked by major hydrological changes comparable to those accompanying the transition from glacial to interglacial periods (Gasse and Van Campo 1994) (Fig. 2.10). Even in temperate and higher latitudes, many lines of evidence point to Holocene changes in both temperature (Bradley 2000) and precipitation/evaporation balance (Bradbury et al. 2000) that lie outside the range of values captured by instrumental records (Fig. 2.11). There is some suggestion that Holocene variability in the North Atlantic region has included a damped continuation of that marked by Dansgaard/Oeschger Cycles during preceding glacial times (e.g., Bond et al. 1997).

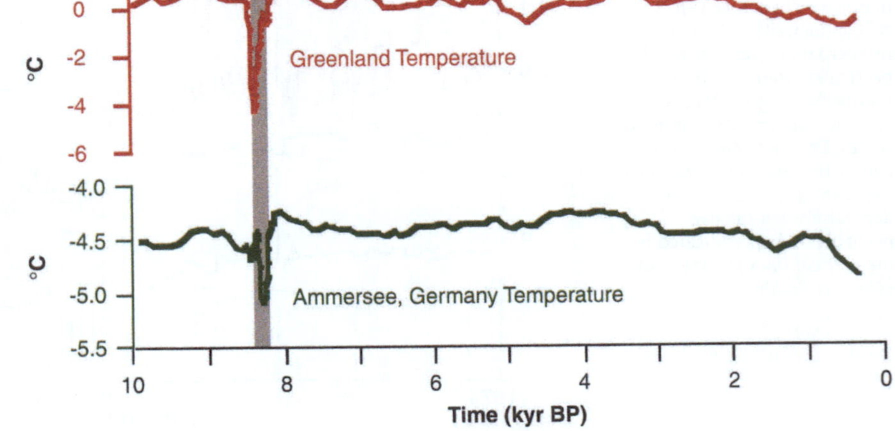

Fig. 2.9. Snow accumulation and isotopically inferred temperature records in the Greenland GISP2 ice core and a temperature record derived from oxygen isotope measurements of fossil shells in the sediments of Lake Ammersee, southern Germany (Von Grafenstein et al. 1998), showing a major climatic instability event that occurred around 8 200 years ago, during the Holocene. The event was large both in magnitude, as reflected by a temperature signal in Greenland of order 5 °C, and in its geographical extent, as indicated by the close correlation of the signal in these two locations. The dramatic event is also seen in the methane record from Greenland, indicating possible major shifts in hydrology and land cover in lower latitudes

Palaeo-climatic evidence demonstrating that strong variability persisted into the second half of the Holocene is particularly important as this is the period that forms the baseline used to assess the significance of major anthropogenic perturbations. The amplitude and frequency of past hydrological variability (e.g., droughts) are especially important because human populations are concentrated in temperate and tropical latitudes. Moreover, in many of the populous areas of the globe, hydrological changes are much more crucial for human survival than are changes in temperature. One important implication of this part of the palaeo-record is that for many parts of the world, and even without any hypothesised anthropogenic greenhouse gas forcing, the future course of climate change is likely to include variability beyond any expectations based on an analysis of the short period of instrumental measurements alone.

Natural Late Holocene variability on sub-decadal to century timescales may be characterised in some cases as shifts in mean temperature that, despite regional differences of detail in timing and expression, appear to show a degree of coherence over a wide area (for example, the so-called Medieval Warm Epoch and Little Ice Age) (Bradley et al. 2003). Variability may also manifest itself as shifts in the spatial teleconnection patterns (teleconnections are defined here as the correlation between specific planetary processes in one region of the

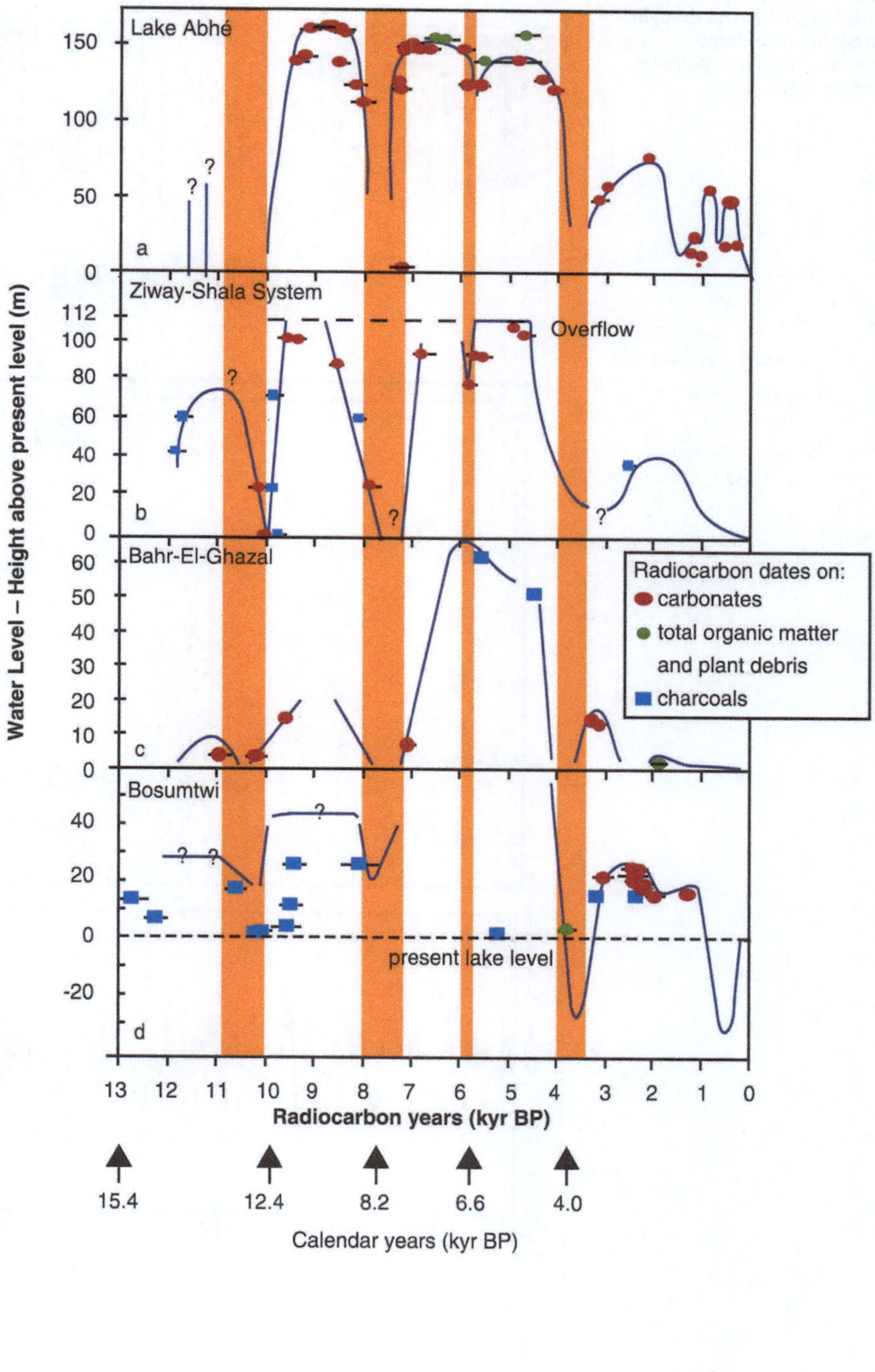

Fig. 2.10. Changes in lake level over the past 15 000 years in an east to west transect of lakes in the northern monsoon domain of Africa (Gasse 2000), showing lake level variations as much as 100 metres, an indication of large changes in the regional hydrological balance

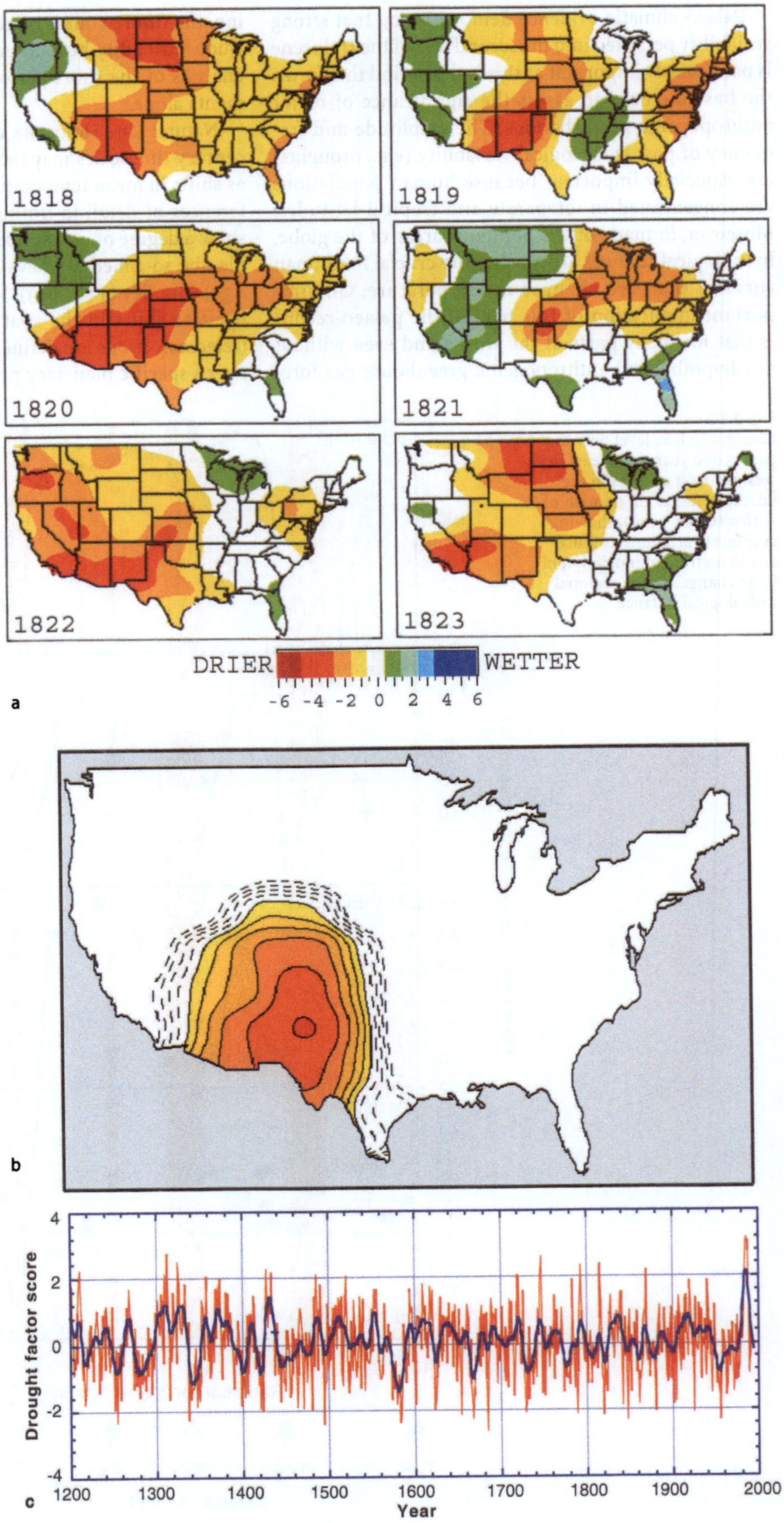

Fig. 2.11. Multiyear drought reconstructed from tree rings: **a** a severe drought through much of the central USA centred around 1820, surpassing in extent and severity the so-called Dust Bowl of the 1930s; **b** Palmer Drought Severity Index grid derived from optimally interpolated tree ring chronologies; and **c** time series of drought as recorded in tree rings for the period 1200–1994, with a correlation with instrumental data of 0.77 on an annual basis and 0.86 on a smoothed, inter-decadal basis over the period 1895–1994 (Cook et al. 1998; NOAA NESDIS drought variability data: *http://www.ngdc.noaa.gov/paleo/drought.html*)

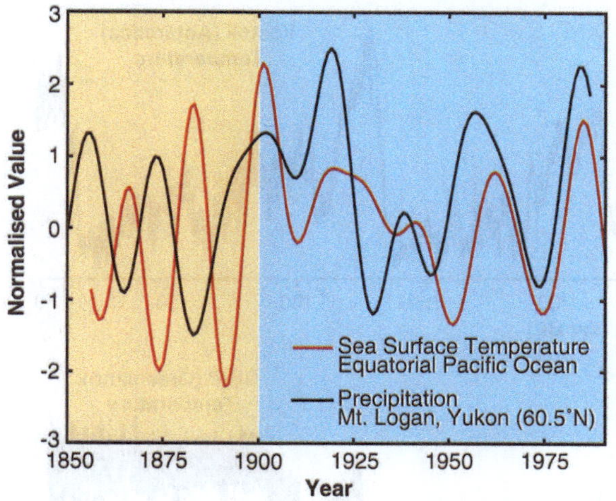

Fig. 2.12. Changes over time in climate teleconnection patterns between the high latitudes and the tropics: records of sea surface temperature and precipitation, showing close correlation after 1900 but anti-correlation prior to 1900 (GWK Moore et al. 2001)

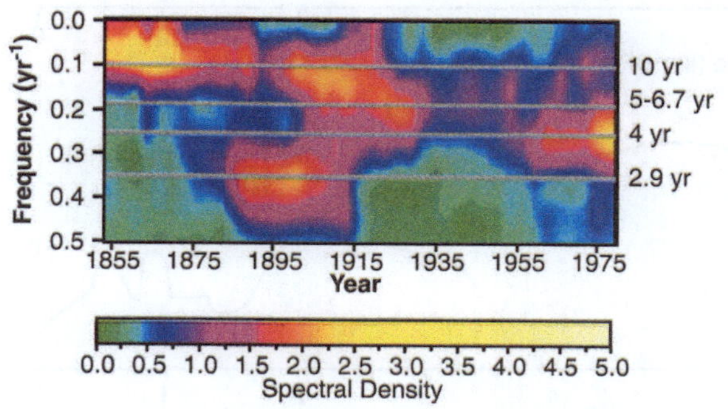

Fig. 2.13a. Evolutionary spectral analysis of sea surface temperature estimates as recorded in a coral from Mariana Atoll in the equatorial Pacific Ocean (Urban et al. 2000). *Colours* show how sea surface temperatures (related to El Niño /La Niña conditions) have varied over time. In recent years this oscillation has been centred on a 4-year period but in the nineteenth century the oscillation occurred only once every 10 years; the shift from low to high frequencies took place early in the twentieth century

world to distant and seemingly unconnected regions elsewhere; see Sect. 2.5.5 and Fig. 2.12) and temporal (frequency and severity) expression of some of the dominant contemporary modes of climate variability. These include ENSO, the El Niño-Southern Oscillation, (Markgraf and Diaz 2000; Labeyrie et al. 2003; Bradley et al. 2003) (Figs. 2.12 and 2.13) and the North Atlantic Oscillation (Hurrell 1995; Appenzeller et al. 1998; Luterbacher et al. 2000; Cullen et al. 2001) on time-scales of decades and centuries, or as changes in the magnitude and frequency of extreme events (Knox 2000; Eden and Page 1998). When current changes are placed in a longer-term context, one important implication is that the pre-anthropogenic baseline is a highly dynamic one. In consequence, no single time interval can serve as a simple, fixed baseline (Millar 2000). Moreover, any anthropogenically forced changes in climate are already interwoven with the pattern of natural variability and the two processes will continue to interact in the future. This adds complexity to both the detection and prediction of current and future climate change.

A highly condensed impression of climate variability on time scales ranging from tens of thousands of years and longer (the so-called Milankovitch frequencies) is given in Fig. 2.14 (Oldfield and Alverson 2003). The diagram illustrates some of the points either explicit or implicit in the foregoing text, namely that:

- variability is not just reflected in temperature; hydrological variability, which is often of much greater importance than temperature to human populations, has been significant on all time scales in the past;
- no single proxy, archive or region truly reflects global variability of climate. A high degree of global coherence may be articulated through immense variations in regional response;

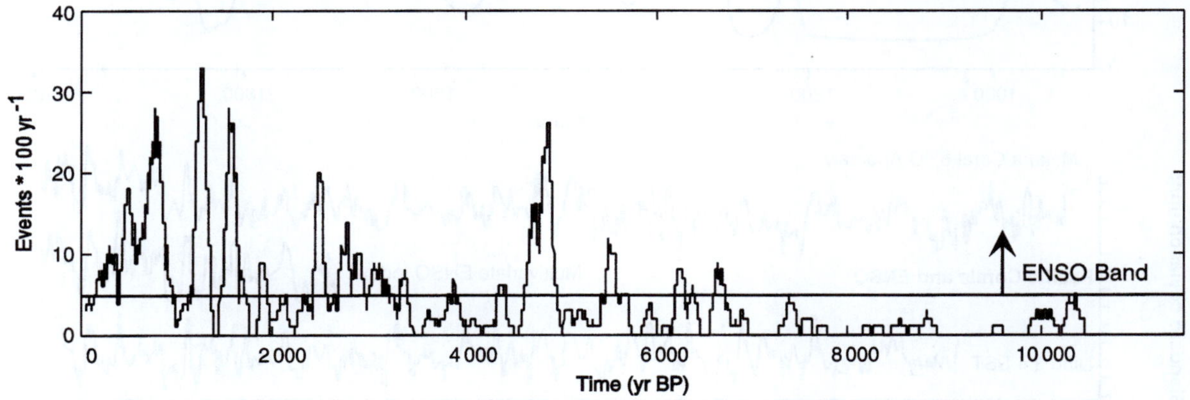

Fig. 2.13b. Long-term ENSO variability during the Holocene (Moy et al. 2002). In this ENSO record in a laminated lake sediment sequence from Laguna Pallcacocha, south Ecuador, spanning the whole of the Holocene, ENSO events have been identified from the presence of light coloured, clastic laminae that, for the recent period, correspond with documented moderate to strong ENSO occurrences. The graph expresses the changing frequency of ENSO as events per overlapping hundred-year window. The *horizontal line* represents the level above which variance in the ENSO frequency band occurs. There is no record of variance within the ENSO band before 7000 calendar years before present, after which an irregular, pulsed increase occurs leading to peak ENSO activity some 1200 years ago

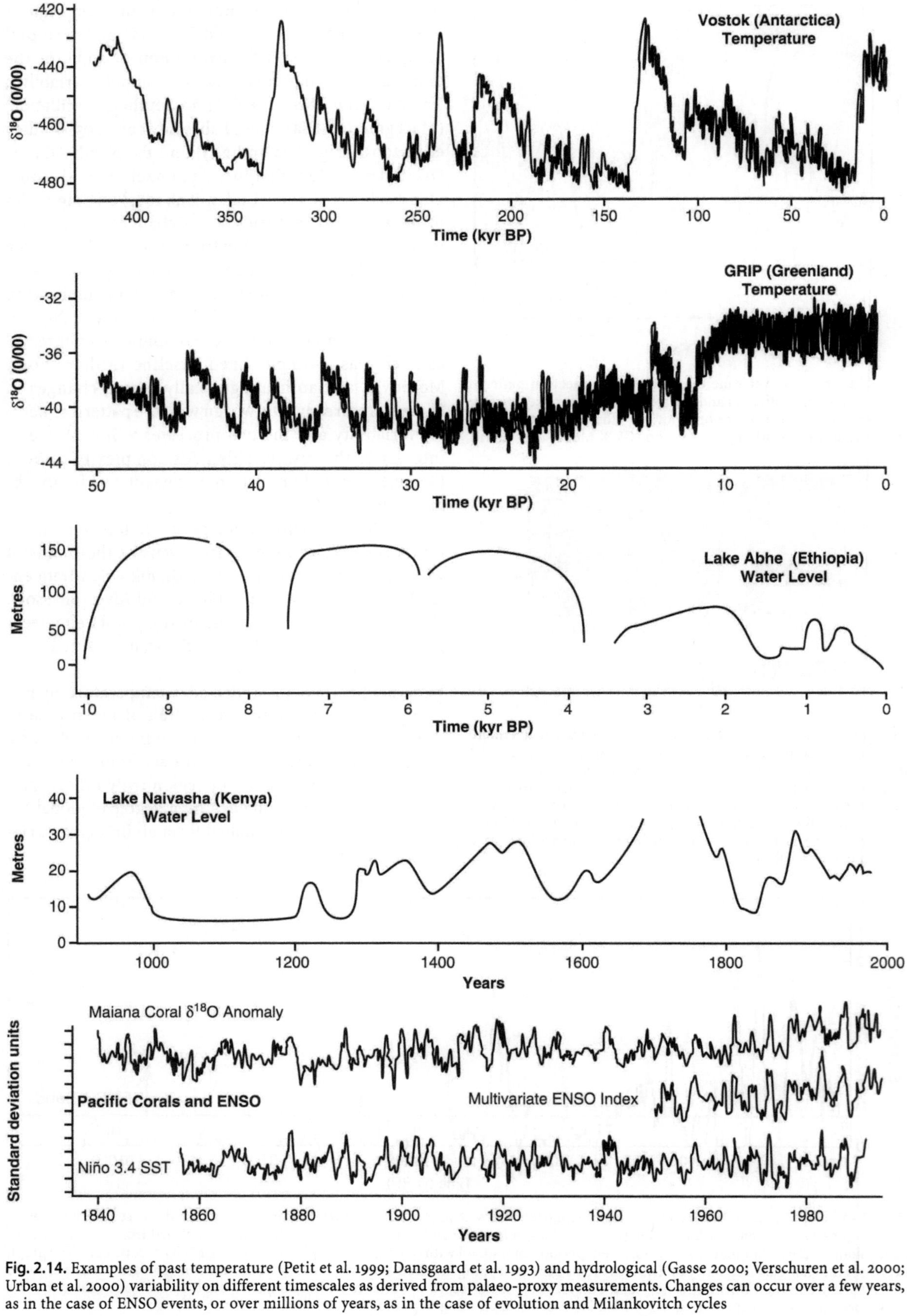

Fig. 2.14. Examples of past temperature (Petit et al. 1999; Dansgaard et al. 1993) and hydrological (Gasse 2000; Verschuren et al. 2000; Urban et al. 2000) variability on different timescales as derived from palaeo-proxy measurements. Changes can occur over a few years, as in the case of ENSO events, or over millions of years, as in the case of evolution and Milankovitch cycles

- although major, abrupt transitions, reflecting reorganisation of the Earth System, are most evident in predominantly cold, glacial periods, they are not absent in the Holocene, especially at lower latitudes; and
- during the late Holocene, when the natural forcings and boundary conditions were similar to those operating today, there is strong evidence that the range of variability significantly exceeded that captured by instrumental records. Relying mainly on the recent period of instrumental records thus gives a false sense of the true variability of the Earth System, even in the late Holocene.

Irrespective of any evaluation of the actual and potential effects of anthropogenic forcing through atmospheric greenhouse gases and aerosols, understanding the expression, causes and consequences of *past* natural variability is of vital concern for developing realistic scenarios of the future. Moreover, the complex interactions between external forcings and internal system dynamics on all timescales implies that at any point in time, the state of the Earth System reflects not only characteristics that are an indication of contemporary processes, but others that are inherited from past influences, all acting on different time scales. The need for an understanding of Earth System functioning that is firmly rooted in knowledge of the past is essential.

2.2.4 Temporal Variability in the Biota

In understanding temporal variability in biota, it is not possible to focus exclusively on the last million years. This is partly due to the slow rate of evolution itself, but also the fact that historical changes are quantified by examining the (very incomplete) fossil record. Living organisms are, however, constantly evolving and, on a geological time scale, organisms appear and disappear. For example, Sepkoski (1992) has estimated that currently about 200 000 species are present in the oceans, but that 7.8 million of the species that evolved in the oceans over the last 600 million years are now extinct.

Changes in global biodiversity are the product of two opposing processes: (*i*) the success of organisms in spreading in local, regional and, ultimately, global ecosystems and (*ii*) changes in the environment that prevent organisms from existing or spreading. Global biodiversity is intrinsically a product of the interaction between biological processes and abiotic environmental/planetary factors.

Although the time scales are different, the record of biodiversity shows considerable variability through time in the same way that atmospheric gases and temperature have. Expansion of biodiversity through history did not proceed at a constant rate. There were periods of rapid expansion, relative stability and rapid decline. In fact, five of the periods of decline were so dramatic that

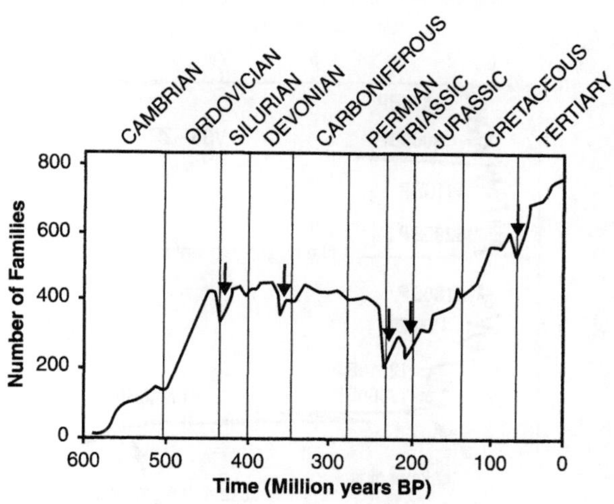

Fig. 2.15. Long-term trend in biodiversity (Wilson 1992). The *arrows* indicate major extinction events. Biodiversity is represented here by families of marine organisms, not total number of species

they are recognised as mass extinction events (Flessa et al. 1986) (Fig. 2.15).

The first two mass extinctions of the past 500 million years, the Ordovician approximately 440 million years ago and the Devonian about 360 million years ago, may have been of less significance than more recent ones and are thought to be the result of glaciation of Gondwanaland, which led to extensive climate cooling and sea level drop (Brenchley et al. 1995; Armstrong 1995, 1996). In the Triassic extinction event that occurred 50 million years after the Permian extinction and lasted for a period of only 10 000 years (Ward et al. 2001), approximately 80% of all species disappeared (Jablonski 1991). The greatest extinction event was the Permian mass extinction (Erwin 1994). This occurred 250 million years ago and killed off over 90% of all marine species, about 70% of terrestrial vertebrate genera, and most land plants (Raup 1979; Sepkoski 1989; Visscher et al. 1996; Ward et al. 2000). It is estimated that this event took place over a period of time ranging from 8 000–500 000 years (Bowring et al. 1998; Rampino et al. 2000). Two hypothesis for the cause of the Permian extinction are bolide impact (either an asteroid or comet) (Rampino and Haggerty 1996) or explosive volcanism (Renne et al. 1995), which then led to massive environmental changes. The last of the five mass extinctions is the Cretaceous-Tertiary (K-T) extinction 65 million years ago, best known for the complete extinction of the dinosaurs. However, it is estimated that between 60–70% of all species became extinct during this period (Stanley 1987). The most widely accepted theory for the K-T extinction is an extraterrestrial impact by either an asteroid or comet (Alvarez et al. 1980; Shukolyukov and Lugmair 1998; Mukhopadhyay et al. 2001). Others argue that the palaeo-evidence points to voluminous volcanism as the trigger for the K-T extinction event (Officer et al. 1987).

Although the rate of extinction is in general several orders of magnitude faster than the rate of evolution and

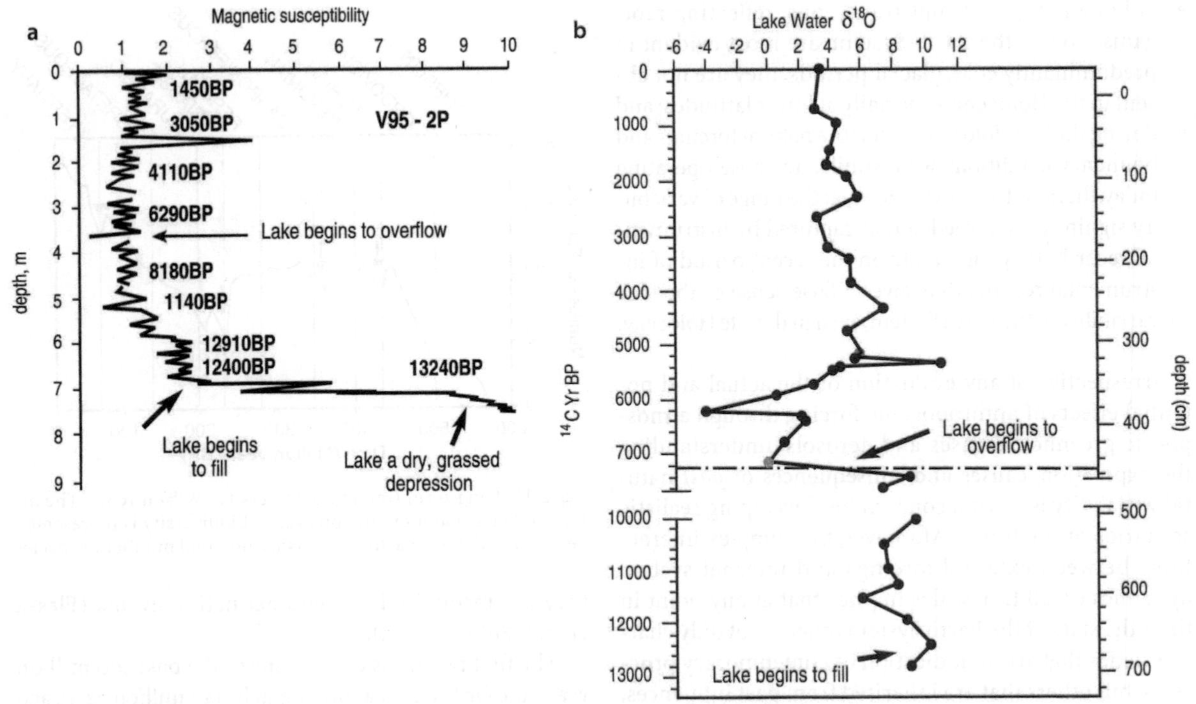

Fig. 2.16. Lake Victoria sediment-core variations in **a** magnetic susceptibility (after Johnson et al. 1996) and **b** lake water $\delta^{18}O$ (Beuning et al. 1998) to illustrate the early Holocene refilling and overflowing of the lake

generation of new species, the response of speciation to environmental change may be dramatic and surprisingly rapid. Before 13 000 years ago Lake Victoria in central Africa was a temporarily dry, grass-covered depression (Johnson et al. 1996) (Fig. 2.16). The lake began to refill about 13 000 years ago and overflowed at around 7 500 years ago (Beuning et al. 1998). Immediately prior to the overflowing, diatom productivity dropped to a minimum (Johnson et al. 1998). In the first 500 years after the lake began to fill about 13 000 years ago, primary production was extremely high, nourished by high input of nutrients from the flooded landscape (Johnson et al. 1998). A few species of cichlids and other fish emerged out of their fluvial refugia to colonise the new lake, apparently generating hundreds of new endemic species over just the last 13 millennia (Johnson et al. 1998). However, a more recent analysis suggests that the current lineage of cichlids in Lake Victoria is much older (~100 000 years), having migrated from the nearby and geologically older Lake Kivu (Verheyen et al. 2003). In general, macroevolution is a slow process in comparison to the one-million-year timeframe of this volume so increases in the overall number of biological species due to evolution can be considered as negligible during this period.

2.3 New Insights into the Role of Biology in Earth System Functioning

Climate has long been recognised to influence soil and biota, but until recently the converse influence of soil and the biota on climate has been viewed as either very slight or non-existent (Walter and Breckle 1985). However, it has now been demonstrated that the Earth's hydrological cycle and climate are strongly modulated by the biological processes of vegetation and phytoplankton and by characteristics of the land surface (Kabat et al. 2004). There is now clear evidence that biota are a critical component of the Earth System and that they respond to and influence the geochemistry and physics of the System in many different ways. Characterisation of biological processes is essential in order to understand climate variability, to predict global weather and climate and to project the functioning of the Earth System into the future (Claussen 2004).

2.3.1 Biogeophysical Processes

Vegetation on land influences water cycling and climate through a variety of mechanisms, including the evaporation and transpiration of water, the reflectance or absorption of incoming solar radiation and the transfer of momentum with atmospheric flows (Box 2.2). Although these mechanisms have been known for some time, the biogeophysical role of vegetation at larger spatial scales, up to the scale of the Earth System itself, has not been appreciated until the past decade or so. A good example of new insights is the importance for global climate of the albedo of different vegetation types in the high latitudes, the so-called taiga-tundra feedback.

Box 2.2. Biogeophysical Effects of Vegetation on Climate

Richard Betts

Vegetation can affect climate through its influence on the physical characteristics of the land surface. These characteristics determine the inputs and outputs of energy, water and momentum to/from the atmosphere at its lower boundary, and hence can exert major influences on both local and global climates.

The fraction of solar radiation reflected away by the surface (the surface albedo) is strongly determined by the nature of the vegetation cover. A forested landscape generally absorbs a greater fraction of incident radiation than open land, partly due to darker tree foliage and partly because the complex canopy structure traps light through multiple reflections within the canopy. This effect is accentuated in snowy conditions – lying snow can entirely cover short vegetation and produce a bright, highly reflective surface, whereas snow falling on a forest may be shed from the canopy, leaving the dark foliage exposed. Even a snow-loaded canopy maintains a low albedo because of multiple internal reflections. Therefore, compared to unforested land, a forested surface will typically be warmed more by a given incident flux of solar radiation.

Water falling as precipitation may be intercepted by a vegetation canopy and re-evaporated, or may reach the surface, infiltrate into the soil profile, be absorbed by plant roots and evaporated through leaves back to the atmosphere (transpiration). Evaporation and transpiration provide a recycling of water back to the atmosphere, and in some continental regions this can be a significant proportion of the supply of moisture for precipitation. Evaporation and transpiration also carry energy away from the surface in the form of latent heat, which would otherwise be available as sensible heat. The extent of this partitioning of the available energy affects temperature near the surface, with greater evaporation and transpiration providing a cooling influence. Moisture in the atmosphere also affects the radiation budget, either both directly through the action of water vapour as a greenhouse gas and indirectly through its influence on cloud formation.

The rates of evaporation and transpiration are highly dependent on the nature of vegetation cover. On the whole, woody vegetation may provide a greater flux of moisture to the atmosphere by capturing more water on the canopy and by accessing soil moisture to a greater depth. In warm, moist regions, the cooling effect of enhanced evaporation and transpiration by forests generally exerts a greater influence on surface temperatures than the warming by decreased surface albedo. Tropical forests therefore cool their local climate and can enhance their local rainfall.

As well as depending on vegetation type and climatic conditions, transpiration can be directly affected by the atmospheric CO_2 concentration. Stomata, the microscopic pores in leaf surfaces which allow CO_2 to diffuse in for photosynthesis, open less wide under elevated CO_2 (Field et al. 1995). This means that less water is lost by transpiration through the stomata, which may lead to a warmer surface climate and reduce moisture recycling.

Vegetation type can also influence the frictional effect of the landscape on low-level air flow. A rougher land surface increases turbulence, which affects the fluxes of heat and moisture and also modifies windspeed. Large-scale changes in surface roughness can modify synoptic-scale atmospheric circulations and hence influence weather patterns.

A change in climate is determined by the forcing (the initial perturbation) and also any feedbacks present in the system. Biophysical effects of vegetation can feature in both of these components of climate change. Feedbacks on climate change forced by the enhanced greenhouse effect can occur from changes in the coverage and distribution of different vegetation types. For example, the onset of past ice ages may have been accelerated by shrinkage of the boreal forests providing a positive feedback on cooling by increasing the surface albedo (Gallimore and Kutzbach 1996). Precipitation reductions in tropical areas could be exacerbated by decreased evaporation and transpiration due to reduced forest cover.

Vegetation may also be involved in direct anthropogenic forcings of climate change. Historical deforestation in the mid-latitudes exerted a direct cooling influence through increased surface albedo (Fig. 2.17), comparable with the climatic forcings exerted by anthropogenic aerosols and changes in greenhouse gases other than carbon dioxide and methane (IPCC 2001). Afforestation and reforestation activities intended to mitigate climate change through carbon sequestration may therefore also exert warming influences through decreased surface albedo (Betts 2000).

A further forcing involving vegetation is the effect of CO_2 on stomatal opening and transpiration. Since CO_2 can warm surface temperatures directly through this mechanism, this means that increasing CO_2 can actually exert two forcings on the climate system: (*i*) as a greenhouse gas, and (*ii*) as an influence on plant physiology (Sellers et al. 1996). This may increase the difficulty of comparing the effects of CO_2 with other greenhouse gases.

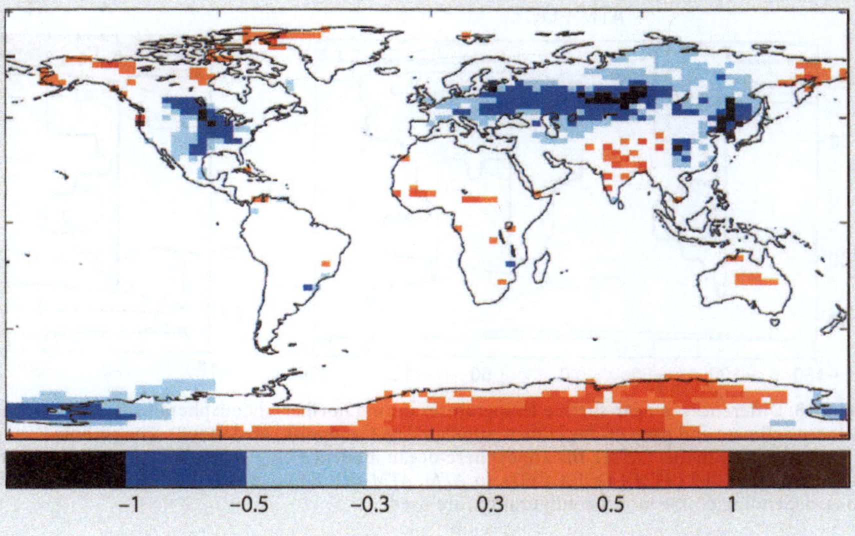

Fig. 2.17.
Simulated changes in near-surface air temperature (°C) due to past deforestation (Betts 2001). Apparent changes over Antarctica are not statistically significant

The albedo of snow-covered vegetation is much lower for tall vegetation, such as boreal forest (taiga), than for low vegetation, such as tundra. In effect, the snow lies beneath the vegetation canopy in a forest, and thus the land surface appears dark to incoming radiation. For tundra the snow completely covers the vegetation, presenting a bright surface to incoming radiation. The measured albedo differs sharply for the two vegetation types, about 0.75 for snow-covered tundra and only 0.2 to 0.4 for boreal forest. The darker forest therefore absorbs more solar energy, heating the ground more strongly than snow-covered tundra. This additional warming favours the growth and expansion of the forest, giving rise to the feedback effect.

The importance of the taiga-tundra feedback effect has been shown in the past behaviour of the Earth System. An example is the so-called biome paradox (Berger 2001). This refers to evidence for warmer winters in the early to mid-Holocene (around 9 000 to 6 000 years ago) in mid- to high latitudes in the northern hemisphere. Boreal forests (taiga) extended north of the modern tree line during that time (Prentice et al. 2000), probably as a result of the strong insolation in the northern hemisphere summer (Harrison et al. 1998). By contrast, insolation during the northern hemisphere winter was weaker than today. Thus the differences in insolation alone would have led to summers that were warmer but winters that were colder than today. However, there is good evidence (e.g. Cheddadi et al. 1997; Prentice et al. 2000; Kaplan et al. 2003) to suggest that both summers *and winters* in many regions of the northern hemisphere were warmer than today. Climate models in which only atmospheric dynamics, driven by the orbital changes, are considered cannot warm the winter climate sufficiently to allow the northward extent of the early to mid-Holocene forests (Texier et al. 1997; Harrison et al. 1998). Only by including the taiga-tundra feedback is the reconstructed winter climate adequately simulated. It is also necessary to include the synergism between the taiga-tundra feedback and the Arctic sea-ice albedo feedback to simulate the amplitude of the winter warming correctly (Ganopolski et al. 1998) (Fig. 2.18).

A second example of the biogeophysical effects of vegetation on the functioning of the Earth System again comes from the mid-Holocene era, in this case from North Africa. Palaeo-climatological reconstructions (e.g. Yu and Harrison 1996; Dupont 1993; Sarntheim 1978) and archaeological evidence (Hoelzmann et al. 2001; Gasse

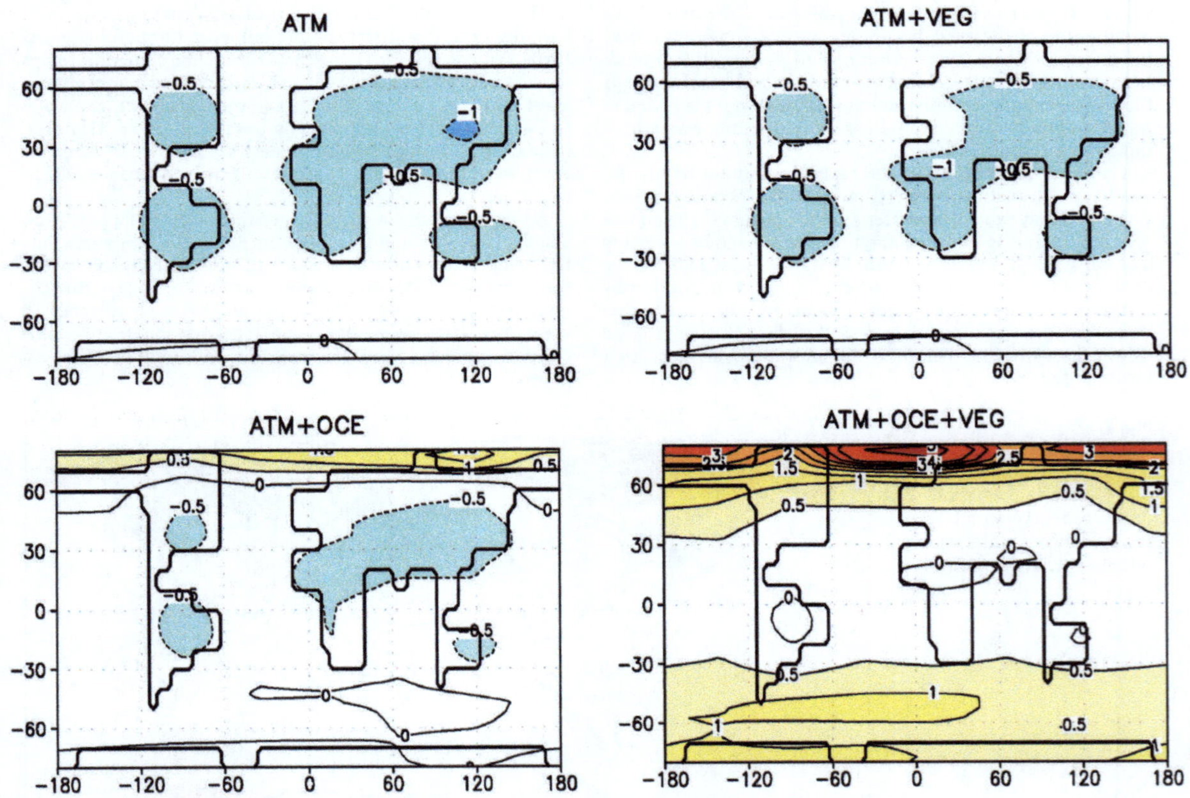

Fig. 2.18. Differences in near-surface temperatures during northern hemisphere winter (December, January, February) between mid-Holocene climate and present-day climate simulated by Ganopolski et al. (1998). The authors used different model configurations: the atmosphere-only model (*ATM*), the atmosphere-ocean model (*ATM + OCE*), the atmosphere-vegetation model (ATM + VEG) and the fully coupled model (*ATM + OCE + VEG*). In *ATM*, *ATM + OCE* and *ATM + VEG*, present-day land-surface and ocean-surface conditions, depending on the model configuration, are used

2000; Crowley and North 1991) indicate that about 6 000 years before present, the Sahel and Sahara regions were wetter than today. Fossil pollen showed that the vegetation limit between the Sahara and the Sahel was at least as far north as 23° N (e.g., Jolly et al. 1998). It has been hypothesised (Kutzbach and Guetter 1986) that the wetter climatic conditions were probably caused by changes in tropical insolation due to changes in the Earth's orbit 6 000 years ago. Numerical simulations using General Circulation Models of the atmosphere showed that these orbital changes led to an amplification of the North African and Indian summer monsoon, which increased the moisture transport into North Africa (Kutzbach and Guetter 1986). However, most simulations produced a smaller climate change than expected on the basis of palaeo-climatic evidence (Joussaume et al. 1999; Kutzbach and Street-Perrott 1985).

The discrepancy between observed and simulated responses to orbital forcing is attributed to the fact that the climate simulations consider only the changed orbital forcing and not changes of both sea-surface temperature and land-surface characteristics. Using a coupled atmosphere-ocean-vegetation model, Claussen and Gayler (1997) were able to simulate the wetter climate of the Sahara 6 000 years ago and the occurrence of vegetation much further north. Claussen and Gayler (1997) and Claussen et al. (1998) concluded that the success of the simulation is due to the inclusion of both (*i*) the interactive heat transfer between the atmosphere and the upper layers of the ocean and (*ii*) the interaction between the high albedo of Saharan sand deserts and the atmospheric circulation as hypothesised by Charney (1975). The latter also included factors related to moisture convergence and the associated convective precipitation and interactive soil moisture. Depending on initial conditions of high or low albedo in the Sahara, the coupled model yields either a desert or a vegetated Sahara. Changes in albedo are found to be more important than changes in other surface parameters. Subsequent simulations support the notion of multiple forcings and feedbacks. Comparisons of simulations by coupled atmosphere-biosphere models of the Green Sahara by de Noblet-Ducoudré et al. (2000) suggest that land-based biogeophysical feedbacks (Broström et al. 1998) along with atmosphere-ocean coupling in the African monsoon regime (Kutzbach and Liu 1997) tend to amplify climate change due to changes in orbital forcing. This amplication, leading to a northward shift of vegetation into the Sahara, in turn leads to a moister climate in this region. Braconnat et al. (1999) show that the vegetation and ocean feedbacks are synergistic.

The formation of clouds high in the troposphere may seem far removed from the metabolism of vegetation on the Earth's surface. Nonetheless even in this case biogeophysical processes may play a strong role in the functioning of the Earth System. It is likely that about half, or perhaps more, of the precipitation that falls on the Amazon Basin is recycled via evapotranspiration from the moist tropical ecosystems themselves (Marengo 2003). In this feedback system enhanced precipitation is important in maintaining the tropical ecosystems themselves. Is emitting water vapour the only role that the vegetation plays in this system? There is some evidence (Andreae et al. 2002; Artaxo et al. 1998) that the vegetation also plays a significant indirect role through the emission of volatile organic compounds. Emission rates of 3 to 15 g C m^{-2} for isoprene and monoterpenes and of 6 to 30 g C m^{-2} for total volatile organic carbon compounds have been measured in the Amazon forests (Artaxo et al. 1998, 2001). About 70–80% of these organic aerosols are water soluble, rendering them efficient as cloud condensation nuclei in the atmosphere. It is estimated that during the wet season the aerosol particle number concentration is about 300 particles cm^3 and the average water droplet size is 15–25 µm. These figures are close to those for tropical ocean environments, leading to the same type of shallow clouds that are found maritime regions (Silva Dias et al. 2002). Such clouds are efficient in producing rainfall and in the case of the Amazon Basin, provides a positive feedback to the tropical vegetation that produces the volatile organic compounds. Substantial evidence also exists to show that water droplet and cloud formation are sensitive to other forms of atmospheric aerosol particle loading and that subtle changes in the amount and type of particles can have significant impacts on the climate system (see later in Chap. 4).

The influence of biota on biophysical processes is not restricted to the land. In the oceans, planktonic coccolithophores produce millions of $CaCO_3$ scales or liths which act as tiny mirrors in the surface water. In large concentrations, these liths reflect incoming solar radiation and so influence the albedo of the water surface.

Marine biota can also have a profound effect on the composition of the atmosphere through the emission of trace gases. Some phytoplankton, including coccolithophorids, release sulphur compounds (particularly DMS – dimethylsulphide) to the atmosphere which, like their volatile organic compound counterparts on land, lead to the formation of condensation nuclei that affect cloudiness and temperature and hence influence climate processes on large scales. Coupled climate-ocean biology models suggest that even a modest change in the rate of DMS emissions, by only a factor of two, can change global mean temperature by up to one degree (Spall et al. 2001). Palaeo-studies reveal that DMS changes in emission of similar magnitude have occurred in the past (Legrand et al. 1991). Similar changes may also occur in the future with varying nitrogen and iron deposition from the atmosphere to the ocean (Turner et al. 1996).

2.3.2 Biogeochemical Processes

Biological processes are fundamental to the cycling of chemical elements on Earth. For example, almost all phytoplankton carry out photosynthesis and fix CO_2. As a by-product of photosynthesis, phytoplankton produce oxygen. This form of photosynthesis evolved in cyanobacteria about 3.5 billion years ago and, through geological time, the oxygen produced by these tiny organisms has led to the oxygen-rich atmosphere that now characterises the Earth System. So dramatic was the impact of the biota in shaping the Earth System that oxygen is now the most abundant element at the Earth's surface.

As phytoplankton remove inorganic carbon from the waters that surround them, they influence the partial pressure of CO_2. This, in turn, influences the equilibrium reaction describing the exchange of CO_2 between water and air. In this way, phytoplankton are important in the transfer of carbon from the atmosphere to the ocean. In addition, through sinking, they are important in transporting carbon from the surface to the intermediate and deep waters as well at to the ocean sediments, i.e., the largest carbon reservoirs in the Earth System. All phytoplankton contribute to this general process of transferring carbon (as well as nitrogen and phosphorus) to the deeper layers of the ocean. However, physiological processes specific to individual groups of phytoplankton can also influence elemental cycling on the global scale.

The scaly $CaCO_3$ liths produced by coccolithophorids remove carbonate from the surrounding seawater and can significantly affect local seawater carbon chemistry. In addition, liths are heavy compared to the surrounding seawater and sink quickly through the water column. They contribute to the accumulation of $CaCO_3$ at the sea floor that leads to the formation of chalk and limestone geological strata. *Emiliania huxleyi* (Fig. 2.19) is the most important coccolithophorid in terms of contributing to the transfer of carbon from surface to deeper layers. Because of the light reflection from liths, blooms of this organism are easily detected from satellites. Approximately 1.4 million km^2 of ocean surface may be influenced annually (Brown and Yoder 1994).

Other phytoplankton groups contribute to the global cycling of specific elements; diatoms, with their silica frustules, are responsible for transferring Si to the inner ocean via the so-called silica pump (Dugdale and Wilkerson 1998). While important for the geochemical cycling of Si, this process is not equally distributed over the world's oceans. It is estimated that about 75% of the deposition of biogenic silica in the world's oceans today occurs in the waters near Antarctica (Ledfield-Hoffman et al. 1986).

Cyanobacteria fix nitrogen and thus mediate the transfer of nitrogen from the atmosphere to the ocean. The primeval ocean was not rich in nitrogen and the majority of the nitrogen found in the oceans today, and upon which ocean productivity depends, originates from this biological fixation. The introduction of nitrogen into the oceans allowed the evolution of nitrifying and denitrifying bacteria. The latter catalyse the conversion of dissolved inorganic nitrogen back to gaseous nitrogen, which can be returned to the atmosphere. In this way biology regulates both the input and the removal of

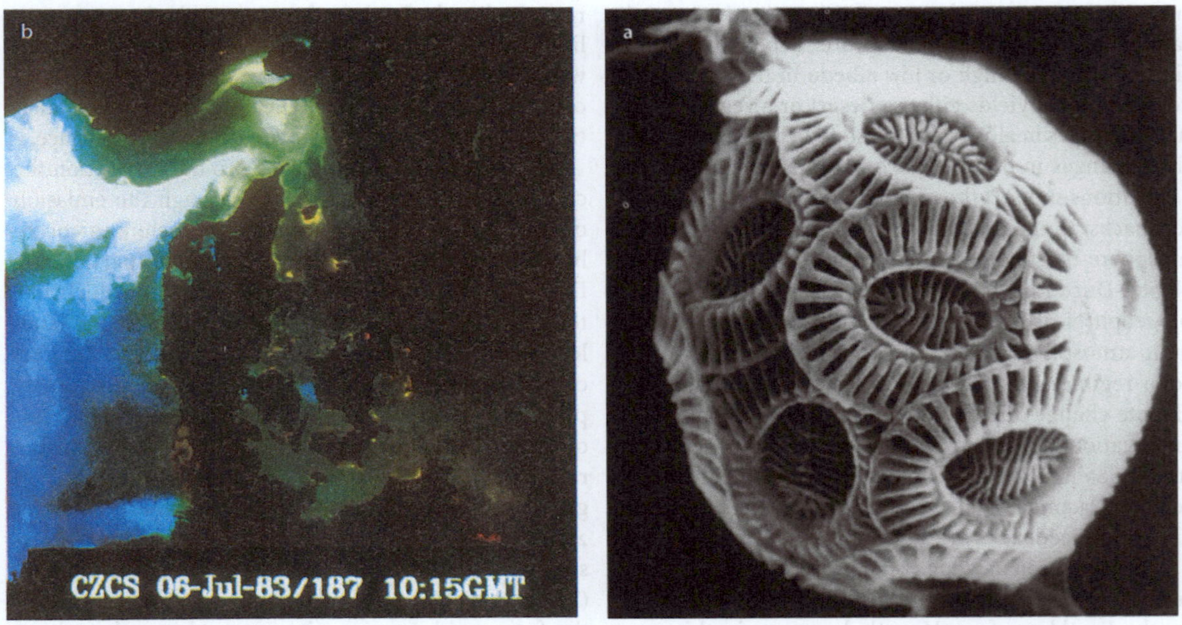

Fig. 2.19. a Electronmicrograph of *E. huxleyi* (*image:* Helge Thomsen, Danish Fisheries Research Institute) and **b** coastal zone color scanner satellite image of a Coccolithophorid bloom (*white area*) in surface waters of the northern North Sea and Skagerrak (*image:* Thorkild Aarup, IOC)

nitrogen in the open (i.e., non-coastal) ocean. The magnitude of present-day atmosphere-to-ocean transfer remains to be quantified; it may be much greater than previously acknowledged (Zehr et al. 2001).

The last decade of global change research has produced many new insights into the role of the terrestrial biosphere in the cycling of elements through the Earth System. An enormous effort has been made to better understand the physiological processes by which terrestrial ecosystems transform elements as they cycle through the Earth System. Of the many new insights that have arisen from this work, two stand out as especially important in terms of Earth System functioning: (*i*) the linkage between the important cycles of carbon, nitrogen and phosphorus; and (*ii*) the significance of indirect ecosystem responses to variability in the abiotic environment.

In describing how the Earth System as a whole works, a common approach is to describe the cycling of key elements individually through the compartments of the System – oceans, land and atmosphere. However, the cycling of one element cannot be understood in isolation from the cycling of others. The availability of one element can interact with the terrestrial biosphere in such a way as to change the availability of other nutrients. For example, most terrestrial ecosystems are limited by nitrogen, and thus the rate of nitrogen mineralisation determines the overall net primary productivity of an ecosystem. In turn, this controls the carbon uptake from the atmosphere as well as phosphorus and sulphur assimilation. When the concentration of nitrogen in tropical grassland and savanna vegetation drops below the 1% threshold, which often occurs during the dry season, then these ecosystems support a lower level of mammalian herbivory. As a result, dry biomass accumulates, which readily leads to fires and consequently to emissions of CO_2 into the atmosphere.

Conversely, the atmospheric concentration of CO_2 can influence the amount of nitrogen taken up by plants that have nitrogen-fixing symbiants in their root structure by altering the biological nitrogen fixation rate. These particular types of plants can utilise increased availability of nitrogen to increase leaf nitrogen, which leads in turn to increased photosynthetic capacity. The biological nitrogen fixation rate, however, is also limited by another element, phosphorus. The CO_2 level can also alter the amount of phosphorus available to plants through stimulation of the production of root exudates, labile carbon compounds that influence the activity of mycorrhizae associated with the plant roots, helping to make phosphorus more available to plants.

Ecosystem physiology is clearly complex (Fig. 2.20). Recent research has begun to unravel some of the indirect responses of terrestrial ecosystems to variability in the abiotic world around them. In some cases, these indirect effects may be more significant for ecosystem functioning than the more obvious direct ones. A good example is the interplay between increasing atmospheric CO_2 concentration, the productivity of the terrestrial biosphere, and the hydrological cycle (see Mooney et al.

Fig. 2.20.
Initial paradigm for ecophysiological research in the Global Change and Terrestrial Ecosystems (GCTE) core project of the IGBP (Steffen et al. 1992)

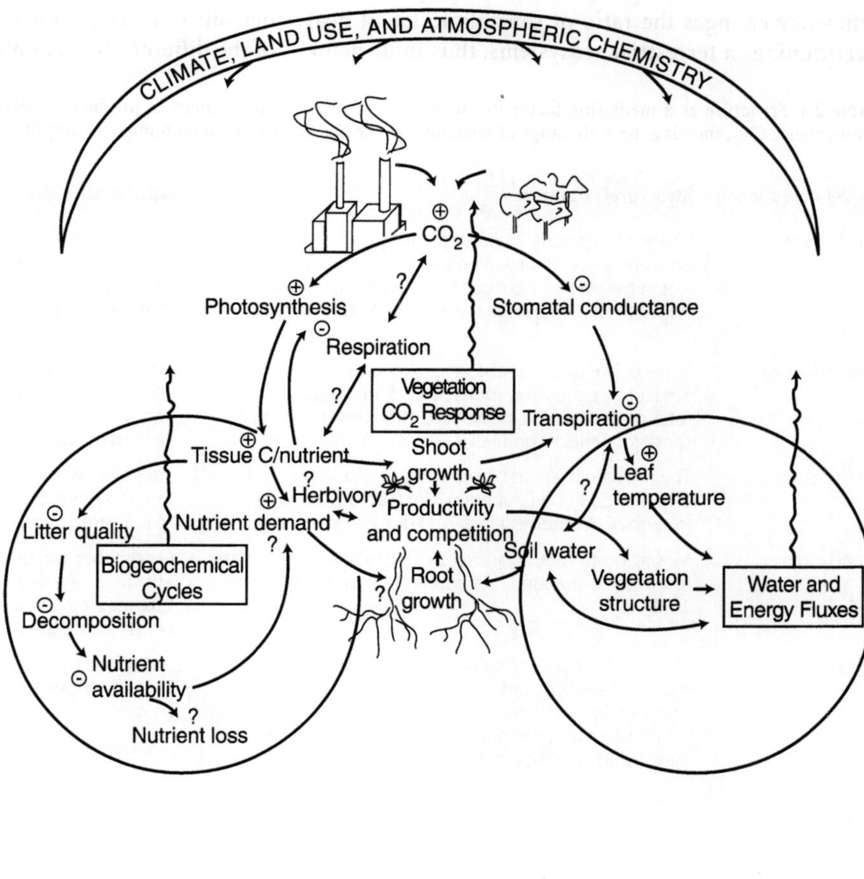

1999 for a review). Atmospheric CO_2 has varied by about 100 ppm between glacial and interglacial states (Fig. 1.3), and terrestrial ecosystems have obviously responded to such changes. The first-order response is the direct effect of increased CO_2 in the interglacial periods on photosynthesis and hence on the productivity of ecosystems. However, experiments exploring the responses of terrestrial ecosystems to a range of CO_2 concentrations show complex behaviour. In addition to the direct increase in productivity, indirect effects acting via changes in the interaction between ecosystems and the hydrological cycle are also important.

Increasing CO_2 concentration causes the stomates on leaves to be open for shorter periods of time; plants can obtain the CO_2 they need more readily. Open stomates not only allow CO_2 from the atmosphere to diffuse into the plant's interior to be transformed by the photosynthetic machinery, but they also allow water vapour from the plant's interior to escape to the atmosphere. The amount of carbohydrate produced per amount of water defines *water use efficiency*. It is to a plant's benefit to maximise this ratio. An indirect effect of changing atmospheric CO_2 concentration is to change the water use efficiency, which is especially important for semi-arid vegetation. Increasing water use efficiency then leads to increased soil moisture, which in turns enhances plant productivity. In many regions of the world, this indirect effect may be comparable to the direct CO_2 enhancement of productivity or perhaps may even surpass it (Owensby et al. 1996; Hungate et al. 1997). In addition, from an Earth System perspective, changing water use efficiency changes the ratio of sensible to latent heat partitioning in terrestrial ecosystems, thus influencing biogeophysical feedbacks to the Earth System (cf. Sect. 2.3.1).

Much global change research in the terrestrial compartment of the Earth System has been, and continues to be, focused on ecosystem physiology. However, the importance for Earth System functioning of the *structure* (composition and spatial pattern) of the terrestrial biosphere, not only its physiology, has become clearer during the past decade. Ten or 15 years ago most treatments of the terrestrial biosphere in GCMs assumed for the entire Earth a static vegetation with a single set of generic physiological properties – the so-called big leaf or green slime models. This treatment was based on an assumption that the effects of structure on physiology were simple and could be readily scaled from a leaf to the globe. However, structure is now known to be a strong mediating factor in the response of terrestrial ecosystems to changes in abiotic forcing functions at every scale (Table 2.1). In fact, in terms of the role of the terrestrial biosphere in the changing functioning of the Earth System, the different physiology of new structural assemblages of vegetation compared to those that they replaced will be at least as important as direct changes in the physiology of existing assemblages (Mooney et al. 1999).

Most simulations of vegetation dynamics at the global scale are now explicitly based on the functional type approach (Smith et al. 1997; Woodward and Cramer 1996), in which vegetation is classified by both structural and functional characteristics. For example, forests are classified into broad-leafed and needle-leafed and into evergreen and deciduous types. Grasses are often classified into C_3 and C_4 types of photosynthetic pathway. Each of the different types of plant has its own characteristic

Table 2.1. Structure as a mediating factor in the ecosystem response to changes in production-related processes under a doubling of atmospheric CO_2, showing the wide range of structural processes that can either attenuate or amplify the process response (Shugart 1997)

Level of response	Structural change	Functional implication
Leaf tissue	Change in stomatal index of plants grown in different CO_2 environments. Effect can be seen in plant material collected before the Industrial Revolution and can be induced under laboratory conditions (Woodward 1987)	Alteration of the stomatal conduction response of the plant. Implication that plant responses historically may be different from present responses (due to change in stomatal index induced by ambient atmospheric conditions)
Individual plant	Rates of leaf photosynthesis can increase on the order of 50% for C_3 plants in response to a doubling of CO_2 but rates of whole plant growth are often less than 20% of those for control conditions (Körner 1993)	Structural considerations including photosynthate allocation and internal interactions (e.g. with nutrients such as nitrogen) can moderate the carbon fixation at the leaf level
Plant stand	The increase in stand biomass is less by a factor of about 0.30 (Shugart and Emanuel 1985) than the increase in growth of the individual plants comprising the stand	Stand interactions (competition, shading, etc.) reduce the stand level biomass increase (or yield) in response to increased growth rates of individual plants
Landscape	Mosaic properties of landscapes alter the stand biomass response to changed conditions (Bormann and Likens 1979)	Landscapes can be thought of as mosaics in different states of recovery from natural disturbances. Changes in plant and stand process are mediated by the local state of disturbance recovery
Region	The terrestrial surface can alternate between being a source to a sink of carbon in the transient response to environmental change – even in cases in which the long-term response to change is similar to the initial condition (Smith and Shugart 1993)	Shifts in vegetation in response to change are delayed by large-scale processes involving dispersal, recovery and other inertial effects

physiological parameters, and combinations of functional types give rise to vegetation assemblages that have integrated sets of biogeophysical parameters (e.g., albedo, roughness) that can interact with the physical climate system. Such dynamic global vegetation models are now being used to simulate changes in vegetation structure and functioning due to climate and atmospheric changes (Cramer et al. 2001) (Fig. 2.21) and are being dynamically coupled to atmosphere-ocean general circulation models to form Earth System simula-

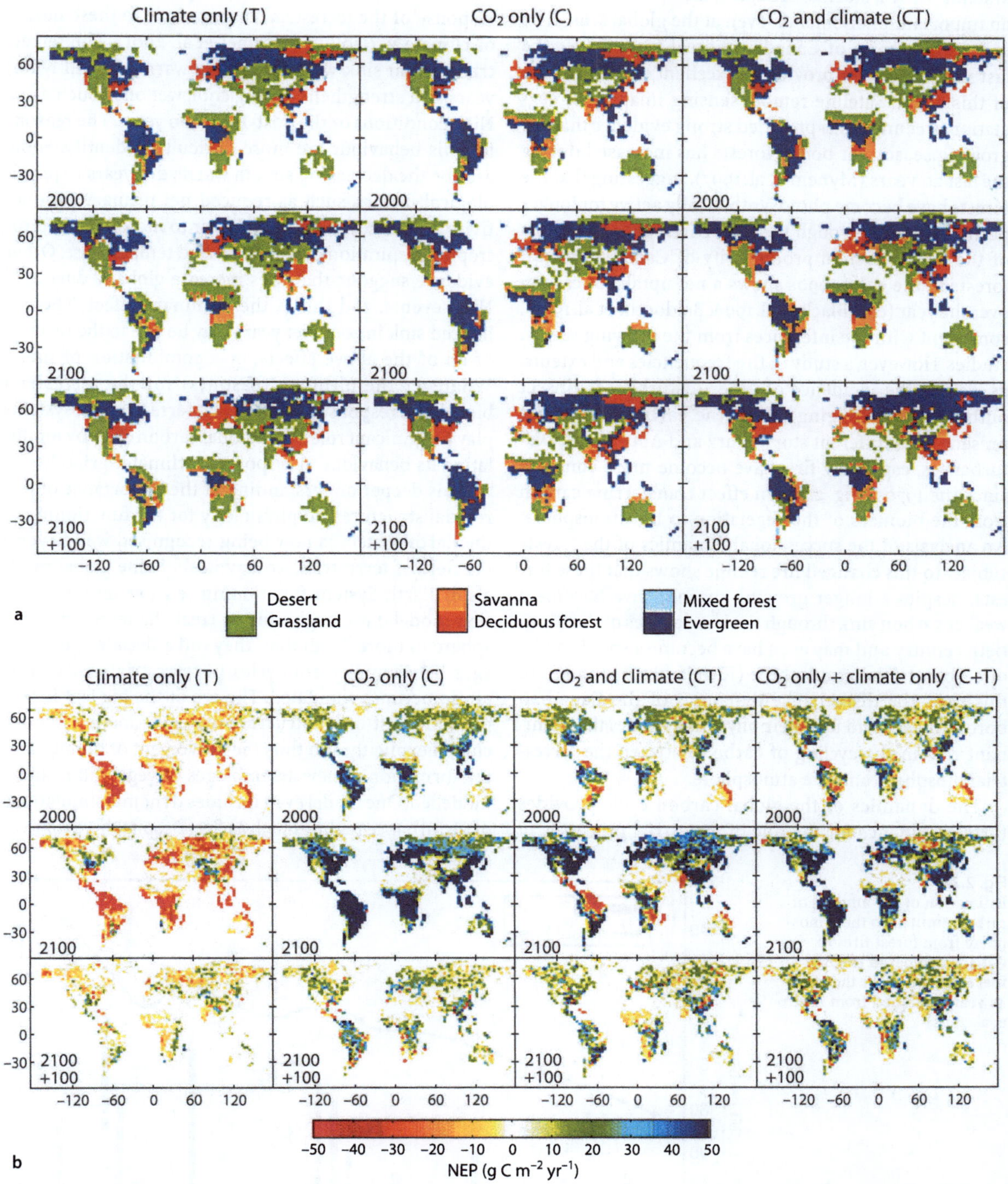

Fig. 2.21. Results of three model experiments based on the averaged output of six dynamic global vegetation models driven by dynamic CO_2 and climate scenarios from 1861 to 2099 with a 100-year stabilisation period (no further changes in CO_2 and climate) beyond 2099 (Cramer et al. 2001): **a** simulated vegetation distribution at the end of the twentieth and the twenty-first centuries and after 100 additional years of stabilised CO_2 and climate; and **b** model average net ecosystem productivity (NEP) at 2000, 2100 and the end of the stabilisation period (2100 + 100) from the three model experiments (mean of last 10 years in each century). The final column shows the sum of NEP from the climate-only and CO_2-only experiments

tors that begin to capture the potentially significant effects of terrestrial ecosystem changes on Earth System functioning (Cox et al. 2000).

Another significant insight from the last decade is the role of the disturbance dynamics of terrestrial ecosystems – fires, storms, droughts, insect outbreaks – as an important phenomenon even at the global scale. The carbon dynamics of Canadian boreal forests over the last several decades provide an excellent demonstration of this effect. Satellite remote sensing imagery of vegetation greenness has provided strong evidence that the growing season for boreal forests has increased during the last 20 years (Myneni et al. 1997), suggesting that the forests have become photosynthetically active for longer periods of time through the year. Direct measurements of the net ecosystem productivity of Canadian boreal forests in the early 1990s shows a net uptake of carbon over the year (e.g., Black et al. 1996; Baldocchi et al. 1997), consistent with the inferences from the growing season studies. However, a study of fire frequencies and extents, as well as the incidence of other disturbances (insect outbreaks and logging), from the early 1990s to the present tell a different story (Kurz and Apps 1999). Disturbances, especially fire, have become more common since the 1970s (Fig. 2.22), in effect transferring carbon from the biomass of the vegetation to the atmosphere. An analysis of the successional dynamics of the forests subject to this changed fire regime shows that these forests, despite a longer growing season, have become a weaker carbon sink through the last decades of the twentieth century and may even have become a small source of carbon to the atmosphere (Kurz and Apps 1999). In this case, the disturbance dynamics of the Canadian boreal forests, and not their physiology, plays the dominant role in the cycling of carbon between the terrestrial biosphere and the atmosphere.

The dynamics of the global carbon cycle provides further evidence for the role of terrestrial processes in Earth System functioning, particularly through the relationship between the inter-annual variation in the atmospheric CO_2 growth rate and major modes of natural variability in the Earth System, such as ENSO or the eruption of Mt Pinatubo (Fig. 2.23). First, there is good evidence that the variation in CO_2 is indeed due to the response of the terrestrial carbon cycle to these modes of climate variability (Schimel et al. 2001). The terrestrial carbon sink weakens during warm, dry (El Niño) years and strengthens during cool, wet ones, such as La Niña conditions or the post-Pinatubo years. The reasons for this behaviour are more difficult to identify. Some ascribe the decreasing sink in warm, dry years to physiological factors, such as reduced net primary productivity due to extensive droughts or to increasing heterotrophic respiration due to increased temperature. Other evidence suggests that fires increase globally during El Niño events, and this is the dominant effect. The enhanced sink in cool, wet years may be due to the reversal of all of the above effects, or a combination of them. Whatever the ultimate reason(s) for the terrestrial biospheric response, it is clear that terrestrial ecosystems play a significant role in the global carbon cycle by modulating its behaviour in response to climate variability.

This deeper understanding of the importance of terrestrial structure and physiology for the functioning of the Earth System is now being recognised well beyond the field of terrestrial ecology itself. Some global models of Earth System functioning (e.g., general circulation models) are beginning to treat the terrestrial biosphere in more detail than they did a decade ago. Linkages between element cycles in terrestrial ecosystems are sometimes simulated. The era of the big leaf is well past; the first-order structure of vegetation is often included explicitly and thus the movement of biomes and the formation of new assemblages of vegetation can be modelled. One model even includes dynamic simulations of disturbance regimes such as fire. Despite this progress,

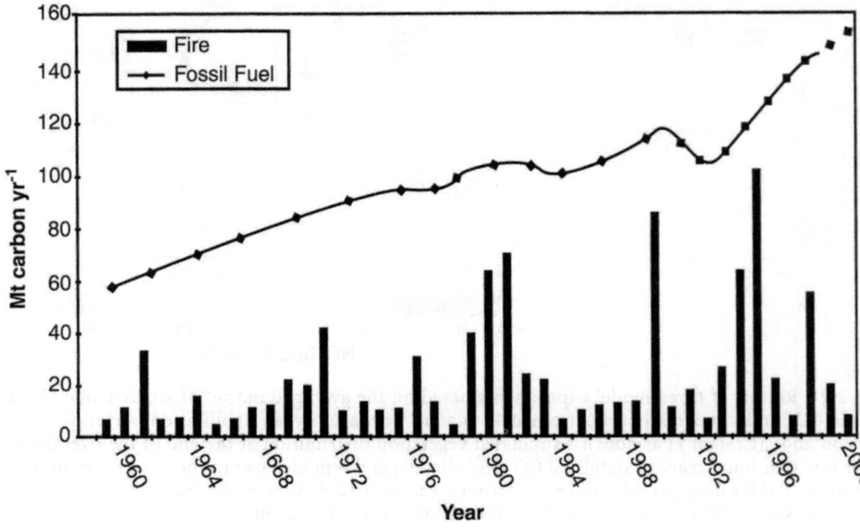

Fig. 2.22.
Estimation of the amount of carbon emitted to the atmosphere from forest fires in Canada compared to fossil fuel emissions over the last 40 years (adapted from Amiro et al. 2003)

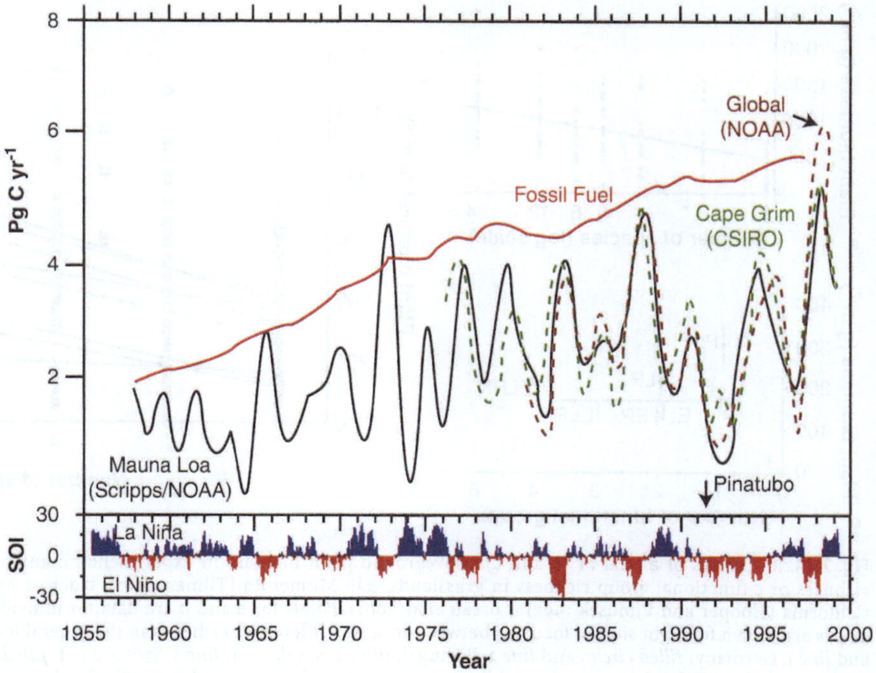

Fig. 2.23. Variability in annual CO_2 growth rate, estimated globally and measured at two observing stations, and global fossil fuel emissions. Also shown is the Southern Oscillation Index (SOI), an indicator of ENSO events (R. J. Francey, pers. comm.)

much is yet to be done to incorporate terrestrial dynamics more realistically in such global models. However, there is one fundamental question about the role of both the terrestrial and marine biospheres in the functioning of the Earth System that is emerging as the most significant for the coming decade at least. How important is biological diversity to the functioning of the Earth System? Do species matter? How much and what types of biological diversity/complexity can be lost before Earth System functioning is impaired?

2.3.3 The Role of Biodiversity

Over the past decade research has begun to unravel some features of the relationship between biological diversity and ecosystem functioning. Experiments that manipulate the biodiversity of terrestrial ecosystems and study the consequences for ecosystem functioning are already showing significant effects. To date much of the experimentation has occurred on temperate grasslands and the primary ecosystem functions studied are primary production and nutrient cycling. Figure 2.24 (Loreau et al. 2001) summarises the main findings, and shows that in general an increase in the number of species in the system increases the plant biomass of the system. Two possible explanations for this effect have been put forward. First, the effect may be due to localised deterministic processes, such as niche differentiation and facilitation, whereby having more species means that more ecological niches are filled (e.g., by species which are able to exploit nutrient poor environments). In this way more synergistic relationships between species (e.g., symbiotic nitrogen fixation) are possible. An alternative explanation is that increasing the number of species in a system simply means that there is a higher probability that a dominant, highly productive species is included in the assemblage (Huston 1997). In reality, these two explanations are at opposite ends of a continuum, and the ultimate explanation likely includes elements of both. Evidence is also emerging that in marine ecosystems the number of species present influences ecosystem functioning. Nutrient turnover rates have been shown to be a function of biodiversity (Emmerson et al. 2001).

Biodiversity may play another role in ecosystems by acting as a buffer or providing insurance against rapid changes in the abiotic environment (Yachi and Loreau 1999). That is, more complex ecosystems may be more resilient or resistant to disturbances or extreme events than simpler systems. For example, species of termites that appear to be completely unimportant in processing litter in semi-arid ecosystems under normal conditions may become very important after a fire when the population of the dominant species of termites is severely reduced. There is some circumstantial evidence for the insurance hypothesis from experimental studies that show decreases in the variability of ecosystem processes with increasing species richness of the system (Fig. 2.25). However, many important questions, such as whether the stabilising effect of increasing diversity saturates at some level, remain to be addressed.

Nearly all of the diversity-functioning experiments and observational studies carried out so far have focussed on temperate grassland ecosystems at small spatial scales. Difficult questions are associated with attempts to scale up this understanding. Do similar principles hold for

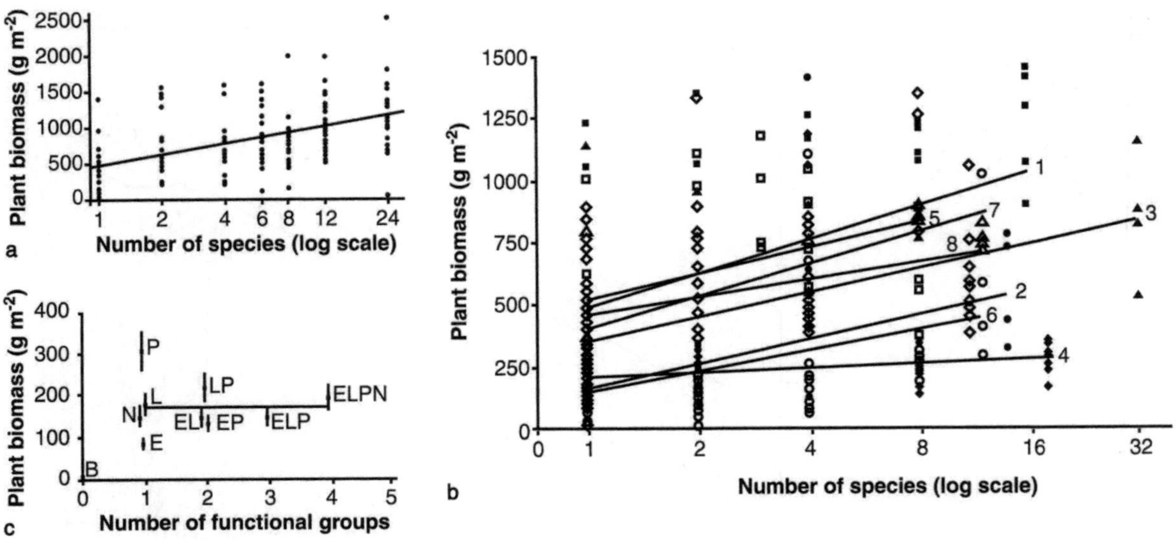

Fig. 2.24. Responses of **a** total or (**b** and **c**) aboveground plant biomass to experimental manipulations of (**a** and **b**) plant species richness or **c** functional group richness in grasslands: **a** in Minnesota (Tilman 2001); **b** across Europe (Hector et al. 1999) and **c** in California (Hooper and Vitousek 1997) (Loreau et al. 2001). *Points* for **a** and **b** are data for individual plots. In **b** different regression slopes are shown for eight sites to focus on between-location differences rather than the general log-linear relationship. *Filled squares* and *line 1*, Germany; *filled circles* and *line 2*, Portugal; *filled triangles* and *line 3*, Switzerland; *filled diamonds* and *line 4*, Greece; *open squares* and *line 5*, Ireland; *open circles* and *line 6*, Sweden; *open diamonds* and *line 7*, Sheffield (UK); *open diamonds* and *line 8*, Silwood Park (UK). Symbols in **c** correspond to functional groups and their combinations: *B*, bare ground; *E*, early-season annuals; *L*, late-season annuals; *P*, perennial bunchgrasses; *N*, nitrogen fixers

other ecosystems, such as forests? How does the diversity-functioning relationship operate at different trophic levels? The latter is exceptionally important in the Earth System context, as small organisms such as viruses, bacteria, archaea, protists and microarthropods drive the bulk of the chemical transformations by which terrestrial ecosystems play a key role in global biogoechemical cycles. In terms of extrapolation to larger spatial scales, an important issue is the effect of landscape fragmentation on the diversity-functioning relationship. More specifically, fragmentation of landscapes may diminish the ability of dominant or key species to migrate and thus be recruited into local ecosystems where they may be essential following a disturbance.

The nexus between diversity and functioning is just as important for marine ecosystems. Differences in sizes, shapes and taxonomic affinity of the organisms that comprise the ocean's vegetation are critical to the cycling of a number of elements in the sea and, ultimately, in the Earth System as a whole. The vegetation of the open ocean is comprised primarily of microscopic plants (phytoplankton). Because of their small size (about 0.2 to 2 mm in length), there is a temptation to consider them as one functional group. Indeed, in many modelling exercises, there is a tendency to treat phytoplankton as a single homogeneous group. In fact, they differ greatly in size, chemical composition and activity and these differences can profoundly influence the functioning of the Earth System. For example, the relative volume difference between the smallest and the largest phytoplankton is at least as great as the relative volume difference between a mouse and an elephant. These, relatively speaking, large size differences between phytoplankton influence both their sinking rates and their susceptibility to different predators. Thus, the probability of sinking to the deeper layers of the ocean differs for different phytoplankton species. Once in the deeper layers of the ocean, the elements contained in the sinking phytoplankton may be out of contact with and unable to react with the atmosphere for centuries or longer, depending on where in the ocean this sinking takes place.

The geological record indicates that *E. huxleyi* has been the most important coccolithophorid in terms of lith transfer to the sediment throughout the Holocene. However, prior to the last glacial period, another coccolithophorid, *Gephyrocapsa*, dominated (Verity and Smetacek 1996). *Gephyrocapsa* is still found today but plays an insignificant role in terms of carbon transfer. No one knows what conditions led to *E. huxleyi* overtaking *Gephyrocapsa* as the dominating bloom former within this taxonomic group. However, the example illustrates again the importance of biodiversity in the Earth System, as a mechanism to respond to change. In this case, it illustrates the importance of functional redundancy, or the importance of having more than one species in a particular functional type (a species or group of species that carry out the same functional role in an ecosystem). Had *Gephyrocapsa* been the only species in its functional type, carbon transfer to the deep ocean via the biological pump would have been significantly impaired with the decline of the species.

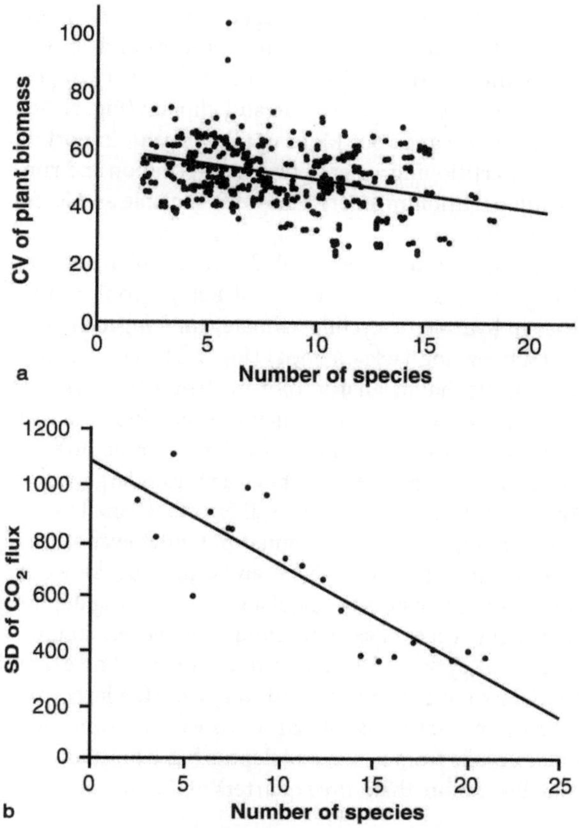

Fig. 2.25. Observed decreases in variability of ecosystem processes as species richness increases (Loreau et al. 2001): **a** adjusted coefficient of temporal variation (*CV*) of annual total plant biomass (g m^{-2}) over 11 years for plots differing in number of species in experimental and natural grasslands in Minnesota (Tilman 1999) (the correlation of variation in soil nitrogen with species richness in these plots precludes the interpretation of increased stability as a pure diversity effect (Huston 1997), although the diversity effect remained significant even after controlling for potential confounding variables (Tilman et al. 1996)); **b** standard deviation (*SD*) of CO_2 flux (in microlitres per 18 hours) from microbial microcosms (McGrady-Steed et al. 1997) (in these data temporal variability in response to diversity is confounded with between-replicate variability)

In addition to differences in functioning due to size, phytoplankton species in general play various roles in many biogeochemical and physical processes; almost all fix CO_2, some producing large amounts of $CaCO_3$ in the process; others are important in the silicon pump (see Box 2.4); still others are important in the oceanic nitrogen cycle. In addition, although not yet well studied, it is clear that other potentially reactive gases are also produced by various phytoplankton (see Sect. 2.4.3). Thus, different phytoplankton species influence global biogeochemical cycling in different ways and the distribution of vegetation in the oceans is as important for the functioning of the Earth System as is diversity in the terrestrial biota.

In its broadest sense this evolving understanding of the role of biological diversity in the functioning of ecosystems represents a significant shift in the way that the functioning of the Earth System itself is conceptualised.

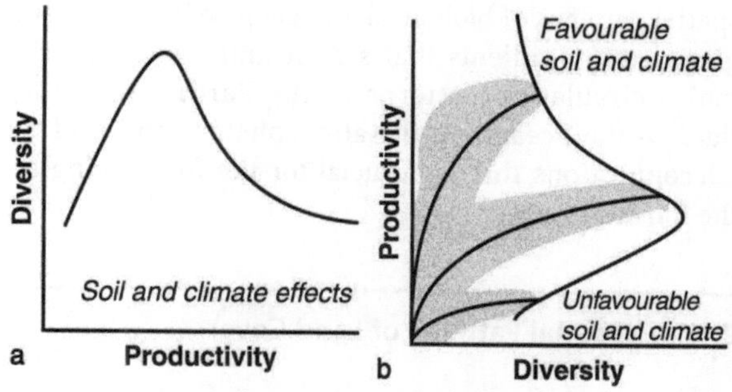

Fig. 2.26. a Hypothesised relationship between diversity-productivity patterns driven by environmental conditions across sites; and **b** the local effect of species diversity on productivity (Loreau et al. 2001). In **a** comparative data often indicate a unimodal relationship between diversity and productivity driven by changes in environmental conditions. That is, the environment drives changes in the productivity of terrestrial ecosystems, which in turn affects the biological diversity of the system. In **b** experimental variation of species richness under three specific sets of environmental conditions (*thin lines*) show increasing productivity with increasing diversity for all environmental conditions. The experiments also produce a pattern of decreasing between-replicate variance and increasing mean response with increasing diversity, as indicated by the *shaded areas* (scatter of responses) around the *thin, curved regression lines*

Until the last decade or two the dominant paradigm has been that the variability of the abiotic environment drives the processes and patterns of the biosphere in a largely one-way fashion. For example, changes in climate and soils drive changes in the productivity of terrestrial ecosystems, which in turn determines the biological diversity of the system (Fig. 2.26a). However, today it is increasingly recognised that changes in the biological diversity of ecosystems can change their functioning in critical ways (Fig. 2.26b), which in turn feeds back to the abiotic environment, as described in the previous two sections. In reality, biological diversity, ecosystem processes, and the abiotic environment are all bound together as a complex system with a web of forcings and feedbacks that give rise to the behaviour of the system as a whole. Biological diversity is not simply the one-way result of abiotic forcings; species matter for the functioning of ecosystems and for their feedbacks to the abioitic environment.

2.4 New Insights into Spatial Variability of the Earth System

Interactions between biota and geochemical and geophysical processes vary and ultimately result in *patterns* of productivity, trace gas exchange, climate, chemistry, and many other processes that vary spatially over the Earth's terrestrial and marine surfaces. The nature of these spatial patterns themselves give further insights into the processes which give rise to them, often leading to enhanced understanding of the interactions among processes. In addition, the spatial heterogeneity of the Earth's physiography coupled with the spatial patterns of biological and geophysical processes give rise to gradients that significantly influence the major circulation patterns of the Earth's two great fluids – the ocean and the atmosphere – and lead to teleconnections that are crucial for the functioning of the Earth System.

2.4.1 Spatial Patterns of Land Cover

The nature of land cover influences the functioning of the Earth System in several ways, as has been described in Sect. 2.2.1 on biogeophysical processes. In addition to the importance of the processes themselves, the spatial pattern of land cover influences the operation of the biogeophysical feedbacks. Figure 2.27 shows how various patterns of bare soil and vegetation in an idealised landscape can influence regional precipitation patterns. The relative amounts of vegetated and bare land cover are important for the regional climate, but also different arrangements of the land cover, in the same overall ratio, produce significantly different rainfall patterns. This effect may become particularly important when the outputs of global climate models are downscaled in attempts to simulate precipitation patterns at finer scales.

In addition to the type of vegetation cover itself, the nature of the vegetation's rooting patterns and activity and characteristics of the soil substrate are very important for the hydrological cycle and climate (Fig. 2.28c). Soil moisture and root water uptake play an important role in partitioning water between evaporation and runoff and partitioning energy between sensible and latent heat fluxes.

Recent analyses have revealed several global patterns that give insights into the role of below-ground processes in hydrologic cycling, climate, and biogeochemistry (Schenk and Jackson 2002) (Fig. 2.28). These analyses indicate that mean 95% rooting depths (the depth of soil which contains 95% of a plant's roots) increase with decreasing latitude from 80° to 30° but showed no clear trend in the tropics, where they vary from large values (deep-rooted systems) in tropical forests to small values in other tropical biomes. Annual potential evapotranspiration, annual precipitation, and length of the warm season are all positively correlated with rooting depths, suggesting a feedback loop among the vegetation, hydrological cycle and local/regional climate. For example, deep roots in tropical forests play a particularly critical role in energy partitioning. In eastern Amazonia water uptake from 2–8 m soil depths has been shown to contribute more than three quarters of the transpiration

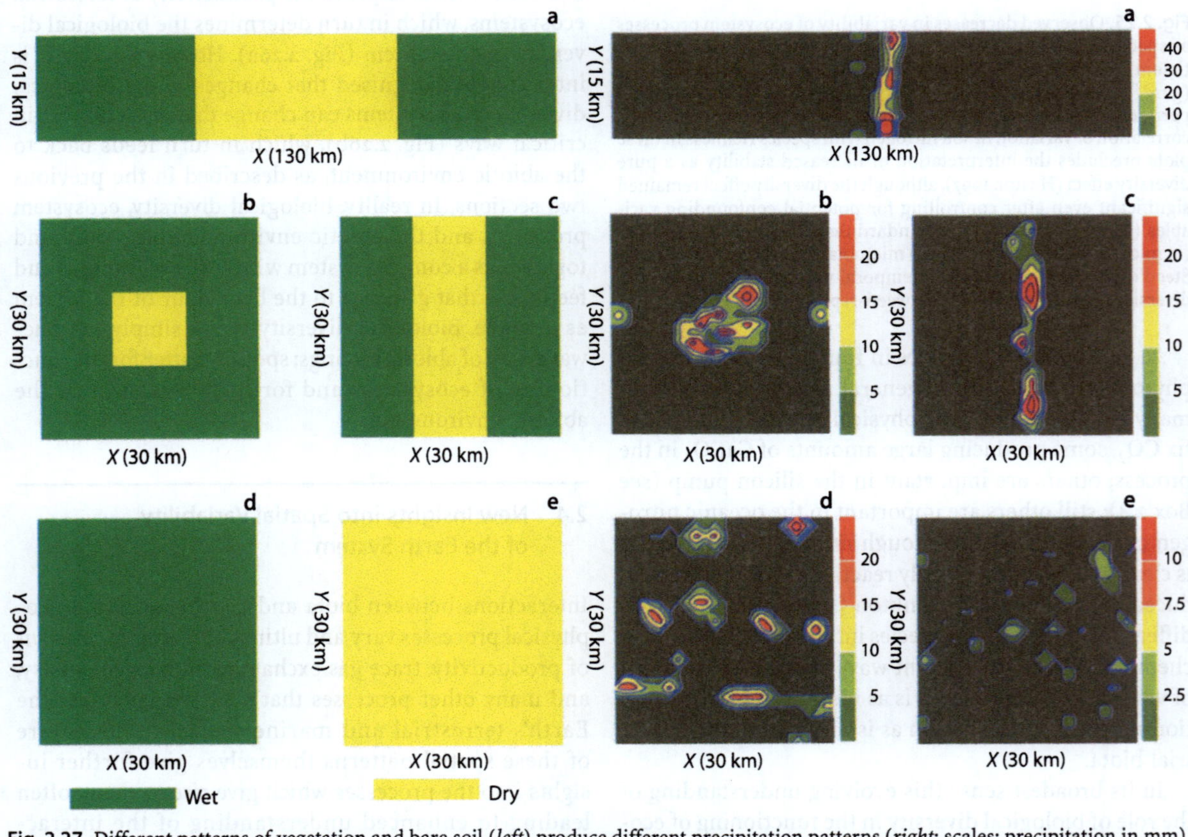

Fig. 2.27. Different patterns of vegetation and bare soil (*left*) produce different precipitation patterns (*right*; scales: precipitation in mm) in these idealised landscapes, based on a model simulation of the relative contribution of turbulence and mesoscale (<200 km) circulation on clouds and precipitation in homogeneous and heterogeneous landscapes (adapted from Avissar and Liu 1996)

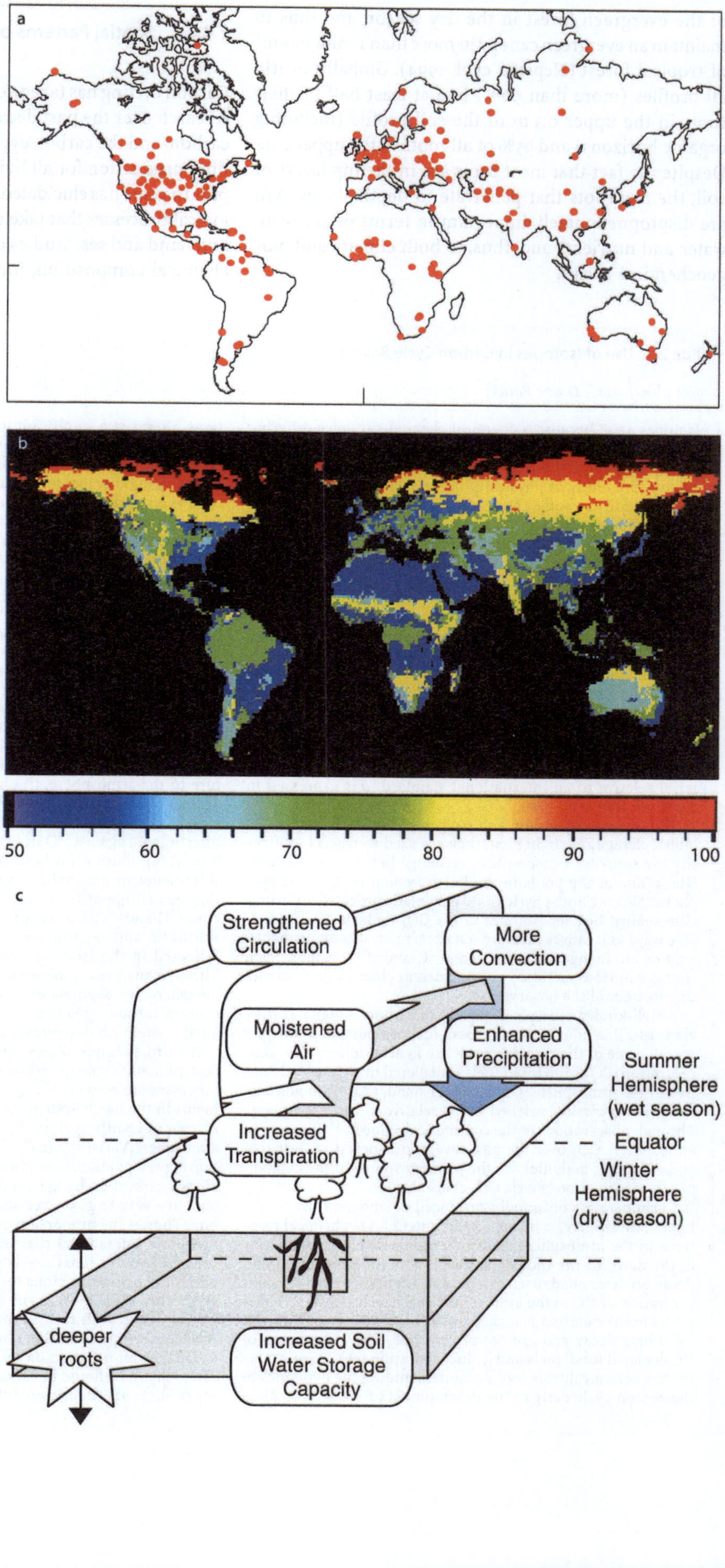

Fig. 2.28.
a Location of soil profiles where rooting depth measurements have been made. **b** Percentage of root biomass in the upper 30 cm of soil (Jackson et al. 1996). **c** Diagrammatic representation of the role of deep roots in the climate system (Kleidon et al. 2000)

of the evergreen forest in the dry season and thus to maintain an evergreen canopy in more than 1 million km^2 of tropical forest (Nepstad et al. 1994). Globally, nearly all profiles (more than 90%) had at least half of their roots in the upper 0.3 m of the soil profile (including organic horizons) and 95% of all roots in the upper 2 m. Despite the fact that most roots are in the top layers of soil, the few roots that penetrate to depths below 2 m are disproportionately important in terms of access to water and nutrients and, thus, to both climate and biogeochemical cycles.

2.4.2 Spatial Patterns of Carbon Sources and Sinks

Carbon cycling has taken centre stage in biogeochemical research over the past decade. Life on Earth is based on carbon, and the carbon cycle is the key to food, fuel, and fibre production for all living things. Research over the past 10 years has elucidated the spatial dimension of ecological processes that take up, store, and emit carbon on both land and sea. Studies of the pattern of atmospheric chemical composition, including isotopes (Box 2.3) of

Box 2.3. Use of Isotopes in Carbon Cycle Research

Jim Ehleringer · Diane Pataki

Isotopes arise because a chemical element can have differing numbers of neutrons in the nucleus of the atom. The atoms are identical chemically but they have different mass and thus can behave differently in physical processes where mass is important. In addition, some (but not all) isotopes are unstable and decay radioactively over time, with a characteristic half-life, into a stable isotope. In CO_2, both carbon and oxygen have multiple isotopes that are useful for global change studies. Carbon has three isotopes of importance. The most common carbon isotope is ^{12}C, with six protons and six neutrons in the nucleus. Carbon-13 is a naturally occurring isotope with one extra neutron; it is also a stable isotope and represents about 1.1% of all carbon in the Earth System. Carbon-14 is also naturally occurring, but it is rare and decays radioactively. The two most abundant oxygen isotopes, oxygen-16 and oxygen-18, are both stable, with the heavier oxygen-18 representing about 0.2% of all oxygen in the Earth System. Isotope abundances are typically presented in δ notation, which describes the ratio of heavy to light isotopes (e.g., $^{13}C/^{12}C$ and $^{18}O/^{16}O$) in a compound or material relative to an international standard. δ is expressed in parts per thousand (‰).

The isotopes of carbon and oxygen have two major uses in global change research. First, they are used as tracers to identify the sources of carbon that exchange between or accumulate in one of the pools in the Earth System (e.g., the atmosphere). Second, both physical and enzymatic processes discriminate against heavier isotopes in the CO_2 molecule in predictable ways as it moves into and out of terrestrial compartments, leading to changes in observed $\delta^{13}C$ and $\delta^{18}O$ values. Such changes in isotopic ratios give important clues as to processes that lie behind the observations.

A well-known example of the use of carbon isotopes as tracers is as a line of evidence for fossil fuel combustion as the primary source of the contemporary rise in atmospheric CO_2 concentration. CO_2 from fossil fuels is depleted in ^{13}C and ^{14}C relative to the atmosphere. On the other hand, CO_2 from volcanic activity is typically enriched in ^{13}C relative to the atmosphere. Through observations of the continued decline in ^{13}C content of atmospheric CO_2 over the past several decades, it is clear that anthropogenic activities are the primary source of the contemporary rise in atmospheric CO_2 concentration.

Carbon-14 is continually produced cosmogenically in the upper atmosphere, is incorporated into CO_2 via chemical reactions in the atmosphere, and then radioactively decays. Prior to the dawn of the Industrial Revolution, the production and decay processes had reached a near steady state so that the concentration of ^{14}C in the atmosphere was nearly constant. However, once the carbon is locked away from the atmosphere, the ^{14}C atoms decay and are not replenished. The combustion of ^{14}C depleted fossil fuels and its incorporation into organic matter has been a valuable tool for understanding the dynamics of the carbon cycle early in the development of modern civilizations. However, a significant increase in aboveground nuclear weapons testing in the late 1950s and early 1960s sharply increased the ^{14}C concentration in the atmosphere, inadvertently providing another tool for tracing the carbon cycle. The ^{14}C concentration peaked at double the steady-state level in 1963 with the implementation of test-ban treaties but has since decayed exponentially as ^{14}C is rapidly incorporated into organic carbon in both oceans and terrestrial reservoirs. Atmospheric $^{14}CO_2$ currently remains about 10% above its pre-testing levels. This so-called bomb ^{14}C signal has been used to constrain global air-sea CO_2 exchange and to track the carbon once it is in the ocean, thereby allowing evaluation of ocean circulation models; it is also used to trace the incorporation, storage, and turnover of carbon in plants and soils.

Photosynthesis and respiration processes in terrestrial ecosystems contribute to large annual fluctuations in the $\delta^{13}C$ and $\delta^{18}O$ values of atmospheric CO_2. Plants with C_3 photosynthesis discriminate against ^{13}C by about 19‰ during the initial steps of photosynthesis, imparting its characteristic isotopic signature to the atmosphere. C_4 photosynthesis discriminates less against ^{13}C but still influences atmospheric isotopic composition. At the same time, photosynthetic activities on land discriminate against ^{18}O in CO_2. Therefore photosynthetic carbon accumulation on land tends to increase the $\delta^{13}C$ and $\delta^{18}O$ values of atmospheric CO_2. On the other hand, respiration has an opposite effect, releasing CO_2 that is depleted in both ^{13}C and ^{18}O. Changes in the relative activities of photosynthesis and respiration over the course of the year are reflected in the isotopic composition of atmospheric CO_2. Through analyses of the isotopic composition of CO_2 in the atmosphere, the separate effects of photosynthesis and respiration on carbon dynamics in the terrestrial component of the Earth System can be unravelled.

A third example shows how knowledge of photosynthetic discrimination among carbon isotopes gives insights into which processes are responsible for fluxes of carbon across compartments in the Earth System. As noted above, C_3 plants discriminate significantly against ^{13}C in their fixation of CO_2 from the atmosphere. Variations in C_3 discrimination appear to be a function of precipitation. Both photosynthesis in C_4 plants and ocean air-sea carbon exchange, on the other hand, do not discriminate to a very large degree against the heavier isotopes of carbon. Thus, if the proportion of C_3 and C_4 plants is known, seasonal and interannual changes in the atmospheric $\delta^{13}C$ value can give insights into the relative importance of terrestrial and oceanic processes as sinks for atmospheric carbon. It is partly on this basis that the large interannual variability in the atmospheric CO_2 growth rate is thought to be primarily due to fluctuations in terrestrial sink strength.

Further information on the use of isotopes in carbon cycle research can be found in Flanagan and Ehleringer (1997), Bowling et al. (2002), Ehleringer et al. (2002), and Pataki et al. (2003).

carbon, oxygen, and other elements, have been combined with surface-based studies of photosynthesis, respiration, carbon storage, and circulation, with remotely sensed data, and with simulation models to refine the understanding of the regulation of carbon cycling, and the magnitude and dynamics of sources and sinks of CO_2 at the global scale. In all of these studies, the spatial pattern of carbon cycle dynamics is crucial to understanding processes and projecting behaviour into the future.

Ocean Cycling of Carbon – Spatial Patterns of Transport, Exchange and Storage

Over the past two decades, intensive observations and process studies, regional integration, and global modelling and remote sensing approaches have dramatically increased the understanding of the biological, physical, and chemical processes controlling the carbon cycle in the ocean (Fasham 2003; Le Quére et al. 2003; Pedersen et al. 2003). Significant new insights into the vertical and lateral transport processes that control ocean dynamics, as well as the role of biological processes in ocean biogeochemistry, result from this work (Box 2.4).

One of the most important achievements of this research has been the synthesis of tens of thousands of individual measurements of air-sea flux of CO_2 in every ocean basin from every season. The pattern of CO_2 air-sea fluxes – the so-called breathing of the oceans – can be seen in Fig. 2.30 (Takahashi et al. 1997). As a result, knowledge about the overall amount of carbon going into oceans at global and regional scales has increased. Likewise, as these data have been synthesised, knowledge about the underlying physical and biological processes that drive carbon exchange has deepened. Pattern and process are intricately linked.

For example, one of the strongest features of the air-sea flux (Fig. 2.30) is the equatorial Pacific bulge, the large source of CO_2 flux that releases up to 1 Gt C per year. The immediate cause is the strong upwelling along the equator, but the flux is also influenced by the strong warming that the upwelling water undergoes, which decreases the solubility of CO_2 in the water and releases dissolved CO_2 to the atmosphere. Additionally, the flux is influenced by the relative lack of phytoplankton activity in the locality. As expected, the upwelling water brings a large source of nutrients to the surface. However, phytoplankton are not able to make use of it because of a lack of sufficient iron and thus do not take up large amounts of carbon in photosynthesis. Thus, the pattern of carbon exchange in this region is related to a combination of physical and biological controls (Watson 1999).

Likewise, biological and physical factors combine to regulate the regions of highest uptake of CO_2. The most intense sink is found in the North Atlantic, where warm water carried north by the Gulf Stream and North Atlantic Drift is rapidly cooled, allowing more CO_2 to dissolve. In addition, biological activity is high due to ample nutrients, including iron transported via the atmosphere to the tropical Atlantic Ocean from the Sahara Desert and then transported northwards by the surface currents.

Further insights into the biological processes that influence the patterns of CO_2 uptake come from estimates of global ocean primary production from remotely sensed patterns of chlorophyll *a* in the surface ocean (Fig. 2.31). The productivity estimates clearly show areas of high biological activity in the North Atlantic, confirming the significant role that the biological pump plays in that region. In addition, there are several coastal areas where biological activity is especially intense, although relating that activity to carbon uptake or loss is not straightforward. Especially interesting are regions of the oceans where upwelling and hence provision of nutrients is known to be strong but where significant biological activity is absent. In addition to the equatorial Pacific Ocean discussed above, the subarctic North Pacific Ocean and parts of the Southern Ocean are so-called high nutrient-low chlorophyll regions. About 30% of the global ocean is classified as such. Patches of higher productivity in the Southern Ocean lying east of the continents appear to be triggered by the atmospheric delivery of iron (see Sect. 2.5.2).

Another important factor controlling the ocean cycling of carbon is the mix of bacteria, phytoplankton, and zooplankton (and ultimately higher trophic levels such as fish) that make up the ecosystem community structure. The community structure can vary by region and by season, but observation of key species can give insights into carbon fluxes. For example, *Trichodesmium* is a cyanobacterium that can fix nitrogen from the dissolved N_2 pool in the oceans. Blooms of *Trichodesmium*, which can be observed from space, are often associated with enhanced primary productivity and enhanced uptake of CO_2 and export of carbon to the deep ocean.

The export of carbon to the deep ocean is an important process that eventually controls the net uptake or release of CO_2 from the oceans but which cannot be readily discerned from observations of surface patterns. For a net carbon sink to be present, carbon must be exported from the surface waters, where most products of planktonic activity (more than 90% on average) are quickly metabolised and the carbon returned to the atmosphere, to the deeper layers of the ocean where it can be stored for hundreds or thousands of years. Several factors influence the export rates. The structure of the ecological community is critical, as communities dominated by larger algae such as diatoms and dinoflagellates that are grazed by large zooplankton produce large, fast-sinking faecal pellets and aggregations of diatoms that export significant amounts of carbon to depth. Often export of carbon occurs in pulsed events associated with storms, the onset of monsoons and the development of vertical temperature gradients in the ocean. In addition,

Box 2.4. The Marine Carbon Cycle

Katherine Richardson · Roger B. Hanson

Carbon in the oceans resides in dissolved organic carbon (DOC), dissolved inorganic carbon (DIC), particulate organic carbon (POC) and in marine biota. Most of the carbon in all the ocean layers is in the form of DIC but less than 1% of this carbon is gaseous carbon dioxide (CO_2). Around 90% of the inorganic form is bicarbonate ions and the rest is in the form of carbonate ions. These three forms are in a pH-dependent equilibrium. The capacity of the ocean's bicarbonate system to buffer changes in CO_2 is limited by the addition of calcium and, to a lesser extent, magnesium from the slow weathering of rocks. Thus, the ability of oceans to absorb excess CO_2 on short time scales is constrained.

Marine biomass makes up less than about 2 Pg C in the oceans. Despite this fact, marine biota are globally as productive as terrestrial systems (Falkowski et al. 2000). However, the biota do not accumulate biomass in carbon-rich structures in the way that land plants do. Primary production in the photic zone, approximately 50 Pg C per year globally, is the major input of organic carbon to the marine ecosystem. Over the past decade much has been learned about primary producers as well as what controls their distribution, rate of carbon uptake and turnover areas.

The carbon fixed at the surface is transported only slowly to deep ocean layers. Most of the carbon in the oceans is in the intermediate and deep waters. Once there, estimated lifetimes are approximately decades to centuries. The higher concentrations of inorganic carbon in these deeper layers in all basins result from the combined influence of two fundamental processes, known as the solubility and the biological pumps (Fasham et al. 2001), which operate at different rates temporally and spatially in diverse regions of the oceans. Thus, the strength and magnitude of these two pumps gives rise to regions of natural sources and sinks of carbon in the oceans.

The solubility pump operates as a result of thermohaline circulation and latitudinal and seasonal changes in ocean ventilation. Being more soluble in cold, more saline waters, CO_2 is sequestered to deeper waters by the formation of cold, dense water at high latitudes. The most intense natural sink region in the world is the North Atlantic, where the Gulf Stream and North Atlantic Drift transport warm water northward. There it cools and sinks, increasing the solubility of CO_2 from the atmosphere and removing it from the surface to deeper waters. Off Scandinavia and Canada, the dissolved CO_2 and fixed carbon in these waters sinks to great deeps over 500 m and is subsequently transported south along the bottom.

The biological pump (Fig. 2.29) operates through the biological assimilation of dissolved CO_2, which enhances the air-sea gas exchange and contributes to the vertical transport of carbon via physical processes. It functions through a complex food web of small plankton, which primarily recycle CO_2 within the photic zone, and larger plankton, which generate most of the particulate and dissolved forms that sink to the deep ocean. Approximately 25% of the carbon fixed by phytoplankton globally in the upper ocean layer sinks to the interior (Falkowski et al. 1998), where it is oxidised and recycled by bacteria and other heterotrophs into dissolved inorganic and organic carbon forms. Only a small percentage (1–2%) reaches the ocean floor. The deep benthos consumes and recycles most of what falls from the interior while the rest is buried in ocean sediments.

The magnitude of vertical flux from the surface layer to the deep ocean depends on the pathway and food web. Food webs dominated by large phytoplankton and macrozooplankton produce the largest vertical export flux to the deeper parts of the ocean (Fasham 2003). Phytoplankton species, which form silicate (opal) or carbonate shells, contribute to what is sometimes referred to as the silicon or carbonate pumps. There is no best way to estimate the magnitude of the biological pumps. However, several global estimates agree quite well and set the transfer rate at about 10 Pg C yr^{-1} (Falkowski et al. 2000, Laws et al. 2000, Schlitzer 2002). Recent findings suggest that the phytoplankton species involved in this process may hold a key to the rate and potential for the flux to change.

Once carbon sinks to the ocean interior via the biological and solubility pumps and then is transported laterally, dissolved CO_2 in the water is effectively prevented from re-equilibrating with the atmosphere until transported back to the surface later. Along the equator, vigorous upwelling occurs, which warms the deep water as it rises to the surface, decreases the solubility of CO_2 and thus releases it to the atmosphere. Along many diverse ocean margins, intense seasonal upwelling occurs and supports a strong phytoplankton and heterotrophic foodweb, referred to as a continental carbon pump. Recent evaluation of the magnitude of this pump indicates that these food webs absorb upwards of 0.2–1 Pg C yr^{-1}, a significant component of the global carbon cycle (Fasham 2003). In many regions, eastern and western ocean boundary currents sweep carbon deep offshore, while in high latitude margins, cold water sinks into the intermediate layer of the open ocean. The magnitude of the lateral and the vertical carbon flux at the seafloor, which is less than 0.2 Pg C yr^{-1} globally, agrees well with benthic respiration, sediment trap and primary production data (Fasham 2003).

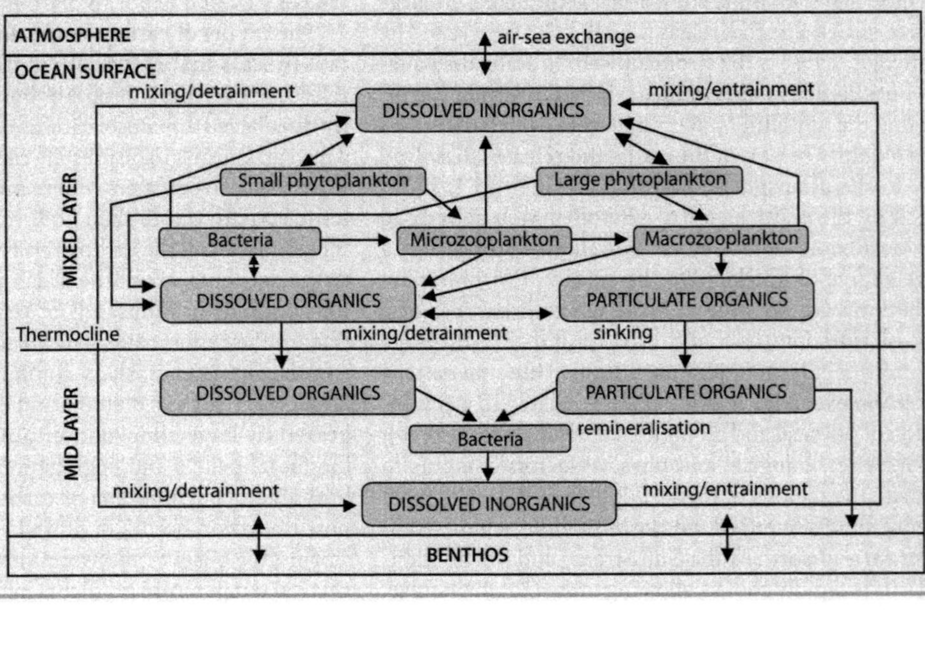

Fig. 2.29. The biological pump (conceptual diagram from the Joint Global Ocean Flux Study (JGOFS) International Project Office, Bergen, Norway)

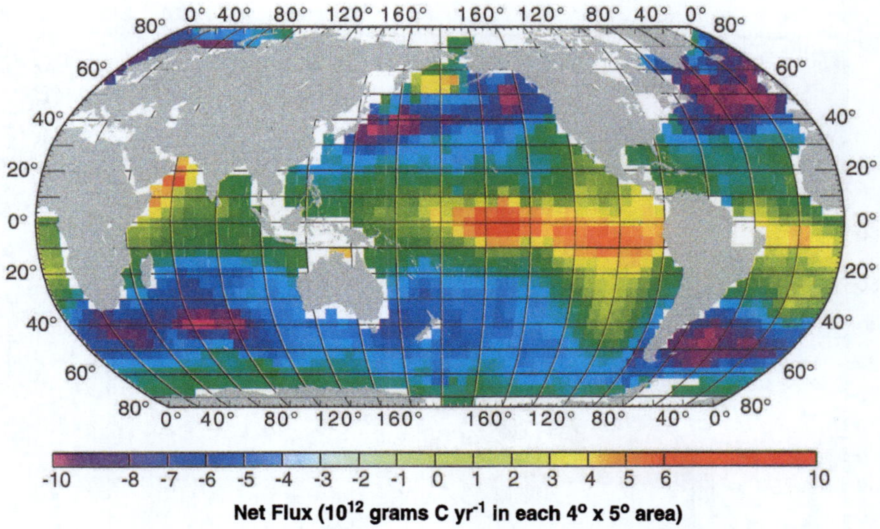

Fig. 2.30.
Mean annual exchange of CO_2 across the sea surface (Takahashi et al. 1997). *Blue and purple colours* denote regions in which the ocean takes up large amounts of CO_2 while *orange and yellow colours* denote areas where significant outgassing of CO_2 occurs

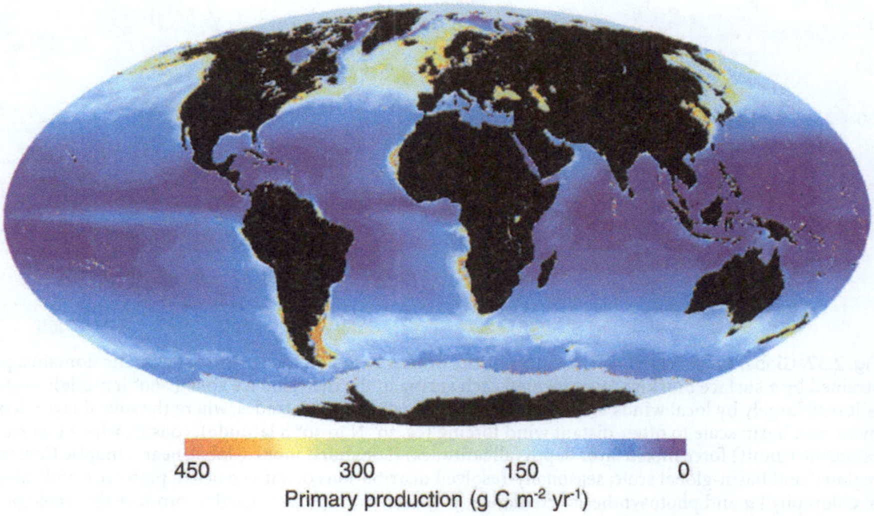

Fig. 2.31.
Global ocean primary production estimated from chlorophyll *a* distribution derived from SeaWiFS data (*image:* Paul Falkowski and Dorota Kolber, Rutgers University, pers. comm.)

recent work (Hansell and Carlson 2001) has demonstrated the importance of dissolved organic carbon instead of particulate organic carbon for export; in some regions the former may be as much as 20% of the flux.

The role of physical and biological processes in determining the spatial pattern of CO_2 uptake and outgassing from the ocean suggests the possibility of an underlying spatial classification of the oceans roughly equivalent to biomes on land. Classifying the ocean geographically is not straightforward given the diffuse and shifting boundaries between regions in the marine environment. Nevertheless, considerable progress has been made in recent years. The notion of ocean biogeochemical provinces was first developed by Platt and Sathyendranath (1988) and was based on remote sensing of ocean colour. The concept has been further developed through more detailed consideration of the properties and processes that define coherent regions (Longhurst 1998; Ducklow 2003; Fig. 2.32). For example, critical biogeochemical processes such as CO_2 exchange, particulate and dissolved organic carbon fluxes to depth, high (or low) bacterial activity and phytoplankton blooms can all be useful, when combined with physically defined zones such as equatorial upwelling areas, ice zones and ocean-coastal margins, in developing the concept of ocean biogeochemical provinces.

The coastal zone, where land, sea, and atmosphere are contiguous and interact, has provided special challenges for the evaluation of carbon cycling and carbon storage at the global scale (Maxwell and Buddemeier 2002). Because of its extreme heterogeneity, caused in part by variation in physiographic characteristics as well as by variations in the delivery of materials from the terrestrial watersheds, both direct observation and modelling of carbon cycling have been difficult. In addition, the coastal zone has been strongly affected by human activities both within the coastal zone itself and in the upstream terrestrial regions. Thus, determining the pattern of carbon sources and sinks in the coastal zone before significant human influence began is very diffi-

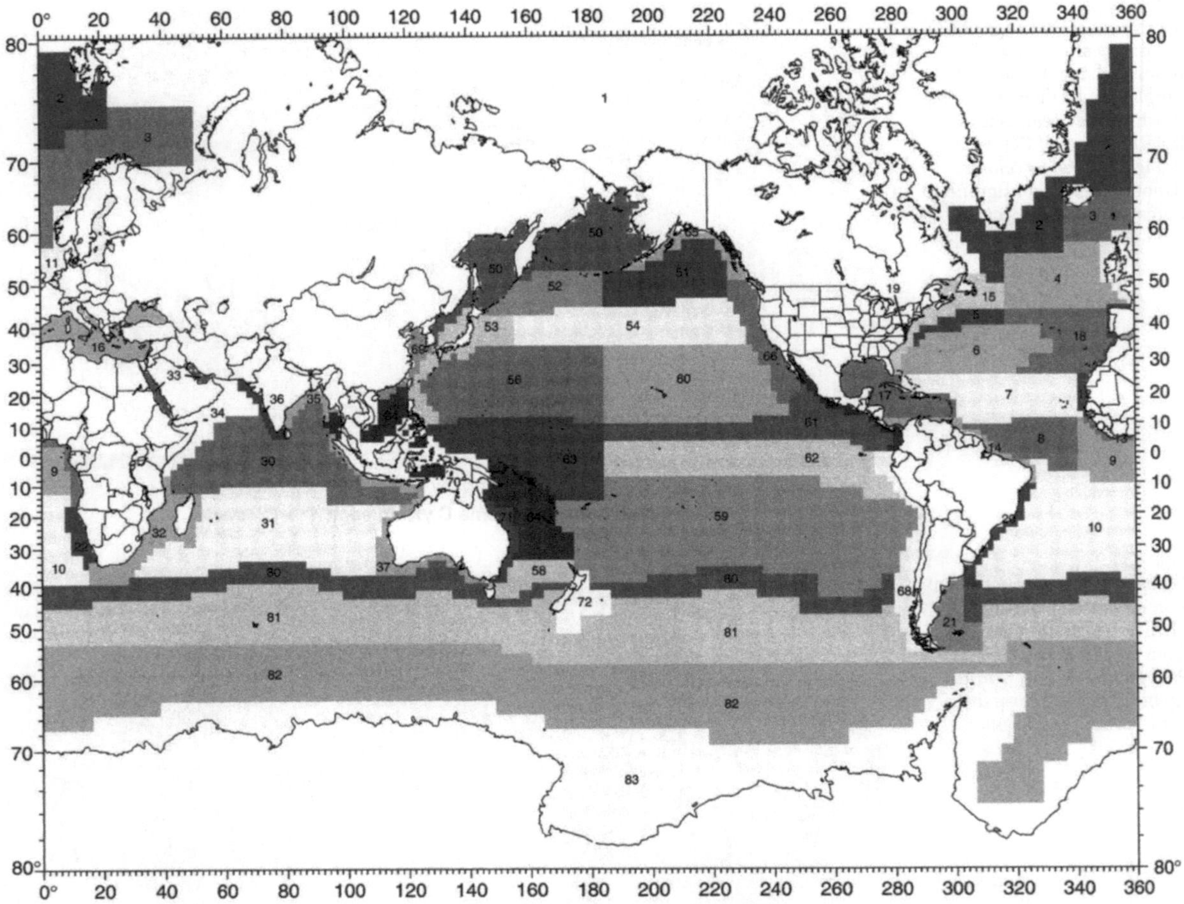

Fig. 2.32. Global biogeochemical provinces of the oceans based on four principal climatic domains: polar, where the mixed layer is constrained by a surface brackish layer formed each spring in the marginal ice zone (>60° latitude); westerlies, where the mixed layer depth is forced largely by local winds and irradiance (ca. 30–60° latitude); trades, where the mixed layer depth is forced by geostrophic adjustment on a basin scale to often-distant wind forcing (ca. 30° N to 30° S latitude); coastal, where diverse coastal process (e.g., tidal mixing, estuarine runoff) force mixed layer depth (all latitudes) (Longhurst 1998). Global, near-synoptic Coastal Zone Colour Scanner data sets on regional and basin-global scale, seasonally-resolved distributions of surface ocean pigments, and extensive data on the vertical structure of chlorophyll *a* and photosynthesis-irradiance (p-I) relationships were used to produce this ecological geography of the sea

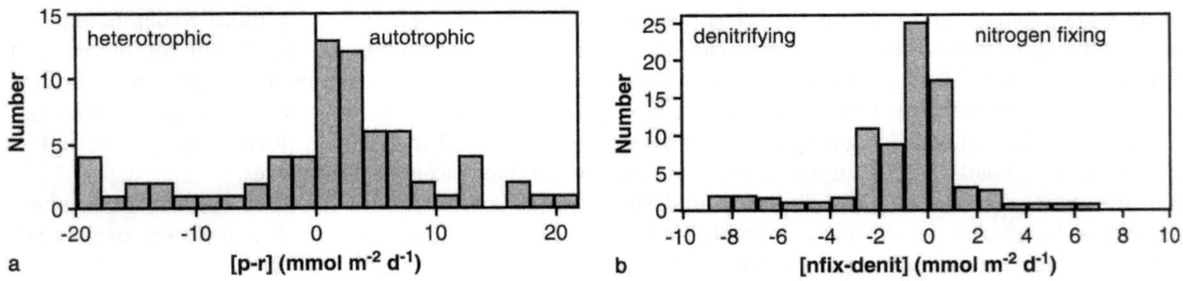

Fig. 2.33. Frequency distributions of $(p-r)$ and $(nfix-dinit)$ at coastal zone biogeochemical budget sites (Smith and LOICZ Modelling Team 2002). The term $(p-r)$ is primary production minus respiration; $(nfix-dinit)$ is nitrogen fixation minus denitrification

cult. Recent research, employing budgets for the delivery and export of dissolved phosphorus and nitrogen to and from several hundred coastal systems, is attempting to estimate the importance of coastal zone ecosystems as sources or sinks for CO_2 in the present era. Early results suggest net carbon uptake (Fig. 2.33) but these estimates include the effects of human perturbations as well as natural processes.

In addition to providing insights into spatial variation in fluxes and processes, measurement of CO_2 fluxes, taken over more than a decade, provide insights into the temporal variability of the spatial patterns of the fluxes. For example, they have demonstrated the strong temporal variability of the equatorial Pacific outgassing. During El Niño events, the upwelling of water in the central Pacific is notably slowed and the outgassing of CO_2

Box 2.5. The Terrestrial Carbon Cycle

R. J. Scholes

The terrestrial carbon cycle is the *chemical engine* that supplies energy and mass to most life on earth, including human life. It is also intimately involved in regulating the composition of the global atmosphere, and thus the energy balance of the climate system.

Plants remove carbon dioxide (CO_2) from the atmosphere and convert it to carbohydrate, through the process of photosynthesis (Fig. 2.34). Some of this carbon is used by the plant to fuel its own metabolism in a process known as autotrophic respiration, resulting in the release of CO_2 back to the atmosphere. The remainder, called net primary production, builds the leaves, stems and roots of the plants. At around 500 Pg C, the global quantity of plant biomass is small relative to the huge amount of carbon stored in the ocean and in fossil fuels, but it is important because it can increase or decrease relatively rapidly in response to climate or management, with strong and immediate effects on the atmospheric carbon pool.

Plant parts eventually die and decay, or are eaten by herbivores, or are consumed by fire. In the first two cases, the carbon is used by microorganisms or animals as an energy source and to build their own biomass, which also eventually passes into the food chain. In these processes most of the carbon is released as CO_2, a pathway known as heterotrophic respiration. Disturbances, such as storms, pest outbreaks, harvesting or clearing accelerate heterotropic respiration. So, in general, does climate warming.

Part of the carbon flowing through the terrestrial ecosystem builds up in the soil as dead plant and animal material, and byproducts of microbial metabolism or fire that are resistant to decomposition. Although the carbon retained in the soil is a small fraction of the flow, cumulatively it amounts to 1 500–2 000 Pg C. Soil organic matter is less easily returned to the atmosphere than is plant biomass, causing the soil to act as a medium-term buffer in the global carbon cycle. Intensive tillage, coupled with removal of most of the crop biomass, reduces the carbon content of the topsoil by up to half, over a period of decades. It can be built up again over a similar period by fallowing or by reduced tillage practices.

Almost all land ecosystems burn at some stage in their history, but the permanently moist ones do so very infrequently. Tropical savannas and boreal forests, on the other hand, emit about 2–5 Pg C yr^{-1} to the atmosphere through fire (Crutzen and Andreae 1990).

There are several other minor mechanisms by which carbon returns to the atmosphere. Plants give off a wide range of carbon-based gases, collectively known as Volatile Organic Compounds (VOC), generally in small quantities. When decomposition occurs in oxygen-starved conditions (such as underwater or in the gut of an animal), carbon is given off as the powerful but short-lived greenhouse gas CH_4. Gases such as the VOCs and CH_4 eventually convert to CO_2 in the atmosphere, but in the process they alter the regional air chemistry, in some cases generating another powerful greenhouse gas, tropospheric ozone.

A very small fraction of the carbon cycling through terrestrial ecosystems leaks into groundwater, rivers, and eventually the ocean, in the form of dissolved or suspended organic carbon (Schlesinger and Melack 1981). It contributes to the organic-rich sediments in lakes and on the coastal shelf, where it eventually decomposes and returns to the atmosphere. This leakage is important in that it results in land ecosystems being small net sinks of carbon, even when the biosphere as a whole is in equilibrium. Another long-term sink mechanism is the formation of black carbon (soot) when biomass burns. Black carbon is virtually inert and is thus removed from an active role in the biosphere for a very long time.

The absolute annual exchange of carbon between land-based ecosystems and the atmosphere is somewhat larger than the exchange between the atmosphere and the oceans (about 120 Pg C yr^{-1} and 90 Pg C yr^{-1} in each direction, respectively). Together these biospheric exchanges dwarf the annual input of carbon to the atmosphere by modern human activities (about 7 Pg C yr^{-1}). In pre-industrial times the exchanges in each direction were approximately in balance. That is no longer true. Between about 1700 and 1960 the land is calculated to have been a net source of carbon to the atmosphere, due to the emissions resulting from the conversion of natural vegetation to croplands. More recently, photosynthesis has exceeded respiration, despite ongoing land conversion in the tropics, and the land has become a net sink for carbon. The terrestrial sink is currently removing about 30% of the additional CO_2 injected into the atmosphere by human activity. Unlike the ocean carbon sink, it cannot continue to do so at the present rate for centuries or millenia. Models project that the sink will peak within the twenty-first century (Cramer et al. 2001).

More details on the terrestrial carbon cycle can be found in Scholes et al. (1999), Bolin et al. (2000) and Prentice et al. (2001).

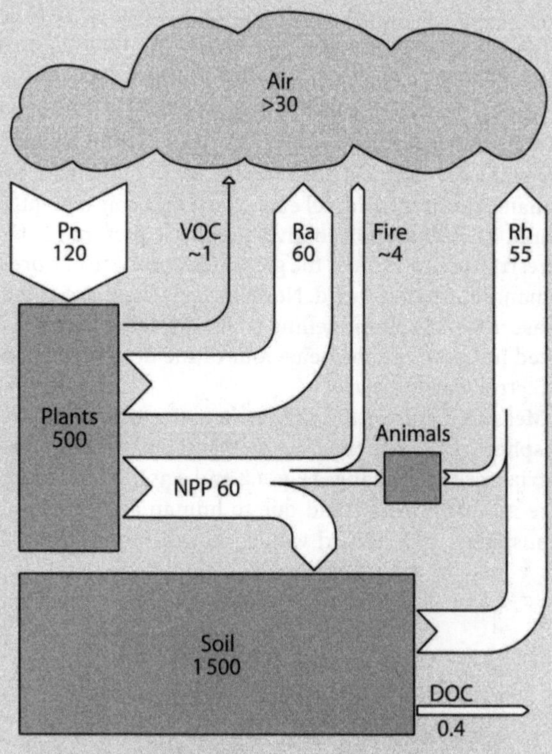

Fig. 2.34. The main elements of the terrestrial carbon cycle. The units are global totals for a pre-industrial world, in Pg C for the pools and Pg C yr^{-1} for the fluxes (Prentice et al. 2001). The rate of biomass formation (net primary productivity, *NPP*), is equal to photosynthesis (*Pn*) minus autotropic respiration (*Ra*) and Volatile Organic Carbon (*VOC*) losses. The rate of carbon uptake or loss from a given patch of land is the *NPP* minus the heterotrophic respiration (*Rh*) and denotes Net Ecosystem Productivity (*NEP*). Over a long period of time, the small but steady loss as dissolved carbon (*DOC*), and the larger intermittent losses due to disturbance, fire or harvesting, become important and must be subtracted to give the Net Biome Productivity (*NBP*) (Schulze and Heimann 1998), the true long-term effect of the land on the atmosphere. Under equilibrium conditions this is a very small number. Under the current human-induced disturbance it is about 1.4 Pg C yr^{-1}

is abruptly shut off for many months. In total, the flux of CO_2 from the ocean to the atmosphere can be reduced by up to 50% compared to a non-El Niño year. These flux data, together with isotopic data, provide spatial and temporal information with which to test ocean carbon-cycle models that incorporate such processes as ocean circulation, temperature-dependent solubility, and biotic uptake and transport to depth (e.g., Orr et al. 2001). Moreover, these data provide a backdrop against which to evaluate change in the future.

Terrestrial Carbon Cycling – Spatial Patterns of Exchange and Storage

Terrestrial ecosystems also exchange CO_2 rapidly with the atmosphere and they store large amounts of carbon in biomass, especially in soils (Box 2.5). Spatial variation in these processes is considerable. Unlike the situation for the oceans, however, hundreds of thousands of measurements of the CO_2 flux across the land-atmosphere boundary from which a Takahashi-type map for terrestrial ecosystems may be constructed are not available. In addition, the current pattern of carbon sources and sinks on land has been significantly influenced by human activities (Schimel et al. 2001) and consequently cannot be relied upon to give a realistic picture of the terrestrial component of the global carbon cycle in a pre-human dominated world. Nevertheless, there are some measurements and modelling techniques that have produced insights into the behaviour of the non-perturbed terrestrial carbon cycle.

Measurements of the change in concentration of atmospheric CO_2 with time suggest spatial patterns of terrestrial carbon exchange (Fig. 1.4 and 2.35). Even if there were no underlying trend due to human activities, the atmospheric CO_2 record would show strong intra-annual variability with a very regular pattern. Concentrations always decrease in the northern hemisphere during the middle of the year and peak during the December–February period; the southern hemisphere shows a similar but smaller and anti-phased pattern. This is evidence of the strong bias of terrestrial ecosystem activity towards the northern hemisphere, where most of the Earth's land mass is found. The photosynthetic activity of northern terrestrial ecosystems peaks during the May–September period whereas heterotrophic respiration reaches its maximum some one to three months later, in phase with the lag of maximum soil temperature from the summer solstice. This temporal imbalance of photosynthesis and respiration is sufficiently strong to leave a prominent imprint on the northern hemispheric record of atmospheric CO_2.

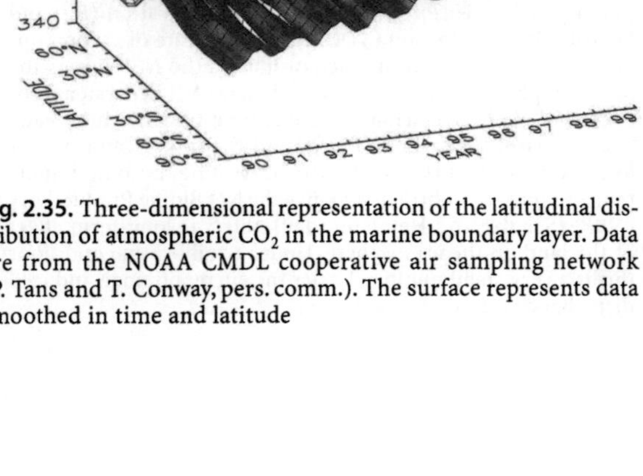

Fig. 2.35. Three-dimensional representation of the latitudinal distribution of atmospheric CO_2 in the marine boundary layer. Data are from the NOAA CMDL cooperative air sampling network (P. Tans and T. Conway, pers. comm.). The surface represents data smoothed in time and latitude

An alternative approach is to use process-based or satellite-driven models of terrestrial productivity to simulate the pattern of carbon uptake and loss based on potential vegetation, that is, on the distribution of ecosystems assuming no significant land-use change. Figure 2.36 shows the global pattern of terrestrial net primary productivity as estimated from an average of such models (Cramer et al. 1999; Kicklighter et al. 1999). Although net primary productivity is not necessarily indicative of the net carbon exchange (Box 2.5), it does give important clues as to the overall pattern of carbon uptake and its relationship to the two most important environmental controlling factors, temperature and precipitation. There is a clear, broad latitudinal gradient in net primary productivity, with the most productive systems lying in the tropics and productivity generally decreasing towards the high latitudes. Superimposed on this is the effect of plant available moisture (a combination of precipitation and the evaporative demand of the atmosphere), with desert and semi-arid areas having low productivity.

To understand more about the processes that regulate carbon fluxes and carbon storage and that give rise to the observed patterns in different kinds of ecosystems and in different places on Earth's surface, especially the processes which release carbon back to the atmosphere via respiration, more localised, detailed bottom-up studies are employed. A network approach is useful to draw out generalisations from the individual studies and to place them in a global context and hence to draw out spatial patterns.

In FLUXNET, a rapidly growing and extensive network focused on understanding spatial and temporal variation in carbon, water and energy fluxes in a variety of terrestrial ecosystems around the globe, over 70 research teams are now measuring such fluxes using micrometeorological approaches (Baldocchi et al. 1996), ecophysiological measurements, and associated measurements of vegetation structure, soils, site history and hy-

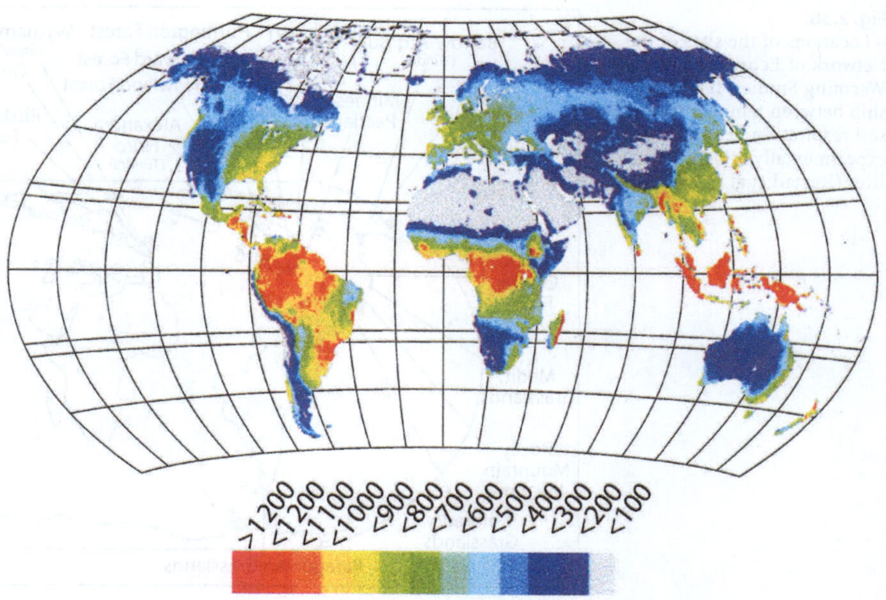

Fig. 2.36. Annual terrestrial net primary production (g C m^{-2}) estimated as the averaged output of 17 terrestrial biogeochemical models (Cramer et al. 1999)

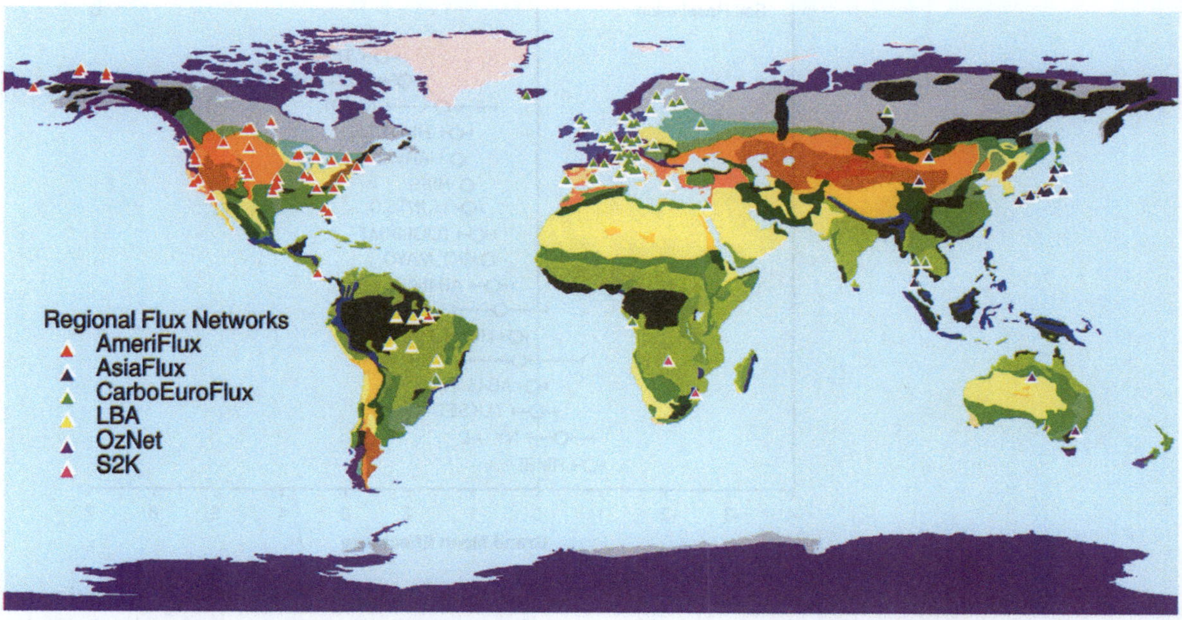

Fig. 2.37. Locations of flux towers that routinely measure water, energy and CO_2 exchange between the land surface and the atmosphere as part of FLUXNET (ORNL DAAC, FLUXNET: *http//public.ornl.gov/fluxnet/ecoregions.cfm*, 10 January 2002)

drologic characteristics (Valentini et al. 2000) (Fig. 2.37). These studies are already showing (*i*) net ecosystem exchange of CO_2 in temperate forests is highly sensitive to growing season length; (*ii*) climate variability is a major driver of variability in CO_2 exchange at a number of scales; and (*iii*) water availability appears to be an important factor in influencing heterotrophic respiration, especially in semi-arid ecosystems (Valentini et al. 2000).

A second network focuses on temperature as another major controlling factor of heterotrophic respiration. The Network of Ecosystem Warming Studies aims to quantify the importance of temperature in controlling heterotrophic respiration, and in influencing nutrient cycling more generally. Early results (Fig. 2.38) show that heterotrophic respiration generally increases with increasing temperature, although there is significant variation amongst sites (Rustad et al. 2001). Especially important is the spatial pattern of responses of ecosystems to warming, as the combination of sensitivity to temperature and large stores of carbon in the soil point to areas where the net exchange of CO_2 may be particularly responsive to climate variability and to longer-term systemic changes in climate.

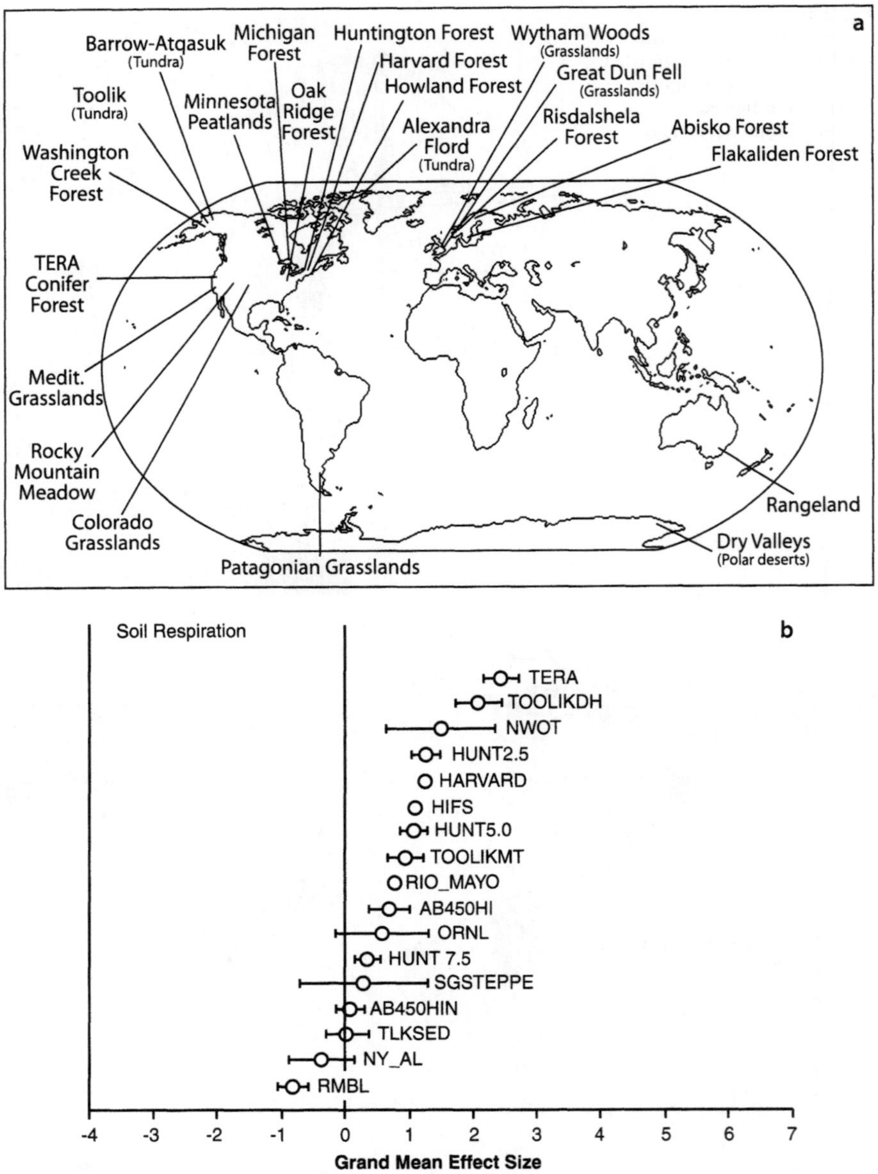

Fig. 2.38.
a Locations of the sites of the Network of Ecosystem Warming Studies; **b** relationship between temperature and soil respiration determined experimentally at network sites (Rustad et al. 2001)

The importance of heterotrophic respiration for the overall carbon exchange between land and atmosphere focuses attention on the soil component of terrestrial ecosystems, where about two-thirds of the carbon in the terrestrial system is stored. Knowledge of the spatial pattern, here in the vertical dimension, of soil organic matter coupled with vegetation distribution yields important insights regarding the processes that store carbon in soil. Recent work suggests that different plant functional types, because of their differences in biomass allocation and root distributions, contribute to different degrees to soil organic matter and carbon distribution, a result that highlights the potential importance of vegetation change and soil organic matter pools for carbon sequestration strategies (Jobbágy and Jackson 2000). Globally, soil organic matter storage in the top three meters of soil is 2 340 Pg C, with 1 500 Pg of this total in the first metre (see also Batjes 1996). Thus, estimates based on a standard one metre rooting depth can be rather misleading. The biomes with the most soil organic matter stored in deeper soil layers, from 1–3 m depth, are tropical evergreen forests (160 Pg C) and tropical grasslands/savannahs (150 Pg C).

Climate is known to be important in controlling the absolute amount of soil organic matter in soils; cold-climate ecosystems generally contain more soil organic matter than those in warm climates. On the other hand, vegetation (plant functional type) is the dominant factor in controlling the vertical distribution of organic matter in the soil profile. For a given climate, organic matter is preferentially found in the deeper layers in shrublands, intermediate in grasslands, and shallower in forests (although the absolute amount in deeper layers can still be large for forests given their large areas, as for tropical evergreen forests). The proportion of soil organic matter in the top 20 cm relative to the first metre is on average 33%, 42%, and 50% for shrublands, grasslands, and forests, respectively. Surprisingly, shrublands, and not forests, consequently store carbon preferentially at the greatest depths.

2.4.3 Trace Gas Exchanges Between Earth's Surface and the Atmosphere

The past decade of research has demonstrated without a doubt that attempts to understand the current atmosphere, or predict the future evolution of the atmosphere, must account for the spatial patterns of biological processes that regulate the production, consumption, and storage of gases. A large number of gases are exchanged between biota and the atmosphere (Mooney et al. 1987) (Fig. 2.39). These include stable, relatively long-lived gases such as CO_2, CH_4, and N_2O, other nitrogen- and sulphur-containing compounds, as well as a wide range of chemically reactive species, including a large suite of volatile organic carbon compounds (Scholes et al. 2003).

Plants produce and emit over 60 volatile organic carbon compounds, including various hydrocarbons, alcohols, carbonyls, fatty acids and ester compounds (Scholes et al. 2003). The amount of carbon released to the atmospheric through these compounds may be substantial; up to 30% of annual accumulation of carbon in an ecosystem can be lost in this way. Volatile organic carbons are often highly reactive in the atmosphere, with lifetimes of hours to weeks, and many of them play important roles in regulating tropospheric ozone, aerosol particle and carbon monoxide production. The spatial pattern of volatile organic carbon emissions (Fig. 2.40) gives some clues about the nature of the terrestrial ecosystems that are especially large emitters of these gases. As a general rule, the production of volatile organic carbon within the chloroplast of plants and their emission to the atmosphere are strongly dependent on temperature. Consequently, the tropics predominate in spatial emission patterns. However, it is important to note that there are many processes by which volatile organic carbon can be produced (chloroplast, metabolic by-products, decaying and drying vegetation, specialised defence mechanisms, plant growth hormones and floral scents) and regional patterns of emissions are controlled by the types of process and their particular environmental controls (Scholes et al. 2003).

Microbial respiration in aerobic soils, sediments, and waters produces CO_2; under anaerobic conditions, the breakdown of organic compounds by microorganisms is a source of methane (CH_4), one of the trace atmospheric carbon gases that plays a critical role as a greenhouse gas and affects the concentration of the hydroxyl radical (OH), thus influencing the oxidising capacity of the atmosphere. The spatial patterns of CH_4 emission are strongly related to areas where anaerobic conditions occur at or near the land or sediment surface. Natural wetlands are the most important natural biogenic source of methane, releasing approximately 110 Tg yr^{-1}. In these systems, CH_4 is produced by methanogens in anaerobic

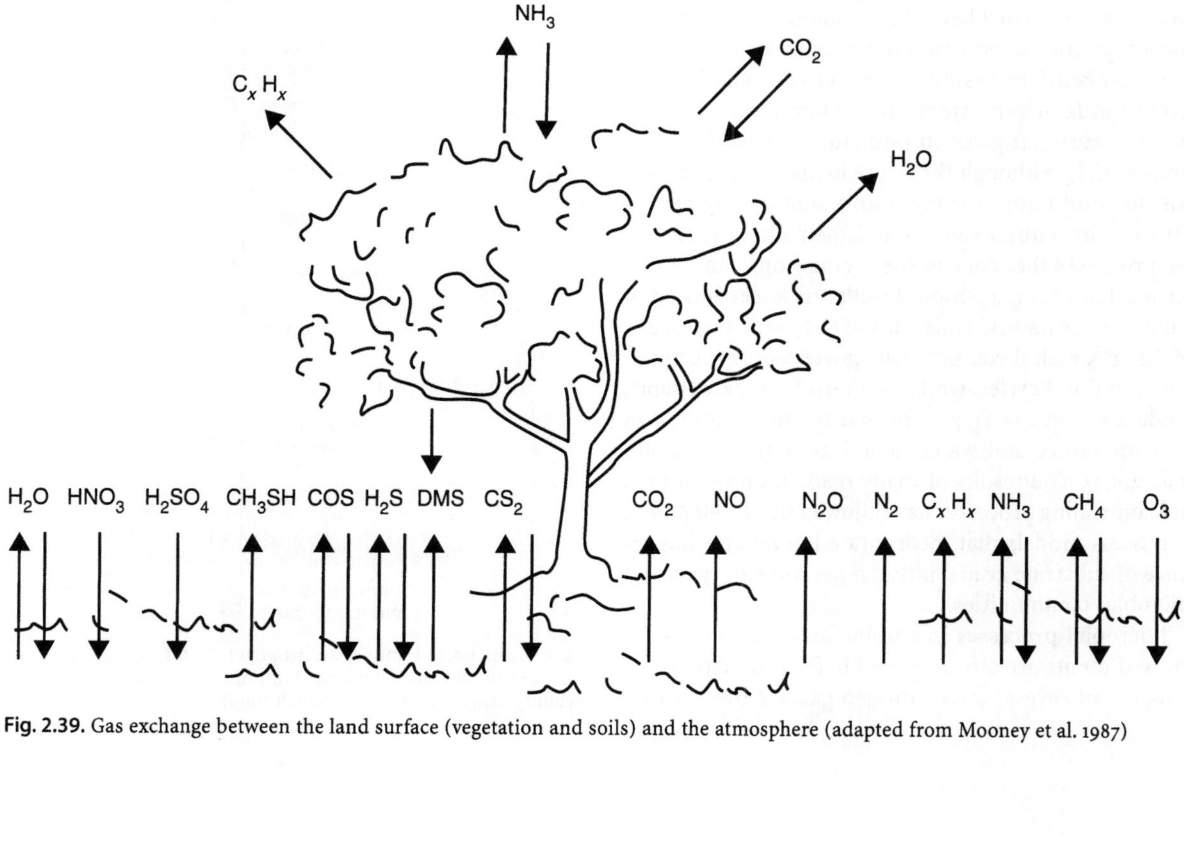

Fig. 2.39. Gas exchange between the land surface (vegetation and soils) and the atmosphere (adapted from Mooney et al. 1987)

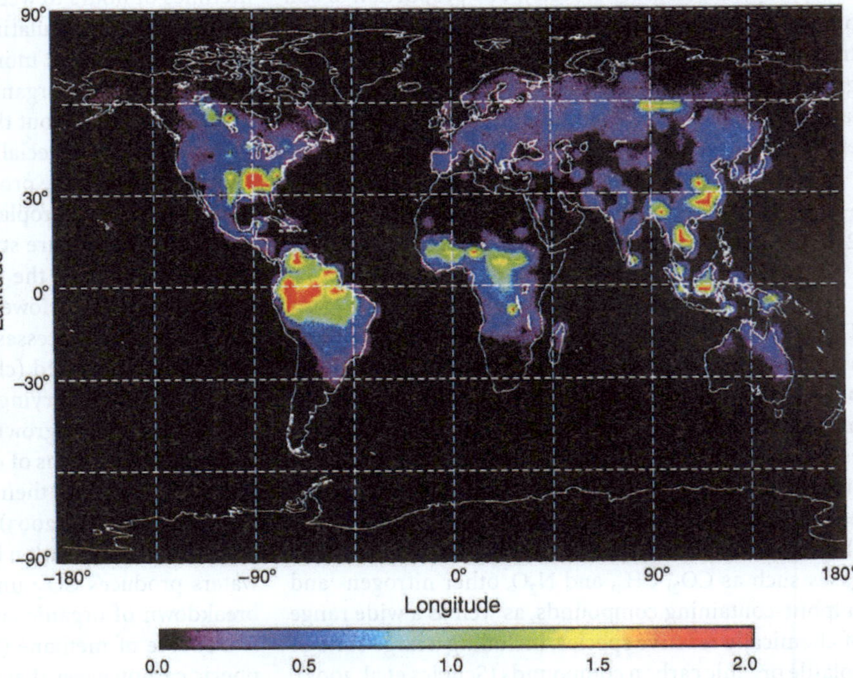

Fig. 2.40. Global distribution of estimated isoprene emissions for July (g C m^{-2} $month^{-1}$) (Guenther et al. 1995)

layers and then diffuses through soil layers to the surface. However, in soils and sediments with aerobic microsites or layers, methanotrophs (methane oxidising bacteria) may consume some of the CH_4, so that the source strength depends on the difference between production and consumption. In wetlands, vegetation can serve as a conduit for CH_4 produced deep in the soil layers, preventing consumption from occurring and thus increasing overall source strength. Research over the past decade has elucidated the regulation of production, consumption, and emission of CH_4, and has developed refined quantitative estimates of source strength in different kinds of ecosystems and different areas of the world. Figure 2.41 gives an estimate of the sources and sinks of CH_4. Although the figure includes both anthropogenic and natural sources and sinks, some spatial patterns for natural sources and their implications for the processes that control these emissions can be discerned. For example, tropical wetlands are estimated to dominate the natural emissions of CH_4 with a release of 66 Tg yr^{-1}, with fluxes generally governed by precipitation and flood cycles, while northern latitude wetlands produce around 38 Tg yr^{-1}, primarily under the control of temperature and water table interactions (Scholes et al. 2003). The results of many field studies on fluxes and controlling processes have allowed the development of process models that incorporate the relative importance of substrate availability, temperature, oxygen, and microbial communities.

Microbial processes in aerobic and anaerobic soils and sediments are also responsible for production and emission of several trace nitrogen gases. Nitrous oxide

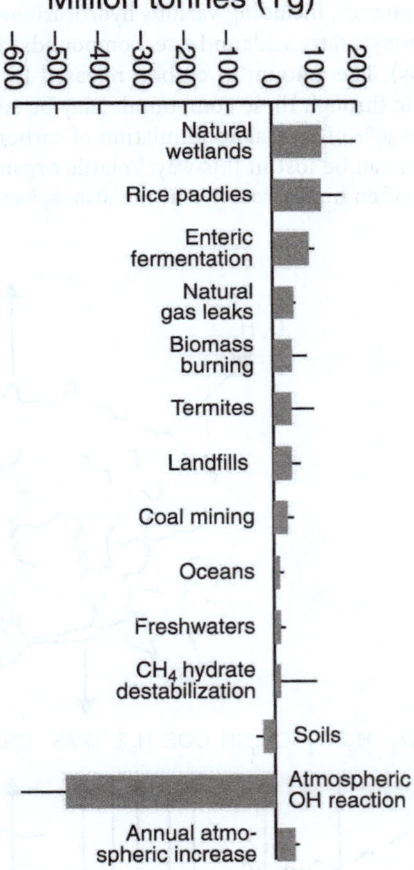

Fig. 2.41. Estimated annual anthropogenic and natural sources and sinks of methane (*thick bars*) in millions of tonnes, with uncertainty ranges (*thin lines*) (Milich 1999)

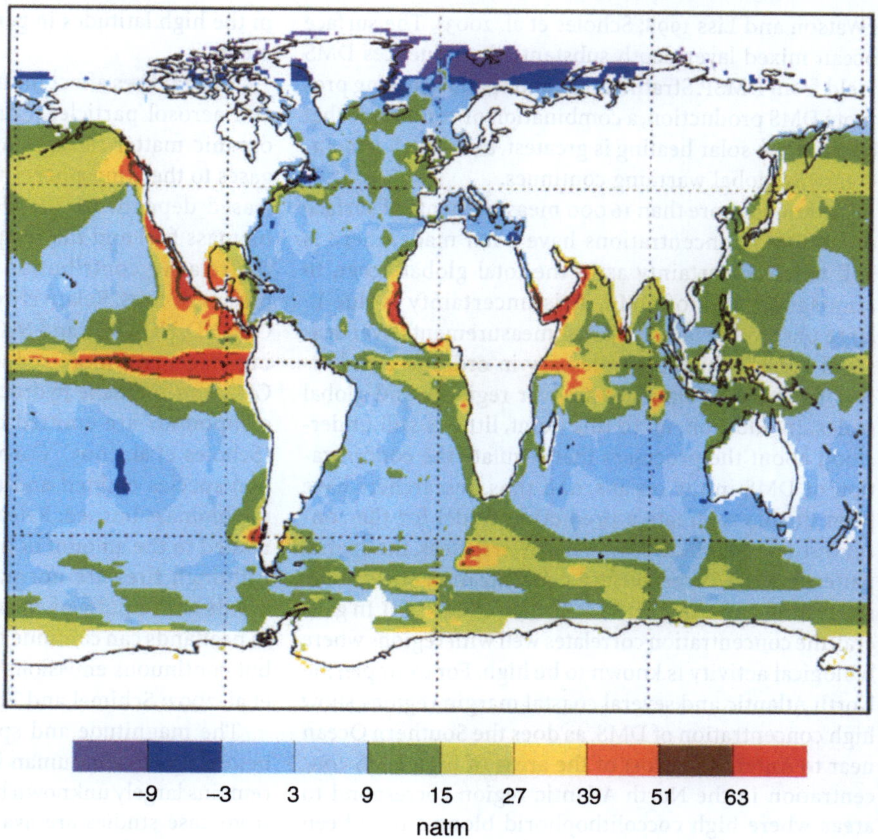

Fig. 2.42.
Annual composite surface difference in partial pressure of N_2O over oceans (in natm = 10^{-9} atmospheres) (Nevison et al. 1995)

(N_2O) and nitric oxide (NO) are both produced as byproducts of microbially-mediated nitrogen transformations in soils, including nitrification (the oxidation of ammonium to nitrate) and denitrification (the reduction of nitrate to N_2O or N_2). It is difficult to determine the spatial variability of these trace gas fluxes in a pre-human dominated world because of the paucity of measurements made on natural ecosystems. Much of the emphasis in the last two decades has been on determining the effects of human perturbations, most notably fossil fuel combustion and the extensive use of fertilisers, on the fluxes of these gases. Box 3.6 gives a quantitative summary of the global nitrogen cycle, both before and after significant human influence.

Prior to the mid 1980s most estimates of natural sources of N_2O were based on only a handful of flux measurements in a few ecosystems, assumed to be constant over large areas. Likewise, prior to the mid 1980s there were fewer than five measurements of NO in natural or managed ecosystems. Studies throughout the late 1980s and 1990s were carried out in sites selected to maximise variation in climate, soil age and type, and disturbance characteristics, all factors known to affect nitrogen cycling in ecosystems (Matson and Ojima 1990). In general, soil emissions (from both natural and managed ecosystems) contribute 70% and 20% of the global budgets of N_2O and NO, respectively. Among natural ecosystems, humid tropical forest soils are the most important source of N_2O, accounting for 20–50% of all the global sources of atmospheric N_2O (Keller et al. 1986; Potter et al. 1996; Verchot et al. 1999), and tropical savannas are the largest natural source of NO (Scholes et al. 2003).

Measurements carried out in marine ecosystems have likewise multiplied in recent years. Data from shipboard measurement programs in many different regions have generally reinforced the view that open ocean upwelling regions along eastern ocean boundaries and in equatorial regions, as well as coastal regions represent major sources of N_2O (Scholes et al. 2003 and Fig. 2.42). The equatorial Pacific Ocean and the coastal regions around Central America are particularly strong sources. This pattern suggests a strong role for temperature as an environmental controller of emissions. The global oceans are thought to contribute about 5 Tg yr^{-1}, or about one-third the global natural source strength.

Microbes in soil, sediments, and waters also produce and consume a suite of sulphur containing gases. In marine systems, specific phytoplankton produce dimethylsulphoniopropionate (DMSP), which is then consumed or released with grazing and microbial processes and diffuses to the atmosphere as dimethylsulphide (DMS). DMS plays a critical role in the creation of aerosol particles that can act as cloud condensation nuclei

(Watson and Liss 1998; Scholes et al. 2003). The surface ocean mixed layer depth substantially influences DMS yield from DMSP. Stratification and shallow mixing promote DMS production, a combination of conditions that occur when solar heating is greatest, and which may increase as global warming continues.

Although more than 16 000 measurements of surface ocean DMS concentrations have been made, there is still much uncertainty as to the total global ocean to atmosphere flux of DMS. This uncertainty is due in large part to the lack of global measurements over seasonal cycles, which are necessary in order to avoid errors from extrapolating data over regional and global scales. In addition, up to this point, little is still understood about the processes that regulate the concentration of DMS in the oceans, and thus, researchers have been unable to create a predictive model for the concentration of DMS in seawater. Nevertheless, the spatial patterns of DMS sea surface concentration show some interesting patterns (Kettle et al. 1999) (Fig. 2.43). In general, the concentration correlates well with regions where biological activity is known to be high. For example, the North Atlantic and several coastal margin regions show high concentration of DMS, as does the Southern Ocean near to Antarctica. Most of the areas of high DMS concentration in the North Atlantic region correspond to areas where high coccolithophorid blooms have been observed (Brown and Yoder 1994). It should be noted, however, that data are sparse in the Southern Ocean and the estimated high DMS concentrations there are more a reflection of the high concentrations thought to exist in the high latitudes in general than to direct measurements.

Natural fires affect the atmosphere by emitting gases and aerosol particles. Combustion of living and dead organic matter releases numerous stable and reactive gases to the atmosphere; just what and how much is released depends on the elemental composition of the biomass fuel and the temperature of the fire, and thus the relative contribution of flaming and smoldering phases of fires. Relatively oxidised compounds, such as CO_2, NO, NO_2, SO_2, and N_2O are emitted during the flaming stage, and more reduced compounds such as CO, CH_4, non-methane hydrocarbons, and reduced sulphur compounds are emitted during the smoldering stage (Scholes et al. 2003). Carbon dioxide is the major carbon species emitted and accounts for over 80–90% of the biomass burned. Release of nitrogen compounds is related to the amount of nitrogen contained in the fuel. Although fires are normally considered to be abrupt events with a pulse of emissions to the atmosphere, fires in peatlands can continue for months or years with small but continuous emissions to the atmosphere (e.g., Page et al. 2002; Schimel and Baker 2002).

The magnitude and spatial pattern of natural fires before significant human influence in the Earth System remains largely unknown because only fragmentary data from case studies are available. Nearly all data on fire frequency and extent comes from the contemporary period. It is clear, however, that wildfires are a natural part of the ecosystem dynamics of two major biomes – boreal forests and savannas – and thus played a significant role

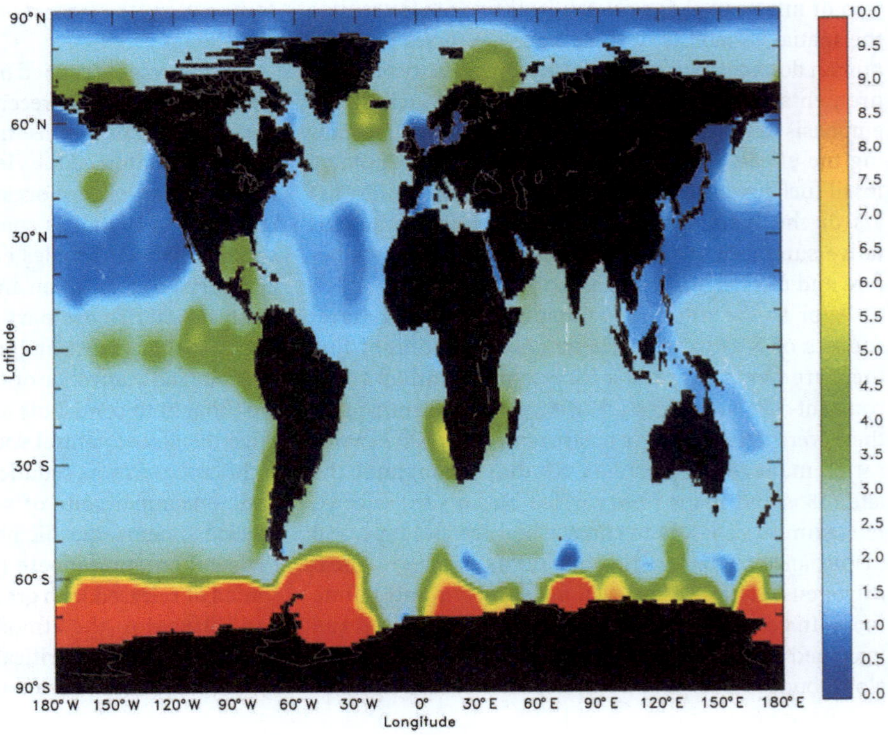

Fig. 2.43. Smoothed field of annual mean DMS (dimethylsulphide) sea surface concentration (10^{-9} mol l^{-1}); the original field was smoothed with an 11-point unweighted filter to remove discontinuities between biogeochemical provinces (Kettle et al. 1999)

in gaseous and aerosol emissions in the pre-human dominated world. The return frequencies of wildfires varied widely, however, between those two biomes, with values of 3–5 years for savannas and values of hundreds of years, perhaps as long as 500, for boreal forests.

2.5 New Insights into the Connectivity of the Earth System over Space and Time

Much of the emphasis in global change research to date has focused strongly on the vertical links between the atmosphere and the Earth's surface. However, components of the global environment, such as ecosystems on land and in the sea, are also connected to each other laterally through the dynamics of the Earth System – through the horizontal movement of water and materials through it, through atmospheric transport and deposition, and through the movement of plants and animals. Moreover, they are connected through energy transfers and through chemical and biological legacies that linger over time. The following sections provide examples of recent global change research that has illustrated the importance of each of these pathways to the functioning of the Earth System.

2.5.1 Connectivity via the Oceans and their Currents

The oceans play a critical role in controlling Earth's environment. They cover 71% of the Earth's surface area and contain over 97% of all water on the planet, (excluding water bound in sediments and sedimentary rocks). While they work in concert with the atmosphere, their capacity for storing and redistributing heat and water around the globe exerts a moderating effect on climate extremes at all time scales. The ocean currents, even though sluggish compared with the winds, share the redistribution task almost equally with the atmosphere. In addition to water and heat, ocean circulation and mixing move materials around the Earth System through horizontal (lateral) transport and vertically through up- and down-welling of ocean water. These transports provide nutrients for marine ecosystems and move particulates and dissolved gases and solids far from their point of origin. In addition, ocean currents play a central role in defining major modes of climate variability, the best known of which is the El Niño-Southern Oscillation phenomenon in the Pacific Ocean. Through climate teleconnections, ENSO impacts are felt in many parts of the world through changed weather patterns. The oceans influence patterns of climate variability even on geological time scales.

The processes driving the circulation of the ocean are complex. They include the external forcing by winds, the effects of the Earth's rotation and the constraints of the topography of the ocean basins and their interconnecting channels. The distribution of density (dependent on salinity, temperature and pressure) within the ocean is also of primary importance. Although these fundamental processes have been known for some time, the last decade or two of global change research has added much detail to understanding of oceanic circulation and has significantly improved knowledge of the role of the ocean in Earth System functioning.

The dominant patterns of surface currents have closed anticyclonic gyres, (clockwise in the northern, anticlockwise in the southern hemisphere) and intensified on their western sides by the influence of the Earth's rotation to form poleward-flowing western boundary currents such as the Gulf Stream and Kuroshio. These gyres meet in a series of complex equatorial currents and counter currents (Fig. 2.44).

The surface gyres dominate the upper ocean above the thermocline that separates warm surface waters from those of the cold abyss. The strength of the surface currents generally decreases with increasing depth. Below the thermocline, the decreasing strength of the wind-driven circulation is matched by currents on which density distribution, the Earth's rotation and bottom topography are the dominant influences. The currents at all levels can be measured directly by drifters, floats and moored current meters and their integrated effects can be seen in the distribution of water properties (for example by the distribution of anomalously salty water from the Mediterranean spreading near 1 000 m depth through the North Atlantic). Superimposed on these clearly-defined current systems is variability on scales of only a few hundred kilometres and a few tens or hundreds of days that is the oceanic equivalent of the changes in weather patterns caused by cyclones and anticyclones in the atmosphere.

A crucial part of the integrated effect of all these current elements is an exchange of water between the surface and deep ocean. In essence it is driven by the cooling of surface water in the polar regions that, coupled with the formation of sea ice and consequent increase in salinity of the remaining unfrozen water, produces water masses dense enough to sink to sub-thermocline levels and in some cases to the ocean floor. The subsequent equatorward flow of these cold deep waters combined with the predominantly wind-driven poleward transport of warm surface water is referred to as the thermohaline circulation. The deep waters migrate slowly back towards the ocean surface driven by the effects of mixing in the ocean interior.

The integrated effects of all of these currents serves to transport the excess solar energy falling on the low latitude oceans towards the poles where at higher latitudes it is lost to the colder atmosphere. The amounts of heat are enormous. In the North Atlantic it reaches a

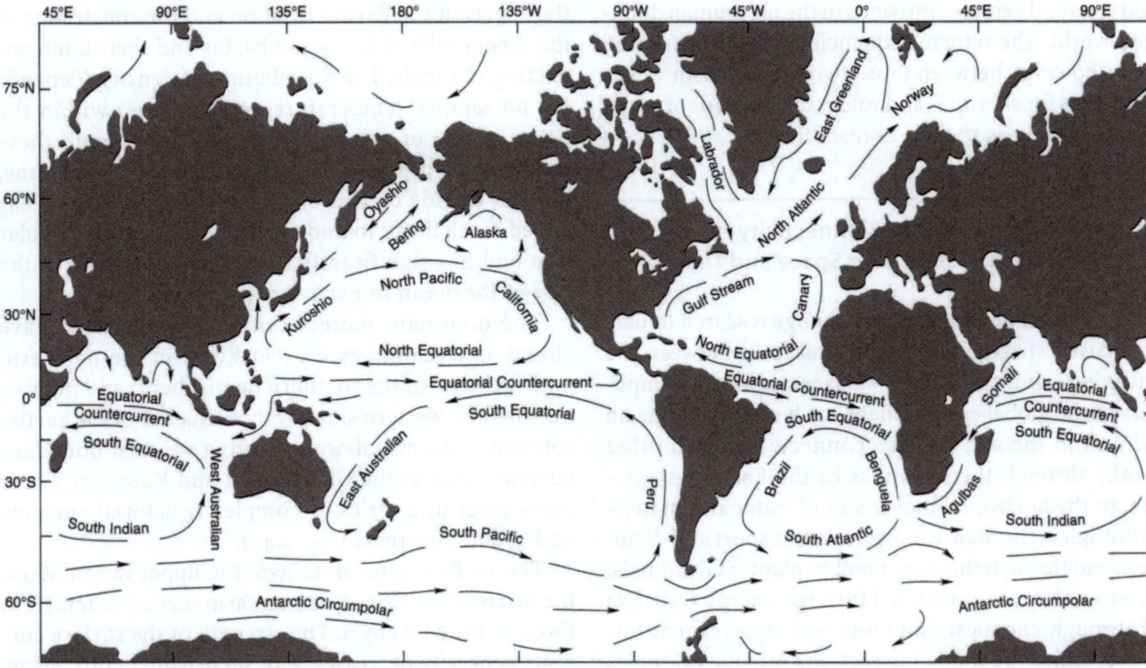

Fig. 2.44. Dominant patterns of surface currents of the world's oceans (adapted from Apel 1987; available on-line at *www.acl.lanl.gov/GCM/currents.html*)

maximum at 24° N of about 1.2 petawatts (1.2 thousand, million megawatts). Changes in the North Atlantic thermohaline circulation and its heat transport have been implicated in the global transitions between glacial and interglacial periods perhaps triggered by changes in freshwater inputs to the high latitude North Atlantic from the atmosphere and the land (see Sect. 5.3.1).

The thermohaline circulation has been likened to a conveyor belt that moves water, heat and other properties and substances around the world's ocean basins (Fig. 2.45). However, this concept is a simplified methaphor rather than a true representation of the complex circulation and transport pathways.

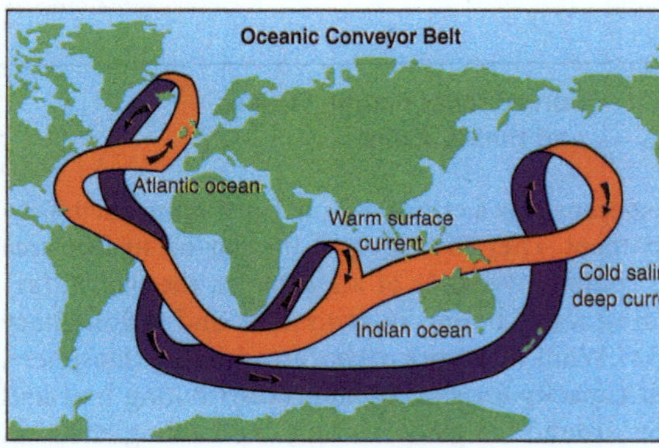

Fig. 2.45. Schematic represention of the thermohaline circulation of the ocean (adapted from Broecker 1991)

2.5.2 Connectivity via Atmospheric Transport

Atmospheric transport is the process by which atmospheric motions carry physical or chemical properties from one region of the Earth to another. Like oceanic transport of energy and matter, atmospheric transport is an exceptionally important feature of Earth System functioning. It transports vast amounts of energy around the planet; atmospheric motions on the planetary scale when aggregated amount to the world's synoptic weather systems, often called the general circulation (and hence the name General Circulation Model for models of the global climate system). In terms of materials, atmospheric transport enables different chemical compounds originating from different parts of the Earth System to interact. Without transport, compounds produced by the terrestrial or marine biospheres, for example, would tend to accumulate in the location of their production until an equilibrium between production and destruction would be achieved. Thus, it is atmospheric transport that maintains the atmosphere in a disequilibrium state and hence develops and maintains regions that are characteristically net sources and net sinks for chemical substances (Wuebbles et al. 2003).

Figure 2.46 shows the main latitudinal patterns of atmospheric circulation in the lower 20 kilometres of the atmosphere, the layer where much of the transport of energy and materials occurs. Atmospheric motions occur on all scales, from planetary, global to synoptic (of the order of 1 000 km), mesoscale (10–500 km) and small scale (below 10 km). From an Earth System functioning perspective, two modes of circulation are espe-

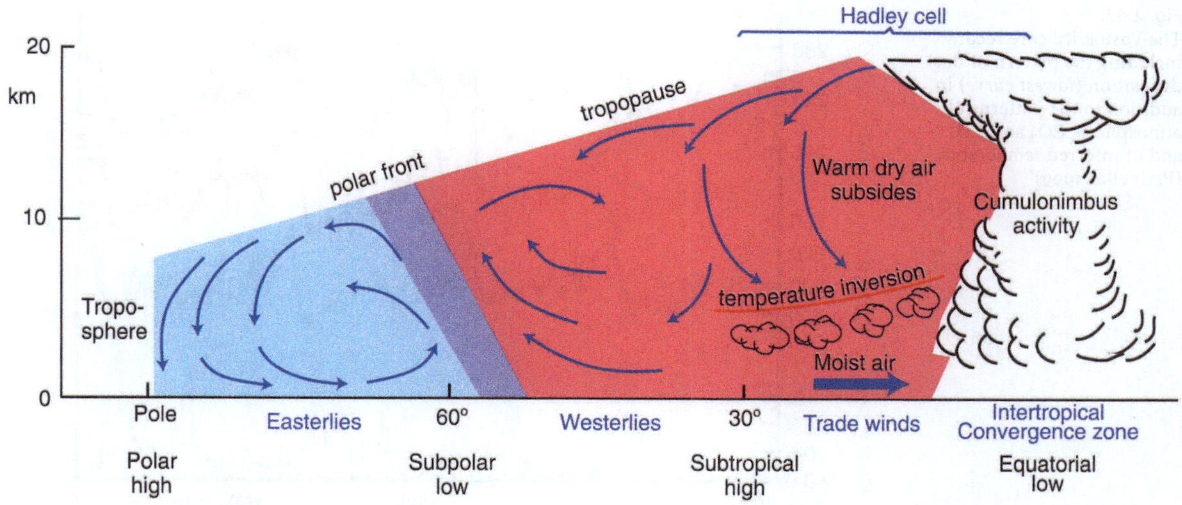

Fig. 2.46a. Schematic view of the latitudinal atmospheric circulation below 20 km altitude, from the equator to the pole, showing the Inter-Tropical Convergence Zone near the equator where strong upward motions take place, tropical clouds form and intense thunderstorm activity occurs, the return flow along the tropopause and the slow air subsidience in the subtropics. The polar front and the polar circulation are also shown (Brown et al. 1989)

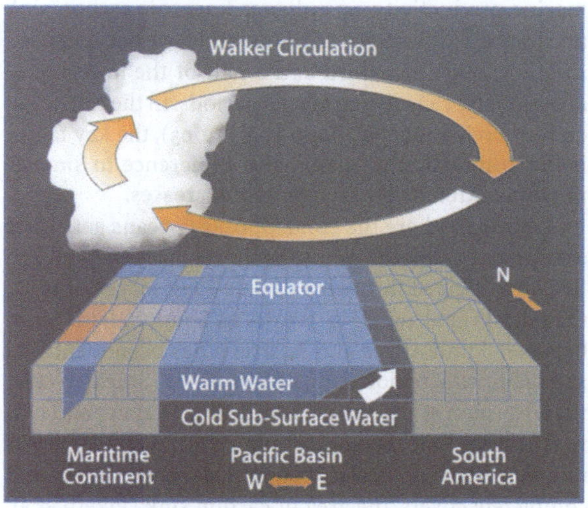

Fig. 2.46b. Walker circulation above the Pacific Ocean, showing the normal form of circulation pattern when El Niño is not active (*image:* redrawn after NASA)

cially important in terms of inter-continental connectivity. First, materials must be transported from their site of production into the atmosphere. The planetary boundary layer is the thin but variable (shallow over land during the night and morning, deeper during the afternoon and evening) layer that directly connects the Earth's surface – land or ocean – with the atmosphere. All materials generated by the biota and all energy and materials deposited from the atmosphere to the surface must pass through the boundary layer. In general, the planetary boundary layer is characterised by weak horizontal winds, strong turbulence, high humidity and high concentrations of compounds emitted at the surface. The linkage of the boundary layer to the free troposphere above is crucial for the movement of materials from their locations of origin but the rate of transfer across this interface remains poorly quantified.

Secondly, once materials are in the free troposphere, large-scale quasi-horizontal advection drives inter-continental transport. This process is responsible both for important biogeochemical linkages between continents and between land and ocean and, in the contemporary period, for trans-boundary air pollution. At mid-latitudes this horizontal flow is predominantly from west to east, but in the tropics and subtropics (within 30° of the equator) the atmospheric flow is dominated by cells consisting of strong convection near the equator, giving rise to fast vertical transport and heavy precipitation and slower downward movement of drier air in the subtropics. Seasonal patterns of atmospheric motions that operate at the regional scales, such as the Asian monsoon, are superimposed on these general patterns. The examples below show how atmospheric transport, especially the passage of materials through the planetary boundary layer and then horizontally for thousands of kilometres before being deposited back to the surface, is an important feature of the Earth System in its pre-human dominated state.

Atmospheric transport is a very effective means of moving large quantities of essential elements around the Earth. For example, ice cores from Greenland (Mayewski et al. 1994, 1997), the Peruvian Andes (Thompson 2000) and Antarctica (Petit et al. 1999 and Fig. 2.47) all show much higher levels of dust deposition during glacial maxima than during the Holocene. Kohfeld and Harrison (2000) suggest that global dust deposition during the Last Glacial Maximum (LGM) was 2–5 times that in the mid-Holocene. The dominant dust source during the LGM in central Greenland was central Asia (Biscaye et al. 1997) and, in Antarctica, it was most prob-

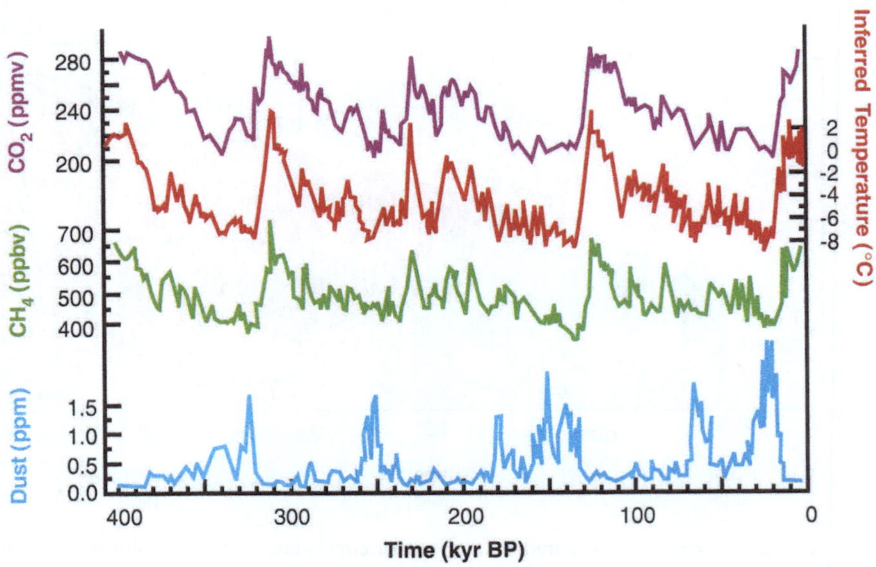

Fig. 2.47. The Vostok ice core record including the pattern of dust deposition (*lowest curve*) in addition to the patterns of atmospheric CO_2 and CH_4 and of inferred temperature (Petit et al. 1999)

ably Patagonia (Delmonte et al. 2002). Records of dust deposition in the western Pacific (Ono and Naruse 1997) at LGM also point to the importance of remote continental sources in central Asia. There appear to be strong links between high latitude dust deposition and the deglaciation sequence in each hemisphere. The high dust flux in Antarctica stopped soon after the LGM, whereas in Greenland the dust flux remained high until the rapid warming at the beginning of the Bölling-Alleröd at about 14 500 years ago. After stopping abruptly at that time, high dust deposition resumed at the onset of the Younger Dryas cold phase and continued until the opening of the Holocene (Raynaud et al. 2003).

Dust is the main source of iron to vast areas of ocean remote from riverine inputs and sources of iron in coastal sediments. Where iron has been the main limiting nutrient in ocean primary productivity, there is a strong *prima facie* case for variations in aeolian dust from remote continental sources to be a significant influence on the carbon budget. Furthermore, iron is known to be an essential ingredient in order for phytoplankton to flourish even in areas of oceans where other nutrients are abundantly available. Some areas, where phytoplankton blooms are known to occur regularly, act as effective sinks of CO_2 (Behrenfeld et al. 1996; Cooper et al. 1996). Controlled experiments have revealed the extent to which iron fertilisation of small areas of surface waters over short periods of time can significantly affect the growth of phytoplankton in waters where iron is otherwise a limiting nutrient (Cooper et al. 1996; Boyd et al. 2000). It is predicted that, over longer time scales, the increased photosynthesis in an iron-enriched marine region acts to enhance the biological CO_2 pump (Behrenfeld et al. 1996; Cooper et al. 1996). Although the demonstration of a stimulation of the biological pump has yet to be achieved in controlled enrichment experiments, several model simulations (Lefévre and Watson 1999; AJ Watson et al. 2000; Bopp et al. 2003) suggest that dust deposition, leading to iron fertilisation, increased marine productivity, an enhanced oceanic carbon sink and hence a draw-down of CO_2, could have been responsible for between a third and a half of the lowering of atmospheric CO_2 during glacial periods. In the full ocean carbon cycle model of Bopp et al. (2003), the key factor is the stimulation of diatoms in preference to smaller phytoplankton as the iron supply increases.

Key areas where the contemporary oceans act either as sources or sinks of atmospheric CO_2 have been identified and mapped through a decade-long survey of pCO_2 (Takahashi et al. 1997; Fig. 2.30). The South Indian Ocean between South Africa and Australia has been identified as a major sink region. Both observations (Sturman et al. 1997; Herman et al. 1997) and modelling (Rayner and Law 1995) reveal that the aerosol plume from southern Africa may reach Australasia on occasions. It crosses, and completely covers, the area of carbon sink (Piketh et al. 2000; Fig. 2.48). Subsidence in the South Indian anticyclone in the vicinity of 70° E causes air, on occasion, to subside to the surface. Given that ocean circulation patterns suggest no significant upwelling, and hence provision of nutrients from the ocean itself, in these sink regions the atmospheric transport of aerosol particles may hold the key to the major carbon uptake.

The quantitative evidence for iron fertilisation by the African aerosols is strong. From measurements made at the high-altitude (3 000 m), baseline site at Ben Macdhui on the edge of the Lesotho Massif (and located beneath the locus of the mean transport plume streaming off South Africa), it has been possible to determine mid-tropospheric mass fluxes of iron-bearing aerosol particles as they are transported towards Australasia. Deposition of particles takes place either as rain-out in wet deposition or as dry deposition. The latter can occur either as dust fallout from an elevated plume or by tra-

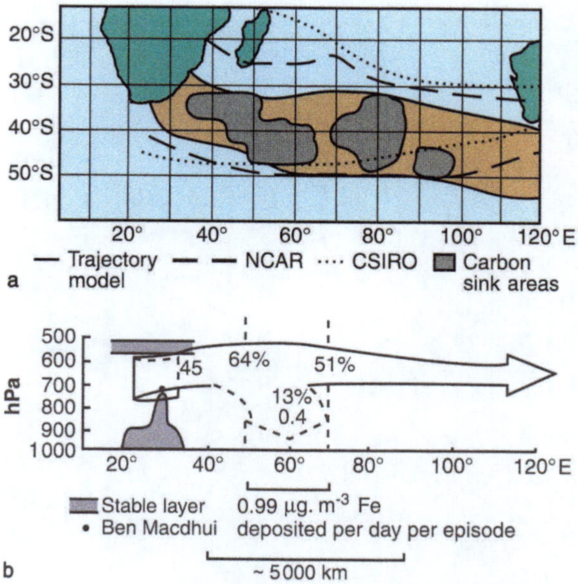

Fig. 2.48. **a** Observed and modelled aerosol particle and trace gas transport plumes passing over areas of enhanced biological productivity and carbon sinks in the South Indian Ocean; and **b** mean daily iron deposition of 0.99 µg m^{-3} from southern Africa into the ocean per iron-transport episode between 50° E and 70° E as estimated from a 5-year air-transport trajectory climatology and PIXE measurements of iron loading in the dust plume passing over Ben Macdhui Mountain at an altitude of 3 000 m in South Africa (Piketh et al. 2000). Approximate annual mass fluxes of total aerosol loading are given as Mt yr^{-1} in *bold*; numbers of trajectories passing given meridians or having subsided to the surface are given as percentages

jectories of air parcels being forced to the surface by atmospheric subsidence. It is deposition by subsidence that is considered here. Over long periods of time, as evidenced in a 5-year climatology, on average 13% of air parcel trajectories leaving South Africa subside to the surface between 50° and 70° E. Iron constitutes 0.7% of the fine particle loading at Ben Macdhui and high-iron transport episodes occur about every 11 days on average. Mean daily deposition of iron into the sea in the central South Indian Ocean following a peak-concentration episode over eastern South Africa is around 0.99 µg m^{-3} (Piketh et al. 2000). Such peak concentrations occur on around 33 days in a year. Given that the average duration of an episode centred on the peak is three days, the number of days a year in which iron fertilisation may be significant appears to be around 100.

The estimated daily episodic concentration of 0.99 µg m^{-3} deposited over the South Indian Ocean from the southern African aerosol plume is broadly similar in magnitude to a concentration of 0.11 µgm^{-3} used in an equatorial Pacific Ocean iron enrichment experiment which produced significant blooming of phytoplankton (Behrenfeld et al. 1996; Cooper et al. 1996). Not all the iron in the plume from southern Africa is likely to be in soluble form; that which is will be available for phytoplankton enrichment.

Strong evidence thus exists to suggest that aeolian transport of aerosol particles from southern Africa, and consequent atmospheric iron fertilisation of marine biota, support enhanced biological productivity and consequently the carbon sink in the South Indian Ocean between South Africa and Australia. Inter-regional nutrient transfers over distances exceeding 5 000 km have been established and links have been demonstrated between continental, terrestrial ecosystems and their remote marine equivalents.

Atmospheric transportation of iron-laden dust can, however, have negative repercussions for other components of the Earth System. Every year roughly one billion tonnes of African dust mainly from the Sahel region is deposited over the Caribbean (Fig. 2.49a). This figure is five times greater than the amount of dust transported atmospherically in the 1970s. The iron-rich dust eventually settles over coral reefs located in this region and help to fertilise algae, which can smother and kill the reefs (USGS 2000). Sahelian dust also carries pathogens that attack *Diadema*, a sea urchin vital to the continuing health of coral reefs. One of *Diadema's* roles is to graze the algae that grow on coral surfaces. Increased algae growth as the result of both iron fertilisation and reduced predation effects, along with introduction of foreign viruses and bacteria that are carried on African dust particles, has led to the proliferation of weakened and dead coral reefs throughout the Caribbean Sea, otherwise known as bleaching events. The rate of change in outbreaks of invader species and diseases in coral reefs in the Caribbean is coincident with the increased dust deposition in the 1970s and 1980s. Each of these bleaching events has been correlated with increases in warm water associated with El Niño weather patterns and peaks in African dust production and transport (USGS 2000) (Fig. 2.49b).

Likewise, terrestrial ecosystems, located on continents or remote islands in the middle of oceans, receive via atmospheric transport rock- and soil-derived nutrients that may be critical to their biogeochemical processes. As soils develop over time, nutrients are gradually lost via weathering of rocks and soil erosion. Thus it would seem that ecosystems would reach an irreversible point of nutrient depletion, yet this is not the case. Instead, Swap et al. (1992) have shown that dust from the African Sahel also settles over the Amazon Basin, serving as a significant source of elements. In addition, recent isotopic studies in the Hawaiian Islands have indicated that dust transported from the Ghobi Desert, over 6 000 km away, provides most of the biologically useful phosphorus necessary to sustain the islands' forests growing on old, highly weathered volcanic substrates (Chadwick et al. 1999).

In addition to particles, chemicals are transported great distances through the atmosphere in gaseous or solution forms, while being chemically modified and eventually affecting downwind air, water, and terrestrial systems. Studies employing aircraft and balloon-based measurements, satellite remote sensing, and ground-based measurements have characterised the transport and chemical processing of gases from point sources and

Fig. 2.49a. Satellite image (NOAA, 28 May 1999) of the southeast United States, Central America and the Amazon region of South America showing a large African dust cloud over the Caribbean

Fig. 2.49b.
Overall increase in African dust reaching the island of Barbados since 1965 (Courtesy of J. Prospero in USGS 2000). Barbados is situated in the Windward Islands, which are hidden beneath the cloud of dust shown in the satellite image in **a**. Peak years for dust deposition were 1983 and 1987, also the years of extensive environmental change on Caribbean coral reefs

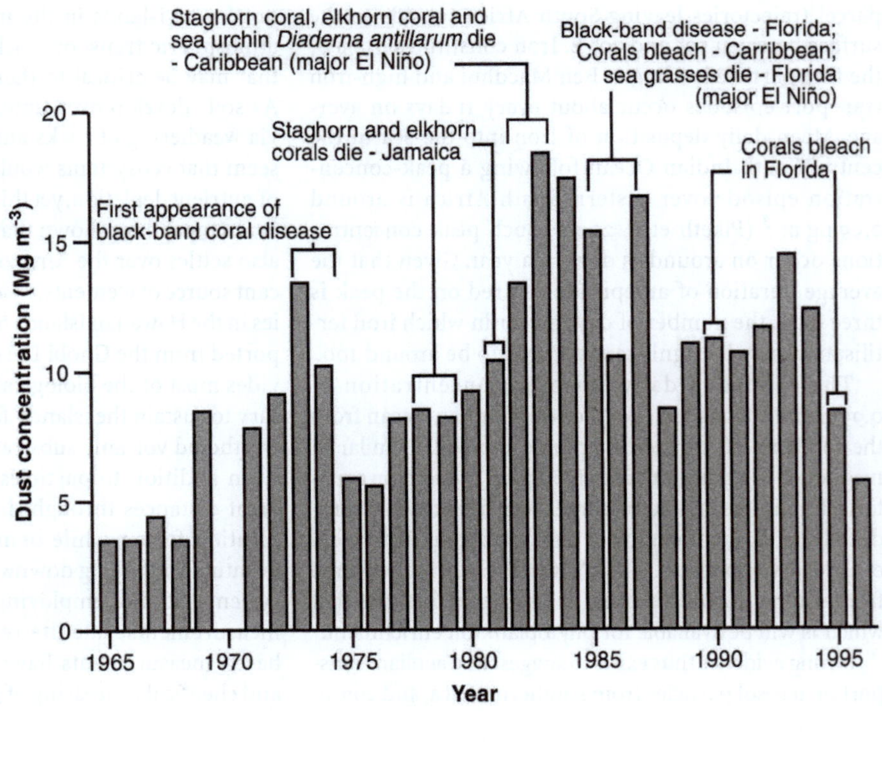

the consequences of those chemicals for downwind ecosystems. For example, it was earlier thought that the emissions from vegetation fires in tropical savannas, a natural part of the dynamics of savanna ecosystems, were largely confined to the region itself due to large-scale subsidence and lack of convection during the dry season when fires occur. However, recent studies (e.g., Chatfield et al. 1996; Thompson et al. 1996; Andreae et al. 2001) have shown that even modest transport by trade winds can move large amounts of smoke into the intertropical convergence zone, where deep convection transports emissions into the middle and upper troposphere and hence into long-range circulation patterns. In general, these and other similar studies are showing that the connectivity between various parts of the planet via atmospheric transport is far greater than earlier acknowledged (Scholes et al. 2003).

2.5.3 Connectivity via Hydrologic Transport

River networks provide conduits that link mountains and coastal areas, land and water ecosystems across landscapes that are regional to subcontinental in scale. Riverine fluxes tie land to the sea, and are a major pathway by which carbon and other materials are transferred from the land to the ocean. The presence of this material in the coastal zone allows unique forms of life to flourish at the interface between land and ocean. There is now evidence that the food resources available in the coastal zone may have been important for the evolution of the human brain (Crawford et al. 2001) and that humans dispersed themselves around the Earth via these coastal zones (Stringer 2000). Thus, riverine transport of material to the coastal zone may actually have created the conditions necessary for human evolution.

Global change research in the past decade has focused on the mobilisation and transport of terrestrial materials to the world's coastal zones, and in particular has focused on the human perturbation to these flows. However, this research raises the question of how hydrologic transport has operated in the past, before human activities have become significant in the functioning of the Earth System. This is not an easy question to answer, as humans are fundamentally reliant on freshwater supplies both directly and indirectly, for example, for agriculture, and thus have extensively modified continental aquatic systems for millennia. Only fragmentary evidence is available to inform about the behaviour of hydrologic transport systems in the past. Nevertheless, a few important generalisations are possible.

Over the past 18 000 years, the architecture of river systems and the rates of water and material transport have changed dramatically (Fig. 2.50). These changes have been related to climate change and its effects on run-off, variable sea level, variation in the distribution and storage of water in ice caps, and the occurrence of giant lake systems at the boundaries of ice caps (Vörösmarty

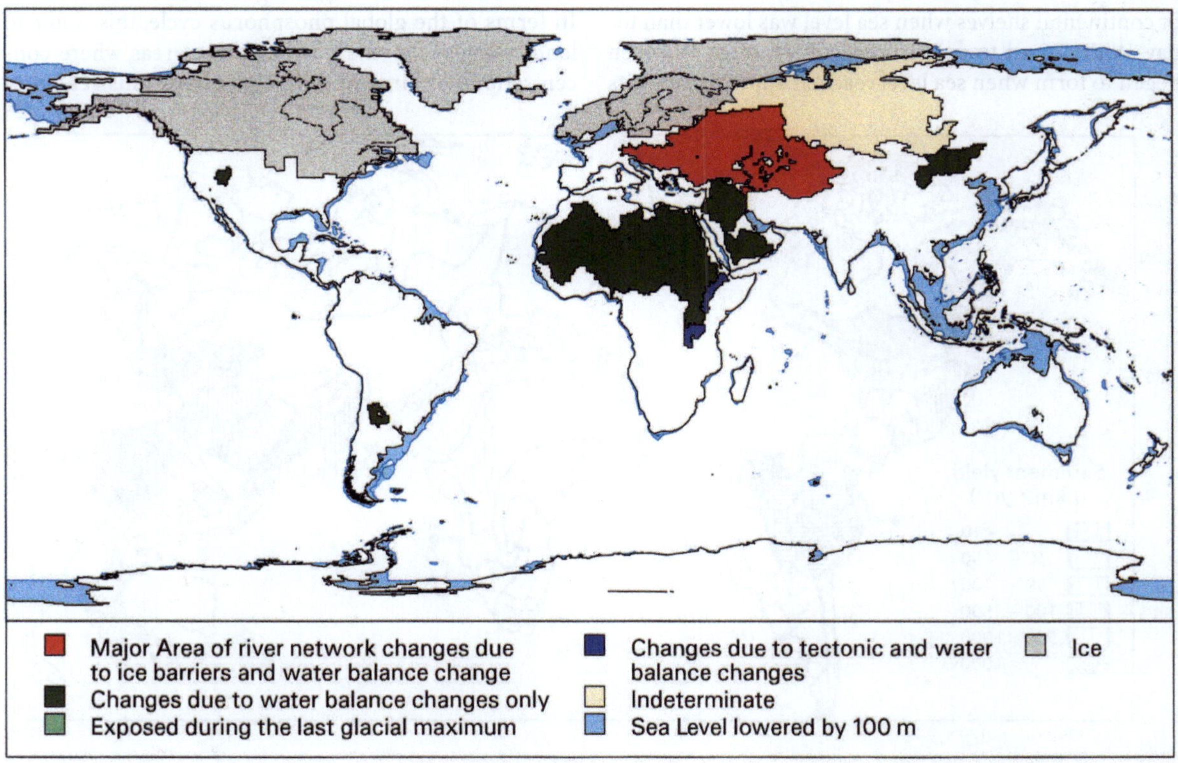

Fig. 2.50. Changes in river networks during the last glacial/interglacial cycle (R. Lammers, pers. comm.)

and Meybeck 1999). Compared to 18 000 years ago about one-third of the present land mass (about 40 million km^2) has been affected. This figure includes the land covered by ice (18.4 million km^2), continental shelf area exposed during the maximum lowering of sea level then (15.9 million km^2) and currently non-flowing rivers discharging during the warm, wet period of the mid-Holocene. These changes were large enough to reverse the flow of Eastern European and West Siberian rivers owing to northern ice barriers. They also created what was potentially the largest drainage system in the world, linking the Black Sea, Caspian Sea and Aral Sea basins through water budget changes in currently dry Central Asia. Other inland water balance changes have affected extended regions in Africa (e.g., Sahara, Okavango-Zambezi Basin), North America (Great Basin, Chihuahua), South America (Mars Chiquita-Parana Basin) and Central Asia (Kerulen-Amur Basin). In addition to these climate-driven changes, recent tectonic activity has also modified river networks, as in the East African Rift, where Lake Victoria and the Kivu, Tanganiika and Turkana Lakes have been connected and/or disconnected from the Zaire and Nile Basins.

Before the development of widespread agriculture, pastoralism and urbanisation, the largest changes of sediment and nutrient flux to the ocean occurred during the major climatic shifts. Deglaciation, for example, exposed large areas of fresh sediment that was moved downstream by meltwater. This so-called periglacial sedimentation has left river terraces in many parts of the world and contributed substantially to the accretion of continental shelves when sea level was lower than today. With regard to modern coastlines, most of which began to form when sea level reached approximately its present position about 6 000 years ago, major changes in sediment flux were driven by the widespread decline in available moisture between ca. 4 000 and 6 000 years ago, especially in lower latitudes. Many rivers delivered less sediment to the coast and delta growth slowed. However, and probably more importantly, concatenations of events – e.g., wet periods, followed by fire, then floods – have driven most changes in sediment flux. Extreme events have been and are very important. These temporal patterns of change have been superimposed onto broad spatial patterns of sediment flux (Fig. 2.51). Catchments with high relief, steep slopes and earthquakes produce much more sediment than low gradient, low relief and aseismic catchments. Before human disturbance, rivers draining the Himalayas, China, the Andes and high lands such as New Guinea dominated the sediment flux to the oceans. They probably also dominated the particulate phosphorus flux.

2.5.4 Connectivity via the Biota

The focus of the previous three sections has been on the abiotic transfer of materials within the Earth System. However, the biosphere also plays a role in the transfer of materials. Phosphorus, for example, is released from the land through the weathering of rocks and transferred from the land to the ocean via the hydrological cycle. Some of this phosphorus is returned to the land by birds and animals migrating between water and land. In terms of the global phosphorus cycle, this water to land transport is small. However, in areas where concentrations of birds or animals is large, this return of

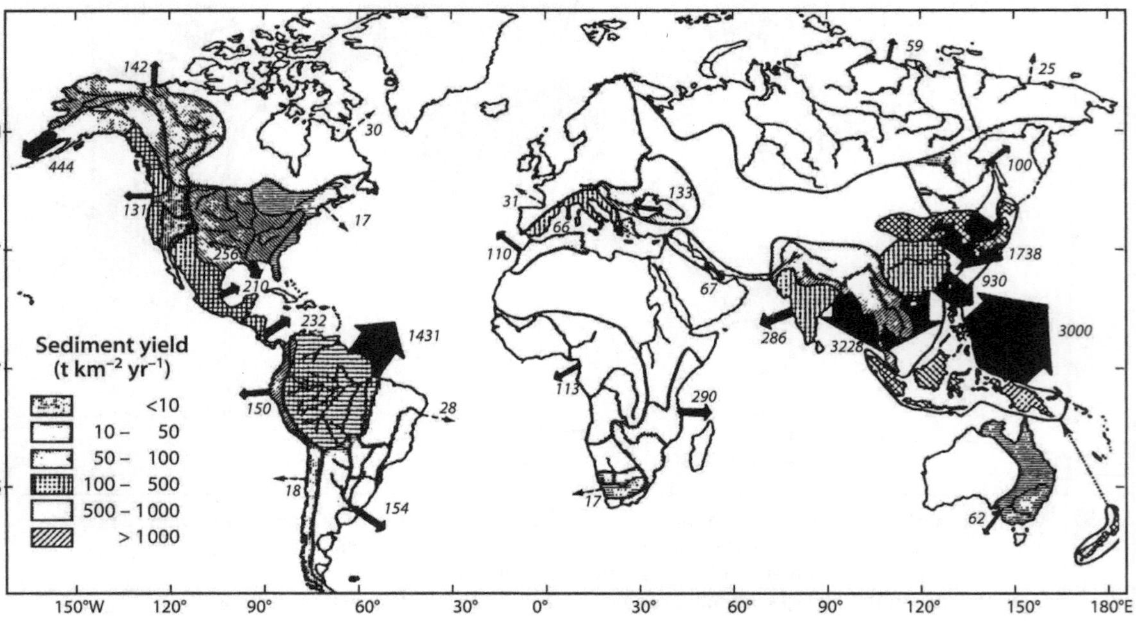

Fig. 2.51. Annual suspended sediment discharge (Milleman and Meade 1983). Relative magnitudes of discharge are indicated by the widths of the *arrows*

phosphorus (and nitrogen) from water to land can influence local vegetation patterns and soil chemistry, which in turn, can influence cycling of other materials. The island of Nauru in the southwestern Pacific Ocean is a good example of how phosphorus transport has been brought about by sea birds. The deposits of phosphorus on the island have become so large that they are now extensively mined for fertiliser production.

One of the more intriguing examples of biogeochemical connectivity via biota occurs in the Amazon Basin, where the rainforest trees have developed a unique way to share phosphorus with their neighbours (Artaxo et al. 2002). In these tropical ecosystems phosphorus is the most limiting nutrient. The phosphorus content of Amazonian rainwater is exceptionally low, about 2 ppb or less. Thus, ensuring the most efficient possible distribution of phosphorus in the ecosystem is important to maximising productivity. One way in which phosphorus is distributed throughout the ecosystem is through the emission of aerosols that contain phosphorus. Figure 2.52 shows the pattern of aerosol concentration in the canopies of Amazonian rainforests.

Two features stand out in the observations. First, there is a bias in the coarse fraction aerosol, which contains most of the phosphorus, towards the lower parts of the canopy airspace. There is no such bias in the fine fraction aerosols. Secondly, the plants preferentially emit the coarse aerosols during the night so that the phosphorus concentrations are even more highly skewed towards the lowest levels of the canopy during the night. This recycling of phosphorus in the ecosystem is important because the Amazonian rainforests lie beneath one of the most highly convective regions of the Earth's atmosphere. During the daytime convection drives air (and gases and materials) from the surface high into the troposphere, where they can be transported far away from their point of origin. During the night atmospheric conditions are highly stable with a thin boundary layer, minimising the risk of particles being transported far from their origins via convection. It appears that the plants emit phosphorus-containing particles primarily in coarse mode and preferentially at night to minimise the risk of losing this key nutrient from the ecosystem, thus sharing it only with their immediate neighbours.

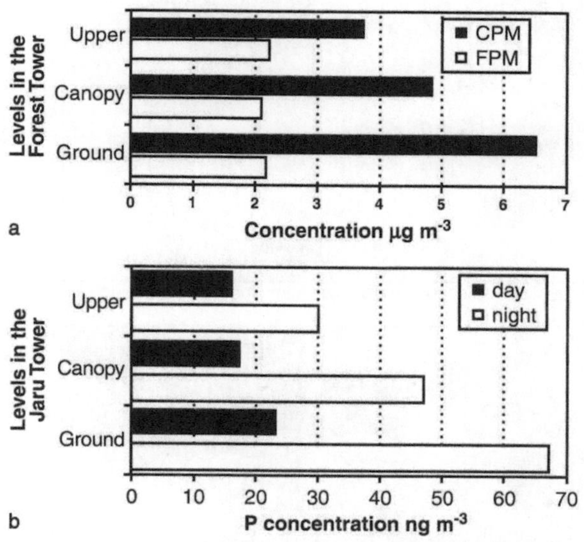

Fig. 2.52. Aerosol and phosphorus concentrations in the canopy of Amazonian rainforest: **a** average aerosol particle concentration in the fine size fraction (FPM) ($dp < 2$ µm) and in the coarse mode size (CPM) ($2 < dp < 10$ µm) for three levels (upper level: 52 m; canopy level: 28 m; ground level: 2 m) of a primary forest tower in Rondônia; and **b** average coarse mode phosphorus concentrations at the three levels of the primary forest tower in Rondônia for daytime and nighttime samples (Artaxo et al. 2001)

Examples of elemental transport by biota are more common in the ocean, where the microscopic animals have much more freedom of movement than many terrestrial organisms. For instance, zooplankton are able, to a much greater extent than phytoplankton, to control their own vertical position in the water column. Zooplankton, both through their feeding and excretion, produce dissolved organic carbon and contribute to turnover of other elements as well. Many zooplankton undergo diurnal and/or seasonal vertical migrations between surface and deeper layers of the ocean. Thus, the migrations of zooplankton may contribute to the transfer of carbon and other elements between depth compartments of the ocean. The transfer of, for example, carbon between ocean compartments can potentially be very important for global carbon cycling as the deeper layers of the ocean are not in direct contact with the atmosphere and cannot react directly with the atmospheric carbon cycle.

In terms of carbon cycling, the seasonal migrations of some zooplankters are potentially more important than the diurnal migrations mentioned above. Some species, for example, *Calanus finmarchicus*, the dominant mesozooplankter in the North Atlantic, overwinter in a specific water mass at depths of over 1 000 m. This organism grows in surface waters, where it eats phytoplankton, and then descends to the depth where this specific water mass is found. Thus, the carbon transported with the *Calanus* to the deeper layers of the ocean is carbon fixed in the surface layer. Water masses at 1 000 m are not in direct contact with the atmosphere. Thus, carbon transferred from surface to deeper waters may be sequestered in the deeper ocean for some time. The importance of carbon removal from surface ocean layers by zooplankton and other marine organisms on atmospheric carbon cycling has yet to be demonstrated. However, considerable research activity is now being directed towards quantification of the biological transfer of carbon in the ocean system.

Recent studies (Heath and Jónasdóttir 1999) of the seasonal migrations of *Calanus finmarchicus* have demonstrated that during the winter, when the organism is in diapause, it is transported with the 1 000 m deep water mass. Climatic influences on the movements and distribution of this water mass thus influence the distribution of this organism. In particular, the North Atlantic Oscillation index has been shown to correlate with the numbers of these organisms reaching their summer surface feeding grounds in the North Sea. In periods, then, this organism is a passive passenger travelling with ocean currents. However, it has evolved to carefully select the currents of which it takes advantage. Its choice of currents allows the organism to migrate between the deep open ocean and surface coastal waters without leaving a specific region of distribution. Here is a clear case of connectivity of biological, hydrological and atmospheric (climatic) components of the Earth System.

2.5.5 Teleconnections in the Climate System

Teleconnections are increasingly recognised as an important feature in the climate system. For example, the ENSO mode of variability across the equatorial Pacific is linked to patterns of floods, drought, and forest fires in areas as widely separated as East Africa, tropical and sub-tropical Australia, low to mid-latitude areas of the North and South American west coasts, and the often arid or semi-arid areas inland from these (Glantz et al. 1991) (Fig. 2.53). In the palaeo-record there are even more dramatic examples. For instance, the sequence of climate changes associated with Dansgaard-Oeschger cycles (Grootes et al. 1993) were propagated over a vast area of the globe (see Sect. 2.2.2). Although there is strong evidence in support of the hypothesis that these oscillations were linked to salinity-driven changes in the North Atlantic that provoked the shutdown of the North Atlantic deep water formation, the precise way in which these changes in the North Atlantic region generated widespread responses in that ocean and surrounding lands – in the record of changing wind-blown particle size assemblages in sections of loess in western China and in the sequence of changes in water stratification and bottom-water redox values in the Santa Barbara Basin off the west coast of California (Behl and Kennett 1996) (Fig. 2.6) – is still not clear. These and other examples have emphasised the importance of apparently regional-scale changes in climate dynamics for the Earth System as a whole. Current experience with ENSO-related teleconnections (Labeyrie et al. 2003) and past evidence for the reverberation of teleconnected climate changes and impacts on civilisations (Chap. 5) argue strongly for the importance of understanding the linkages between regional and global change.

In reality, these teleconnections operate through the transport of energy by the two great fluids of the Earth System, the atmosphere and the oceans, and by their

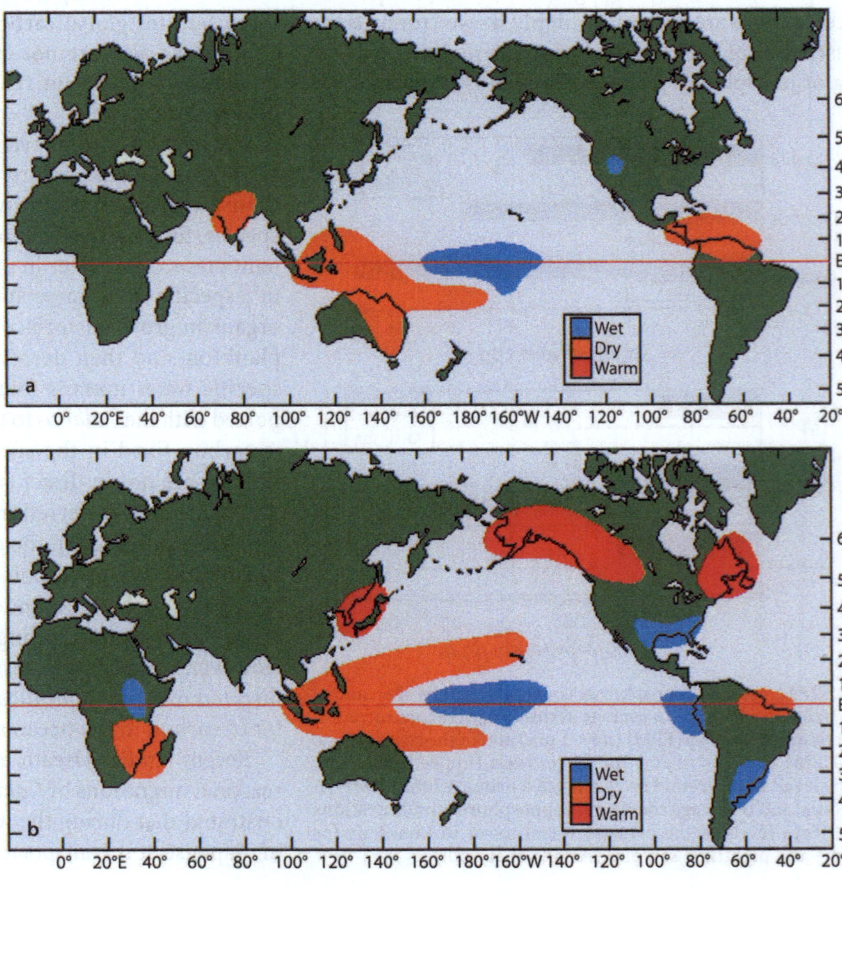

Fig. 2.53. Impacts of ENSO on different regions of the world. **a** Northern hemisphere summer; **b** northern hemisphere winter (*source:* NOAA, *http://www.pmel.noaa.gov/tao/elnino/*)

coupling. The ENSO phenomenon (Fig. 2.54) appears to originate in changes in the equatorial Pacific Ocean trade winds, which in turn cause changes in the surface ocean circulation in the equatorial Pacific. These variations give rise to a change in sea surface temperatures (SST) that feed back to the atmosphere to induce further changes in the surface wind field. This cyclical ocean-atmosphere interaction builds until a vast shift in warm waters occurs across the Pacific with their movement eastward towards the South American coast. The SST in the eastern Pacific can rise by as much as 2–4 °C with an El Niño event. A change of this magnitude in oceanic and atmospheric circulation cannot be confined to the tropical Pacific region only. Global atmospheric circulation must readjust to the changes in the Pacific Basin, and

a December–February Normal Conditions

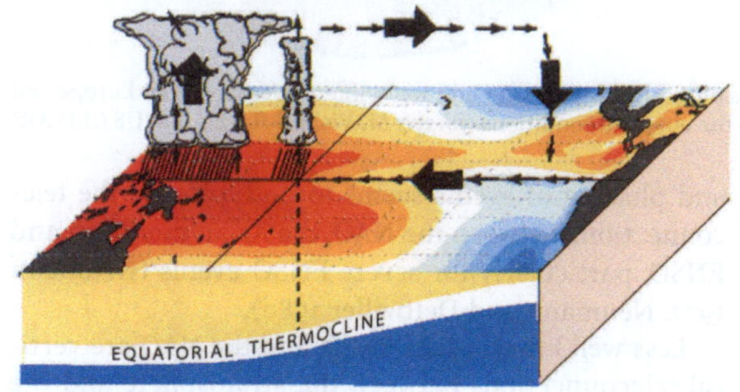

b December–February El Niño Conditions

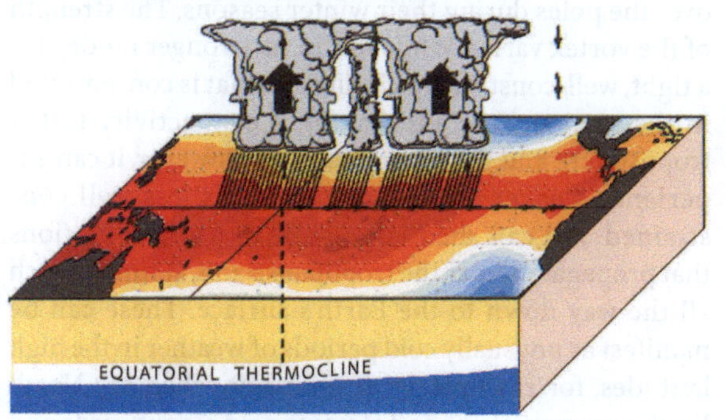

c December–February La Niña Conditions

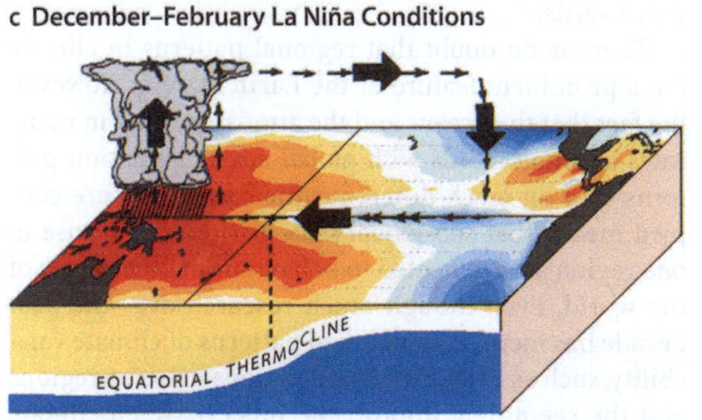

Fig. 2.54. Schematic view of sea surface temperature (*shaded, blue-cold; orange-warm*) and tropical rainfall during **a** normal years, **b** El Niño years and **c** La Niña years (*image:* modified from NOAA, *http://iri.columbia.edu/climate/ENSO/background/basics.html*)

this shifts the patterns of precipitation as well. Torrential rains occur along the coast of Ecuador and Peru, areas that are normally semi-arid, the western part of North America also becomes wetter while the East dries, and severe drought comes to Northeast Brazil, Australia, Indonesia and Southern Africa.

The Asian monsoon is a regional climate system of enormous importance to humans, as about 60% of humanity lives within its direct influence. Much of the intrinsic variability of this system is driven by the dynamics of the system within the region itself. Although the Asian monsoon may appear to be a relatively stable and predictable regional climate system, there is evidence from ice cores in the Himalaya (Thompson et al. 2000) that it has exhibited substantial variability in the past, with devastating consequences for the human societies that depend upon it. The prolonged drought of 1790–1796, clearly recorded in the ice core evidence, is believed to have led to the death of some 600 000 people in northern India (Bradley et al. 2003). In addition, there is some statistical evidence that the behaviour of the Asian monsoon has, at times, been linked to the ENSO phenomenon. For example, the severe fires on the Indonesian islands of Sumatra and Kalimantan in 1997/1998 were likely due to a combination of excessive land clearing coupled with an unusually weak monsoon that was almost surely linked to the strong El Niño event during that period. Interestingly, the link between ENSO and the Indian component of the Asian monsoon has virtually disappeared from 1976 onwards (Kumar et al. 1999a,b), reinforcing the importance of placing current teleconnections in a longer time perspective.

Another major mode of climate variability is the North Atlantic Oscillation (NAO) (Fig. 2.55), which also appears to be driven by changes in ocean-atmosphere coupled dynamics with teleconnections that go beyond the North Atlantic region (Busalacchi 2002). The NAO is a large-scale alteration of atmospheric mass that is manifested as a pressure difference between the Icelandic low-pressure region in the north and the Azores high-pressure region in the south. It is most pronounced during the winter period and seems to have evolved over the last century from a biennial periodicity to its present predominantly decadal periodicity. Despite its direct manifestation in the North Atlantic region, it is associated with fluctuations further afield, in particular with SST fluctuations and surface air temperature seesaws between northern Eurasia and eastern Canada and Greenland, and between Mediterranean Eurasia/Africa and the eastern United States. Transitions in the NAO have also been associated with changes in patterns of precipitation as far away as Northeast Brazil and the western Sahel region of Africa, as well as with storminess over the Atlantic Ocean and adjacent land areas and with shifts in the intensities and pathways of severe hurricanes. There is also strong evidence for a link between

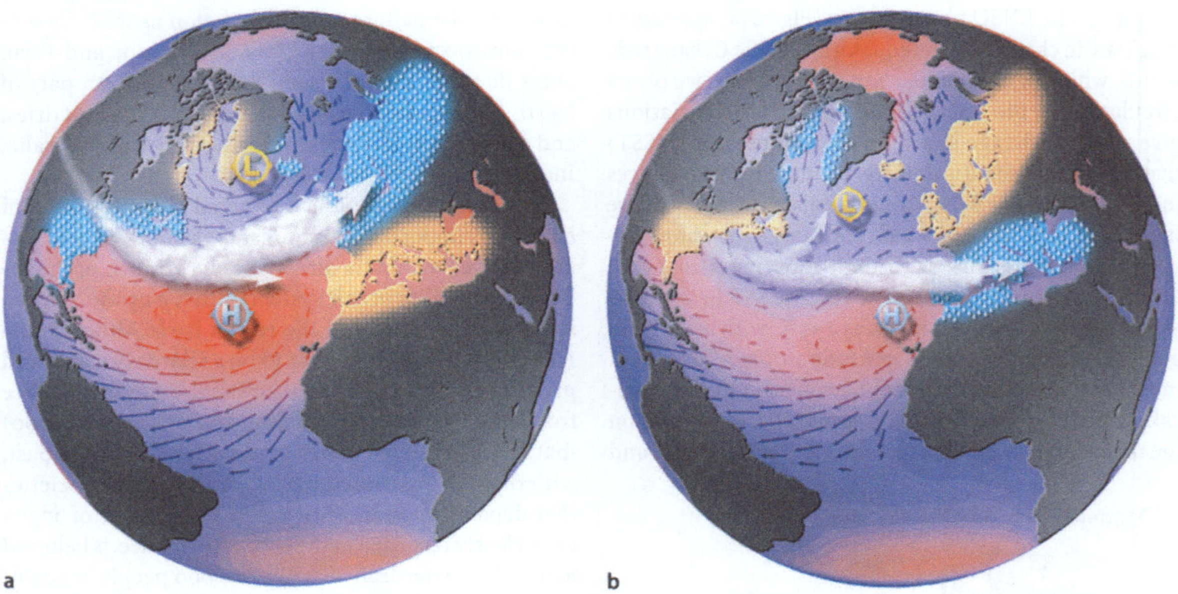

a b

Fig. 2.55. The two phases of the North Atlantic Oscillation: **a** positive phase, leading to mild and wet winters over Northern Europe and dry conditions in the Mediterranean; and **b** negative phase, leading to the opposite conditions (*image:* M. Visbeck/LDEO and US CLIVAR)

the NAO and the so-called Arctic Oscillation (Thompson and Wallace 1998), although the precise nature of the link and the mechanisms responsible for it are not well understood.

Given the global nature of atmospheric circulation, it is likely that all major modes of climate variability are linked at some level. A recent analysis hypothesises that coupling of the North Atlantic Oscillation, the Asian monsoon and the El Niño-Southern Oscillation phenomenon during the 1788–1795 period led to severe climate anomalies around the globe, in turn contributing to social disturbance and upheaval (Grove 2003). During that period there were prolonged drought and crop failures in many parts of central Europe, in the West Indies and Egypt, on St Helena and other islands in the South Atlantic and in India. The climate-induced stress may have contributed to social upheaval in France and Catalonia. Severe drought at the same time nearly led to the collapse of the new British colony in Sydney Cove, Australia. In addition, unusually mild winter conditions were observed on the British Isles and in North America, leading to epidemics of vector-borne diseases in the latter. It appears that these anomalies were due to teleconnections triggered initially in 1788 by a weak phase of the North Atlantic Oscillation, which led to a very cold winter, persistent high pressure over Europe, a cold, wet spring and then a summer drought. This sequence of changes was followed by monsoonal blocking and drought in South and Southeast Asia. About a year after the early stages of the event were observable in southern India, the first stages of an ENSO event were noticed as a warm current along the Peruvian coast. More recent evidence confirms that this multi-year sequence and phasing of events seems to be typical of the teleconnections between the NAO, the Asian monsoon and ENSO, particularly for severe ENSO events (Neumann 1977; Neumann and Dettwiller 1980).

Less well known than ENSO and the NAO are vertical teleconnections between the stratosphere and the Earth's surface. A strong vortex forms in the stratosphere over the poles during their winter seasons. The strength of the vortex varies. While it is in its stronger mode, it is a tight, well-constrained circulation that is concentrated in the stratosphere and has little connectivity to the troposphere and the Earth's surface. However, it can experience weaker modes in which it is less well constrained and can shed large-scale dynamical motions that propagate across the tropopause and, at times, reach all the way down to the Earth's surface. These can be manifest as unusually cold periods of weather in the high latitudes, for example, in Scandinavia when the North Pole's vortex weakens and streams cold air outwards and downwards.

There is no doubt that regional patterns in climate are a prominent feature of the Earth System. However, the fact that the oceans and the atmosphere act in many ways as single fluids with global-scale circulation patterns (but on much different time scales) and are coupled means that sub-global scale patterns that arise in one region are connected to behaviour in other parts of the world. Even though much research over the past decade has focused on regional patterns of climate variability, such as ENSO, teleconnnections between regions and the seemingly improbable links between climate events in far distant corners of the planet are a characteristic feature of the Earth System.

2.5.6 Connectivity through Time: The Legacy of Past Disturbance

Earth System processes are also linked over time. The existence of oxygen in the atmosphere is a legacy of the activities of earlier forms of life, beginning with the appearance of photosynthetically active organisms in the ocean 3.8 billion years ago. Biogeochemical and ecological processes occurring today reflect the evolution of landscapes over millions of years, and the responses of ecological systems over decades to hundreds of years. Biosphere-atmosphere interactions have a memory, and concentrations and content of the atmosphere today are influenced by what they were yesterday, as well as by interactions with the biosphere on much longer time scales. Likewise, ecological systems have memories, and their states and functioning at any given time reflect the state and functioning of that system as it was decades, centuries, and even millennia before. Moreover, their responses to various stresses or forcings are determined or at least tempered by that history. In the past decade much progress has been made in identifying some of the connections between past and present. Three examples taken from terrestrial ecology demonstrate these temporal links; they illustrate the importance of the past in understanding the current dynamics of the Earth System.

The current functioning of the coastal plain of eastern Siberia reflects processes that occurred thousands of years ago (Chapin et al. 2003). During the Pleistocene this region was a steppe-like dry tundra grazed by large herds of bison, mammoths, horses, and caribou, and was underlain by deep and organic rich soils. Due to the cold climate, ground ice occupied 50–70% of the soil volume (Zimov et al. 1995). At the beginning of the Holocene about 11 600 years ago the animals disappeared, due to some combination of human hunting and climatic change, and the steppe-like dry tundra was converted to the moss-dominated tundra that currently occupies the region. The warmer Holocene climate melted much of the ice-rich soil, forming lakes with organic rich sediments that are now important sources of methane to the atmosphere. Thus, the carbon processing in these lakes today, as observed by the ^{14}C content of the CH_4 emitted, carries the imprint of changes that occurred at the beginning of the Holocene. In a similar vein, the fact that tropical forests are the single most important source of nitrous oxide today can be attributed to soil development that has been taking place for millions of years. In these soils, long-term, uninterrupted leaching of water and elements through soil has led to a general depletion of rock-derived nutrients like phosphorus and calcium. As these rock-derived elements are lost and thus become limiting to plant growth, atmospherically derived nutrients such as nitrogen cycle in excess and are easily lost. Connections through time matter.

Disturbance history also influences the behaviour of ecosystems. For example, research over the past decade has shown that, in many of the regrowing forests of the eastern USA and parts of Europe, ecosystem functioning today reflects the severity, frequency, and type of disturbance experienced hundreds of years ago (Foster 1992; Foster et al. 1992; Fuller et al. 1998). Ecologists have long recognised that ecosystem structure and function change during succession, as species replace one another over time. What has only recently become evident is that the actual rate and sequence that succession takes, and the changes in rates of carbon exchange and accumulation, nutrient cycling and trace gas, energy and water exchange that accompany that succession, are all affected by the characteristics of the disturbance. Figure 2.56 shows the pattern of sudden, episodic loss of carbon from a forest due to disturbance and the slow regrowth and associated carbon uptake of the forest. Depending on the time period over which the carbon flux is measured and the position in the successional cycle of this period, a forest can appear to be a strong sink, carbon neutral, or even a carbon source. The long term temporal dynamics of forest succession likely play an important role in terrestrial carbon dynamics at the global scale. Recent analyses (Schimel et al. 2001; Prentice et al. 2001; McGuire et al. 2001) point to rebound from changed disturbance regimes (especially past land use) as a significant contributor to the observed terrestrial sink, along with physiological processes such as fertilisation by increasing atmospheric CO_2 and/or nitrogen deposition. It is still not possible, however, to attribute relative magnitudes to the various processes that are contributing to the terrestrial carbon sink.

A third example shows very long timescale linkages triggered by human activities at a time when humans were an insignificant force in the functioning of the Earth System. Centuries ago in southern Africa pastoralists penned their cattle in the centre of the village at night to protect them from predators. Over the approximately 50-year life of the village, dung from the animals fertilised the site. When humans left the area, the nutrient

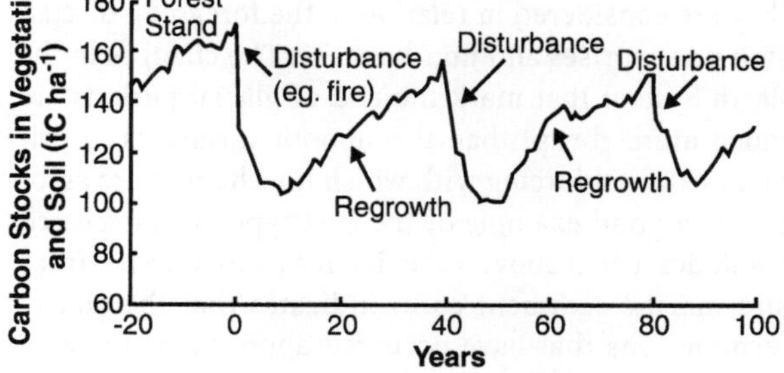

Fig. 2.56. The time-evolution of annual-average on-site carbon stocks for a forest ecosystem subject to periodic disturbances (adapted from RT Watson et al. 2000)

enrichment shifted the species composition of the patch towards sweet, nitrogen rich, as opposed to the surrounding less palatable sour, nitrogen-poor, veld. Wild animals continued to preferentially graze these patches, maintaining a high level and turnover of nutrients. Even today, seven centuries later, the ancient pattern of nutrient rich patches remains, uninterpretable without an understanding of the grazing and settlement patterns of early humans in the area (Blackmore et al. 1990).

2.6 New Insights into Non-linearities, Surprises and Thresholds in the Earth System

Strictly speaking, surprises can only be defined in relation to expectations. Hence, they are always subjective and, in part at least, a comment on the current lack of understanding, or at least the inadequacies of conceptual models of what may be expected. By contrast, non-linearities should be definable in objective, quantitative and, ideally, mathematical terms. In reality, very few processes in nature are linear, although linear approximations often have to serve in modelling. Three kinds of past surprise findings that were unexpected given the state of knowledge of the time are especially relevant:

- examples of situations where there is a disproportionate response to a given forcing;
- cases where outcomes from a given set of apparent causal factors are counter-intuitive according to the present state of knowledge; and
- situations where occurrences that are well documented in the past lie beyond the range of expectations based on direct observation and instrumental records.

2.6.1 Glacial Terminations

In preceding sections some of the dramatic changes in the Earth System during the geologically recent past have been considered first as reflections of temporal variability (Sect. 2.2.2), then in terms of the spatial teleconnections that they demonstrate (Sect. 2.5.5). Here they are considered in relation to the foregoing discussion on surprises and non-linearity. The changes in the Earth System that mark the end of glacial periods are much more abrupt than the smooth variations in primary external forcing with which the changes are associated, a good example of the first type of unexpected result described above. All evidence from both polar ice and marine sediment cores indicates that the glacial terminations that have occurred approximately every 100 000 years over at least the last 600 000 years are much more rapid than glacial onsets (e.g., Bradley 1999). Furthermore, the amplitude of the shift from glacial to interglacial conditions seems massively out of proportion to the small changes in solar radiation to which Milankovitch forcing has given rise on 100 000 year intervals through changes in orbital eccentricity. Both the pace and the magnitude of change are thus truly surprising and must reflect strong feedback mechanisms operating within the Earth System. Raynaud et al. (2003) set out the sequence for each of the last four transitions as revealed by the latest studies on the Vostok core (Petit et al. 1999; Pépin et al. 2002) when they pointed out that changing the orbital parameters initiated the glacial-interglacial climatic changes, then the greenhouse gases amplified the weak orbital signal, accompanied several thousand years later by the effect of decreasing albedo during the retreat of the northern hemisphere ice sheets.

The glacial termination with the most information available is the last one, beginning some 16 000 years ago and ending around 11 600 years before present with the initiation of the Holocene (Fig. 2.57). Two features demonstrate without a doubt that the termination was not a single, rapid, globally coherent, monotonic process. At the very beginning of the warming trend from glacial to Holocene, the record from Antarctica leads that from Greenland. Moreover, during the period of warming, the sharp fluctuations in temperature that mark the Greenland record in particular and the northern hemisphere more generally are largely out of phase with the changes in Antarctica. Thus, any concept of a rapid shift associated with glacial terminations has to be qualified by the realisation that the rate of change is not uniform over the whole globe and that the oscillations that are superimposed on the trend are not globally parallel in sign. The rapid oscillations that are superimposed on the warming trend and which tend to emphasise the dramatic nature of the termination in many parts of the northern hemisphere especially are, as noted in Sect. 2.2.2, the last of the Dansgaard-Oeschger cycles. These themselves are examples of major surprises, so much so that the first evidence for them in the Greenland Camp Century ice core record (Dansgaard et al. 1969) needed independent confirmation before becoming generally accepted.

2.6.2 Heinrich and Dansgaard-Oeschger Events

A vast range of climate indicators in ice, marine and continental archives point to rapid, high amplitude switches between full glacial and relatively mild interglacial conditions between about 70 000 years and 11 000 years ago (see, for example, Fig. 2.6). Many of these switches appear to have been well over half the amplitude of the temperature change from the glacial maximum to peak Holocene warmth. Moreover, the events themselves seem to be grouped in cycles (Bond and Lotti 1995; Bond et al. 1997) beginning with the most severe and gradually declining in amplitude until the next multiple cycle is initiated. In the North Atlantic

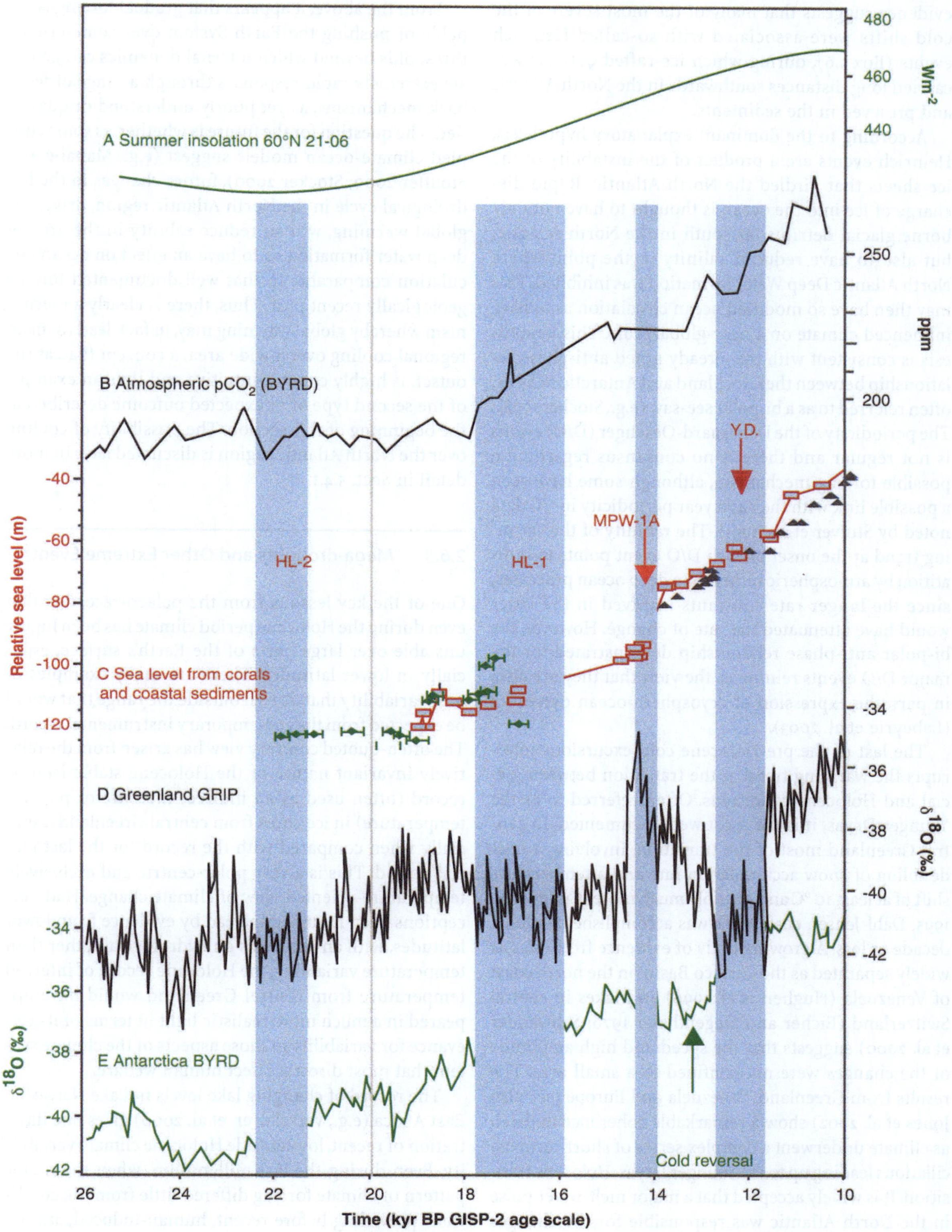

Fig. 2.57. Last deglaciation (25 000–10 000 years ago) through various proxies: **A** northern summer insolation (based on Berger 1977); **B** atmospheric pCO_2 from the Byrd ice core (Blunier et al. 1998); **C** relative sea level from Barbados coral (*squares*) (Fairbanks 1989), Tahiti coral (*triangles*) (Bard et al. 1996), and sedimentary facies (*diamonds*) from northwest Australian shelf (Yokohama et al. 2001) and Sunda shelf (Hanebuth et al. 2000); **D** GRIP ice $\delta^{18}O$ (Johnsen et al. 1992); and **E** Byrd ice $\delta^{18}O$ (Blunier et al. 1997). The GRIP and Byrd records are shown here on the GISP-2 time scale based on annual layer counting (Grootes et al. 1993; Meese et al. 1994; Stuiver et al. 1995; Grootes and Stuiver 1997). Correlation of the GISP-2 and GRIP $\delta^{18}O$ signals was done with Analyseries (Paillard et al. 1996) and transferred to the Byrd CH_4 stratigraphy of Blunier et al. (1997, 1998)

evidence suggests that many of the most severe of the cold shifts were associated with so-called Heinrich events (Box 5.6), during which ice-rafted detritus was carried long distances southwards in the North Atlantic and preseved in the sediments.

According to the dominant explanatory hypothesis, Heinrich events are a product of the instability of the ice sheets that girdled the North Atlantic. Rapid discharge of ice into the ocean is thought to have not only borne glacial detritus far south in the North Atlantic, but also to have reduced salinity to the point where North Atlantic Deep Water formation was inhibited. This may then have so modified ocean circulation as to have influenced climate on a near-global scale. This hypothesis is consistent with the already noted anti-phase relationship between the Greenland and Antarctic records, often referred to as a bi-polar see-saw (e.g., Stocker 1998). The periodicity of the Dansgaard-Oeschger (D/O) events is not regular and there is no consensus regarding a possible forcing mechanism, although some have seen a possible link with the 1 470 year periodicity in ^{14}C data noted by Stuiver et al. (1991). The rapidity of the warming trend at the onset of each D/O event points to inititition by atmospheric rather than deep ocean processes, since the longer rate constants involved in the latter would have attenuated the rate of change. However, the bi-polar anti-phase relationship demonstrated for the major D/O events reinforces the view that they are also, in part, the expression of cryosphere-ocean dynamics (Labeyrie et al. 2003).

The last of the pre-Holocene cold excursions interrupts the warming trend at the transition between glacial and Holocene conditions. Often referred to as the Younger Dryas, it is the most well documented. In central Greenland most of the transition, involving a near doubling of snow accumulation rate and a temperature shift of at least 10 °C and possibly much more (Cuffey et al. 1995; Dahl-Jensen et al. 1998), was accomplished within a decade or less. A growing body of evidence from sites as widely separated as the Cariaco Basin on the north coast of Venezuela (Hughen et al. 1996) and lakes in central Switzerland (Eicher and Siegenthaler 1976; Schwander et al. 2000) suggests that the speed and high amplitude of the changes were not confined to a small area. The results from Greenland, Venezuela and Europe (see also Jones et al. 2002) show a remarkable coherence in detail as climate underwent a complex series of short-term oscillations leading up to the Younger Dryas-Holocene transition. It is widely accepted that a major melt-water pulse in the North Atlantic was responsible for the changes in climate and CO_2 associated with the Younger Dryas (Marchal et al. 1999). As in the case of the preceding Dansgaard-Oeschger cycles, this would have given rise to major changes in thermohaline circulation. However, no credible *external* forcing mechanism has been established for these widespread and dramatic shifts.

From the above, it appears that gradual forcing is capable of pushing the Earth System over some crucial thresholds beyond which internal dynamics can generate extremely rapid responses through a range of feedback mechanisms, as yet poorly understood or quantified. The question for the future is whether, as some coupled climate-ocean models suggest (e.g., Manabe and Stouffer 2000; Stocker 2000), future changes in the hydrological cycle in the North Atlantic region, driven by global warming, will so reduce salinity in the area of deep water formation as to have an effect on ocean circulation comparable to that well documented for the geologically recent past. Thus, there is clearly a mechanism whereby global warming may, in fact, lead to mean regional cooling over a wide area, a concept that, at the outset, is highly counter-intuitive and thus an example of the second type of unexpected outcome described at the beginning of this section. The possibility of cooling over the North Atlantic region is discussed later in more detail in Sect. 5.4.1.

2.6.3 Mega-droughts and Other Extreme Events

One of the key lessons from the palaeo-record is that even during the Holocene period climate has been highly unstable over large parts of the Earth's surface, especially in lower latitudes. There are many examples of past variability that lie well outside the range that would be expected from the contemporary instrumental record. The often-quoted contrary view has arisen from the relatively invariant nature of the Holocene stable isotope record (often used as an indirect measure of past air temperature) in ice cores from central Greenland, especially when compared with the record for the last glacial period. This is a very polar-centric and exclusively temperature-oriented view of climate change. Had perceptions been moulded instead by evidence from lower latitudes, with an emphasis on hydrological rather than temperature variability, the Holocene record of inferred temperature from central Greenland would have appeared in a much more realistic light in terms of its relevance for variability in those aspects of the climate system that most directly affect human welfare.

The record of changing lake levels in Lake Naivasha, East Africa (e.g., Verschuren et al. 2000) gives one illustration of recent, low latitude Holocene climate variability. Even during the last millennium, when the major pattern of climate forcing differed little from the conditions prevailing before recent, human-induced, atmospheric greenhouse gas increases, there were at least three marked periods of extended drought. In each case, the amplitude and duration greatly exceeded the span and severity of any droughts that have occurred during the short, recent period for which instrumental records exist. Bradley et al. (2003) show that these changes closely parallel changes in inferred trade wind strength on the other side of the Atlantic, in the Cariaco Basin, off the northern coast of Venezuela (Fig. 2.58). Both records show strong coherence with variations in the ^{14}C concentration in atmospheric CO_2. If these variations reflect changes in solar radiation, the Lake Naivasha record illustrates not only the wide amplitude of recent variability at a regional scale but also the disproportionate regional response to what are very slight variations in incoming radiation of the order of 0.24%. The possible link between changing atmospheric ^{14}C and solar variability remains controversial however, since atmospheric ^{14}C is also influenced by the rate of ocean turnover, which could be governed by processes unrelated to solar variability.

Dramatic evidence for recent climate extremes well beyond the range of the instrumental record exists in regions outside the low latitudes also. In the central and western United States even the Dust Bowl of the 1930s pales into insignificance alongside the widespread, extreme and unrelenting drought during the fifteenth century (Cook et al. 1999 and Fig. 2.11). Indeed, as already noted above, the story is similar in many parts of the world whether in terms of the duration of severe droughts, the changing magnitude-frequency relationships of extreme events such as floods, or switches in the periodicity and amplitude of dominant modes of variability such as ENSO. During the second half of the Holocene, when external forcing as well as global sea level and ice extent were broadly similar to those of today, climate and hydrological variability greatly exceeded expectations derived from the short, recent instrumental record alone.

In summary, the palaeo-record is replete with instances that show how strongly the Earth System has been influenced by internal feedback mechanisms that have served as amplifiers, generating non-linear responses to even modest external perturbations.

2.6.4 The Browning of the Sahara

The story of the green Sahara in Sect. 2.3.1 has a sequel that demonstrates other features of the Earth System (Claussen et al. 1999; deMenocal et al. 2000). As noted in Sect. 2.3.1, about 6 000 years ago the climate in the Sahel-Sahara region was much more humid than today with vegetation cover resembling that of a modern-day African savanna. About 5 500 years ago, an abrupt change in the regional climate occurred, triggering a rapid conversion of the Sahara into its present desert condition (Fig. 2.59).

The ultimate cause was a small, subtle change in Earth's orbit, leading to a small change in the distribution of solar radiation on Earth's surface (Fig. 2.60a). Model simulations (Claussen et al. 1999) suggest that this small change nudged the Earth System across a threshold that triggered a series of biophysical feedbacks that led, in turn, to a drying climate (Fig. 2.60b). The rate at which precipitation decreased, however, was much more rapid than the change in solar radiation. The vegetation changed even more sharply in response to changing rainfall (Fig. 2.60c) and the region became the present-day desert. Given the changes in vegetation, the Earth System model used in this study then predicted the resulting increase in wind erosion and deposition of sand off the West African coast. The agreement of the model re-

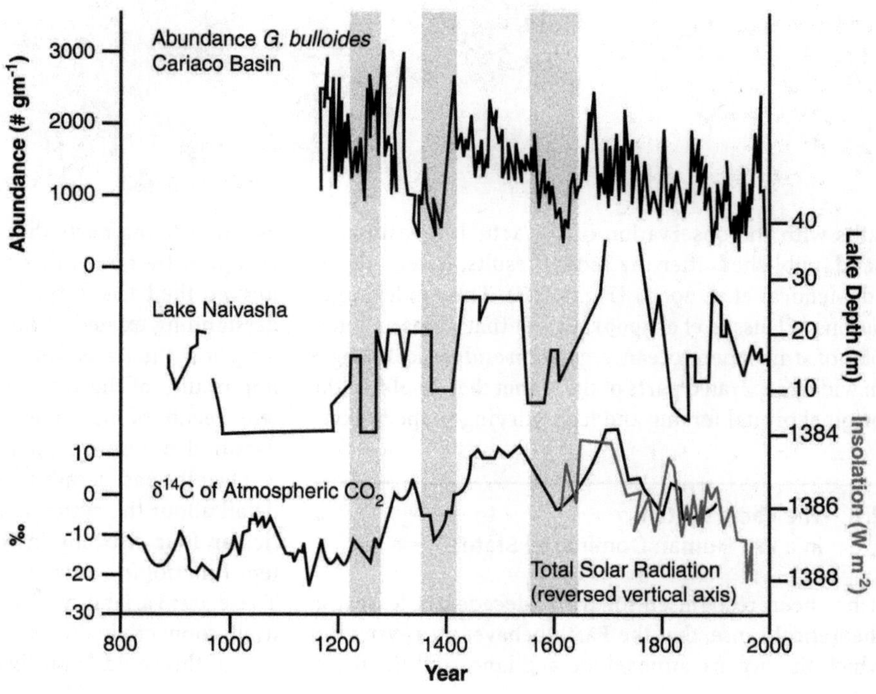

Fig. 2.58. Comparison of records of North Atlantic trade-wind strength (inferred from *G. bulloides* abundance at the Cariaco Basin (Black et al. 1999), Lake Naivasha level (inferred from sedimentological indicators; Verschuren et al. 2000) and solar radiation (inferred from the $\Delta\delta^{14}C$ of atmospheric CO_2 by Stuiver and Reimer (1993) and for the past 400 years from a reconstruction by Lean et al. (1995)). Several of the multidecadal changes in these records are coincident (*highlighted by grey bars*), suggesting the possibility of a common response to radiative forcing on this time scale (Bradley et al. 2003)

Fig. 2.59.
a Much of North Africa is currently a vegetation-less desert (*photo:* IGBP Image Archive); and **b** during the mid-Holocene, about 9 000 to 6 000 years ago, this same region was a savanna, carrying significant populations of large animals and humans (*photo:* D. Parsons)

sults with the observation of the actual deposition of sand, published after the model results, is remarkable (deMenocal et al. 2000) (Fig. 2.60d). The model simulations (Claussen et al. 1999) suggest that it was an interplay of atmosphere, ocean, vegetation and sea ice changes in widely separated parts of the planet that amplified the original orbital forcing and led to drying in the region.

2.7 The Earth System in a Pre-Human Dominated State

It has been recognised for many decades, at least in a theoretical sense, that the Earth behaves as a system in which the oceans, atmosphere and land, and the living and non-living parts therein, are all connected. While accepted by many, this working hypothesis seldom formed the basis for global change research. Little understanding existed of how the Earth worked as a system, how the parts were connected, or even about the importance of the various component parts of the system. Feedback mechanisms were not always clearly understood, nor were the dynamics controlling the system.

Over the past decade much has been learned in greater detail about the systemic nature of the planet. In particular, four critically important features of Earth System functioning have come into much sharper focus. These are: variability in space and time, the role of life itself, connectivity in space and time and the consideration of thresholds, non-linearities and abrupt change.

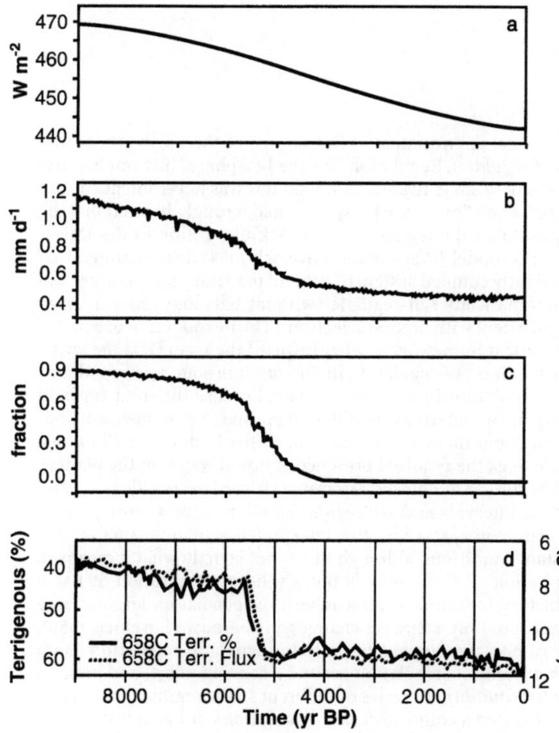

Fig. 2.60. The abrupt change from savanna to desert in North Africa during the mid-Holocene: **a** change in the regional flux of solar radiation at Earth's surface due to a small change in the Earth's orbit; **b** simulated change in rainfall ultimately triggered by the change in incident solar radiation; **c** change in fraction vegetation cover in response to changing rainfall; and **d** model predictions (*solid line*) and observations (*broken line*) of the resulting increase in wind erosion and deposition of sand off the West African coast (Claussen et al. 1999; deMenocal et al. 2000)

Variability in space and time. Human perceptions of the nature of the Earth System are strongly biased by the short period of documentary records and quantitative observations. Human civilisations have developed during the Holocene; the Industrial Revolution began scarcely two centuries ago; and the period of instrumental measurements on which much knowledge of the global environment is based is even shorter. These facts have given human societies a highly misleading picture of the variability of the planetary life support system upon which they depend. Variability is a predominant feature of the Earth System at almost every space and time scale. Many of these modes of natural variability are beyond human experience, with amplitudes that are not captured in the modern instrumental record. Antarctic and Greenland ice cores show that major shifts in the climate system have occurred in relatively short time frames, such as the sub-millennium scale. Changes in the hydrological cycle have occurred over even shorter periods of time and have been observed during the late Holocene. For example, the droughts experienced during the past several centuries have been much less severe than those that have occurred during the past 2 000 years.

Many of these natural modes of variability of the Earth System have characteristic patterns, limits and time scales that provide insights into the combinations of forcings and feedbacks that give rise to them. Understanding Earth's natural variability also provides a critical baseline upon which to evaluate the impacts of the growing human influence on the planetary machinery. Finally, a knowledge of Earth's behaviour in the past gives warnings about what could happen if human activities inadvertently trigger modes of natural variability that have not operated during the period of human civilisation but would have devastating consequences if they occur in the near future (Chap. 5).

The role of life itself. Biological processes interact strongly with physical and chemical processes to create the planetary environment, but biology plays a much stronger role than previously thought in the functioning of the Earth System. Biological processes in the sea and on land strongly influence the atmospheric concentration of CO_2. The rooting patterns and activity of vegetation affect both carbon and water storage and fluxes between the land and the atmosphere. The biological diversity of terrestrial and marine ecosystems influences the magnitude of key ecosystem processes such as productivity, and plays a role in the long-term stability of ecosystem functioning in the face of environmental change.

Much more, however, needs to be understood about the role of biological processes in the functioning of the Earth System. Two questions stand out. The first is focused on biodiversity. The current debate about biodiversity cannot be considered as an aesthetic, cultural or economic question alone, important as these aspects may be. Biodiversity is now becoming a central feature in research, discussion and debate about the future of Earth's life support system. How important is biological diversity for the functioning of the planet? Do species really matter? If so, which ones and why? How much diversity or complexity in ecological systems can be lost before critical Earth System functions are impaired?

The second, related, question focuses more strongly on the nature of the feedbacks from ecological systems to the abiotic environment. There are many examples, such as the emission by marine and terrestrial plants of aerosols that enhance precipitation by acting as cloud condensation nuclei, where biological systems act in ways that help to control the abiotic environment around them. These examples lie at the heart of the Gaia theory (Lovelock 1979) (Box 2.6). Yet other examples can be found, such as the browning of the Sahara (Sect. 2.6.4), where vegetation appears, in a non-Gaian way, to amplify a deleterious change triggered by a very modest change in an abiotic forcing factor. While there is no doubt that biology plays a crucial role in the function-

Box 2.6. Gaia

Tim Lenton

The Gaia concept arose in the 1960s when James Lovelock was employed by NASA to help design means of detecting life on Mars. Lovelock took a general approach to the problem of life detection and suggested that if a planet harbours abundant life, it will have an atmosphere shifted many orders of magnitude away from thermodynamic equilibrium (Lovelock 1965). Earth has a remarkable atmosphere, whilst Mars and Venus have atmospheres more modestly shifted from equilibrium by the predictable action of photochemistry (Hitchcock and Lovelock 1967). Earth's atmospheric composition is also stable, in the sense that the concentrations of major gases have varied relatively little over time periods many times the residence time of their constituent molecules. This indicated to Lovelock that the composition of Earth's atmosphere was being regulated and he suggested that life was doing the regulating. The fact that life has persisted on Earth for at least 3.5 billion years, despite a ~25% increase in the luminosity of the Sun, led him to propose that the climate has also been regulated, in concert with the atmospheric composition. The concept and the system to which it refers were named Gaia (Lovelock 1972) after the Greek Goddess of the Earth.

The microbiologist Lynn Margulis played a key role in developing the Gaia concept and framing it as a hypothesis of "... atmospheric homeostasis by and for the biosphere" (Lovelock and Margulis 1974). The Gaia hypothesis postulated that "... the climate and chemical composition of the Earth's surface are kept in homeostasis at an optimum by and for the biosphere" (Lovelock and Watson 1982). A number of observations were given in support, for example, the concentration of oxygen (the product of past photosynthesis), at 21% of the atmosphere by volume, is sufficient to support the activities of large animals but below the threshold as which frequent fires threaten the persistence of slowly regenerating forests, and it has remained within these bounds for at least the last ~350 million years. Gaia had already proved successful as a hypothesis generator; Lovelock proposed that the biologically important elements sulphur and iodine would have gaseous forms produced by marine life and he then discovered dimethylsulphide (DMS) (Lovelock et al. 1972) and methyl iodide (Lovelock et al. 1973). In the late 1970s, the Gaia hypothesis was extended to encompass regulation of aspects of the composition of the ocean, including its salinity (Lovelock 1979) and the concentration of some biologically important elements (Whitfield 1981), including phosphorus and nitrogen, which had been discussed prior to Gaia (Redfield 1958).

Once the Gaia hypothesis had appeared in book form (Lovelock 1979), it attracted strong and constructive criticisms (Doolittle 1981; Dawkins 1983). These led to some of its tenets being altered or abandoned, and in the late 1980s, the original hypothesis was replaced with what Lovelock named the Gaia theory. Unfortunately, the changes are often overlooked and the outdated Gaia hypothesis continues to draw new critiques (Crutzen 2002). Hence the key changes are spelt out here:

- The notion of regulation "by and for the biosphere" was rejected with the realisation that it is the *whole system* of life and its material environment at the surface of the Earth (Gaia) that *self-regulates*. Regulation "for the biosphere" had implied teleology to some readers, although this was never intended, and regulation "by ... the biosphere" had wrongly implied that life was doing the regulating. Lovelock's invention of the Daisyworld model (Watson and Lovelock 1983) demonstrated that a tightly coupled system of life and planetary temperature can automatically self-regulate (without teleology) in a manner consistent with natural selection. (The introduction of a mathematical framework is what inspired the term Gaia *theory*.)
- The notion of regulation in an *optimum* state was broadened to regulation in a *habitable* state, because different types of organism can have very different optima (e.g., temperature optima), but there are shared habitability bounds for all known life (e.g., the required presence of liquid water on the planet).
- The notion of *homeostasis* was restricted to specific cases and time intervals and *self-regulation* adopted as a more general term. Homeostasis is often interpreted as maintenance of constant conditions, although this is not strictly what it means in physiology. Constancy is not a valid generalisation of Earth history, for example, atmospheric oxygen has undergone large and possibly stepwise changes in the past (Lovelock 1988), prior to the current broadly homeostatic regulation. Both *homeorhesis* (regulation around a moving point) and *punctuated equilibria* (a series of different stable regimes) were early suggested as more accurate descriptions of Earth history, but neither term caught on. Homeostasis has also often been misunderstood as implying a system dominated by negative feedback. In fact, as in cybernetics, engineering and physiology, it was always intended to imply a system with a combination of positive and negative feedbacks.

With these changes, the Gaia theory was presented as a framework for understanding the Earth as a system and its development over time (Lovelock 1988). The effects of life within the system continued to be emphasised, but the focus shifted to understanding emergent properties of the whole system, including self-regulation. Four properties of life were identified as the basis for generating biota-environment feedbacks: First, all organisms alter their environment. Second, organisms grow and multiply, potentially exponentially (a positive feedback intrinsic to life). Third, the state of the environment constrains the growth of life (for each environmental variable, there is a level or range at which growth of a particular organism is maximum and constraints outside of which it cannot grow). Finally, natural selection determines that the types of life that leave the most descendants come to dominate their environment.

The different types of biotic feedback that can emerge from this life-environment coupling have been classified (Lenton 1998) and new examples of planetary-scale feedback mechanisms continue to be identified and quantified (Lenton and Watson 2000a, 2000b). Amongst the best established now is the proposal (Lovelock and Watson 1982; Lovelock and Whitfield 1982) that biological amplification of silicate rock weathering is part of a respon-

ing of the Earth System, knowledge of the overall nature of this role remains elusive. On balance, does the biosphere work in ways that influence the abiotic environment in directions of benefit to the biosphere? Does the biosphere provide an overall stabilising influence on the dynamics of the planetary machinery? Is the Earth System a strongly Gaian world or not (Boxes 2.6 and 2.7)?

Connectivity in space and time. Processes in far distant corners of the planet are linked in ways that could hardly be conceived of 10–20 years ago. The great conveyor belt of the ocean moves energy and nutrients in a slow waltz around the planet. Atmospheric circulation picks up dust from the continents and deposits it on the ocean surface thousands of kilometres away, triggering planktonic blooms and thus directly linking terrestrial with marine ecosystems. The combination of climate change and human hunting over 10 000 years ago has stamped northern Eurasian ecosystems with a signature still readable today in the patterns of trace gas emissions.

◄

sive negative feedback on long-term carbon dioxide and global temperature change (Schwartzman and Volk 1989). Much ongoing research was generated by the proposal (Charlson et al. 1987) that emission of dimethylsulphide gas from surface ocean ecosystems generates sulphate aerosol, an increased number density of cloud condensation nuclei and hence more reflective (higher albedo) clouds, with a consequent reduction in insolation and cooling at the surface of the ocean. It was suggested that this could generate negative feedback on the growth of dimethylsulphide producers. It is now understood that above the temperature at which the surface ocean stratifies (~10 °C) the sign of feedback is likely to switch to positive (warming leading to less dimethylsulphide emission thus amplifying the change). This and other biogenic positive feedbacks can amplify changes in the system, for example, at the termination of ice ages. Such behaviour is taken to indicate that at present, over-arching negative feedbacks are nearing the limits of their operation and the system may be approaching a major transition/reorganisation (Lovelock 1991; Lovelock and Kump 1994).

From the outset Gaia has referred (implicitly or explicitly) to a system. Initially this system was described as a quasi-living organism (Lovelock 1979), and later as a super-organism (Lovelock 1988). Both descriptions generated understandable resistance from biologists, whilst appealing to a general readership. The term *Earth System* arose later (NASA 1986) and it is worth asking whether it refers to the same thing? The Gaia system has been defined (Lenton and van Oijen 2002) as the thermodynamically open system at Earth's surface comprising life (the biota), atmosphere, hydrosphere (ocean, ice, freshwater), dead organic matter, soils, sediments and those parts of the lithosphere (crust) and mantle that interact with surface processes (including sedimentary rocks and rocks subject to weathering). The Earth's internal heat source is considered (like the Sun) to be outside of the influence of the system. In practice, the *surface* Earth System as studied by the IGBP and illustrated in the Bretherton diagram (NASA 1986) is the Gaia system minus some of the slowest processes. In contrast, geologists sometimes consider the Earth System to include the entire interior of the planet, and to include states without life, including those before the origin of life, whereas Gaia refers to a system with abundant life. Thus Gaia may be considered a sub-system of the *geologic* Earth System, with a shorter life span.

The Amsterdam Declaration (Box 6.11) has as its first bullet point: "The Earth System behaves as a single, self-regulating system comprised of physical, chemical, biological and human components" (B. Moore et al. 2001). This adopts the central notion of a "self-regulating system" from the Gaia theory. It doesn't say what is regulated, at what level, or what is meant by self-regulation, but these issues continue to be explored in depth in the Gaia literature (Lenton 2002) and in this volume. Thus a common theoretical framework for understanding the system now seems to be emerging, whatever we choose to call it.

Much research on global change and the nature of the Earth System has focused on the vertical fluxes of water, energy and materials between the land and the atmosphere and between the ocean and the atmosphere. While such fluxes are undoubtedly important, there is a growing awareness of the absolutely essential role that lateral flows and transformations play in the functioning of the Earth System. The central importance of the planetary circulation of the two great fluids – the atmosphere and the oceans – is obvious, although much remains to be understood about the nature of these circulations and their coupling. More recent is the appreciation of the critical connectivity of the land to the ocean through riverine systems for global biogeochemical cycling. Even more recent is the evidence that connectivity between land and ocean and between continents via the atmosphere is a surprisingly important part of the natural Earth System. Every place on Earth is downwind of someplace else, and almost every point on the continents is downstream of someplace else.

Thresholds, non-linearities and abrupt change. The behaviour of the Earth System is typified not by stable equilibria but by strong non-linearities, whereby relatively small changes in a forcing function can push the System across a threshold and lead to abrupt changes in key functions. Some of the modes of variability noted above contain the potential for very sharp, sudden changes that are unexpected given the relatively small forcing that triggers such changes. For example, the speed and amplitude of the movement of the Earth System from its glacial to its interglacial state is massively out of proportion to the small change in the distribution of incoming solar radiation that set the transition in motion. Even more dramatic are the sudden swings in temperature in the northern hemisphere, oscillations of up to 10 °C in only a decade, that appear to be associated with changes in the salinity of the surface waters in the North Atlantic Ocean sufficient to trigger major changes in deep water formation and ocean circulation.

The potential for abrupt change is a characteristic that is extremely important for understanding the nature of the Earth System. The existence of such changes has been convincingly demonstrated by palaeo-evidence accumulated during the past decade. Unravelling the triggers of such changes and the internal dynamics of the Earth System that connect the trigger to the outcome is one of the most pressing challenges to improving understanding of the planetary machinery.

Although much more is now known about the functioning of the Earth System in its state prior to significant human influence, much of this knowledge is still fragmentary. There are many examples of individual features of Earth System dynamics and many case studies showing feedbacks and other interactions. Yet, the patterns revealed by the Vostok ice core (Fig. 1.3) demonstrate without a doubt that the Earth behaves as a single, interlinked, self-regulating system in which all of these processes and connections work together to define the behaviour of the System. Thus, in terms of understanding the Earth System in its pre-human dominated form, one of the biggest challenges is to explain the Vostok patterns. Many attempts have been made (see Prentice et al. 2001 for a summary) but none has achieved a convincing explanation; the challenge remains.

Box 2.7. Anti-Gaia

Paul Crutzen

A recent review of the Gaia theory (Crutzen 2002), largely building on arguments presented before by others, summarises the case against the radical versions of Gaia: *Optimising Gaia* which states in James Lovelock's own words that "climate and chemical composition of the Earth are kept in homeostasis at an optimum by and for the biosphere", and its somewhat weaker form *Homeostatic Gaia*: "The notion of the biosphere as an active and adaptive control system, able to maintain the Earth in homeostasis we are calling the Gaia hypothesis". The problem with these *healing Gaia* versions is that they rest on the false assumption that the influence of the biosphere in the Earth System is dominated by negative (healing) feedbacks, counteracting cosmic disturbances and those imposed by the non-living physical parts of the Earth System. Major examples to the contrary can be given, such as the *Snowball Earth* episodes some 2 200 and again 600–750 million years ago when on several occasions most or all of the Earth was covered by glaciers (Hoffmann and Schrag 1999), thereby wiping out much of the then existing life forms. Another example of anti-Gaian behaviour is that of the atmospheric concentrations of the greenhouse gases CO_2 and CH_4 during the glacial and inter-glacial periods of the quarternary.

A further interesting example goes squarely against one which is often presented by the Gaia proponents (Box 2.6) in support of healing Gaia, stating that the constancy of the atmospheric oxygen content at about 21% over the past 350 million years is regulated by the frequency of forest fires: if the atmospheric oxygen level gets too high, more frequent forest fires will re-establish equilibrium. However, integrated over geological timescales actually the contrary is the case: while between fires the forests re-grow, recurrent fires will produce and accumulate increasing amounts of long-lived elemental carbon (charcoal), thereby actually increasing the oxygen and lowering the atmospheric CO_2 content. This positive feedback on the atmospheric oxygen content should be further explored (Kuhlbusch and Crutzen 1995). (It is interesting to note that the impact of the huge meteoritic collision at the K-T boundary some 65 million years ago may have been particularly severe because for the first time large amounts of sunlight absorbing black carbon particles were produced by the extensive forest fires, causing global darkness and subfreezing surface temperatures.)

With many arguments against it, it is good that the healing Gaia concept is now abandoned as a general principle about the role of the biosphere in Earth System, being replaced by versions such as *habitable Gaia*, building on "shared habitability bounds for all known life" (quotes from Box 2.6), a notion which can be further condensed into the questions of why there is this great variety of life on Earth, how it could have co-emerged under such diverse and variable environmental and climatic conditions, and how stable it is. In the new Gaia version, "the effects of life within the system continued to be emphasised, but the focus shifted to understanding emergent properties of the whole system, including self-regulation in the meaning that the whole system of life and its material environment at the Earth surface self regulates". Throughout Earth's history, it appears that this has indeed been the case in stepwise fashion, interrupted by disastrous breakdowns of the existing *equilibria* and adapted biospheres resulting from extraterrestrial causes and internal destabilising forces in the highly chaotic Earth System. If this is the new Gaia, it is closer to what is known now about the Earth System, except maybe for the notion of self-regulation, but it is difficult to see what is actually new.

Self-regulation is a term which is now used instead of *homeostasis* (Box 2.6) and which is also often used in the IGBP community, as in this book and in the Amsterdam Declaration (Box 6.11), whose first sentence is grossly misleading when it is taken out of the context of the rest of the declaration. Those who misinterpret this statement could incorrectly assume that the Earth System as a whole, including the human component, will self-regulate, whatever this vague term means. Such a misinterpretation on self-regulation trivialises the already severe, and strongly growing, impacts by humans on the environment on all scales.

To address the goals of the IGBP to understand, detect and predict major disturbances and instability regimes in the Earth System on timescales of centuries up to millennia, there is little that the new Gaia has to offer as a guiding principle. Also, regarding the creation of new science, the contribution of old Gaia has been modest, contrary to what is often claimed by its supporters. The important role of dimethylsulphide (DMS) in producing cloud condensation nuclei over the oceans (Charlson et al. 1987) can qualify as a Gaia discovery, but whether DMS acts in a *hurting or healing Gaia* fashion is an open question. Measurements are too few to yield a definitive answer. For many researchers in the field of atmospheric chemistry and climate, the greatest stimulus for improved understanding of the processes which determine (note, not regulate) the Earth System has come from the need to understand, predict, and help to avert the ever growing influence of human activities in the *Anthropocene*, the main objective of the IGBP and the other global change programmes.

There is no doubt about the great importance of biological and life processes within the Earth System on all scales and going back far in the past to the miraculous emergence of life from non-living matter. Whether there will ever be something like a unified theory on the evolving role of the biosphere in the Earth System, however, is difficult to say and may well not be possible. Lovelock bravely attempted this but did not succeed. Reaching this goal will be much more difficult than arriving at the unified field theory in the physical world, which, even after its discovery, will be understood by only very few.

About 70 years ago, the Russian geologist Vernadsky (1930) made some prophetic statements which relate closely to the goals of the IGBP and global change research. He stated "Thus the ubiquity and pressure of life can be considered as the expression of the natural principle governing the biogenic migration of elements in the biosphere" and "In an insignificant time the biogenic migration has been increased by the use of man's skill to a degree far greater than to be expected from the whole mass of living matter". Further, "The surface of the Earth has been transformed unrecognizably, and no doubt far greater changes will come", and finally, "We are confronted with a new form of biogenic migration resulting from the activity of the human reason".

Understanding the natural rhythms and patterns of Earth System functioning is an essential prerequisite to understanding the impacts and consequences of global change. Such understanding lays out the field on which the recent changes to the global environment are being played out. The next chapter of this book focuses on the source of these changes to the Earth System – human beings and their societies – and examines in detail how this one species has become a force of global significance over the last few hundred years. Chapters 4 and 5 then place this new force back into the Earth System itself, drawing on the background understanding of Earth System functioning described in this chapter, and ask two fundamental questions: How much are human activities changing the functioning of the Earth System and what are the consequences?

References

Allen DJ, Pickering KE, Thompson A, Stenchikov G, Kondo Y (2000) A three-dimensional total odd nitrogen (NO_y) simulation during SONEX using a stretched-grid chemical transport model. J Geophys Res 105:3851–3876

Alley RB (2000) The Younger Dryas cold interval as viewed from central Greenland. Quaternary Sci Rev 19:213–226

Alley RB, Meese DA, Schuman CA, Gow AJ, Taylor KC, Grootes PM, White JWC, Ram M, Waddington ED, Mayewski PA, Zielinski GA (1993) Abrupt increase in Greenland snow accumulation at the end of the Younger Dryas event. Nature 362:527–529

Alvarez JW, Alvarez W, Asaro F, Michel HV (1980) Extraterrestrial cause for the Cretaceous-Tertiary extinction. Science 208:1105–1107

Alverson K, Oldfield F (2000) Abrupt climate change. Past Global Changes, Bern, Switzerland. PAGES:CLIVAR Newsletter 8(1):7–8

Amiro BD, MacPherson JI, Desjardins RL, Chen JM, Liu J (2003) Post-fire carbon dioxide fluxes in the western Canadian boreal forest: Evidence from towers, aircraft and remote sensing. Agr Forest Meteorol, in press

Amman B, Oldfield F (2000) Biotic responses to rapid climatic changes around the Younger Dryas. Palaeogeogr Palaeocl 159:3–4

Andreae MO, Artaxo P, Fischer H, Freitas SR, Grégoire J-M, Hansel A, Hoor P, Kormann R, Krecji R, Lange L, Lelieveld J, Lindinger W, Longo K, Peters W, de Reus M, Scheeren HA, Silva Dias MAF, Ström J, van Velthoven PFJ, Williams J (2001) Transport of biomass burning smoke to the upper troposphere by deep convection in the equatorial region. Geophys Res Lett 28:951–954

Andreae MO, Artaxo P, Brandão C, Carswell FE, Ciccioli P, da Costa AL, Culf AD, Esteves JL, Gash JHC, Grace J, Kabat P, Lelieveld J, Malhi Y, Manzi AO, Meixner FX, Nobre AD, Nobre C, Ruivo LP, Silva-Dias MA, Stefani P, Valentini R, von Jouanne J, Waterloo MJ (2002) Biogeochemical cycling of carbon, water, energy, trace gases and aerosols in Amazonia: The LBA-EUSTACH experiments. J Geophys Res 107(D20):LBA33-1–LBA33-25

Anklin M, Schwander J, Stauffer B, Tschumi J, Fuchs A, Barnola JM, Raynaud D (1997) CO_2 record between 40 and 8 kyr BP from the Greenland Ice Core Project ice core. J Geophys Res 102: 26539–26546

Apel JR (1987) Principles of ocean physics. Academic Press, New York

Appenzeller C, Stocker TF, Anklin M (1998) North Atlantic Oscillation dynamics recorded in Greenland ice cores. Science 282: 446–449

Armstrong HA (1995) High resolution biostratigraphy (conodonts and graptolites) of the Upper Ordovician and Lower Silurian-evaluation of the late Ordovician mass extinction. Modern Geology 20:41–68

Armstrong HA (1996) Biotic recovery after mass extinction – the role of climate and ocean-state in the post-glacial (Upper Ordovician-Lower Silurian) recovery of the conodonts. In: Hart MB (ed) Biotic recovery from mass extinction events. Geological Society London Special Publication 102:105–117

Artaxo P, Fernandes ET, Martins JV, Yamasoe MA, Hobbs PV, Maenhaut W, Longo KM, Castanho A (1998) Large-scale aerosol apportionment in Amazonia. J Geophys Res 103:31827–31847

Artaxo P, Andreae MO, Guenther A, Rosenfeld D (2001) Unveiling the lively atmosphere-biosphere interactions in the Amazon. Global Change Newsletter No. 45, International Geosphere Biosphere Programme, Stockholm, Sweden, pp 12–15

Artaxo P, Martins JV, Vanderlei J, Yamasoe MA, Procópio AS, Pauliquevis TM, Andreae MO, Guyon P, Gatti LV, Leal AMC (2002) Physical and chemical properties of aerosols in the wet and dry season in Rondônia, Amazonia. J Geophys Res 107(D20):LBA49-1–LBA49-14

Avissar R, Lui Y (1996) A three-dimensional numerical study of shallow convective clouds and precipitation induced by land-surface forcing. J Geophys Res 101:7499–7518

Baldocchi D, Valentini R, Running S, Oechel W, Dahlman R (1996) Strategies for measuring and modeling carbon dioxide and water vapour fluxes over terrestrial ecosystems. Global Change Biol 2:159–168

Baldocchi DD, Vogel CA, Hall B (1997) Seasonal variation of carbon dioxide exchange rates above and below a boreal jack pine forest. Agr Forest Meteorol 83:147–170

Bard E, Hamelin B, Arnold M, Montaggioni L, Cabioch G, Faure G, Rougerie F (1996) Deglacial sea-level record from Tahiti corals and the timing of global meltwater discharge. Nature 382: 241–244

Batjes NH (1996) Total carbon and nitrogen in the soils of the world. Eur J Soil Sci 47:151–163

Battle M, Bender M, Sowers T, Tans PP, Buttler JH, Elkins JW, Ellis JT, Conway T, Zhang N, Lang P, Clarke AD (1996) Atmospheric gas concentrations over the past century measured in air from fern at the South Pole. Nature 383:231–235

Behl RJ, Kennett JP (1996) Brief interstadial events in the Santa Barbara Basin, NE Pacific, during the past 60 kyr. Nature 379: 243–246

Behrenfeld MJ, Bale AJ, Kolber ZS, Aiken J, Falkowski PG (1996) Confirmation of iron limitation of phytoplankton photosynthesis in the equatorial Pacific Ocean. *Nature* 383:508–511

Bender M (2002) Orbital tuning chronology for the Vostok climate record supported by trapped gas composition. Earth Planet Sc Lett 204:275–289

Berger A (1977) Long-term variations of the earth's orbital elements. Celestial Mech 15:53–74

Berger A (2001) The role of CO_2, sea-level and vegetation during the Milankovitch-forced glacial-interglacial cycles. In: Bengtsson L (ed) Proceedings 'Geosphere-Biosphere Interactions and Climate'. Pontifical Academy of Sciences, 9–13 November 1998, Vatican City

Betts RA (2000) Offset of the potential carbon sink from boreal forestation by decreases in surface albedo. Nature 408:187–190

Betts RA (2001) Biogeophysical impacts of land use on present-day climate: Near-surface temperature change and radiative forcing. Atmospheric Science Letters doi:1006/asle.2001.0023

Beuning KRM, Kelts K, Stager JC (1998) Abrupt climatic changes associated with the Younger Dryas interval in Africa. In: Lehman JT (ed) Environmental change and response in East African lakes. Kluwer Academic Publishers, Dordrecht, pp 147–156

Birks HH, Battarbee RW, Birks HJB (2000) The development of the aquatic ecosystems at Krakenes Lake, western Norway, during the late-glacial and early Holocene – a synthesis. J Paleolimnol 23:91–114

Biscaye PE, Grousset FE, Revel M, VanderGaast S, Zielinski GA, Vaars A, Kukla G (1997) Asian provenance of glacial dust (stage 2) in the Greenland Ice Sheet Project 2 ice core, Summit, Greenland. J Geophys Res 102:26765–26782

Black DE, Peterson LC, Overpeck JT, Kaplan A, Evans MN, Kashgarian M (1999) Eight centuries of North Atlantic Ocean atmosphere variability. Science 286:1709–1713

Black TA, den Hartog G, Neumann HH, Blanken PD, Yang PC, Russell C, Nesic Z, Lee X, Chen SG, Staebler R, Novak MD (1996) Annual cycles of water vapour and carbon dioxide fluxes in and above a boreal aspen forest. Global Change Biol 2:219–229

Blackmore AC, Mentis MT, Scholes RJ (1990) The origin and extent of nutrient-enriched patches within a nutrient-poor savanna in South Africa. J Biogeogr 17:463–470

Blunier T, Schwander J, Stauffer B, Stocker T, Dällenbach A, Indermuhle A (1997) Timing of the Antarctic cold reversal and the atmospheric CO_2 increase with respect to the Younger Dryas event. Geophys Res Lett 24: 2683–2686

Blunier T, Chappellaz J, Schwander J, Dälenbach A, Stauffer B, Stocker TF, Raynaud D, Jouzel J, Clausen HB, Hammer CU, Johnsen SJ (1998) Asynchrony of Antarctic and Greenland climate change during the last glacial period. Nature 394: 739–743

Bolin B, Sukumar R, Ciais P, Cramer W, Jarvis P, Khesghi H, Nobre C, Semenov S, Steffen W (2000) Global perspective. In: Watson RT, Noble IR, Bolin B, Ravindranath NH, Verardo DJ, Dokken DJ (eds) Land use, land use change and forestry: A special report of the IPCC. Cambridge University Press, Cambridge, UK, pp 23–51

Bond G, Lotti R (1995) Iceberg discharges into the North Atlantic on millenial time scales during the last glaciation. Science 267:1005–1010

Bond G, Showers W, Cheseby M, Lotti R, Almasi P, deMenocal P, Priore P, Cullen H, Hadjas I, Bonani G (1997) A pervasive millennial-scale cycle in North Atlantic Holocene and glacial climates. Science 278:1257–1266

Bopp L, Kohfeld KE, Le Quéré C, Aumont O (2003) Dust impact on marine biota and atmospheric CO_2 during glacial periods. Paleoceanography, in press

Borman FH, Likens GL (1979) Pattern and process in a forested ecosystem. Springer-Verlag, New York

Bowling DR, McDowell NG, Bond BJ, Law BE, Ehleringer JR (2002) ^{13}C content of ecosystem respiration is linked to precipitation and vapor pressure deficit. Oecologia 131:113–124

Bowring SA, Erwin DH, Jin YG, Martin MW, Davidek K, Wang W (1998) U/Pb zircon geochronology and tempo of the End-Permian mass extinction. Science 280:1039–1045

Boyd PW, Watson A, Law CS, Abraham E, Trull T, Murdoch R, Bakker DCE, Bowie AR, Buesseler K, Chang H, Charette M, Croot P, Downing K, Frew R, Gall M, Hadfield M, Hall J, Harvey M, Jameson G, La Roche J, Liddicoat M, Ling R, Maldonado M, McKay RM, Nodder S, Pickmere S, Pridmore R, Rintoul S, Safi K, Sutton P, Strzepek R, Tanneberger K, Turner S, Waite A, Zeldis J (2000) A mesoscale phytoplankton bloom in the polar Southern Ocean stimulated by iron fertilization. Nature 407:695–702

Braconnot P, Joussaume S, Marti O, de Noblet N (1999) Synergistic feedbacks from ocean and vegetation on the African monsoon response to mid-Holocene insolation. Geophys Res Lett 26:2481–2484

Bradbury JP, Grosjean M, Stine S, Sylvestre F (2000) Full and late Glacial lake records along the PEP 1 Transect: Their role in developing interhemispheric paleoclimate interactions. In: Markgraf V (ed) Interhemispheric climate linkages. Academic Press, San Diego, USA, pp 265–289

Bradley RS (1999) Palaeoclimates: Reconstructing climates of the quaternary, 2nd Edition. Academic Press, San Diego, USA

Bradley RS (2000) Past global changes and their significance for the future. Quaternary Sci Rev 19:391–402

Bradley RS, Alverson K, Pedersen TF (2003) Challenges of a changing earth: Past perspectives, future concerns. In: Alverson K, Bradley R, Pedersen T (eds) Paleoclimate, global change and the future. IGBP Global Change Series. Springer-Verlag, Berlin Heidelberg New York, pp 163–167

Brasseur GP, Prinn RG, Pszenny AAP (eds) (2003) The changing atmosphere: An integration and synthesis of a decade of tropospheric chemistry research. IGBP Global Change Series. Springer-Verlag, Berlin Heidelberg New York

Brenchley PJ, Marshall JD, Carden GAF, Robertson DBR, Long DG, Meidla T, Hints L, Anderson TF (1995) Bathymetric and isotopic evidence for a short-lived late Ordovician glaciation in a greenhouse period. Geology 22:295–298

Broecker WS (1991) The great ocean conveyor. Oceanography 4:79–89

Broström A, Coe MT, Harrison SP, Gallimore R, Kutzbach JE, Foley J, Prentice IC, Behling P (1998) Land surface feedbacks and palaeomonsoons in northern Africa. Geophys Res Lett 25:3615–3618

Brown CW, Yoder JA (1994) Coccolithophorid blooms in the global ocean. J Geophys Res 99:7467–7482

Brown J, Colling A, Park D, Phillips J, Rothery D, Wright J (1989) Ocean circulation. Open University/Pergamon Press, New York

Busalacchi A (2002) The coupled climate system: Variability and predictability. In: Steffen W, Jäger J, Carson D, Bradshaw C (eds) Challenges of a changing earth: Proceedings of the Global Change Open Science Conference. Amsterdam, The Netherlands, 10–13 July 2001. IGBP Global Change Series. Springer-Verlag, Berlin Heidelberg New York, pp 117–120

Caillon N, Severinghaus JP, Jouzel J, Barnola J-M, Kang J, Lipenkov VY (2003) Timing of atmospheric CO_2 and Antarctic temperature changes across termination III. Science 299:1728–1731

Chadwick OA, Derry LA, Vitousek PM, Huebert BJ, Hedin LO (1999) Changing sources of nutrients during four million years of ecosystem development. Nature 397:491–497

Chapin FS III, Matson PA, Mooney HA (2003) Principles of terrestrial ecosystem ecology. Springer-Verlag, Berlin Heidelberg New York, in press.

Charlson RJ, Lovelock JE, Andreae MO, Warren SG (1987) Oceanic phytoplankton, atmospheric sulphur, cloud albedo and climate. Nature 326:655–661

Charney JG (1975) Dynamics of deserts and drought in the Sahel. Q J Roy Meteor Soc 101:193–202

Chatfield RB, Vastano HBSJA, Sachse G (1996) A general model of how fire emissions and chemistry produce African/Oceanic plumes (O_3, CO, PAN, smoke) in TRACE-A. J Geophys Res 101: 24279–24306

Cheddadi R, Yu G, Guiot J, Harrison SP, Prentice IC (1997) The climate of Europe 6000 years ago. Clim Dynam 13:1–9

Chen FH, Bloemendal J, Wang JM, Li JJ, Oldfield F (1997) High-resolution multi-proxy climate records from Chinese loess: Evidence for rapid climatic changes over the last 75 kyr. Palaeogeogr Palaeocl 130:323–335

Claussen M (2004) Does land surface matter in weather and climate? In: Kabat P, Claussen M, Dirmeyer PA, Gash JHC, Bravo de Guenni L, Meybeck M, Pielke Sr, RA, Vörösmarty CJ, Hutjes RWA, Lütkemeier S (eds) Vegetation, water, humans and climate: A new perspective on an interactive system. IGBP Global Change Series. Springer-Verlag Berlin Heidelberg New York

Claussen M, Gayler V (1997) The greening of the Sahara during the mid-Holocene: results of an interactive atmosphere-biome model. Global Ecol Biogeogr 6:369–377

Claussen M, Brovkin V, Ganopolski A, Kubatzki C, Petoukhov V (1998) Modelling global terrestrial vegetation-climate interaction. Philos T Roy Soc B 353:53–63

Claussen M, Kubatzki C, Brovkin V, Ganopolski A, Hoelzmann P, Pachur HJ (1999) Simulation of an abrupt change in Saharan vegetation at the end of the mid-Holocene. Geophys Res Lett 24:2037–2040

Cook ER, D'Arrigo RD, Briffa KR (1998) A reconstruction of the North Atlantic Oscillation using tree-ring chronologies from North America and Europe. Holocene 8:9–17, *http://www.ngdc.noaa.gov/paleo/drought.html*

Cook ER, Meko DM, Stahle DW, Cleaveland MK (1999) Drought reconstructions for the continental United States. J Climate 12: 1145–1162

Cooper DJ, Watson AJ, Nightingale PD (1996) Large decrease in ocean-surface CO_2 fugacity in response to in situ iron fertilization. *Nature* 383:511–513

Cortijo E, Lehman S, Keigwin L, Chapman M, Paillard D, Labeyrie L (1999) Changes in meridional temperature and salinity gradients in the North Atlantic Ocean (30° to 72° N) during the last interglacial period. Paleoceanography 14:23–33

Cox PM, Betts RA, Jones CD, Spall SA, Totterdell IJ (2000) Acceleration of global warming due to carbon-cycle feedbacks in a coupled model. Nature 408:184–187

Craig H, Horibe Y, Sowers T (1988) Gravitational separation of gases and isotopes in polar ice caps. Science 242:1675–1678

Cramer W, Kicklighter DW, Bondeau A, Moore III B, Churkina G, Nemry B, Ruimy A, Schloss AL, The participants of the Potsdam NPP Model Intercomparison (1999) Comparing global models of terrestrial net primary productivity (NPP): Overview and key results. Global Change Biol 5(1):1–15

Cramer W, Bondeau A, Woodward FI, Prentice IC, Betts RA, Brovkin V, Cox PM, Fisher V, Foley JA, Friend AD, Kucharik C, Lomas MR, Ramankutty N, Sitch S, Smith B, White A, Young-Molling C (2001) Global response of terrestrial ecosystem structure and function to CO_2 and climate change: Results from six dynamic global vegetation models. Global Change Biol 7:357–373

Crawford MA, Bloom M, Cunnane S, Holmsen M, Ghebremeskel K, Parkington J, Schmidt W, Sinclair AJ, Leigh Broadhurst C (2001) Docosahexaenoic acid and cerebral evolution. In: Hamazaki T, Okuyama H (eds) Fatty acids and lipids – new findings. World Rev Nutr Diet 88:6–17

Crowley T, North G (1991) Paleoclimatology. Oxford Monographs on Geology and Geophysics No. 18. Oxford University Press, New York, USA

Crutzen PJ (2002) A critical analysis of the Gaia hypothesis as a model for climate/biosphere interactions. Gaia 11:96–103

Crutzen PJ, Andreae MO (1990) Biomass burning in the tropics: Impacts on atmospheric chemistry and biogeochemical cycles. Science 250:1669–1678

Cuffey KM, Clow GD, Alley RB, Stuiver M, Waddington ED, Saltus RW (1995) Large arctic temperature change at the Wisconsin-Holocene glacial transition. Science 270:455–458

Cullen HM, D'Arrigo RD, Cook ER, Mann ME (2001) Multiproxy reconstructions of the North Atlantic Oscillation. Paleoceanography 16:27–39

Dahl-Jensen D, Mosegaard K, Gunderstrup N, Clow GO, Johnsen SJ, Hansen AW, Balling N (1998) Past temperature directly from the Greenland Ice sheet. Science 252:268–271

Dansgaard W, Johnsen SJ, Moller J, Langway Jr CC (1969) One thousand centuries of climatic record from Camp century on the Greenland ice sheet. Science 166:377–381

Dansgaard W, Johnsen SJ, Clausen HB, Dahl-Jensen D, Gundestrup NS, Hammer CU, Hvidberg CS, Steffensen JP, Sveinbjornsdottir AE, Jouzel J, Bond G (1993) Evidence for general instability of past climate from a 250-kyr ice-core record. Nature 364:218–220

Dawkins R (1983) The extended phenotype. Oxford University Press, Oxford, UK

Delmas RJ (1993) A natural artefact in Greenland ice-core CO_2 measurements. Tellus 45B:391–396

Delmonte B, Petit J-R, Maggi V (2002) Glacial to Holocene implications of the new 27000 year dust record from the EPICA Dome C (East Antarctica) ice core. Clim Dynam 18:647–660

deMenocal PB, Ortiz J, Guilderson T, Adkins J, Sarnthein M, Baker L, Yarusinki M (2000) Abrupt onset and termination of the African Humid Period: Rapid climate response to gradual insolation forcing. Quaternary Sci Rev 19:347–361

de Noblet-Ducoudré N, Claussen M, Prentice IC (2000) Mid-Holocene greening of the Sahara: First results of the GAIM 6000 year BP experiment with two asynchronously coupled atmosphere/biome models. Clim Dynam 16:643–659

Doolittle WF (1981) Is nature really motherly? CoEvolution Quarterly 29:58–63

Ducklow HW (2003) Biogeochemical provinces. Towards a JGOFS synthesis. In: Fasham MJR (ed) Ocean biogeochemistry: The role of the ocean carbon cycle in global change. IGBP Global Change Series. Springer-Verlag, Berlin Heidelberg New York, pp 3–17

Dugdale RC, Wilkerson FP (1998) Understanding the eastern equatorial Pacific as a continuous new production system regulating on silicate. Nature 391:270–273

Dupont LM (1993) Vegetation zones in NW Africa during the Brunhes chron reconstructed from marine palynological data. Quaternary Sci Rev 12:189–202

Eden DN, Page MP (1998) Palaeoclimatic implications of a storm erosion record from late Holocene lake sediments, North Island, New Zealand. Palaeogeogr Palaeocl 139:37–58

Ehleringer JR, Bowling D, Fessenden J, Flanagan LB, Helliker BR, Martinelli LA, Ometto JP (2002) Stable isotopes and carbon cycle processes in forest and grasslands. Plant Biology 4:181–193

Eicher U, Siegenthaler U (1976) Palynological and oxygen isotope investigations on Late-Glacial sediment cores from Swiss lakes. Boreas 5:109–117

Emmerson MC, Solan M, Emes C, Paterson DM, Raffaelli D (2001) Consistent patterns and the idiosyncratic effects of biodiversity in marine ecosystems. Nature 411:73–77

Erwin DH (1994) The Permo-Triassic extinction. Nature 367: 231–235

Etheridge DM, Steele LP, Langenfields RL, Francey RJ, Barnola J-M, Morgan VI (1996) Natural and anthropogenic changes in atmospheric CO_2 over the last 1000 years from air in Antarctic ice and fern. J Geophys Res 101:4115–4128

Fairbanks RG (1989) A 17000-year glacio-eustatic sea level record: Influence of glacial melting rates on the Younger Dryas event and deep ocean circulation. Nature 342:637–642

Falkowski PG, Barber RT, Smetacek V (1998) Biogeochemical controls and feedbacks on ocean primary production. Science 281: 200–206

Falkowski P, Scholes RJ, Boyle E, Canadell J, Canfield D, Elser J, Gruber N, Hibbard K, Högberg P, Linder S, Mackenzie FT, Moore III B, Pedersen T, Rosenthal Y, Seitzinger S, Smetacek V, Steffen W (2000) The global carbon cycle: A test of our knowledge of Earth as a system. Science 290:291–296

Fang XM, Ono Y, Fukusawa H, Pan BT, Li J, Guan DH, Oi K, Tsukamoto S, Torii M, Mishima T (1999) Asian summer monsoon instability during the past 60000 years: Magnetic susceptibility and pedogenic evidence from the western Chinese Loess Plateau. Earth Planet Sc Lett 168:219–232

Fasham MJR (ed) (2003) Ocean biogeochemistry: The role of the ocean carbon cycle in global change. IGBP Global Change Series. Springer-Verlag, Berlin Heidelberg New York

Fasham MJR, Baliño BM, Bowles MC (eds) (2001) A new vision of ocean biogeochemistry after a decade of the Joint Global Ocean Flux Study (JGOFS). Ambio Special Report 10

Field CB, Jackson RB, Mooney HA (1995) Stomatal responses to increased CO_2: Implications from the plant to the global scale. Plant Cell Environ 18:1214–1225

Flanagan LB, Ehleringer JR (1997) Ecosystem-atmosphere CO_2 exchange: Interpreting signals of change using stable isotope ratios. Trends Ecol Evol 13:10–14

Flessa KW, Erben HK, Hallam A, Hsü KJ, Hüssner HM, Jablonski D, Raup DM, Seponski JJ Jr., Soulé ME, Stinnesbeck W, Vermeij GJ (1986) Causes and consequensces of extinction. In: Raup DM, Jablonski D (eds) Patterns and process in the history of life. Springer-Verlag, Berlin Heidelberg New York, pp 235–257

Flückiger J, Dällenbach A, Blunier T, Stauffer B, Stocker TF, Raynaud D, Barnola J-M (1999) Variations in atmospheric N_2O concentration during abrupt climatic changes. Science 285:227–230

Flückiger J, Monnin E, Stauffer B, Schwander J, Stocker TF, Chappellaz J, Raynaud D, Barnola JM (2002) High resolution Holocene N_2O ice core record and its relationship with CH_4 and CO_2. Global Biogeochem Cy 10.1029/2001GB001417

Foster DR (1992) Land-use history (1730–1990) and vegetation dynamics in central New England, USA. J Ecol 80:753–772

Foster DR, Zebryk T, Schoonmaker P, Lezberg A (1992) Post-settlement history of human land-use and vegetation dynamics of a *Tsuga canadensis* (hemlock) woodlot in central New England. J Ecol 80:773–786

Fuller JL, Foster DR, McLachlan JS, Drake N (1998) Impact of human activity on regional forest composition and dynamics in central New England. Ecosystems 1:76–95

Gallimore RG, Kutzbach JE (1996) Role of orbitally induced changes in tundra area in the onset of glaciation. Nature 381:503

Ganopolski A, Kubatzki C, Claussen M, Brovkin V, Petoukhov V (1998) The influence of vegetation-atmosphere-ocean interaction on climate during the mid-Holocene. Science 280:1916–1919

Gasse F (2000) Hydrological changes in the African tropics since the Last Glacial Maximum. Quaternary Sci Rev 19:189–211

Gasse F, Van Campo E (1994) Abrupt post-glacial events in West Asia and north Africa. Earth Planet Sc Lett 126:453–456

Glantz MH, Katz RW, Nicholls N (1991) Teleconnections linking worldwide climatic anomalies. Cambridge University Press, Cambridge

Grootes PM, Stuiver M (1997) Oxygen 18/16 variability in Greenland snow and ice with 133 to 105-year resolution. J Geophys Res 102:26455–26470

Grootes PM, Stuiver M, White JWC, Johnsen S, Jouzel J (1993) Comparison of oxygen isotope records from the GISP2 and GRIP Greenland ice cores. Nature 366:552–554

Grove R (2003) Revolutionary weather: The climate and economic crisis of 1788–1795 and the discovery of El Niño. In: Proceedings of the Conference 'Climate and Culture', November 2002. Australian Academy of Science, Canberra, in press

Guenther A, Newitt CN, Erickson D, Fall R, Geron C, Graedel T, Harley P, Klinger L, Lerdau M, McKay W, Pierce T, Scholes B, Steinbrecher R, Tallamraju R, Taylor J, Zimmermann P (1995) A global model of natural volatile organic compound emissions. J Geophys Res 100:8873–8892

Guiot J, Pons A, de Beaulieu J-L, Reille M (1989) A 140000 years continental climate reconstruction from two European pollen records. Nature 338:309–313

Haan D, Raynaud D (1998) Ice core record of CO variations during the last two millennia: Atmospheric implications and chemical interactions within the Greenland ice. Tellus 50B:253–262

Hanebuth T, Stattegger K, Grootes PM (2000) Rapid flooding of the Sunda shelf: A late-Glacial Sea-Level record. Science 288:1033–1035

Hansell DA, Carlson CA (2001) Biogeochemistry of total organic carbon and nitrogen in the Sargasso Sea: Control by convective overturn. Deep-Sea Res Pt II 48:1649–1667

Harrison SP, Jolly D, Laarif F, Abe-Ouchi A, Dong B, Herterich K, Hewitt C, Joussaume S, Kutzbach JE, Mitchell J, de Noblet N, Valdes P (1998) Intercomparison of simulated global vegetation distributions in response to 6 ky BP orbital forcing. J Climate 11:2721–2742

Heath MR, Jónasdóttir SH (1999) Distribution and abundance of overwintering *Calanus finmarchicus* in the Faroe-Shetland Channel. Fish Oceanogr 8(1):40–60
Hector A, Schmid B, Beierkuhnlein C, Caldeira MC, Diemer M, Dimitrakopoulos PG, Finn JA, Freitas H, Giller PS, Good J, Harris R, Högberg P, Huss-Danell K, Joshi J, Jumpponen A, Körner C, Leadley PW, Loreau M, Minns A, Mulder CPH, O'Donovan G, Otway SJ, Pereira JS, Prinz A, Read DJ, Scherer-Lorenzen M, Schulze E-D, Siamantziouras A-SD, Spehn EM, Terry AC, Troumbis AY, Woodward FI, Yachi S, Lawton JH (1999) Plant diversity and productivity experiments in European Grasslands. Science 286:1123–1127
Herman JR, Bhartia PB, Torres O, Hsu C, Seftor C, Celarier E (1997) Global distribution of UV-absorbing aerosols from Nimbus-7/TOMS data. J Geophys Res 102:16749–16759
Hitchcock DR, Lovelock JE (1967) Life detection by atmospheric analysis. Icarus 7:149–159
Hoelzmann P, Keding B, Berke H, Kröpelin S, Kruse H-J (2001) Environmental change and archaeology: Lake evolution and human occupation in the Eastern Sahara during the Holocene. Palaeogeogr Palaeocl 169:193–217
Hoffmann PF, Schrag DP (1999) The snowball earth. *http://eps.harvard.edu/people/faculty/hoffmann/snowball-paper.html*
Hooper DU, Vitousek PM (1997) The effects of plant composition and diversity on ecosystem processes. Science 277:1302–1305
Hughen KA, Overpeck JT, Peterson LC, Trumbore S (1996) Rapid climate changes in the tropical Atlantic region during the last deglaciation. Nature 380:51–54
Hungate B, Chapin FS III, Zhong H, Holland EA, Field CB (1997) Stimulation of grassland nitrogen cycling under elevated CO_2. Oecologia 109:149–153
Hurrell JW (1995) Decadal trends in the North Atlantic Oscillation: Regional temperatures and precipitation. Science 269: 676–679
Huston MA (1997) Hidden treatments in ecological experiments: Re-evaluating the ecosystem function of biodiversity. Oecologia 110:449–460
Indermuhle A, Stocker TF, Joos F, Fischer H, Smith HJ, Wahlen M, Deck B, Mastroianni D, Tschumi J, Blunier T, Meyer R, Stauffer B (1999) Holocene carbon-cycle dynamics based on CO_2 trapped in ice at Taylor Dome, Antarctica. Nature 398:121–126.
IPCC (2001) Climate change 2001: The scientific basis. Contribution of Working Group I to the Third Assessment Report of the Intergovernmental Panel on Climate Change. Houghton JT, Ding Y, Griggs DJ, Noguer M, van der Linden PJ, Dai X, Maskell K, Johnson CA (eds), Cambridge University Press, Cambridge New York
Jablonski D (1991) Extinctions: A paleontological perspective. Science 253:754–756
Jackson RB, Canadell J, Ehleringer JR, Mooney HA, Sala OE, Schulze ED (1996) A global analysis of root distributions for terrestrial biomes. Oecologia 108:389–411
Jacobson M, Charlson RJ, Rodhe H, Gordon OH (2000) Earth system science: From biogeochemical cycles to glacial change. Academic Press, London, UK
Jobbágy EG, Jackson RB (2000) The vertical distribution of soil organic carbon and its relation to climate and vegetation. Ecol Appl 10:423–436
Johnsen SJ, Clausen HB, Dansgaard W, Fuhrer K, Gundestrup N, Hammer CU, Iversen P, Jouzel J, Stauffer B, Steffensen JP (1992) Irregular glacial interstadials recorded in a new Greenland ice core. Nature 359:311–313
Johnson TC, Scholz CA, Talbot MR, Kelts K, Ssemanda I, McGill JW (1996) Late Pleistocene desiccation of Lake Victoria and rapid evolution of cichlid fishes. Science 273:1091–1093
Johnson TÇ, Chan Y, Beuning KRM, Kelts K, Ngobi G, Verschuren D (1998) Biogenic silica profiles in Holocene cores from Lake Victoria: Implications for lake level history and initiation of the Victoria Nile. In: Lehman JT (ed) Environmental change and response in East African lakes. Kluwer Academic Publishers, Dordrecht, pp 75–88
Jolly D, Harrison SP, Damnati B, Bonnefille R (1998) Simulated climate and biomes of Africa during the late Quaternary: Comparison with pollen and lake status data. Quaternary Sci Rev 17:629–657
Jones PD, Briffa KR, Osborn TJ, Bergstrom H, Moberg A (2002) Relationships between circulation strength and the variability of growing season and cold season climate in northern and central Europe. Holocene 12:643–656
Joussaume S, Taylor KE, Braconnot P, Mitchell JFB, Kutzbach JE, Harrison SP, Prentice IC, Broccoli AJ, Abe-Ouchi A, Bartlein PF, Bonfils C, Dong B, Guiot J, Herterich K, Hewitt CD, Jolly D, Kim JW, Kislov A, Kitoh A, Loutre MF, Masson V, McAvaney B, McFarlane N, de Noblet N, Peltier WR, Peterschmitt JY, Pollard D, Rind D, Royer JF, Schlesinger ME, Syktus J, Thompson S, Valdes P, Vettoretti G, Webb RS, Wyputta U (1999) Monsoon changes for 6 000 years ago: Results of 18 simulations from the Paleoclimate Modeling Intercomparison Project (PMIP). Geophys Res Lett 26:859–862
Kabat P, Claussen M, Dirmeyer PA, Gash JHC, Bravo de Guenni L, Meybeck M, Pielke Sr. RA, Vörösmarty CJ, Hutjes RWA, Lütkemeier S (eds) (2004) Vegetation, water, humans and climate: A new perspective on an interactive system. IGBP Global Change Series. Springer-Verlag, Berlin Heidelberg New York
Kaplan JO, Bigelow NH, Prentice IC, Harrison SP, Bartlein PJ, Christensen TR, Cramer W, Matveyeva AD, McGuire AD, Murray DF, Razzhivin VY, Smith B, Walker DA, Anderson PM, Andreev AA, Brubaker LB, Edwards ME, Lozhkin AV (2003) Climate change and Arctic ecosystems II: Modeling, palaeodata-model comparisons, and future projections. J Geophys Res, in press
Keller M, Kaplan WA, Wofsy SC (1986) Emissions of N_2O, CH_4 and CO_2 from tropical forest soils. J Geophys Res 91:11791–11802
Kettle AJ, Andreae MO, Amouroux D, Andreae TW, Bates TS, Berresheim H, Bingemer H, Boniforti R, Curran MAJ, DiTullio GR, Helas G, Jones GB, Keller MD, Kiene RP, Leck C, Levasseur M, Maspero M, Matrai P, McTaggart AR, Mihalopoulos N, Nguyen BC, Novo A, Putaud JP, Rapsomanikis S, Roberts G, Schebeske G, Sharma S, Simo R, Staubes R, Turner S, Uher G (1999) A global database of sea surface dimethylsulphide (DMS) measurements and a simple model to predict sea surface DMS as a function of latitude, longitude and month. Global Biogeochem Cy 13:399–444
Khodri M, Leclainche Y, Ramstein G, Braconnot P, Marti O, Cortijo E (2001) Simulating the amplification of orbital forcing by ocean feedbacks in the last glaciation. Nature 410:570–574
Kicklighter DW, Bruno M, Dönges S, Esser G, Heimann M, Helfrich J, Ift F, Joos F, Kadku J, Kohlmaier GH, McGuire AD, Melillo JM, Meyer R, Moore B III, Nadler A, Prentice IC, Sauf W, Schloss AL, Sitch S, Wittenberg U, Würth G (1999) A first order analysis of the potential of CO_2 fertilization to affect the global carbon budget: A comparison of four terrestrial biosphere models. Tellus 51B:343–366
Kleidon A, Fraedrich K, Heimann M (2000) A green planet versus a desert world: Estimating the maximum effect of vegetation on the land surface climate. Climatic Change 44:471–493
Knox JC (2000) Sensitivity of modern and Holocene floods to climate change. Quaternary Sci Rev 19:439–458
Körner Ch (1993) CO_2 fertilization: The great uncertainty in future vegetation development. In: Solomon AM, Shugart HH (eds) Vegetation dynamics and global change. Chapman & Hall, New York, pp 53–70
Kohfeld KE, Harrison SP (2000) How well can we simulate past climates? Evaluating the models using global palaeoenvironmental datasets. Quaternary Sci Rev 19:321–346
Kuhlbusch TAJ, Crutzen PJ (1995) Toward a global estimate of black carbon in residues of vegetation fires representing a sink of atmospheric CO_2 and a source of O_2. Global Biogeochem Cy 4: 491–501
Kumar KK, Rajagopalan B, Cane MA (1999a) On the weakening relationship between the Indian monsoon and ENSO. Science 284:2156–2159
Kumar KK, Kleeman R, Cane MA, Rajagopalan B (1999b) Epochal changes in Indian monsoon-ENSO precursors. Geophys Res Lett 26:75–78
Kurz WA, Apps MJ (1999) A 70-year retrospective analysis of carbon fluxes in the Canadian forest sector. Ecol Appl 9:526–547
Kutzbach JE, Guetter PJ (1986) The influence of changing orbital parameters and surface boundary conditions on climate simulations for the past 18 000 years. J Atmos Sci 43:1726–1759

Kutzbach JE, Liu Z (1997) Response of the African monsoon to orbital forcing and ocean feedbacks in the middle Holocene. Science 278:440–443
Kutzbach JE, Street-Perrott FA (1985) Milankovitch forcing of fluctuations in the level of tropical lakes from 18 to 0 kyr BP. Nature 317:130–134
Labeyrie L, Cole J, Alverson K, Stocker T (2003) The history of climate dynamics in the late Quaternary. In: Alverson K, Bradley R, Pedersen T (eds) Paleoclimate, global change and the future. IGBP Global Change Series. Springer-Verlag, Berlin Heidelberg New York, pp 33–61
Laws EA, Falkowski PG, Smith Jr. WO, Ducklow HW, McCarthy JJ (2000) Temperature effects on export production in the open ocean. Global Biogeochem Cy 14:1231–1246
Le Quéré C, Aumont O, Bopp L, Bousquet P, Ciais P, Francey R, Heimann M, Keeling CD, Keeling RF, Kheshgi H, Peylin P, Piper SC, Prentice IC, Rayner P (2003) Two decades of ocean CO_2 sink and variability. Tellus, in press
Lean J, Beer J, Bradley RS (1995) Reconstruction of solar irradiance since 1610: Implications for climate change. Geophys Res Lett 22:3195–3198
Ledfield-Hoffman PA, DeMaster DJ, Nittrouer CA (1986) Biogenic silica in the Ross Sea and the importance of Antarctic continental shelf deposits in the marine silica budget. Geochim Cosmochim Ac 50:2099–2110
Lefévre N, Watson AJ (1999) Modeling the geochemical cycle of iron in the oceans and its impact on atmospheric CO_2 concentrations. Global Biogeochem Cy 13:727–736
Legrand M, Feniet-Saigne C, Saltzman ES, Germain C, Barkov NI, Petrov VN (1991) Ice-core record of oceanic emissions of dimethylsulphide during the last climate cycle. Nature 350:144–146
Lenton TM (1998) Gaia and natural selection. Nature 394:439–447
Lenton TM (2002) Testing Gaia: The effect of life on Earth's habitability and regulation. Climatic Change 52:409–422
Lenton TM, van Oijen M (2002) Gaia as a complex adaptive system. Philos T Roy Soc B 357:683–695
Lenton TM, Watson AJ (2000a) Redfield revisited: 1. Regulation of nitrate, phosphate and oxygen in the ocean. Global Biogeochem Cy 14:225–248
Lenton TM, Watson AJ (2000b) Redfield revisited: 2. What regulates the oxygen content of the atmosphere? Global Biogeochem Cy 14:249–268
Leuenberger M, Lang C, Schwander J (1999) $\delta^{15}N$ measurements as a calibration tool for the paleothermometer and gas-ice age differences. A case study for the 8200 BP event on GRIP ice. J Geophys Res 104:22163–22169
Longhurst A (1998) Ecological geography of the sea. Academic Press, San Diego, USA
Loreau M, Naeem S, Inchausti P, Bengtsson J, Grime JP, Hector A, Hooper DU, Huston MA, Raffaelli D, Schmid B, Tilman D, Wardle DA (2001) Biodiversity and ecosystem functioning: Current knowledge and future challenges. Science 294:804–808
Lovelock JE (1965) A physical basis for life detection experiments. Nature 207:568–570
Lovelock JE (1972) Gaia as seen through the atmosphere. Atmos Environ 6:579–580
Lovelock JE (1979) Gaia: A new look at life on Earth. Oxford University Press, Oxford, UK
Lovelock JE (1988) The ages of Gaia: A biography of our living earth. W. W. Norton & Co., New York, USA
Lovelock JE (1991) Gaia: The practical science of planetary medicine. Gaia Books, London, UK
Lovelock JE, Kump LR (1994) Failure of climate regulation in a geophysiological model. Nature 369:732–734
Lovelock JE, Margulis LM (1974) Atmospheric homeostasis by and for the biosphere: The Gaia Hypothesis. Tellus 26:2–10
Lovelock JE, Watson AJ (1982) The regulation of carbon dioxide and climate: Gaia or geochemistry? Planet Space Sci 30:795–802
Lovelock JE, Whitfield M (1982) Life span of the biosphere. Nature 296:561–563
Lovelock JE, Maggs RJ, Rasmussen RA (1972) Atmospheric dimethylsulphide and the natural sulphur cycle. Nature 237:452–453
Lovelock JE, Maggs RJ, Wade RJ (1973) Halogenated hydrocarbons in and over the Atlantic. Nature 241:194–196
Luterbacher J, Schmutz D, Gyalistras D, Jones PD, Davies TD, Wanner H, Xoplaki E (2000) Reconstruction of highly resolved NAO and EU indices back to AD 1500. Geophysical Research Abstracts 2:OA34
Manabe S, Stouffer RJ (2000) Study of abrupt climate change by a coupled ocean-atmosphere model. Quaternary Sci Rev 19:285–299
Marchal O, Stocker TF, Joos F, Indermühle A, Blunier T, Tschumi J (1999) Modelling the concentration of atmospheric CO_2 during the Younger Dryas climate event. Clim Dynam 15:341–354
Marengo JA (2003) On the characteristics and spatial-temporal variability of the water balance in the Amazon basin. Clim Dynam, submitted
Markgraf V, Diaz HF (2000) The ENSO record: A synthesis. In: Diaz HF, Markgraf V (eds) El Niño and the Southern Oscillation: Multiscale variability, global and regional impacts. Cambridge University Press, Cambridge, UK, pp 465–488
Matson PA, Ojima DS (1990) IGBP Report No. 13: Terrestrial biosphere exchange with global atmospheric chemistry. International Geosphere Biosphere Program, Stockholm, Sweden
Maxwell BA, Buddemeier RW (2002) Coastal typology development with heterogeneous data sets. Springer-Verlag, Berlin Heidelberg New York
Mayewski PA, Meeker LD, Whitlow S, Twickler MS, Morrison MC, Bloomfield P, Bond GC, Alley RB, Gow AJ, Grootes PM, Meese DA, Ram M, Taylor KC, Wumkes W (1994) Changes in atmospheric circulation and ocean ice cover over the North Atlantic during the last 41000 years. Science 263:1747–1751
Mayewski PA, Meeker LD, Twickler MS, Whitlow S, Yang Q, Lyons WB, Prentice M (1997) Major features and forcing of high-latitude northern hemisphere atmospheric circulation using a 110 000-year-long glaciochemical series. J Geophys Res 102:26345–26366
McGrady-Steed J, Harris PM, Morin PJ (1997) Biodiversity regulates ecosystem predictability. Nature 390:162–165
McGuire AD, Sitch S, Clein JS, Dargaville R, Esser G, Foley J, Heimann M, Joos F, Kaplan J, Kicklighter DW, Meier RA, Melillo JM, Moore III B, Prentice IC, Ramankutty N, Reichenau T, Schloss A, Tian H, Williams LJ, Wittenberg U (2001) Carbon balance of the terrestrial biosphere in the twentieth century: Analyses of CO_2, climate and land-use effects with four process-based ecosystem models. Global Biogeochem Cy 15:183–206
McManus JF, Oppo DW, Cullen JL (1999) A 0.5 million year record of millennial scale climate variability in the North Atlantic. Science 283:971–97
Meese DA, Gow AJ, Grootes P, Mayewski PA, Ram M, Stuiver M, Taylor KC, Waddington ED, Zielinski GA (1994) The accumulation record from the GISP2 core as an indicator of climate change throughout the Holocene. Science 266:1680–1682
Milich L (1999) The role of methane in global warming: Where might mitigation strategies be focused? Global Environ Chang 9:179–201
Millar CI (2000) Evolution and biogeography of *Pinus radiata* with a proposed revision of its Quaternary history. New Zeal J For Sci 29:335–365
Milliman JD, Meade RH (1983) World-wide delivery of river sediment to the oceans. J Geol 91:1–21
Monnin E, Indermuhle A, Dallenbach A, Flückiger J, Stauffer B, Stocker TF, Raynaud D, Barnola JM (2001) Atmospheric CO_2 concentrations over the last glacial termination. Science 291:112–114
Mooney H, Canadell J, Chapin FS, Ehleringer J, Körner Ch, McMurtrie R, Parton W, Pitelka L, Schulze D-E (1999) Ecosystem physiology responses to global change. In: Walker BH, Steffen W, Canadell J, Ingram J (eds) The terrestrial biosphere and global change: Implications for natural and managed ecosystems. IGBP Book Series No. 4, Cambridge University Press, Cambridge, pp 141–189
Mooney HA, Vitousek PM, Matson PA (1987) Exchange of materials between the biosphere and atmosphere. Science 238:926–932
Moore B, Underdal A, Lemke P, Loreau M (2001) The Amsterdam Declaration on Global Change. In: Steffen W, Jäger J, Carson D, Bradshaw C (eds) Challenges of a changing earth: Proceedings of the Global Change Open Science Conference. Amsterdam, The Netherlands, 10–13 July 2001. IGBP Global Change Series. Springer-Verlag, Berlin Heidelberg New York, pp 207–208, *http://www.sciconf.igbp.kva.se/fr.html*

Moore GWK, Holdsworth G, Alverson K (2001) Extra-tropical response to ENSO 1736–1985 as expressed in an ice core from the Saint Elias mountain range in northwestern North America. Geophys Res Lett 28:3457–3461
Moy CM, Seltzer GO, Rodbell DT, Anderson DM (2002) Variability of El Niño/Southern Oscillation activity at millennial timescales during the Holocene epoch. Nature 420:162–165
Mukhopadhyay S, Farley KA, Montanari A (2001) A short duration of the Cretaceous-Tertiary boundary event: Evidence from extraterrestrial helium-3. Science 291:1952–1955
Myneni RB, Keeling CD, Tucker CJ, Asrar G, Nemani RR (1997) Increased plant growth in the northern high latitudes from 1981–1991. Nature 386:698–702
NASA Earth System Science Committee (1986) Earth system science: A closer view. National Aeronautics and Space Administration, Washington, USA
Nepstad DC, de Carvalho CR, Davidson EA, Jipp PH, Lefebvre PA, Negreiros GH, da Silva ED, Stone TA, Trumbore SE, Veira S (1994) The role of deep roots in the hydrological and carbon cycles of Amazonian forests and pastures. Nature 372:666–669
Neumann J (1977) Great historical events that were significantly affected by the weather: Part 2. The year leading to the Revolution of 1789 in France. B Am Meteorol Soc 58:163–168
Neumann J, Dettwiller J (1980) Great historical events that were significantly affected by the weather: Part 9. The year leading to the Revolution of 1789 in France (II). B Am Meteorol Soc 71: 33–41
Nevison C, Weiss R, Erickson DJ III (1995) Global oceanic emissions of nitrous oxide. J Geophys Res 100:15809–15820
Officer CB, Hallam A, Drake CL, Devine JD (1987) Late Cretaceous and paroxysmal Cretaceous/Tertiary extinctions. Nature 326:143–149
Oldfield F, Alverson K (2003) The societal relevance of paleoenvironmental research. In: Alverson K, Bradley R, Pedersen T (eds) Paleoclimate, global change and the future. IGBP Global Change Series. Springer-Verlag, Berlin Heidelberg New York, pp 1–11
Ono Y, Naruse T (1997) Snowline elevation and eolian dust flux in the Japanese islands during isotope stages 2 and 4. Quatern Int 37:45–54
Orr J, Maier-Reimer E, Mikolajewicz U, Monfray P, Sarmiento JL, Toggweiler JR, Taylor NK, Palmer J, Gruber N, Sabine CL, Le Quéré C, Key RM, Boutin J (2001) Estimates of anthropogenic carbon uptake from four 3-D global ocean models. Global Biogeochem Cy 15:43–60
Owensby CE, Ham JM, Knapp A, Rice CW, Coyne PI, Auen LM (1996) Ecosystem-level responses of tallgrass prairie to elevated CO_2. In: Koch GW, Mooney HA (eds) Carbon dioxide and terrestrial ecosystems. Academic Press, San Diego, USA, pp 147–162
Page SE, Siegert F, Rieley JO, Boehm H-DV, Jaya A, Limn S (2002) The amount of carbon release from peat and forest fires in Indonesia during 1997. Nature 420:61–65
Paillard D, Labeyrie LD, Yiou P (1996) AnalySeries 1.0:a Macintosh software for the analysis of geophysical time-series. EOS T Am Geophys Un 77:379
Pataki DE, Ehleringer JR, Flanagan LB, Yakir D, Bowling DR, Still CJ, Buchmann N, Kaplan JO, Berry JA (2003) The application and interpretation of Keeling plots in terrestrial carbon cycle research. Global Biogeochem Cy, in press
Pedersen TF, Francois R, Francois L, Alverson K, McManus J (2003) The late Quaternary history of biogeochemical cycling of carbon. In: Alverson K, Bradley R, Pedersen T (eds) Paleoclimate, global change and the future. IGBP Global Change Series. Springer-Verlag, Berlin Heidelberg New York, pp 63–79
Pépin L, Raynaud D, Barnola J-M, Loutre MF (2002) Hemispheric roles of climate forcings during glacial-interglacial transitions as deduced from the Vostok record and LLN-2D model experiments. J Geophys Res 106:31885–31892
Petit JR, Jouzel J, Raynaud D, Barkov NI, Barnola J-M, Basile I, Bender M, Chappellaz J, Davis M, Delaygue G, Delmotte M, Kotlyakov VM, Legrand M, Lipenkov VY, Lorius C, Pépin L, Ritz C, Saltzman E, Stievenard M (1999) Climate and atmospheric history of the past 420 000 years from the Vostok ice core, Antarctica. Nature 399:429–436
Piketh SJ, Tyson PD, Steffen W (2000) Aeolian transport from southern Africa and iron fertilisation of marine biota in the South Indian Ocean. S Afr J Sci 96:244–246
Platt T, Sathyendranath S (1988) Oceanic primary production: Estimation by remote sensing at local and regional scales. Science 241:1613–1620
Potter CS, Matson PA, Vitousek PM, Davidson EA (1996) Process modelling of controls on nitrogen gas emissions from soils worldwide. J Geophys Res 101:1361–1377
Prentice IC, Jolly D, BIOME 6000 members (2000) Mid-Holocene and glacial-maximum vegetation geography of the northern continents and Africa. J Biogeogr 27:507–519
Prentice IC, Farquhar GD, Fasham MJR, Goulden ML, Heimann M, Jaramillo VJ, Kheshgi HS, Le Quéré C, Scholes RJ, Wallace DWR (2001) The carbon cycle and atmospheric carbon dioxide. In: Houghton JT, Ding Y, Griggs DJ, Noguer M, van der Linden PJ, Dai X, Maskell K, Johnson CA (eds) Climate change 2001: The scientific basis. Contribution of Working Group I to the Third Assessment Report of the Intergovernmental Panel on Climate Change. Cambridge University Press, Cambridge and New York
Rampino MR, Haggerty BM (1996) Hazards due to asteroids. University of Arizona Press
Rampino MR, Porokoph A, Adler AC (2000) Tempo of the end-Permian event: High-resolution cyclostratigraphy at the Permian-Triassic boundary. Geology 28:643–646
Raup DM (1979) Nature of the Permo-Triassic bottleneck and its evolutionary implications. Science 206:217–218
Raynaud D, Jouzel J, Barnola J-M, Chappellaz J, Delmas RJ, Lorius C (1993) The ice record of greenhouse gases. Science 259: 926–933
Raynaud D, Barnola JM, Chappellaz J, Blunier T, Indermuhle A, Stauffer B (2000) The ice record of greenhouse gases: A view in the context of changes. Quaternary Sci Rev 19:9–17
Raynaud D, Blunier T, Ono Y, Delmas RJ (2003) The Late Quaternary history of atmospheric trace gases and aerosols: Interactions between climate and biogeochemical cycles. In: Alverson K, Bradley R, Pedersen T (eds) Paleoclimate, global change and the future. IGBP Global Change Series. Springer-Verlag, Berlin Heidelberg New York, pp 13–31
Rayner PJ, Law RM (1995) A comparison of modelled responses to prescribed CO_2 sources. CSIRO Australia, Division of Atmospheric Research, Technical Paper No. 36
Redfield AC (1958) The biological control of chemical factors in the environment. Am Sci 46:205–221
Renne PR, Zichao Z, Richards MA, Black MT, Basu AR (1995) Synchrony and causal relations between Permian-Triassic boundary crises and Siberian flood volcanism. Science 269:1413–1416
Rustad LE, Campbell JL, Marion GM, Norby RJ, Mitchell MJ, Hartle AE, Cornelissen JHC, Gurevitch J, GCTE-NEWS (2001) A meta-analysis of the response of soil respiration, net nitrogen mineralization, and aboveground plant growth to experimental ecosystem warming. Oecologia 126:543–562
Sarnthein M (1978) Sand deserts during glacial maximum and climatic optimum. Nature 272:43–46
Schenk HJ, Jackson RB (2002) The global biogeography of roots. Ecol Monogr 72:311–328
Schimel D, Baker D (2002) Carbon cycle: The wildfire factor. Nature 420:29–30
Schimel DS, Enting I, Heimann M, Wigley T, Raynaud D, Alves D, Siegenthaler U (1995) CO_2 and the carbon cycle. In: Houghton JT, Meira Filho LG, Bruce J, Hoesung Lee, Callander BA, Haites E, Harris N, Maskell K (eds) Climate change 1994: radiative forcing of climate change and evaluation of the IPCC IS92 emission scenarios. Cambridge University Press, Cambridge, UK, pp 35–71
Schimel DS, House JI, Hubbarde KA, Bousquet P, Ciais P, Peylin P, Braswell BH, Apps MJ, Baker D, Bondeau A, Canadell J, Churkina G, Cramer W, Denning AS, Field CB, Friedlingstein P, Goodale C, Heimann M, Houghton RA, Melillo JM, Moore III B, Murdiyarso D, Noble I, Pacala SW, Prentice IC, Raupach MR, Rayner PJ, Scholes RJ, Steffen WL, Wirth C (2001) Recent patterns and mechanism of carbon exchange by terrestrial ecosystems. Nature 414:169–172

Schlesinger WH (1999) Carbon sequestration in soils. Science 284:2095
Schlesinger WH, Melack JM (1981) Transport of organic carbon in the world's rivers. Tellus 33:172–187
Schlitzer R (2002) Carbon export fluxes in the Southern Ocean: Results from inverse modelling and comparison with satellite based estimates. Deep-Sea Res Pt II 49:1623–1644
Scholes MC, Matrai PA, Andreae MO, Smith KA, Manning MR (2003) Biosphere-atmosphere interactions. In: Brasseur GP, Prinn RG, Pszenny AAP (eds) The changing atmosphere: An integration and synthesis of a decade of tropospheric chemistry research. IGBP Global Change Series. Springer-Verlag, Berlin Heidelberg New York
Scholes RJ, Schulze ED, Pitelka LF, Hall DO (1999) Bioeochemistry of terrestrial ecosystems. In: Walker BH, Steffen W, Canadell J, Ingram J (eds) The terrestrial biosphere and global change: Implications for natural and managed ecosystems. IGBP Book Series No. 4, Cambridge University Press, Cambridge, UK, pp 271–303
Schulze E-D, Heimann M (1998) Carbon and water exchange of terrestrial ecosystems. In: Galloway JN, Melillo JM (eds) Asian change in the context of global change. IGBP Book Series No. 3, Cambridge University Press, Cambridge, UK, pp 145–161
Schwander J (1989) The transformation of snow to ice and the occlusion of gases. In: Oeschger H, Langway CC Jr. (eds) The Environmental record in glaciers and ice sheets. John Wiley, New York, USA, pp 53–67
Schwander J, Stauffer B (1984) Age difference between polar ice and the air trapped in its bubbles. Nature 311:45–47
Schwander J, Eicher U, Ammann B (2000) Oxygen isotopes of Lake Marl at Gerzensee and Leysin (Switzerland), covering the Younger Dryas and two minor oscillations, and their correlation to the GRIP ice core. Palaeogeogr Palaeocl 159:203–214
Schwartzman DW, Volk T (1989) Biotic enhancement of weathering and the habitability of Earth. Nature 340:457–460
Sellers PJ, Bounoua L, Collatz GJ, Randall DA, Dazlich DA, Los SO, Berry JA, Fung I, Tucker CJ, Field CB, Jensen TG (1996) Comparison of radiative and physiological effects of doubled atmospheric CO_2 on climate. Science 271:1402–1406
Sepkoski JJ (1989) Perdicity in extinction and the problem of catastrophism in the history of life. J Geol Soc London 146:7–19
Sepkoski JJ (1992) Phylogenetic and ecologic patterns in the Phanerozoic history of marine biodiversity. In: Eldredge N (ed) Systematics, ecology, and biodiversity crisis. Columbia University Press, New York, USA, pp 77–100
Severinghaus JP, Sowers T, Brook EJ, Alley RB, Bender M (1998) Timing of abrupt climate change at the end of the Younger Dryas interval from thermally fractionated gases in polar ice. Nature 391:141–146
Shugart HH (1997) Terrestrial ecosystems in changing environments. Cambridge University Press, Cambridge, UK
Shugart HH, Emanuel WR (1985) Carbon dioxide increase: the implications at the ecosystem level. Plant Cell Environ 8:381–386
Shukolyukov A, Lugmair GW (1998) Isotopic evidence for the Cretaceous-Tertiary impactor and its type. Science 282:927–929
Silva Dias MA, Rutledge S, Kabat P, Silva Dias PL, Nobre C, Fisch G, Dolman AJ, Zipser E, Garstang M, Manzi A, Fuentes JD, Rocha H, Marengo J, Plana-Fattori A, Sá L, Alvalá R, Andreae MO, Artaxo P, Gielow R, Gatti L (2002) Clouds and rain processes in a biosphere-atmosphere interaction context in the Amazon Region. J Geophys Res 107:10.1029/2001JD000335
Smil V (2001) Enriching the earth: Fritz Haber, Carl Bosch, and the transformation of world food production. MIT Press, Cambridge, USA
Smith TM, Shugart HH (1993) The transient response of terrestrial carbon storage to a perturbed climate. Nature 361:523–526
Smith TM, Shugart HH, Woodward FI (eds) (1997) Plant functional types: Their relevance to ecosystem properties and global change. IGBP Book Series No. 1, Cambridge University Press, Cambridge, UK
Smith SV, LOICZ Modelling Team (2002) Carbon-nitrogen-phosphorus fluxes in the coastal zone: The global approach. International Geosphere Biosphere Programme, Global Change Newsletter No. 49. Stockholm, Sweden, pp 7–11
Sowers T (2001) The N_2O record spanning the penultimate deglaciation from the Vostok ice core. J Geophys Res 106:31903–31914
Spall SA, Jones A., Roberts DL, Woodage MJ, Anderson TR (2001) Simulating DMS feedbacks on climate with a 3D coupled atmosphere/ocean climate model. Abstract IM01-11, Joint Oceanographic Assembly of IAPSO and IABO, Mar del Plata, Argentina, October 2001
Stanley S (1987) Extinction. Scientific American books, Inc.
Steffen WL, Walker BH, Ingram JSI, Koch GW (eds) (1992) Global change and terrestrial ecosystems: The operational plan. The International Geosphere Biosphere Programme, IGBP Report No. 21, Stockholm, Sweden
Stocker TF (1998) The seesaw effect. Science 282:61–62
Stocker TF (2000) Past and future reorganisations in the climate system. Quaternary Sci Rev 19:301–319
Stringer C (2000) Coasting out of Africa. Nature 405:24–27
Stuiver M, Reimer PJ (1993) Extended ^{14}C data base and revised Calib 3.0 ^{14}C age calibration program. Radiocarbon 35:215–230
Stuiver M, Braziunas TF, Becker B, Kromer B (1991) Climatic, solar, oceanic and geomagnetic influences on late glacial and Holocene atmospheric $^{14}C/^{12}C$ change. Quaternary Res 35:1–24
Stuiver M, Grootes PM, Braziunas TF (1995) The GISP2 ^{18}O climate record of the past 16 500 years and the role of sun, ocean, and volcanoes. Quaternary Res 44:341–354
Sturman AS, Tyson PD, D'Abreton PC (1997) Transport of air from Africa and Australia to New Zealand. J Roy Soc New Zeal 27: 485–498
Swap R, Garstang M, Greco S, Talbot R, Kallbert P (1992) Saharan dust in the Amazon Basin. Tellus 44B:133–149
Takahashi T, Feely RA, Weiss RF, Wanninkhof RH, Chipman DW, Sutherland SC, Takahashi TT (1997) Global air-sea flux of CO_2: An estimate based on measurements of sea-air pCO_2 difference. P Natl Acad Sci USA 94:8292–8299
Texier D, de Noblet N, Harrison SP, Haxeltine A, Jolly D, Joussaume S, Laarif F, Prentice IC, Tarasov P (1997) Quantifying the role of biosphere-atmosphere feed-backs in climate change: Coupled model simulations for 6 000 years BP and comparison with palaeodata for northern Eurasia and northern Africa. Clim Dynam 13:865–882
Thompson AM, Pickering KE, McNamara DP, Schoeberl MR, Hudson RD, Kim J-H, Browell EV, Kirchhoff VWJH, Nganga D (1996) Where did tropospheric ozone over southern Africa and the tropical Atlantic come from in October 1992? Insights from TOMS, GTE TRACE-A, and SAFARI 1992. J Geophys Res 101: 24251–24278
Thompson DWJ, Wallace JM (1998) The Arctic Oscillation signature in the wintertime geopotential height and temperature fields. Geophys Res Lett 25:1297–1300
Thompson LG (2000) Ice core evidence for climate change in the tropics: Implications for our future. Quaternary Sci Rev 19: 19–35
Thompson LG, Yao T, Mosley-Thompson E, Davis ME, Henderson KA, Lin P-N (2000) A high-resolution millennial record of the South Asian monsoon from Himalayan ice cores. Science 289: 1916–1919
Tilman D (1999) The ecological consequences of changes in biodiversity: A search for general principles. The Robert H. MacArthur Award Lecture. Ecology 80:1455–1474
Tilman D (2001) Effects of diversity and composition on grassland stability and productivity. In: Press MC, Huntly NJ, Levin SA (eds) Ecology: Achievement and challenge. British Ecological Society Symposium, Vol. 41. Blackwell Science, Oxford, UK, pp 183–207
Tilman D, Wedin D, Knops J (1996) Productivity and sustainability influenced by biodiversity in grassland ecosystems. Nature 379:718–720
Turner SM, Nightingale PD, Spokes LJ, Liddicoat MI, Liss PS (1996) Increased dimethylsulphide concentrations in sea water from in situ iron enrichment. Nature 383:513–517
Tyson P, Fuchs R, Fu C, Lebel L, Mitra AP, Odada E, Perry J, Steffen W, Virji H (eds) (2002) Global-regional linkages in the earth system. The IGBP Global Change Series. Springer-Verlag, Berlin Heidelberg New York
Urban FE, Cole JE, Overpeck JT (2000) Influence of mean climate change on climate variability from a 155-year tropical Pacific coral record. Nature 407:989–993

USGS (2000) African dust causes widespread environmental distress. United States Geological Survey Information sheet. *http://www.coastal.er.usgs.gov/african_dust/*, 31 Oct 2002
Valentini R, Matteucci A, Dolman AJ, Schulze ED (2000) Respiration as the main determinant of carbon balance of European forests. Nature 404:861–865
van Kreveld S, Sarnthein M, Erlenkeuser H, Grootes P, Jung S, Nadeau MJ, Pflaumann U, Voelker A (2000) Potential links between surging ice sheets, circulation changes, and the Dansgaard-Oeschger cycles in the Irminger sea, 60–18 kyr. Paleoceanography 15:425–442
Verchot LV, Davidson EA, Cattanio JH, Ackerman IL, Erickson HE, Keller M (1999) Land use change and biogeochemical controls of nitrogen oxide emissions from soils in eastern Amazon. Global Biogeochem Cy 13:31–46
Verheyen E, Salzburger W, Snoeks J, Meyer A (2003) Origin of the superflock of Cichlid fishes from Lake Victoria, East Africa. Science 300:325–329
Verity PG, Smetacek V (1996) Organism life cycles, predation and the structure of marine pelagic ecosystems. Mar Ecol-Prog Ser 130:277–293
Vernadsky V (1930) The biosphere. Translated and reprinted by Synergetic Press, Oracle, USA
Verschuren D, Laird KR, Cumming BF (2000) Rainfall and drought in equatorial East Africa during the past 1100 years. Nature 403:410–414
Visscher H, Brinkhuis H, Dilcher DL, Elsik WC, Eshet Y, Looy CV, Rampino MR, Traverse A (1996) The terminal Paleozoic fungal event: Evidence of terrestrial ecosystem destabilization and collapse. *P Natl Acad Sci USA* 93:2155–2158
Vitousek PM, Mooney HA, Lubchenco J, Melillo JM (1997) Human domination of earth's ecosystems. Science 277:494–499
Von Grafenstein U, Erlenkeuser H, Muller J, Jouzel J, Johnsen S (1998) The cold event 8200 years ago documented in oxygen isotope records of precipitation in Europe and Greenland. Clim Dynam 14:73–81
Vörösmarty CJ, Meybeck MM (1999) Riverine transport and its alteration by human activities. International Geosphere Biosphere Programme, Global Change NewsLetter No. 39. Stockholm, Sweden, pp 24–29
Walter H, Breckle S (1985) Ecological systems of the geobiosphere. Springer-Verlag, Berlin Heidelberg New York
Wang YJ, Cheng H, Edwards RL, An Z, Wu J, Shen C-C, Dorale JA (2001) A high-resolution absolute-dated late Pleistocene monsoon record from Hulu Cave, China. Science 294:2345–2348
Ward PD, Montgomery DR, Smith R (2000) Altered river morphology in South Africa related to the Permian-Triassic extinction. Science 289:1740–1743
Ward PD, Haggart JW, Carter ES, Wilbur D, Tipper HW, Evans T (2001) Sudden productivity collapse associated with the Triassic-Jurassic boundary mass extinction. Science 292:1148
Watson AJ (1999) Iron in the oceans: Influences on biology, geochemistry and climate. Progress in Environmental Science 1:345–370
Watson AJ, Liss PS (1998) Interactions between the marine biota and climate via the carbon and sulphur biogeochemical cycles. Philos T Roy Soc B 353:41–51
Watson AJ, Lovelock JE (1983) Biological homeostasis of the global environment: The parable of Daisyworld. Tellus 35B:284–289
Watson AJ, Bakker DCE, Ridgwell AJ, Boyd PW, Law CS (2000) Effect of iron supply on Southern Ocean CO_2 uptake and implications for glacial atmospheric CO_2. Nature 407:730–733
Watson RT, Noble IR, Bolin B, Ravindranath NH, Verardo DJ, Dokken DJ (eds) (2000) Land use, land use change and forestry: A special report of the IPCC. Cambridge University Press, Cambridge, UK
Whitfield M (1981) The world ocean: Mechanism or machination? Interdiscipl Sci Rev 6:12–35
Wilson EO (1992) The diversity of life. Allen Lane, the Penguin Press
Woodward FI (1987) Stomatal numbers are sensitive to increase of CO_2 from pre-industrial levels. Nature 327:617–618
Woodward FI, Cramer W (eds) (1996) Plant functional types and climate change. Opulus Press, Uppsala, Sweden
Wuebbles DJ, Brasseur GP, Rodhe H (2003) Changes in the chemical composition of the atmosphere and potential impacts. In: Brasseur GP, Prinn RG, Pszenny AAP (eds) The changing atmosphere: An integration and synthesis of a decade of tropospheric chemistry research. The IGBP Global Change Series. Springer-Verlag, Berlin Heidelberg New York
Yachi S, Loreau M (1999) Biodiversity and ecosystem productivity in a fluctuating environment: The insurance hypothesis. P Natl Acad Sci USA 96:1463–1468
Yokohama Y, Deckker PD, Lambeck K, Johnston P, Fifield LK (2001) Sea-level at the last glacial maximum: Evidence from northwestern Australia to constrain ice volumes for oxygen isotope stage 2. Palaeogeogr Palaeocl 165:281–297
Yu G, Harrison S (1996) An evaluation of the simulated water balance of Eurasia and northern Africa at 6000 yr BP using lake status data. Clim Dynam 12:723–735
Zehr JP, Waterbury JB, Turner PJ, Montoya JP, Omoregie E, Steward GF, Hansen A, Karl DM (2001) Unicellular cyanobacteria fix N_2 in the subtropical North Pacific. Nature 412:635–638
Zimov SA, Chuprynin VI, Oreshko AP, Chapin FS III, Reynolds JF, Chapin MC (1995) Steppe-tundra transition: A herbivore-driven biome shift at the end of the Pleistocene. American Naturalist 146:765–794

Chapter 3

The Anthropocene Era: How Humans are Changing the Earth System

The planet is now dominated by human activities. Human changes to the Earth System are multiple, complex, interacting, often exponential in rate and globally significant in magnitude. They affect every Earth System component – land, coastal zone, atmosphere and oceans. The human driving forces for these changes – both proximate and ultimate – are equally complex, interactive and frequently teleconnected across the globe. The magnitude, spatial scale, and pace of human-induced change are unprecedented. Today, humankind has begun to match and even exceed some of the great forces of nature in changing the biosphere and impacting other facets of Earth System functioning. In terms of fundamental element cycles and some climatic parameters, human-driven changes are pushing the Earth System well outside of its normal operating range. In addition, the structures of the terrestrial and marine biospheres have been significantly altered directly by human activities. There is no evidence that the Earth System has previously experienced these types, scales, and rates of change; the Earth System is now in a no-analogue situation, best referred to as a new era in the geological history of Earth, the Anthropocene.

3.1 A Human-Dominated Planet?

Until very recently in the history of Earth, humans and their activities have been an insignificant force in the dynamics of the Earth System. Human-induced environmental change has been, for most of history, highly localised and at times regional. Only very occasionally could the human impact be considered global, as in the loss of Quaternary megafauna or the colonial impacts on biota (Martin and Klein 1984; Crosby 1986; Turner and Butzer 1992; Alroy 2001; Roberts et al. 2001). These qualities of the human imprint on the Earth System changed with the Industrial Revolution in the late eighteenth century. With the advent of fossil fuel-based energy systems, the very structure of human existence changed and, with it, the human capacity to affect the planet. The new order technology increased societal capacity to extract, consume, and produce (Grübler 1998), facilitating an enormous rise in the global population, from just under one billion people in 1800 to a projected nine billion or more by 2050 (UN Population Division 2000) (Fig. 3.1), and a shift to market economies and information flows that have raised lifestyle expectations everywhere (Dicken 1992) (Fig. 3.2). The result is a global escalation in both total and per capita demands for Earth's resources, including fish stocks, inert materials, freshwater, and prime agricultural soils (Tolba et al. 1992), and in the environmental consequences of production and consumption for the functioning of the Earth System.

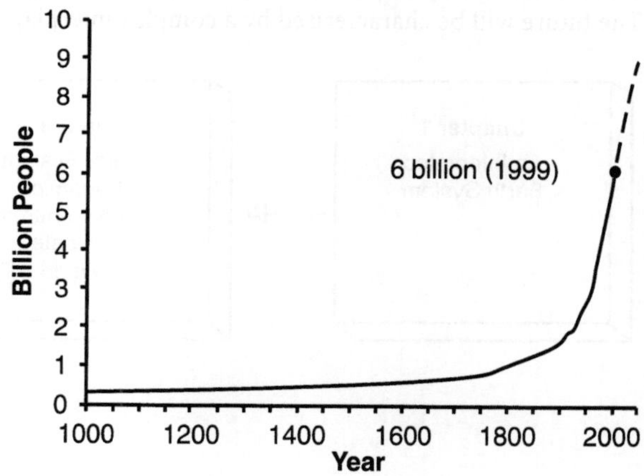

Fig. 3.1. Population growth from 1000 to 2000 (data from UN Population Division 2000; US Bureau of the Census 2000)

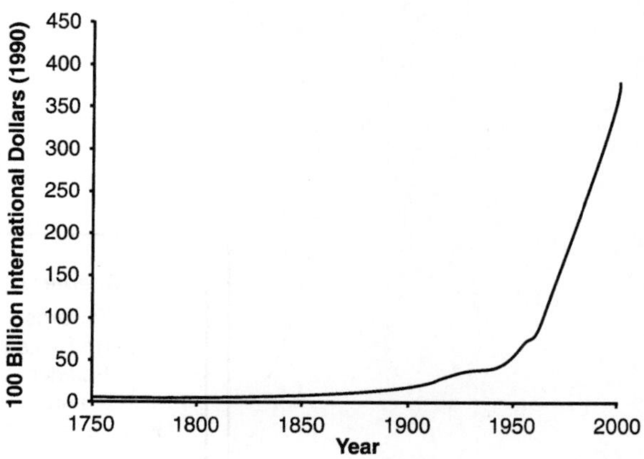

Fig. 3.2. Total world Real Gross Domestic Product reflecting the increased consumption from both an increasing population and intensified consumer lifestyle (data from Nordhaus 1997)

Much of this history of human impact on environmental systems is concerned with those aspects of global change that are cumulative rather than systemic (Turner et al. 1990), that is, changes that occur on a local scale but are so ubiquitous around the planet that when aggregated have a global-scale effect. Such changes have, so far, had a much greater impact on hydrology, atmosphere, soils, vegetation and renewable resource systems generally over the last 150 years than have systemic changes in atmospheric composition and climate. This relative importance of cumulative change is underscored by the fact that the climatic warming from the end of the Little Ice Age (~1850) through to the last decade of the twentieth century has already been, at least in terms of northern hemisphere temperatures, one of the greatest climate (systemic) changes recorded during the late Holocene (Mann et al. 1999; Bradley 2000). Cumulative change continues, registered in the large-scale land changes worldwide and in the current mass extinction of species (Myers 1997). On the other hand, as the systemic aspects of global change accelerate throughout this century, as they are projected to do, they will become a more important feature of global change. The future will be characterised by a complex interplay of systemic and cumulative changes interacting in different ways with multiple modes of natural variability.

In this chapter, the ways in which human activities have evolved from insignificance in terms of Earth System functioning to becoming a force equal to, and in some cases dominant over, the great forces of nature are described. Human-driven changes go far beyond the well-known increase in greenhouse gases due to fossil fuel combustion; they affect every major Earth System compartment and influence a wide range of fundamental Earth System processes. The chapter is organised into three main parts:

- A discussion of the nature of human activities that are now altering the global system in myriad ways. Although many different approaches to explaining these activities are possible, two broad lines of reasoning are used: drivers linked in hierarchical sets and drivers conceptualised as human-environment conditions (Sect. 3.2).
- A description of the changes due to human activities that have occurred in the recent past, primarily over the past few centuries, to major components of the Earth System (Sect. 3.3).

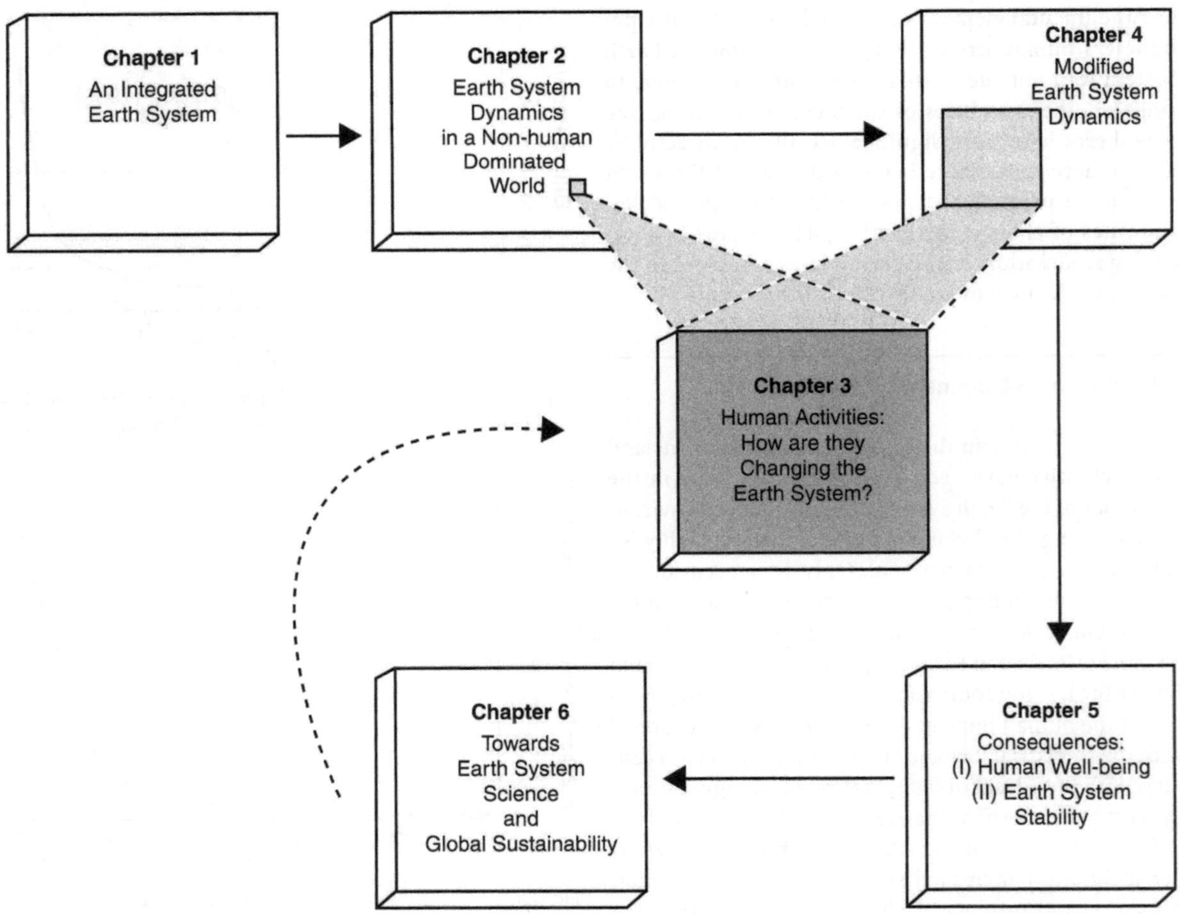

Fig. 3.3. Chapter 3 focus: the nature, magnitude and rate of human-driven changes to the Earth System

- After dealing with the changes themselves and their proximate and ultimate drivers in turn, the changes are then brought back together into an Earth System perspective at the end of the chapter (Sect. 3.4).

This chapter focuses exclusively on the nature of human-driven changes to the Earth System. It does not carry the discussion through to the impacts and consequences of these changes, as many readers will be accustomed to seeing in other treatments of global change. Rather, the responses to these human-driven changes in terms of Earth System functioning are discussed in Chap. 4, and their consequences for humans – in terms of resources and the planetary life support system – are discussed in Chap. 5 (Fig. 3.3).

3.2 Drivers of Change

Over the past two centuries the interactions among population, technology and socio-political organisation have changed dramatically. As a result, the scope and degree of human alteration of the Earth System have also changed. Resource consumption has escalated to the point that the impacts influence the state and functioning of the Earth System (Turner et al. 1990). This consumption is reflected in agriculture and food production, industry, energy production, and urbanisation as well as less widely recognised influences such as recreation and international commerce (WCED 1987; Vitousek et al. 1997a; NRC 1999). These enterprises and others are transforming Earth's land surface, altering its biogeochemical and hydrologic cycles, and adding and deleting species, destroying and modifying ecosystems, and ultimately changing climate and biological diversity (Fig. 3.4). In the following sections, current trends in these human endeavours are discussed, followed by several perspectives on how their influence is played out in a heterogeneous world.

3.2.1 Sectoral Activities as Drivers of Change

The global population at the beginning of the twenty-first century is around six billion people. Over the past decade approximately 80 million people were added to the population each year. There is no doubt that the growth of the human population over the last few cen-

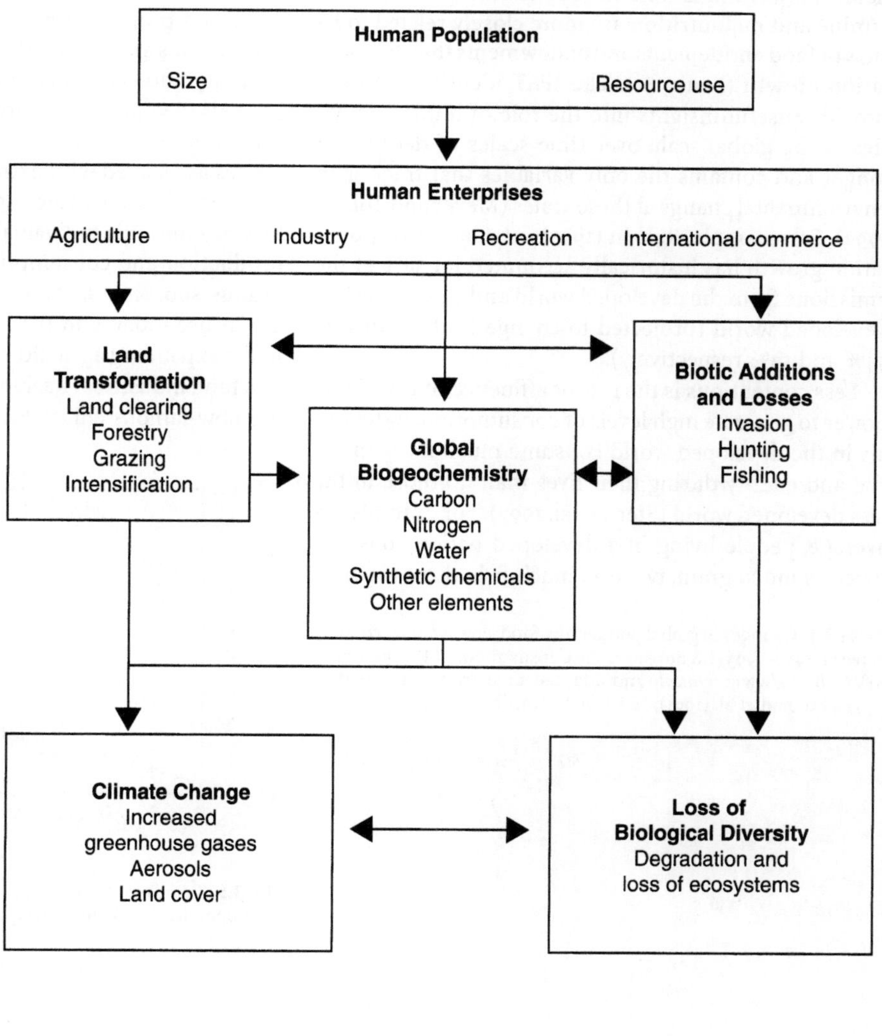

Fig. 3.4. The scope, degree and interactive nature of human alteration of the Earth System (adapted from Vitousek et al. 1997a)

turies has been substantial, and that this growth correlates with many global environmental changes at the global scale and over long time scales. While it is inaccurate to translate this correlation to a simple cause-effect relationship, there are a number of basic human needs that are related to the number of the people on the planet – a demand for food, water, shelter/housing, community health and employment – the basics of life. The ways in which society meets those needs are critical determinants of the environmental consequences that result at all scales. This relationship is grossly captured by the IPAT identity, in which the environmental impact of resource use (I) is a function of population (P), affluence (A) and technology (T) to give

$$I = P \times A \times T$$

(Ehrlich and Holdren 1971; Holdren and Ehrlich 1974).

It is more difficult to apply this simplified framework at small space and time scales (i.e., for case studies); for example, an emphasis on the population variable can have the effect of blaming the victims (as in high fertility rates among economically marginal households in the tropical world) for consequences such as tropical deforestation and famine-malnutrition. In fact, modern famine and malnutrition are more closely related to issues of food entitlements and endowments than to population growth (Sen 1981). The IPAT identity, however, provides useful insights into the role of human activities at the global scale over time scales of decades or longer and contains the only variables that track with environmental change at these scales (Meyer and Turner 1992). For example, Raskin (1995) calculates that population growth has historically accounted for 38% of the emissions from the developed world and 22% in the less developed world (projected to change in the future to 34% and 53% respectively).

Less contentious is the role of affluence, at least in its power to generate high levels of consumption. Individuals in the developed world consume much more material and energy during their lives than do those in the less developed world (Stern et al. 1997). For example, on average, people living in a developed nation consume twice as much grain, twice as much fish, three times as much meat, nine times as much paper and eleven times as much gasoline as people living in a developing nation (Laureti 1999). They do so because affluence increases expectations and underpins material lifestyles, especially with regard to consumption for entertainment, mobility, communication, and a broad range of goods and services that are beyond the reach of those in the less developed world. Affluence, however, also correlates with technological efficiency in production and the capacity to reduce waste on a per unit of production basis. It may also be associated with an environmental Kuznets curve, which shows a decoupling of consumption and deleterious environmental consequences at certain inflection points in affluence. Nevertheless, as global affluence has increased between 1970 and 1997, for example, the global consumption of energy rose from 217 to 400 million kJ, and consumption of materials, although more difficult to globally aggregate, likewise increased dramatically (WRI 1999) (Table 3.1).

Table 3.1. Changes in global population, food, energy, and economy between 1950–1993 (US Bureau of the Census 2000; USDA 1996, 1991; RIVM *http://www.rivm.nl*; and adapted data from Darmstadter (1971); Etemad et al. (1991); IEA (1998); Nordhaus (1997))

	Growth factor
Population	2.2
Grain	2.7
Energy	2.0
Total World Real GDP	6.0

Industry

With dramatically increasing consumption, opportunities for industry have grown and will continue to do so. The availability of inexpensive fuel and the development and proliferation of transportation, communication systems and other industries have increased the connectivity and economic development of the world, but have also led to large environmental impacts, including the emissions of air and water pollutants and other wastes associated with extraction, production and consumption of goods (Fig. 3.5). Such releases include the loss of non-toxic but valuable materials wasted during production and consumption, and also toxic and hazardous substances. Over 100 000 industrial chemicals are in use today, and the number is increasing rapidly in the expanding agriculture, metals, electronics, textiles and food industries (Raskin et al. 1996). Among these are the now-famous chlorofluorocarbons (CFCs) that have

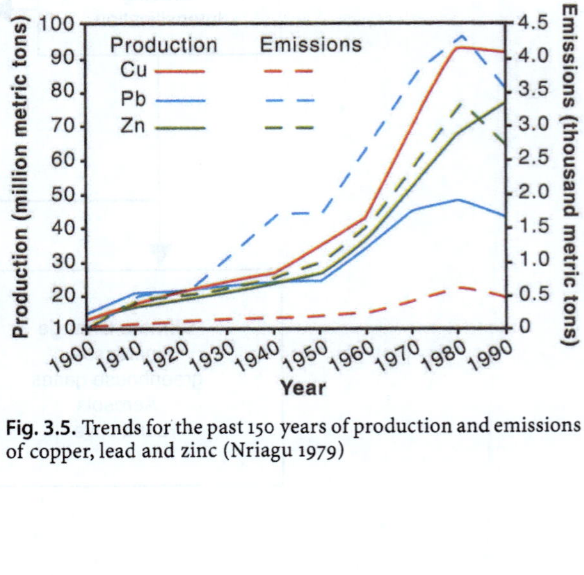

Fig. 3.5. Trends for the past 150 years of production and emissions of copper, lead and zinc (Nriagu 1979)

led to depletion of stratospheric ozone (WMO 1999) and various other trace gases that pollute the troposphere and damage human and plant life over large regions (Chameides et al. 1994). Others, such as heavy metals and some chlorinated organic compounds (Fig. 3.6 and 3.7), pose threats to the environment and human health due to their persistence and bio-accumulation through the food chain.

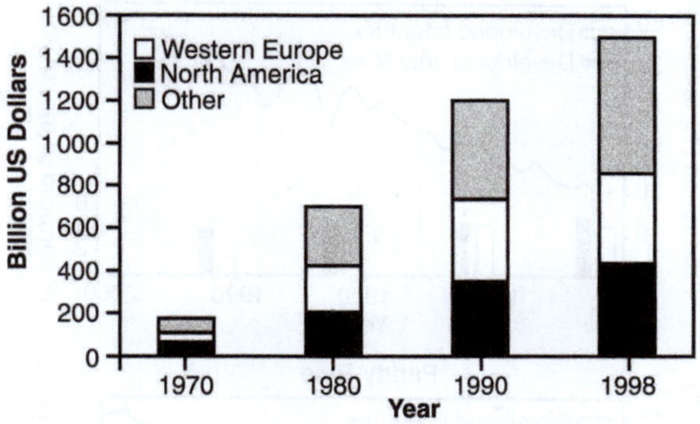

Fig. 3.6. Global production of chemicals by region (OECD 2001)

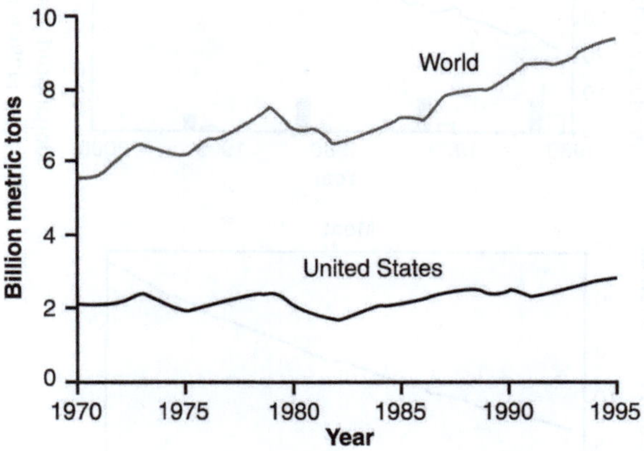

Fig. 3.7. World and United States consumption of metals from 1970–1995 (Matos and Wagner 1998, based on data from UN 1967, 1976, 1985, 1994; FAOSTAT 2002)

Energy and Transport

Energy is needed for most activities of industrialised and industrialising countries. As noted in Table 3.1, worldwide energy use doubled in the last half of the twentieth century, and it will be likely to increase more as the growing human population develops. Today, most energy is derived from the combustion of fossil fuels (Fig. 3.8). In the past, concerns about sparse reserves of coal, oil and gas loomed large (Fig. 3.9). However, now the major drawback to fossil fuel use has to do with its environmental effects – combustion is the major source of critical air pollution problems at local, regional, and global scales – and more than any other human activity drives anthropogenic climate change. Transport comprises 25% of the world's energy use and 50% of the world's oil production, with motor vehicles accounting for 80% of transport-related energy. The number of motor vehicles has increased from 40 million after World War II to 676.2 million in 1996 (UNEP 2000). When put in the context of projections estimating 1 000 million motor vehicles by 2025, the importance of transport as an energy consumer is evident.

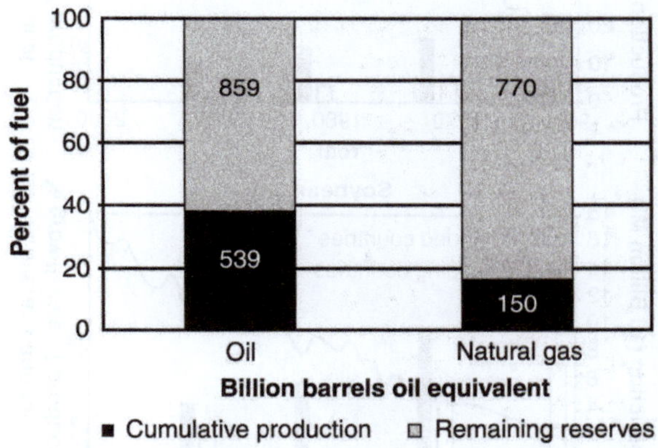

Fig. 3.9. Fraction of known oil and natural gas reserves consumed to date (USGS 2000)

Fig. 3.8. World energy use from 1971 to 1998 (OECD/IEA 2000)

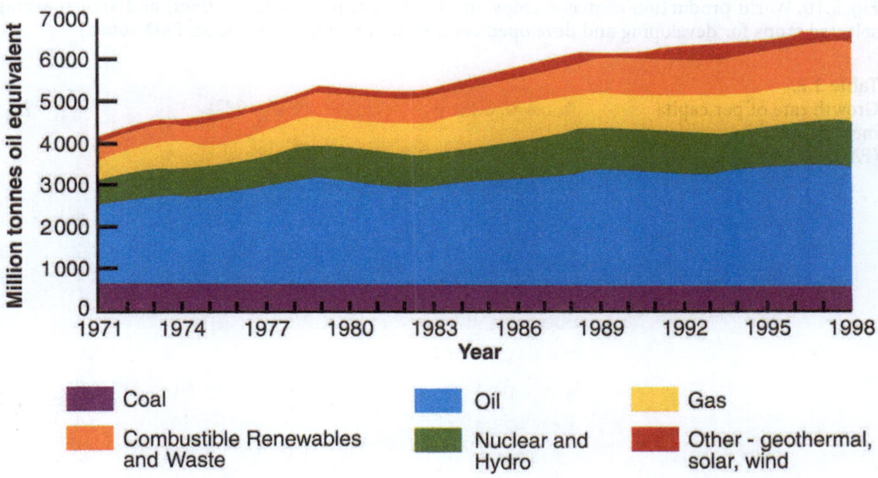

Provision of Food

The provision of food for a rapidly expanding population is a major driver of change in both the terrestrial and marine biospheres. World cereal production has more than doubled since 1961, keeping abreast of world population growth (Fig. 3.10). World wheat, maize and rice production have all either tripled or almost tripled since 1961, while soybean production has increased almost eight-fold (Fig. 3.10). In addition, world meat demand and production have also increased dramatically, tripling since 1961 (Fig. 3.10, Table 3.2). These impressive increases have come largely through intensification of agricultural production rather than expansion of croplands. World

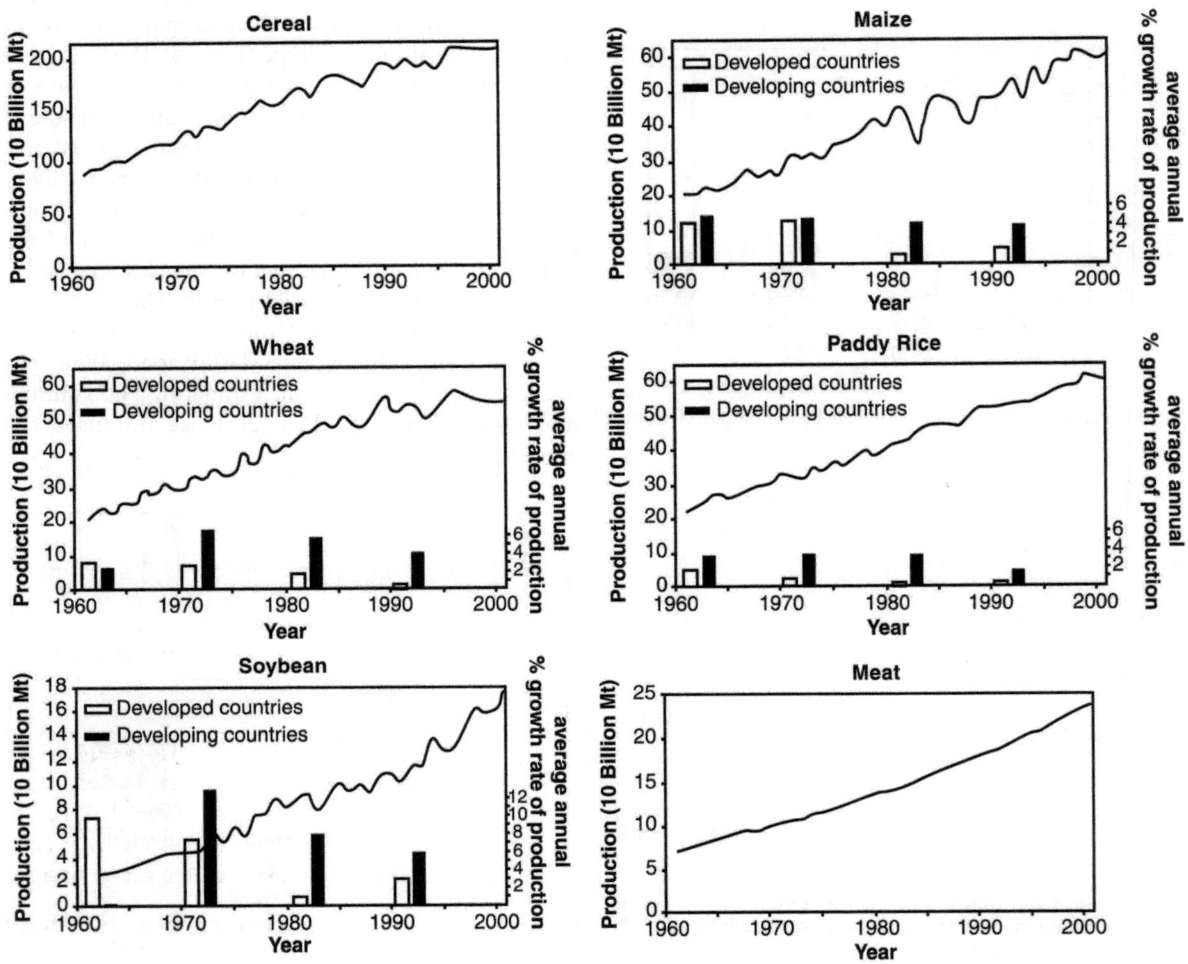

Fig. 3.10. World production of major crops and meat from 1960 to 2000 (*lines*) and annual average growth rates of production for selected crops for developing and developed countries (*bars*) (FAOSTAT 2002; FAO 2000a)

Table 3.2.
Growth rate of per capita meat demand by region (FAOSTAT 2002)

Region/country	1967–1982	1982–1990	1990–1997
Former Soviet Union	1.8	2.2	-8.5
Eastern Europe	2.1	1.3	-2.4
United States	0.5	0.8	0.9
EU15	1.9	0.9	0.2
Latin America	1.4	0.9	3.5
Sub-Saharan Africa	0.2	-0.8	0.2
West Asia/North Africa	3.4	0.0	0.9
Developing Asia	2.4	6.0	7.1
Developing World	2.0	3.4	5.2
Developed World	1.5	1.2	-1.1
World	1.1	1.5	1.4

fisheries and aquaculture have also experienced astounding increases in the past 50 years, increasing from 20 million tons to 120 million tons of production in the late 1990s (Fig. 3.65). On average, people around the world rely on fish for 16% of the protein that they consume (Table 3.3).

Agricultural trade has also increased dramatically since 1961 (Fig. 3.11), expanding at a rate faster than food production. Many developing countries depend largely on agricultural trade for export earnings (FAO 2000a). However, prices for most agricultural commodities have fallen over the last 15 years, forcing developing countries to export more in order to maintain earnings (WRI 1996). Despite this dependency on agricultural exports, the developing countries' share of world agricultural exports dropped from 40 to 27% between the 1960s and the present (FAO 2000a).

The demand for food is the ultimate driver of other environmental changes in addition to land cover itself. For example, since 1950 the amount of irrigated cropland has increased from 90 million hectares to 270 million hectares (FAO 2000a) (Fig. 3.12). It is estimated that irrigation is responsible for 50% of the global food production increase between the mid-1960s and mid-1980s. Currently, irrigation accounts for a third of total production (WRI 1996). Between 1950 and 2000 fertiliser use increased 10-fold, from 14 million tons to 141 million tons (FAO 2000a; Brown et al. 2001) (Fig. 3.13). Fertiliser consumption in developing regions is projected to continue increasing through the year 2010 (WRI 1996). Pesticide applications have also increased dramatically since the 1950s. Consumption grew by more than 10% per year until the 1980s, when it slowed to about 3% growth per year (Yudelman et al. 1998).

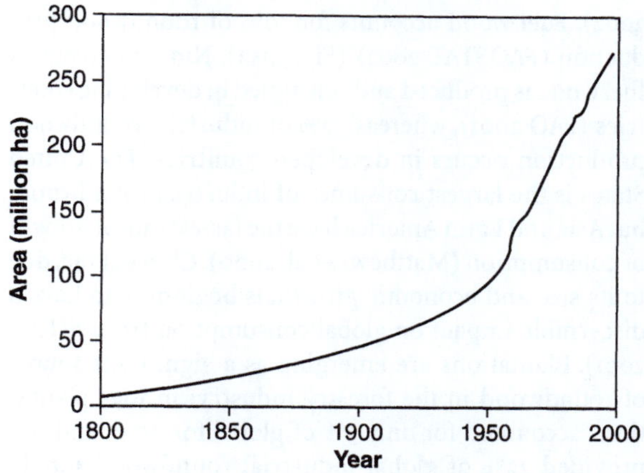

Fig. 3.12. World agricultural area under irrigation (FAOSTAT 2002)

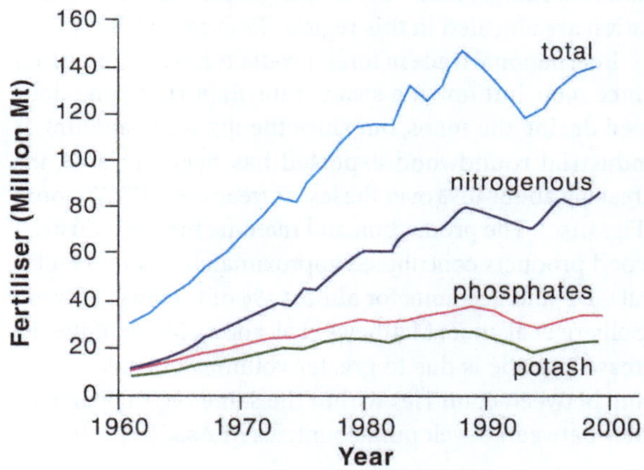

Fig. 3.13. Global fertiliser consumption (FAOSTAT 2002)

Table 3.3. Contribution of fish to diet, as a percentage of overall protein consumption (FAO 1993)

Region	Fish as share of animal protein consumed (%)
North America	6.6
Western Europe	9.7
Africa	21.1
Latin America and Caribbean	8.2
Near East	7.8
Far East	27.8
Asian centrally planned economies	21.7
World	16

Fig. 3.11.
World agricultural exports, 1961–1998 (FAO 2000a)

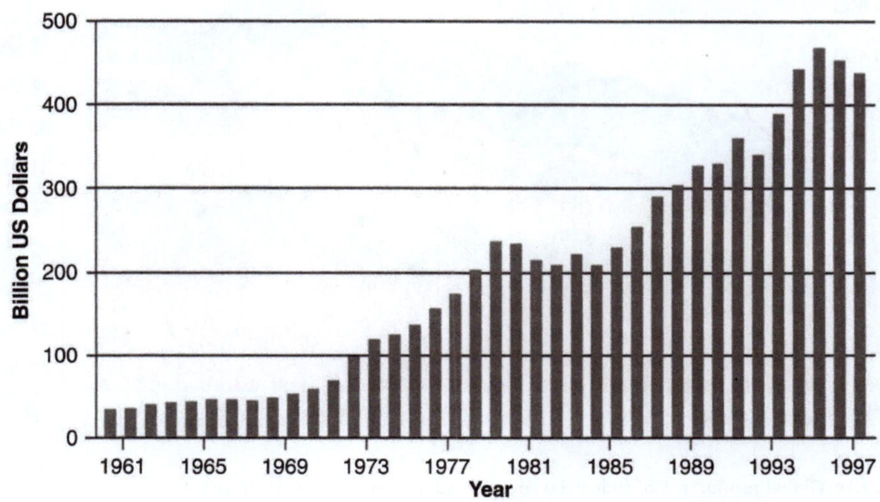

Forestry

Like agricultural production, wood production has increased over the last half century, albeit at a slower rate. Global production of industrial roundwood has increased by almost 50% since 1960 (Fig. 3.14). Roundwood consumption includes consumption of sawlogs, veneer logs, pulpwood, fuel wood and industrial roundwood (FAOSTAT 2002). Fuel wood accounts for 50% of roundwood production (FAOSTAT 2002) (Fig. 3.15a). Ninety percent of fuel wood is produced and consumed in developing countries (FAO 2001), whereas 79% of industrial roundwood production occurs in developed countries. The United States is the largest consumer of industrial roundwood, but Asia and Latin America have the fastest rate of growth of consumption (Matthews et al. 2000). China alone, due to its size and economic growth, is beginning to have a discernible impact on global consumption trends (FAO 2001). Plantations are emerging as a significant source of roundwood in the forestry industry; in 1995 plantations accounted for only 3% of global forested land but provided 22% of global industrial roundwood supply (Brown 1999). Asia has recently begun to dominate plantation holdings, with 62% of the global industrial plantation area located in this region (FAO 2001) (Fig. 3.15b).

International trade in forest products has also increased since 1960, but not at a steady rate. Export volume doubled during the 1960s, but since the 1970s the amount of industrial roundwood exported has been variable, increasing about 30% over the last 30 years (FAOSTAT 2002) (Fig. 3.15c). The production and manufacture of industrial wood products contributed approximately 2% to the global GDP and accounts for almost 3% of total world trade (Solberg et al. 1996; Matthews et al. 2000). Much of the increase in trade is due to greater volumes of trade occurring between countries within the same region, particularly between developing countries in Asia (FAO 2001).

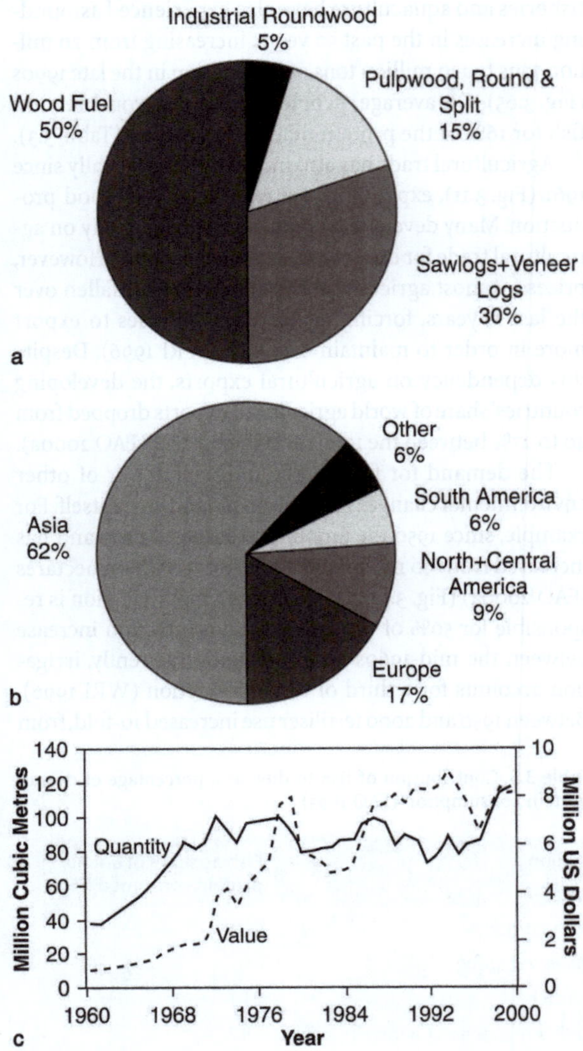

Fig. 3.15. a Roundwood production by use, 2001 (FAOSTAT 2002); b distribution of forest plantation area by region, 2000 (FAO 2001); c world export of industrial roundwood, 1961–2000 (FAOSTAT 2002)

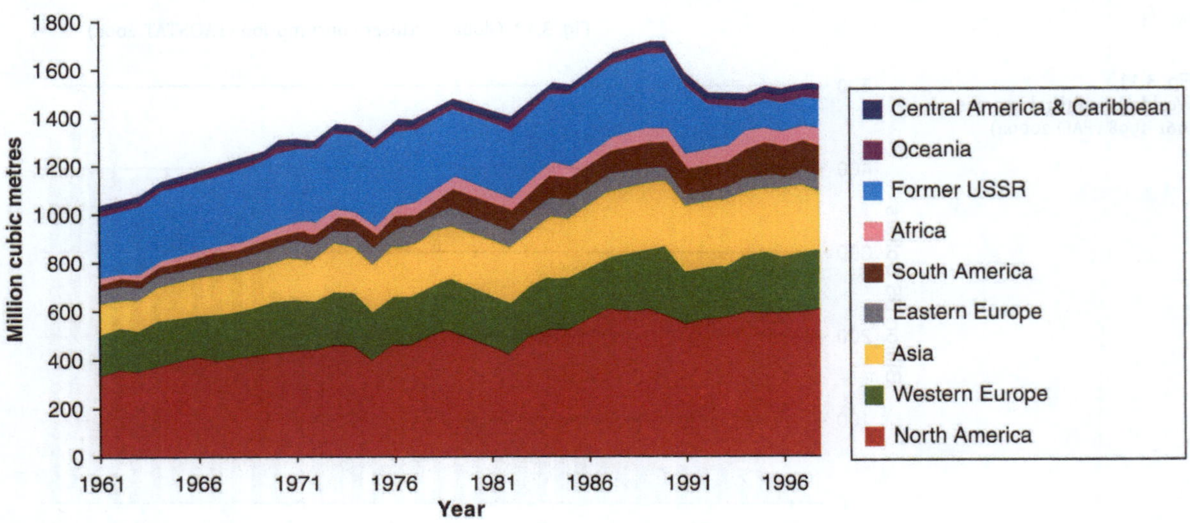

Fig. 3.14. Global production of industrial roundwood, 1961–1998 (FAOSTAT 2002)

Recreation and Tourism

Travel and tourism, currently a minor contributor to environmental change, is one of the fastest growing industries in the world. From being a very small industry at the beginning of the twentieth century, it had grown by 1999 to account for 11.7% of the world's GDP, or about USD 3 550 billion worth of economic activity. Travel and tourism now account for almost 200 million jobs worldwide, or about 8% of the world's total employment. The rate at which the industry is growing is consistently outpacing baseline economic growth. Using annual trends in arrivals as an indicator, international arrivals grew at an annual average rate of about 4.2% during the 1990s, despite slow economic growth and high unemployment in major industrialised countries during the first part of the decade and the Asian financial crisis later in the decade (World Tourism Organization 2001) (Fig. 3.16).

There are some interesting regional trends in the industry. For example, the East Asia-Pacific region has consistently experienced the highest growth rate of international visitors since 1970, with its share of the world market growing from 3.2% in 1970 to 13.9% in 1998. For the 1990s, the region's growth rate was 6.7%, substantially higher than the global average despite the Asian financial crisis. Strong growth seems set to continue for the foreseeable future (World Tourism Organization 2001).

Trends in Sectoral Drivers

As the twenty-first century unfolds, population, affluence and technology are interacting in ways that will have important implications for the nature of sectoral drivers of change. Although the global population will increase for several decades to come, there is now strong evidence that fertility rates are dropping in most parts of the world and that the stabilisation of the global human population may be realised during this century (UN Population Division 2000). In many economies in the developed world rapid technological change is leading to increases in efficiency, reduction in wastes and dematerialisation (producing more with less materials), which in turn are leading to a decoupling between economic growth in major sectors and the resource usage and emissions into the environment. This is becoming evident in the energy, industry and food provision sectors. The energy intensity in most modern economies is increasing, that is, there is increasing economic output for the same or less amount of energy expended. Industry is producing more for the use of the same or fewer input materials (cf. Fig. 3.5). Precision farming is leading to enhanced agricultural production with less fertiliser and water usage.

To date all of these technological gains in efficiency and reductions in impacts are almost exclusively available to high-income sectors of developed countries only. These sectors are so economically powerful, however, that they can transfer environmental consequences elsewhere, as in the case of large-scale tropical deforestation in Indonesia to provide pulp for international consumption (Jeppson et al. 2001). Thus, perhaps the biggest unknown in projecting future trends in sectoral drivers, and hence their implications for the Earth System, is the future relationship between the wealthy and the poor. Present trends suggest that the gap between the wealthy and the poor is increasing almost universally, both within countries and between countries. For example, as noted in Chap. 1, the ratio of income of the world's richest 20% to the poorest 20% increased from 30:1 in 1960 to 78:1 in 1997 (World Bank 2002; UNDP 2000). Given these trends, the majority of the world's population will not readily be able to benefit from technological advances. Access to new technologies and methodologies, rather than the development of them,

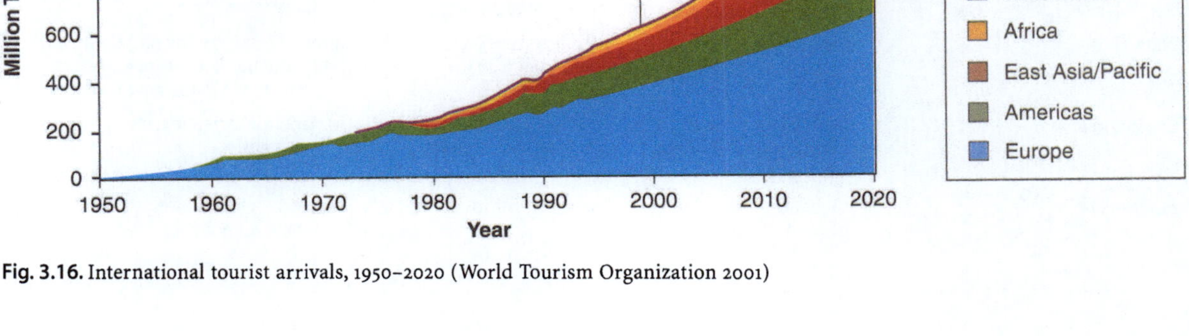

Fig. 3.16. International tourist arrivals, 1950–2020 (World Tourism Organization 2001)

thus may well become the limiting factor in terms of changing the future trajectories of sectoral activity on a global scale. Wealth differences, however, can be more complex and are often linked to different political economies and their effect on the ability of countries and locales to protect resources or enforce rules in their use. Wealth differences between countries have been shown to have significant impacts on resource use. For example, far eastern Siberia is currently experiencing rapid deforestation that is being driven by markets for wood and pulp in China (Turner 2002).

3.2.2 A Systems Approach to the Drivers of Earth System Change

The preceding section presented the human drivers of change to the Earth System on a sectoral basis in the context of production and consumption linked to a population-affluence-technology perspective. Other, more systems-oriented approaches can also be used to describe human-driven change to the global environment. One such approach is to view human activities as a complex hierarchical cascade of drivers from proximate to ultimate, or underlying, drivers (Table 3.4). Proximate drivers are the immediate human activities that cause the environmental change in question: examples include the burning of forests by small-holders for cultivation; household choice to use an automobile; corporate design of an industrial facility; or state or public choice of waste treatment. Ultimate drivers are the human demands that shape consumption expectations and patterns, for example, the demands for and access to food, consumer products, lifestyle, mobility, recreation, employment, comfort and safety. In reality, however, there is rarely a simple cause-effect relationship or causal chain between the ultimate causes and the proximate drivers of change. The proximate drivers are more typically driven by complex interactions between trends in human population growth, consumption by that population, its distribution in cities vs. rural settings, national goals or economic status, technological capacity, political and resource institutions, cultural values, markets, and so on.

Cause-effect relationships and other such simplifications help to draw attention to human-environment problems and their potential drivers, but they also help entrench and calcify public attitudes and even policy directed towards problem resolution. One classical example of the importance of understanding complex hierarchical cascades of drivers is found in the case of tropical deforestation. Population pressure and poverty have often been cited as the primary causes of tropical deforestation. However, a careful analysis of a large number of case studies across the tropics suggests that a more complex array of drivers including market and policy failures and terms of trade and debt are likely influences on the patterns and trajectories of land-use change in the tropics (Geist and Lambin 2002) (Fig. 3.17). As noted in one extensive review of the literature, forests fall because it is profitable to someone or some group (Angelsen and Kaimowitz 1999). However, the simplicity of this statement belies the complexity of the interlinked causes underlying this profitability.

These various factors come together in different ways in the three major humid tropical areas – Latin America, Central Africa and Southeast Asia – to produce patterns of land-use change characteristic of each region. Thus, tropical deforestation in Africa is driven primarily by impoverished farmers seeking to expand subsistence farming; in Latin America by medium to large-scale holders, facilitated by government policies, seeking to transform forest to pasture; and in Southeast Asia, by a mix of corporate logging and plantation development as well as policy-driven resettlement schemes (e.g., Barraclough and Ghimire 1995; Angelsen and Kaimowitz 1999; Lambin et al. 2001).

Table 3.4. Proximate and underlying drivers of human transformation of Earth

Compartment/cycle transformed	Proximate driver	Underlying driver
Land	Clearing (cutting forest and burning), agricultural practices (e.g. tillage, fertilisation, irrigation, pest control, high yielding crops, etc.), abandonment	Demand for food (and dietary preferences), recreation, other ecosystem goods and services
Atmosphere	Fossil fuel burning, land-use change (e.g., agricultural practices), biomass burning, industrial technology	Demand for mobility, consumer products, food
Water	Dams, impoundments, reticulation systems, waste disposal techniques, management practices	Demand for water (direct human use), food (irrigation), consumer products (water usage in industrial processes)
Coastal/marine	Land-cover conversion, groundwater removal, fishing intensity and technique, coastal building patterns, sewage treatment technology, urbanisation	Demand for recreation, lifestyle, food, employment
Biodiversity	Clearing of forest/natural ecosystems; introduction of alien species	Demand for food, safety, comfort, landscape amenity

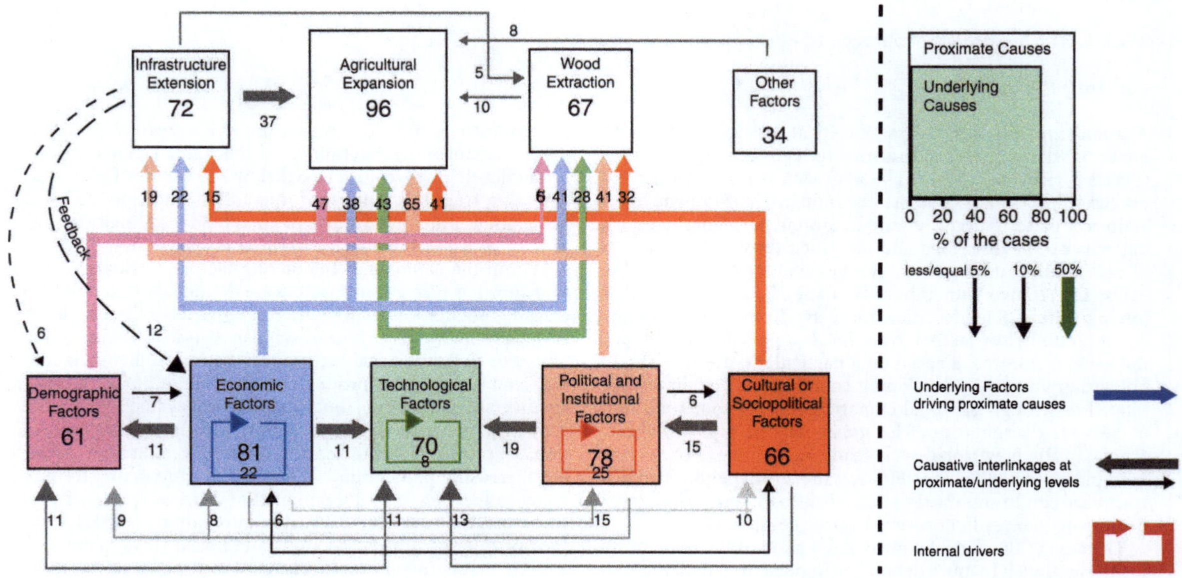

Fig. 3.17. The causative patterns of tropical deforestation from 1850 to 1997 based on a synthesis of 152 case studies examining both proximate and underlying causes (Geist and Lambin 2002)

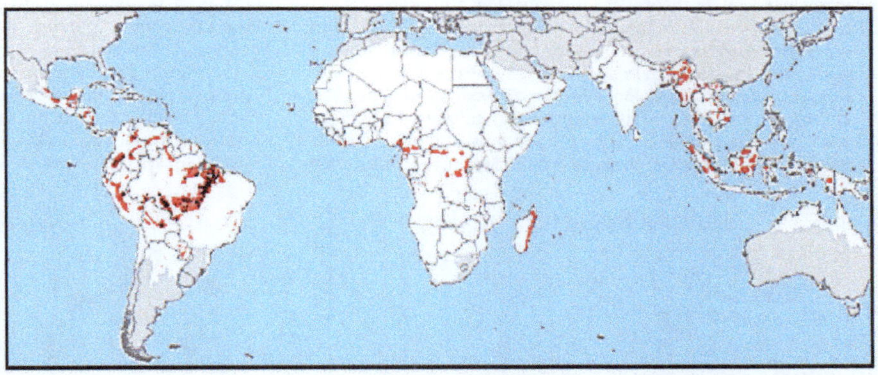

Fig. 3.18. Locations of rapid tropical deforestation over approximately the last 20 years (Lepers et al. 2003). The map indicates the number of times each 0.1° grid was identified as undergoing rapid deforestation. Data derived from Achard et al. (2002), DeFries et al. (2002) and Landsat Pathfinder for the Amazon Basin. *Pink colours* indicate that only one of the data sets identified the grid as undergoing rapid deforestation, *red* indicates two data sets and *black* indicates three

The fact that a variety of forces and processes can come together in specific geographical locations and conditions, giving rise to specific environmental outcomes, suggests another approach to understanding human drivers and impacts. Place-based synthesis of this kind permits comparative assessments from which common human-environment conditions can be drawn – those combinations of factors tending toward similar environmental impacts or trajectories of change (NRC 1999). Several examples of such an approach exist, some directed toward the identification of hot spots of deforestation (Fig. 3.18) or loss of biotic diversity (Achard et al. 1998), others involving regions at risk – places (human-environment conditions) in which livelihoods and human health were threatened by environmental degradation (Cutter 2001; Kasperson et al. 1995). Attempts have also been made to delineate and classify specific types of human-environment systems in terms of their vulnerability towards degradation (Box 3.1).

Systematic assessments of differing kinds of human-environment conditions and their environmental consequences worldwide do not exist, but research is mounting at sub-global levels providing strong clues about them. It is common to find the driving forces of land and ecosystem change to be geographically external to the location in which the change is taking place. This spatial separation is obvious in cases such as acid rain impacts on forest health where the pollutant source lies well upwind of the deposition. It may not be so obvious when the linked causes are less direct: international timber markets, co-joined corporate and state interests, concessions to cut tropical forests, abandoned logging roads, marginalised populations in adjacent regions, emigration following the roads and maintenance of forest clearance for small holder cultivation are a common combination of causes in Southeast Asia (e.g., Kummer 1992; Brookfield et al. 1995). Such placed-based approaches can highlight the connectivity between the biophysical and socio-economic conditions of the particular location undergoing change, and between that location and others.

Box 3.1. The Syndromes Approach to Place-Based Assessment

Gerhard Petschel-Held

A promising approach to assess critical developments in the Earth System identifies and models syndromes of change (Petschel-Held et al. 1999). The approach is based on the hypothesis that similar processes of human-environment interactions or patterns occur repeatedly around the planet, that such patterns can be described and identified through the analysis of case studies, and that they can be qualitatively modelled to assess the relative vulnerabilities of specific places to various forcing factors. The identification of syndromes takes into account a number of factors, from local to global. For example, the Sahel Syndrome is based on a particular pattern of smallholder agriculture in developing countries, but might also be related to strongly seasonal climatic patterns, poor soils and/or semi-arid vegetation. The *upper part* of Fig. 3.19 shows schematically how global-scale features combine with specific local places to yield characteristic patterns. Depending on distinct local conditions these patterns can evolve into critical syndromes or into resilient co-evolutions (homeostases).

The case of the Sahel Syndrome is a good example to examine the approach in more detail. It is important to note that this pattern does not solely refer to the Sahel region, but sees the region as a showcase for an archetypical interaction between humans and nature. The central feature is to develop a wiring diagram for the dynamics of the system that is based on case studies as well as theoretical insights (Fig. 3.19). For this pattern of human-nature interaction the critical relationships involve the growth rate of population; the allocation of labour to agricultural and off-farm activities; the nexus among wages, total income and the income relative to basic needs; the availability and adoption of agricultural technology and management practices; yields of crops and animals; and ultimately the quality of the resource base. In particular, in the end the approach must be able to simulate the well-known impoverishment – resource degradation spiral that threatens many parts of the developing world. The Sahel Syndrome itself is seen as emerging from this pattern of interaction, which as such can bring about other syndromes of smallholder agriculture, e.g., The Rural-Exodus Syndrome (WBGU 1997).

Within the complex of interacting factors, qualitative relationships between various factors are defined. For example, increasing population leads to increasing availability of labour. However, the modelling approach can also identify inflection points or thresholds. An increase in the availability of labour may lead to increasing productivity for a while, but as the society diversifies, off-farm labour may become available and more attractive. In addition, factors can interact. Increasing population can place increasing demands on scarce natural resources, but increasing population can also lead to increasing technological effectiveness and learning, thus decreasing pressure on resources. With a potentially vast array of potential interactions theoretically possible, case studies are used to supplement a generic modelling approach that captures the large variety of contexts and places. This generic character is achieved by implementing a modelling technique that formally describes interactions by qualitative statements such as: "Without improvements in agricultural technologies, an intensification of agriculture induces an increase in soil degradation". This sign-oriented technique, originated in the computer science field of artificial intelligence, does not make use of numbers but directly gives results for the trends, e.g., increasing or decreasing soil degradation. Therefore, regions that may vary by numbers but not by sign might all be described by the same generic qualitative model.

Based on the model, conditions for the emergence of syndromes or homeostases can be assessed. These conditions yield a global

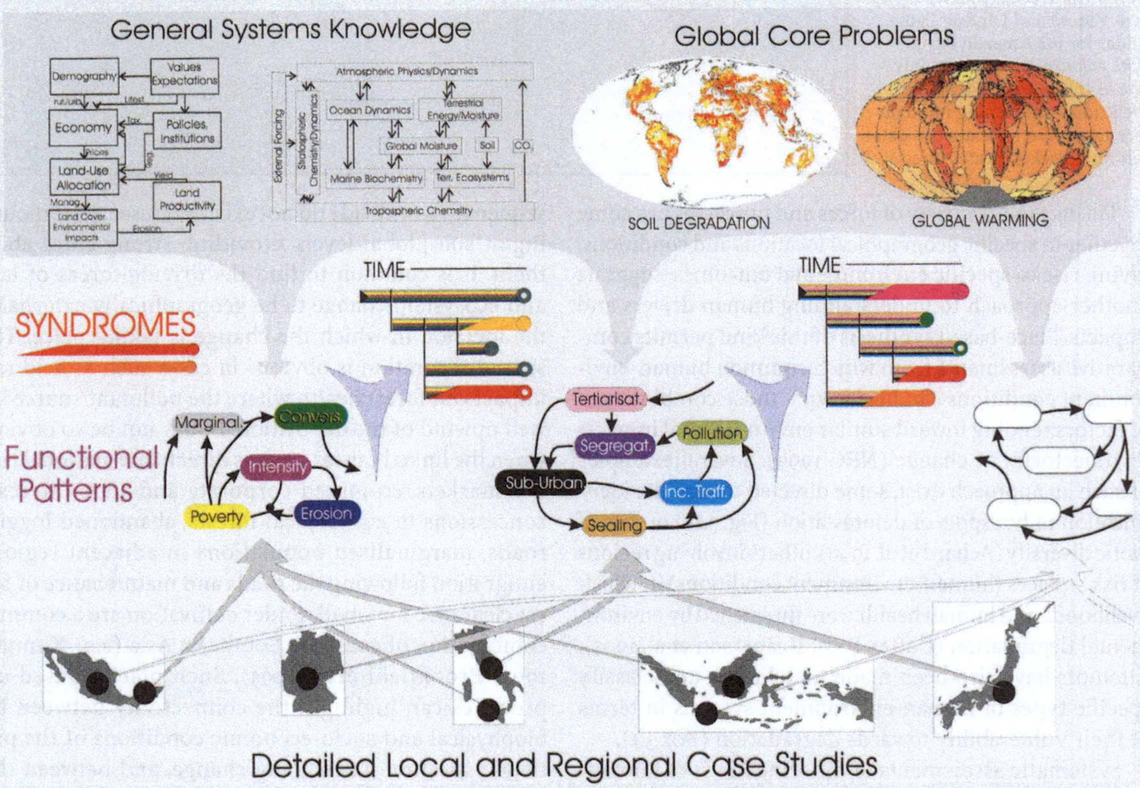

Fig. 3.19. Grouping of case studies and Earth System properties into typical functional patterns (Petschel-Held 2001)

◀

picture of relative susceptibilities, or vulnerabilities, to a particular syndrome. Based on data from the early 1990s, Fig. 3.20 shows the relative disposition of various regions of the world to the Sahel Syndrome, ranging from 0 to 1, that is, from no vulnerability to highest, respectively. In addition to the Sahel region itself, wide regions in southwest and east Africa show high vulnerability. Other areas in Asia and South America could also be subject to the downward spiral of increasing poverty and resource degradation.

How can the syndrome approach be used to assess global change impacts and vulnerabilities to particular forces (cf. Sect. 5.2.2)? Regions can be considered to be highly vulnerable if climate or global change prove to be the straw that breaks the camel's back. The syndrome approach can help to identify these regions, as it takes into account the multiple factors for stress. It is in these regions that fine scale analysis of global change impacts should be undertaken.

Fig. 3.20. Geographical distribution of regions with various levels of vulnerability (*0* denotes low vulnerability, *1* highest risk) in the Sahel Syndrome (Lüdeke et al. 1999)

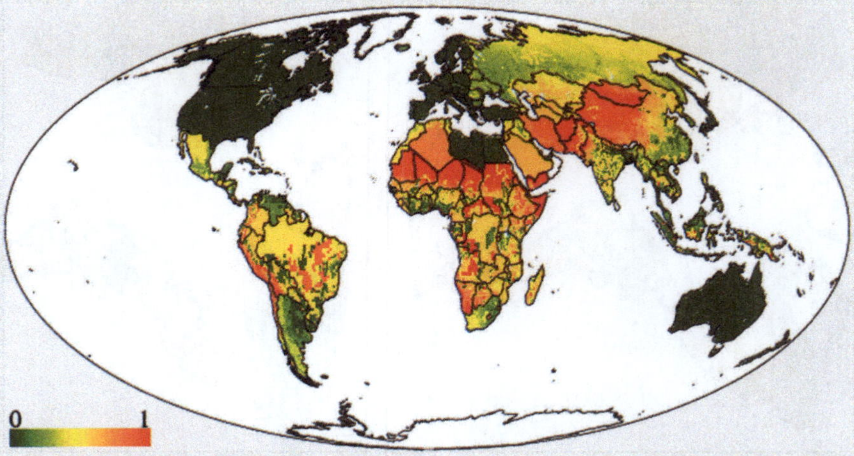

3.3 Characterisation of Changes in the Earth System

No component of the Earth System has been wholly unaffected by the human endeavours discussed above. Earth's atmosphere, freshwater and marine systems, land systems, biota, and the biogeochemical cycles that link them, all reflect the influence of humans. There is little doubt that Earth's land cover has undergone the longest and most pervasive changes as a result of numerous human endeavours, acting singly or in concert, but other components of the Earth System have been changed as well. The following sections identify and characterise some of the immediate changes experienced by different components of the Earth System.

Fig. 3.21. Transformation of natural ecosystems to rice paddies, such as those pictured here in the terraced landscape of the Philippines (*image:* FAO photo, R. Faidutti 1990)

3.3.1 The Earth's Land as Transformed by Human Activities

Humans are a terrestrial species. The very existence of large numbers of people and their development into complex societal organisation requires an ever-increasing use and management of the Earth's land surface. It is no surprise, therefore, that human-induced environmental changes with the longest history and most pervasive global imprint, until recently, are those involving land use and land cover (Fig. 3.21). Elements of this imprint can be found early in human prehistory, but they have subsequently enlarged in kind and scale and with increasing intensity (e.g., Meyer and Turner 1994; Thomas 1956). While the geographic focus of land change has shifted through time, there is little reason to believe that its global magnitude will be reduced in the near future.

Early people were hunter-gatherers and had few advantages in the hunt, other than their social organisation and technological inventiveness. The first such management invention with large-scale environmental impacts was the control of fire (Pyne 1991). Burning eased the hunt and led to abundance in slaughter beyond the consumption capacity of the hunters (Insenberg 2000) but, more

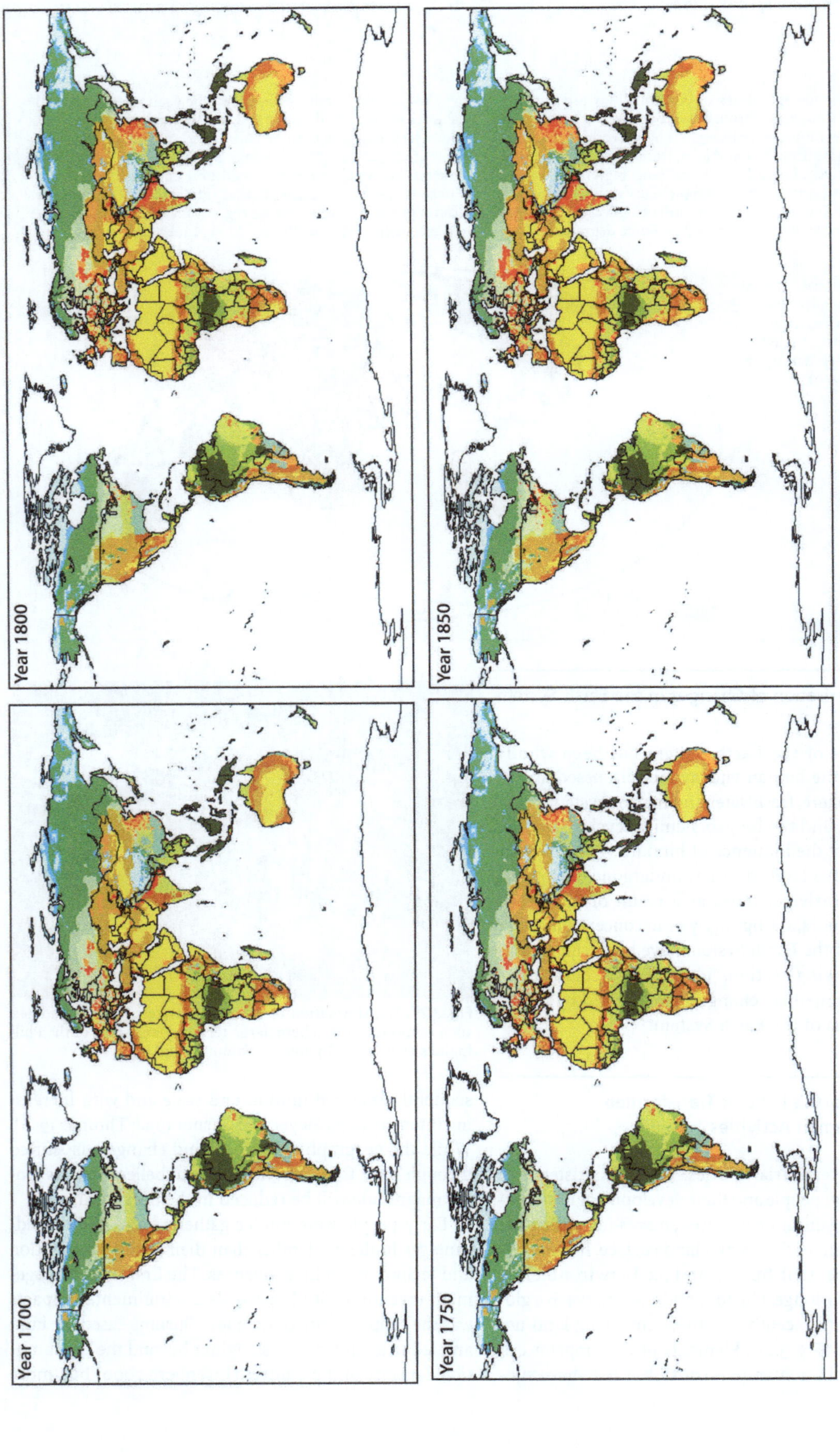

Fig. 3.22a. Transformation of the Earth's land surface by human activities over the past 300 years (Ramankutty and Foley 1998; Klein Goldewijk 2001)

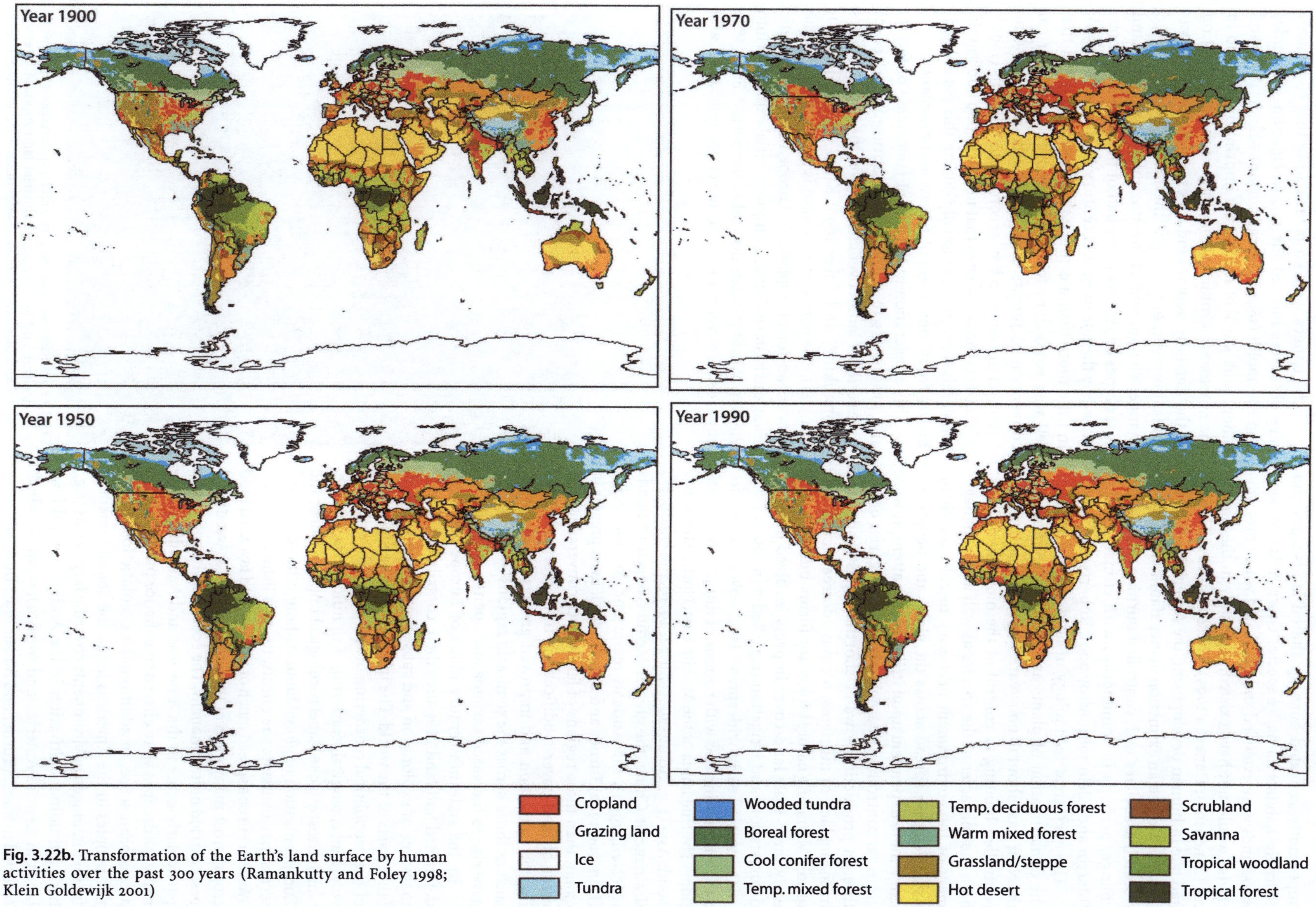

Fig. 3.22b. Transformation of the Earth's land surface by human activities over the past 300 years (Ramankutty and Foley 1998; Klein Goldewijk 2001)

importantly, repeated burning altered landscapes and hence the habitats for native species. Grasslands and open woodlands were extended and/or created where they otherwise would not have existed, especially in the Americas (Sauer 1961; Denevan 1995). Those continents inhabited late in human development, the Americas and Australia, witnessed an extinction in megafauna which some research associates with controlled burning and habitat change, perhaps in conjunction with climate change (Martin and Klein 1984; Alroy 2001; Roberts et al. 2001).

The next great technology of land change involved the domestication of plants and animals beginning in the Near East about 10 000 years ago, and the move from controlled hunting to control of the biota themselves, both the character of the species as well as the management of the environment necessary to cultivate them. As agriculture spread across the globe and people became more dependent upon it, increasing attempts were aimed at controlling or ameliorating the vagaries of nature on production – water shortfalls, pest and weed invasion, lack of nutrients, biomass decreases and loss of soil. Managing nature took many forms, but those with early, large-scale land-cover impacts involved burning, deforestation, and irrigation (e.g., Redman 1999).

Centres of early widespread land cover change for agriculture coincided with regions of initial state formation – riverine lands of the Middle East, North Africa, and South Asia (e.g., Adams 1965; Butzer 1976), the Huang Ho drainage of China, the greater Mediterranean rimlands (McNeill 1992), the Andean realm and Mesoamerica (Denevan 1992; Turner and Butzer 1992). Even greater in extent than these regions of land-cover conversion were areas of land-cover modifications from repeated burning for cultivation and improved grasses for livestock, such as those across the semi-arid regions of Africa that gave rise to increased savanna and open woodlands.

By the sixteenth century much of Europe had been deforested and land pressures there contributed in part to the age of exploration and subsequent colonisation of other parts of the world. The impacts went beyond those of the geopolitical reordering of the world; they also involved an ecological reordering of significant magnitude, which Crosby (1986) labels ecological imperialism or the Europeanisation of global biota. Although these emotive terms may overstate the effects, there is little doubt that Western Europe spread much of its biota and favoured food crops across many parts of the world, as well as weeds and pests carried unintentionally. The ecological impacts were particularly acute in the Americas and Australia, where many landscapes were altered to enhance production in a European sense. In addition, the dramatic loss of native peoples in the Americas led to the abandonment of further managed environments covering large areas (e.g., high altitudinal cultivation in the Andes; tropical lowlands in Central America and Mexico), some of which reverted to forest (Turner and Butzer 1992; Denevan 1992).

Such changes were so far reaching that contemporary observers noticed the transformation. Alexander von Humboldt (1849), George Perkins Marsh (1864) and others took stock of the human imprint on the land in the nineteenth century, remarking on the modification of the Earth by humankind. These and other observations, however, only caught the leading wave of land change. The amount of forest converted to other land cover types by 1850 was only half of that which was converted by the late twentieth century (Turner et al. 1990); in little more than a century the amount of forest that fell was equivalent to the entire previous historical conversion of forests over thousands of years, some 3 200 000 to 3 900 000 km^2 according to Williams (1990). Richards (1990) estimates that nearly 10 000 000 km^2 in croplands were gained globally during this period, primarily from conversion of forests and grasslands.

This acceleration in the rate of conversion of forests and grasslands to croplands is captured in a recent time series of land-cover change over the past 300 years (Ramankutty and Foley 1998; Klein Goldewijk 2001; Fig. 3.22). The period 1700 to 1850 is characterised by the slow expansion of croplands in Europe, the Indo-Gangetic Plain and eastern China, with some initial expansion into the New World towards the end of the period. From 1850 onwards, however, the pace of conversion quickened, and

Fig. 3.23. Transformation of temperate and tropical regions of the world by agriculture and forestry into industrial landscapes of intensive cropping and plantation forests (*photo:* Brian Prechtel, ARS-USDA)

particularly from 1900 to 1970 the expansion of croplands was dramatic in many parts of the world.

By the final quarter of the twentieth century, the transformation of the Earth's land had become staggering in scope (Fig. 3.23), drawing attention and worldwide concern in terms of their implications for food production, biogeochemical flows, biodiversity and ecosystem services (Vitousek 1994) (see also discussion to follow in Chap. 4 and 5). The current state of land cover globally is indicated in Table 3.5 and Fig. 3.24. The magnitude of human modification of the land can be measured in several ways, highlighted below as four different aspects of land-use/cover change that give a sense of the near-complete human domination of the Earth's land surface.

Table 3.5. Global estimates of land cover and its change

Land cover	Year	Area (10^6 km^2)	Change (%)	Method	Source
Forest/woodland	1990	41.50	−29	Expert opinion	Klein-Goldewijk 2001
	1992	43.90	−20	Inventory	Ramankutty and Foley 1999
	1994	41.72		State reported	FAOSTAT 2002
	1997	33.36		Adjusted UN	Bryant et al. 1997
	2000	41.70			IPCC 2001
	2000	45.38		State reports	UNEP-WCMC 2000
Steppe/savanna/grassland	1990	17.50	−49	Expert opinion	Klein-Goldewijk 2001
	1992	26.70	−20	Inventory	Ramankutty and Foley 1999
	2000	35.00			IPCC 2001
	2000	53.66		Satallite imagery	White et al. 2000
Cropland	1990	14.70	+444	Expert opinion	Klein-Goldewijk 2001
	1992	20.30	+408	Inventory	Ramankutty and Foley 1999
	1994	15.10		State reported	FAOSTAT 2002
	2000	16.00			IPCC 2001
Pasture	1990	3.10		Expert opinion	Klein-Goldewijk 2001
	1994	34.65		State reported	FAOSTAT 2002
	2000	3.50			IPCC 2001

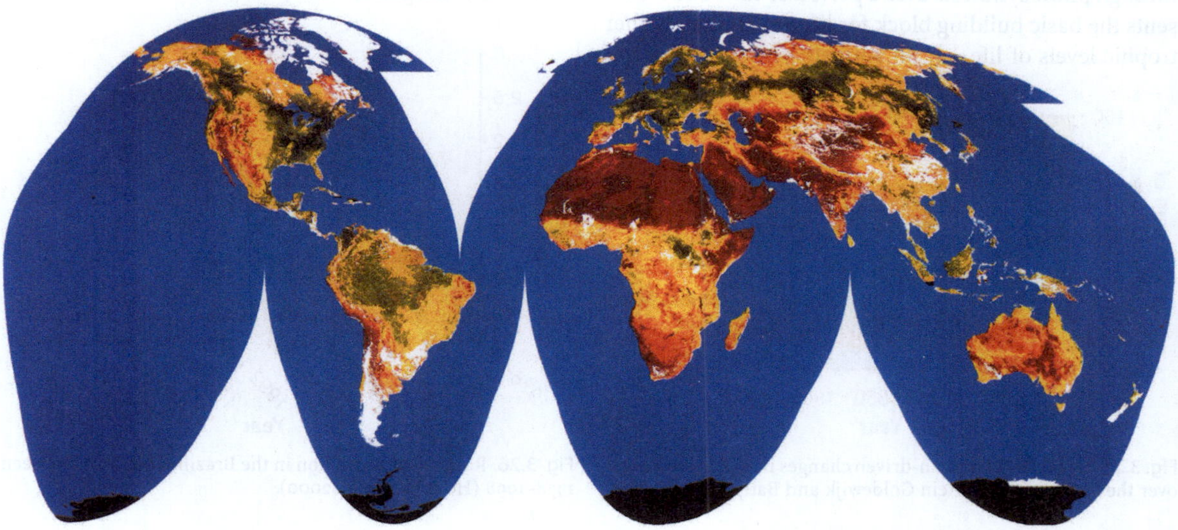

▲
Fig. 3.24a.
High resolution (1 km) global map of present-day land cover based on data from the Advanced Very High Resolution Radiometer in 1992–1993 (Loveland and Belward 1997)

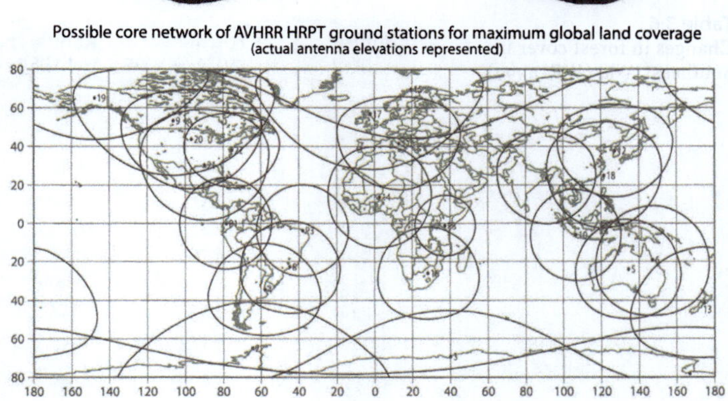

Fig. 3.24b. ▶
AVHRR ground stations (Loveland and Belward 1997)

Area of Land Converted

Most estimates of the total degree of land-cover change suggest that ~50% of the ice-free land surface of the Earth has been converted or substantially modified by humankind over the past 10 millennia or so (e.g., Fig. 3.25). Virtually no region or land-class remains untouched by human-induced impacts of some kind, and direct use of the land is undertaken everywhere, except for the most extreme environments (cold or dry). The current foci of large-scale land-uses involve metroplexes (megacities and their surrounds) along the world's coastal zones and major waterways; prime cultivation lands characterised by soil quality, relatively stable rainfall regimes and/or water sources for irrigation and terrain suitable for mechanised production; and more marginal agricultural lands used for stocking or, in the case of the tropical world, for small-holder cultivation.

Use of Terrestrial Net Primary Productivity

Another measure of the degree of human domination of the land is through the estimate of terrestrial net primary productivity that is used, co-opted, or otherwise diverted from its natural metabolic pathways by human activity. Net primary productivity is a measure of the net amount of carbon assimilated by terrestrial vegetation through photosynthesis over a period of time and represents the basic building block for life on land. All higher trophic levels of life, i.e., all animals, ultimately depend on this process for their existence. An estimate about a decade ago suggested that ~40% of global net primary productivity is used or co-opted by humans, primarily through land-uses (Vitousek et al. 1986). More recent work has greatly expanded the range of the estimates to 5–50% (DeFries et al. 1999; Rojstaczer et al. 2001).

Rates of Tropical Deforestation

The highest rates of deforestation in the world currently take place in the tropics (FAO 1999), where rates can sometimes be as high as 4% per year for extended periods (see Table 3.6; WRI 1999). Like many natural and human processes in the context of the Earth System, though, deforestation in the tropics is neither a constant nor a linear process. Regional ebbs and flows of deforestation occur across the globe, linked to major policy shifts and international economic conditions. Economic downturns and the absence of subsidies that lead to deforestation can trigger a cessation in deforestation, and subsequent land abandonment and forest regrowth. For example, the rates of deforestation have fluctuated considerably in Amazonia (Houghton and Skole 1990) (Fig. 3.26) and rates as high as 4.5% have been estimated between 1980–1990 in southern Mexico (Mas Caussel 1996). For a number of reasons, determining a global average rate of tropical deforestation is not straightforward (Box 3.2).

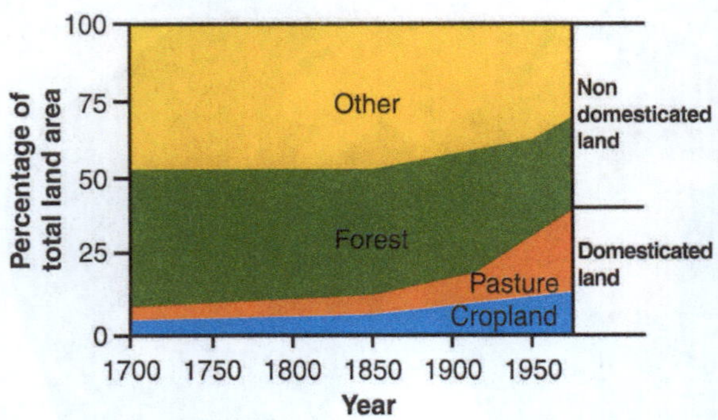

Fig. 3.25. Overview of human-driven changes in global land cover over the last 300 years (Klein Goldewijk and Battjes 1997)

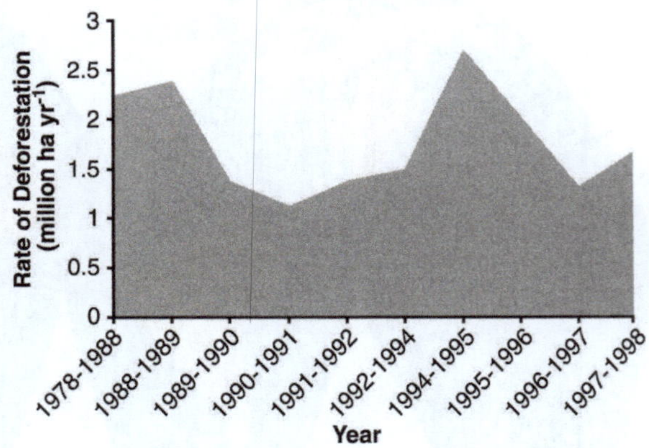

Fig. 3.26. Rates of deforestation in the Brazilian Amazon between 1978–1998 (Houghton et al. 2000)

Table 3.6.
Changes in forest cover in Southeast Asia (WRI 1999)

Country	Forest cover (10^3 ha) 1995	Annual change (%) 1980–1995	Frontier forest cover as % of orig. forest 1996	Present forest as % of orig. forest 1996
Cambodia	9830	−2.71	10.3	65.1
Indonesia	109791	−1.18	28.5	64.6
Lao PDR	12435	−1.41	2.1	30.0
Malaysia	15471	−2.83	14.5	63.8
Myanmar	27151	−1.75	0.0	40.6
Philippines	6766	−3.96	0.0	6.0
Singapore	4	0.00	0.0	3.1
Thailand	11630	−3.58	4.9	22.2
Vietnam	9117	−1.45	1.9	17.2
Total for SE Asia	202195	−1.44		

Box 3.2. Determining Rates of Tropical Deforestation

R. DeFries

Felling trees to obtain timber and make way for settlements and croplands is as old as the first attempts by humans at sedentary lifestyles. In past centuries expansion of cropland in the temperate forests of Eurasia and North America led to massive loss of forests. In more recent decades, the focus of deforestation has shifted to the tropical belt as clearing has expanded into remaining forests tracts of Latin America, central Africa, and Southeast Asia. The forces driving the tropical clearing are varied, in some locations resulting from resource extraction to satisfy timber demands in distant parts of the world, in other locations from expansion of commercial agriculture, and still in others from clearing by local populations for subsistence farming (Geist and Lambin 2002).

Quantifying the rates of deforestation and identifying locations undergoing rapid change are key to understanding the driving forces and consequences for the Earth System. Yet, knowledge of the changes occurring in tropical forests is surprisingly limited. Several factors contribute to this lack of consensus. First, there is no globally accepted definition of *forest* (UNEP 2001). Some definitions include only primary forest and some include only economically productive forests. Definitions vary according to tree canopy density. For example, the United Nations Food and Agriculture Organization (FAO) defines forest cover as greater than 10% canopy cover (FAO 2000c), while the IGBP defines it as greater than 60% (Loveland et al. 2000). These variations in definitions create difficulties for comparing deforestation estimates from various sources.

Another factor contributing to lack of consensus about rates of tropical deforestation is a paucity of observations spanning the entire tropical belt at multiple time periods and analysed with consistent methods. The FAO Forest Resource Assessments are the most widely-used source of information on forest area and decadal changes (FAO 1995, 2000c). The information, however, is aggregated at a national level and questions remain about varying methods and definitions used by different countries for data collection (Matthews 2001). Several studies based on analysis of satellite data indicate that the FAO Forest Resource Assessment overestimates the rate of deforestation in some countries (Steininger et al. 2001; Achard et al. 2002; DeFries et al. 2002).

Unlike national assessments and ground-based surveys, satellite data inherently provide consistent observations over time and, if consistent methods are used to interpret the data, provide useful estimates of deforestation rates. Data from the Landsat sensor, at a spatial resolution of 30 m, have provided estimates for some regions, primarily the Amazon basin (Skole and Tucker 1993; Houghton et al. 2000; Steininger et al. 2001), and in some *hot spots* of deforestation defined from expert opinion (Achard et al. 2002) (Fig. 3.22). Cloud coverage, lack of past acquisitions, and limitations in abilities to handle very large volumes of data have impeded Landsat-derived analyses throughout the entire tropical belt. Coarser resolution data from sensors such as the Advanced Very High Resolution Radiometer (AVHRR) span the entire land area with daily coverage and have provided insights into locations of rapid change and rates of deforestation (Malingreau et al. 1995; DeFries et al. 2002; Hansen and DeFries 2003), though the resolution of 1 km is much coarser than the patchy scale at which deforestation generally occurs. More recently launched sensors, such as SPOT VEGETATION and Moderate Resolution Imaging Spectroradiometer (MODIS), combined with improved approaches to characterise forest cover (Hansen et al. 2002), are promising for monitoring deforestation in the future.

An additional issue confounding estimates of tropical deforestation rates is that forest cover is dynamic, resulting both from natural disturbances such as fire and human management practices of abandonment and regrowth. Except in the most extreme forest conversion for permanent commercial agriculture or settlements, locations which appear deforested at one time are likely to grow back and possibly become deforested again. Deforestation rates derived from comparisons of static snapshots of forest cover at discrete time periods do not capture these dynamic processes and lead to misleading conclusions. For Earth System science questions such as the implications of deforestation for carbon fluxes and biological diversity, processes are fundamentally different in locations undergoing deforestation or regrowth. Spatially-disaggregated information on these locations, combined with other information such as biomass or species habitat, are needed to address these questions.

A few data sets are available to compare estimates of rates of deforestation and net changes in forest area at the continental scale. Achard et al. (2002) analysed Landsat data for the 1990s in locations where expert opinion indicates that change has occurred. DeFries et al. (2002) analysed the full extent of the tropical domain from 1982–2000 but at a coarser spatial resolution of 8 km with AVHRR data, interpreted in combination with higher resolution information where available. These two satellite-derived estimates, one in selected locations at higher resolution and the other over the full area at coarser resolution, provide complementary approaches for comparison with the FAO Forest Resource Assessment country reports. They converge on similar conclusions. The comparisons suggest that the reported FAO rates of net change (deforestation minus regrowth area) appear larger than the estimates from both satellite-based studies. However, the DeFries et al. (2002) analysis indicates an increasing overall tropical rate of deforestation from the 1980s to 1990s, particularly in Southeast Asia. In contrast, the FAO reports indicate a downward trajectory in the rate.

Accurate assessments in the overall rate of deforestation, the locations of the change and the trajectories of deforestation rates underlie analyses of the driving forces and consequences for the Earth System. Such analyses require not only reduced uncertainty in the area undergoing conversion, but an improved understanding of the dynamic nature of deforestation and regrowth and the interactions between human management and natural processes of changes in tropical forests.

Table 3.7. Estimates of region-wide changes in forest area (10^6 ha yr^{-1}) from three sources: analysis of 8 km AVHRR data from 1982–2000 (DeFries et al. 2002), analysis of Landsat data for 1990s (Achard et al. 2002), and FAO country data from the Forest Resources Assessment (FAO 1995, 2000c). The latter provides only estimates of net changes in forest area. The estimates are not directly comparable because Achard et al. (2002) include countries only in the humid tropics while DeFries et al. (2002) include the full tropical domain. Net change is deforestation minus regrowth area ([a] for countries comparable with DeFries et al. data set; [b] for countries comparable with Achard et al. data set)

	Latin America		Southeast Asia		Africa	
Mean annual change (1980–1990)						
Deforestation: DeFries et al.	4.4	(2.9–5.1)	2.2	(1.5–2.9)	1.5	(1.0–1.5)
Net change: DeFries et al.	3.6	(1.9–4.5)	1.2	(0.1–2.3)	0.3	(0–0.6)
Net change: FAO country report	7.1		2.4		3.9	
Mean annual change (1990–2000)						
Deforestation: DeFries et al.	4.0	(2.6–4.6)	2.7	(1.9–3.7)	1.3	(0.9–1.3)
Net change: DeFries et al.	3.2	(1.7–4.0)	2.0	(0.8–3.2)	0.4	(0–0.6)
Net change: FAO country report[a]	4.4		2.4		5.2	
Deforestation: Achard et al.	2.5	±1.4	2.5	±0.8	0.85	±0.3
Net change: Achard et al.	2.2	±1.2	2.0	±0.8	0.8	±0.3
Net change: FAO country report[b]	2.7		2.5		1.2	

Reforestation in Europe and North America

In contrast to these trends and trajectories of continuing land-cover conversion or modification from natural towards domesticated systems, some counter-trends have emerged or are emerging. For example, considerable reforestation has taken place in Western Europe and eastern North America over the last few decades as local economies have shifted to industrial-service sector orientations whose economic wealth can import agricultural goods from elsewhere and can outbid farmers (often on marginal lands) for local lands (Fig. 3.27). Suburbanised forests or densely settled forest landscapes follow. Likewise, the economics of farming in developed countries is driving production towards high yielding systems on favoured lands (Waggoner and Ausubel 2001) and away from more marginal lands. The overall result is the abandonment of agricultural land and its reversion to more natural ecosystems, usually forest or scrubland. Thus, while total world agricultural area has increased by nearly 10% since 1960, agricultural area in the USA and the European Union (12 original member states) has decreased by nearly 15% and 7% respectively over the same period, freeing 19 million and 30 million hectares respectively from agricultural use (FAOSTAT 2002).

Given the trends of land-cover change in the recent past and the complexities of forces that drive land-use and land-cover change, what are the projections of land-cover change for the coming decades? Land changes have been and remain intimately tied to global and regional political economies and to technological change. Inasmuch as the broader strokes of these facets of society can be understood, so can that for land change. Pending some unforeseen technological breakthroughs, and assuming current trends in population and the continuation of market economies, most forecasts indicate a shrinkage of cropland in the developed world, an intensification of cultivation on prime lands everywhere, and growth in urbanisation everywhere which, in turn, will mean that megacities will outbid agriculture for land (Turner 2002). These trajectories are less certain in the developing world, especially in the tropical forest frontiers, although increasingly agricultural land uses will move from subsistence to market production. One projection of land-cover change suggests that conversion of tropical forests will diminish in the near future in both South America and Southeast Asia, but in Central Africa conversion will accelerate and continue unabated until most of the forests have been converted to agricultural systems (Fig. 3.28). Whatever projection for future land cover is adopted, there is little doubt that nearly all of the Earth's land surface will be formally managed, including the increasing array of parks, reserves and protected forests.

3.3.2 The Atmosphere as Transformed by Human Activities

The composition of the Earth's atmosphere has changed significantly over the history of the planet. A clean and slowly varying atmosphere was essential for the origins of life, and life in turn influences the composition of the atmosphere. Human activities are now having a profound influence on the atmospheric system; in fact, they have been influencing the atmosphere for nearly as long as they have been altering the landscape (Fig. 3.29). Traces of widespread discernible impacts of human activities on atmospheric chemistry began with the early days of extensive metal smelting (Nriagu 1984, 1996). By the time of the Greek and Roman Empires, clear signs of enhanced concentrations of lead, copper and other trace metals can be found in ice cores from Greenland (Hong et al. 1994) and in lake sediments (Renberg et al. 1994) and peats (Lee and Tallis 1973; Livett et al. 1979; Shotyk et al. 1996, 1998; Martínez-Cortizas et al. 1999) from Europe. From this time onwards, human impacts on atmospheric composition remain measurable and from early medieval times onwards, they increase.

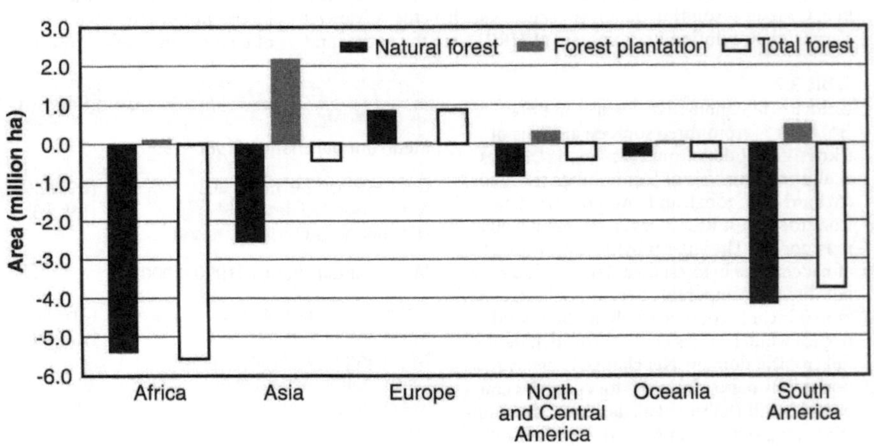

Fig. 3.27. Annual net change in forest area by region, 1990–2000 (FAO 2001). Aggregation of North and Central America hides an overall positive net change of 388 000 ha in forest for North America

Human influence on atmospheric composition has risen sharply over the past two centuries, especially over the last 100 years. Today's atmosphere bears a much greater chemical burden, even in the most remote areas, than in the recent past. The rate at which these changes in atmospheric composition are occurring is astonishingly fast compared to the natural rhythms of change in the atmosphere. The human-induced changes are driven by a combination of wide-scale industrialisation, growth of the transport sector, and the effects of intensification and extensification of agriculture, which include the application of fertilisers, biomass burning, deforestation and clearance of land for cropping and grazing. Against this overall trend of increasing human impact on the atmosphere, the realisation of the deleterious effects of deteriorating air quality at the local and regional scale has led to some lessening of the human impact at a regional scale. The best-known case was the acid rain episode in Western Europe and eastern North America. Although the first clear evidence for damage to ecosystems in these areas was published in the late 1960s (e.g. Odén 1968), it took many more years before the industries responsible for most of the harmful atmospheric emissions, especially of sulphur dioxide, acknowledged the growing body of science evidence and took effective measures to reduce them. Emissions of sulphur dioxide peaked in Western Europe and eastern North America in the late 1970s and 1980s and have been reduced substantially since then.

The record in environmental archives has confirmed the trend towards an amelioration of air quality impacts in many parts of the world (Nriagu 1990; Boutron et al. 1991; Candelone et al. 1995; Schwikowski et al. 1999). Nevertheless, there are signs that even in the developed countries, air quality problems persist, but with a different character. Elsewhere, factors such as the presence of major economic constraints, the legacy of past political systems, continuing dependence on fossil fuel as the main energy source, population expansion and focus on economic development lead to a continued increase in human impact on the atmosphere at regional and global levels.

This section describes in more detail the ways in which Earth's atmosphere has been significantly impacted by the human activities discussed earlier in this chapter. Four important types of atmospheric constituents, namely greenhouse gases, photo-oxidants, aerosols and other industrial pollutants and trace metals, will be considered.

Greenhouse Gases

The atmospheric concentrations of the greenhouse gases CO_2, CH_4 and N_2O have increased from their pre-industrial values to levels unprecedented in the last 420 000 years (Fig. 3.30). The CO_2 concentration is currently nearing 370 ppm; prior to the seventeenth century, CO_2 concentrations fluctuated in the range 275–285 ppm for a millennium and varied between about 180 and 280 ppm throughout the last four glacial-interglacial cycles. There is no doubt that fossil fuel combustion over the last 200 years, and especially during the last decades of the twentieth century, is the main process responsible for these elevated atmospheric CO_2 concentrations. Forest clearance and increased biomass burning contribute significantly to atmospheric concentrations, although they represent only about one-fifth of the total flux (IPCC 2001).

Increases in atmospheric CH_4 concentrations have been even more rapid and dramatic than those for CO_2 over the last few decades, having risen from a pre-industrial level of 670 ppb to concentrations greater than 1 700 ppb at the present. The sources of the CH_4 emissions are also largely human-related, being primarily domestic ruminants, rice paddies, human-induced fires, landfills,

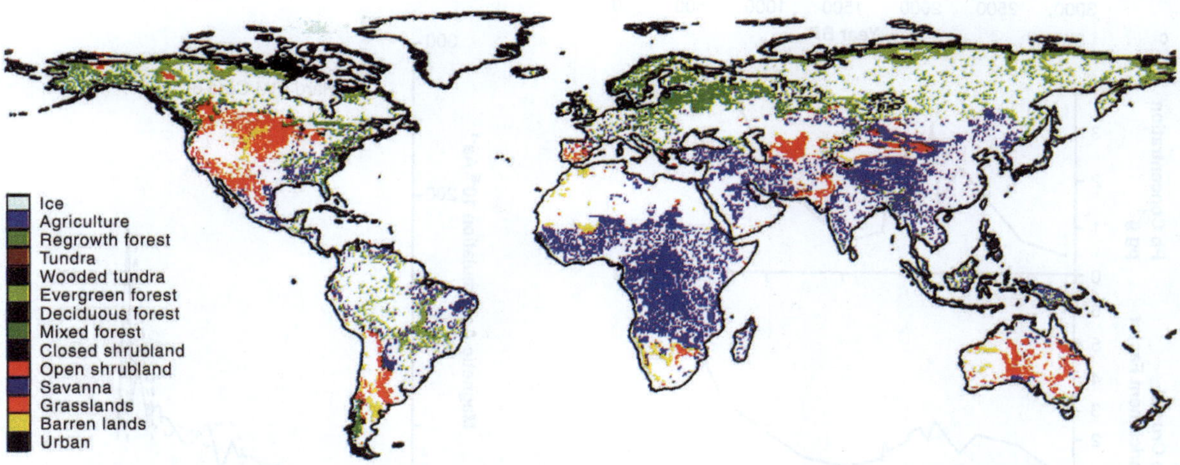

Fig. 3.28. Projected global land-cover change from 1990–2090 using the IMAGE 2.1 model (modified by J. Landgridge from Alcamo et al. 1996). *Coloured areas* depict regions projected to change from one cover type to the type indicated

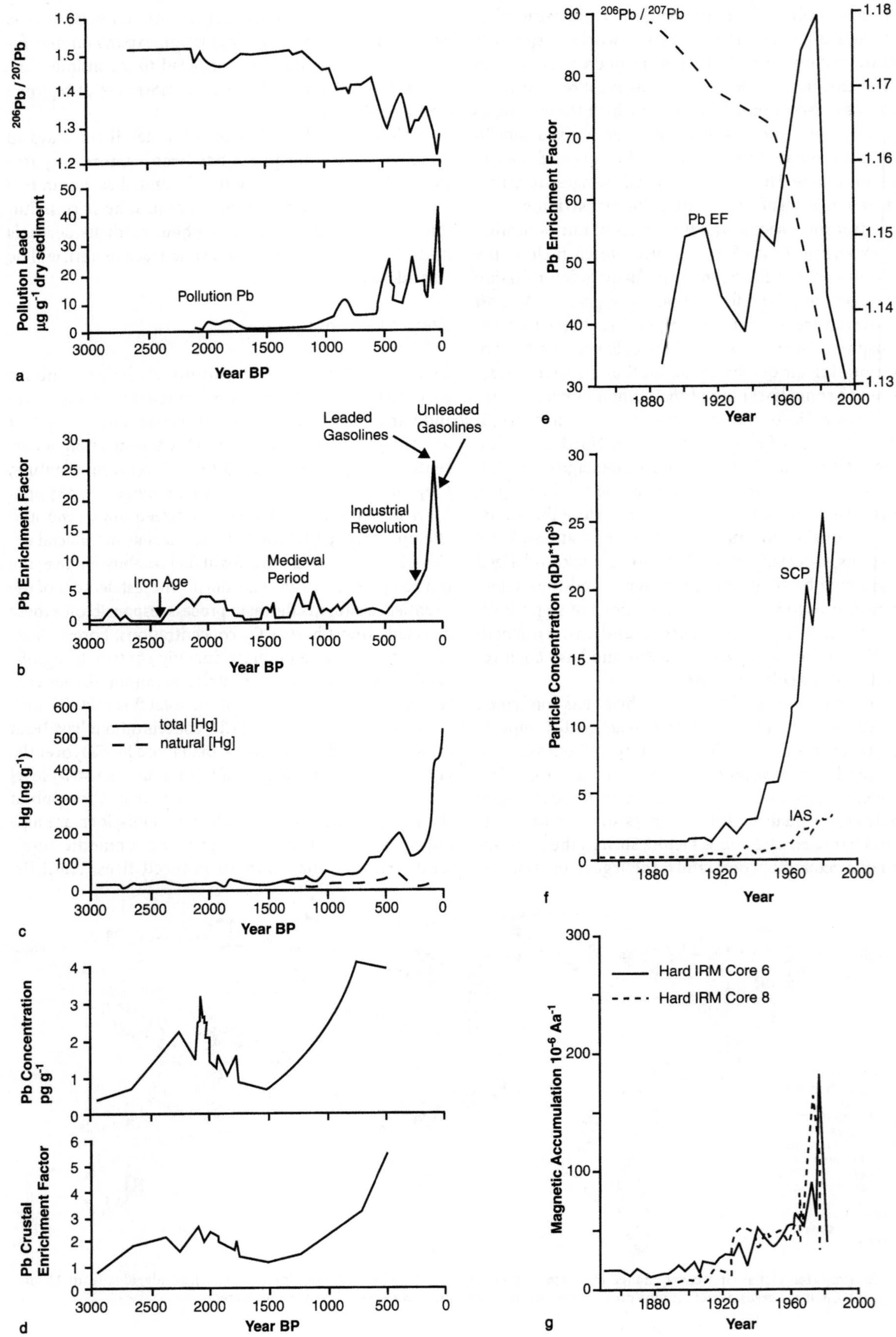

206Pb / 207Pb
Pollution Lead µg g-1 dry sediment
Pollution Pb
Year BP
a
Pb Enrichment Factor
Leaded Gasolines
Unleaded Gasolines
Industrial Revolution
Medieval Period
Iron Age
Year BP
b
Hg (ng g-1)
total [Hg]
natural [Hg]
Year BP
c
Pb Concentration pg g-1
Pb Crustal Enrichment Factor
Year BP
d
206Pb / 207Pb
Pb Enrichment Factor
Pb EF
206Pb / 207Pb
Year
e
Particle Concentration (qDu*10³)
SCP
IAS
Year
f
Magnetic Accumulation 10-6 Aa-1
Hard IRM Core 6
Hard IRM Core 8
Year
g

◀ **Fig. 3.29.** Sedimentary histories of trace metal and industrially generated particulate deposition: **a** pollution lead concentrations and Pb^{206}/Pb^{207} ratios as recorded in Koltjärn, a small lake in southern Sweden (Renberg et al. 1994, 2000); **b** total lead concentration record from Penido Vello, a peat profile in northern Spain (Martínez-Cortizas et al. 1997), set against a series of historical events and cultural stages from 3000 BP onwards; **c** total (*solid line*) and natural (*dashed line*) mercury (Hg) concentration record at the same site as **b** set against cultural and technological changes (Martínez-Cortizas et al. 1999); **d** total lead concentrations and crustal enrichment values from Greenland ice for the period 3000 to 500 years ago (Hong et al. 1994); **e** a short-term record of total lead deposition and lead stable isotope ratios from La Tourbiere des Genevez, an ombortrophic (precipitation-dependent) peat bog in Switzerland (Weiss et al. 1999) (*EF* denotes the Pb enrichment factor. The twentieth century pattern is comparable to that at Koltjärn in **a**); and **f** indicators of industrially generated atmospheric particulate deposition as recorded in the sediments of a small lake in northwestern Scotland remote from industrial sources (Rose 1994). (The inorganic ash spheres (*IAS*) are mostly fly-ash derived from coal-fired power stations, the spherical carbonaceous particles (*SCP*) are derived from both coal-fired and oil-fired power plants, but mainly the latter); **g** magnetic accumulation rates 10^{-6} A a^{-1} used as a proxy for industrial particulate at Big Moose Lake in the northeastern USA (Oldfield and Dearing 2003)

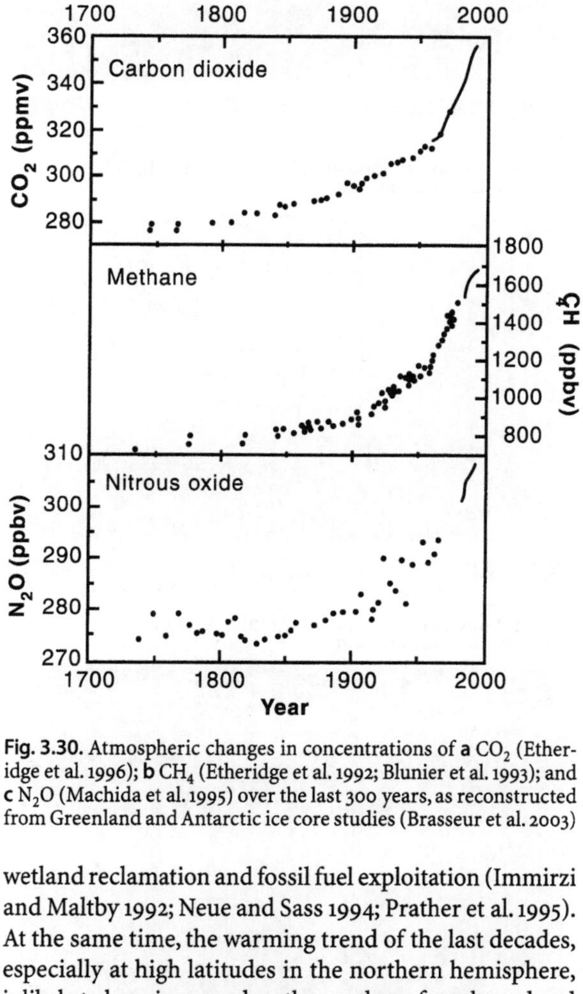

Fig. 3.30. Atmospheric changes in concentrations of **a** CO_2 (Etheridge et al. 1996); **b** CH_4 (Etheridge et al. 1992; Blunier et al. 1993); and **c** N_2O (Machida et al. 1995) over the last 300 years, as reconstructed from Greenland and Antarctic ice core studies (Brasseur et al. 2003)

wetland reclamation and fossil fuel exploitation (Immirzi and Maltby 1992; Neue and Sass 1994; Prather et al. 1995). At the same time, the warming trend of the last decades, especially at high latitudes in the northern hemisphere, is likely to have increased methane release from boreal and sub-arctic peatlands and organic soils through acceleration of the chemical reactions involved (Conrad 1989; Dise 1993; Kirschbaum 1995; Granberg et al. 1997), changes in hydrology (Knowles and Moore 1989; Moore et al. 1990; Laine and Vanha-Majamaa 1992; Moore and Roulet 1993; Laine and Minkkinen 1996) and thawing permafrost (Post 1990; Vourlitis et al. 1993; Waelbroeck et al. 1997).

The concentration of N_2O has increased from 285 to more than 310 ppbv since the pre-industrial era. The main sources are the increase in industrial agricultural fertilisation, although the introduction of legumes in crop rotation and use of animal manure also contribute (Matson et al. 1997; Vitousek et al. 1997b; Kroeze et al. 1999).

Photo-oxidants

While much of the attention in recent decades has been focussed on greenhouse gases, the chemical composition of the atmosphere has been changing in other significant ways. In contrast to the concentration of ozone (O_3) in the stratosphere (see below), the concentration of O_3 has been increasing in the troposphere (Fig. 3.31), where it can be harmful, not beneficial. Ozone is usually not produced directly by human activities, but rather is the result of chemical reactions involving precursors that are emitted by human activities. Observations at the Earth's surface at mountain sites in Europe suggest that tropospheric ozone has increased by a factor of 4 or more during the past century (WMO 1999). This increase is believed to be mainly due to human-driven emissions of nitrogen oxides NO_x (Fig. 3.32), hydrocarbons and CO, which interact with solar radiation to produce O_3. Tropospheric O_3 plays a critical role in determining the oxidising and cleansing efficiency of the atmosphere, as discussed in more detail in Chap. 5.

Only a few data are presently available on the concentration of CO, another significant photo-oxidant, from the past. An increase of about 20 ppb is observed between 1800 and 1950 for this compound from the ice core from Summit, Greenland, where the pre-industrial level is 90 ppb (and 55–60 ppb in the Antarctic). There is some evidence that the CO concentration may have increased in the higher latitudes of the northern hemisphere since about 1850 but not in the southern hemisphere (Haan et al. 1996) (Fig. 3.33).

Aerosols

Aerosols, atmospheric mixtures containing liquid or solid particulates of various sizes and compositions suspended in carrier gases (Box 3.3), have been recognised as critical factors in both climate dynamics and heterogeneous atmospheric chemistry. Perhaps the most well known, and certainly among the most important, are sulphate aerosol particles. Sulphate as well as nitrate aerosol particles from fossil fuel burning and smelting began increasing in northern hemisphere polar snow and ice at the start of the twentieth century (Mayewski et al. 1990; Legrand et al. 1997). In comparison to a pre-industrial level of 26 ng g^{-1}, sulphate concentrations

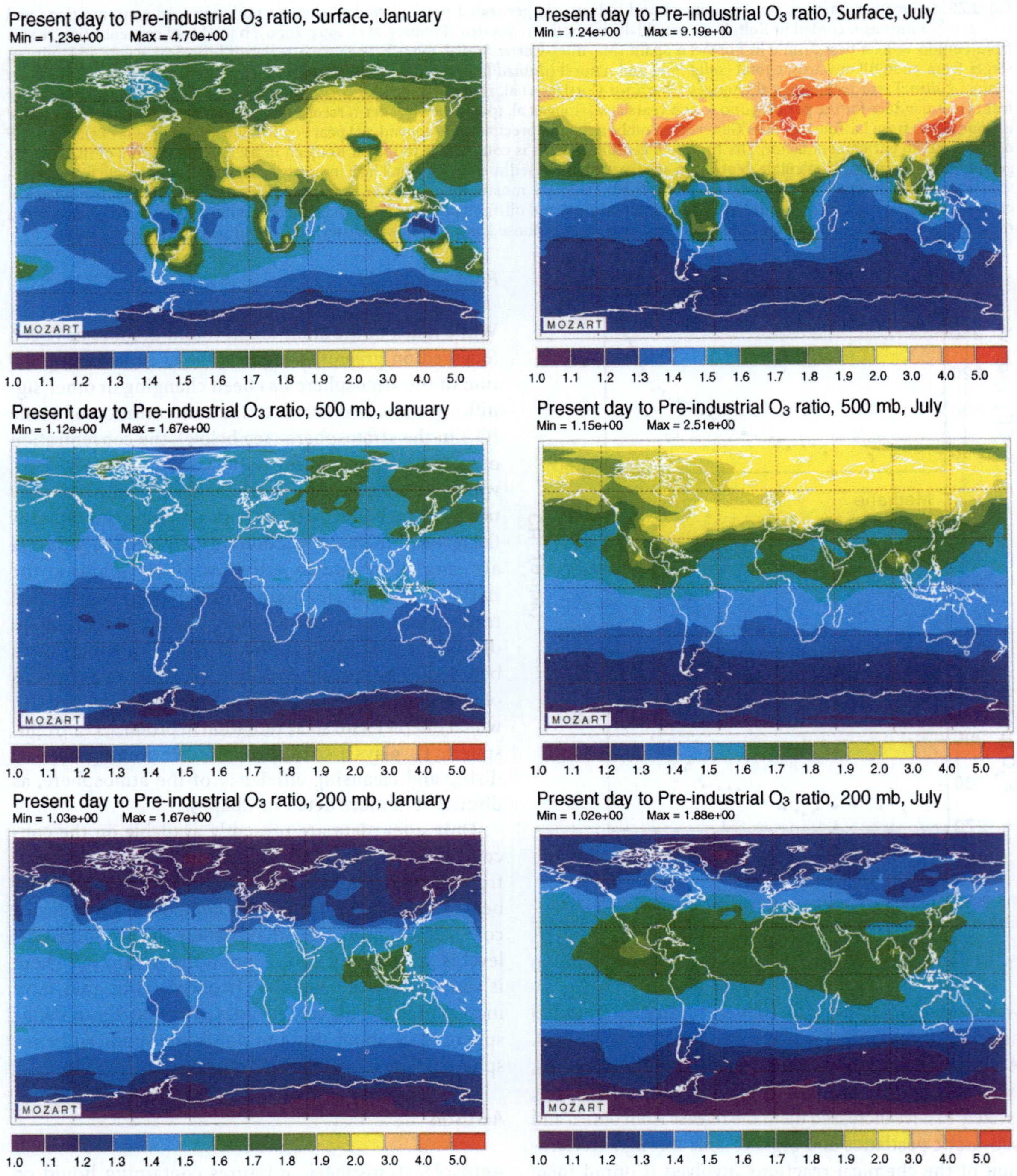

Fig. 3.31. Estimated changes in the ozone concentration since the pre-industrial era at three different altitudes (surface, 500 mb, 200 mb) in January (*left*) and July (*right*) (Hauglustaine and Brasseur 2001). The changes are expressed as the ratio of the present-day to the concentration estimated for the year 1850

peaked at 110 ng g^{-1} at the end of the 1960s. Thereafter, decreasing anthropogenic sulphur emissions are clearly reflected in Greenland sulphate profiles, which exhibit a steady decreasing trend since about 1980. Recent instrumental records of the composition of the atmosphere also show a decline in SO_2 in eastern North America and Europe, although N_2O has not shown similar declines. The concentrations of both gases are increasing in Southeast, South, and Temperate East Asia, however (Fu et al. 2002). Globally, anthropogenic sources of oxidised compounds of sulphur are still more than double the natural sources, indicative of the massive human influence on the atmosphere's aerosol particle burden.

The particles produced from the combustion of vegetation (biomass burning) and also from fossil fuels represent another major source of particles in

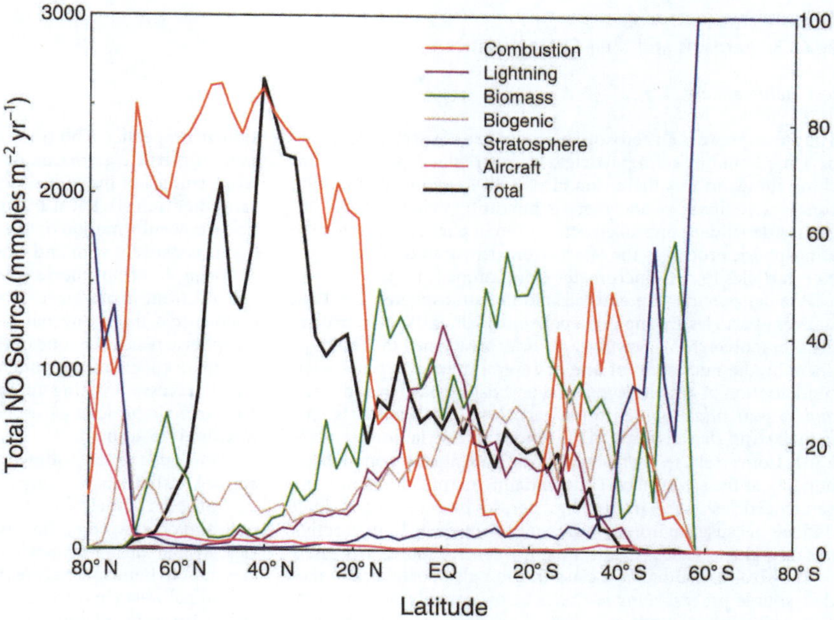

Fig. 3.32.
Latitudinal distribution of zonally-averaged NO_x emissions (*image:* H. Levy II, NOAA/GFDL)

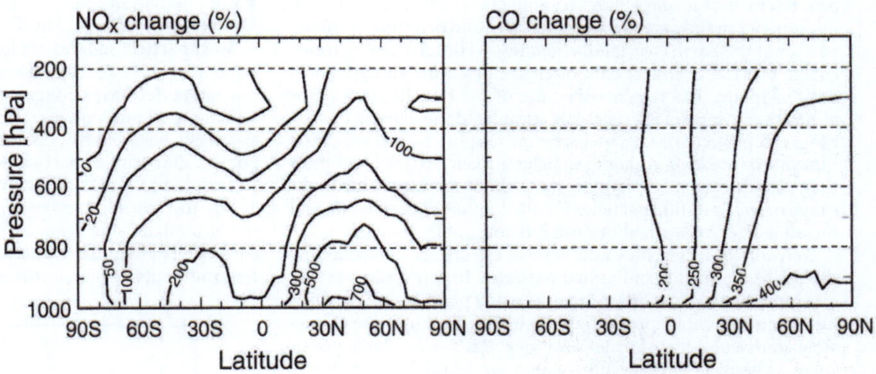

Fig. 3.33.
Percentage change in the yearly and zonally averaged concentration of NO_x and CO from pre-industrial times to present (Wang and Jacob 1998)

the atmosphere. These plumes (smoke), which can be transported long distances and thus affect atmospheric composition far from the source of the burning, consist primarily of a large range of partially burned organic particles. As noted in Chap. 2, biomass burning is a natural part of the ecosystem dynamics of some biomes, especially tropical savannas and boreal forests, and thus the pre-human atmosphere contained particles from these fires. However, through deforestation and the increased use of fuel wood by humans, the magnitude of biomass burning is estimated to have risen by 30–50% over the last century or two (Scholes et al. 2003). The overall effect of biomass burning and fossil fuel combustion on particle loading is difficult to quantify. From pre-industrial to modern times, for example, carbonaceous particle concentrations at a high-Alpine glacier site have increased by factors of 3.7, 3.0, 2.5 and 2.6 for black, elemental, organic and total carbon respectively (Lavanchy et al. 1999). Globally, the amount of soot in the atmosphere due to human activities is estimated to have increased by 10-fold over natural sources (Brasseur et al. 2003).

A third important type of aerosol particle is mineral dust, which is generated in the arid parts of continents and transported long distances. As shown in the Vostok ice core record (Petit et al. 1999), the amount of mineral dust in the natural atmosphere varies widely. Human activities can affect the loading of dust through land-cover and land-use change, especially in those areas where land degradation has occurred. Recent examples include the generation of large dust storms in degraded lands in western China and their transport across the Pacific Ocean. However, the globally averaged increase in dust loading due to human activities is difficult to quantify; recent estimates suggest that the anthropogenically-derived fraction of the total dust loading could be 30–50% (Heintzenberg et al. 2003).

Global distributions of aerosol particle loading as characterised by the aerosol optical depth (Box 3.3) are shown in Fig. 3.38. These figures, derived from satellite measurements (for oceans) and models coupled with observed meteorology, show the effects of both natural and anthropogenic aerosol particles. Unfortunately, a comparable global map showing the distribution of

Box 3.3. Aerosols and their Characteristics

Jost Heintzenberg

The atmosphere is an aerosol, i.e., a mixture of carrier gases suspending liquid or solid particles. There are good reasons to include clouds in this definition also. Both systems often exhibit comparable lifetimes and there is mounting evidence for a humidity-dependent unbroken continuum in particle sizes, related atmospheric processes and effects from dehydrated sub-micrometer particles to decamicrometer cloud drops (Fig. 3.34).

Primary particles are emitted into the atmosphere from their source as particles. Examples are pollen, sea salt, and soot from combustion sources. New, *secondary particles* are formed in the atmosphere by the nucleation of one or several precursor gases while condensation of gas phase material on existing particles adds *secondary particulate matter* to the particles. *Crustal particles* refer to their wind-driven origin in the eroded terrestrial surface of the Earth. Conversely, *sea salt particles* originate mostly from bubbles bursting at the sea surface. Differentiating natural and anthropogenic particles in the atmospheric aerosol becomes increasingly difficult because environmental changes caused by human activities on a global scale also affect natural aerosol source processes.

The size-related particle classification also correlates to particle source process. *Fine particles* below roughly one micrometer are mainly secondary in nature or result from combustion processes whereas *coarse particles* with diameters larger than one micrometer mainly are primary in nature and originate from the surface of the Earth (sea, land, vegetation). For technical reasons, coarse particle chemical data often are given for an upper size limit of 10 µm. The regular structure of the size distribution of atmospheric aerosol particles has stimulated the naming of subranges in particle size. Particles below roughly 20 nm are called *ultrafine* or *nucleation-mode* particles. *Aitken particles* are larger than ultrafine particles, ranging up to about 100 nm. A third subrange or *mode* of fine particles is called *accumulation mode* and includes sizes between about 100 nm and 1 µm.

Aerosol characteristics and aerosol effects are controlled by the properties of the individual particles. In any given particle size interval a volume of air contains many particles, all of which have their individual properties because of the multitude of possible source and transformation processes. The available physical and chemical particle information on the other hand mostly concerns bulk properties of a rather large number of particles as required for the quantification in specific analytical instruments after sampling from the atmosphere.

The distribution of particle properties amongst a population of particles with a common sorting property such as their electrical mobility or mass is called *state of mixture*. This state varies between the extreme possibility of an *external mixture*, in which essential particle properties such as their light absorption differ between particles of a given sorting property and an *internal mixture*, in which all particles of a given size range have very similar properties. The traditional methods of quantifying this state of mixture are based on topochemical or electron microscope studies of individual deposited particles. With recent advancements in electrical and mass spectroscopy of airborne particles, new information on the state of mixture has been derived that is consistent with and extends microscopic data.

Owing to atmospheric transformation processes, there is a general trend from more externally mixed particles in source regions to a more internal mixture towards the end of the atmospheric residence time of the particles. The main reasons are Brownian coagulation, condensation from the gas phase and in-cloud processes evening out inter-particle differences with time. To date no large scale aerosol dynamics model takes the state of mixture into account.

Knowledge of the 3-dimensional distribution of atmospheric aerosol particles is not complete. Satellite data and global chemical transport models show that crustal particles are concentrated over and downwind of the deserts of southern and northern Africa, Arabia and Eurasia whereas the maxima of sea salt particles in both hemispheres are found in the stormy mid- and high-latitudinal west wind zones. Anthropogenic aerosols are concentrated (in order of importance) around and downwind of the major population centres of East Asia, the Indian sub-continent, Africa, North America and Europe.

Most particle sources are located at or near the ground. Consequently, most aerosol characteristics related to particle concentration decrease strongly with altitude. Known exceptions are total number and possibly particle surface, which show secondary peaks near the tropopause where both natural and anthropogenic (air traffic) particle sources have local maxima. In the stratosphere there is a perennial particle layer between 15 and 25 km, the so-called Junge-layer. Global coverage with size distribution data is far from satisfactory. Over the oceans, surface coverage is about 28% whereas no comparable value exists for the continents or upper atmospheric layers.

Fig. 3.34. A typical example of an atmospheric aerosol

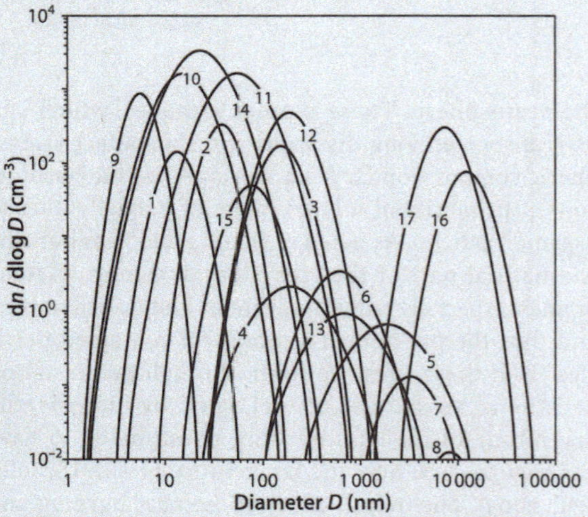

Reservoir	No.		
Arctic	1	Anthropogenic, fine	9
Ocean, fine	2	Anthropogenic, fine	10
Ocean, fine	3	Anthropogenic, fine	11
Sea salt (film)	4	Anthropogenic, fine	12
Sea salt (jet)	5	Anthropogenic, coarse	13
Crustal, mode 1	6	Upper troposphere	14
Crustal, mode 2	7	Stratosphere	15
Crustal, mode 3	8	Marine stratus	16
		Continental stratus	17

Fig. 3.35. Particle size distribution within an atmospheric aerosol

◀

Including non-precipitating clouds, the sizes of atmospheric aerosol particles are distributed over more than four orders of magnitude, from 1 nm to about 100 µm. This distribution of particle numbers over particle diameter *D* is conveniently approximated by a superposition of a number of log-normal distributions. Figure 3.35 illustrates the shape of the different particle size distributions by atmospheric reservoir and particle source. Size position and total number of the sub-micrometer modes are controlled by the strengths of the related gas-to-particle sources combined with the effects of aerosol dynamics and liquid phase chemical reactions. The different modes depicted in Fig. 3.35 will never be present at the same time and place in the atmosphere. Numerical models of atmospheric aerosol dynamics predict the transformations between the depicted modes in relation to the source and sink processes controlling a particular setting. Despite the very wide range of median particle sizes, the geometric standard deviations of the distributions vary between only 1.4 and 2.

Chemical composition is a major factor that controls the atmospheric effects of aerosol particles. The many possible sources coupled with physical and chemical atmospheric transformation processes lead to high variability in particle composition. This variability combined with very low mass mixing ratios of particulate matter in the atmosphere (mostly much less than parts per million) put very high demands on chemical analyses. Consequently, global coverage and chemical completeness of aerosol analyses are far from satisfactory for most aerosol-related issues; rather, typical examples from larger aerosol experiments of the polluted Central Europe and from low pollution marine regions are presented.

Figure 3.36 summarises the chemical mass balances over central Europe for summer and winter. Ammonium sulphate and carbonaceous components, both largely due to combustion sources, dominate fine particle composition. The higher mass fraction of elemental carbon in winter results from increased combustion of fossil fuels during the cold season.

The marine aerosol data shown in Fig. 3.37 are aggregated from recent cruises in the North and South Atlantic Ocean and in the Indian Ocean, from which the first chemical mass closures of size-segregated aerosol particles were reported. The significant contribution of continental dust to sub-micrometer particle composition and the even stronger contribution to coarse particle composition is due to the ship passing through the Saharan plume over the Atlantic. Super-micrometer composition in Fig. 3.37b indicates that the two major possible contributors are sea salt and dust with their actual proportion controlled by local wind speed and the distance to the nearest continental crustal source. The relatively high mass fraction of elemental carbon is due to the African biomass-burning plume reaching out over the tropical Atlantic. Nearly 90% of the particulate mass is in the coarse particle range.

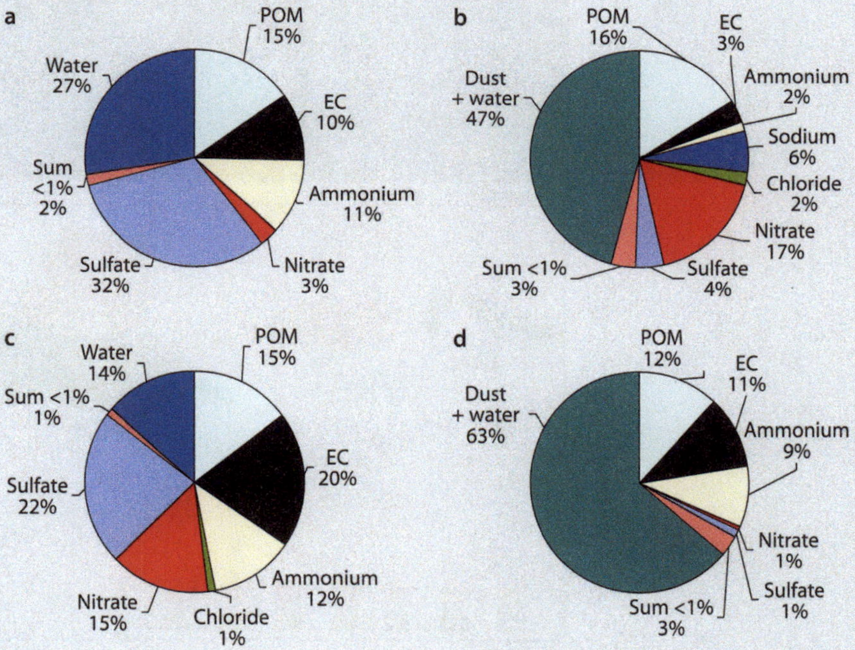

Fig. 3.36. Particle composition by mass of an aerosol over central Europe in summer and in winter; **a** fine particles, summer; **b** coarse particles, summer; **c** fine particles, winter; **d** coarse particles, winter

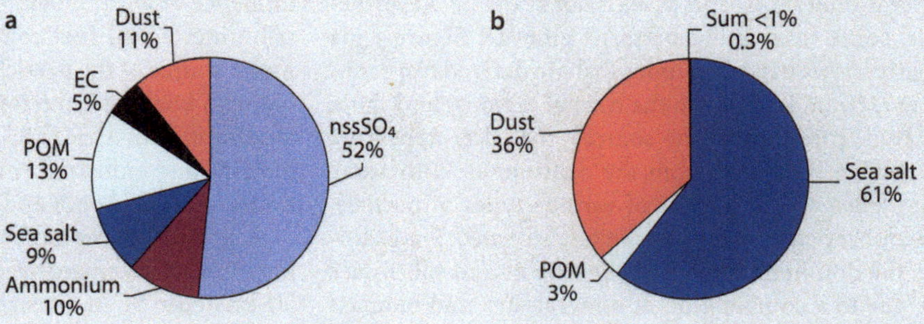

Fig. 3.37. Particle composition by mass in marine aerosols over the Atlantic and Indian Oceans: **a** fine fraction; **b** coarse fraction

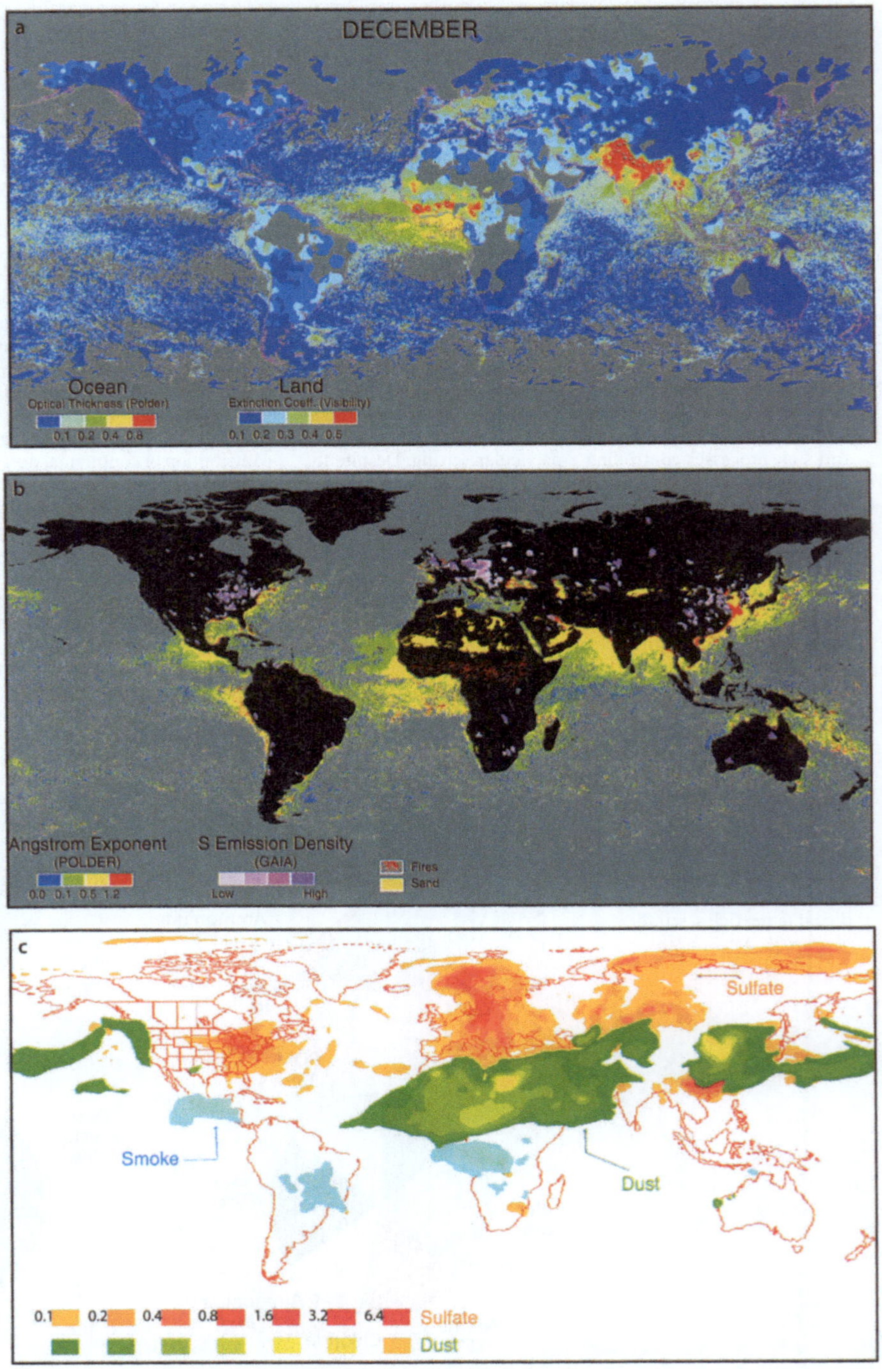

Fig. 3.38. Global horizontal patterns for **a** December 1996 and **b** June 1997 of aerosol optical thickness and Ångström exponents over the oceans from the POLDER instrument aboard the ADEOS satellite and extinction coefficients over land areas from visibility data (reproduced with permission of LOA, LSCE and CNES, France; and NASDA, Japan). **c** Global fields of mineral dust, sulphate and smoke, calculated with a chemical transport model using actual meteorology (*http://www.nrlmty.navy.mil/aerosol/*; Heintzenberg et al. 2003)

natural-only aerosol particles is not available. Nevertheless, some interesting patterns emerge. Figure 3.38a clearly shows particle plumes that are derived from continents (found mainly in the coastal regions) and those derived from the marine sources, the latter appearing as patches detached from the continents. With some knowledge of the sources of various types of particle, the anthropogenic imprint can be discerned. For example, the dominant plumes in the tropics and subtropics are due to a combination of mineral dust and biomass burning emissions. Figure 3.38c more clearly shows the influence of anthropogenic activities on the global distribution. Fossil fuel combustion is undoubtedly the major source of the particles in the high latitudes, while biomass burning is a strong contributor to the particle loading in both Africa and South America. Both biomass burning and industrial emissions contribute to the particle loading in South and Southeast Asia.

A striking example of sharp increase in particle loading of the atmosphere over the past century is the regular existence of an extensive haze in the January–April period over most of the Indian Subcontinent, the tropi-

Fig. 3.39.
Aerosol particle loading in the dry season (January–April) over and south of the Indian subcontinent (NASA, *http://www.earthobservatory.nasa.gov/*, 2001)

cal Indian Ocean north of 5° S, over much of Southeast Asia and into the western Pacific Ocean (Fig. 3.39). This 3-km deep layer of haze, sometimes called the Asian Brown Cloud (Box 3.4), consists of a mixture (by mass) of sulphates (32%), ammonium (8%), organic molecules (26%), nitrates (10%), black carbon (14%), fly ash (5%), dust (10%) and potassium (2%). Average aerosol optical depths can be as high as 0.3, compared to a globally and annually averaged mean value of 0.12. Similar, but deeper, aerosol clouds are to be observed over southern Africa and the adjacent Indian Ocean (Tyson et al. 2002). The impacts of increased aerosol loading in the atmosphere are significant and are discussed in Box 3.4 and further in Chap. 4.

Other Industrial Contaminants and Trace Metals

Apparently harmless anthropogenically produced gases can have surprisingly drastic impacts on the atmosphere. One of the most dramatic changes in atmospheric composition observed during the past decades is the depletion of ozone in the stratosphere due to the emission of chlorine- and bromine-containing gases primarily of industrial origin. Chlorofluorocarbons were first produced in the 1950s for use as refrigerants, propellants, solvents, etc., and they were heralded for their non-reactivity in the lower atmosphere. Their tropospheric concentrations increased dramatically throughout the second half of the twentieth century (Fig. 3.42). The re-

Box 3.4. The Asian Brown Cloud

A. Jayaraman · A. P. Mitra

The findings of the Indian Ocean Experiment – INDOEX (Ramanathan et al. 2001a) on the extended haze spread over the South Asian and the tropical Indian Ocean regions and its radiative impact have led to a major international effort, called the Asian Brown Cloud (ABC) Project (Ramanathan and Crutzen 2001). Haze is a common feature of industrialised areas, though not necessarily of large regions. The Asian haze covers an area of a few million square kilometres and has a vertical extent of about three kilometres. Ship cruise observations over the tropical Indian Ocean show a large latitudinal gradient in aerosol particle amount from coastal India to the pristine ocean (Fig. 3.40) where atmospheric dynamics play a significant role in the dispersal of particles (Jayaraman et al. 1998; Krishnamurti et al. 1998). The INDOEX field experiments, involving ships, island stations and aircraft, conducted over the tropical Indian Ocean during the winter months of 1998 and 1999, helped in documenting the phenomenon and its possible impact on climate (Ramanathan et al. 2001a).

Every year from December to April the low-level northeast trade wind blowing from the Southeast Asian region transports a large amount of anthropogenic particles derived mainly from fossil fuel and biomass burning to the tropical Indian Ocean. The high concentration of black carbon, of the order of 14% in the fine particle mass (Lelieveld et al. 2001), contributes to absorption of solar radiation which gives a brownish tinge to the haze. The north to south dispersion of particles is, however, limited by the Inter-Tropical Convergence Zone (ITCZ), where the surface trade winds from the two hemispheres meet and is also a region of strong convective activity. The ITCZ exhibits an annual oscillation with respect to the equator, and for the whole globe the maximum excursion is found over the Indian Ocean, from about 10° S in January to about 20° N in July. Yearly variation in the strength of the ITCZ and its progression greatly influence the spatial and temporal distribution of aerosol particle loading over the Indian Ocean (Rajeev et al. 2000; Li and Ramanathan 2002). Based on the chemical composition it is estimated that about 80% (±10%) of the particles found in the northern Indian Ocean during winter is anthropogenic in nature (Ramanathan et al. 2001a). The chemical properties of the particles determine their refractive index and hence the scattering and absorbing efficiency of the aerosol.

Sulphate particles scatter light and help to cool the Earth's atmospheric system (negative aerosol forcing) while soot particles also absorb the radiation and thus contribute to atmospheric warming (positive forcing). The single scattering albedo, which is a measure of the scattering ability of the aerosol mixture, is found to be very low, of the order of 0.9 in the haze region compared to nearly 0.99 over the pristine ocean region south of the ITCZ (Ramachandran and Jayaraman 2002). The dominant effect of the haze is in decreasing the solar flux reaching the Earth's surface by as much as 23 (±2) $W m^{-2}$ of which only 7 (±1) $W m^{-2}$ is backscattered to space and the remaining about 16 (±2) $W m^{-2}$ is trapped within the atmosphere (Fig. 3.41). This three-fold difference between the surface and top of the atmosphere forcing is mainly due to absorption of solar radiation by soot particles within the first few kilometres of the atmosphere (Satheesh and Ramanathan 2000; Ramanathan et al. 2001a).

Such *hot spots* of high aerosol forcing are bound to have an effect on the global climate system (Menon et al. 2002; Chung et al. 2002). However, several questions still remain. For example, during summer months the southwest monsoon is active over most of the South Asian region and the associated deep convection lifts particles to great heights while inhibiting large scale horizontal transport. The well known effects of aerosol particles on the modification of microphysical properties of clouds (e.g., increase in cloud lifetime associated with a reduction in precipitation efficiency) become important in assessing the overall aerosol forcing on climate (Ackerman et al. 2000). It is an observational challenge to quantify the aerosol radiative forcing in the presence of clouds as the forcing is very sensitive to the position of the aerosol layer, whether it is below or above the cloud (Podgorny and Ramanathan 2001). A detailed study is required to quantify the impact of air pollution as a global climate forcing (Ramanathan et al. 2001b; Hansen 2002). The recently initiated ABC Project will address these issues and make impact analyses on climate, agriculture and health with the support of participating countries in establishing additional observatories and model simulations.

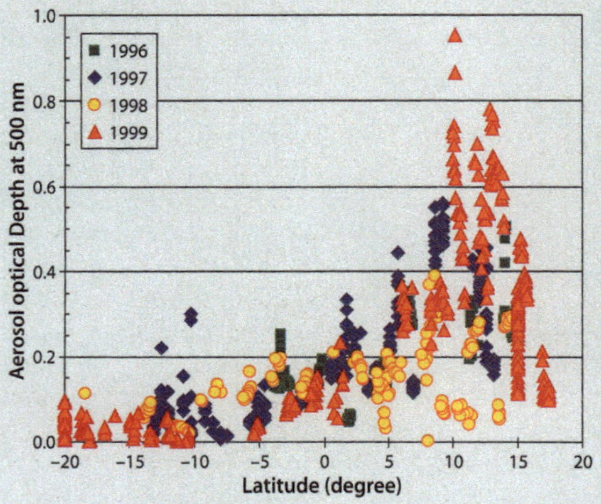

Fig. 3.40. The latitudinal variation of columnar aerosol optical depth at 500 nm over the tropical Indian Ocean, measured using a sun-photometer onboard the Indian research vessel Sagar Kanya during 1996 to 1999. The measurement accuracy is ±0.03

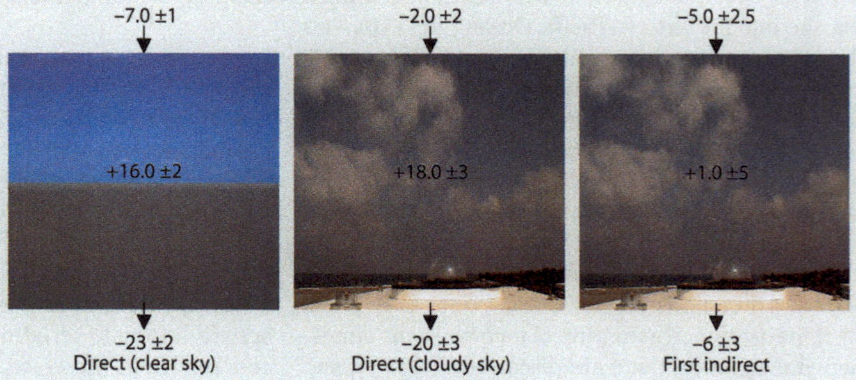

Fig. 3.41. Aerosol radiative forcing for the North Indian Ocean Region (0° N to 20° N and 40° E to 100° E) (Ramanathan et al. 2001a). The values on the *top*, *middle* and *bottom* represent forcing at the top of the atmosphere, within the atmosphere and at the surface, respectively. The values include the effects due to both natural and anthropogenic aerosols

verberations of these changes were noted in the late 1970s, when measurements of total column ozone indicated pronounced ozone depletion at high southern latitudes during the spring months. Peak ozone loss has approached 60% at the South Pole; ozone loss in the stratosphere over the mid-latitudes, although much less, is still noticeable. Total column ozone during the period 1997–2001 was 3% less than the pre-1980 value for the northern mid-latitudes and 6% less for the southern mid-latitudes. While there is no seasonality in the southern hemisphere decreases, the northern hemisphere experiences increased loss (about 4%) in winter-spring and smaller decreases (about 2%) in summer-autumn. There is no discernible change to total column ozone over the tropics (UNEP/WMO 2002). Concentrations of CFCs in the atmosphere are now in decline, due to emission reductions agreed in the Montreal Protocol and subsequent agreements.

Human activities have had a marked impact on the global atmospheric cycles of several trace metals as shown by their polar snow records, in particular in Greenland (Candelone et al. 1995). The ratio between pre-industrial level and the last 20 years is found to be significantly higher for metals than for sulphate or nitrate: for example, factors of 2.3, 2.7 and 9 are found for zinc, copper and cadmium, respectively. The figure is considerably higher (250) for lead due to the use of organic lead derivatives in gasoline. Global increase of mercury in the atmosphere is a good example, as its deposition rate is estimated to exceed the pre-industrial value by a factor of two worldwide but up to five or ten times in North America and northern Europe (Bergan et al. 1999). This increased deposition of mercury from the atmosphere, coupled with increased leaching of mercury from acidified soils (ultimately due to atmospheric deposition of nitrogen and sulphur compounds), has led to a bioaccumulation of organic mercury in some fish, resulting in toxic concentrations in some lakes in Scandinavia and Canada (Lindqvist et al. 1991).

In summary, the imprint of human activities on the global atmosphere is unmistakable and profound. The concentrations of greenhouse gases have risen well above their pre-industrial levels and continue to rise. The concentrations of many oxidising gases, such as O_3, have also risen in the troposphere, but the direct human influence on their concentrations is not easy to determine as they react with other species in the atmosphere, such as CH_4. Some human perturbations of atmospheric composition compensate for each other in terms of their effects. On the other hand, the increase in aerosol particle loading in the atmosphere is dramatic and is undoubtedly the dominant change in atmospheric composition that has occurred over the last decade or two.

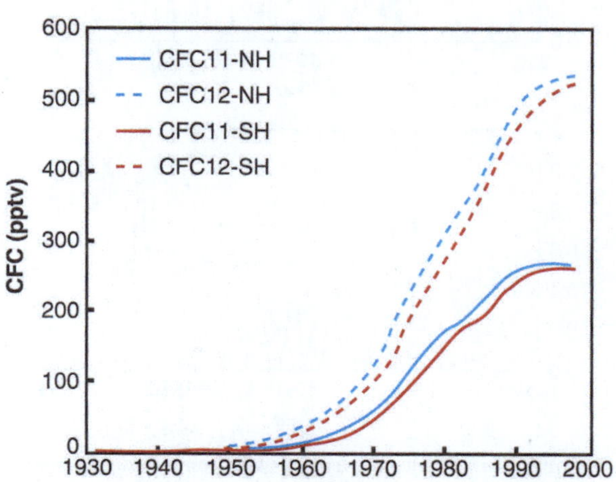

Fig. 3.42. The observed evolution of atmospheric CFC-11 and CFC-12 concentration for the northern and southern hemispheres (Walker et al. 2000)

3.3.3 The Hydrological Cycle as Transformed by Human Activities

Given the central importance to humans of water for domestic use, industry and agriculture, the hydrological cycle, especially its terrestrial component, has been one of the most impacted components of the Earth System. Major human influence on the hydrological cycle started about 4 000 years ago with water engineering in association with agricultural development. This influence has accelerated sharply in the last century, and during the last 50 years it has likely exceeded natural forcings of continental aquatic systems in many parts of the world (Meybeck and Vörösmarty 2004).

Humans are prolific engineers of continental aquatic systems (Box 3.5). Dams, reservoirs and flow diversions have been in existence for thousands of years but only over the past 50–75 years have they become massive in size and pandemic in scope (Fig. 3.43). More than 45 000 dams over 15 m high are registered in the world and many hundreds of thousands of smaller dams have been built on rivers and on farms. In the northern hemisphere only 23% of the flow in 139 of the largest rivers is unaffected by reservoirs (Dynesius and Nilsson 1994). During the 1940 to 1990 period, withdrawals of water for

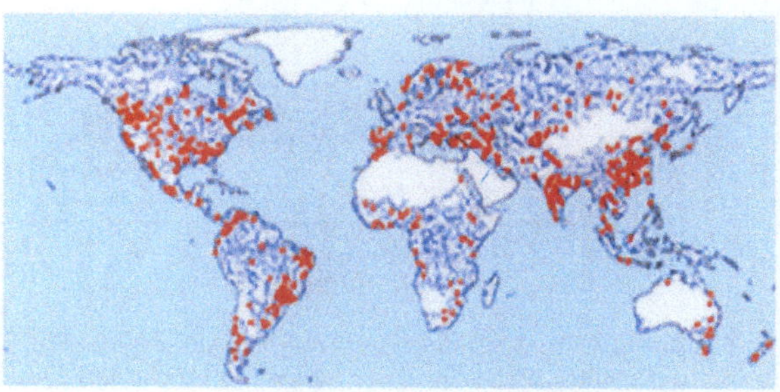

Fig. 3.43. Global distribution of 622 large reservoirs, classified as those with a maximum capacity greater than 0.5 cubic kilometres (Vörösmarty et al. 1997)

Box 3.5. Human-Driven Changes to Continental Aquatic Systems

Michel Meybeck · Charles Vörösmarty

Human activities have had and continue to have significant direct impacts on continental aquatic systems (Table 3.8). Water engineering, in the form of damming, withdrawal and consumption, diversion and channelisation, can dramatically change basin water balances, flow characteristics, sediment and nutrient transport as well as aquatic habitat. This spectrum of changes is driven by the forces of population growth, agriculture, economic development, and industrialisation. The objective of this box is to describe these direct human-driven changes to water systems and their immediate effects. These changes reverberate through the Earth System and are described in Chap. 4. The multiple effects of global change and local and regional influences on water quality are discussed in Chap. 5.

Damming, water extraction and redirection of flows are the three most important ways in which engineering directly affects the water system. Figure 3.44 illustrates these three types of flow distortion by examples of water engineering works in three heavily-regulated rivers. The Nile River time series (Fig. 3.44a) for discharge just below the Aswan High Dam demonstrates an enormous change. It is not difficult to identify when the dam was constructed and Lake Nasser reservoir filled. The post-impoundment Nile shows reduced overall discharge, truncated peak flows, higher low flows and a seasonal shift in the timing of the natural hydrograph (Fig. 3.44b). Progressive losses of discharge for the Syr-Darya River (Fig. 3.44c) are associated with expanding water use for irrigation and the much-publicised contraction of the Aral Sea. Increases in discharge for the Burntwood River in Manitoba, Canada, (Fig. 3.44d) are typical of inter-basin transfer schemes, here to optimise hydroelectricity production between the Churchill and Nelson Rivers, which empty into Hudson Bay.

These are but a few examples of the impact of hydraulic control on river systems, which is sure to expand along with the water services necessary to support a growing and economically developing human population. Such changes represent a virtually instantaneous reconstitution of the behavior of rivers and connections to the coastal zone. These changes are pandemic and collectively impart an important human signature

Table 3.8.
Impacts of human activities on water systems

Water system state change	Immediate impacts
River damming and channelisation	Nutrient and carbon retention; retention of particulates; loss of longitudinal and lateral connectivity; creation of new wetlands; flow stabilisation and habitat fragmentation, including impedence of fish migration
Irrigation and water transfer	Partial to complete loss of river fluxes; salinisation through evaporation
Release of industrial and mining wastes	Heavy metal increase in water system; acidification of surface waters
Release of urban and domestic wastes	Enhance disease vectors; persistent organic pollutants; eutrophication

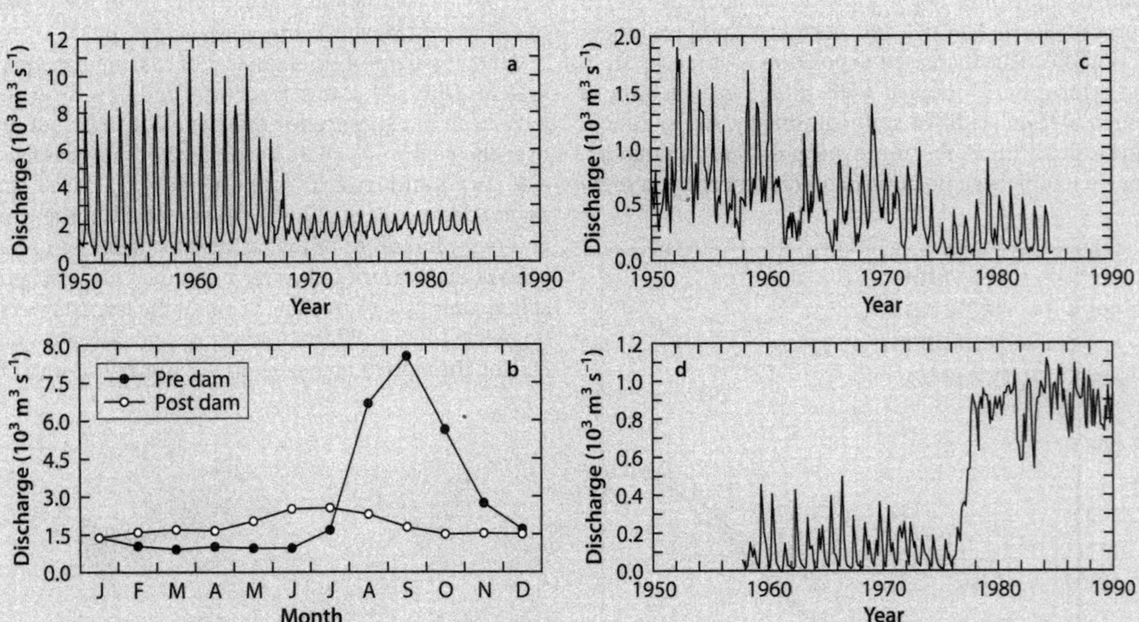

Fig. 3.44. The history of flow distortion by water engineering works in three heavily regulated rivers (adapted from Vörösmarty and Sahagian 2000). **a** The Nile River time series for discharge just below the Aswan High Dam; **b** pre- and post-impoundment seasonal discharge of the Nile River; **c** progressive losses of discharge for the Syr-Darya River associated with expanding water use for irrigation and the much-publicised contraction of the Aral Sea; **d** increases in discharge for the Burntwood River in Manitoba (Canada) owing to inter-basin transfer schemes between the Churchill and Nelson Rivers, which empty into Hudson Bay

◄

on the water cycle. Is the proliferation of still waters in otherwise free-flowing rivers of global significance?

Several quantitative measures demonstrate the extent to which human activities have changed continental aquatic systems. One such measure is the 700% increase – relative to the natural state – in the standing stock (still waters) of river channel water owing to impoundments (Vörösmarty et al. 1997). Another is the estimate that in the northern hemisphere, more than three-quarters of mean natural flow is highly regulated and fragmented by humans (Dynesius and Nilsson 1994). Yet another is the aging or the change in the residence time of continental runoff held by river systems in response to large reservoirs (Fig. 3.45). The increase in residence time varies with the nature of the river system and the engineering works, but the global mean natural residence time of two weeks has been significantly increased to a global average of four months, and in some modified river systems up to a year. The largest increases in residence time occur in the largest river systems, such as the Nile, the Colorado and the Rio Grande (Vörösmarty et al. 2000).

The consequences of water aging for associated material transport are to trap a substantial proportion of the incoming suspended sediments and to modify the concentration of dissolved compounds of nitrogen, phosphorus and silicon. Many of the world's largest river basins show nearly complete sediment retention from large reservoirs alone (i.e. >0.5 km^3 storage capacity). It has been estimated that the impact of the 45 000 largest registered reservoirs is to trap nearly 30% of global sediment flux destined for the ocean (Vörösmarty et al. 2003). This estimate is undoubtedly low and will rise further with inclusion of the approximately 800 000 smaller impoundments (McCully 1996). The amount of sediment retention will increase further in future through continued dam construction.

With expanding demands for freshwater with population growth and economic development, there will be increasing pressure to regulate the continental water cycle. Thus, for the next several decades at least, the already profound human intervention in this important element of the Earth System will intensify further.

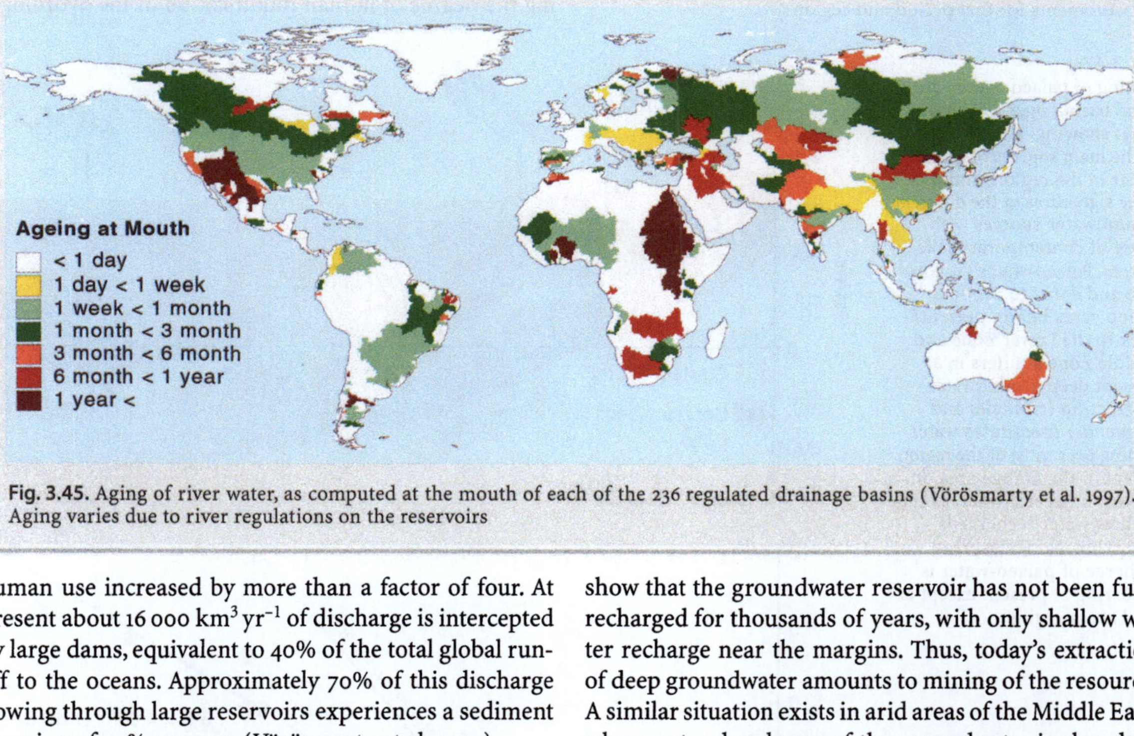

Fig. 3.45. Aging of river water, as computed at the mouth of each of the 236 regulated drainage basins (Vörösmarty et al. 1997). Aging varies due to river regulations on the reservoirs

human use increased by more than a factor of four. At present about 16 000 km^3 yr^{-1} of discharge is intercepted by large dams, equivalent to 40% of the total global runoff to the oceans. Approximately 70% of this discharge flowing through large reservoirs experiences a sediment trapping of 50% or more (Vörösmarty et al. 2003).

Human demands on water affect other aspects of the hydrologic system. Groundwater pumping in excess of recharge, with resulting land subsidence or, in coastal areas, contamination due to salt-water intrusion, is occurring in several parts of the world, particularly in arid and semi-arid areas. In the Great Plains area of the United States, pumping of groundwater from the large Ogallala aquifer for agricultural use has led to decreasing water levels in the aquifer and to abandonment of agricultural land. Figure 3.46 shows the annual average rates of water level change by region for the past 30 years. In the Chad Basin, Nigeria, palaeo-records (Fig. 3.47) show that the groundwater reservoir has not been fully recharged for thousands of years, with only shallow water recharge near the margins. Thus, today's extraction of deep groundwater amounts to mining of the resource. A similar situation exists in arid areas of the Middle East, where natural recharge of the groundwater is also close to zero so that pumping of groundwater for irrigation of agriculture amounts to the mining of fossil water.

Nearly all of the research related to human influence on the hydrological cycle has understandably focused on liquid water. Another way to measure the impacts of humans on the global water cycle is to estimate the diversion of vertical water vapour flows from land to atmosphere as a result of direct human impacts on the hydrological cycle and indirectly through land-use change. A continental scale analysis for Australia has been carried out and has yielded interesting results (Gordon 2000). Australia is the driest inhabited continent on Earth and

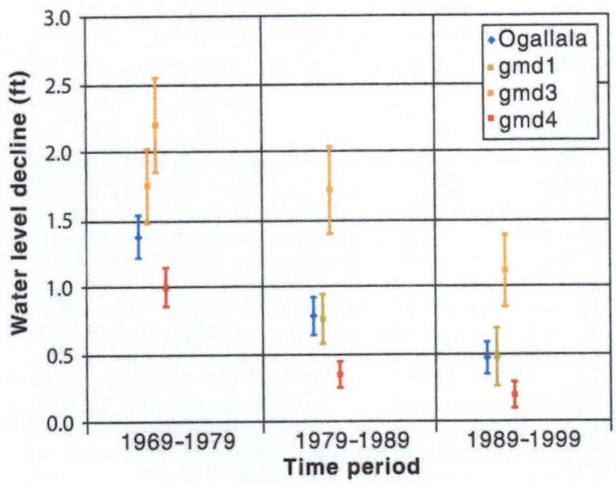

Fig. 3.46. Average annual rate of water level change for four aquifers in the High Plains region, USA over the past three decades (Woods et al. 2000). The *central points* are the average values and the *vertical lines* represent an uncertainty of one standard deviation – the range that can be expected to include about 66% of all measurements for that period and region

has the most skewed distribution of evapotranspiration and runoff. About 90% of all precipitation is returned to the atmosphere via evapotranspiration; only 10% returns to the oceans as runoff. Given Australia's relatively low human population and on a continental scale relatively modest conversion of land to intensive agriculture, an insignificant human effect on evapotranspiration was expected. The analysis showed, however, that the continent's natural water vapour flows have been reduced by about 10% due to human activities, namely, the replacement of native woody vegetation by annual crops and grasslands. On a regional basis, the figure would be much higher as much of Australia's land-cover conversion has occurred in a narrow strip along the eastern coastline and in the southwestern corner of Western Australia.

A global-scale analysis based on the methodology used for Australia has not been completed, but it is likely that the degree of water vapour flow diversion globally will be larger. Estimates such as these are necessary for assessing the degree of human modification of the hydrologi-

Fig. 3.47.
Mining of palaeo-waters of the Chad basin, Nigeria (Edmunds 1999) showing: **a** the location of the main sources of groundwater in the region; and **b** isotopic signatures of the different groundwater sources with those of contemporary lake waters. Palaeo-water (*squares* in **b** and dated to 18 000 to 25 000 years before present) refer to the Lower Zone and Middle Zone aquifers in **a**. Present deep groundwater abstraction (*triangles* and *diamonds*) constitutes water mining over most of the region except in the Manga area of northern Nigeria where some shallow water recharge is occurring. No widespread recharge of palaeo-water is occurring at present except at the Lake margins

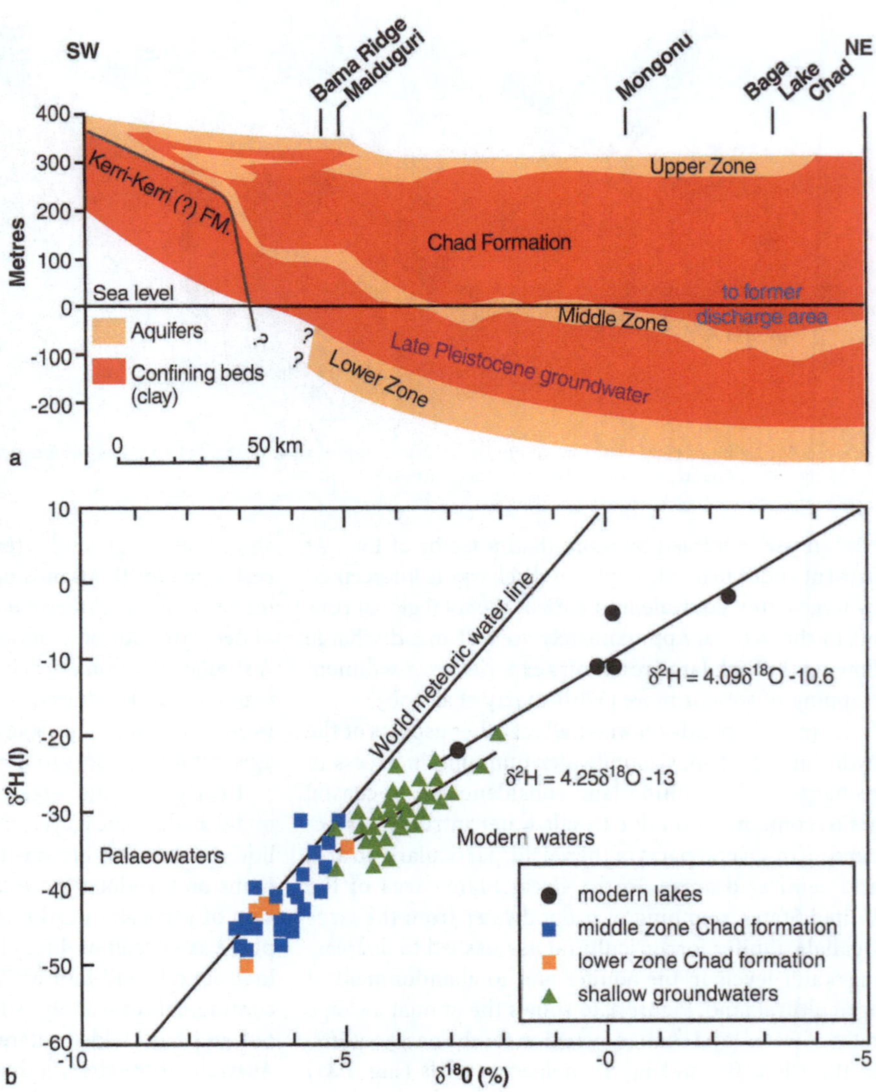

cal cycle from an Earth System perspective. It is through interception, storage and diversion of the lateral flows of water between land and ocean and through changes to the amount and pattern of vertical water vapour flows that human activities have become a global-scale force in the functioning of continental aquatic systems.

3.3.4 Coastal and Marine Environments as Transformed by Human Activities

More than 50% of the human population reside within 100 km of a coast (Kremer and Crossland 2002), and consequently coastal margins are strongly influenced by human activities. In terms of direct impacts, humans alter the geomorphology of the coastal zone through the construction of shoreline engineering structures, ports and urban developments. They also impact coastal ecosystems directly by converting them from their natural state to modified systems designed for the direct production of food. A prominent example is the conversion of mangrove forests to prawn farms. Globally, approximately 50% of mangrove systems have been converted to other uses as a result of human activities (Kelleher et al. 1995; WRI 1996, 2000; Naylor et al. 2000), and about 29% of coastal land (excluding Antarctica) is classified as altered or semi-altered in some way. Coastal ecosystems are also affected by the direct or inadvertent introduction of non-indigenous species, which can alter the structure and functioning of coastal ecosystems.

As the coastal zone is an interface between the land and the ocean, the health of coastal ecosystems depends

Fig. 3.48. MODIS image (12 April 2002) of the Yangtze River, showing the heavy load of sediments carried by the Yangtze River into the East China Sea (*image:* Jacques Descloitres, MODIS Land Rapid Response Team, NASA GFSC 2002)

on the through-flow of water and suspended and dissolved materials from upstream areas through the coastal zone to the continental shelves. As described in the previous section, human activities have significantly affected these flows both in terms of the amount of water transported to the coastal zone (due to water use for urban development, industry and agriculture) and the timing of these flows (due to flood/wave mitigation, reservoir storage or entire water diversion schemes). The amount of sediments delivered to the coastal zone by these water flows, which are important to coastal geomorphology as the source material for river deltas, has also been significantly altered by human activities, in two opposing directions. There have been regional increases in some areas in the delivery of sediments to the coastal zone through increased soil erosion driven by agriculture, construction, mining and forestry (Fig. 3.48). In other areas, however, delivery of sediment to the coastal zone as been decreased through sediment trapping within reservoirs and other pondages upstream.

The situation for nutrients is clearer than that for sediments. The amount of nutrients moving through and transformed within the coastal zone has increased dramatically due to human activities and is predicted to continue to do so into the future (Fig. 3.49). Some of the increased nutrient loading in coastal waters is due to direct human injection within the coastal zone itself, for example, from urban sewage treatment plants or from industries in coastal cities. Additional loadings come from upstream human activities, primarily from agriculture. It is estimated that the increase in nitrogen delivery via rivers entering the North Atlantic has increased by a factor of between 3 and 20 (Howarth et al. 1996). Increased nutrient delivery to coastal regions occurs also through atmospheric deposition of agricultural, industrial, and transport emissions of NO_x (Jaworski et al. 1997; Howarth et al. 1996). In addition to nutrients, contaminants such as heavy metals, persistent organic pollutants, various other synthetic chemicals, radioactive materials, bacteria and slowly degrading solid waste like plastics are transported from land to the coastal regions. More open marine areas are also subject to contaminants from offshore activities (e.g. oil pollution from shipping and drilling activities) and to nutrient deposition from the atmosphere (Galloway et al. 1994; Paerl and Whitall 1999).

The most direct human impact on marine ecosystems is through fisheries. Humans rely on the ocean fisheries for 16% of their animal protein (FAO 1993). One recent estimate suggested that humans ultimately harvest 8% of the primary production of the oceans, with much greater percentages for the upwelling and continental shelf areas (Pauly and Christensen 1995). Among the major marine fish stocks or groups of stocks for which information is available, about 47–50% of stocks are fully exploited, 15–18% are overexploited and 9–10% have been depleted or are recovering from depletion (FAO 2000b). In addition to influencing the size and abundance of target species, commercial fisheries may influence population characteristics of species incidentally caught in the fishery. An equivalent to 25% of the annual production of marine fisheries is discarded as bycatch each year (FAO 2000b).

As for land change, there are some counter trends, almost entirely in the developed world, where the environmental state of the coastal zone has been improved through human activities. Some rivers and harbour areas of major

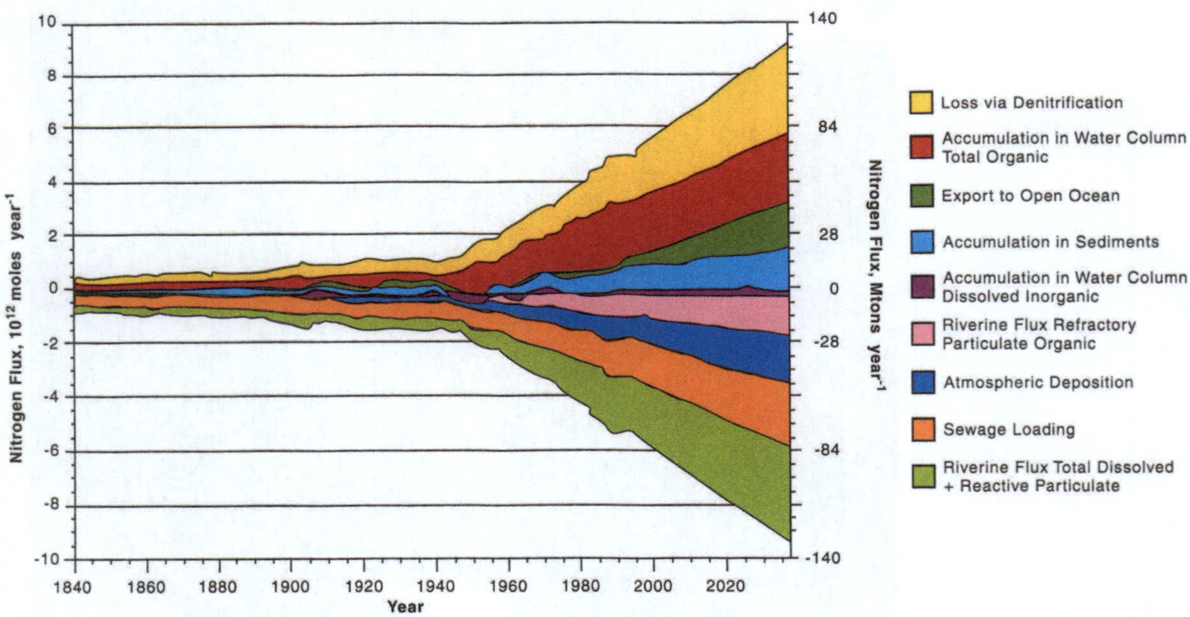

Fig. 3.49. Model-calculated partitioning of the human-induced nitrogen perturbation fluxes in the global coastal margin for the period since 1850 to the present (2000) and projected to 2035 under a business-as-usual scenario (Mackenzie et al. 2002)

western coastal cities have been significantly cleaned of pollutants (e.g., the Thames in London), and the transport of excess nutrients and other pollutants through the coastal zone has been diminished in many cases. However, from the perspective of the Earth System, it is the functioning of the entire coastal zone around the planet that is important. A typology approach has been used to identify coastal regions of the world that are relatively pristine or undisturbed. Data filters were applied to exclude coastal regions with populations densities of >10 persons km^{-2} and >5% cropland in the adjacent terrestrial regions. The result is startling (Fig. 3.50). Virtually no large stretches of coastal area outside of Greenland, northern Canada and Siberia and remote areas of South America and Australia are now without significant human influence.

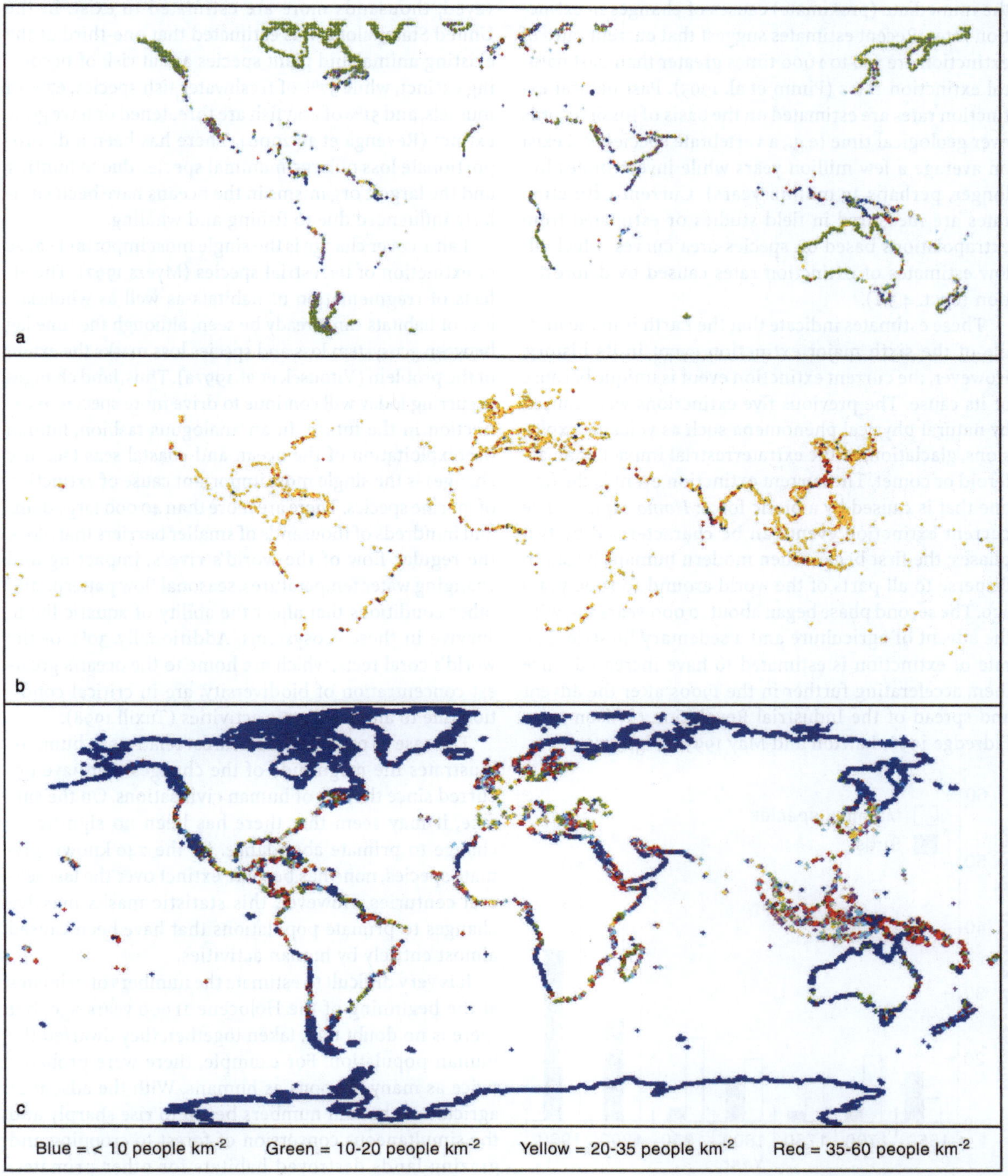

Fig. 3.50. Global estimates of relatively pristine or undisturbed coastal regions. Polar regions have been cropped and data filtered for population density >10 persons km^{-2} and land cover >5% cropland (LOICZ 2002)

3.3.5 Biological Diversity as Transformed by Human Activities

As was shown in the previous chapter, biological extinction is a natural process. Here it is necessary to examine the proposition that human activities are causing changes in the biological diversity of Earth beyond the rates of natural changes. It is also necessary to consider the immediate (proximate) causes of changes in extinction rates. Recent estimates suggest that current rates of extinction are 100 to 1 000 times greater than past natural extinction rates (Pimm et al. 1995). Past natural extinction rates are estimated on the basis of fossil records over geological time (e.g., a vertebrate species will exist on average a few million years while invertebrates live longer, perhaps 10 million years). Current extinction rates are measured in field studies or estimated from extrapolations based on species-area curves, which allow estimates of extinction rates caused by deforestation (Sect. 4.3.2).

These estimates indicate that the Earth is in the middle of the sixth major extinction event in its history. However, the current extinction event is unique because of its cause. The previous five extinctions were caused by natural physical phenomena such as volcanic explosions, glaciation, or the extraterrestrial impact of an asteroid or comet. The current extinction event is the first one that is caused by a biotic force: *Homo sapiens*. The current extinction event can be characterised by two phases; the first began when modern humans began to disperse to all parts of the world around 100 000 years ago. The second phase began about 10 000 years ago with the advent of agriculture and a sedentary lifestyle. The rate of extinction is estimated to have increased since then, accelerating further in the 1900s after the advent and spread of the Industrial Revolution (Wilson 1992; Eldredge 1998; Lawton and May 1995) (Fig. 3.51).

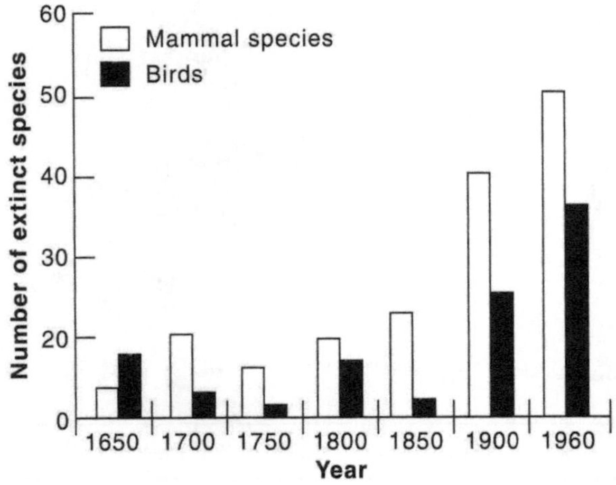

Fig. 3.51. Global rates of extinction for birds and mammals over the last 400 years (Reid and Miller 1989)

Rates of decline for mammals, fish, birds, amphibians and reptiles are high (Vitousek et al. 1997a; Tuxill 1998). Of the birds surveyed, 20% are threatened by extinction, 39% of mammals and fish are threatened, while 26% of reptiles and 30% of amphibians are threatened (Baillie and Groombridge 1996). Furthermore, between 22 and 47% of all plant species are in danger of becoming extinct (Pitman and Jorgensen 2002). These estimates include only those species that have been surveyed; thousands more are estimated to exist. In the United States alone, it is estimated that one-third of the existing animal and plant species are at risk of becoming extinct, while 37% of freshwater fish species, 67% of mussels, and 51% of crayfish are threatened or have gone extinct (Revenga et al. 2000). There has been a disproportionate loss of large mammal species due to hunting and the largest organisms in the oceans have been similarly influenced due to fishing and whaling.

Land-cover change is the single most important cause of extinction of terrestrial species (Myers 1997). The effects of fragmentation of habitats as well as wholesale loss of habitats can already be seen, although the time lag between ecosystem loss and species loss masks the extent of the problem (Vitousek et al. 1997a). Thus, land changes occurring today will continue to drive more species to extinction in the future. In an analogous fashion, human overexploitation of the ocean and coastal seas (sea-use change) is the single most important cause of extinction of marine species. There are more than 40 000 large dams and hundreds of thousands of smaller barriers that block the regular flow of the world's rivers, impacting and changing water temperatures, seasonal flow patterns and other conditions that affect the ability of aquatic life to survive in these ecosystems. Additionally, 30% of the world's coral reefs, which are home to the ocean's greatest concentration of biodiversity, are in critical condition due to anthropogenic activities (Tuxill 1998).

The case of primates, the nearest relatives to humans, illustrates the magnitude of the changes that have occurred since the rise of human civilisations. On the surface, it may seem that there has been no significant change to primate abundance. Of the 240 known primate species, none has become extinct over the last several centuries. However, this statistic masks massive changes to primate populations that have been caused almost entirely by human activities.

It is very difficult to estimate the numbers of primates at the beginning of the Holocene 11 600 years ago, but there is no doubt that, taken together, they dwarfed the human population. For example, there were probably twice as many baboons as humans. With the advent of agriculture, human numbers began to rise sharply and the simultaneous conversion of forest to cropping and grazing lands destroyed habitats for other primates. Around 2 000 years ago it is estimated that *Homo sapiens*, with a population of about 300 million, became the most abundant of all of the primates. However, in the last 50–100 years the changes have been even more startling. Around 1930 the human population of about 2 billion likely outnumbered the population of all other primates combined. Now, with a population of 6.1 billion, humans are massively outpacing the capability of their nearest relatives, the great apes (e.g., gorillas, chimpanzees, bonobos), to keep up. The total of 300 000 humans that are born every day almost equals the entire population of less than 400 000 great apes on Earth.

The International Union for the Conservation of Nature has summarised the current situation for primates in terms of their likelihood of extinction. Of the 240 known species, 19 are critically endangered, having experienced extreme and rapid reductions in their populations. They are in danger of extinction in a 10–20 year timeframe. There are 46 species at the endangered level, and an additional 51 listed as vulnerable. This gives a total of 116, or nearly half of the total number of primate species, that are in danger of becoming extinct within this century if present trends continue. All of these totals have risen from their estimates of a decade ago, suggesting that the situation is worsening and thus is still a long way from stabilisation or improvement.

Similar sets of statistics could be presented for other groups of species. All hide the true impact of the current changes to life on Earth. Many species that are not yet extinct, and may indeed survive into the future, have nevertheless been so reduced in numbers that they have lost their ecological function and thus have become *ecologically extinct*. The case of primates described above is a good illustration of the point. Although only one subspecies may have gone extinct, it is likely that dozens of species have become so few in number that they have lost their ecological role and many more are likely to be driven into this state in the coming decades.

Although the primary cause of species extinctions is habitat alteration and/or destruction, the introduction of non-native species into an ecosystem can also have negative impacts on the population sizes of native species. Non-native species are introduced by one of three means: accidental introduction; species imported for a limited purpose, but then escape; and deliberate introductions on a large-scale (Levine 1989). The magnitude of the transport of species (biological invasion) is enormous and becoming ever more frequent due to the increase in global commerce and international travel (Figs. 3.16 and 3.66). In many continental areas, 20% or more of the plant species are non-indigenous; on many islands, that number is 50% or more (Rejmanek and Randall 1994). Many biological invasions are effectively irreversible – once established in an environment, it is difficult and often prohibitively expensive to remove them. Some of these degrade human health and the health of native species; others cause huge economic losses; still others alter the structure and functioning of whole ecosystems and drive losses of species diversity in those systems.

An example is that of the brown tree snake. This is a mildly venomous one- to three-metre-long tree dweller and nocturnal hunter that was inadvertently introduced to Guam from New Guinea. A fairly innocuous reptile, its presence wasn't detected until nearly a decade later in the late 1960s when biologists observed that the southern third of the island was completely devoid of the usual bird song. Although scientists had identified the cause of the disappearance of birds, as well as that of some lizards and mammals, they were unable to contain the brown snake by that time, and by the 1990s its population had exploded to more than 10 times its normal population density and had caused the extinction of 9 of Guam's 13 native forest birds (Baskin 2002).

The problem of species relocation is not confined to terrestrial systems. Relative species abundance in marine waters can also be influenced by the introduction of alien or exotic species. In some cases, such introductions are intentional for the purposes of improving or establishing fisheries (i.e., the introduction of the King Crab in the 1960s in the Barents Sea), or in connection with aquaculture/sea ranching activities or as an aid in the process of land reclamation. Often, however, introductions occur unintentionally via transport in ballast water or as passengers on the bottoms of ships. It is estimated that ballast waters are responsible for transporting between 3 000 and 10 000 species daily on a global scale. Sometimes the dumping of ballast water causes serious disruption of local ecosystems by adding an exotic predator species. It was ballast waters that brought the zebra mussel to the Great Lakes, northern Pacific seastars to southern Australia, and comb jellies to the Black Sea.

Non-intentional introductions can also occur as secondary introductions along with organisms intentionally introduced. In most cases, introductions of alien species have little effect in the ecosystem to which they are introduced but there have been several unintentional introductions that have dramatically altered ecosystems (e.g. the introduction of the comb jelly *Mnemiopsis* to the Black Sea which destroyed the anchovy fishery by consuming fish larvae) and/or had serious economic consequences. Combating the zebra mussel that was introduced to the Great Lakes in North America, is, for example, estimated to cost approximately US$100 billion annually (OTA 1993).

3.3.6 Alteration of Carbon, Nitrogen, Phosphorus and Sulphur Fluxes

Undoubtedly the best-known impact of human activities on the Earth System is the increase of CO_2 in the atmosphere due to fossil fuel combustion. In a broader context, this is just one of the many ways in which human activities are altering natural biogeochemical cycles. Others include intensification of agriculture, which often involves the use of fertilisers and irrigation; changes to lateral biogeochemical fluxes through dams and impoundments; the burning of biomass associated with clearing of forests and modification and management of rangelands; and the nature of forestry management practices. Many of these individual effects have been addressed in the previous discussions of change in Earth System components. What has not been addressed, however, is the *extent* to which the overall cycles of carbon, nitrogen, phosphorus and sulphur have been altered by human activities.

Human alteration of Earth's biogeochemical cycles is complex and consists of two major components: (*i*) perturbations of and additions to natural fluxes within biogeochemical cycles and (*ii*) the responses of the Earth System to these changed fluxes and feedbacks to biogeochemical cycling as a whole. Thus,

- Alteration of fluxes = human perturbations of or additions to natural fluxes
- Changed biogeochemical cycling = human perturbations/additions + Earth System responses

This section deals only with human alterations of biogeochemical fluxes, the first equation above. It examines two aspects of the alterations: (*i*) the nature and magnitude of the direct human perturbation of or addition to natural fluxes, and (*ii*) the compartments of the Earth System from which and into which the altered or added flux flows. Chapter 4 deals with the reverberations of these altered fluxes through the Earth System – the responses of the System to the altered fluxes and the feedbacks to biogeochemical cycling. This discussion is required before undertaking an examination, in an integrated way, of changes to Earth's biogeochemical cycling, the second equation above.

Figure 3.52 illustrates the two components of changed biogeochemical cycling described in the second equation above. The *top* curve in the figure shows the amount of annual CO_2 emissions from fossil fuel combustion; this is discussed below and represents a new, additional flux due to human activities (the first component on the right-hand side of the second equation above – *human perturbations/additions*). The *bottom*, more variable curve represents *changed biogeochemical cycling* in terms of the change in atmospheric CO_2 concentration (the left-hand side of the equation). The difference between these two curves (and the reason that not all of the fossil fuel CO_2 emissions stay in the atmosphere) is due to the *responses of the Earth System* to this new flux (the second term on the right-hand side of the equation). These latter two are dealt with in Chap. 4.

For the carbon cycle the magnitude of the human alteration of and addition to fluxes is also shown in diagrammatic form (Fig. 3.53), superimposed on the natural fluxes within the global carbon cycle. For the other cycles (N, P, S), there still remain large uncertainties in many of the natural fluxes between Earth System compartments and a consensus global budget for these elements has not yet been achieved (for examples of global budgets see Galloway et al. 1995; Charlson et al. 2000; Jahnke 2000; Reeburgh 1997). The discussions below for N, P and S describe the nature and magnitude of the human alterations of fluxes in comparison with the appropriate natural flux only.

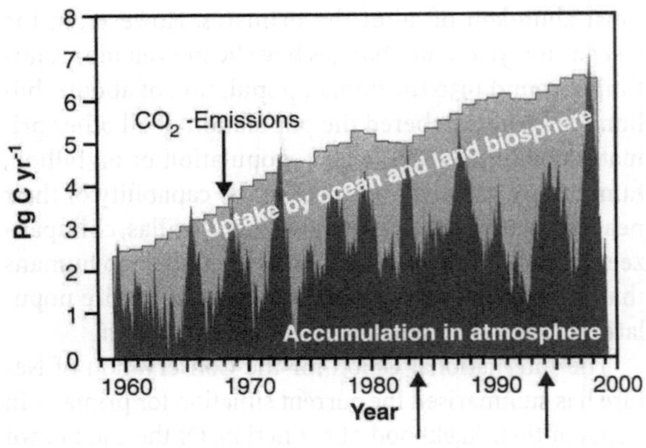

Fig. 3.52. Anthropogenic emissions of CO_2 and changed carbon cycling (adapted from IPCC 2001). The stepwise increase shown by the *upper curve* represents human-driven emissions of CO_2 (Marland et al. 2002); the *dark grey curve* shows the measured rise of CO_2 in the atmosphere for each year (Keeling and Whorf 2000), the difference between the curves being the amount of CO_2 taken up by land or ocean, that is, the response of the Earth System to the human perturbation. Large interannual changes in the uptake of CO_2 by land or ocean appear to be related to the El Niño events indicated by *arrows* on the horizontal axis

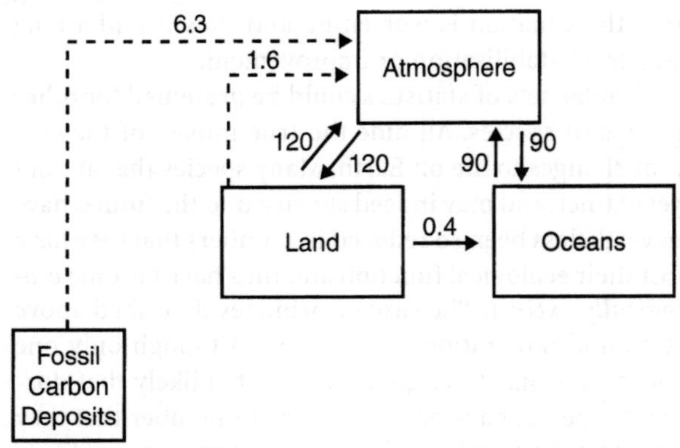

Fig. 3.53. Natural fluxes of carbon between the land, atmosphere, oceans (*solid arrows*) plus the anthropogenic addition of carbon to the atmosphere from fossil carbon deposits and from land-use change (*broken arrows*) (based on data from Prentice et al. 2001)

Carbon Cycle

The largest human-driven flux in the carbon cycle is the emission of CO_2 from the burning of fossil fuels. Currently this flux is about 6.3 Pg C yr^{-1} and continues to rise. This carbon flows from the fossil stores of carbon deep underground and is deposited in the atmosphere. The natural flux of carbon between the fossil store and the atmosphere is virtually zero so the human-driven flux is a big perturbation. In effect, the combustion of fossil fuel removes carbon from a store that is isolated from the atmosphere and inserts it in the active land-ocean-atmosphere cycle at the surface of the Earth.

The second significant human perturbation to the carbon cycle is the removal of carbon from terrestrial ecosystems through land-cover and land-use change. For the decades of the 1980s and 1990s this flux was estimated to be about 1.5 Pg C yr^{-1}, but there is considerable uncertainty surrounding the estimate. The most important land-use/cover change in terms of the carbon cycle is tropical deforestation and conversion to agriculture, which removes carbon from both the biomass of the trees through burning of the slash and from the soil through subsequent oxidation and erosion. Biomass burning is a flux of carbon from terrestrial ecosystems to the atmosphere while loss of soil carbon is a flux from terrestrial ecosystems to both the atmosphere (via oxidation) and eventually to the ocean (via erosion). The relative magnitudes of these two fluxes are still a matter for debate and further research.

A few important characteristics of the anthropogenically driven fluxes should be noted. First, the two are fundamentally different in that the land-cover/use change driven flux is a perturbation of natural land-atmosphere and land-ocean fluxes while fossil fuel combustion is essentially a new flux that is absent from the natural world. This difference is often neglected in the formulation of policy responses to climate change issues. Second, although the land-cover/use change flux is small compared to the gross fluxes of carbon between the atmosphere and terrestrial ecosystems, it is significant in the context of the *net* natural flux between these stores, which has been virtually zero during the Holocene. Third, there is considerable variability in natural and anthropogenic carbon fluxes both spatially and temporally (Sect. 2.4.2).

Nitrogen Cycle

Human-driven changes to the global nitrogen cycle have been profound (Box 3.6). The most important human-driven nitrogen flux is the industrial reduction (fixation) of atmospheric N_2 to ammonia (NH_3), primarily for use as a fertiliser. This represents a flux from the atmosphere to terrestrial ecosystems and is currently around 80 Tg N yr^{-1}. A second deliberate human flux is that due to the cultivation of soybeans, alfalfa and other leguminous crops, which biologically fix around 40 Tg N yr^{-1}. Again, this is a flux from the atmosphere to terrestrial ecosystems. A third perturbation is related to fossil fuel combustion, which inadvertently oxidises N_2 in the atmosphere to NO and NO_2. Although this is not a net flux between compartments, it does have implications for the functioning of the nitrogen cycle (Sect. 4.5.2). About 20 Tg N yr^{-1} is fixed by fossil fuel combustion. The fourth significant human-driven flux is the emission of N-gases, such as NO_x and N_2O, to the atmosphere as a result of biomass burning. Some of these emissions are associated with natural cycles of burning. Anthropogenically driven burning beyond the natural cycles is estimated to add about 10 Tg N yr^{-1} to the atmosphere (Jaffe 2000). Since the source of this nitrogen is from vegetation, this represents a flux from terrestrial ecosystems to the atmosphere. Taken together, these fluxes are now greater in magnitude than the natural fixation of nitrogen by all of Earth's terrestrial ecosystems (about 90–130 Tg N yr^{-1}). Rates of biological nitrogen fixation in oceans are less certain, but perhaps as much as for terrestrial ecosystems (Vitousek et al. 1998).

Phosphorus Cycle

The human perturbation of and additions to the phosphorus cycle are simpler in that the phosphorus cycle does not have a significant atmospheric component. The only human perturbation is the mining of phosphate deposits, their conversion to fertilisers or detergents and their application to terrestrial-based systems. However, this flux is estimated to be 12.5 Tg P yr^{-1} (Jahnke 2000) and represents a massive increase of the net mobilisation of phosphorus above the natural rate of about 2.2 Tg yr^{-1} due to weathering of phosphate rock and the activity of the roots of terrestrial biota (Reeburgh 1997).

Sulphur Cycle

Fossil fuel combustion is by far the largest anthropogenic source of oxidised sulphur gases in the atmosphere, with nearly all of the emissions in the form of SO_2. Like the emission of CO_2 from fossil fuel combustion, this is a flux to the atmosphere from a store isolated from land-ocean-atmosphere cycling. Anthropogenic emissions of sulphur to the atmosphere are estimated to be about 80 Tg S yr^{-1}, which is substantially larger than the natural fluxes of similar sulphur compounds of about 25 Tg S yr^{-1} (Charlson et al. 2000).

As noted above, this analysis simply documents the nature and size of the human-driven fluxes of carbon, nitrogen, phosphorus and sulphur; it does not yet include the responses of the Earth System to these new and perturbed fluxes. This will be done in Sect. 4.5, where the integrated responses of the Earth System to human forcing are discussed.

Box 3.6. The Global Nitrogen Cycle: Past, Present and Future[1]

James Galloway

In the late-twentieth century human activities surpassed natural terrestrial processes in converting unreactive N_2 to reactive N (Nr)[2]. This fact has great significance because not only is N the very stuff of life, but the lack of Nr is often limiting to the productivity of ecosystems and too much Nr contributes to most environmental issues of the day.

Nr creation requires the breaking of the triple bonds within the N_2 molecule, a reaction that requires energy. In nature N_2 is converted to Nr mainly by biological nitrogen fixation (BNF), performed by certain unique microorganisms that have developed the special metabolic machinery necessary to produce biologically active reduced forms of nitrogen such as ammonia, amines, and amino acids, the structural constituents of proteins and nucleic acids (Vitousek et al. 2002). These specialized organisms include a few free-living bacteria and blue-green algae, and also certain symbiotic bacteria that have developed special metabolic relationships with the roots of leguminous crop plants such as soybeans, clover, and N-fixing trees such as alder. In addition, N_2 can also be converted to Nr during lightning strikes. In the pre-human world BNF was the dominant means by which new Nr was made available to living organisms. Humans create Nr in three ways: cultivation-induced BNF (e.g., legume production), Haber-Bosch process (industrial fixation of N_2 to NH_3), and fossil fuel combustion (high temperature conversion of N_2 and fossil organic N to NO_x). By about 1970 anthropogenic processes overtook natural terrestrial process in Nr creation on a global scale.

The impact of this significant increase in the rate of Nr creation on the global N cycle is illustrated by contrasting Nr creation and distribution from 1890 to 1990. The former is an appropriate starting point for examination of the N cycle since at that time limited Nr was created by human activities. Although the global population was ~25% of the current number, the world was primarily agrarian and produced only 2% of the energy and 10% of the grain produced today. Most energy (75%) was provided by biomass fuels; coal provided most of the rest (Smil 1994). Petroleum and natural gas production was limited and was of little consequence relative to the global supply of energy and the creation of Nr as NO_x through combustion. In total, fossil fuel combustion created only about 0.6 Tg N yr^{-1} in 1890, through production of NO_x (Fig. 3.54).

Crop production was primarily sustained by recycling crop residue and manure on the same land where food was raised. Since the Haber-Bosch process was not yet invented, the only new Nr created by human activities was by legume and rice cultivation (the latter promotes Nr creation because rice cultivation creates an anaerobic environment which enhances nitrogen fixation). While estimates are not available for 1890, Smil (1999) estimates that in 1900 cultivation-induced Nr creation was on the order of 15 Tg N yr^{-1}. Additional Nr was mined from guano (~0.02 Tg N yr^{-1}) and nitrate deposits (~0.13 Tg N yr^{-1}) (Smil 2000).

Thus in 1890 the total anthropogenic Nr creation rate was ~15 Tg N yr^{-1}, almost entirely for food production. In contrast, the natural rate of Nr creation was on the order of 220 Tg N yr^{-1}. Terrestrial ecosystems created ~100 Tg N yr^{-1} and marine ecosystems created ~120 Tg N yr^{-1} (D. Capone, personal communication). An additional ~5 Tg N yr^{-1} was fixed by lightning. On a relative basis for the globe, human activities created about 5% of the total Nr fixed and about 13% when only terrestrial systems are considered.

One century later the world's population had increased by a factor of ~3.5, from about 1.5 to about 5.3 billion, but the global food and energy production increased about 7-fold and 90-fold, respectively. Just as was the case in 1890, in 1990 (and now) food production accounts for most of the new Nr created. What changed most since 1890 was the magnitude of Nr created by humans. Smil (1999) estimated that in the mid-1990s cultivation-induced Nr production was ~33 Tg N yr^{-1}. The Haber-Bosch process, which did not exist in 1890, created an additional ~85 Tg N yr^{-1} in 1990, mostly for fertilizer (~78 Tg N yr^{-1}) and the remainder in support of industrial activities such as the manufacture of synthetic fibers, refrigerants, explosives, rocket fuels, nitroparaffins, etc.

For energy production, during the period 1890 to 1990, much of the world was transformed from a bio-fuel to a fossil-fuel economy. The increase in energy production by fossil fuels resulted in increased NO_x emissions from ~0.6 Tg N yr^{-1} in 1890 to ~21 Tg N yr^{-1} in 1990. By 1990 over 90% of energy production resulted in the creation of new Nr, in contrast to 1890 where very little energy production caused Nr creation.

By 1990 Nr created by anthropogenic activities was ~140 Tg N yr^{-1}, a ~9-fold increase over 1890, contrasted to a ~3.5-fold increase in global population. Coupled with the increase in Nr creation by human activities was a decrease in natural terrestrial N fixation because of conversion of natural grasslands and forests to croplands, etc., from ~100 Tg N yr^{-1} to ~89 Tg N yr^{-1} (Cory Cleveland, pers. comm.).

The fate of anthropogenic Nr for the three anthropogenic sources is clear: NO_x from fossil fuel combustion is emitted directly into the atmosphere; RNH_2 from rice and legume cultivation is incorporated into biomass; NH_3 from the Haber-Bosch process is primarily converted into commercial fertiliser, which is applied to agroecosystems to produce food. However, little of the fertiliser N actually enters the human mouth in the form of food; most is in fact ultimately released to environmental systems.

In 1890 both the creation and fate of Nr was dominated by natural processes (Fig. 3.54a). There were limited Nr transfers via atmospheric and hydrologic pathways relative to the amount of Nr created. For terrestrial systems, of the ~115 Tg Nr yr^{-1} created, only about ~15 Tg N yr^{-1} were emitted to the atmosphere as either NH_3 or NO_x. There was limited connection between terrestrial and marine ecosystems; only about 5 Tg N yr^{-1} of dissolved inorganic nitrogen was transferred via rivers into coastal ecosystems in 1890 and only about 17 Tg N yr^{-1} were deposited to the ocean surface.

In 1990 by contrast, when creation of Nr was dominated by human activities (Fig. 3.54b), there were also significant changes in Nr distribution. NH_3 emissions increased from ~9 Tg N yr^{-1} to ~43 Tg N yr^{-1} as a consequence of food production; NO_x emissions increased from ~7 Tg N yr^{-1} to ~34 Tg N yr^{-1} from both energy and food production. The increased emissions resulted in widespread distribution of Nr to downwind ecosystems. Transfer of Nr to marine systems also increased. By 1990 riverine fluxes of dissolved inorganic nitrogen to the coastal ocean had increased to 20 Tg N yr^{-1} and atmospheric N deposition to marine regions had increased to 27 Tg N yr^{-1}. While evidence suggests that most of the riverine N is denitrified in coastal and shelf environments (Seitzinger and Giblin 1996), most of the atmospheric flux is deposited directly to the open ocean, although a portion of the 27 Tg N yr^{-1} is deposited to coastal ocean and shelf regions, with significant ecological consequences (Rabalais 2002).

A key component missing from Fig. 3.54b is the ultimate fate of the ~140 Tg N yr^{-1} Nr created by human action in 1990. On a global basis, Nr created by human action is either accumulated (stored) or is denitrified. Unfortunately, it is not possible to estimate the relative importance of these two processes. This in-

[1] Adapted primarily from Galloway and Cowling (2002).

[2] The term reactive nitrogen (Nr) as used in this box includes all biologically active, photochemically reactive, and radiatively active N compounds in the atmosphere and biosphere of the Earth. Thus Nr includes inorganic reduced forms of N (e.g., NH_3, NH_4^+), inorganic oxidised forms (e.g., NO_x, HNO_3, N_2O, NO_3^-), and organic compounds (e.g., urea, amines, proteins).

◀

ability represents one of the largest uncertainties in the understanding of the nitrogen budget at any scale.

There are thus large uncertainties regarding the rates of Nr accumulation in various reservoirs. This limits the ability to determine the temporal and spatial distribution of environmental effects. These uncertainties are even more significant because of the sequential nature of the effects of Nr on environmental processes. This sequence of transfers, transformations and environmental effects is referred to as the nitrogen cascade (Galloway et al. 2003). A single atom of newly created Nr (as either NH_x or NO_x) can alter a wide array of biogeochemical processes and exchanges among environmental reservoirs. For example, a molecule of NO emitted to the atmosphere during fossil fuel combustion can, in sequence, increase ozone concentrations in the troposphere, decrease atmospheric visibility and increase concentrations of PM2.5 particles, increase precipitation acidity, increase soil acidity, increase or decrease forest productivity, increase surface water acidity, increase hypoxia in coastal waters, increase greenhouse warming, and decrease stratospheric ozone.

A principle feature of the cascade is the accumulation rate of Nr in environmental systems. This is one of the most important research questions associated with the impact of humans on the nitrogen cycle. Human creation of Nr will continue to increase in the future as population grows. Even after population has peaked Nr creation is still likely continue to increase due to growth in per capita resource use. How high will the Nr creation rate go? In 1990 it was ~140 Tg N yr^{-1}, and the average per capita Nr creation rate was ~24 kg N $person^{-1}$ yr^{-1}, ranging from ~7 kg N $person^{-1}$ yr^{-1} in Africa to ~100 kg N $person^{-1}$ yr^{-1} in North America. If the global population peaks at ~8.9 billion people and if all people had the same per capita Nr creation rate from food and energy production as North America in 1990 (~100 kg N $person^{-1}$ yr^{-1}), then the total Nr creation rate would be ~900 Tg N yr^{-1}, with about half occurring in Asia. Given the environmental concerns about Nr, it is unlikely that this value will be reached. What the final maximum Nr creation rate turns out to be, however, will depend to a very large extent on how the world manages its use of nitrogen for food production and its control of N in energy production in future.

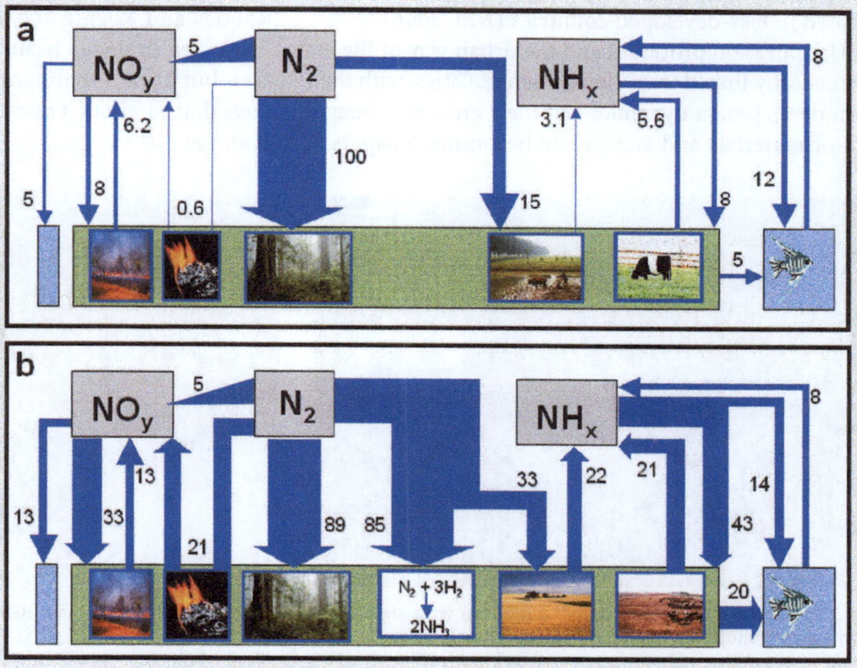

Fig. 3.54. Global terrestrial nitrogen budget for **a** 1890 and **b** 1990 in Tg N yr^{-1}. The emissions to the *NO_y box* from the *coal* reflect fossil fuel combustion. Those from the *vegetation* include agricultural and natural soil emissions and combustion of biofuel, biomass (savanna and forests) and agricultural waste. The emissions to the *NH_x box* from the *agricultural field* include emissions from agricultural land and combustion of biofuel, biomass (savanna and forests) and agricultural waste. The NH_x emissions from the *cow* and *feedlot* reflect emissions from animal waste. The transfers to the *fish box* represent the lateral flow of dissolved inorganic nitrogen from terrestrial systems to the coastal seas

3.4 Putting Human-Driven Changes into an Earth System Perspective

Most of this chapter has considered the ways in which human activities are changing the Earth System by examining a large number of detailed pieces – the human driving forces and the resulting changes in the Earth System. It should be clear from the foregoing discussion that many of the system changes occur due to interactions among a large number of human activities, aggregated globally over long periods of time. Several properties of these interactions emerge as important features of an altered Earth System.

3.4.1 Socioeconomic and Cultural Teleconnections

Just as connections in the biophysical part of the Earth System link processes across long distances (see Sect. 2.5), socioeconomic and cultural connections link human activities in widely separated regions of the planet. Two of the most important of these are urbanisation and globalisation. Together they are linking and moving people, processes and products across the Earth. Any understanding of the evolving human role in the functioning of the Earth System must take into account this accelerating human planetary network that is transforming the anthroposphere.

Urbanisation

As the world's human population has grown, its distribution, among nations and between rural and urban areas, has changed dramatically. Urban populations are currently growing more rapidly than rural populations all over the world, particularly in developing countries (Figs. 3.55 and 3.56). It is estimated that more than half of the world's population now live in urban areas. More than 90% of future population growth will be concentrated in cities in developing countries (e.g., Fig. 3.57) and a large percentage of this population will be poor (Sanchez 2002). The number of large cities has also grown significantly. In 1950 there were 81 cities with a population of between 1 and 10 million and by 1990 there were 270 cities of this size. In 1950 there were only two megacities (London and New York), by 1990 there were 21 cities of this size (greater than 10 million) and by 2015 there are predicted to be 33 megacities, most of them located in less-developed countries (NRC 1999).

Urbanisation processes and the urban way of life are intrinsically linked with global change. Cities, with their high development dynamics and their growing throughput of materials and energy, are becoming hotspots of change in terms of both resource demand and environmental impact. Cities are also a focus of concern with regard to the complex interlinkages between socioeconomic development, disparities in wealth, environmental change and public health (Caldwell and McMichael 2002).

While urban land use, in the form of built-up or paved areas, occupies less than 2% of the land area of the Earth (Gubler 1994; Lambin et al. 2001), the services provided by ecosystems that cover much larger areas are required to support urban populations. A useful concept to show the impact of cities on the Earth's resources is the calculation of the ecosystem services (for example, provision of clean water, food, building materials, energy and waste disposal) that a city requires that are essentially imported from other regions (Box 3.7). For example, it has been estimated that Hong Kong requires ecosystems over an area that is 2 200 times the built-up area of the city in order to support its inhabitants with essential ecosystem goods and services (Warren-Rhodes and Koenig 2001). The 29 largest cities in the Baltic Sea drainage basin cover only 0.1% of the land area but their inhabitants appropriate an ecosystem area that is about 1 000 times the urban area (Folke et al. 1997).

Fig. 3.55. Contrasts in urbanisation: **a** modern with sophisticated infrastructure; and (*photo:* IGBP Photo Archive) **b** impoverished with unsafe conditions (*photo:* R. Sanchez 2002)

Fig. 3.56. Distribution of densely populated regions of the world as shown by night lights (*image:* NASA, *http://www.gsfc.nasa.gov*)

Urbanisation has also been shown to have significant environmental impacts at all scales (Berry 1990). Locally, urban areas change the character of the Earth's surface by replacing the natural surface of soil and vegetation with brick, concrete, glass, metal etc. at different levels above the ground. This changes the local reflection and radiation characteristics of the Earth's surface, altering heat exchange patterns as well as the aerodynamic roughness of the surface. Regionally, urban areas generate large amounts of heat and alter the composition of the atmosphere, resulting in significant pollution due to sulphate aerosols, other particulates and tropospheric ozone. Urban surfaces also have major effects on the regional water balance and water quality; the paving alters peak-flow characteristics and total runoff. Berry (1990) estimates that peak discharge after rainfall is six times higher in urban areas than in non-urbanised areas. Globally, cities are major contributors to increased concentrations of greenhouse gases to the atmosphere, as well as reactive gases. For example, Dhakal et al. (2002) estimate that between 1970 and 2002, CO_2 emissions from energy use in Tokyo more than doubled, and for 1990–1998 the annual average growth rates of CO_2 emissions for Tokyo and Seoul were 1.7% and 1.63% respectively.

Studying patterns and underlying driving factors of urbanisation has become one of the main contributions to the discussions of future sustainable development (IHDP 2002). Urbanisation relieves some environmental pressures by concentrating populations in small geographical areas and thus providing at least the potential for more efficient provision of services and treatment of wastes. On the other hand, urban populations everywhere tend to command higher standards of living than rural populations. Consumption levels are higher in urban areas (Parikh 2002).

In summary, urban areas are hot spots of global change. They act as magnets for resources, drawing in materials and energy from vast distances and appropriating goods and services from ecosystems many times their areas. They emit a large range of effluents in highly concentrated fluxes into local airsheds and watersheds, effluents that then spread out into the Earth System, affecting regions and processes far removed from the urban areas themselves.

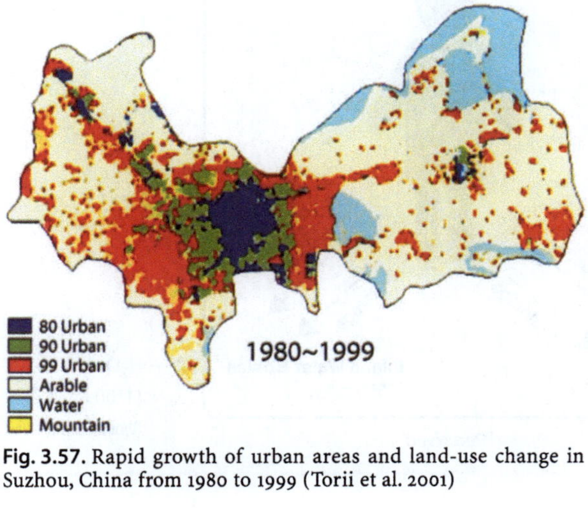

Fig. 3.57. Rapid growth of urban areas and land-use change in Suzhou, China from 1980 to 1999 (Torii et al. 2001)

Globalisation

One of the most obvious changes experienced by societies over the past decade has been termed globalisation. Globalisation can be defined as the worldwide interconnectivity of places and people through global markets, communication systems and information flows, the flows of capital and commerce, and international laws and conventions. Globalisation results in the disconnection of locations of demand from those for production of food and other commodities. Globalisation has numerous consequences for understanding and interpreting contemporary global change and for projecting the future evolution of the Earth System.

Globalisation integrates not only the economy but also culture, technology and governance (Tong et al. 2002). Electronic communications in particular are causing major changes. For example, worldwide the number of mobile phone users increased from 0 in 1978 to 305 million in 1998, while the number of internet users increased from 0 in 1985 to 180 million in 1998 (UNDP 1999; Tong et al. 2002). Such changes have not only linked regions and people more than before, but they have also widened the availability of information about the environment. Without doubt, enhanced availability of information has enhanced the spread of Western consumption patterns. Globalisation has also led to the spread of viruses and pests worldwide.

Economic growth has been stimulated by globalisation, but the process has brought with it both positive and negative environmental implications. The challenge is to decouple economic growth from environmental pressures (Vellinga and Herb 2000; Azar et al. 2002; Ruffing 2002). Southeast Asia offers an excellent example. Recent trends show how global socioeconomic activities impact regional development and drive regional change. Global forces, especially through industrialisation and the commercialisation of agriculture and forestry, have changed the Southeast Asia landscape (Lim 1995; Parnwell and Bryant 1996; Lebel 2000) (Fig. 3.60a,b). The extensive, biodiversity-rich tropical forests of the region are undergoing rapid transformation, primarily to mono-culture plantation forestry or to agriculture. Deforestation has been driven primarily by international timber markets and national timber concessions granted to timber companies to log forests (Kummer 1992; Brookfield et al. 1995; Than 1998; King 1998; Jeppson et al. 2001). Investment from, and trade with, Japanese trans-national companies (Dauvergne 1997), and to a lesser extent, those in the USA and Europe, has been central in diversifying and integrating Southeast Asia's production into the global economy (Rock 2000; Knight 1998).

Box 3.7. Ecological Footprints

Åsa Jansson

The concept of *ecological footprints* (Wackernagel and Rees 1996) is useful for estimating, in a semi-quantitative way, the dependence of human populations on the natural world around them. The idea of showing human dependence on nature by quantifying ecosystem areas is not new. In 1967 Borgström introduced the concept of *ghost acreage* to describe a population's need for agricultural products. Since then several similar approaches have been put forward. Odum (1975, 1989) coined the term *shadow area*, using solar energy as the measuring unit and Jansson and Zucchetto (1978), estimating energy flows, showed that the offshore fisheries of the island of Gotland in the Baltic Sea require extensive marine areas.

The ecological footprint approach can be applied in different ways. In some cases the notion of a global steady state is used; in others the world is regarded as dynamic and complex (Folke et al. 1997; Jansson et al. 1999). A dynamic and complex view acknowledges that the ecological footprint approach does not provide information on the resilience of the system, or how close to thresholds the support capacity might be. Furthermore, the dynamic view to ecological footprints does not reduce the work of nature to a single dimension to be used as an operational indicator of ecological carrying capacity, sustainability or as a basis for discussions on equity.

The appropriation of ecosystem goods by a human population, in terms of food and timber, can be quantified by the areas of land and water (sea areas included for marine-derived food) required to produce the amount of these goods consumed by a defined human population. The amount of land or sea needed to absorb wastes is more difficult to estimate. In general, the approach is to first identify potential sink ecosystems within a defined area, followed by an estimation of the amount of land or sea needed to sequester the emission of the substance or compound of interest. Thus, estimating the area needed to sequester the CO_2 emitted annually from human activities often entails estimating forested land, as many forest types are known to provide a significant sink for CO_2 (Dixon et al. 1994). Furthermore, the amount of agricultural land needed to absorb phosphorus (as sewage sludge) excreted by a human population can be estimated, as well as the amount of wetlands needed to retain and denitrify the nitrogen compounds emitted directly by humans. These estimates are conservative. They do not account, for example, for P and N emitted by food processing, household waste, car emissions, etc., nor do they account for the amount of land and inland waters needed to absorb and process the aerosols and dissolved reactive gases emitted by cities and eventually rained out of the atmosphere.

Freshwater is essential for both humans and ecosystems. However, the concomitant use of freshwater by appropriated ecosystems such as forests, wetlands and lakes is almost always neglected. As shown in Fig. 3.58, this water usage is substantial and dwarfs the amount of water that is consumed directly for personal and industrial use, in this case by the average Baltic region dweller. These indirect water footprints show that the trade off between alternative uses of freshwater needs to be explicitly addressed in both renewable freshwater management and assessment, as well as in ecosystem management.

In only a few years the majority of the world's population will live in cities. In response there is a growing interest in investigating the relationship between nature and urbanisation. The ecological footprint approach has an important role to play in this context. Urban areas require a wide range of ecosystem goods and services for their existence – food, water, raw materials for industry and the capacity to absorb and process the wastes that are generated by urban areas. These services tend to be out of sight and mind to most urban dwellers. They are also usually neglected in economic analyses of the benefits and costs of urbanisation. Yet these ecosystem services are a fundamental underpinning of the social and economic development of cities.

Using the ecological footprint approach based on the notion of a dynamic and complex world, Folke et al. (1997) estimated the ecological footprint of the urban areas in the Baltic Sea drainage basin (Fig. 3.59). In the Baltic region the average urban resident requires between 60 000 and 115 000 m^2 of land for appropriation of ecosystem goods and services. When the demands of all 29 major cities in the basin were aggregated, an area corresponding to 75 to 150% of the entire Baltic Sea drainage basin is required, even though the urban areas themselves occupy only 0.1% of the area of the basin. Extrapolation of the methodology to the global scale leads to even more interesting results. For ex-

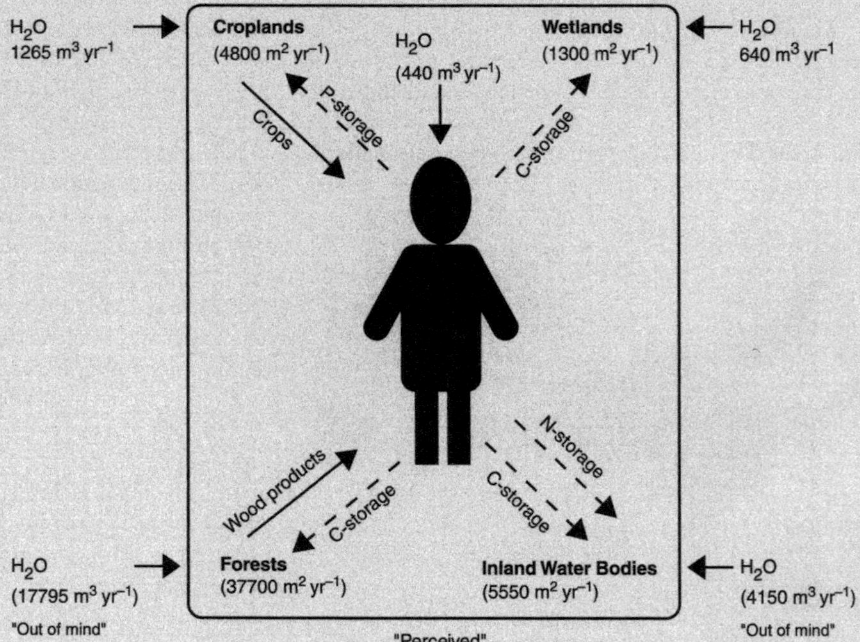

Fig. 3.58. Goods and services appropriated by a Baltic Sea region dweller: estimated ecosystem areas of forest, croplands, wetlands and inland water bodies annually appropriated by an average person living in the Baltic Sea drainage basin (adapted from Jansson et al. 1999). Services shown include timber and terrestrial food production and the absorption and processing of carbon dioxide (CO_2), phosphorus (P) and nitrogen (N). Also included are the freshwater flows required to generate these services

◀

ample, the world's 744 largest cities contain 20% of the human population and are responsible for 32% of the global emissions of CO_2 from fossil fuel combustion. The area of forest required to absorb these emissions, even taking into account that a significant fraction would be absorbed by the oceans, is, at a minimum, equal to the entire current global C sink capacity of forests and at a maximum would be three times this area.

The ecological footprint approach does not imply that all of the area required to provide ecosystem goods and services for a city lies in a contiguous area surrounding the city itself. Through world trade, urban areas now appropriate significant amounts of goods and services from far distant places around the Earth. Likewise, through atmospheric and hydrological transport, the effluents from cities move far away from their sources before they are absorbed and reprocessed. The fact that many of these effluents are building up in the global environment is evidence that the overall absorptive capacity of the Earth System is already being exceeded. In many ways urbanisation is a global change phenomenon.

The ecological footprint approach provides a powerful tool for illuminating the dependence of city dwellers on vast ecosystem areas. However, being able to provide only static snapshots, it does not have the capability to answer questions on whether present levels of resource use are sustainable or not (Deutsch et al. 2000).

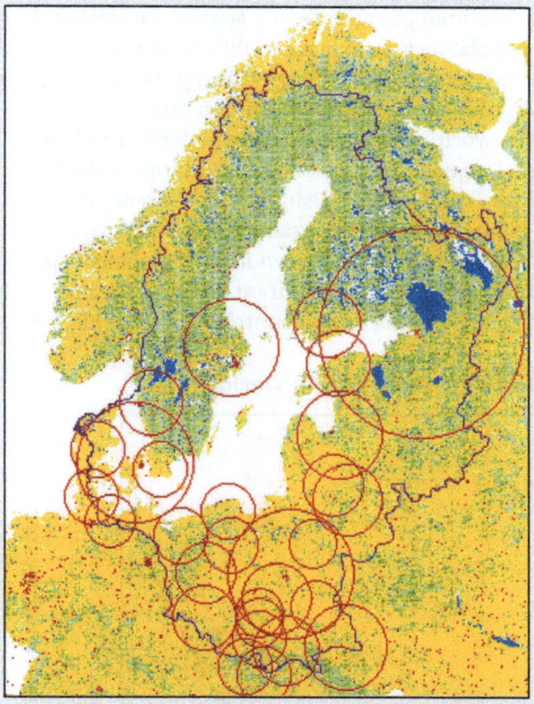

Fig. 3.59. Ecological footprints for Baltic region city dwellers: the location of the 29 largest cities in the Baltic Sea drainage basin and their hidden demand for ecosystem support (Folke et al. 1997). The area of hidden demand is illustrated by the *circles* around each city. The *circles* represent the aggregated average footprints needed to support timber and terrestrial and marine food consumption, carbon sequestration from energy production and phosphorus and nitrogen retention. The figure does not imply that the cities appropriate the actual area within the circles, only that they demand this area. Due to trade appropriation may take place elsewhere on Earth

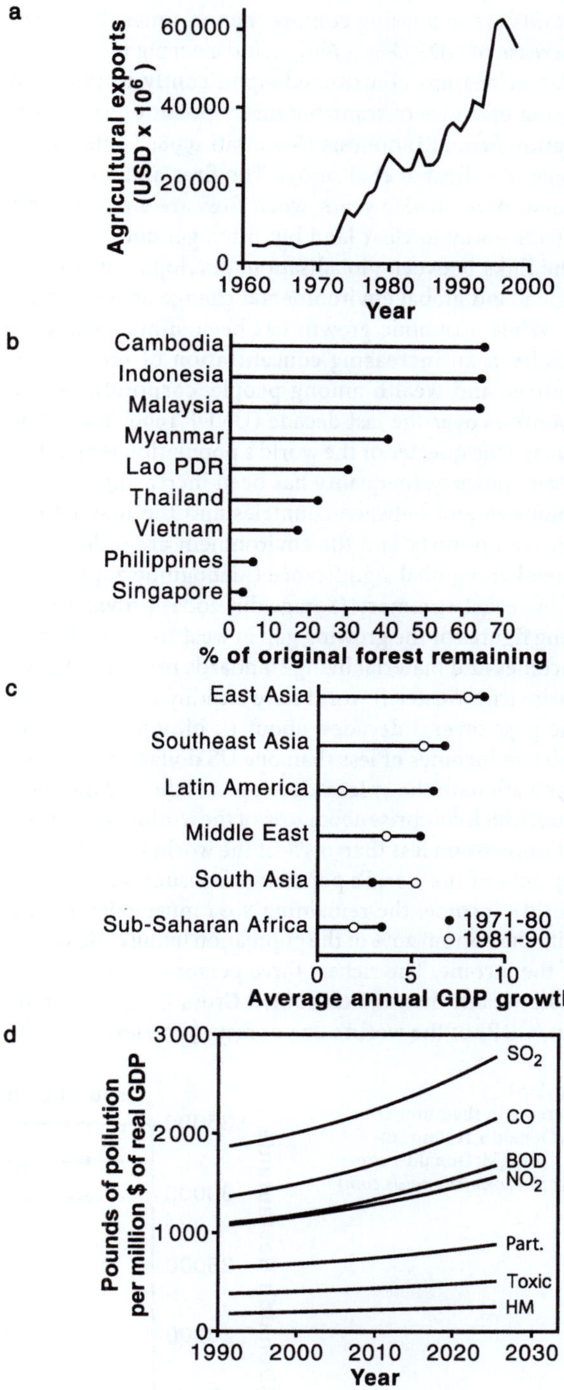

Fig. 3.60. Globalisation and environmental trends in Southeast Asia: **a** value of agricultural exports from ASEAN countries 1960–1998 (FAOSTAT 2002); **b** 1996 forest cover as a percentage of original forest in ASEAN countries (WRI 1999); **c** average annual GDP growth rate of various developing regions of the world (Stallings 1995); and **d** predicted pollution intensity of Thai GDP for 1992–2025 by pollutant (*Part.:* particulate; *HM:* heavy metals)(Angel and Rock 2000)

The environmental consequences of this development include high rates of deforestation (Fig. 3.60b) and accelerating regional air and water pollution problems. In Thailand, for instance, the concentrations of many pollutants are increasing at more than 10 times the rate of increase of GDP (Fig. 3.60d). Land clearing for oil palm plantations has contributed significantly to the most recent episodes of trans-boundary pollution from vegetation fires in Indonesia (Siscawati 1998; Tomich et al. 1998; Murdiyarso et al. 2003). The fires and smoke are worst in dry ENSO years, when fires are a particularly effective way to clear land but often get out of control. The links between globalisation, development, and regional and global environmental change are evident.

While economic growth has been stimulated, there has been an increasing concentration of income, resources and wealth among people, corporations and countries over the last decade (UNDP 1999; Tong et al. 2002). One quarter of the world's population remains in severe poverty. Inequality has been increasing in many countries and between countries and the interactions between poverty and the environment are of local, regional and global significance (Mabogunje 2000).

A recent summary (Serageldin 2002) provides startling figures of the growing gap in wealth. While the real incomes and material living standards in the developed (primarily Western) world have steadily increased over the past several decades, about 1.2 billion people still exist on incomes of less than one US dollar per day. The populations of the 47 least developed of the world's countries, which comprise about 10% of the world's total population, exist on less than 0.5% of the world's income. The top 20% of the world's population consumes 85% of the world's income; the remaining 80% must exist on 15% with the bottom 20% of the population living on just 1.3% of the income. The richest three persons on Earth have more wealth than the combined Gross Domestic Product (GDP) of the world's 47 poorest countries. The richest 15 persons on Earth have more wealth than the combined GDP of all of the countries in sub-Sahara Africa, which have a total population of 550 million. These disparities appear to be growing with further globalisation of the world's economy (Serageldin 2002).

As a result of these disparities in wealth, significant portions of the world's population still face severe development challenges (Serageldin 2002):

- more than 2 billion people do not have access to adequate sanitation;
- 1 billion people do not have access to clean water;
- 1.3 billion people, mostly living in cities in the developing world, are breathing air below the standards set by the World Health Organization; and
- 700 million people, mostly women and children, suffer from indoor air pollution due to the burning of biomass in stoves.

Finally, the phenomenon of globalisation is profoundly affecting the connectivity between the cultures of the world through massive international migration, education, communication and corporate outreach. For example, immigrants currently account for about 65% of the population growth in most OECD countries, compared to 45% less than a decade ago (during the 1990–1995 period). Globally there are 150 million migrants, about 3% of the world's population. At least half of the world's 4 000 languages are predicted to disappear during the twenty-first century. Businesses are rapidly becoming more international. For example, the McDonald's chain of hamburger restaurants, an icon of globalisation, has expanded from one restaurant in the USA in 1954 to 29 000 restaurants in 121 countries in 2000 (Fig. 3.61).

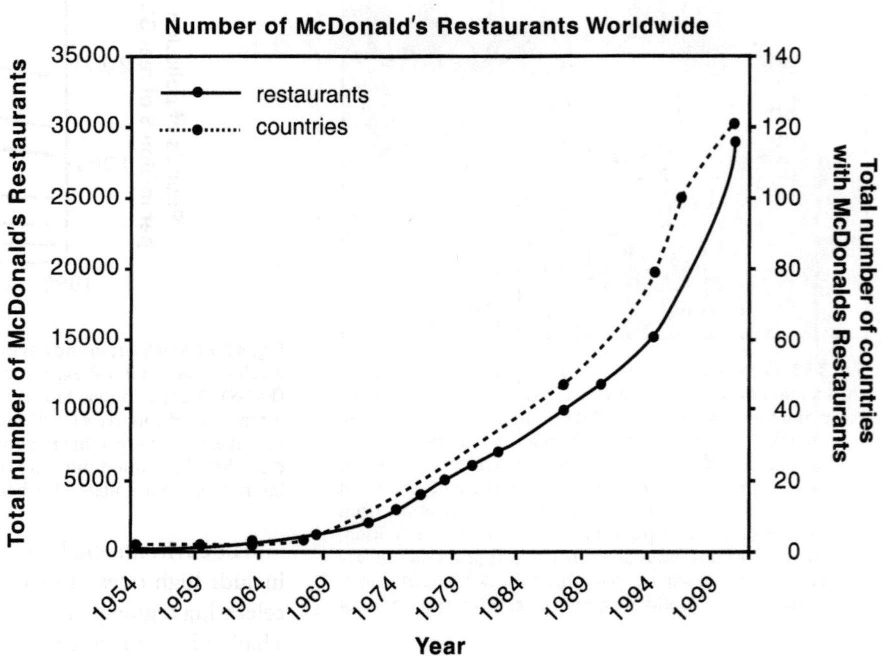

Fig. 3.61. Increase in the number of McDonald's restaurants globally (McDonald's 2002: *http://www.mcdonalds.com*)

At the same time there is growing resistance in many ways to this increasing cultural connectivity as individual core cultures seek to retain their identity in the face of increasing interaction from external influences. The speed and the breadth and depth of interconnectivity of cultures, particularly the penetration of Western consumption-based culture into more traditional cultures around the globe, is almost surely unprecedented in the history of human civilisations.

3.4.2 Interaction Between Human-Driven Changes and Natural Variability

The human activities described throughout this chapter do not operate in isolation, nor do they operate on a static or equilibrium global environment. The Earth System shows a wide range of natural variability in its non-human dominated state, sometimes of large magnitude and on rapid time scales. Any contemporary human influences are superimposed on and interact with these natural patterns of variation. It is thus of paramount importance to disentangle human impacts from natural variability in order to understand their relative importance as drivers of change in Earth System functioning. Moreover, within each set of influences, natural and human, there is a range of processes, the impacts of which need to be disentangled. However, it is of equal, perhaps even greater, importance to increase the understanding of their interactions, especially in those situations where their effects are mutually reinforcing or their combined impact is to drive systems over critical thresholds into modes of non-linear change.

Two important examples illustrate cases where it is commonly assumed that human activities have had a direct and demonstrable effect on the global or regional environment. The first is greenhouse gas emissions and climate change; the second is over-fishing and the decline in fish populations. In each case, however, the situation is much more complex and can only be explained by a strong interaction between natural variability and increasing human pressure.

Greenhouse Gases and Climate Change

One of the central questions in global change science is the extent to which human activities, especially the emissions of greenhouse gases into the atmosphere, are influencing the global climate. Given that the climate is subject to a wide range of natural modes of variability, how can the recent observed changes in climate be attributed unequivocally, at least in part, to human influence?

Climate models are good tools for integrating the various forcing factors on climate within a single framework and estimating the overall effect on the climate system. Figure 3.62 shows the results of a climate model

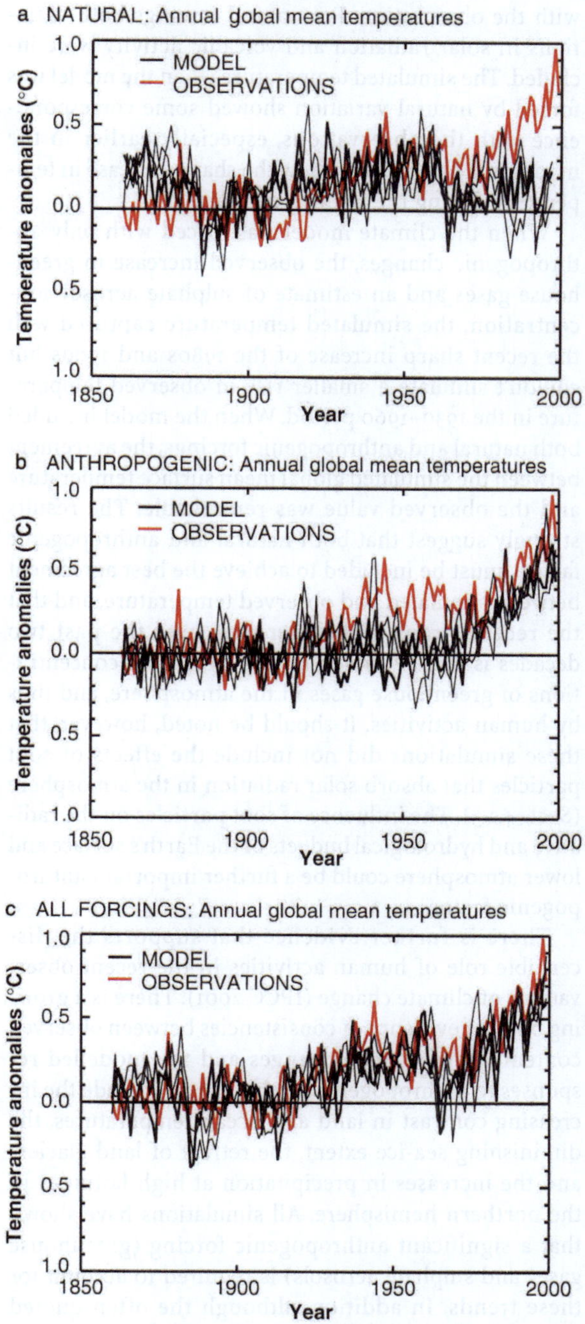

Fig. 3.62. Global annual mean surface temperature anomalies (°C) defined as the difference between the observed 1880–1920 mean and modelled temperatures determined from ensembles of four simulations with a coupled ocean-atmosphere climate model forced **a** with solar and volcanic forcing only; **b** with anthropogenic forcing only including well-mixed greenhouse gases, changes in stratospheric and tropospheric ozone and the direct and indirect effects of sulphate aerosols; and **c** with all forcings, both natural and anthropogenic (IPCC 2001). The model data are only sampled at the locations where there are observations

forced with natural drivers of the climate system and anthropogenic drivers separately, and then both types of forcing together. For each simulation the model projection of global mean surface temperature is compared with the observation. For natural forcings both variations in solar radiation and volcanic activity were included. The simulated temperature when the model was forced by natural variation showed some correspondence with the observations, especially earlier in the model run, but failed to show the sharp increase in temperature during the last two decades.

When the climate model was forced with only anthropogenic changes, the observed increase in greenhouse gases and an estimate of sulphate aerosol concentration, the simulated temperature captured well the recent sharp increase of the 1980s and 1990s but couldn't simulate a smaller rise in observed temperature in the 1930–1960 period. When the model included both natural and anthropogenic forcings, the agreement between the simulated global mean surface temperature and the observed value was remarkable. The results strongly suggest that both natural and anthropogenic factors must be included to achieve the best agreement between simulated and observed temperature, and that the recent strong global warming over the past two decades is largely driven by the increasing concentrations of greenhouse gases in the atmosphere, and thus by human activities. It should be noted, however, that these simulations did not include the effects of soot particles that absorb solar radiation in the atmosphere (Sect. 4.2.3). The influence of soot particles on the radiative and hydrological budgets of the Earth's surface and lower atmosphere could be a further important anthropogenic factor.

There is further evidence that supports the discernible role of human activities in the recent observations of climate change (IPCC 2001). There is a growing body of evidence of consistencies between observed contemporary climate changes and the modelled responses to anthropogenic forcing. These include the increasing contrast in land and ocean temperatures, the diminishing sea-ice extent, the retreat of land glaciers and the increases in precipitation at high latitudes in the northern hemisphere. All simulations have shown that a significant anthropogenic forcing (greenhouse gases and sulphate aerosols) is required to account for these trends. In addition, although the often-quoted discrepancy between the modelled and observed vertical profile of temperature in the troposphere has not yet been fully resolved, it has been significantly reduced through the use of more realistic forcing histories in the climate models.

In summary, the most comprehensive and thorough analysis of contemporary changes to the climate system (IPCC 2001) highlights the reality of anthropogenic greenhouse gas-driven climate change, clearly evident beyond natural patterns of variability. Thus, although global change is much more than climate change, it is beyond doubt that the global changes unfolding through the twenty-first century will have a strong climate dimension.

Fluctuating Fish Stocks: Over-fishing or Natural Variability?

The crashes of the stocks of some commercial fish species, such as the Japanese sardine, are often cited as evidence of over-fishing (Fig. 3.63). For some of the commercially important fish stocks there are now relatively good catch data covering the last 100 years or longer. These data give important insights into the difficult question of what causes large fluctuations in fish stocks. Although catch data are not a perfect proxy for the abundance of fish stocks, their fluctuations are large enough over a century time scale that they cannot reflect changes in fishing activity alone and must reflect, to a large extent, real changes in species abundances. A noteworthy feature of the data is that fish stocks from different parts of the world often fluctuate in synchrony. This suggests some kind of teleconnection between these stocks or, in other words, that these stocks must be responding to the same signals in the Earth System.

Evidence suggests that climate may be driving the observed similar changes in geographically separated stocks (Klyashtorin 1998). Just how the abundance of commercially important fish stocks and other marine organisms are related to major atmospheric processes influencing climate remains to be elucidated, however.

Notwithstanding natural variability, it is possible to discern the impact of human activities (i.e., fishing) on fish stock abundance. Intensive fishing of Californian sardine in the 1940s, when the stocks were naturally low due to climate variability, led to a stock collapse, the concomitant collapse of the industry, and a slowed recovery when the climate became more favourable. With Greenland cod, stocks varied with natural variability during most of the twentieth century, but not during the latter decades of the century (Fig. 3.64). Owing to heavy fishing pressure, the stock may have become too small during the latter part of the century to capitalise on the conditions that, in earlier years when more eggs were produced, allowed the abundance of the stock to increase once favourable conditions were re-established.

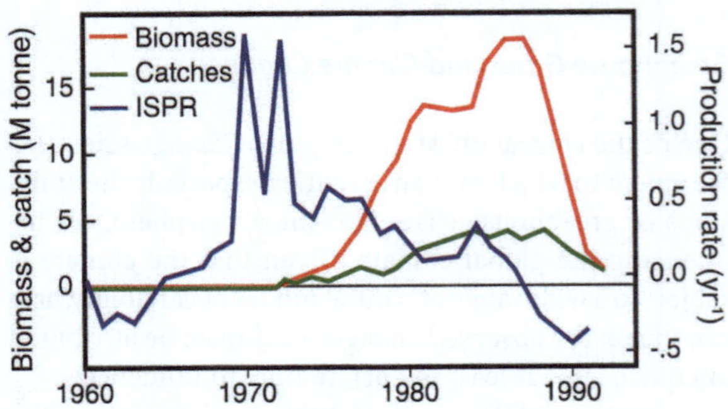

Fig. 3.63. Variations in Instantaneous Surplus Production Rate (ISPR) of Japanese sardine stocks, indicating the complex interplay between natural variability and human pressures and showing the stock collapse due to over-fishing around 1990 (Jacobson et al. 2001)

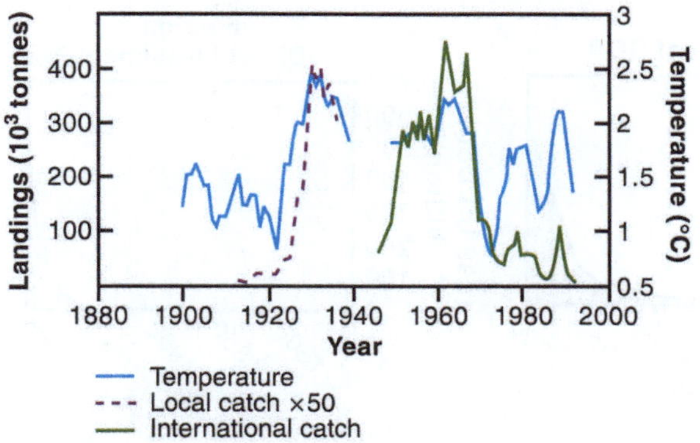

Fig. 3.64. Variations in Greenland Cod catches with temperature from 1900 to 2000 (Barange 2002)

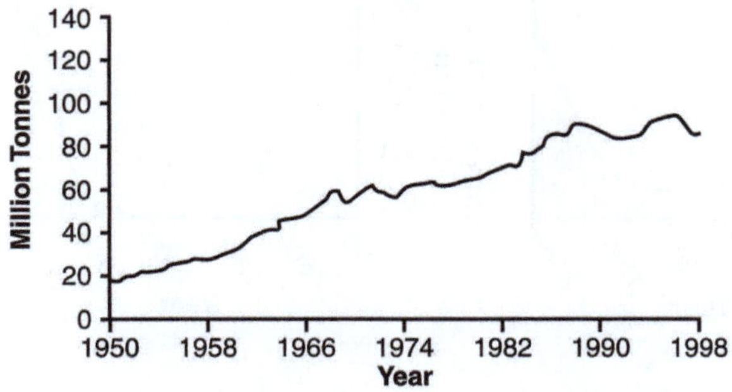

Fig. 3.65. Global fish catch from 1950 to 1998 (adapted from FAO 2000b)

After increasing dramatically throughout most of the twentieth century, fish catches stabilised in the late 1980s and remained more or less at the same level throughout the 1990s (Fig. 3.65). This stabilisation is believed to represent a human factor, that most commercially valuable species are now fully exploited or over-exploited (FAO 2000b).

3.4.3 Global Change: Magnitudes, Rates and Significance of Human Changes

What is the ultimate significance for the functioning of the Earth System of complex, interacting drivers of change originating in the burgeoning human enterprise? It is a surprisingly difficult question to address. It is not easy to assess change and its implications while standing in the midst of it (Kates et al. 1990). The evidence and interpretations in this chapter might be interpreted as highly negative and alarmist, at least in the sense that environmental damage, even threats to the functioning of the Earth System as a whole, loom large on the horizon and are the apparent outcome of changing human-environment relationships. These threats are real and are serious. Historically, however, humankind has responded positively to such threats as much as it has failed to respond or responded negatively. Society has employed technology and institutions to reduce or slow environmental changes and/or find substitutes for the nature reduced or lost, and there are projections of similar responses to this challenge (e.g., Lomborg 2001; Ausubel 2002). Indeed, there are some signs that a societal transformation is growing. Significant amounts of land in the developed world are being returned to natural ecosystems as agricultural efficiency rises, although land elsewhere often provides the substitute for production. More economic activity is occurring for less energy expended. Globally, per capita water withdrawals have declined over the past two decades as water use efficiency has risen to meet increasing water demand (UN 1992; Shiklomanov 1993; Gleich 1998). Local and regional pollution problems in many parts of the world have been significantly ameliorated in response to demands for a cleaner environment and through improving technologies.

Like many historical improvements in environmental quality, these positive trends are, up to now, limited in scale and only partly address the most pressing of the environmental problems. What makes this moment different from those in the past, however, is the *global* character of the problem – human activity *anywhere* affects *everywhere* through systemic change. Past change largely affected the condition of the environment at specific locations. Change now affects global-scale biogeochemical flows, threatening the functioning of the Earth System as a whole in ways not yet fully understood and leading to increasing chances of surprises (Schneider et al. 1998; NRC 2002).

It is important to understand the magnitude and rate of the evolving human enterprise and its impact on the Earth System. Figure 3.66 shows examples of changes in the Anthroposphere over the past few hundred years; it is an attempt to define a few key indicators that capture the changing nature of human societies at this pivotal time in the development of the human-environment relationship. All of the trends shown are global and mask important regional differences. Nevertheless, at the level of the Earth System, global-scale indicators are appropriate and important.

One feature stands out as remarkable. The second half of the twentieth century is unique in the entire history of human existence on Earth. Many human activities reached take-off points sometime in the twentieth century and have accelerated sharply towards the end of the century. The last 50 years have without doubt seen the most rapid transformation of the human relationship with the natural world in the history of humankind.

Figure 3.67 shows that the impacts of these accelerating human changes are now clearly discernible at the level of the Earth System as a whole. Many key indicators of the functioning of the Earth System are now showing responses that are, at least in part, driven by the changing human imprint on the planet. All components of the global environment – oceans, coastal zone,

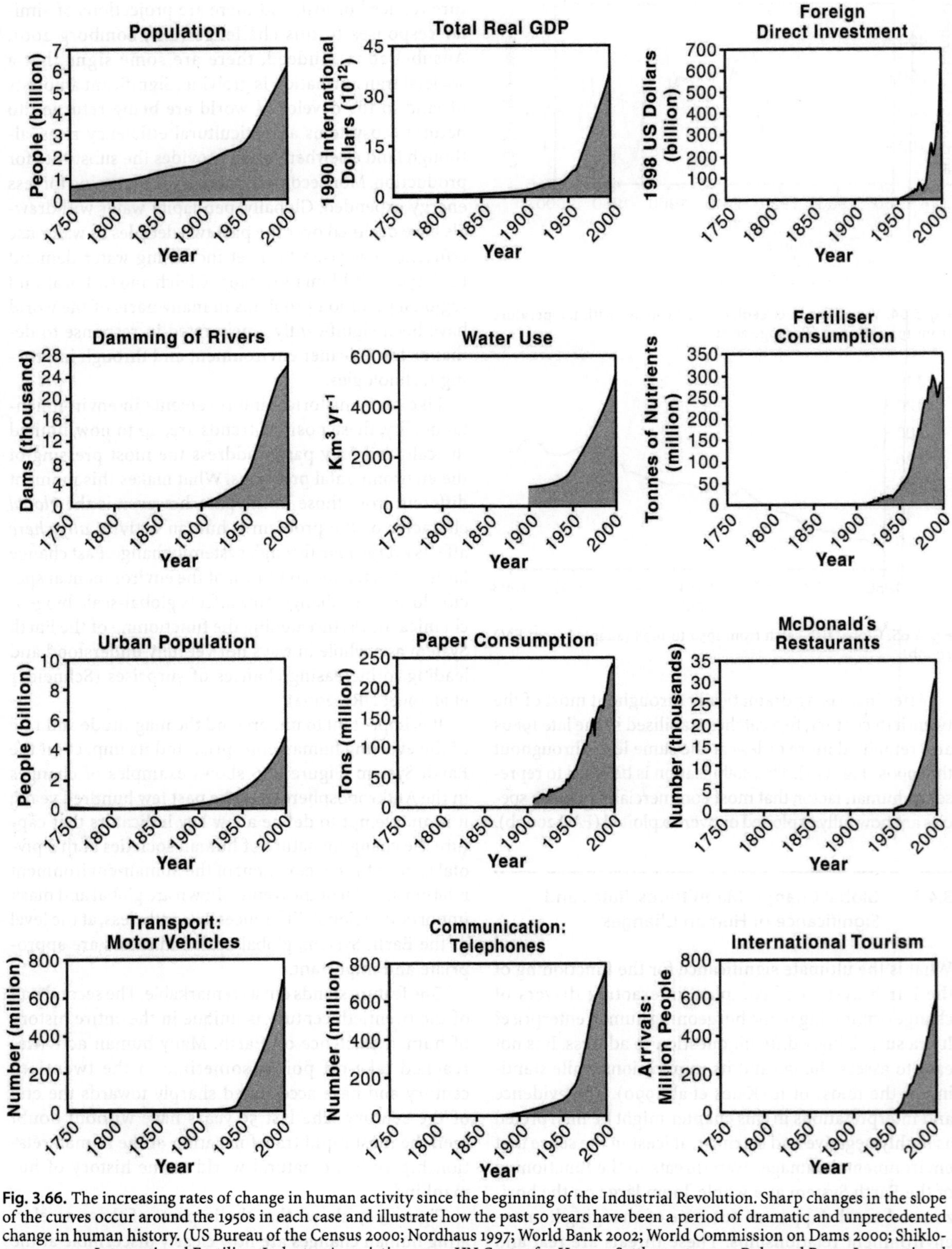

Fig. 3.66. The increasing rates of change in human activity since the beginning of the Industrial Revolution. Sharp changes in the slope of the curves occur around the 1950s in each case and illustrate how the past 50 years have been a period of dramatic and unprecedented change in human history. (US Bureau of the Census 2000; Nordhaus 1997; World Bank 2002; World Commission on Dams 2000; Shiklomanov 1990; International Fertilizer Industry Association 2002; UN Center for Human Settlements 2001; Pulp and Paper International 1993; McDonald's 2002; UNEP 2000; Canning 2001; World Tourism Organization 2001)

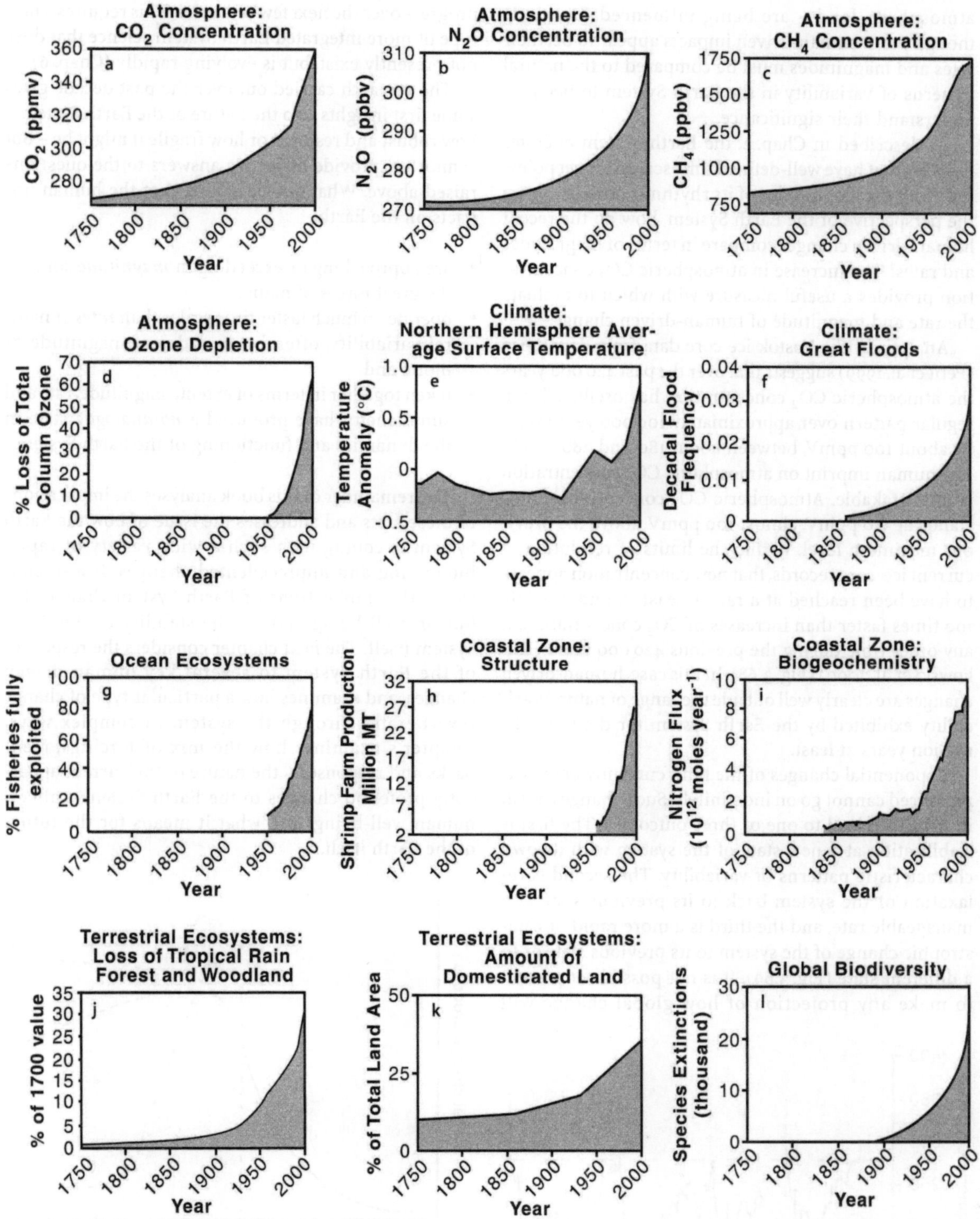

Fig. 3.67. Global-scale changes in the Earth System as a result of the dramatic increase in human activity: **a** atmospheric CO_2 concentration (Etheridge et al. 1996); **b** atmospheric N_2O concentration (Machida et al. 1995); **c** atmospheric CH_4 concentration (Blunier et al. 1993); **d** percentage total column ozone loss, using the average annual total column ozone, 330, as a base (*image:* J. D. Shanklin, British Antarctic Survey); **e** northern hemisphere average surface temperature anomalies (Mann et al. 1999); **f** decadal frequency of great floods (one-in-100-year events) after 1860 for basins larger than 200 000 km^2 with observations that span at least 30 years (Milly et al. 2002); **g** percentage of global fisheries either fully exploited, overfished or collapsed (FAOSTAT 2002); **h** annual shrimp production as a proxy for coastal zone alteration (WRI 2003; FAOSTAT 2002); **i** model-calculated partitioning of the human-induced nitrogen perturbation fluxes in the global coastal margin for the period since 1850 (Mackenzie et al. 2002); **j** loss of tropical rainforest and woodland, as estimated for tropical Africa, Latin America and South and Southeast Asia (Richards 1990; WRI 1990); **k** amount of land converted to pasture and cropland (Klein Goldewijk and Battjes 1997); and **l** mathematically calculated rate of extinction (based on Wilson 1992)

atmosphere, land – are being influenced. Dramatic though these human-driven impacts appear to be, their rates and magnitudes must be compared to the natural patterns of variability in the Earth System to begin to understand their significance.

As described in Chap. 2, the Earth System operates in cycles that have well-defined time scales and set points that limit the magnitudes of its rhythmic changes. From the perspective of the Earth System, how do the recent human-driven changes compare in terms of magnitudes and rates? The increase in atmospheric CO_2 concentration provides a useful measure with which to evaluate the rate and magnitude of human-driven change.

Analysis of the Vostok ice core data from Antarctica (Petit et al. 1999) suggests that over the past 420 000 years the atmospheric CO_2 concentration has oscillated in a regular pattern over approximately 100 000 year cycles by about 100 ppmV, between about 180 and 280 ppmV. The human imprint on atmospheric CO_2 concentration is unmistakable. Atmospheric CO_2 concentration now stands at 370 ppmV, almost 100 ppmV above the previous maximum level. Within the limits of resolution of current ice-core records, that new concentration appears to have been reached at a rate at least 10 and possibly 100 times faster than increases of CO_2 concentration at any other time during the previous 420 000 years (Falkowski et al. 2000) (Fig. 3.68). In this case, human-driven changes are clearly well outside the range of natural variability exhibited by the Earth System for the last half-million years at least.

Exponential changes of the kind currently being experienced cannot go on indefinitely. Such changes must, in principle, lead to one of three outcomes. The first is stabilisation at a new state of the system with its own characteristic patterns of variability. The second is relaxation of the system back to its previous state at a manageable rate, and the third is a more rapid or catastrophic change of the system to its previous state or to a different state (Fig. 3.69). It is not possible at present to make any projection of how global change will progress over the next few centuries. This requires a new type of more integrated Earth System science that does not presently exist, but is evolving rapidly (Chap. 6).

The research carried out over the past decade gives some first insights into the nature of the Earth System – how robust and resilient or how fragile it might be – but cannot yet provide definitive answers to the questions raised above. What can be said is that the human impacts on the Earth:

- are approaching or exceeding in *magnitude* some of the great forces of nature;
- operate on much faster time scales than *rates* of natural variability, often by an order of magnitude or more; and
- taken together in terms of extent, magnitude, rate and simultaneity, have produced a *no-analogue* state in the dynamics and functioning of the Earth System.

The remainder of this book analyses the implications of these facts and addresses the issue of how the Earth System is coping with such a wide variety of rapid, interacting and unprecedented changes. It also considers the implications of Earth System changes for human well-being and for the stability of the Earth System itself. The next chapter considers the responses of the Earth System to several key human-driven changes, and examines how a particular type of change reverberates through the system in complex ways. Chapter 5 examines how the mix of forcings, feedbacks and responses – the nature of the current apparently profound changes to the Earth System – affects human well-being, and what it means for the future of the Earth itself.

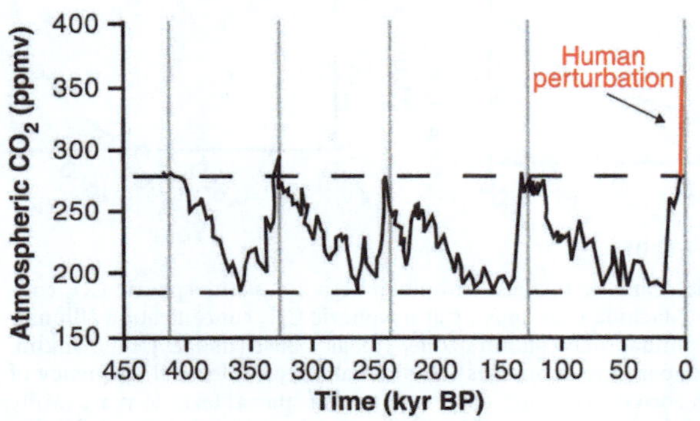

Fig. 3.68. Atmospheric CO_2 concentration over past 420 000 years from the Vostok ice core with the recent human perturbation superimposed (Petit et al. 1999; Keeling and Whorf 2000)

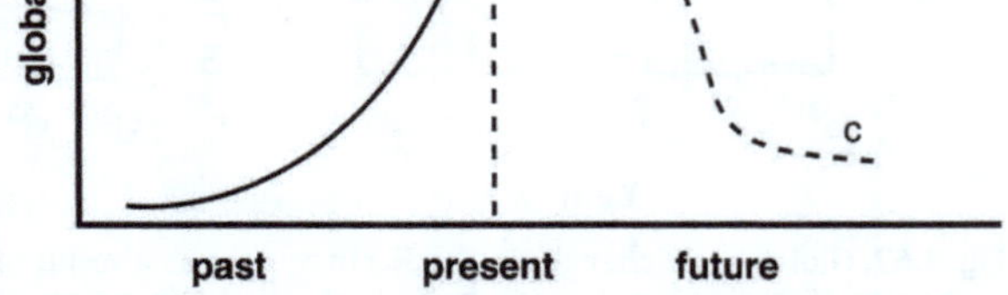

Fig. 3.69. Schematic diagram showing possible future trajectories of current exponential changes: **a** stabilisation at a new state of the system; **b** relaxation of the system to its previous state at a manageable rate; and **c** a rapid or catastrophic change of the system to its previous state or to a different state. Which trajectory will be followed for a given process depends on both human responses to global change and the perturbed dynamics of the Earth System

References

Achard F, Eva H, Glinni A, Mayaux P, Richards T, Stibig HJ (eds) (1998) Identification of deforestation hot spot areas in the humid tropics. Trees Publ. Series B. Research Report No. 4. Space Application Institute, Global Vegetation Monitoring Unit. Joint Research Centre, European Commission, Brussels

Achard F, Eva H, Stibig HJ, Mayaux P, Gallego J, Richards T (2002) Determination of deforestation rates of the world's humid tropical forests. Science 297:999–1002

Ackerman AS, Toon OB, Stevens DE, Heymsfield AJ, Ramanathan V, Welton EJ (2000) Reduction of tropical cloudiness by soot. Science 288:1042–1047

Adams RMc (1965) Land behind Bagdhad: History of settlement on the Diyala Plains. University of Chicago Press, Chicago

Alcamo J, Kreileman GJJ, Bollen JC, van den Born GJ, Gerlagh R, Krol MS, Toet AMC, de Vries HJM (1996) Baseline scenarios of global environmental change. Global Environ Chang 6(4): 261–303

Alroy J (2001) A multispecies overkill simulation of the End-Pleistocene megafaunal mass extinction. Science 292:1893–1896

Angel DP, Rock MT (2000) Asia's clean revolution industry, growth and the environment. Sheffield, Greenleaf

Angelsen A, Kaimowitz D (1999) Rethinking the causes of deforestation: Lessons from economic models. Word Bank Res Obser 14:73–98

Ausubel J (2002) Maglevs and the vision of St. Hubert. In: Steffen W, Jäger J, Carson D, Bradshaw C (eds) Challenges of a changing earth: Proceedings of the Global Change Open Science Conference. Amsterdam: The Netherlands, 10–13 July 2001. Springer-Verlag, Berlin Heidelberg New York

Azar C, Holmberg J, Karlsson S (2002) Decoupling – past trends and prospects for the future. A report for the Swedish Environmental Advisory Council. Ministry of the Environment, Stockholm, Sweden

Baillie J, Groombridge B (eds) (1996) 1996 IUCN red list of threatened animals. World Conservation Union, Gland, Switzerland

Barange M (2002) Influence of climate variability and change on the structure, dynamics and exploitation of marine ecosystems. In: Hester RE, Harrison RM (eds) Global environmental change. Royal Society of Chemistry, pp 57–82

Barraclough SL, Ghimire KB (1995) Forests and livelihoods: The social dynamics of deforestation in developing countries. St. Martins Press, New York

Baskin Y (2002) A plague of rats and rubber-vines: The growing threat of species invasions. A SCOPE-GISP Project, Island Press, Washington

Bergan T, Gallardo L, Rodhe H (1999) Mercury in the global troposphere: A three-dimensional model study. Atmos Environ 33: 1575–1585

Berry BJL (1990) Urbanization. In: Turner BL II, Clark WC, Kates RW, Richards JF, Mathews JT, Meyer WB (eds) The Earth as transformed by human action. Cambridge University Press, Cambridge

Blunier T, Chappellaz J, Schwander J, Barnola J-M, Desperts T, Stauffer B, Raynaud D (1993) Atmospheric methane record from a Greenland ice core over the last 1000 years. J Geophys Res 20:2219–2222

Borgström G (1967) The hungry planet. MacMillan, New York

Boutron CF, Gorlach U, Candelone J-P, Bolshov MA, Delmas RJ (1991) Decrease in anthropogenic lead, cadmium and zinc in Greenland snows since the late 1960s. Nature 353:153–156

Bradley RS (2000) Past global changes and their significance for the future. Quaternary Sci Rev 19:391–402

Brasseur GP, Artaxo P, Barrie LA, Delmas RJ, Galbally I, Hao WM, Harriss RC, Isaksen ISA, Jacob DJ, Kolb CE, Prather M, Rodhe H, Schwela D, Steffen W, Wuebbles DJ (2003) An integrated view of the causes and impacts of atmospheric changes. In: Brasseur GP, Prinn RG, Pszenny AAP (eds) The changing atmosphere: An integration and synthesis of a decade of tropospheric chemistry research. The International Global Atmospheric Chemistry Project (IGAC). The IGBP Global Change series. Springer-Verlag, Berlin Heidelberg New York

Brookfield HC, Potter L, Byron Y (1995) In place of the forest. Environmental and socio-economic tranformation in Borneo and the Eastern Malay Peninsula. UNU press, Tokyo

Brown C (1999) Global forest products outlook study: Thematic study on plantations. Working Paper No. GFPOS/WP/03 (draft). Food and Agriculture Organization of the United Nations, Rome

Brown L, Flavin C, French H, Abramovitz JN, Dunn S, Gardner G, Mastny L, Mattoon A, Roodman D, Sampat P, Sheehan MO, Starke L (2001) State of the world 2001. The Worldwatch Institute. Norton and Company, New York

Bryant D, Nielsen D, Tangley L (1997) Last frontier forests: Ecosystems and economies on the edge. World Resources Institute, 42 pp

Butzer K (1976) Early hydraulic civilization in Egypt: A study in cultural ecology. University of Chicago Press, Chicago

Caldwell B, McMichael T (2002) Cities: Are they good for health? International Human Dimensions Programme, Global Environmental Change Newsletter 3/2002:4–5

Candelone J-P, Hong S, Pellone C, Boutron C (1995) Post-industrial revolution changes in large scale atmospheric pollution of the northern hemisphere by heavy metals as documented in central Greenland snow and ice. J Geophys Res 100:16605–16616

Canning D (2001) World Bank: A database of world infrastructure stocks, 1950–95. World Bank, Washington DC

Chameides WL, Kasibhatla PS, Yienger J, Levy H II (1994) Growth in continental-scale metro-agroplexes: Regional ozone pollution and world food production. Science 264:74–77

Charlson RJ, Anderson TL, McDuff RE (2000) The sulfur cycle. In: Jacobson MC, Charlson RJ, Rodhe H, Orians GH (eds) Earth system science: From biogeochemical cycles to global change. Academic Press, London, pp 343–359

Chung CE, Ramanathan V, Kiehl JT (2002) Effects of the South-Asian absorbing haze on the Northeast monsoon and surface-air heat exchange. J Climate 15:2462–2476

Conrad R (1989) Control of methane production in terrestrial ecosystems. In: Andreae MO, Schimel DS (eds) Exchange of trace gases between terrestrial ecosystems and the atmosphere. John Wiley & Sons Ltd., New York, pp 39–58

Crosby AW (1986) Biological imperialism: The biological expansion of Europe, 900–1900. Cambridge University Press, New York

Cutter SL (2001) American hazardscapes: The regionalization of hazards and disasters. John Henry press, Washington DC

Darmstadter J (1971) Energy in the world economy: A statistical review of trends in output, trade, and consumption since 1925. John Hopkins Press, Baltimore

Dauvergne P (1997) Shadows in the forest: Japan and the politics of timber in Southeast Asia. MIT Press, Cambridge MA

DeFries RS, Field CB, Fung I, Collatz GJ, Bounuoa L (1999) Combining satellite data and biogeochemical models to estimate global effects of human-induced land cover change on carbon emissions and primary productivity. Global Biogeochem Cy 13:803–815

DeFries R, Houghton RA, Hansen M, Field C, Skole DL, Townshend J (2002) Carbon emissions from tropical deforestation and regrowth based on satellite observations for the 1980s and 90s. P Natl Acad Sci USA 99(22):14256–14261

Denevan DW (1995) Prehistoric agricultural methods as models for sustainability. Advanced Plant Pathology II:21–43

Denevan WM (1992) The pristine myth: The landscape of the Americas in 1492. Ann Assoc Am Geogr 82:426–443

Deutsch L, Jansson Å, Troell M, Rönnbäck P, Folke C, Kautsky N (2000) The 'ecological footprint': Communicating human dependence on nature's work. Ecol Econ 32:351–355

Dhakal S, Kaneko S, Imura H (2002) CO_2 emissions from energy use in Asian mega-cities: An analysis on driving factors for Tokyo and Seoul. Submitted for publication to Environmental Systems Research Volume 30, Japan Society of Civil Engineers, Japan

Dicken P (1992) Global shift: The internationalization of economic activity, 2nd ed. Guilford Press, New York

Dise NB (1993) Methane emissions from Minnesota peatlands: Spatial and seasonal variability. Global Biogeochem Cy 7:371–384

Dixon RK, Brown S, Houghton RA, Solomon AM, Trexler MC, Wisniewski J (1994) Carbon pools and flux of global forest ecosystems. Science 263:185–190

Dynesius M, Nilsson C (1994) Fragmentation and flow regulation of river systems in the northern third of the world. Science 266:753–762

Edmunds WM (1999) Groundwater: A renewable resource? Focus in Sahara and Sahel. Brochure for the GASPAL (Groundwater as Palaeoindicator) project (EC ENRICH, ENV4-CT97-0591); available from the author

Ehrlich PR, Holdren JP (1971) Impact of population growth. Science 171:1212–1217

Eldredge N (1998) Life in the balance: Humanity and the biodiversity crisis. Princeton University Press, Princeton New Jersey

Etemad B, Bairoch P, Luciani J, Toutain J-C (1991) World energy production 1800–1985. Libraire Droz, Geneve, pp 272

Etheridge DM, Steele LP, Langenfelds RL, Francey RJ, Barnola J-M, Morgan VI (1996) Natural and anthropogenic changes in atmospheric CO_2 over the last 1000 years from air in Antarctic ice and firn. J Geophys Res 101:4115–4128

Etheridge DM, Pearman GI, Fraser PJ (1992) Changes in tropospheric methane between 1841 and 1978 from a high accumulation-rate Antarctic ice core. Tellus 44B: 282–294

Falkowski P, Scholes RJ, Boyle E, Canadell J, Canfield D, Elser J, Gruber N, Hibbard K, Högberg P, Linder S, Mackenzie FT, Moore B III, Pedersen T, Rosenthal Y, Seitzinger S, Smetacek V, Steffen W (2000) The global carbon cycle: A test of knowledge of Earth as a system. Science 290: 291–296

Folke C, Jansson Å, Larsson J, Costanza R (1997) Ecosystem appropriation by cities. Ambio 26:167–172

FAO (1993) Marine fisheries and the law of the sea: A decade of change. Food and Agricultural Organization of the United Nations, Fisheries Circular No. 853, Rome

FAO (1995) Forest resource assessment 1990: Global synthesis. Food and Agriculture Organization of the United Nations, Rome

FAO (1999) The state of the world's forest 1999. Food and Agricultural Organization of the United Nations, Rome

FAO (2000a) The state of food and agriculture 2000. Food and Agricultural Organization of the United Nations, Rome. *http://www.fao.org/decrep/x4400e/x4400e09.htm*, 31 July 2002

FAO (2000b) The state of world fisheries and aquaculture. Food and Agricultural Organization of the United Nations, Rome. *http://www.fao.org/sof/sofia/index_en.htm*, 30 April 2002

FAO (2000c) Global forest resources assessment 2000. Food and Agriculture Organization of the United Nations, Rome

FAO (2001) State of the world's forest 2001. Food and Agriculture Organization of the United Nations, Rome. *http://www.fao.org/forestry/fo/sofo/sofo-e.stm*, 12 August 2002

FAOSTAT (2002) Statistical databases. Food and Agriculture Organization of the United Nations, Rome. *http://www.apps.fao.org*, 12.08.2002

Fu C, Harasawa H, Kasyanov V, Kim J-W, Ojima D, Wan Z, Zhao S (2002) Regional-global interactions in East Asia. In: Tyson P, Fuchs R, Fu C, Lebel L, Mitra AP, Odada E, Perry J, Steffen W, Virji H (eds) Global-regional linkages in the earth system. IGBP Global Change Series, Springer-Verlag, Berlin Heidelberg New York, pp 109–149

Galloway JN, Cowling EB (2002) Reactive nitrogen and the world: Two hundred years of change. Ambio 31: 64–71

Galloway JN, Aber JD, Erisman JW, Seitzinger SP, Howarth RH, Cowling EB and Cosby BJ (2003) The nitrogen cascade. Bioscience, in press

Galloway J, Levy H, Kasibhatia P (1994) Consequences of population growth and development on deposition of oxidized nitrogen, year 2020. Ambio 23:120–123

Galloway JN, Schlesinger WH, Levy H II, Michaels A, Schnoor JL (1995) Nitrogen fixation: Anthropogenic enhancement-environmental response. Global Biogeochem Cy 9(2):235–252

Geist HJ, Lambin EF (2002) Proximate causes and underlying forces of tropical deforestation. Bioscience 52(2):143–150

Gleich PH (1998) The world's water: The biennial report on freshwater resources. Island Press, Washington DC

Gordon I (2000) Extent and impacts of dryland salinity in Queensland. National Land and Water Resources Audit Dryland Salinity Project Report, Canberra

Granberg G, Mikkelä C, Sundh I, Svensson BH, Nilsson M (1997) Sources of spatial variation in methane emission from mires in northern Sweden: A mechanistic approach in statistical modelling. Global Biogeochem Cy 11:135–150

Grübler A (1998) Technology and global change. Cambridge University Press, Cambridge

Gubler DJ (1994) Perspectives on the prevention and control of dengue haemorrhagic fever. Kaohsiung J Med Sci 10:15–18

Haan D, Martinerie P, Reynaud D (1996) Ice core data of carbon monoxide over Antarctica and Greenland during the last 200 years. Geophys Res Lett 23:2235–2238

Hansen JE (2002) Proceedings of 'Air Pollution as a Climate Forcing: a Workshop'. *http://www.giss.nasa.gov/meetings/pollution02/*, 22 May 2003

Hansen M, DeFries R (2003) Detecting long term forest change using continuous fields of tree cover maps from 8 km AVHRR data for the years 1982–1999. Ecosystems, in press

Hansen MC, DeFries RS, Townshend JRG, Sohlberg R, DiMiceli C, Carroll M (2002) Towards an operational MODIS continuous field of percent tree cover algorithm: Examples using AVHRR and MODIS data. Remote Sens Environ 83(1–2):303–319

Hauglustaine DA, Brasseur GP (2001) Evolution of tropospheric ozone under anthropogenic activities and associated radiative forcing of climate. J Geophys Res 106:32337–32360

Heintzenberg J, Raes F, Schwartz SE, Ackermann I, Artaxo P, Bates TS, Benkovitz C, Bigg K, Bond T, Brenguier J-L, Eisele FL, Feichter J, Flossman AI, Fuzzi S, Graf H-F, Hales JM, Herrmann H, Hoffman T, Huebert B, Husar RB, Jaenicke R, Kärcher B, Kaufman Y, Kent GS, Kulmala M, Leck C, Liousse C, Lohmann U, Marticorena B, McMurry P, Noone K, O'Dowd C, Penner JE, Pszenny A, Putaud J-P, Quinn PK, Schurath U, Seinfeld JH, Sievering H, Snider J, Sokolik I, Stratmann F, van Dingenen R, Westphal D, Wexler AS, Wiedensohler A, Winker DM, Wilson J (2003) Tropospheric aerosols. In: Brasseur GP, Prinn RG, Pszenny AAP (eds) The changing atmosphere: An integration and synthesis of a decade of tropospheric chemistry research. The International Global Atmospheric Chemistry Project (IGAC). IGBP Global Change Series, Springer-Verlag, Berlin Heidelberg New York

Hong S, Candelone JP, Patterson CC, Boutron CF (1994) Greenland ice evidence of hemispheric lead pollination two millennia ago by Greek and Roman civilization. Science 265:1841–1843

Holdren JP, Ehrlich PR (1974) Human population and the global environment. Am Sci 62:282–292

Houghton FA, Skole DL, Nobre CA, Hackler JL, Lawrence KT, Chomentowski WH (2000) Annual fluxes of carbon from deforestation and regrowth in the Brazilian Amazon. Nature 403:301–304

Houghton RA, Skole DL (1990) Carbon. In: Turner BL II, Clark WC, Kates RW, Richards JF, Mathews JT, Meyer WB (eds) The Earth as transformed by human action: Global and regional changes in the biosphere over the past 300 years. Cambridge University Press, Cambridge New York, pp 393–408

Howarth RW, Billen G, Swaney D, Townsend A, Jaworski N, Lajtha K, Downing JA, Elmgren R, Caraco N, Jordan T, Berendse F, Freney J, Kudeyarov V, Murdoch P, Zhao-Liang Z (1996) Regional nitrogen budgets and riverine N and P fluxes for the drainages to the North Atlantic Ocean: Natural and human influences. In Howarth RW (ed) Nitrogen cycling in the North Atlantic Ocean and its watersheds. Kluwer, Dordrecht 304 pp, reprinted from Biogeochemistry 35(1)

IHDP (2002) IHDP update 03/2002. International Human Dimensions Programme on Global Environmental Change, Bonn

Immirzi CP, Maltby E (1992) The global status of peatlands and their role in carbon cycling. Friends of the Earth, London

Insenberg AC (2000) The destruction of the bison: An environmental history, 1750–1920. Cambridge University Press, Cambridge

IPCC (2001) Climate change 2001: The scientific basis. Contribution of Working Group I to the Third Assessment Report of the Intergovernmental Panel on Climate Change. Houghton JT, Ding Y, Griggs DJ, Noguer M, van der Linden PJ, Dai X, Maskell K, Johnson CA (eds) Cambridge University Press, Cambridge New York

IEA (1998) International energy agency database beyond 20/20. Paris

International Fertilizer Industry Association (2002) Fertilizer indicators. *http://www.fertilizer.org/ifa/statistics/indicators/ind_cn_world.asp*, 25 Oct 2002

Jacobson AD, DeOliveira JAA, Barange M, Cisneros-Mata MA, Felix-Uraga R, Hunter JR, Kim JY, Matsuura Y, Niquen M, Porteiro C, Rothschild B, Sanchez RP, Serra R, Uriarte A, Wada T (2001) Surplus production, variability, and climate change in the great sardine and anchovy fisheries. Can J Fish Aquat Sci 58:1891–1903

Jaffe DA (2000) The nitrogen cycle. In: Jacobson MC, Charlson RJ, Rodhe H, Orians GH (eds) Earth system science: From biogeochemical cycles to global change. Academic Press, London, pp 322–342
Jahnke RA (2000) The phosphorus cycle. In: Jacobson MC, Charlson RJ, Rodhe H, Orians GH (eds) Earth system science: From biogeochemical cycles to global change. Academic Press, London, pp 360–376
Jansson Å, Zucchetto J (1978) Man, nature and energy flow on the island of Gotland. Ambio 7:140–149
Jansson Å, Folke C, Rockström J, Gordon L (1999) Linking freshwater flows and ecosystem services appropriated by people: The case of the Baltic Sea drainage basin. Ecosystems 2:351–366
Jaworski NA, Howarth RW, Hetling LJ (1997) Atmospheric deposition of nitrogen oxides onto the landscape contributes to coastal eutrophication in the northeast United States. Environ Sci Technol 31:1995–2004
Jayaraman A, Lubin D, Ramachandran S, Ramanathan V, Woodbridge E, Collins WD, Zalpuri KS (1998) Direct observations of aerosol radiative forcing over the tropical Indian Ocean during the Jan-Feb 1996 Pre-INDOEX cruise. J Geophys Res 103:13827–13836
Jeppson P, Jarvine JK, MacKinnon K, Monk KA (2001) The end of Indonesia's lowland forests? Science 292:859–861
Kasperson JX, Kasperson RE, Turner BL II (1995) Regions at risk: Comparisons of threatened environments. United Nations University, Tokyo
Kates RW, Turner BL II, Clark WC (1990) The great transformation. In: Turner BL II, Clark WC, Kates RW, Richards JF, Mathews JT, Meyer WB (eds) The Earth as transformed by human action. Cambridge University Press, Cambridge
Keeling CD, Whorf TP (2000) Atmospheric CO_2 records from sites in the SIO air sampling network. In: Trends: A Compendium of Data on Global Change. Carbon Dioxide Information Analysis Center, Oak Ridge National Laboratory, US Department of Energy, Oak Ridge, TN, USA
Kelleher G, Bleakley C, Wells S (1995). A global representative system of marine protected areas. Volume 1. World Bank, Washington, DC
King VT (1998) Environmental challenges in Southeast Asia. Curzon Press, Surrey
Kirschbaum MUF (1995) The temperature dependence of soil organic matter decomposition, and the effect of global warming on soil organic C storage. Soil Biol Biochem 27(6):753–760
Klein Goldewijk K (2001) Estimating global land use change over the past 300 years: The HYDE database. Global Biogeochem Cy 15:417–434
Klein Goldewijk K, Battjes JJ (1997) One hundred year database for integrated environmental assessments. National Institute for Public Health and the Environment (RIVM), Bilthoven, Netherlands
Klyashtorin LB (1998) Long-term climate change and main commercial fish production in the Atlantic and Pacific. Fish Res 37:115–125
Knight M (1998) Developing countries and the globalization of financial markets. World Dev 26:1185
Knowles TR, Moore R (1989) The influence of water table levels on methane and carbon dioxide levels from peatland soils. Can J Soil Sci 69(1):33–38
Kremer H, Crossland C (2002) Coastal change and the "Anthropocene": Past and future perspectives of the IGBP-LOICZ project. In: Deutsches Nationalkomitee für das Internationale Hydrologische Programm (IHP) der UNESCO und das Operationelle Hydrologische Programm (OHP) der WMO, Herausgeber, Koblenz, Germany. IHP/OHP Berichte, Sonderheft 13
Krishnamurti TN, Jha B, Prospero J, Jayaraman A, Ramanathan V (1998) Aerosol and pollutant transport and its impact on radiative forcing over the tropical Indian Ocean during the Jan.-Feb. 1996 Pre-INDOEX cruise. Tellus 50B(5):521–542
Kroeze C, Mosier AR, Bouwman L (1999) Closing the global N_2O budget: A retrospective analysis 1500–1994. Global Biogeochem Cy 13:1–8
Kummer DM (1992) Deforestation in the postwar Philippines. Geography Research Paper No. 234, University of Chicago Press, Chicago
Laine J, Minkkinen K (1996) Effect of forest drainage on the carbon balance of a mire: A case study. Scand J Forest Res 11:307–312
Laine J, Vanha-Majamaa I (1992) Vegetation ecology along a trophic gradient on drained pine mires in southern Finland. Ann Bot Fenn 29:213–233
Lambin EF, Turner BL II, Geist H, Agbola S, Angelsen A, Bruce JW, Coomes O, Dirzo R, Fischer G, Folke C, George PS, Homewood K, Imbernon J, Leemans R, Li X, Moran EF, Mortimore M, Ramakrishnan PS, Richards JF, Skånes H, Steffen W, Stone GD, Svedin U, Veldkamp T, Vogel C, Xu J (2001) The causes of land-use and -cover change: Moving beyond the myths. Global Environ Chang 11(4)5–13
Laureti E (1999) Fish and fishery products: World and apparent consumption statistics based on food balance sheets. FAO Fisheries Circular No. 821, Revision 5. Food and Agriculture Organization of the United Nations, Rome
Lavanchy VMH, Gäggeler HW, Schotterer U, Schwikowski M, Baltensperger U (1999) Historical record of carbonaceous particle concentrations from a European high-alpine glacier (Colle Gnifetti, Switzerland). J Geophys Res B:21227–21236
Lawton JH, May RM (1995) Extinction rates. Oxford University Press, Oxford, UK
Lebel L (2000) Global change and development in Southeast Asia. Southeast Asia Regional Committee for START, Bangkok
Lee JA, Tallis JH (1973) Regional and historical aspects of lead pollution in Britain. Nature 245:216–218
Legrand M, Hammer C, Deangelis M, Savarino J, Delmas R, Clausen H, Johnsen SJ (1997) Sulfur-containing species (methanesulfonate and SO_4) over the last climatic cycle in the Greenland Ice Core Project (central Greenland) ice core. J Geophys Res 102(C12): 26663–26679
Lelieveld J, Crutzen PJ, Ramanathan V, Andreae MO, Brenninkmeijer CAM, Campos T, Cass GR, Dickerson RR, Fischer H, de Gouw JA, Hansel A, Jefferson A, Kley D, de Laat ATJ, Lal S, Lawrence MG, Lobert JM, Mayol-Bracero OL, Mitra AP, Novakov T, Oltmans SJ, Prather KA, Reiner T, Rodhe H, Scheeren HA, Sikka D, Williams J (2001) The Indian Ocean experiment: Widespread air pollution from South and Southeast Asia. Science 291:1031–1036
Lepers E, Lambin E, DeFries R, Janetos A, et al. (2003) Areas of rapid land-cover change of the world. MEA Rep., Millenium Ecosystem Assessment, Penang, Malaysia
Levine SA (1989) Analysis of risk for invasions and control programmes. In: Drake JA, Mooney HA, de Castri F, Groves RH, Kruger FJ, Rejmanek M, Williamson M (eds) Biological invasions: A global perspective. Scope 37. John Wiley and Sons
Li F, Ramanathan V (2002) Winter to summer monsoon variation of aerosol optical depth over the tropical Indian Ocean. J Geophys Res 107(D16):10.1029/2001JD000949
Lim LYC (1995) Southeast Asia: Success through international openness. In: Stallings B (ed) Global change, regional response. Cambridge University Press, Cambridge, pp 238–271
Lindqvist O, Johansson K, Aastrup M, Anderson A, Bringmark L, Hovsenius G, Håkanson L, Iverfeldt Å, Meili M, Timm B (1991) Mercury in the Swedish Environment: Recent research on causes, consequences and corrective measures. Water Air Soil Poll 55:1–261
Livett EA, Lee JA, Tallis JH (1979) Lead, zinc and copper analyses of British blanket peats. J Ecol 67:865–891
LOICZ (2002) Coastal typology project of the IGBP/LOICZ project. Land-Ocean Interactions in the Coastal Zone International Project Office, Texel, Netherlands, *http://www.nioz.nl/loicz/res.htm*
Lomborg B (2001) The skeptical environmentalist: Measuring the real state of the world. Cambridge University Press, Cambridge
Loveland TR, Belward AS (1997) The IGBP-DIS global 1-km land cover data set, DISCover: First results. Int J Remote Sens 18:3289–3295
Loveland TR, Reed BC, Brown JF, Ohlen DO, Zhu Z, Yang L, Merchant JW (2000) Development of a global land cover characteristics database and IGBP DISCover from 1-km AVHRR data. Int J Remote Sens 21:1303–1330
Lüdeke MKB, Moldenhauer O, Petschel-Held G (1999) Rural poverty driven soil degradation under climate change: The sensitivity of the disposition towards the Sahel syndrome with respect to climate. Environ Model Assess 4(4):315–326

Mabogunje AL (2002) Poverty and environmental degradation: Challenges within the global economy. Environment 44(1):8–30
Machida T, Nakazawa T, Fujii Y, Aoki S, Watanabe O (1995) Increase in the atmospheric nitrous oxide concentration during the last 250 years. Geophys Res Lett 22:2921–2924
Mackenzie FT, Ver LM, Lerman A (2002) Century-scale nitrogen and phosphorus controls of the carbon cycle. Chem Geol 190:13–32
Malingreau JP, Achard F, D'souza G, Stibig HJ, D'souza J, Estreguil C, Eva H (1995) AVHRR for global tropical forest monitoring: The lessons of the TREES project. Remote Sensing Reviews 12: 29–40
Mann ME, Bradley RS, Hughes MK (1999) Northern hemisphere temperatures during the past millennium: Inferences, uncertainties, and limitations. Geophys Res Lett 26(6):759–762
Marland G, Boden TA, Andres RJ (2002) Global, regional, and national CO_2 emissions. In: Trends: A compendium of data on global change. Carbon Dioxide Information Analysis Center, Oak Ridge National Laboratory, US Department of Energy, Oak Ridge, TN, USA
Marsh GP (1965) Man and nature; Or, the Earth as modified by human action (orig. 1864). Belknap Press of Harvard University Press, Cambridge, MA
Martin PS, Klein RG (1984) Quaternary extinctions: A prehistoric revolution. University of Arizona Press, Tucson
Martínez-Cortizas A, Pontevedra-Pombal X, Novoa-Munoz JC, Garcia-Rodeja E (1997) Four thousand years of atmospheric Pb, Cd and Zn deposition recorded by the ombrotrophic peat bog of Penido Vello (northwestern Spain). Water Air Soil Poll 100:387–403
Martínez-Cortizas A, Pontevedra-Pombal X, García-Rodeja E, Nóvoa-Muñoz JD, Shotyk W (1999) Mercury in a Spanish peat bog: Archive of climate change and atmospheric metal deposition. Science 284:939–942
Mas Caussel J (1996) Estimación preliminar de las tasas de deforestación en el estado de Campeche. Jaina 7:5–6
Matos G, Wagner L (1998) Consumption of materials in the United States, 1900–1995. Annu Rev Energ Env 23:107–122
Matson PA, Parton WJ, Power AG, Swift MJ (1997) Agricultural intensification and ecosystem properties. Science 277:504–509
Matthews E (2001) Understanding the FRA 2000: Forest briefing No. 1. World Resources Institute, Washington DC
Matthews E, Payne R, Rohweder M, Murray S (2000) Pilot analysis of global ecosystem. Forest ecosystems. World Resources Institute. Washington DC, *http://www.wri.org/wr2000*, 12 August 2002
Mayewski PA, Lyons WB, Spencer MJ, Twickler MS, Buck CF, Whitlow SI (1990) An ice core record of atmospheric response to anthropogenic sulphate and nitrate. Nature 346:554–556
McCully P (1996) Silenced rivers: The ecology and politics of large dams. Zed Books, London
McDonald's (2002) Homepage. *http://www.mcdonalds.com*, 28 Oct. 2002
McNeill JR (1992) The mountains of the Mediterranean world: An environmental history. Cambridge University Press, Cambridge
Menon S, Hansen J, Nazarenko L, Luo Y (2002) Climate effects of black carbon aerosols in China and India. Science 297:2250–2253
Meybeck M, Vörösmarty CJ (2004) The integrity of river and drainage basin systems: challenges from environmental change. In: Kabat P, Claussen M, Dirmeyer PA, Gash JHC, de Guenni LB, Meybeck M, Pielke RA Sr., Vörösmarty C, Hutjes RWA, Luetkemeier S (eds) Vegetation, water, humans and the climate: A new perspective on an interactive system; Part D. IGBP Global Change Series. Springer-Verlag, Berlin Heidelberg New York, pp 297–480
Meyer WB, Turner BL II (1992) Human population growth and global land-use/cover change. Annu Rev Ecol Syst 23:39–61
Meyer WB, Turner BL II (eds) (1994) Changes in land use and land cover: A global perspective. Cambridge University Press, Cambridge
Milly PCD, Wetherald RT, Dunne KA, Delworth TL (2002) Increasing risk of great floods in a changing climate. Nature 415: 514–517
Moore T, Roulet N, Knowles R (1990) Spatial and temporal variations of methane flux from subarctic/northern boreal fens. Global Biogeochem Cy 4(1):29–46
Moore TR, Roulet NT (1993) Methane flux: Water table relations in northern wetlands. Geophys Res Lett 20:587–590
Murdiyarso D, Lebel L, Gintings AN, Tampubolon SMH, Heil A, Wasson M (2003) Policy responses to complex environmental problems: Insights from a science-policy activity on transboundary haze from vegetation fires in Southeast Asia. Agr Ecosyst Environ, in press
Myers N (1997) Mass extinction and evolution. Science 278:597–598
Naylor RL, Goldburg RJ, Mooney H, Beveridge M, Clay J, Folke N, Kautsky N, Lubcheno J, Primavera J, Williams M (2000) Nature's subsidies to shrimp and salmon farming. Science 282:883–884
Neue HU, Sass RL (1994) Rice cultivation and trace gas exchange. In: Prinn RG (ed) Global atmospheric-biospheric chemistry. Plenum Press, New York, pp 119–147
Nordhaus W (1997) Do real wage and output series capture reality? The history of lighting suggests not. In: Bresnahan T, Gordon R (eds) The economics of new goods. University of Chicago Press, Chicago
NRC (2002) Abrupt climate change: Inevitable surprises. National Research Council, National Academy Press, Washington DC
NRC (1999) Our common journey: A transition toward sustainability. Board on Sustainable Development, National Research Council, National Academy Press, Washington DC
Nriagu JO (1996) A history of global metal pollution. Science 272:223–224
Nriagu JO (1979) Global inventory of natural and anthropogenic emissions of trace metals to the atmosphere. Nature 279:409–411
Nriagu JO (ed) (1984) Changing metal cycles and human health. Springer-Verlag, Berlin Heidelberg New York
Nriagu JO (1990) Global metal pollution. Poisoning the biosphere. Environment 32:7–33
Odén S (1968) The acidification of air and precipitation and its consequences in the natural environment. Swedish National Science Research Council, Stockholm
Odum EP (1975) Ecology: The link between the natural and the social sciences, second edition. Holt-Saunders, Japan
Odum EP (1989) Ecology and our endangered life-support systems. Sinauer, Sunderland, Massachusetts
Oldfield F, Dearing JA (2003) The role of human activities in past environmental change. In: Alverson K, Bradley R, Pedersen T (eds) Paleoclimate, global change and the future. IGBP Global Change Series, Springer-Verlag, Berlin Heidelberg New York
OECD (2001) Environmental outlook for the chemicals industry. Organisation for Economic Co-operation and Development. Available at *http://www.oecd.org/ehs*, 22 May 2003
OECD/IEA (2000) Key world energy statistics from the IEA. Organisation for Economic Co-operation and Development / International Energy Agency. *http://www.iea.org/statist/keyworld/keystats.htm*, 14 May 2003
OTA (1993) Harmful non-indigenous species in the United States. U.S. Congress, Office of Technology Assessment, U.S. Government Printing Office, Washington DC
Paerl HW, Whitall DR (1999) Anthropogenically-derived atmospheric nitrogen deposition, marine eutrophication and harmful algal bloom expansion: Is there a link? Ambio 28:307–311
Parikh J (2002) Consumption patterns: Economic and demographic change. In: Munn T (ed) Encyclopedia of global environmental change, vol. 3. John Wiley & Sons Ltd., pp 249–252
Parnwell MJG, Bryant RL (eds) (1996) Environmental change in South-east Asia: People, politics and sustainable development. Routledge, London
Pauly D, Christensen V (1995) Primary production required to sustain global fisheries. Nature 374:255–257
Petit JR, Jouzel J, Raynaud D, Barkov NI, Barnola J-M, Basile I, Bender M, Chappellaz J, Davis M, Delaygue G, Delmotte M, Kotlyakov VM, Legrand M, Lipenkov VY, Lorius C, Pépin L, Ritz C, Saltzman E, Stievenard M (1999) Climate and atmospheric history of the past 420 000 years from the Vostok ice core, Antarctica. Nature 399:429–436
Petschel-Held G (2001) Actors and their environment – Syndromes of land-use change in developing countries. International Geosphere Biosphere Programme. Global Change NewsLetter 48:27, Stockholm, Sweden
Petschel-Held G, Schellnhuber HJ, Bruckner T, Toth F, Hasselmann K (1999) The tolerable windows approach: Theoretical and methodological foundations. Climatic Change 41(3–4):303–331

Pimm SL, Russell GJ, Gittleman JL, Brooks TM (1995) The future of biodiversity. Science 269:347–350
Pitman NCA, Jorgensen PM (2002) Estimating the size of the World's threatened flora. Science 298:989
Podgorny IA, Ramanathan V (2001) A modeling study of the direct effect of aerosols over the tropical Indian Ocean. J Geophys Res 106:24097–24105
Post WM (ed) (1990) Report of a workshop on climate feedbacks and the role of peatlands, tundra, and boreal ecosystems in the global carbon cycle. ORNL/TM-11457, Oak Ridge National Laboratory, Oak Ridge, TN
Prather M, Derwent R, Ehhalt D, Fraser P, Sanhueza E, Zhou X (1995) Other trace gases and atmospheric chemistry. In: Houghton JT, Meira Filho LG, Bruce J, Lee H, Callander BA, Haites E, Harris N, Maskell K (eds) Climate change 1994: Radiative forcing of climate change and an evaluation of the IPCC IS92 emission scenarios. Cambridge University Press, Cambridge New York, pp 73–126
Prentice IC, Farquhar GD, Fasham MJR, Goulden ML, Heimann M, Jaramillo VJ, Khesghi HS, Le Quéré C, Scholes RJ, Wallace DWR (2001) The carbon cycle and atmospheric carbon dioxide. In: Houghton JT, Ding Y, Griggs DJ, Noguer M, van der Linden PJ, Dai X, Maskell K, Johnson CA (eds) Climate change 2001: The scientific basis. Contribution of Working Group I to the Third Assessment Report of the Intergovernmental Panel on Climate Change. Cambridge University Press, Cambridge New York
Pulp and Paper International (1993) PPI's international fact and price book. In: FAO forest product yearbook 1960–1991. Food and Agriculture Organization of the United Nations, Rome
Pyne SJ (1991) Burning bush: A fire history of Australia. New York, Holt
Rabalais N (2002) Nitrogen in aquatic ecosystems. Ambio 31:102–112
Rajeev K, Ramanathan V, Meywerk J (2000) Regional aerosol distribution and its long-range transport over the Indian Ocean. J Geophys Res 105:2029–2043
Ramachandran S, Jayaraman A (2002) Pre-monsoon aerosol loadings and size distributions over the Arabian Sea and the Tropical Indian Ocean. J Geophys Res 107(D24):4738,doi:10.1029/2002JD002386
Ramanathan V, Crutzen PJ (2001) *www.asianbrowncloud.ucsd.edu/ABCconceptFinal23May01.pdf*
Ramanathan V, Crutzen PJ, Lelieveld J, Mitra AP, Althausen D, Anderson J, Andreae MO, Cantrell W, Cass GR, Chung CE, Clarke AD, Coakley JA, Collins WD, Conant WC, Dulac F, Heintzenberg J, Heymsfield AJ, Holben B, Howell S, Hudson J, Jayaraman A, Kiehl JT, Krishnamurti TN, Lubin D, McFarquhar G, Novakov T, Ogren JA, Podgorny IA, Prather K, Priestley K, Prospero JM, Quinn PK, Rajeev K, Rasch P, Rupert S, Sadourny R, Satheesh SK, Shaw GE, Sheridan P, Valero FPJ (2001a) The Indian Ocean experiment: An integrated assessment of the climate forcing and effects of the great Indo-Asian haze. J Geophys Res 106:28371–28398
Ramanathan V, Crutzen PJ, Kiehl JT, Rosenfeld D (2001b) Aerosols, climate and the hydrological cycle. Science 294:2119–2124
Ramankutty N, Foley JA (1999) Estimating historical changes in global land cover: croplands from 1700 to 1992. Global Biogeochem Cy 13:997–1027
Ramankutty N, Foley JA (1998) Characterising patterns of global land use: An analysis of global cropland data. Global Biogeochem Cy 12:667–685
Raskin P, Chadwick M, Jackson T, Leach G (1996) The sustainability transition: Beyond conventional development. Polestar Series 1. Stockholm Environment Institute, Stockholm
Raskin PD (1995) Methods for estimating the population contribution to environmental change. Ecol Econ 15:225–233
Redman CL (1999) Human impact on ancient environments. University of Arizona Press, Tucson
Reeburgh WS (1997) Figures summarizing the global cycles of biogeochemically important elements. Bulletin of Ecological Society of America 78(4):260–267, *http://www.ess.uci.edu/~~reeburgh*, 23 Oct 2002
Reid WVC, Miller KR (1989) The scientific basis for the conservation of biodiversity. World Resources Institute, Washington DC
Rejmanek M, Randall JR (1994) Invasive alien plants in California: 1993 summary and comparison with other areas in North America. Madroño 41(3):161–177
Renberg I, Wik-Persson M, Emteryd O (1994) Pre-industrial atmospheric lead contamination detected in Swedish lake sediments. Nature 368:323–326
Renberg I, Brannvall M-L, Bindler R, Emteryd O (2000) Atmospheric lead pollution history during four millennia (2000 BC to AD 2000) in Sweden. Ambio 29(3):150–156
Revenga C, Brunner J, Henninger N, Kassem K, Payne R (2000) Pilot analysis of global ecosystems: Freshwater systems. World Resources Institute, Washington DC
Richards J (1990) Land transformation. In: Turner BL II, Clark WC, Kates RW, Richards JF, Mathews JT, Meyer WB (eds) The Earth as transformed by human action: Global and regional changes in the biosphere over the past 300 years. Cambridge University Press, Cambridge, pp 163–201
Roberts RG, Flannery TF, Ayliffe LK, Yoshida H, Olley JM, Prideaux GJ, Laslett GM, Baynes A, Smith MA, Jones R, Smith BL (2001) New ages for the last Australian megafauna: Continent-wide extinction about 46 000 years ago. Science 292:1888–1892
Rock MT (2000) Globalisation and sustainable industrial development in the second tier Southeast Asia newly industrialised countries. In preparation
Rojstaczer S, Sterling SM, Moore NJ (2001) Human appropriation of photosynthesis products. Science 294:2549–2552
Rose NL (1994) Characterization of carbonaceous particles from lake sediments. Hydrobiologia 274:127–132
Ruffing KG (2002) Indicators of decoupling. Paper presented at the Science Forum, World Summit on Sustainable Development, Johannesburg, August 31, 2002
Sanchez R (2002) Cities and global environmental change. International Human Dimensions Programme. Global Environ Chang News 3/2002:1–3
Satheesh SK, Ramanathan V (2000) Large differences in tropical aerosol forcing at the top of the atmosphere and Earth's surface. Nature 405:60–63
Sauer CO (1961) Fire and man. Mitteilungen zur Kulturkunde 7:399–407
Scholes MC, Matrai PA, Andreae MO, Smith KA, Manning MR (2003) Biosphere-atmosphere interactions. In: Brasseur GP, Prinn RG, Pszenny AAP (eds) The changing atmosphere: An integration and synthesis of a decade of tropospheric chemistry research. The International Global Atmospheric Chemistry Project (IGAC), IGBP Global Change series. Springer-Verlag, Berlin Heidelberg New York
Schneider S, Turner BL II, Morehouse Garriga H (1998) Imaginable surprise in global change science. J Risk Res 1(2):165–185
Schwikowski M, Brütsch S, Gäggeler HW, Schotterer U (1999) A high resolution air chemistry record from an Alpine ice core (Fiescherhorn Glacier, Swiss Alps). J Geophys Res 104:13709–13720
Seitzinger SP, Giblin AE (1996) Estimating denitrification in North Atlantic continental shelf sediments. Biogeochemistry 35:235–260
Sen A (1981) Poverty and famine: An essay on entitlements and deprivation. Oxford University Press, Oxford
Serageldin I (2002) World poverty and hunger – The challenge for science. Science 296:54–58
Shiklomanov IA (1993) World fresh water resources. In: Gleick PH (ed) Water in crisis: A guide to the world's fresh water resources. Oxford University Press, New York
Shiklomanov IA (1990) Global water resources. Nature Resour 26(3)
Shotyk W, Cheburkin AK, Appleby PG, Fankhauser A, Kramers JD (1996) Two thousand years of atmospheric arsenic, antimony, and lead deposition recorded in an ombrotrophic peat bog profile, Jura Mountains, Switzerland. Earth Planet Sc Lett 145(1–4):E1–E7
Shotyk W, Weiss D, Appleby PG, Cheburkin AK, Frei R, Gloor M, Kramers JD, Reese J, Van Der Knaap WO (1998) History of atmospheric lead deposition since 12 370 ^{14}C yr BP from a peat bog, Jura Mountains, Switzerland. Science 271:1635–1640
Siscawati M (1998) Underlying causes of deforestation and forest degradation in Indonesia: A case study on forest fire. In: Proceedings of IGES International Workshop on Forest Conservation Strategies for the Asia and Pacific Region, 21–23 July 1998, Hayama, Kanagawa, Institute for Global Environmental Strategies, Japan, pp 44–57

Skole D, Tucker C (1993) Tropical deforestation and habitat fragmentation in the Amazon: Satellite data from 1978 to 1988. Science 260:1905–1910
Smil V (1994) Energy in world history. Westview Press, Boulder, CO
Smil V (1999) Nitrogen in crop production: An account of global flows. Global Biogeochem Cy 13:647–662
Smil V (2000) Energy in the twentieth century: Resources, conversions, costs, uses, and consequences. Annu Rev Energ Env 25:21–51
Solberg B, Brooks D, Pajuoja H, Peck TJ, Wardle PA (1996) Long-term trends and prospects in world supply and demand for wood and implications for sustainable forest management. Research Report 6, European Forest Institute (EFI), Joensuu, Finland
Stallings B (ed) (1995) Global change, regional response: The new international context of development. Cambridge University Press, Cambridge
Steininger MK, Tucker CJ, Townshend JRG, Killeen TJ, Desch A, Bell V, Ersts P (2001) Tropical deforestation in the Bolivian Amazon. Environ Conserv 28(2):127–134
Stern PC, Dietz T, Ruttan VW, Socolow RH, Sweeney JL (eds) (1997) Environmentally significant consumption. National Academy Press, Washington DC
Than M (1998) Introductory overview: Development strategies, agricultural policies and agricultural development in Southeast Asia. ASEAN Economic Bulletin 1:1
Thomas WLJ (1956) Man's role in changing the face of the Earth. University of Chicago Press, Chicago
Tolba MK, El-Kholy OA, El-Hinnawi E, Holdgate ME, McMichael DF, Munn RE (1992) The world environment 1972–1992: Two decades of challenge. United Nations, Chapman & Hall, London
Tomich TP, Fagi AM, de Foresta H, Michon G, Murdiyarso D, Stolle F, van Noordwijk M (1998) Indonesia's fires: Smoke as problem, smoke as a symptom. Agroforestry Today January-March pp 4–7
Tong S, Gerber R, Wolff R, Verrall K (2002) Population, health, environment and economic development. Global Change and Human Health 3:36–41
Torii K, Mori Y, Ji Z, Sato Y, Otsubo K, Yo S (2001) Observing the expansion of the built-up areas of regional capital cities in Yangtze River Delta by satellite images. Paper presented at the 22nd Asian Conference on Remote Sensing, 5–9 November 2001, Singapore. Centre for Remote Imaging, Sensing and Processing (CRISP), National University of Singapore; Singapore Institute of Surveyors and Valuers (SISV); Asian Association on Remote Sensing (AARS)
Turner BL II (2002) Land-use and land-cover change: Advances in 1.5 decades of sustained international research. In: Steffen W, Jäger J, Carson D, Bradshaw C (eds) Challenges of a changing earth: Proceedings of the Global Change Open Science Conference, Amsterdam, The Netherlands, 10–13 July 2001. Springer-Verlag, Berlin Heidelberg New York
Turner BL II, Clark WC, Kates RW, Richards JF, Mathews JT, Meyer WB (1990) The Earth as transformed by human action: Global and regional changes in the biosphere over the past 300 years. Cambridge University Press, Cambridge
Turner BL II, Butzer KW (1992) The Columbian encounter and land-use change. Environment 43(8):16–20,37–44
Tuxill J (1998) Losing strands in the web of life: Vertebrate declines and the conservation of biological diversity. Worldwatch Paper 41. Worldwatch Insitute, Washington DC
Tyson P, Fuchs R, Fu C, Lebel L, Mitra AP, Odada E, Perry J, Steffen W, Virji H (eds) (2002) Global-regional linkages in the Earth System. IGBP Global Change Series. Springer-Verlag, Berlin Heidelberg New York
UN (1967, 1976, 1985, 1994) United Nations Indicators Commodity Statistics Yearbook 1967, 1976, 1985, 1994. United Nations, New York
UN (1992) Protection of the quality and supply of freshwater resources: Application of integrated approaches to the development, management and use of water resources. Chapter 18 of Agenda 21. United Nations, New York
UN Center for Human Settlements (2001) The state of the world's cities, 2001. United Nations, *http://www.unchs.org*, 4 Oct 2002
UNDP (1999) Human development report 1999. United Nations Development Program. Oxford University Press, New York
UNDP (2000) Economic reforms, globalization, poverty and the environment. Reed D, Rosa H (eds). United Nations Development Programme, Oxford University Press, New York
UNEP (2000) Global environmental outlook 2000. Clarke R (ed), United Nations Environment Programme
UNEP (2001) An assessment of the status of the world's remaining closed forests. UNEP/DEWA/TR 01-2. Nairobi, Kenya, Division of Early Warning Assessment (DEWA)
UNEP-WCMC (World Conservation Monitoring Centre) (2000) Statistical analysis of global forest conservation. Global overview. *http://www.wcmc.org.uk/forest/data/cdrom2/stat1.htm*
UNEP/WMO (2002) Scientific assessment of ozone depletion. Prepared by the scientific assessment panel of the Montreal Protocol on substances that deplete the ozone hole. United Nations Environment Programme/World Meteorological Organization
UN Population Division (2000) World population prospects: The 2000 revision. United Nations, Department of Economic and Social Affairs.
US Bureau of the Census (2000) International Database. *http://www.census.gov/ipc/www/worldpop.htm*, updated 10 May 2000
USDA (1991) World grain database. Unpublished printout. United States Department of Agriculture, Washington DC
USDA (1996) Grain: World markets and trade. United States Department of Agriculture Foreign Agricultural Service, Washington DC
USGS (2000) U.S. Geological Survey World Petroleum Assessment 2000. U.S. Geological Survey World Energy Assessment Team, United States Geological Survey, available at *http://www.greenwood.cr.usgs.gov/energy/WorldEnergy/DDS-60/*, 2 April 2002
Vellinga P, Herb N (eds) (2000) Industrial transformation. International Human Dimensions Programme on Global Environmental Change, Bonn
Vitousek PM, Hedin LO, Matson PA, Fownes JH, Neff J (1998) Within-system element cycles, input-output budgets, and nutrient limitations. In: Pace ML, Groffman PM (eds) Successes, limitations, and frontiers in ecosystem science. Springer-Verlag, Berlin Heidelberg New York
Vitousek PM (1994) Beyond global warming: Ecology and global change. Ecology 75:1861–1876
Vitousek PM, Ehrlich PR, Ehrlich AH, Matson PA (1986) Human appropriation of the products of photosynthesis. Bioscience 36:368–373
Vitousek PM, Hattenschwiler S, Olander L, Allison S (2002) Nitrogen and nature. Ambio 31:97–101
Vitousek PM, Mooney HA, Lubchenco J, Melillo JM (1997a) Human domination of Earth's ecosystems. Science 277:494–499
Vitousek PM, Aber JD, Howarth RW, Likens GE, Matson PA, Schindler DW, Schlesinger WH, Tilman D (1997b) Human alteration of the global nitrogen cycle: Sources and consequences. Ecol Appl 7:737–750
von Humboldt A (1849) Ansichten der Natur, 3rd ed. J. G. Cotta, Stuttgart and Tübingen, Reprint 1969, P. Reclam, Stuttgart
Vourlitis GL, Oechel WC, Hastings SJ, Jenkins MA (1993) The effect of soil moisture and thaw depth on CH_4 flux from wet coastal tundra ecosystems on the north slope of Alaska. Chemosphere 26:329–337
Vörösmarty CJ, Sahagian D (2000) Anthropogenic disturbance of the terrestrial water cycle. Bioscience 50:753–765
Vörösmarty CJ, Fekete BM, Meybeck M, Lammers R (2000) A simulated topological network representing the global system of rivers at 30-minute spatial resolution (STN-30). Global Biogeochem Cy 14:599–621
Vörösmarty CJ, Meybeck M, Fekete B, Sharma K, Green P, Syvitksi J (2003) Anthropogenic sediment retention: Major global-scale impact from the population of registered impoundments. Global Planet Change, in press
Vörösmarty CJ, Sharma K, Fekete B, Copeland AH, Holden J, Marble J, Lough JA (1997) The storage and aging of continental runoff in large reservoir systems of the world. Ambio 26:210–219

Wackernagel M, Rees W (1996) Our ecological footprint: Reducing human impact on earth. New Society Publishers, Gabriola Island, BC
Waelbroeck C, Monfray P, Oechel WC, Hastings S, Vourlitis G (1997) The impact of permafrost thawing on the carbon dynamics of tundra. Geophys Res Lett 24:229–232
Waggoner PE, Ausubel JH (2001) How much will feeding more and wealthier people encroach on forests? Popul Dev Rev 27:239–257
Walker SJ, Weiss RF, Salameh PK (2000) Reconstructed histories of the annual mean atmospheric mole fractions for the halocarbons CFC-11, CFC-12, CFC-113, and carbon tetrachloride. J Geophys Res 105:14285–14296
Wang Y, Jacob DJ (1998) Anthropogenic forcing on tropospheric ozone and OH since pre-industrial times. J Geophys Res 103:31123–31135
Warren-Rhodes K, Koenig A (2001) Ecosystem appropriation by Hong Kong and its implications for sustainable development. Ecol Econ 39:347–359
WBGU (1997) World in transition. The research challenge. German Advisory Council on Global Change, Springer-Verlag, Berlin Heidelberg New York
WCED (1987) Our common future. World Commission on Environment and Development. Oxford University Press, Oxford
Weiss D, Shotyk W, Appleby PG, Kramers JG, Cherbukin AK (1999) Atmospheric Pb deposition since the industrial revolution recorded by five Swiss peat profiles: Enrichment factors, fluxes, isotopic composition and sources. Environ Sci Technol 33: 1340–1352
White R, Murray S, Rohweder M (2000) Pilot analysis of global ecosystems: Grassland ecosystems. World Resources Institute, 100 pp
Williams M (1990) Forests. In: Turner BL II, Clark WC, Kates RW, Richards JF, Mathews JT, Meyer WB (eds) The Earth as transformed by human action: Global and regional changes in the biosphere over the past 300 years. Cambridge University Press, Cambridge, pp 163–201
Wilson EO (1992) The diversity of life. Allen Lane, the Penguin Press
WMO (1999) Scientific assessment of ozone depletion: 1998. Global Ozone Research and Monitoring Project – Report No. 44, World Meteorological Organization, Geneva, Switzerland
Woods JJ, Schloss JA, Mosteller J, Buddemeier RW, Maxwell BA, Bartley JD, Whittemore DO (2000) Water level decline in the Ogallala Aquifer. A report on KWO-KGS Contract 99-132, Kansas Geological Survey Open-file Report 2000-29B (v2.0), *http://www.kgs.ukans.edu/HighPlains/2000-29B/Decdir.htm*
World Bank (2002) Data and statistics. *http://www1.worldbank.org/economicpolicy/globalization/data.html*, 4 Oct 2002
World Commission on Dams (2000) Dams and development: A new framework for decision-making. The Report of the World Commission on Dams, Earthscan Publications Ltd, London and Sterling, VA
World Tourism Organization (2001) Tourism industry trends. Industry Science Resources, *http://www.world-tourism.org*, 22 Oct 2002
WRI (2003) A guide to world resources 2002–2004: Decisions for the Earth. A joint publication with UN Development Program, UN Environmental Program, World Bank and World Resources Institute, Washington DC
WRI (1990) Forest and rangelands. In: A guide to the global environment. World Resources Institute, Washington DC, pp 101–120
WRI (1996) World resources: A guide to the global environment – The urban environment. World Resources Institute. Oxford University Press, Oxford New York
WRI (1999) World resources 1998–1999: A guide to the global environment. World Resources Institute, Oxford University Press, Oxford New York
WRI (2000) World resources: People and ecosystems: The fraying web of life. World Resources Institute, Oxford University Press, Oxford New York
Yudelman M, Ratta A, Nygaard D (1998) Pest management and food production: Looking to the future. 2020 Vision, Food, Agriculture and the Environment. Discussion paper 25. International Food Policy Research Institute, Washington DC

Chapter 4

Reverberations of Change: The Responses of the Earth System to Human Activities

Although human activities are, in the context of the Earth System, now equal in magnitude to some of the great forces of nature, it is often difficult to determine the extent to which an observed change in Earth System functioning is due to the anthropogenic forcing or is part of the Earth System's natural variability. Human impacts on the Earth System do not operate in separate, simple cause-effect responses. A single type of human-driven change, for example deforestation or fossil fuel combustion, triggers a large number of responses in the Earth System, which themselves reverberate or cascade through the System, often merging with patterns of natural variability. The responses seldom follow linear chains, but more usually interact with each other, sometimes damping the effects of the original human forcing and at other times amplifying them. Responses become feedbacks, which in turn can lead to further forcings in the System. The sequence of forcings-responses-feedbacks-forcings thus forms loops that can alter overall patterns of behaviour of the Earth System.

4.1 Reverberations of Change

From what has been presented thus far, it is obvious that human activities play a significant role in modulating the Earth System and in promoting and accelerating global change. The question to be addressed in this chapter is the manner in which the Earth System responds to human forcing of one kind or another. The responses are many and varied and typically human-induced changes cascade through the system to bring about further multiple collateral and interacting changes, which in turn produce a myriad further responses. Some systems may be well buffered and show resilience against change; others may not be and in some cases a threshold of resilience may be exceeded and the system may pass into another state, stable or unstable.

The ways in which perturbations are processed by complex systems is described in some detail in extended Box 4.1. This theoretical underpinning is essential to understanding the nature of natural variability of the Earth System (Chap. 2). It is just as important to comprehending the ways in which human forcings of the Earth System are actually manifested in changes in the behaviour of the System – why some apparently severe forcings produce no apparent change in behaviour while other relatively minor forcings can cause catastrophic responses; why some trigger immediate responses while others experience long lags before any responses are observed. Some examples drawn from the impacts of fishing on marine ecosystems illustrate this complexity (Box 4.2) and highlight the general theme of cascading effects to be expanded in more detail in the following sections.

The focus of this chapter (Fig. 4.1) is on the nature of the responses themselves of the Earth System to human forcing, regardless of the consequences for human well-being. The latter will be dealt with in the next chapter. To discuss all the Earth System responses induced by human activities is not feasible and falls beyond the scope of the chapter. Rather, the central point to be emphasised here is that Earth System functioning does not respond to anthropogenic forcing in simplified cause-effect relationships, as implied in statements such as the one that greenhouse gas emissions cause global warming. The nature of the Earth System's responses to the increasing anthropogenic forcing is more complex. These notions will be examined in more detail in the next sections in connection with the responses of the Earth System to fossil fuel combustion and to land use changes followed by intensification of agriculture.

4.2 Responses of the Earth System to Fossil Fuel Combustion

The work of the IPCC Working Group I (1992, 1996, 2001a) has significantly improved understanding of the effects on the climate system of CO_2 emissions from fossil fuel combustion. This chapter draws on that work, but puts it into the broader context of global change rather than climate change. Thus, this chapter will examine the wide range of direct and indirect responses of the Earth System to the suite of gases and aerosols that arise from fossil fuel combustion. Although in reality these responses interact with each other and with other stresses and ameliorating factors, here they are treated somewhat in isolation. The intent is to show how a single forcing fac-

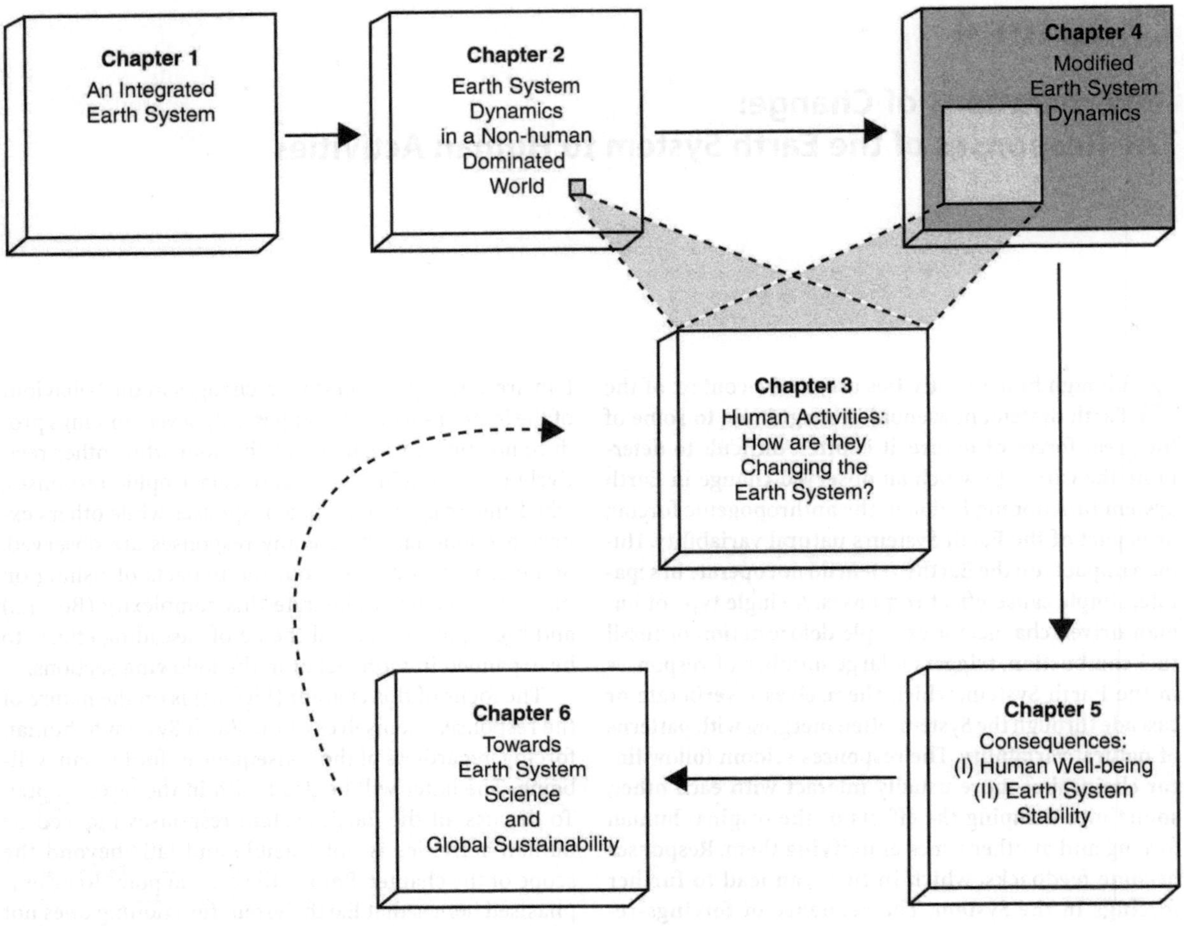

Fig. 4.1. Chapter 4 focus: the responses of the Earth System to human activities

tor – fossil fuel combustion – triggers a large number of responses that reverberate through the Earth System in a wide variety of chains or pathways. The focus will be primarily on the responses of ecosystems, both terrestrial and marine. Not all the responses are deleterious for ecosystems; some are positive. Together they consist of complex combinations of radiation, fertilisation, toxicity and marine chemistry effects. Figure 4.3 shows the wide range of pathways by which fossil fuel combustion can affect the functioning of the Earth System and provides the framework around which the following discussion is organised. It is important to remember that this chapter focuses only on the ways in which anthropogenic forcing causes responses that cascade through the Earth System, regardless of the ultimate consequences for human well-being. The latter is dealt with in Chap. 5. Also, in this chapter Earth System responses to anthropogenic forcing are treated individually, in turn. For example, effects on the climate system appear in several different points throughout the chapter. As in previous chapters, at the end of this chapter the individual responses are brought together again to build an integrated picture of changes to major Earth System functioning, such as the responses of the climate system as a whole.

4.2.1 Non-Reactive Gases

The illustrative gas that will be used is CO_2. For many scientists, and virtually all the public and those within the policy sector, the role of CO_2 in the Earth System is associated only with its direct enhancement of the radiative forcing of the atmosphere and hence with climate change. However, CO_2 has many other direct and indirect impacts on ecosystems, some of which are at least as important quantitatively as projected climate change impacts.

Climate

The best known response of the Earth System to the rising concentration of CO_2 concentration is its contribution to climate change. The Vostok ice core record (Petit et al. 1999) shows how CO_2 has varied naturally within stable limits over the last half million years (see Fig. 1.3). Modern measurements of the atmospheric concentration of the gas reveal that the present rate of increase is unprecedented and without analogue over the last half million years. The atmospheric response to the CO_2 in-

Box 4.1. Evolution of Perturbations in Complex Systems

H. Held · H. J. Schellnhuber

Genuinely complex systems consist of huge numbers of distinct components that interact with each other in a non-linear way. If the current macro-state of such a system is disturbed by a (small external) perturbation, this interference may either kick the system into an entirely new macro-state, fade away with almost no discernible impact, or generate some effect in between these two extreme outcomes. The prediction of the spatio-temporal evolution of perturbations in complex systems is actually one of the most difficult problems in contemporary science. This is illustrated by the notorious difficulties arising in hurricane trajectory forecasting as opposed to the straightforward calculation of the ballistic curve of a cannon ball.

In spite of the intricacies involved, the physical sciences have come quite a long way towards reliable perturbation analysis, even when strong non-linearities play a role. By way of contrast, the same type of analysis for complex ecological or even socio-economic systems is still in its infancy. Unfortunately, the Earth System consists of millions of complex sub-systems from all those categories. Therefore, it may seem impossible to provide a no-nonsense answer to the famous question about "dangerous anthropogenic interference with the climate system" as formulated in the UN-Framework Convention on Climate Change (FCCC 1999). And yet this question must be addressed for the sake of global sustainable development. A number of cognitive approaches are available to provide ways of addressing the issues involved.

Non-specialists are often confused by popular accounts of *chaos theory* maintaining that innocuous butterflies can create devastating storms on the far side of the planet. This notion is not entirely wrong, since tiny perturbations can indeed be amplified to tremendous events in complex systems, but *only if* a significant sub-set of system components are appropriately tuned by chance at the disturbance point in time. *In general*, however, the machinery works the other way round: even strong shocks are normally dissipated into gentle breezes by the intrinsic systems interactions.

Linear Warm-up

Standard perturbation theory as developed by modern physics starts from the simplest systems and conditions conceivable and tries to proceed stepwise towards ever-increasing levels of complexity. The analytic point of departure is defined by media that are governed by linear equations of motion and therefore do not represent truly complex systems in the sense introduced above. In spite of this fact, it is highly instructive to study how local disturbances excite various propagation phenomena in such media. These excitations typically decay quasi-monotonously in space and time, or travel outwards as wave packets that slowly get dispersed. The response to an idealised, point-like perturbation is a Green's function (or fundamental solution), and the response of the linear medium to an arbitrary perturbation can actually be obtained as an appropriate superposition of such Greens's functions (Jackson 1999). Thus the evolution of all disturbances can be regarded as perfectly predictable for this systems type – roughly speaking, the response amplitude will be proportional to the perturbation amplitude, as in linear optics.

Slightly more intricate phenomena occur in linear media when a perpetual external energy source excites the so-called wave trains. Furthermore, if the exciting source moves faster than the pertinent wave velocity supported by the medium, then a *Mach cone* with potentially destructive impacts on the system's boundaries can build up. This effect is well-known from supersonic air travel. But even if the energy source is stationary, something similar may happen under the precondition of sufficiently weak dissipation: the exited wave trains reach the boundaries, are reflected back into the medium, interfere with each other, and generate complicated patterns of dramatically enhanced response amplitude. These high-intensity domains are called *scars* and can be best explained by the methods developed in quantum chaos theory (Heller 1984). This methodology has been suggested for assessing, for instance, the seismic instability of the ground on which Mexico City is built (Mateos et al. 1993; Lomnitz et al. 1999).

Homogenous Non-linear Media

A much higher degree of complexity is present in extended homogeneous media that obey rather simple, mildly non-linear laws (typically expressed in terms of ordinary or partial differential equations). One of the most important laws of this type for Earth System science is summarised in the Navier-Stokes equations that describe fluid dynamics within a wide range of physical situations. Meteorology is particularly interested in shear flows, where small oscillatory perturbations may decay immediately and thus do not challenge the current macro-state of the medium. Under certain conditions, however, such perturbations keep on growing and transform the system's state completely. For instance, in case of baroclinic instability, a tiny initial disturbance can develop into a huge baroclinic cyclone (Holton 1992).

Even water waves cannot always be understood in terms of linear approximations. A beautiful phenomenon, calling for a genuinely non-linear explanation, was observed in 1834 by Scott-Russell: a boat being rapidly drawn along a narrow channel was suddenly stopped. In front of it, a mighty wave packet had built up which kept on propagating down the channel at more than 10 km hr^{-1}. While preserving its shape, it moved forward for more than 2 km! This phenomenon can be modelled by the Korteweg-de Vries equation. It captures a non-linear mechanism that counteracts the usual dispersion of wave packets due to different velocities of the packet's components. Thus the wave packet here becomes a *soliton* (Cherednik 1996) that preserves its form. Solitons are, in fact, very peculiar entities since they are able to collide and penetrate each other without any major deformation.

Granular Non-linear Systems

While the Navier-Stokes equations and similar mathematical objects describing homogeneous media have been studied for centuries now, the investigation of granular non-linear systems, composed of many discrete (identical or dissimilar) pieces, gathered momentum only in the last decades of the twentieth century. Many of these systems are characterised by short-range interactions between the constituents that can cooperate to generate *phase transitions*, i.e., complete transformations in systems state quality. An important example is provided by super-cooled liquids that can be turned into solid lumps by a single small dust particle. Similarly it has been shown that ferromagnetic materials can enter a state of self-organised magnetisation below a well-defined *critical* temperature (Huang 1987). At the critical point these materials become *infinitely susceptible* to external perturbations: the tiniest disturbance can flip the magnetisation vector of the total macroscopic sample. The formal analysis of these phenomena is provided by catastrophe theory (Thom 1975; Zeeman 1976) that explains – in quite general terms – how abrupt systems changes can be triggered by smooth variations of key parameters.

Even if a complex system with many degrees of freedom is far away from its critical states, its reaction to small perturbations may be highly non-trivial. This was demonstrated by Fermi, Pasta and Ulam (1974) in a famous numerical experiment. They investigated how the equilibrium state is approached in a large sample of non-linearly coupled harmonic oscillators. Their expectation was that the interactions would cause a rapid *dissipation* of an initial energy disturbance across all accessible eigenmodes of the aggregate, i.e., a spontaneous thermalisation. The calculations

◄

showed, however, that energy pulses were not distributed evenly over the harmonic modes; in fact, the entire system returned to its initial state after a finite period of time.

This is reminiscent of the soliton behaviour described above; indeed there is a subtle formal relationship between both phenomena where energy inputs are processed through truly dissipative systems (where *friction* prevails) in a very different way. The most prominent example is provided by the *Kolmogorov cascade* for atmospheric turbulence (Kolmogorov 1941; Panofsky 1984), which perceives the energy as trickling down a hierarchy of eddies of all sizes (self-similar or fractal dynamics).

Pertinent insights into the evolution of disturbances can also be gained from the study of reaction-diffusion systems (Field and Burgers 1985) and cellular automata (Haussels et al. 2001), which model a wide range of complicated natural phenomena. Of particular interest within the global change context are the so-called excitation waves: if the system considered is initially in a steady state, then a local perturbation exceeding a certain *critical threshold* can trigger such a wave (Steinbock 1998). This is not entirely surprising since threshold behaviour can actually be seen as a hallmark of non-linear systems. The excitation waves may assume spectacular shapes like oscillating circles or spirals, and even more involved topologies can be produced by the intrinsic forces.

Over the last decade, a similar systems type with non-linear local dynamics called *coupled-map lattices* (Kaneko 1993) has been investigated with increasing intensity. It turns out that most of the non-equilibrium states in these systems exhibit spatio-temporal (i.e., high-dimensional) chaos (Cross and Hohenberg 1993). Most of the typical phenomena can be encountered and studied in a simple representative of the entire system family, namely the *logistic lattice* (Kaneko 1989). Here one finds that the lattice macro-state can assume qualitatively different phases like *frozen random pattern, Brownian motion of defects, spatio-temporal intermittency, pattern selection*, and *fully developed spatio-temporal chaos* when tuned by the two key systems parameters of non-linearity and map-coupling strength. In the chaotic phase, any perturbation of a homogeneous initial state will create a hierarchical domain structure. The number of possible domain structures is actually huge and probably increases exponentially with the system size. Local disturbances can flip the system from one domain structure to another one quite easily, so the whole complex is extremely volatile (Kaneko and Tsuda 2001).

Natural Environmental Systems

Non-linear perturbation theory helps to explain regional (and possibly global) palaeoclimatic variability – like the Dansgaard-Oeschger oscillations of Arctic temperature (Sect. 2.6.2) as recorded in the Greenland ice-core archives (Grootes et al. 1993). An Earth System model of intermediate complexity (Petoukhov et al. 2000; Ganopolski et al. 2001) (Box 6.7) recently demonstrated that the North Atlantic thermohaline circulation, which largely determines the Arctic environmental conditions, could assume three different quasi-equilibrium states under the astrogeophysical conditions of the last glaciation period. Simulation runs and available data sets suggest that the Dansgaard-Oeschger oscillations were caused by *stochastic resonance* (Ganopolski and Rahmstorf 2002) (Box 5.6); a faint background signal of period 1 500 years (probably provided by solar variation) was amplified by erratic intrinsic climate fluctuations to flip the thermohaline circulation between its two upper (i.e. warmer) quasi-stationary states at time intervals that constitute random multiples of 1 500 years.

Another perturbation phenomenon can be observed in models for groundwater contamination dynamics, where steady flows penetrate a porous medium which can adsorb toxic substances like heavy-metal ions (Gruber 1990). This adsorption effect explains, at least partly, the surprisingly high cleansing capacity of groundwater systems. Massive problems arise, however, if detergents are added to the flow via standard washing machine use; the disturbing substances introduced into the pore water compete with the metal ions for adhesion to the porous medium's surface and when present in high enough concentrations expel many of them back into the groundwater. The same happens when the heavy metal ions are pulled off the surfaces by the formation of chemical complexes; the majority of the heavy metal ions in solution are bound into soluble chemical complexes with the detergent. This reduces the concentration of the free (uncomplexed) metal ions. Since the adsorbed concentration is a function of the free (uncomplexed) concentration, a reduction of the latter reduces the concentration on the surfaces as well. In fact, one droplet of detergent generates a wave packet (see above) of ever-increasing heavy-metal infestation downstream causing devastating ecological effects.

A quantum step to a higher level of complexity analysis is made when biology is added to geophysics to capture fully-fledged ecosystems. The response to disturbances within these systems is difficult to describe in terms of well-defined phenomena revealed by traditional perturbation theory – and even more difficult to track down by robust observations. Nevertheless, there is increasing evidence that ecosystems sometimes behave like *catastrophe machines* (Zeeman 1976), i.e., they undergo drastic switches to contrasting states when forced by gradual and smooth changes in environmental conditions (like climate, nutrient loading, habitat fragmentation or biotic exploitation – see Box 4.2) (Scheffer et al. 2001). Conspicuous illustrations of this behaviour, which can be explained within the framework of mathematical catastrophe theory, are the collapse and hysteretic recovery of shallow-lake vegetation as driven by insiduous phosporous concentration change, and the abrupt desertification of the western Sahara some 5 000 years ago due to slowly decreasing insolation (Claussen et al. 1999) (Sect. 2.6.4). A wide range of ecosystems, including coral reefs and woodlands, has been shown to exhibit drastic state shifts.

Catastrophes like such first-order phase transitions (Huang 1987) belong to the simplest non-linear responses to disturbances. Unfortunately, an advanced perturbation analysis of ecosystems, identifying phenomena like the ones described above for granular complex systems in physics, is not yet available. There exist a number of promising approaches, such as the Lotka-Volterra formalism for prey-predator dynamics (Dixon 2000), the graph-theoretical analysis of intricate trophic webs (Yodzis 1989), and other concepts of theoretical ecology that will allow the investigation of disturbance evolution. Such investigations are urgently needed – for instance, assessments of the stability of ecosystem composition (i.e., maintenance of biodiversity) in response to perturbations by, say, invasive species.

Socio-economic Systems

Little is known about the ways in which perturbations propagate across the layers of the anthroposphere (Schellnhuber 1999) and how their character is transformed by complex human response. In the context of climate-change assessment, for instance, one would like to employ a rigorous mathematical analysis of the impacts of, say, three consecutive crop failures on the socio-economic dynamics of a developing country like Burkina Faso. Such an analysis is evidently not available, let alone well-founded theories of *social phase transitions* (revolutions?) as caused by individual, local action or *social tsunamis* (mass hysteria, genocide), i.e., devastating amplification of initially small perturbations as caused, for example, by significant changes in boundary conditions.

One of the few disciplines to have ventured into this *quasi terra incognita* is economics: a number of authors have tried to tran-

◄

scend the neo-classical canon by investigating the effect of shocks (like the 1970s oil crises or the emergence of revolutionary technologies) (Edenhofer and Jaeger 1996; Edenhofer 1999). This analysis can also be inverted by looking for *butterfly economics* behaviour (Allen 1994; Nelson and Winter 1982; Silverberg et al. 1988), where tiny socio-political fluctuations can be non-linearly amplified to dramatic events by the system itself.

The *science of disasters*, i.e., the analysis of disruptive phenomena in socio-economic systems, has been reviewed (Bunde et al. 2002). An example considered is the development of stock market crashes and potential ways of forecasting them in spite of the subtle co-psychological effects involved (Sornette et al. 2002). The approach employed is based on the fundamental assumption that an initial *pessimistic disturbance* somewhere in the market drives stock prices down just as tiny cracks spread out in solids. For example, application of a statistical physics model of the economy suggested a slow oscillatory downward slide of the global stock market by mid-2003 (Hogan 2002). In another example Helbing et al. (2002) attempt to explain the onset and characteristics of panic dynamics within restricted areas with specific outlet geometries.

Inverting the Task: Vulnerability Analysis

In order to understand global change and the complex reverberations of human interference with the Earth System, many of the concepts and techniques discussed so far may be employed. Such is the case with cascading impacts of well-defined anthropogenic perturbations of highly sensitive environmental entities (such as the Amazon rainforests or semi-arid land use systems). The danger is that the powerful analytic tools necessary for truly conclusive systemic (i.e., transdisciplinary) investigations may not be developed fast enough to cope with the current and future transformations of the planetary environment caused by socio-economic driving forces.

An alternative approach, which basically inverts the cognitive task, is worth considering. Instead of investigating cascading impacts caused by a single environmental perturbation within an affected region or the total Earth System, attention can be focussed on specific valuable entities and critical combinations of disturbance factors that jeopardise the entity at stake can be identified (see, e.g., JB Smith et al. 2001). Such a *vulnerability analysis* actually constitutes the methodological cornerstone of *sustainability science* (Kates et al. 2001) (Box 6.10).

The exploration of systemic thresholds (phase transition points/lines) is an important instrument of vulnerability analysis as exemplified by the *tolerable windows approach* to integrated climate change assessment (Toth 2003). Under favorable circumstances, the *fatal* disturbance synergies that potentially push the system in question across a borderline or threshold, as well as the intervention strategies necessary for defusing these synergies, can be determined.

Finally, one may move even further away from the conventional perturbations perspective by searching the *Earth System possibility space* for *dangerous attractors* of nature–society interactions dynamics, i.e., entangled causal patterns onto which environmental degradation processes tend to lock in irrespective of the precise starting conditions. This concept is illustrated, for instance, by the so-called syndromes analysis of global change (Petschel-Held et al. 1999; Petschel-Held and Lüdeke 2001) (Box 3.1) and should come to full fruition once a network of integrated regional studies – as pioneered by LBA (The Large-scale Biosphere-Atmosphere Experiment in Amazonia) (Box 6.6) – become available. In the long run, the major research challenge is to find those intervention portfolios that help to lock up the degradation syndromes currently pervading our planet or lurking around the corner.

crease will continue for a long time even with the most stringent abatement scenarios for future emissions (IPCC 2001a) (Fig. 4.4). The evidence that the contemporary increase in atmospheric CO_2 concentration, in which fossil fuel combustion now plays a dominant role, is directly leading to changes in climate continues to strengthen and is now contested by few (Mann et al. 1999; IPCC 2001a).

Changes in climate are triggering a vast number of further reverberations of change throughout the Earth System, for example, significant changes in the cryosphere (Box 4.3). A few additional important reverberations of change are described below. The aim is to be illustrative rather than exhaustive; the IPCC reports provide many more examples in more detail of the ways in which climate change is influencing other aspects of Earth System functioning. The point is that changes in climate are cascading through the Earth System in many different ways.

Terrestrial and marine ecosystems. Given the complex role that biological systems play in the functioning of the Earth System (Chap. 2), it is no surprise that climate variability and change result in important and readily discernible reverberations through ecosystems, with consequent feedbacks to Earth System functioning. As shown in Fig. 4.3, these reverberations can, to a significant extent, ultimately be traced to the emission of CO_2 to the atmosphere through human activities.

Responses of ecosystems to changing climate were first evaluated on the basis of simple, first-order physiological effects (e.g., enhanced growth from enhanced precipitation) or simple shifts of intact biomes in response to a climate scenario. More recently it has been realised that multiple, interacting stresses on particular ecosystems must be considered instead of cause-effect chains with single drivers. It is also necessary to consider the effects of changes in structure as opposed to changes in physiology alone. For example, in Canadian boreal forests remote sensing studies and direct flux measurements suggest enhanced growth due to climate change and thus enhanced carbon uptake, whereas an evaluation of longer-term structural changes due to changing disturbance regimes suggests that Canada's boreal forests have actually shown a net loss of carbon to the atmosphere despite enhanced growth (Walker et al. 1999). Ecosystem responses as a consequence of extreme events and sequences of events are very different from those associated with monotonic changes in mean conditions such as when prolonged droughts for several centuries produce a collapse or near collapse of a system as a direct consequence of climatic change.

The dynamics of marine ecosystems are closely related to climate variability and thus will be affected significantly by any changes in climate. Figure 4.6 shows the correlation between the landings of several commercially important fish species and the Atmospheric Cir-

Box 4.2. Reverberations of Change Through Marine Ecosystems

Katherine Richardson

When fisheries are considered, images of commercially important fin fishes usually come to mind. In fact, the focus of many early fisheries was on large mammals or reptiles. Examining the ecosystem responses to the fisheries on these animals can illuminate the potential ecosystem responses to the fin fisheries that are currently practised and illustrate more generally the cascade of responses that can result from human-driven changes.

Whales, manatees, dugongs, monk seals, sea otters, crocodiles, turtles and other large vertebrates are now entirely extinct or ecologically extinct in many marine coastal ecosystems. An organism is considered to be ecologically extinct when it has been reduced to such low numbers that the function that these organisms carried out in the ecosystem as a whole can no longer be recorded. An example of such a function is grazing or predation on other organisms. The immediate response within an ecosystem of the ecological or actual extinction of an organism is often that another organism at the same trophic level takes over the function of the missing organism. Thus, there is often a lag period before a structural ecosystem response to the extinction can be observed. This lag can make it difficult to identify the original cause of what usually becomes a more extensive chain or cascade of responses through the ecosystem.

A contemporary example of ecosystem responses to removal of large vertebrate predators is that of the kelp forests on the west and east coasts of North America (Jackson et al. 2001) (Fig. 4.2). Prior to heavy fishing pressure on large fish and mammals, the sea urchin populations on the Alaskan/Californian coasts were held in check by sea otters and sheephead. In the Gulf of Maine, the sea urchin populations were held down by the predators sea mink and cod. In both regions, the sea urchin population feeds on kelp. With the predation pressure on sea urchins from the mammals and large fish, the sea urchin population was kept at a level where it was unable to eradicate the large kelp forests. However, after human fishing activities reduced or eliminated the larger predators, the sea urchin population responded by expanding and the increased grazing pressure of the sea urchins on kelp caused a response in the kelp population in the form of a dramatic population decline.

It is also believed that removal through overfishing of the large oyster beds that were, for example, found in regions such as Chesapeake Bay, may have altered the response of these ecosystems to eutrophication. Oysters survive by filtering phytoplankton. Thus, they remove large quantities of organic material from the water column. If these large populations of filter feeders were still in place, it is suggested that the ecosystem response to eutrophication would be minimised in that there would be less organic material available for bacterial degradation (Jackson et al. 2001).

In commercial fin fisheries at present, the changes in the structure of the associated ecosystems in terms of the size of the organisms found within them or the species they contain have been recorded (Rice and Gislason 1996; Pauly et al. 1998). At least some of these changes are believed to be a direct response to fishing pressure. These changes indicate an ecosystem response in the food web structure, similar to that of the sea urchin example above, where the relative abundance of organisms changes and elemental and energy flow is altered within the ecosystem. Such fishing pressures can also have long-term evolutionary impacts on the selection of genotypes within species. Recent experiments (Conover and Munch 2002) suggest that heavy harvesting produces high yields initially but quickly evolves to lower yields than the control while light harvesting produces the reverse effect. These results were caused by the selection of genotypes with slower growth rates under heavy harvesting and the opposite under light harvesting.

In addition to responding to grazing pressure, the abundance of organisms and their geographic distribution changes in response to changes in the physical environment. This has been seen clearly, for example, in coastal waters where salinity changes have occurred as a result of human-induced changes in water flow (i.e. dam construction) and species shifts from saline to freshwater tolerant fish species have been noted. An increase in global temperatures will elicit a response in ocean temperature distributions as well as in density fields and current systems. Changes in density fields will result in changes in elemental nutrient distributions that will, in turn, elicit a response in the plankton food web and ocean productivity (Bopp et al. 2001). This, then, can be expected to induce responses in commercially important fish stocks. However, it is not yet possible to predict what these responses will be.

Changes in ice cover resulting from increasing global temperatures can also be expected to influence the distribution of marine organisms and their food webs. In the Antarctic, for example, many species (including a number of birds and whales) are dependent on the availability of krill (euphausids) for their survival. An important food source for these krill is phytoplankton that accumulate in high concentrations immediately below the ice margins of the continent (Le Fèvre et al. 1998). Thus, as Antarctic temperatures increase and the ice margins diminish in extent, the food availability and, as a result, the numbers of krill can also be expected to diminish (Anon. 1998). A reduction in krill abundance can be predicted to elicit a response throughout the food web dependent on these organisms. Thus, warming in Antarctic regions can be predicted to influence the whale and bird populations dependent upon krill as a food source.

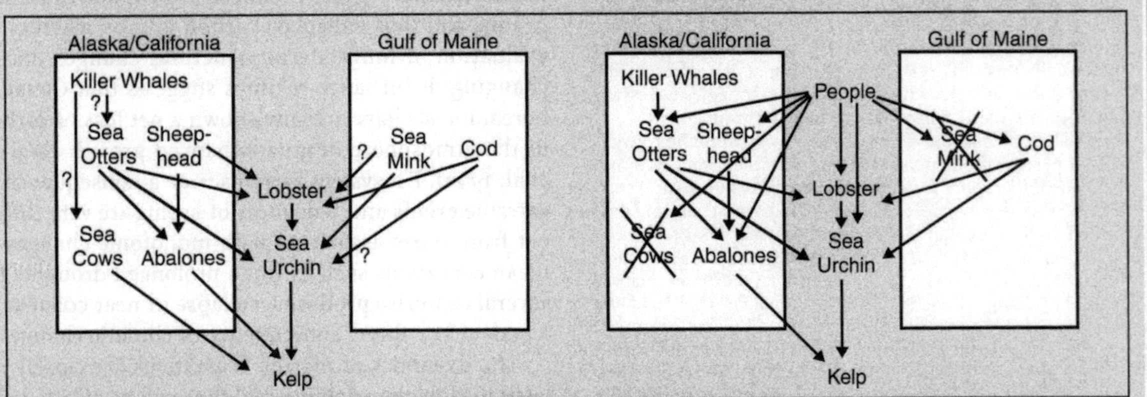

Fig. 4.2. Simplified coastal food webs in kelp forests showing changes in some of the important top-down interactions due to overfishing: before fishing (*left*) and after (*right*) (Jackson et al. 2001)

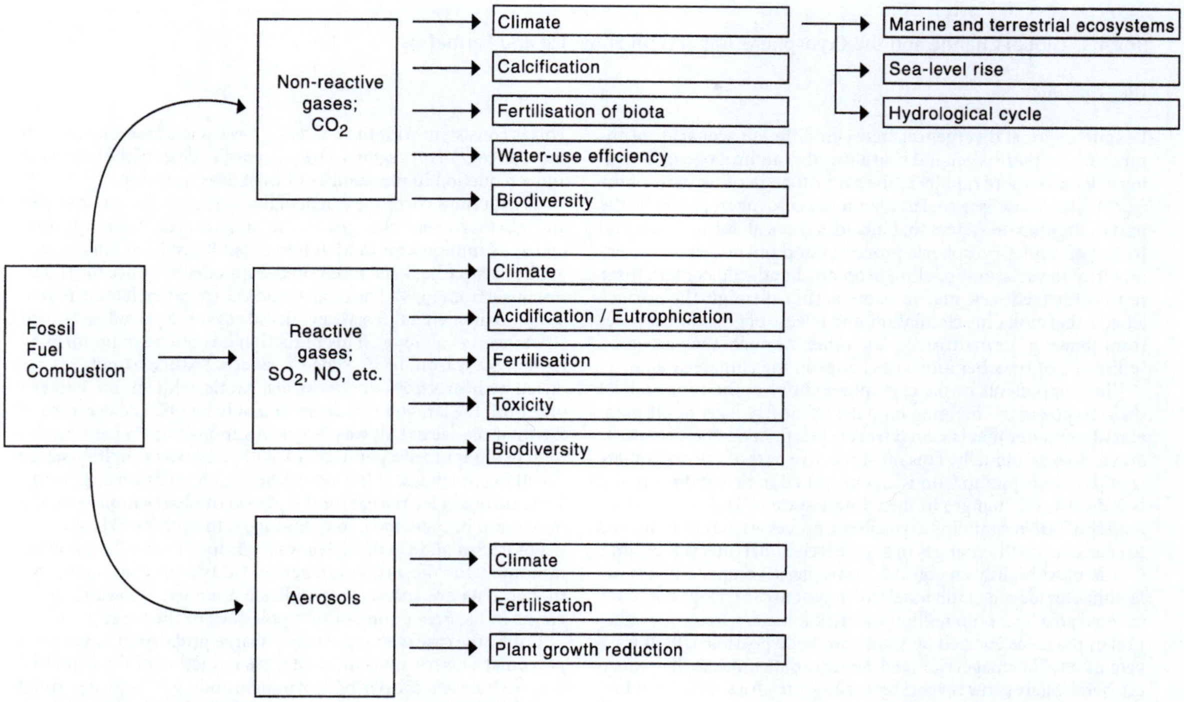

Fig. 4.3. Reverberations of the effects of fossil fuel combustion through the Earth System

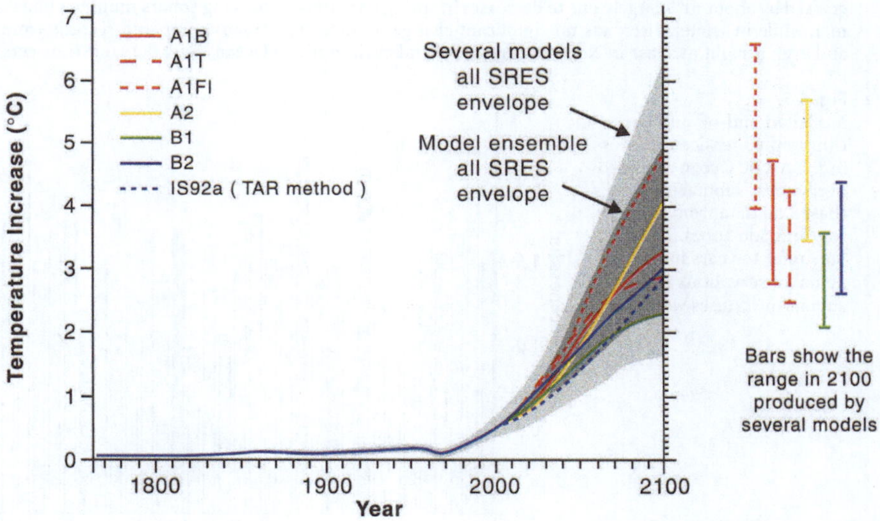

Fig. 4.4. Historical anthropogenic global mean temperature change and future changes for the six illustrative IPCC-SRES scenarios using a simple climate model tuned to seven atmosphere-ocean general circulation models (IPCC 2001a). For comparison, following the same method, results are shown for the IS92a scenario. The *dark grey shading* represents the envelope of the full set of thirty-five SRES scenarios using the simple model ensemble mean results

culation Index, a measure of climate variability. The mechanisms which govern this relationship are not well known, although they almost surely operate further down the food web than the commercially valuable fish populations themselves. Recent results (Heath et al. 1999) suggest that the rate at which the climate system is now changing may be breaking this historic linkage between climate variability and fish populations. Up to 1996 the abundance of the copepod *C. finmarchicus*, a key species in the marine food web in the North Atlantic region, was directly related to annual fluctuations in the North Atlantic Oscillation (Fig. 4.7). However, after 1996 the population has remained at a low level despite the change in the North Atlantic Oscillation. It appears that this shift was triggered by a sharp reduction in available overwintering habitat. Recent observations of the reduction in the formation of Norwegian deep water and the associated overwintering habitat may be responsible. This may be another example of the complex reverberations through the Earth System ultimately attributable, at least in part, to anthropogenic forcings.

Recent reviews of the responses of ecosystems to changing climate shows a coherent and clearly discernible trend across many ecosystems in phenology (the seasonal timing of the activities of plants and animals),

Box 4.3. Global Change and the Cryosphere: Impacts on Snow, Ice and Permafrost

Oleg Anisimov

Despite regional divergences between different scenarios of future climate, there is general consensus that anthropogenic warming will occur more rapidly in the Arctic than in other parts of the world. The Arctic is associated with the cryosphere, which is the part of the climate system that includes snow, all forms of ice, and frozen ground. Cryospheric processes and phenomena are very sensitive to variations of climate on one hand, and contain three important feedback mechanisms acting through the surface albedo, thermohaline circulation, and release of greenhouse gases from thawing permafrost, on the other. As such, they are good indicators of reverberations of change in the climate system.

The components of the cryosphere differ in the time scale of their response to changing climate. Of nearly 80% of all planetary freshwater that is stored frozen, less than one percent melts and refreezes annually. This most reactive part of the cryosphere is in dynamic equilibrium with current climatic conditions and is indicative of changes in the climate system. The observed intensity of fast-responding cryospheric processes may thus be used to characterise the changes over relatively short intervals of time.

The most rapidly changeable cryospheric component is snow. Within a few days an individual storm system can change the snow-covered area by 0.1–1.0 million square kilometres. Routine evaluation of the areas covered by snow was made possible with the advent of satellite imagery in 1966. Satellite data indicate that northern hemisphere snow cover extent changes from 42–50 million km^2 in January to 1–4 million km^2 in August, with large interannual variations of both minimum and maximum (Lawrimore et al. 2001). In the last 30 years annual-mean snow-cover extent has decreased by about 10%, largely due to decreases in spring and summer, while in winter there was no significant change in Eurasia and even a slight increase in North America (Folland et al. 2001). This is consistent with the tendency toward increases in seasonal snowfall over many regions, a trend to earlier dates of the last freeze and a reduction in the number of frost days in spring.

As with snow cover, the extent of floating ice in the Arctic Ocean and northern seas changes in the annual cycle from approximately 15 million km^2 in March to about 8.5 million km^2 in September. Sea ice has been monitored from satellites since 1973 (continuously from 1978). The total Arctic sea-ice extent has decreased by approximately 2.8 ±0.3% per decade over the period 1978–1996 (Parkinson et al. 1999). This reduction has not been uniform by region, by season, or by type of sea ice. Sea-ice extent was reduced by nearly 20% in the eastern Arctic, while in the western part until the late 1990s it decreased only by 5%. The decrease of the total sea-ice extent was larger in summer (4.5% per decade) than in winter (2.2% per decade), with the area of multi-year ice shrinking much faster (7% per decade) (Johannessen et al. 1999). The rate of sea ice retreat for the period of observations was at a maximum in 1990, 1993, 1995, and 1998. In 1998 ice extent in the Beaufort Sea and Chukchi Sea was 25% lower than the previous minimum for the 1953–1997 period (Maslanik et al. 1999). Although data are sparse, they indicate a tendency toward an increase in the area of open water poleward of the ice edge.

Unlike the case with snow cover, a large proportion of sea ice is perennial. Observations indicate a 9% reduction of the perennial ice area between the periods 1979–1989 and 1990–1995. Perennial or multi-year ice responds to seasonal climatic forcing through changes of its thickness, which generally have an amplitude of 0.5–1.0 m. Most robust data on ice thickness come from upward looking sonars mounted beneath the sea surface. Rothrock et al. (1999) compared records from submarines cruising the Arctic Ocean along the same transects and concluded that sea-ice thick-

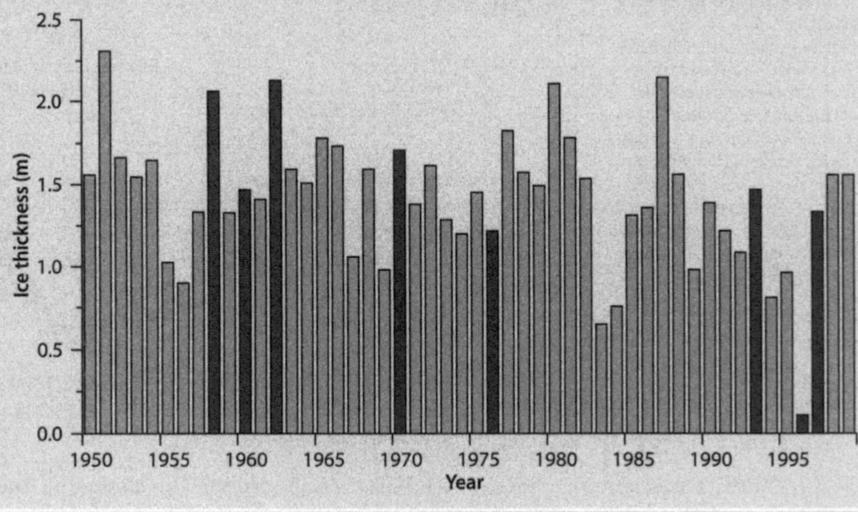

Fig. 4.5. Modelled end-of-summer (minimum) sea-ice thickness in the Arctic Ocean averaged over repeat submarine routes (based on data from Holloway and Sou 2002). The *dark bars* refer to years in which actual observations from submarine cruises were made

distribution and functioning, all strong evidence that the functioning of the biological components of the Earth System is already responding clearly to changing climate (Walther et al. 2002). A similar study of flowering plants in the British Isles showed that 16% flowered significantly earlier in the 1990s compared to the previous four decades and only 3% flowered significantly later (Fitter and Fitter 2002). In terms of phenology, typical changes are normally seen in the spring rather than the autumn end of seasonality and include the first singing or earlier breeding of birds, earlier arrival of migratory birds, earlier shooting and flowering of plants, earlier appearance of butterflies and earlier spawning in amphibians. Figure 4.8 shows how several phenological phases in northern Europe are correlated well with both natural modes of climate variability and with the underlying trend towards a warmer world. More geographically comprehensive analyses confirm that these trends are also discernible and significant at the global scale (Parmesan and Yohe 2003; Root et al. 2003).

◄

ness has decreased by almost 43% in the period from 1960 to 1990. Recent modelling results put this figure into question. Although simulated submarine cruises at the specific years 1958, 1960, 1962, 1970, 1976, and 1993, 1996 and 1997 gave similar figures, if the five earlier cruises each occurred one year earlier, and each of the three later occurred one year later, the result would have been no significant change in sea-ice thickness (Fig. 4.5). Total calculated loss of ice volume in the Arctic Ocean over the 30-year long period is about 12%, of which 9% is due to enhanced ice export and only 3% due to changes in growth/melt ratio and thus to changes in the thickness of sea ice (Holloway and Sou 2002).

Among the most sensitive indicators of climate change are glaciers. They respond to the interplay of changing temperatures and precipitation by changing their thickness and length. Data show that many glaciers receded dramatically in the twentieth century, although a few increased in length. As was indicated by modelling 44 glaciers in different parts of the world, this is consistent with the 0.66 ±0.20 °C global warming over the past 100 years (Church and Gregory 2001).

Large changes have taken place in permafrost regions. In the last century permafrost has become warmer and in many locations retreated poleward. Measurements in northern Alaska indicate a 2–4 °C warming of permafrost from the beginning of the century to the mid-1980s (Lachenbruch and Marshall 1986) and further warming of about 3 °C since the late 1980s. Less pronounced warming, up to nearly 2 °C over the last decade, was detected in Canadian permafrost regions. Warming of frozen ground averaging 1 °C or less has been measured in northwestern Siberia in recent years and in mountainous areas of Europe (Nelson 2003). Although warming of permafrost is evident, due to the large latent heat involved in the phase change of water from solid to liquid, the actual retreat of permafrost is slow.

The immediate response of the permafrost system to climatic warming is an increase in the uppermost layer of seasonal thawing. In ice-rich permafrost it may cause substantial subsidence of the ground surface. Such changes may lead to severe distortions of the terrain and are particularly detrimental to the infrastructure built upon permafrost. Some of the examples from central Alaska show that thaw-induced subsidence over the past two centuries has been as much as 2.5 m (Nelson 2003). Thawing of permafrost and associated topographic changes are occurring at an accelerating rate and have noticeable impacts on hydrology, landscape, and terrestrial ecosystem structure, causing transformation of some of the northern birch forests to fens and bogs. One of the most important biogeochemical implications is the release of greenhouse gases from thawing permafrost. In many regions changes in permafrost condition over the past few decades have already shifted the balance between the accumulation and release of soil carbon and converted much of the tundra from a sink to a net source (Oechel et al. 1993).

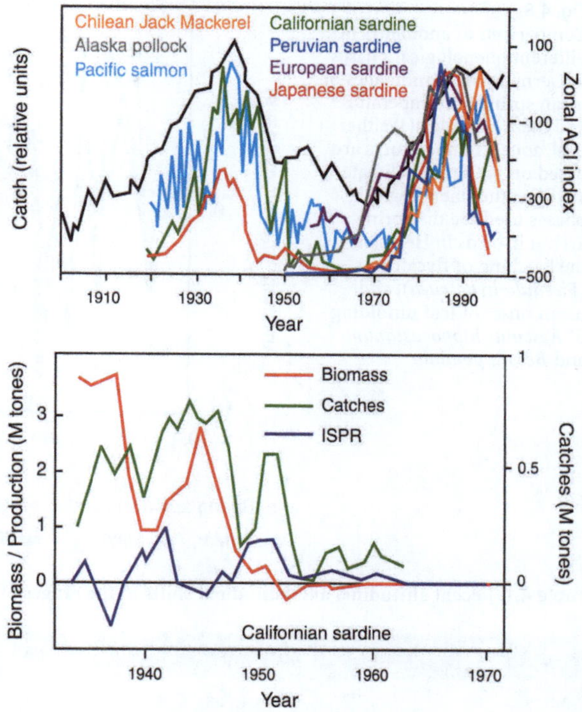

Fig. 4.6. Correlation between catch trends in some major commercial fish species and the dynamics of the Atmospheric Circulation Index (ACI), a measure of climate variability (Klyashtorin 2001)

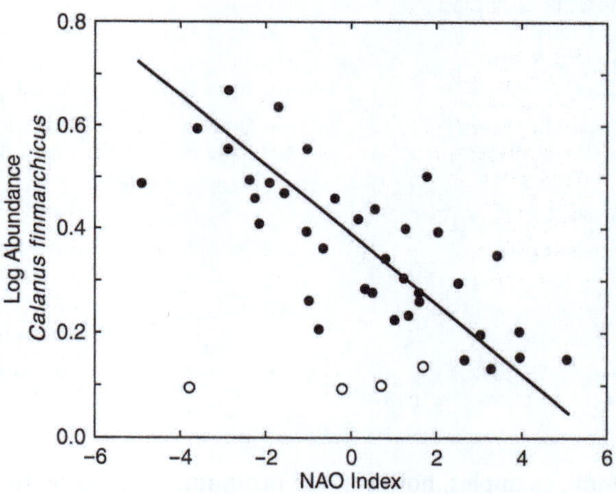

Fig. 4.7. Changing response of the copepod *Calanus finmarchicus* to North Atlantic climate (adapted from Fromentin and Planque 1996). *Black dots* indicate years 1958–1995; *open dots* show years 1996–1999

One of the most commonly cited effects of climate change on terrestrial ecosystems is the poleward and upward shifts of species ranges in response to higher temperatures. Table 4.1 shows that this pattern is now clearly evident at many places around the world and for a wide range of species. Given the fact that species are responding individually to a changing climate, changes in community composition are also expected due to the asymmetry in individual responses. This phenomenon is most evident in ecosystems already near the limits of their ranges. Antarctic terrestrial ecosystems are showing increases in diversity, especially in soil fauna, with warming while coral species in tropical waters, already near their upper temperature limit, are experiencing decreases in abundance and possibly also in diversity with the sharply increased incidence of bleaching events in recent decades.

The overall effects on ecosystem functioning of these changes in phenology, range and community composition are much more difficult to observe and project.

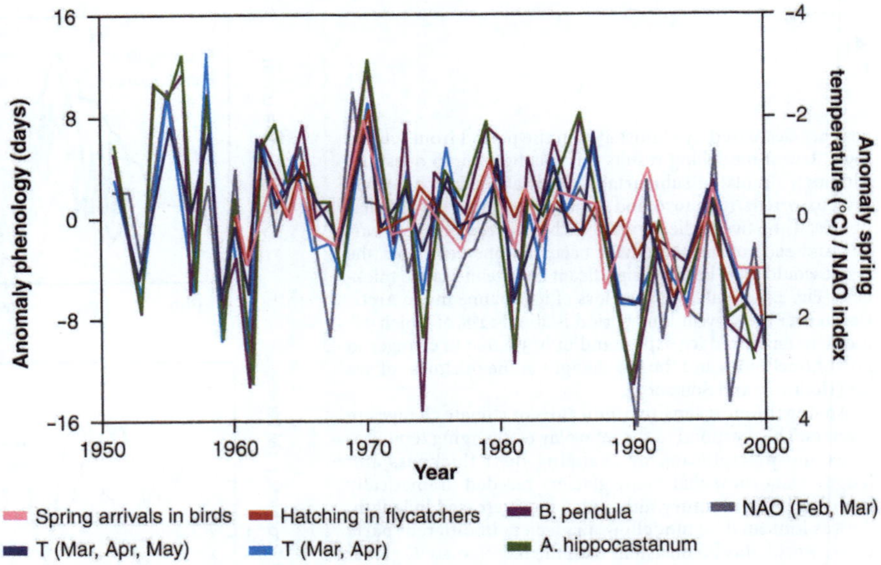

Fig. 4.8. Comparison of anomalies of different phenological phases in Germany with anomalies of mean spring air temperature and the NAO index (Walther et al. 2002). Temperatures are based on 35 German climate stations; the phenological phases used are the spring arrival in birds in Helgoland, the hatching of flycatchers (*Ficedula hypoleuca*), and mean onset of leaf unfolding of *Aesculus hippocastanum* and *Betula pendula*

Table 4.1. Recent altitudinal and latitudinal shifts in the ranges of a number of biological species (Walther et al. 2002)

Species	Location	Observed change	Climate link
Treeline	Europe, New Zealand	Advancement towards higher altitudes	General warming
Arctic shrub vegetation	Alaska	Expansion of shrubs in previously shrub-free areas	Environmental warming
Alpine plants	European Alps	Elevational shift of 1–4 m per decade	General warming
Antarctic plants and invertebrates	Antarctica	Distribution changes	Liquid water availability and increased temperature
Zooplankton, intertidal invertebrate and fish communities	Californian coast, North Atlantic	Increasing abundance of warm-water species	Warmer shoreline ocean temperature
39 butterfly species	North America and Europe	Northward range shifts up to 200 km over 27 years	Increased temperatures
Such as Edith's Checkerspot butterfly (*Euphydryas editha*)	Western United States	124 m upward and 92 km northward shift since the beginning of the twentieth century	Increased temperatures
Lowland birds	Costa Rica	Extension of distribution from lower mountain slopes to higher areas	Dry season mist frequency
12 bird species	Britain	18.9 km average range movement northwards over a 20 year period	Winter temperatures
Red fox (*Vulpes vulpes*), Arctic fox (*Alopex lagopus*)	Canada	Northward expansion of red fox range and simultaneous retreat of Arctic fox range	General warming

Some examples, however, are beginning to emerge. In Britain newts (*Triturus* spp.) are entering ponds earlier in the spring whereas frogs (*Rama emporaria*) have not yet altered their reproductive phenology significantly. This means that larvae and embryos of the frogs are experiencing higher levels of predation by newts (Beebee 1995). Many other such interactions, often much more complex, are no doubt occurring as a result of the reverberation of climate-driven change through terrestrial and marine ecosystems. The consequent feedbacks of such reverberations in ecological systems to the functioning of the Earth System are even more difficult to project but will become increasing important as global change continues.

Sea-level rise. Depending on the climate change scenario used, mean global sea-level rise by 2100 could vary from about 20 cm to nearly 1 m (Fig. 4.9; Box 4.4). As measured at any given point, sea level is influenced by many factors including land movement, steric changes reflecting the thermal expansion of the near-surface waters of the oceans, the effects of modes of ocean-atmosphere variability such as ENSO, the melting of low-latitude and temperate ice caps and changes in the complex mass balance of the major ice caps. The latter are among the hardest to quantify and include strongly time-lagged response processes that are difficult to measure, to model or to extrapolate with any confidence. There is thus no clear consensus on the probability of major ice

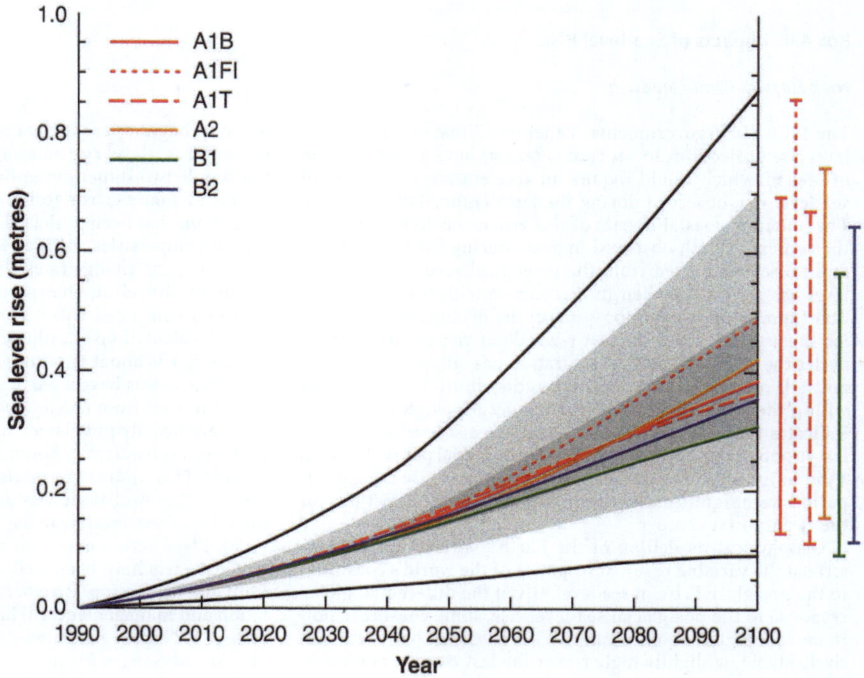

Fig. 4.9. Global average sea-level rise 1990–2100 for the SRES scenarios (IPCC 2001a). Thermal expansion and land ice changes were calculated using a simple climate model calibrated separately for each for seven atmosphere-ocean general circulation models (AOGCMs), and contributions from changes in permafrost, the effect of sediment deposition and the long-term adjustment of the ice sheets to past climate change were added. Each of the six *lines* appearing in the key is the average of AOGCMs for one of the six illustrative scenarios. The region in *dark shading* shows the range of the average of AOGCMs for all 35 SRES scenarios. The region in *light shading* shows the range of all AOGCMs for all 35 scenarios. The region delimited by the *outermost lines* shows the range of all AOGCMs and scenarios including the uncertainty in land ice changes, permafrost changes and sediment deposition

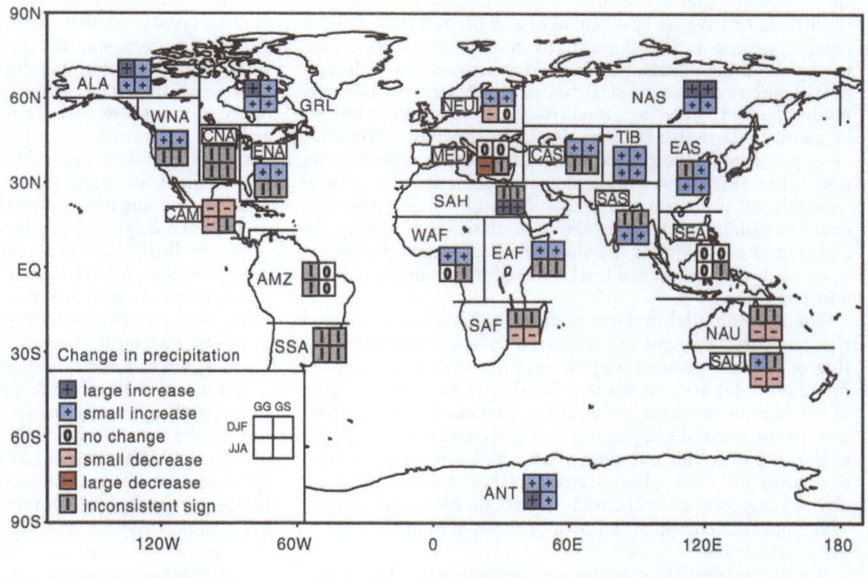

Fig. 4.10. Analysis of inter-model consistency in regional precipitation change (IPCC 2001a). Regions are classified as showing either agreement on increase with an average change of greater than 20% (large increase), agreement on increase with an average change between 5 and 20% (small increase), agreement on a change between −5 and +5% or agreement with an average change between −5 and +5% (no change), agreement on decrease with an average change between −5 and −20% (small decrease), agreement on decrease with an average change of less than −20% (large decrease), or disagreement (inconsistent sign). *GG* is the greenhouse gas only case and *GS* is the greenhouse gas with increased sulphate case

sheet collapse, though its consequences would be truly catastrophic in coastal regions throughout the world.

Hydrological cycle. Given its central role in the climate system, the hydrological cycle is intimately intertwined with any changes in climate. Changes in precipitation patterns, in both means and extremes, are an obvious manifestation of climate change. Precipitation patterns are much more difficult to simulate with reasonable accuracy in climate models than are temperatures. As a consequence, they are much more difficult to predict at all spatial scales. Figure 4.10 shows recent predictions of changes in global precipitation patterns. Although the changes for any particular region may not appear to be very large, especially when compared against temperature changes, the changes in surface runoff will probably be more pronounced, especially in semi-arid areas (Fig. 4.11). The results shown in Fig. 4.11 are for two climate scenarios only. However, certain features in projections of runoff changes appear to be robust across a wide range of emission scenarios and models: increases in runoff in the high latitudes and in South Asia; decreases in runoff in central Asia, the Mediterranean region, southern Africa and Australia. The consequences for these and other changes in the hydrological cycle for human well-being are potentially severe, as discussed in Sect. 5.3.2.

Box 4.4. Impacts of Sea-level Rise

Nick Harvey · Ian Goodwin

The IPCC (Intergovernmental Panel on Climate Change) sea-level rise projections to AD 2100 (Fig. 4.9) have a central value of 0.48 m, which would require an acceleration in the rate of sea level rise observed during the last century (IPCC 2001a). The potential coastal impact of this rise needs to be placed in the context of both observed impacts during the past century and those interpreted from the geological record. Global sea level has risen and fallen in accordance with the quasi-periodic (approximately 100 000-year) cycles of climate change and ice sheet growth over the last 1–2 million years. During these cycles the global, coastal configuration has adjusted to the rise and fall of sea level through the redistribution of sediment within the coastal and inner shelf zones. Geological evidence indicates that the most rapid changes in sea-level during the last 50 000 years occurred during the postglacial period between 8 500 to 12 500 years ago. The rates of sea-level rise during this period were significantly higher than any IPCC projections for the twenty-first century.

Geophysical modelling of the Earth's surface has demonstrated the variable regional response of the world's coastlines to the postglacial rise in sea level. Given the differential global response to the postglacial sea-level rise, some coasts are now in morphological equilibrium to a sea level that has been relatively stable or slightly higher over the last 6 000–7 000 years. Other coasts are still adjusting to the sea-level rise through local or regional glacio-isostatic movements of the lithosphere (known as relative sea-level change) and the redistribution of marine sediment. The global pattern of post-glacial hydro-isostatic coastal adjustment is well documented through field investigations and geophysical modelling (Peltier 2001). Hence, modern and predicted sea-level rise will have a greater impact on coasts that are still experiencing hydro-isostatic subsidence as opposed to those coasts in equilibrium or experiencing uplift. For this reason modern measurements of sea-level change need to have a global isostatic adjustment applied in order to obtain the true secular rate of sea-level change. In addition, the stability of the shoreline is mutually dependent upon both changes in sea level and sediment supply (Cowell and Thom 1994).

The key potential impacts of climate change and sea-level rise on coastal systems are increased coastal erosion; inhibition of primary production processes; more extensive coastal inundation; higher storm-surge flooding; landward intrusion of seawater in estuaries and aquifers; changes in surface water quality and groundwater characteristics; changes in the distribution of pathogenic microorganisms; higher sea-surface temperatures; and reduced sea-ice cover (IPCC 2001b). The most obvious impact is coastal land loss through sustained erosion. This impact is compounded by global population increase in the coastal zone.

The Second and Third Assessment Reports of the IPCC (1996, 2001a) together with the relevant contributions from IPCC Working Group II (Watson et al. 1998; IPCC 2001b) outline the key coastal areas which will be subject to the greatest impact from accelerated sea-level rise such as low-lying coral islands, deltaic and coastal plains, sand beaches, barrier coasts, coastal wetlands and lagoons. In addition, these reports note the potential impact on gravel beaches and barriers, unlithified cliff coasts and ice-rich cliff coasts. Whilst the impact of slow rates of sea-level change on the coast are difficult to identify and often masked by sediment budget variations, the impact is clearly visible on a locally subsiding coast. There are a number of coasts around the world that have been subsiding in recent decades; they provide a useful basis for predicting the effects of sea level rise on these and other coasts.

About 70% of the world's sandy coasts, which occupy about a fifth of the global coastline, have been retreating over the last century and it has been argued (Bird 1993) that sea-level rise will now begin to erode the currently stable sandy coasts. The impact of sea-level rise on sandy coasts is often analysed using the simple two-dimensional Bruun (1962) rule that relates the retreat of sandy coasts to the magnitude of sea-level rise, although this has been criticised because it does not include geological complexities, such as sediment budget changes. However, shoreline change rates along the US east coast produced ratios of shoreline change to sea-level rise varying from 110 to 181 compared with the ratios of 50 to 200 according to Bruun's calculations (Leatherman 2001). The average shoreline change rate is about 150 times the rate of sea-level rise.

Deltaic coasts have a particular significance in terms of potential impact from accelerated sea-level rise because many of these are heavily populated, and are already susceptible to inundation, subsidence, shoreline recession and sediment starvation (IPCC 2001b). An increased subsidence due to groundwater withdrawal alone was attributed to a 17 mm yr^{-1} rise in the relative sea-level near Bangkok (Sabhasri and Suwarnarat 1996). Elsewhere, impacts of sediment starvation on eroding deltaic coasts have been well documented for the Nile, Indus, Ebro and Mississippi Rivers. It is also possible that river regulation and management will have greater impacts than climate change for highly regulated river deltas such as the Nile, Mackenzie and Ganges Rivers.

The impact of a rising sea on wetland coasts needs to be placed in the context of the already significant human impact on these coasts. Extensive areas of mangroves have been cleared globally for firewood, charcoal or to make way for coastal development such as aquaculture. Subsiding mangrove coasts provide a good analogue for response to a rising sea with evidence of mangrove advance inland unless locally impeded by artificial structures.

Coastal marshes respond to sea-level rise by landward horizontal colonization in a similar manner to mangroves but have difficulty keeping up with sea-level rise on subsiding coasts such as southern England, northwest France or Venice Lagoon. As noted by Bird (1996), sea-level rise is likely to cause rapid recession of seaward margins of marshes and mangrove swamps unless they have sufficient peat accumulation or sedimentation rates. On a global scale, Nicholls et al. (1999) estimate that almost one quarter of the world's wetlands could be lost by the 2080s as a result of sea-level rise.

Coral reef coasts have demonstrated geologically an ability to respond to a rising sea-level with vertical Holocene accretion on the Great Barrier Reef of up to 26 m, with growth rates of 6 mm yr^{-1} (Harvey 1986) compared to even higher rates from Atlantic reefs where a vertical Holocene reef accretion of 33 m has been recorded (MacIntyre et al. 1977). Consequently, it has been suggested that healthy reefs with an estimated upper growth limit of 10 mm yr^{-1}, will be able to keep up with projected rates of sea-level rise (Buddemeier and Smith 1988, Schlager 1999). A more significant climate change impact on coral reefs is likely to be the prospect of increased sea surface temperatures causing an increase in coral bleaching events together with a reduction in reef calcification. It has been predicted that increased CO_2 will reduce reef calcification to the extent that this effect should be measurable toward the end of this century (Kleypas et al. 1999). While there is geological evidence that corals have responded to rapid sea-level rise in the past, this ability will be reduced in heavily populated coastal locations because of differential human impacts, such as pollution and increased sedimentation.

High-latitude coasts are particularly susceptible to increased periods of ice thaw causing a reduction of sea-ice and creating greater wave exposure, as well as exposing unlithified coastal sediments. Evidence of this is beginning to emerge from the rapidly eroding sandy coasts in the Gulf of St Lawrence, where severe erosion in recent years has been linked to warmer win-

◄

ters (Forbes et al. 1997). Another impact of global warming is the potential for thawing of sea ice to increase areas of open water in high latitudes and create a longer fetch for wave generation. This type of impact will be exacerbated for high-latitude low energy mud and sand coasts such as in the Canadian Arctic Archipelago (Forbes and Taylor 1994).

A potential increased intensity of coastal storms together with relative sea-level change will reduce the average recurrence interval of storm events used for coastal planning and management purposes, such as storms with an average recurrence interval of 50 to 100 years. This collectively increases the risk to coastal populations, although locally it will depend on the coastal resilience, which has ecological, geomorphic and socio-economic components.

IPCC (2001b) discuss the natural system's physical susceptibility to sea-level rise and define its response capacity in terms of resilience and resistance. Human activities can affect the physical response through the capacity of a society, including its technical, institutional, economic and cultural ability to cope with the impacts of sea-level rise and climate change. This could include any of the IPCC options of protection, accommodation, and retreat (IPCC 1992). However, there has been a mixed success for the various global and regional approaches to coastal vulnerability assessment. Differences between countries in terms of physical susceptibility to sea-level rise and socioeconomic resilience are highlighted by the vulnerability of small island states, located mainly in the tropics and subtropics.

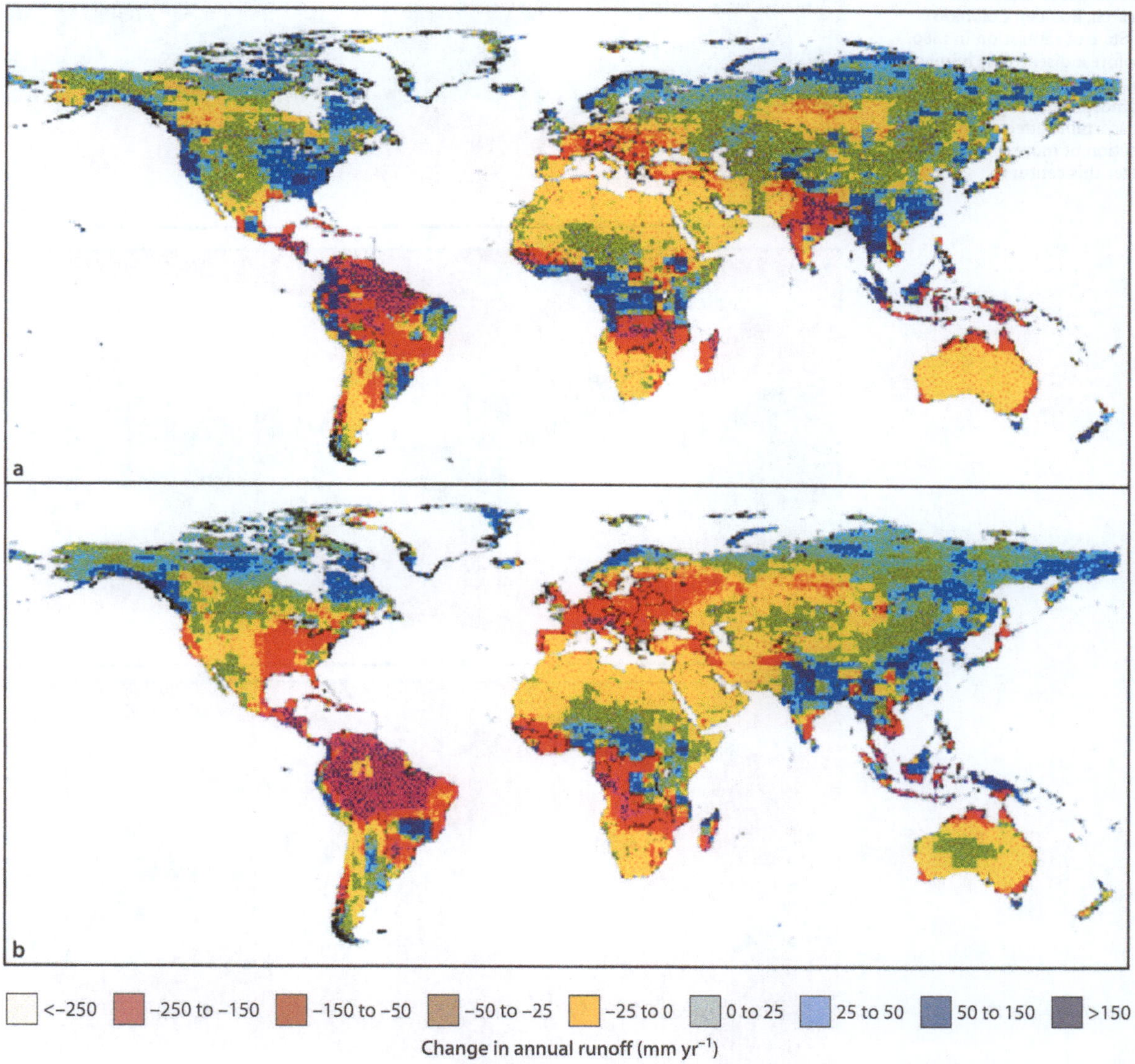

Fig. 4.11. Projected changes in average annual water runoff by 2050, relative to average runoff for 1961–1990, largely follow projected changes in precipitation (Arnell 1999). Changes in runoff are calculated with a hydrologic model using as inputs climate projections from two versions of the Hadley Centre atmosphere-ocean general circulation model for a scenario of 1% per annum increase in effective carbon dioxide concentration in the atmosphere: **a** HadCM2 ensemble mean and **b** HadCM3

Calcification

Returning to the direct impacts of anthropogenic emissions of CO_2 on the Earth System, a potentially important but seldom considered effect concerns the chemistry of the surface waters of the ocean. Rising atmospheric CO_2 concentration leads to a larger pCO_2 difference across the air-sea interface and hence to increased dissolution of CO_2 in the surface waters. This changes the chemical equilibria that control the pH of the surface waters resulting in a decrease in alkalinity. This, in turn, reduces the calcium carbonate saturation state of seawater, which leads to decreased rates of calcification (growth) of the organisms which form coral reefs. The effect may reduce calcification rates by as much as 30% with a doubling of atmospheric CO_2 (Kleypas et al. 1999). The most immediate implication of changes on surface water chemistry may be on coral reefs (Fig. 4.12) (Sect. 4.4.2).

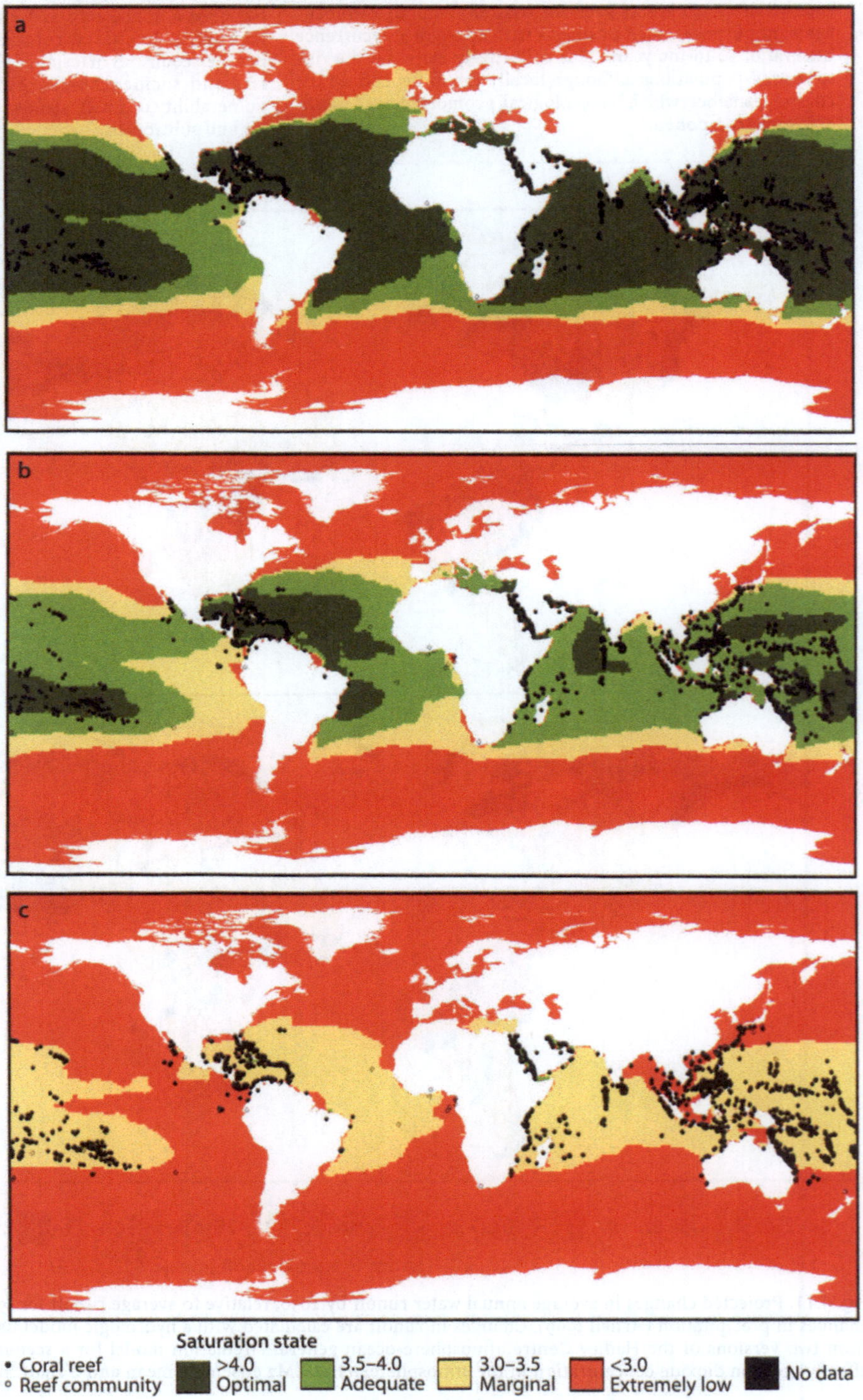

Fig. 4.12. The surface ocean aragonite (calcium carbonate) saturation state and its implications for the growth of coral reefs (R. Buddemeier, pers. comm., project based on datasets supplied by J. A. Kleypas, National Center for Atmospheric Research, Boulder, Colorado). **a** State of saturation in 1880, before a discernible human impact; **b** current levels of atmospheric CO_2, showing a discernible effect; and **c** projection of more severe effects later this century

Fertilisation of Biota

The sharp increase in atmospheric CO_2 concentration over the past two centuries is expected to have significant impacts on the functioning of terrestrial ecosystems. Carbon is the building block of all terrestrial life on Earth. Thus, the direct effects on gross photosynthesis, carrying through to increases in productivity and biomass, the so-called CO_2 fertilisation effect, is likely to be significant. An equivalent effect is unlikely in marine ecosystems, as their productivity is not limited by the amount of dissolved inorganic carbon in the surface ocean and does not respond to increased concentrations.

Over the past two decades, a large number of experiments, ranging from single-plants-in-pots laboratory work to *in situ* field Free-Air Carbon Dioxide Enrichment (FACE) experiments, have been carried out to determine the effects of rising CO_2 on terrestrial ecosystems. A recent synthesis of the results of these experiments (Mooney et al. 1999) indicates that herbaceous ecosystems show a mean increase in above-ground biomass of about 16% with a doubling of CO_2 (Fig. 4.13). Much variation exists around this mean. Some ecosystems show a near-doubled biomass; others show a decrease. In addition, cold ecosystems are less responsive to elevated CO_2 than warmer temperate or tropical ones. Woody species appear to respond most strongly as seedlings or juveniles, but longer-term effects on mature trees are more difficult to judge due to fewer experiments on closed forests (Box 4.5). The most well-documented FACE experiment on an *in situ* forest, that being carried out in North Carolina, USA, on a 15-year-old pine plantation (Hendrey et al. 1999), has shown a sustained increase in woody biomass of 20–25% with enhanced atmospheric CO_2 (200 ppm above ambient) (DeLucia et al. 1999; Hamilton et al. 2002). In agricultural systems, an enhancement of yields of 10–15% is expected with a doubling of atmospheric CO_2 concentration under normal field conditions (Gregory et al. 1999).

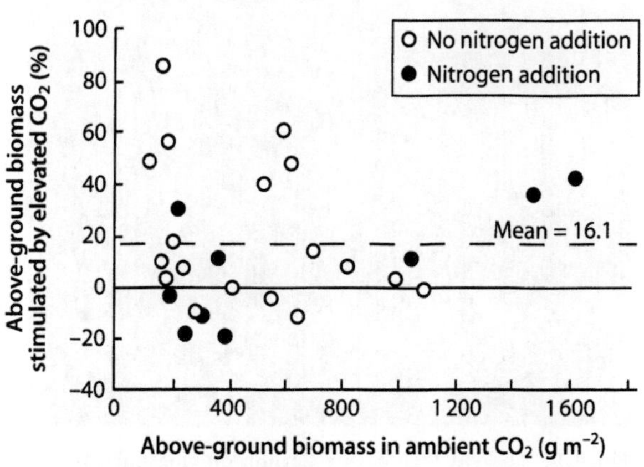

Fig. 4.13. Stimulation of above-ground biomass of herbaceous systems by elevated CO_2 with and without addition of nitrogen (Mooney et al. 1999)

Nearly all experiments use a step change in CO_2 concentration and a single elevated concentration. This is far from realistic in several ways. First, the response of gross photosynthesis is nonlinear, with the largest proportional responses expected at lower CO_2 concentrations. Secondly, terrestrial ecosystems have been subjected to a slow continuous change over a much longer period of time, not a sudden step-change. Third, the response of *in situ* terrestrial ecosystems will be influenced by many other factors that limit growth, especially nutrients. Given these complexities, it is perhaps not surprising that when more than one concentration of CO_2 elevation is used, strong non-linearity in the response can result (Fig. 4.15). This is a sobering reminder that the wealth of information being generated on the responses of ecosystems to a single, elevated CO_2 concentration is difficult to translate to long-term responses in the field.

A complementary approach to experimentation is to analyse the effects of the historical increase in CO_2 concentration from 280 to ca. 370 ppm, which should have already had a significant effect on terrestrial ecosystem productivity. A variety of evidence suggests that over the past decades the terrestrial biosphere has indeed become an increasing net sink for atmospheric CO_2. Apportioning that sink to causal factors, such as the CO_2 fertilisation effect, has proved to be difficult, however.

Water Use Efficiency

The first generation of elevated CO_2 experiments has shown that indirect responses to elevated CO_2 are just as important as the direct impacts on productivity described above (Mooney et al. 1999). The most important of these indirect effects is on water use efficiency. A common response of plants to elevated CO_2 is to reduce stomatal opening as the increased ambient CO_2 allows the plant to maintain the same internal CO_2 concentration with less contact with the atmosphere. This has the side effect of reducing the amount of water vapour that plants transpire through their stomates, thus, in effect, producing more biomass for less water lost. In semi-arid ecosystems the effect is especially pronounced as the increased water use efficiency leads to increased soil moisture and hence to increased productivity. In these systems this indirect effect on productivity often outweighs the direct CO_2-induced increase. Figure 4.16 shows how the leaf-level change in water-use efficiency reverberates through larger scales in terrestrial ecosystems and interacts with other CO_2- and climate-driven changes.

Biodiversity

One common feature of all studies of terrestrial ecosystem responses to elevated CO_2 is that no two species have been found to respond identically (Mooney et al. 1999). Thus, the impact of elevated CO_2 has become a biodiversity issue. Increasing atmospheric CO_2 will likely lead to changes

Box 4.5. Forest Responses to a Future CO_2-Enriched Atmosphere

Richard J. Norby

The importance of forests worldwide to the cycling of carbon and the overall health of the planet demands an understanding of how the structure and function of forests will change in response to the increasing concentration of CO_2 in the atmosphere. The size and longevity of trees preclude direct experimentation on intact forest ecosystems over the time scales at which many ecosystem processes operate, but past and ongoing experiments provide information and understanding about the responses of critical forest processes that can then inform the ecosystem models that predict the structure and function of forests in the future.

Forests in the future will most probably exhibit a higher rate of photosynthesis and gross primary productivity as the CO_2 concentration rises. The initial interaction between a tree and CO_2 in the atmosphere is not different from that of other C_3 plants: photosynthesis at the leaf level is stimulated, averaging 66% higher in field-grown trees grown in atmospheres with 600–700 ppm CO_2 (Norby et al. 1999). Most evidence indicates that the stimulation of photosynthesis in trees will be sustained over time with little down-regulation of photosynthetic capacity, although partial loss of response to CO_2 of Rubisco activity and assimilation has been observed in older foliage (Rogers and Ellsworth 2002). The gross primary productivity of a forest depends, however, not just on leaf-level photosynthesis but on the integration of photosynthesis over the entire forest canopy and over the entire growing season. Since leaf area index of forests (or tree assemblages) does not seem to be affected by CO_2 once the forest has completed its space-filling phase and the canopy has closed, measurements and models both indicate that forest gross primary productivity is higher in a CO_2-enriched atmosphere (Luo et al. 2001, Norby et al. 2002). Plant respiration is not usually much affected by elevated CO_2, so increased gross primary productivity generally translates directly into increased net primary productivity, although the response usually is somewhat attenuated compared to gross primary productivity. Over decadal time frames, nitrogen availability may constrain the capacity of forest ecosystems to respond to an increasing CO_2 concentration, but current evidence has not demonstrated effects of elevated CO_2 on nitrogen cycling (Finzi et al. 2002) or the litter decomposition pathway (Norby et al. 2001).

A more difficult question to address has been the fate of the additional carbon absorbed by forests in a CO_2-enriched atmosphere. The question of whether the increase in net primary productivity will lead to an increase in plant biomass or, alternatively, to an increased rate of cycling of carbon through the ecosystem (Strain and Bazzaz 1983) was posed two decades ago. It has profound implications for the feedback between the terrestrial biosphere and the atmosphere and, therefore, to the trajectory of atmospheric CO_2 concentration and the greenhouse forcing that attends it. Research in the 1980s and 1990s was largely focused toward identifying physiological mechanisms of tree response and the responses of critical ecosystem components. Only recently has it been possible to apply that knowledge at the scale of a forest stand using free-air CO_2 enrichment (FACE) technology. The FACE experiment in Oak Ridge, Tennessee, USA, was constructed in a stand of the broadleaf deciduous tree, *Liquidambar styraciflua* (sweetgum) (Fig. 4.14). Although young and not yet reproductive, the stand is fully occupying the site, the canopy is closed, and the trees are in a linear growth phase. During five years of exposure to an atmospheric CO_2 concentration of about 540 ppm, net primary productivity in this stand has been enhanced by about 21% compared to the forest plots in current ambient CO_2 (Table 4.2), a response that is consistent with projections from ecosystem models, as well as with suppositions based on the results of earlier experiments with smaller, individual trees. The increased net primary productivity can be traced to a sustained increase in photosynthesis at the leaf scale, which averaged 46% higher in elevated CO_2 (Gunderson et al. 2002).

Increased net primary productivity in this sweetgum stand has not led to more plant biomass, but rather an increase in C cycling in elevated CO_2. The additional C taken up in photosynthesis is allocated primarily to the production of fine roots (Table 4.2), which turn over rapidly and do not contribute to a sustained increase in tree biomass or aboveground carbon sequestration. The fine root C does enter the soil, and there is the potential for increased C sequestration into soil organic matter pools. However, detecting small changes in soil C is difficult. The sweetgum responses contrast with the North Carolina FACE experiment in a *Pinus taeda* plantation of similar stature. There, a sustained increase in net primary productivity in response to CO_2 enrichment, similar in magnitude to that of the sweetgum forest, was associated with increased aboveground wood incre-

Fig. 4.14. The FACE (Free-Air Carbon dioxide Enrichment) experiment in a stand of *Liquidambar styraciflua* (sweetgum) trees at Oak Ridge National Laboratory, USA

◄

ment (Hamilton et al. 2002). Inherent differences in carbon allocation patterns (much greater fine root productivity in the sweetgum stand) may explain these results. The challenge in describing future global forest responses is accentuated by the difference in response of these two relatively similar experiments in the southeastern United States.

Carbon sequestration is not the only response of forests that will matter in the future. Much is made of the effect of CO_2 on water-use efficiency of plants, often with an assumption that improved water-use efficiency will mean less transpiration or greater drought tolerance. The primary response to CO_2 of stomatal conductance seems to vary with species or conditions; conductance is lower in elevated CO_2 in the sweetgum FACE experiment but is not affected in the pine FACE. Nevertheless, whole-stand transpiration is hardly affected at all in either stand and may not be an important aspect of forest response to elevated CO_2 (Wullschelger et al. 2002; Schafer et al. 2002). Other forest responses to future increases in atmospheric CO_2 might include changes in the herbaceous or woody species composition of forest stands, interactions with forest pests and pathogens, or alteration of the response to climatic variables or tropospheric ozone (Percy et al. 2002). There is evidence for many such responses in short-term experiments with small trees, but little basis for extending those results to intact forests.

Any prediction about forests of the future will be fraught with uncertainty, and this surely includes predictions of responses to rising atmospheric CO_2. Although it is likely that elevated CO_2 will cause increased carbon uptake from the atmosphere, this response may not be detectable if other co-occurring environmental factors obscure or counteract the CO_2 signal. Likewise, the lack of clear evidence of CO_2 fertilisation in tree-ring records (Jacoby and D'Arrigo 1997) does not indicate that there has been no response to past increases in CO_2, occurring, as they were, in combination with other fluctuating environmental resources. The fate of increased carbon absorbed by forests from the atmosphere could include accumulation in woody biomass, accumulation in soil organic matter pools, or simply faster C cycling through the ecosystem, depending on the characteristics of the ecosystem. Higher-order interactions with CO_2, such as alteration of insect herbivory or nutrient cycles, could have profound effects on the structure and function of forests, but this remains largely speculative. The inevitable and unavoidable constraints that derive from the fundamental nature of forests will not permit us to replicate a future forest in a future atmosphere, and exact prediction of responses will be unattainable. Identifying a range of possible futures, however, based on experimental observations and well-informed ecosystem models is a worthwhile goal that ongoing research programmes are approaching.

Table 4.2. Relative response (%) to a CO_2-enriched atmosphere (545 ppm) of processes at different scales in a FACE experiment in a *Liquidambar styraciflua* (sweetgum) stand. All responses are from the third season of CO_2 exposure (2000) except for gross primary productivity, stomatal conductance, and stand transpiration (1999)

Process	Percentage increase	Reference
Leaf photosynthesis	46	Gunderson et al. 2002
Leaf respiration	0	Tissue et al. 2002
Stem respiration	33	Edwards et al. 2002
Gross primary production	27	Norby et al. 2002
Net primary production	21	Norby et al. 2002
Wood increment	7	Norby et al. 2002
Fine root production	112	Norby et al. 2002
Nitrogen uptake	23	D. W. Johnson (unpublished)
Soil respiration	17	P. J. Hanson (unpublished)
Stomatal conductance	−22	Wullschleger et al. 2002
Stand transpiration	−10	Wullschleger et al. 2002

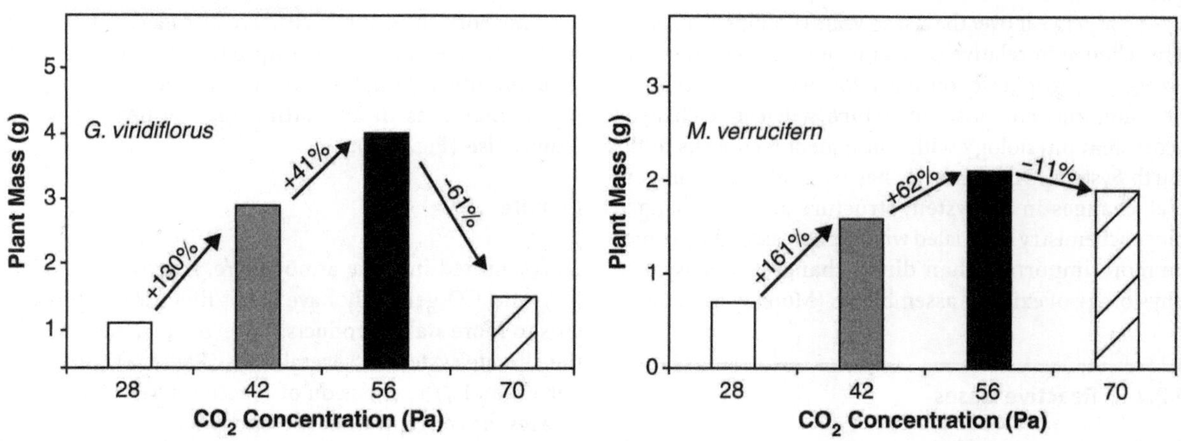

Fig. 4.15. The growth response of two tropical liana species (vine) grown in deep shade at several elevated CO_2 concentrations (Granados and Körner 2002). The responses are highly nonlinear, with the initial stimulation of growth reversed as CO_2 is raised beyond 560 ppm

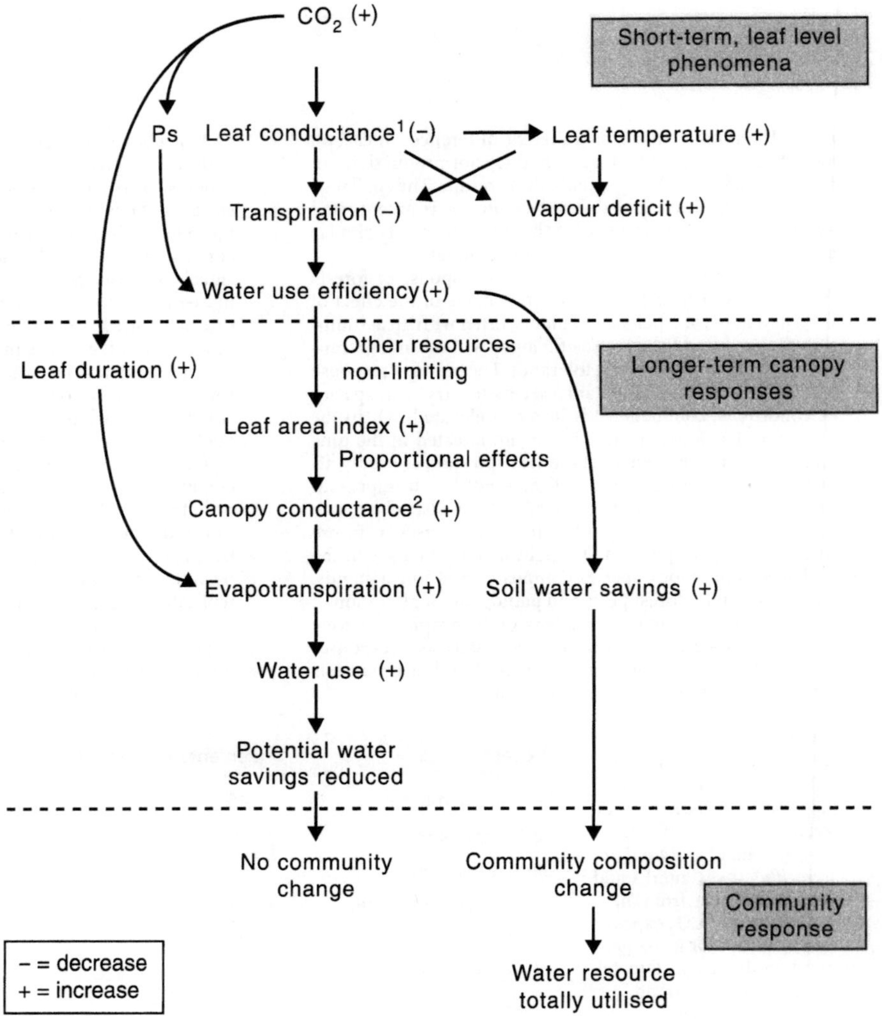

Fig. 4.16.
Water-mediated responses to elevated CO_2 at the leaf-, canopy- and community-levels (Mooney et al. 1999). *1* = Short term sensitivity to water stress enhanced under high CO_2; *2* = sensitivity of canopy conductance to evapotranspiration depends on community roughness

in the competitive abilities of plants, in turn altering species abundances and community composition. For example, a recent study has shown that Amazonian tropical forests are experiencing a strong increase in the density, basal area and mean size of lianas (woody climbing plants) with the dominance of lianas to trees increasing by 1.7–4.6% per year over the last 20 years (Phillips et al. 2002). This change in relative growth is presumably due to increasing atmospheric concentration of CO_2. Such changes in community composition, in turn, will lead to changed ecosystem physiology with consequent feedbacks to the Earth System. One synthesis has concluded that, in general, changes in ecosystem structure and the changed biogeochemistry associated with the new assemblages may be more important than direct changes in ecosystem physiology of existing assemblages (Mooney et al. 1999).

4.2.2 Reactive Gases

In addition to CO_2, the combustion of fossil fuels produces a suite of reactive gases, the precise mix depending on the nature of the fuel and the conditions of combustion. Reactive gases affect the climate system in a number of ways, but they have a larger number of direct effects on the biosphere than does CO_2. In addition to fertilisation, the deposition of reactive gases can lead to acidification, eutrophication and toxicity. The issue of deposition of reactive nitrogen compounds will increase sharply in importance, especially in Asia and other developing regions of the world, if projected increases in deposition for the next 20 years materialise (Fig. 4.17).

Climate

Once emitted into the atmosphere, reactive gases like NO_x and CO generally have short lifetimes as precursors to more stable products. Nevertheless, they impact the climate system in several ways (Fig. 4.18) (Box 4.6). For example, the emission of reactive gases usually increases the concentration of tropospheric O_3, which acts as a greenhouse gas in the troposphere but acts to cool the climate in the stratosphere. The effect of human ac-

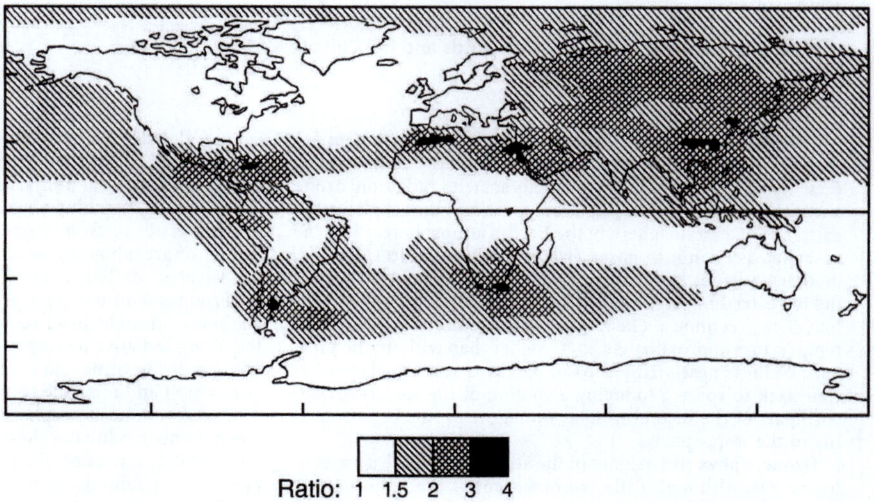

Fig. 4.17.
Ratio of the estimated deposition of oxidised (reactive) forms of nitrogen to ocean and land surfaces in 2020 relative to 1980 (adapted from Galloway et al. 1994 and Watson 1997)

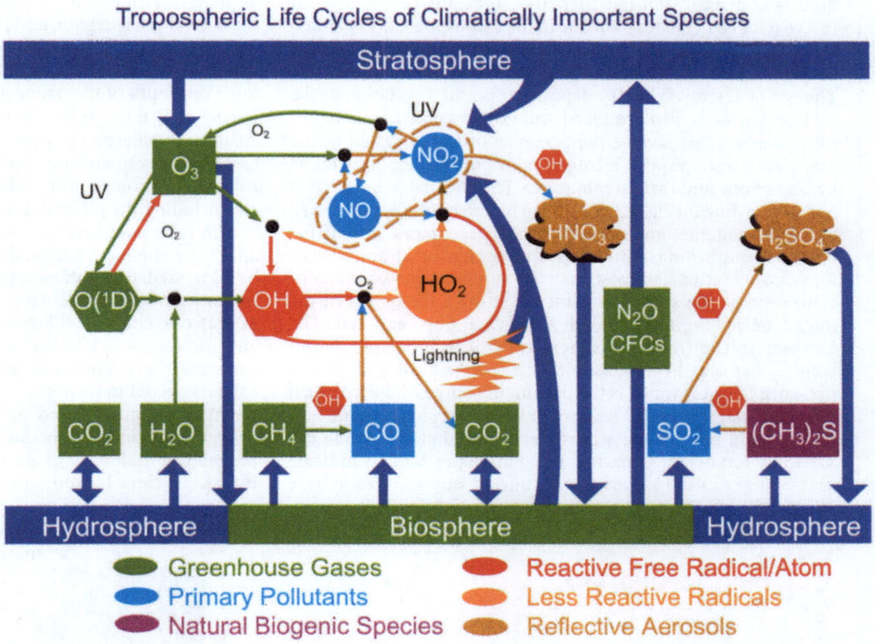

Fig. 4.18.
Tropospheric life cycles of climatically important species (adapted from Prinn 1994)

tivities has been to decrease the concentration of O_3 in the stratosphere (Sect. 5.4.1) (Box 5.5) while increasing it in the troposphere to give a net warming effect in both cases.

A second major effect of reactive gases on climate is their influence on the concentration of oxidising species like the OH radical, which in turn can have a significant indirect influence through their reactivity with the important greenhouse gas CH_4; the largest sink for CH_4 is through oxidation in the atmosphere. Any significant changes in the oxidation chemistry that controls this sink would significantly influence the atmospheric lifetime of CH_4 and thus its impact on climate. Thus, taken together, the impacts of reactive gases on the climate system are important and chemistry-climate coupling (see also discussion on aerosols below) is receiving increasing attention in research to improve understanding of the climate system.

Acidification and Eutrophication

With oxidation of NO_x and SO_x to acids that dissolve in cloud droplets and raindrops, acid rain is produced. Both wet and dry deposition of acidic compounds on terrestrial and marine systems produce ecosystem responses. These range from acidification of lakes, with consequences for the biota and ecosystem functioning, to acid deposition on soils and vegetation. The process of soil acidification was widespread in Europe and North America during the 1970s and 1980s before more stringent emission controls and affected large areas of forest, agriculture and grasslands. The acidification problem is now becoming widespread in Asia, especially China. As soils become acidified from atmospheric deposition, the transport of strong acids from the soils into groundwater is accompanied by an equimolar transport of cations.

Box 4.6. Atmospheric Reactive Compounds and the Climate System

Guy P. Brasseur

Interactions between atmospheric chemistry and climate are complex because they involve numerous feedback processes. First, many chemical compounds interact with radiation: ozone absorbs solar ultraviolet radiation, a process that explains the existence of a stratosphere in the Earth's atmosphere, while the so-called greenhouse gases (water vapour, carbon dioxide, methane, nitrous oxides, ozone, halocarbons, etc.) trap part of the terrestrial infrared radiation in the Earth-atmosphere system. The greenhouse effect maintains the Earth's surface at a temperature approximately 30 °C higher than without the presence of these gases. Tiny aerosol particles scatter solar radiation back to space, producing a cooling of the surface, or absorb part of the incoming solar radiation, producing a warming of the troposphere.

Ozone, a powerful oxidant of the atmosphere, is also a greenhouse gas. Although little is known about the tropospheric abundance of this gas in the pre-industrial era, it is believed that its concentration has increased substantially in the last 100 years. The presence of ozone in the troposphere is explained by downward transport of ozone-rich stratospheric air masses in the vicinity of tropopause foldings, cut-off lows and storms. The gas is destroyed by dry deposition on the vegetation at the surface. Ozone is also produced and destroyed by photochemical processes that involve the presence of the so-called ozone precursors: water vapour, nitrogen oxides, methane, nonmethane hydrocarbons and carbon monoxide. The chemistry involved is highly non-linear (Fig. 4.19), and can be substantially perturbed by anthropogenic emissions of ozone precursors (fossil fuel combustion, biomass burning and agricultural activities). Model calculations suggest that the summertime ozone concentration at the surface has increased by a factor 2–3 in the industrialised regions of North America, Europe and Asia. The increase in the free troposphere (northern hemisphere and tropics) has also been substantial. As ozone photolysis in the presence of water vapor is the dominant source of the hydroxyl (OH) radical, a powerful oxidant in the atmosphere, ozone pollution could modify the atmospheric abundance of some climatically important gases including methane. Models indicate that changes in the tropospheric concentration of ozone have led to a globally and annually averaged direct radiative forcing of approximately 0.4 $W m^{-2}$ since the pre-industrial era, with strong geographical and seasonal variability (Fig. 4.20) and regional peaks reaching 1 $W m^{-2}$. Indirect radiative effects such as those resulting from ozone-induced changes in the concentration of greenhouse gases like methane need to be added to the direct effect. The radiative forcing of ozone, which is rather uncertain due to our poor estimate of pre-industrial concentrations of this chemical compound, has to be compared with the direct radiative forcing resulting from the human-induced increase in the atmospheric concentrations of carbon dioxide (1.4 $W m^{-2}$) and methane (0.5 $W m^{-2}$).

Changes in the atmospheric concentrations of chemical species not only modify the global energy balance in the Earth System, but they also affect the atmospheric circulation when temperature gradients are generated. This effect is probably most significant in the stratosphere where, for example, the formation of the springtime ozone hole in Antarctica produces a cooling of the austral polar region, and hence a strengthening of the polar vortex. This process generates a positive feedback since a stronger dynamical isolation of the Antarctic stratosphere tends to reinforce the depth of the ozone hole. In the troposphere, the direct impact of changes in the trace species concentrations on the circulation is believed to be generally small, except perhaps in the case of water vapour. More important are the changes in the aerosol load and composition, which can affect the hydrological cycle, including the precipitation regime.

The reverse effects of how climate changes can potentially affect the chemical composition of the atmosphere must also be addressed. Here different processes must be considered. First, in a warmer and wetter troposphere, the rates of reactions are expected to change, with potential modifications in the budget and lifetimes of chemical species. With higher water vapour concentrations in the atmosphere, the level of the hydroxyl radical is expected to be higher, with an intensification of key oxidising mechanisms, and probably a reduction in the tropospheric abundance of ozone and other chemical compounds. Biogenic emissions of gases, such as isoprene by vegetation and nitrogen oxides by soils, are affected by temperature changes

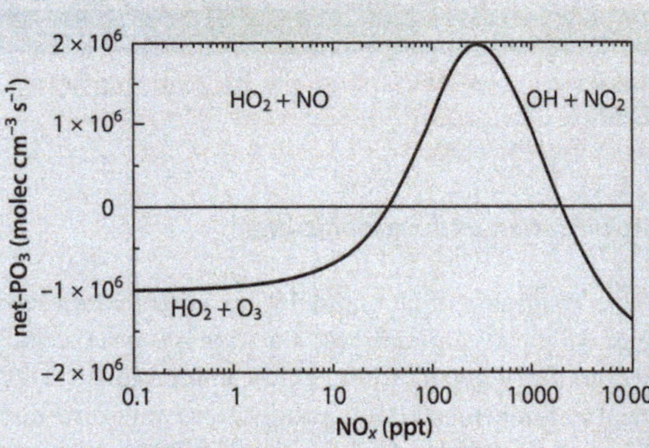

Fig. 4.19. Variation of the net ozone production (molec. $cm^{-3} s^{-1}$) as a function of the atmospheric NO_x ($NO + NO_2$) mixing ratio (pptv) calculated at the Earth's surface for summer time conditions. Under low NO_x levels in clean remote areas, ozone is photochemically destroyed, while in moderately polluted regions, ozone is produced. The threshold value between these two chemical regimes corresponds to an NO_x mixing ratio of approximately 40 pptv. Under the conditions depicted in the figure, the net production reaches a maximum for a NO_x mixing ratio of 300 pptv. Above this level, NO_x becomes sufficiently abundant to titrate ozone molecules. The major mechanisms involved under the different regimes are shown here

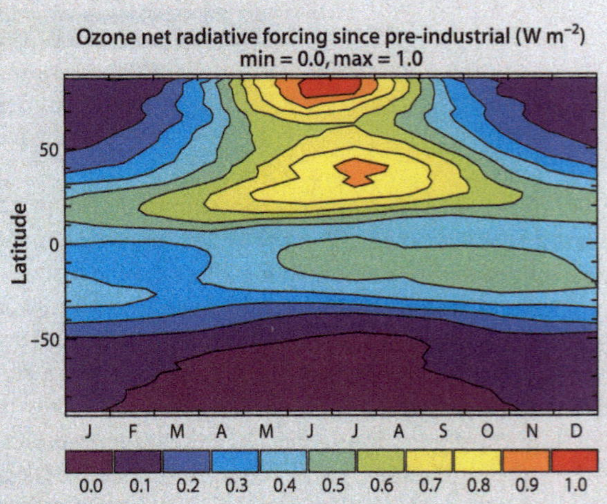

Fig. 4.20. Zonally averaged net radiative forcing ($W m^{-2}$) produced by changes in tropospheric ozone since the pre-industrial era, as estimated by the model of Hauglustaine and Brasseur (2001). The maximum forcing occurs during the summer in the northern hemisphere, where human-induced changes in the ozone concentration have been the largest. Radiative forcing is also significant in the tropics during the entire year

◀

and could therefore be substantially modified under a warming scenario. It is estimated that the emission of isoprene, for example, is enhanced by 10 percent for a 1 °C temperature increase. Tropical sources of chemical compounds could be substantially perturbed by climate changes if the frequency of occurrence and the location of wildfires (biomass burning), and if the distribution and frequency of lightning (which generates nitrogen oxides in the upper levels of the troposphere), are modified in a modified climate. Changes in the tropical ozone budget should result from these perturbations.

Changes could also occur in the exchange rates of gases between the atmosphere and the ocean, with potential effects on the atmospheric aerosol load (sulphates, sea-salt). Wet scavenging of soluble species could be modified as a result of perturbations in rainfall intensity and distribution. Dry deposition of gases and aerosol particles on the surface could be altered by climate-driven changes in land cover and human-induced changes in land use. Finally, changes in the atmospheric circulation from anticipated climate perturbations (large-scale advection, boundary layer ventilation, convective activity, cross-tropopause exchanges) could have a large impact on the distribution and budgets of chemical compounds in the atmosphere. The importance of all these potential effects remains to be quantified.

Finally, climate-chemistry interactions can occur indirectly though disturbances in biogeochemical cycles. An example is provided by the interactions between the carbon and nitrogen cycles. Nitrogen oxides emitted, for example, from fossil fuel combustion are transported and transformed in the atmosphere before being eventually deposited as nitrates on the surface. The deposition of nitrates, which fertilises land and the ocean, should modify the fluxes of carbon dioxide to the atmosphere, with impacts on atmospheric radiative forcing. The fertilisation of the ocean by iron-containing dust particles mobilised in arid regions is also believed to affect the carbon cycle and hence the climate system. Many of these more system-level effects are dealt with in more detail elsewhere in this volume.

Thus, soil acidification is a nonreversible change over anything other than very long time scales. Likewise, acidification of inland and coastal waters from wet and dry deposition of nitrogen compounds may occur with a similar range of ecosystem responses resulting.

Eutrophication of an aquatic ecosystem is defined as an increase in the organic carbon available within the ecosystem (Nixon 1995). For most aquatic ecosystems, the primary source of organic carbon is through photosynthesis, the magnitude of which is dependent on light and nutrient availability. Thus, changes in nutrient availability through, for example, direct discharge to aquatic ecosystems or diffuse input via runoff from agricultural land or atmospheric deposition can lead to an increase in photosynthesis and eutrophication. Ecosystems respond to eutrophication in different ways, depending both on the degree of the eutrophication and the physical characteristics of the system. In some systems, mild eutrophication can result in greater overall productivity in the system and be beneficial for commercial fisheries, for example. More often, however, the ecosystem responses to eutrophication are less desirable. The most obvious response is an increase in phytoplankton biomass but other ecosystem responses including changes in species and/or a reduction in the number of species present, changes in water clarity, and a reduction in oxygen concentration resulting from the bacterial breakdown of organic matter in the system can also occur. In extreme cases, the ecosystem response to eutrophication is anoxia (absence of oxygen) that leads to mass mortality of organisms requiring oxygen. Eutrophication as an ecosystem response is best documented in freshwater systems, where limited water exchange allows only minimal dilution of introduced nutrients. However, in recent decades, it has become obvious that many coastal marine ecosystems have also exhibited eutrophication (e.g., Cooper and Brush 1991, 1993; Jørgensen and Richardson 1996; Andren et al. 1999, 2000; Goolsby 2000).

Fertilisation

In regions where the productivity of terrestrial ecosystems is likely limited by nitrogen, there is some evidence that the deposition of fixed nitrogen compounds (NO_3^-, NH_4^+) may have a fertilising effect. The recent rates of regrowth of some European forests, more vigorous than in the historical past, have been attributed to this effect with atmospheric deposition providing a substantial portion of the required nutrients (Mund 1996). Figure 4.21 shows that the growth rate of young stands of *Picea abies* in Germany and of *Pinus sylvestris* in Siberia was much slower one hundred years ago than in 1996. In addition, the growth rate of old stands has increased at a time when one would expect an age-related decrease in whole stand production; with N deposition the old stand reaches growth rates that are equivalent to those of young stands. A comparison of growth rates with forest inventory yield tables indicates that the young stands grow presently at a higher yield class than the old stands did at the time of their establishment, and that the old stands have changed yield classes and presently approach the same yield class as young stands. However, the more vigorous regrowth of European forests is most likely due to an interacting suite of factors, including better silvicultural practices and probably also CO_2 fertilisation as will as N deposition. At present a definitive, quantitative attribution of enhanced growth to nitrogen deposition is not possible, although it is likely that nitrogen deposition has been significant in driving the changes in growth curves shown in Fig. 4.21. In the future, nitrogen deposition will likely have diminishing effects on growth as systems saturate with nitrogen and other factors become limiting (Scholes et al. 2003).

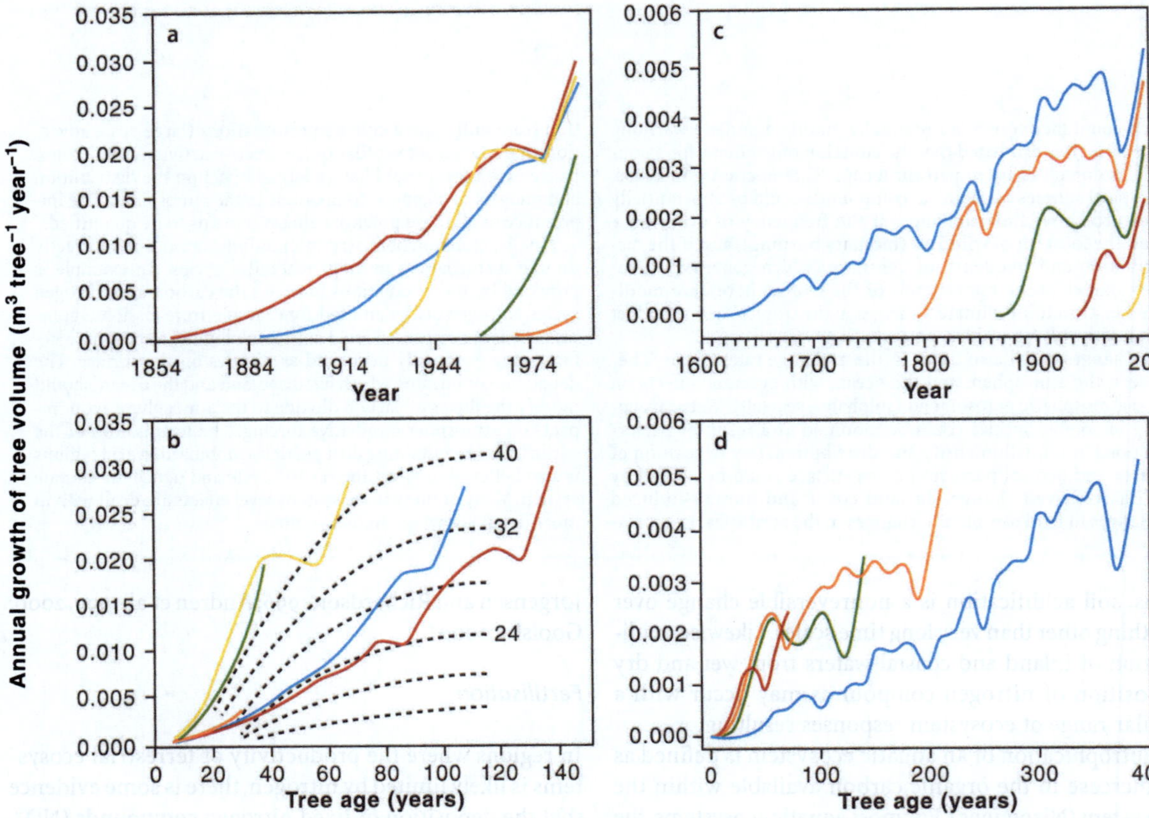

Fig. 4.21. Growth of a chronosequence of *Picea abies* stands in Germany and of *Pinus sylvestris* in Siberia (Mooney et al. 1999)

Toxicity

In high concentrations, the same fixed nitrogen compounds that fertilise ecosystems in lower doses can become toxic. This is particularly likely when other nutrients are strongly limiting and nitrogen compounds are present in excess. This toxicity effect was proposed as one of the factors affecting central European forests during the dieback episodes of the 1970s (Schulze 1995). Tropospheric O_3 can also be toxic to terrestrial vegetation. In some polluted regions there is evidence for strong O_3 impacts, including visible leaf damage on sensitive plant species and even reductions in the growth of some forest tree species (e.g., beech). There is some evidence that increased tropospheric O_3 in the northern mid-latitudes is also leading to significant reductions in crop yields. For example, it is estimated that for wheat there is a 30% yield reduction for a seasonal 7-hour O_3 daily mean of 80 ppb, a concentration level that has been found in parts of the US and Europe (Fuehrer 1996). It is difficult to generalise, however, as responses vary among species, and even between varieties in a single species. In addition, there is some evidence that increasing atmospheric CO_2 may partially ameliorate the O_3 damage through an increase in stomatal closure.

Biodiversity

Manipulative experiments on herbaceous ecosystems have shown that the addition of nitrogen compounds can trigger changes in plant species composition and an overall reduction in diversity. Similar effects have been noted in the field in Europe, where increasing dominance by grasses is occurring in many heath, meadow and forest ecosystems. Some heath species may even have been eliminated from heathlands in the Netherlands (Berendse and Elberse 1990). In general, the impacts are strongest in nutrient-poor systems, where deposition of nitrogen compounds enhances growth of the most responsive species. These then out-compete and eliminate rare species that occupy N-deficient habitats (Mooney et al. 1999; Sala et al. 1999).

4.2.3 Aerosols

Although aerosol particles (discussed earlier in Sect. 3.3.2 and Box 3.3) arise from a wide range of sources, including some natural sources (e.g., dust from arid regions), a significant part of the contemporary increase in particle abundance in the atmosphere is due to fossil fuel combustion. The primary fossil-fuel derived aerosol particle is sulphate, and about 81% of anthropogenic sulphur emissions are from fossil fuel combustion (Heintzenberg et al. 2003). The responses of the Earth System to increased particle loading in the atmosphere occur both in direct and indirect climate forcing and in a range of effects, both beneficial and harmful, on the biosphere.

Climate

In addition to the effects of both non-reactive and reactive gases on climate described above, other changes in the composition of the atmosphere can lead to changes in climate. The distribution of aerosol particles has complex effects on the climate as well as on atmospheric chemistry more generally. The most well-known of the impacts on climate is the so-called direct effect, which is due to the ability of particles to scatter and absorb radiation. This effect reduces the amount of solar radiation reaching the Earth's surface and thus acts to cool it, in opposition to the direct effect of greenhouse gases. However, the direct climate forcing of aerosol particles is not as straightforward as that of greenhouse gases; even the direction of forcing can change depending on the nature of the particle. The best-known effect is due to sulphate aerosol particles, which scatter incoming solar radiation and thus lead to cooling. Carbonaceous particles (black carbon), on the other hand, absorb both solar and infrared radiation and can lead to warming at the surface, partially counteracting the effects of sulphate aerosol cooling. At the same time fossil fuel burning also produces organic carbon particles, which act to cool the Earth's surface.

At the regional scale the effects of aerosol particles on radiative forcing can be highly significant. The INDOEX campaign has shown that over the dry season in South Asia (January to April) the haze layer due mainly to anthropogenically produced particles (Box 3.4) can overshadow the effect of greenhouse gases (Lelieveld et al. 2001; Ramanathan et al. 2001b). During this period the haze layer can cause a net cooling effect at the Earth's surface of ca. 14 W m^{-2}. In contrast, the greenhouse forcing at the Earth's surface is estimated to be 2–3 W m^{-2} at present. Figure 4.22 shows the reduction in absorbed solar radiation at the surface in the South Asia region due to particle loading in the atmosphere.

Aerosol particles also have at least two important indirect effects on the Earth System (Ramanathan et al. 2001a). Figure 4.23 is a simplified representation of a

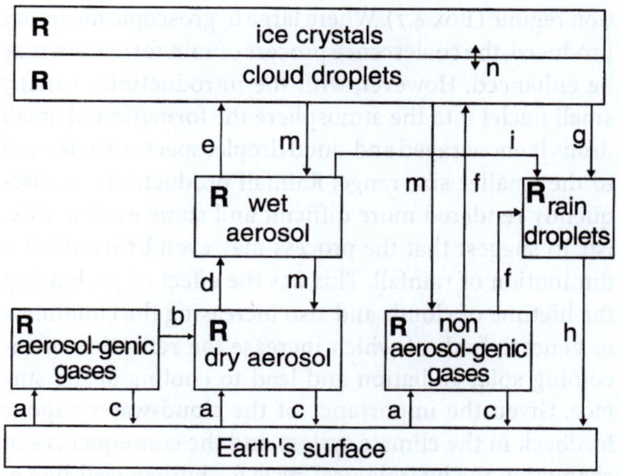

Fig. 4.23. Representation of the multiphase atmospheric system (Facchini 2002). The *arrows* represent physical and chemical processes leading to mass exchange between the different phases: *a* emission, *b* gas-to-particle conversion, *c* dry deposition, *d* condensation, *e* nucleation, *f* dissolution, *g* formation of precipitation, *h* wet deposition, *i* aerosol capture by falling droplets, *m* evaporation, and *n* freezing/melting. Chemical reactions can occur in all phases of the system

Fig. 4.22. Reduction in surface solar radiation absorption due to the Indo-Asian haze effects (measured January to April from 1996–1999) (Ramanathan et al. 2001a)

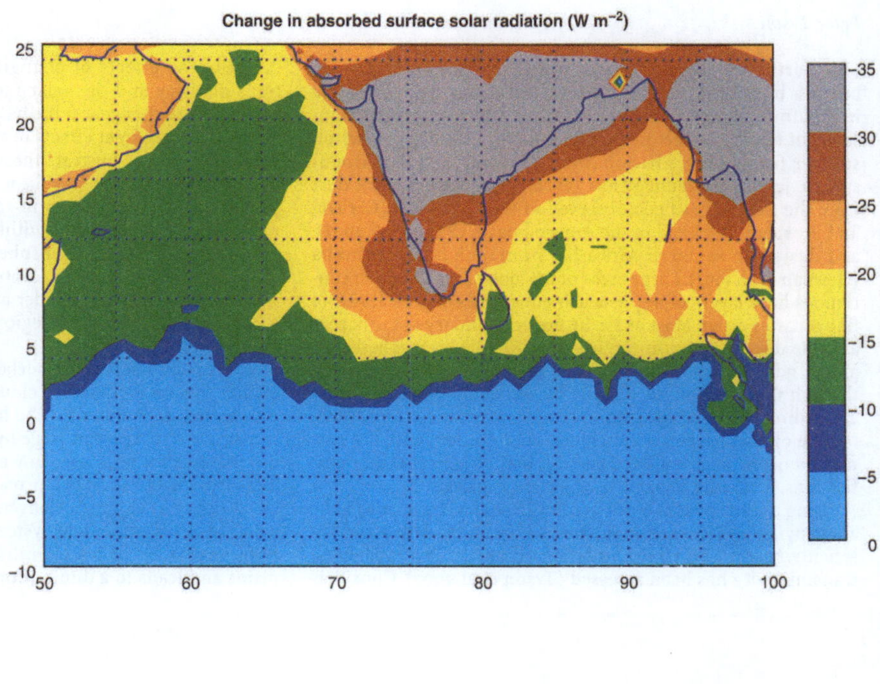

multiphase atmospheric system that captures the indirect aerosol particle effects and puts them into the context of the atmospheric chemical system as a whole. The *arrow* marked *e* is especially important for the Earth System as it shows the role of aerosol particles in acting as condensation nuclei for cloud formation, which gives rise to both indirect effects. The number and nature of the particles is crucial for this process. An increase in aerosol particles can cause an increase in the cloud droplet number concentration, which in turn leads to an increase in reflection to space of solar radiation. This has a cooling effect, and is the first indirect effect of aerosol particles on radiative forcing.

The second indirect effect is related to the precipitation regime (Box 4.7). Where large hygroscopic nuclei are produced, the coalescence process of rain formation may be enhanced. However, with the introduction of many small nuclei into the atmosphere the formation of small drops is encouraged and cloud droplet spectra are skewed to the smaller size range. Rainfall production is consequently rendered more difficult and some evidence exists to suggest that the process may even bring about a diminution of rainfall. This has the effect of prolonging the lifetime of clouds and also increasing the cloudiness in general, both of which increase the reflection of incoming solar radiation and lead to cooling at the surface. Given the importance of the cloud-water vapour feedback in the climate system and the consequences of changes in the hydrological cycle for human well-being, the role of aerosol particles in enhancing or diminishing cloud production and precipitation is looming as one of the biggest questions to be addressed in Earth System science over the next decade.

A further indirect effect on climate via the hydrological cycle may be caused by the increasing aerosol particle loadings. Incoming solar radiation is balanced at the Earth's surface by evapotranspiration and sensible heat flux, with about 70% of the absorbed solar radiation being balanced by evapotranspiration on a global average. If incoming solar radiation is being diminished by increasing aerosol particles in the atmosphere, then evapotranspiration should also be diminished (Box 4.8). This, in turn, would be accompanied by a reduction in precipitation, amounting to an overall spinning down of the hydrological cycle, acting in opposition to the effect of warming on the hydrological cycle. It is not clear, either from model studies or from observations, how important this particular aerosol particle effect might be. However, diminishing subtropical rainfall might be due to it, at least in part. It is noteworthy that rainfall has been decreasing in many parts of East Asia, which has one of the highest aerosol particle loadings in the world.

Fertilisation

There is evidence to suggest that transport and remote deposition of nutrient-laden aerosol particles may be of significance as a source of nutrients to marine and terrestrial ecosystems (Sect. 2.5.2). In this case aerosol particles act in a similar fashion to the nitrogen compounds discussed above. There is some evidence for the long-range transport of nutrients from the Sahara to the Caribbean and Brazil (Formenti et al. 2001) and from southern Africa to the central Indian Ocean (Piketh et al. 2000) via aerosol particles. A similar phenomenon has

Box 4.7. Some Impacts of Aerosols on Rainfall

Peter Tyson

Fine particles derived from the burning of fossil fuels, from biomass burning or naturally from mineral dust typically reside in the atmosphere for up to weeks and may be transported thousands of kilometres before being removed. Industrially derived sulphur from South Africa has been measured at mid-tropospheric levels on Mount Kenya (Tyson and Gatebe 2001); dust from the Sahara has been observed in the Caribbean (Carlson and Prospero 1972) and in the Amazon basin (Swap et al. 1992); and dust from southern Africa has been found over Australia (Herman et al. 1997). Large areas of the globe are covered in contiguous haze layers of air in which enhanced concentrations of fine aerosols occur. Most often the upper boundary of the haze is a stable discontinuity in the lower or middle atmosphere. Regional haze layers reduce the flux of solar radiation passing through the atmosphere, directly by scattering and absorption and indirectly by acting as cloud condensation nuclei and affecting the optical properties of clouds. As a consequence the surface energy balance is altered and this may have important ramifications, including impacts on the hydrological cycle.

Using a simple basin model, the hydrological balance of Lake Tanganyika and its catchment area has been modelled and the sensitivity of modelled precipitation to changes in atmospheric transmissivity has been assessed (Tyson et al. 1997). Considering only the effects of changing aerosol loadings on solar transmissivity and the solar radiation balance, and the feedback of these changes on the basin hydrology, has enabled estimates of the indirect effects of transmissivity on rainfall to be made. The model suggests that a 10% increase in the attenuation of solar radiation over the region of Lake Tanganyika might produce a 15% diminution in rainfall. Such a change in solar radiation attenuation falls within the observed range of interannual variability in atmospheric transmissivity during the second half of the twentieth century in southern Africa. Changes in transmissivity of this order are possible in future if current GCM scenarios for a drier regional future with global warming are realised.

Apart from altering the surface energy balance, aerosols act to alter the microphysics of clouds and may enhance or diminish rainfall depending on the hygroscopic nature and size of aerosols. An increase in large hygroscopic aerosols accelerates the coalescence processes in warm tropical convective clouds, increases the duration of storms and hence the amount of rainfall (Mather et al. 1997). Conversely the addition of numerous small aerosols into cloud systems shifts cloud droplet spectra towards smaller drop sizes, inhibits rainfall-producing mechanisms and leads to a diminution in precipitation.

Box 4.8. The Pan Evaporation Paradox

Michael L. Roderick · Graham D. Farquhar

Imagine a dish full of water sitting inside a laboratory where the relative humidity of the (still) air was less than 100%. Now, imagine that the temperature of the air was increased, while all else was held constant. Under this scenario, the rate at which water evaporated from the dish would increase when the air temperature was increased. Based on reasoning of this sort, there has been an implicit assumption in many disciplines (e.g. ecology, hydrology, agriculture, forestry) that a warming of the terrestrial surface should be concurrent with a general drying out of the surface, all else being equal (IPCC 2001a). Hence, it came as a big surprise when first reported in 1995 that over the previous 50 years pan evaporation (Fig. 4.24) had in general been decreasing over the USA and the former Soviet Union (Peterson et al. 1995). Those data were presented as averages over a great many pans, so there would have been increases at some pans, but the overall trend was a decrease. The surprise was that this had occurred, despite a general increase in surface air temperature in those areas (Folland et al. 2001). Since then, subsequent reports have confirmed these trends by showing that while pan evaporation has increased in some areas, e.g. parts of China (Thomas 2000) and Israel (Cohen et al. 2002), the general trend has been a decrease over the last 50 years, e.g. India (Chattopadhyay and Hulme 1997), parts of China (Thomas 2000) and Italy (Moonen et al. 2002). The fundamental importance of the pan evaporation paradox was recently highlighted by Moonen and colleagues (Moonen et al. 2002), who noted that ecological and hydrological models have generally assumed (or calculated) that pan evaporation should have been increasing. The seemingly divergent trend between surface air temperature and pan evaporation has since become known as the pan evaporation paradox.

The first exploration of this paradox was a proposal, based on the Bouchet hypothesis, that the decrease in pan evaporation could be explained by assuming that the evaporation from the environment surrounding the pan had increased (Brutsaert and Parlange 1998). The underlying basis is that in water-limited environments, the air over the pan is usually cooler and more humid when evaporation from the adjacent environment is high. Consequently, there is a reduction in the vapour pressure deficit, and a resulting reduction in evaporation from the pan. Analysis of rainfall and stream flow data from water-limited environments in both the former Soviet Union and the USA does appear to show an increase in evaporation from the environment (Golubev et al. 2001, Szilagyi et al. 2001) in support of this argument. However, there are two problems with the humidity-based argument. First, it does not make any prediction about changes in evaporation in wet environments, but in these regions, evaporation from both the pan and surrounding environment have decreased (Golubev et al. 2001). The second difficulty is that over the USA, the average vapour pressure deficit has remained virtually constant over the last 50 years (Szilagyi et al. 2001). This shows that a humidity-based mechanism is unlikely to be the most important one. The net result is that one paradox has now led to another – how is it possible for the average air temperature to increase but the average vapour pressure deficit to remain constant?

The key to resolving this second paradox is to consider the definition of the vapour pressure deficit, and from that, determine how this could happen. A previous investigation of that question (Roderick and Farquhar 2002) found that this can only happen for typical surface temperatures (e.g. 5–45 °C) if the increase in dew point temperature is about double the increase in temperature. That is very important, because global observations for the last 50 years show that on average, night time temperatures (~0.2 °C per decade) have been increasing about twice as fast as day time temperatures (~0.1 °C per decade) (Karl et al. 1993; Easterling et al. 1997). This implies that the increase in dew point temperature should be roughly double the increase in day time temperature (Roderick and Farquhar 2002) and that proposition is generally consistent with the observed increase in dew point temperatures in the USA (Gaffen and Ross 1999; Robinson 2000). Hence, the fact that the average vapour pressure deficit in the USA has not changed over the last 50 years is not a paradox – based on the above analysis that is what would be expected. While an increase in dew point temperatures of about double the increase in day time temperature has yet to be verified in other places, it should be at least roughly correct given that the general trend has been for increasing absolute vapour pressures, increasing average temperatures, and declining diurnal temperature ranges (Folland et al. 2001).

Assuming that the average vapour pressure deficit has remained constant and that the changes in wind have been small, the only other explanation for a decrease in pan evaporation is a decrease in global solar irradiance, as has been previously suggested (Stanhill and Cohen 2001; Cohen et al. 2002). A decrease in pan evaporation due to a decrease in sunlight is also consistent with the previously noted observation that evaporation from both the pan and the environment had decreased in wet environments. By a stroke of luck, one of the few places where long term regional scale measurements of sunlight are available is the north-west corner of the former Soviet Union (Abakumova et al. 1996), which coincides with some of the sites used in the original work on the paradox (Peterson et al. 1995; Golubev et al. 2001). The sunlight data from that region show a decrease of 2–4% per decade (1960–1990). From calculations based on accepted physical principles (Penman 1948; Linacre 1993a,b), a change in sunlight of this magnitude would result in a decrease in the annual pan evaporation of ~90–155 mm in the last 30 years. In comparison, the observed decrease in annual pan evaporation from this region varied over the different pans, but was on average ~110 mm, consistent with the estimate based on the reduction in sunlight (Roderick and Farquhar 2002).

On that basis the observed decrease in pan evaporation is not a paradox. On the contrary, it is exactly what is expected based on the observations. The key to resolving the paradox has been to consider the change in sunlight.

Fig. 4.24. Class A pan at Canberra Airport, Australia. In Australia Class A pans are usually equipped with a bird-guard (as shown). This design is not necessarily used elsewhere and care is needed when comparing pan evaporation measurements from different places (see Ohmura and Wild 2002) (*photo:* Jeff Wilson)

been observed in southern Africa. In the water-deficient parts of the ecosystems of the Okavango Delta wetlands, aerosol particles supply up to 52% of phosphates and 30% of nitrates needed by the vegetation (Garstang et al. 1998) (Box 4.9).

Plant Growth Reduction

Increasing aerosol particle loading might also have indirect, negative impacts on vegetation through a reduction of incident solar radiation. Most of the assessments of this effect have so far been carried out on regions with predominantly agricultural ecosystems (Chameides et al. 1999) but the effect would apply equally to more natural systems. In general, vegetation productivity is directly related to incoming solar radiation so, all other things being equal, reductions in solar radiation lead to concomitant decreases in production. The estimates for China suggest that in some areas incident solar radiation has decreased by as much as 30%. Reductions of this magnitude imply significant impacts on food production.

There are also suggestions that an increase in aerosol particle loadings in the atmosphere can actually increase plant productivity through an indirect effect (Roderick et al. 2001). The effect is due to an increase in diffuse as opposed to direct radiation. Diffuse radiation allows leaves lower down in the vegetation canopy, which are shaded from direct radiation by the leaves above them, to photosynthesise and thus increase the overall uptake of carbon by the ecosystem.

The sections above have shown how a single type of anthropogenic forcing – the combustion of fossil fuels – can reverberate through and impact Earth System functioning in very many different ways and in distant places on the planet. To stress the point, a specific illustration of how the use of an air conditioner in a midwestern American home on a hot summer day may influence the ability of an African farmer in the Sahel to grow food for his family is suggested. Box 4.10 shows how this implausible connection might be made.

4.3 Response of the Earth System to Land-Use and Land-Cover Change

Just as for fossil fuel combustion, land-use changes almost never produce responses in the Earth System that operate in simple cause-effect relationships. A single land-use change usually triggers a large number of responses that reverberate and cascade throughout the System, often interacting in complicated ways with natural variability to become feedbacks for yet further change.

Furthermore, it is important to realise that many natural land covers are, in fact, co-evolved landscapes, the result of hundreds and even thousands of years of human-environment interactions. Several examples suffice. The large savannas of Africa have co-evolved

Box 4.9. Impacts of Aerosol Transport on Freshwater Systems: The Okavango Delta Wetland System

Peter Tyson

Annual horizontal mass fluxes of combined natural and anthropogenic aerosols have been estimated for various localities over southern Africa (Garstang et al. 1998; Piketh et al. 1999a,b). Over the central subcontinent, ~12 Mt yr^{-1} are transported over Zimbabwe and Botswana in the direction of the Atlantic Ocean. The deposition of particulate nutrients from the plume being transported may have significant implications for the nutrient balance of ecosystems. Such is the case for the 12 000 km^2 inland Okavango Delta wetlands of Botswana, where annual flooding of the ecosystem from late-summer rains in the Angolan highlands occurs during the dry season (July-August), when aerosol loading and dry deposition are greatest (Garstang et al. 1998) (Fig. 4.25). The general absence of chemical weathering in the Okavango and upstream environment results in a low dissolved-solids load in the water of the river, most of which consists of silica and calcium and magnesium bicarbonates. Nutrient species occur at low concentrations: nitrogen (as nitrate or ammonium ions) is typically 0.08 ppm, phosphorus 0.04 ppm and potassium 2.9 ppm. Atmospheric aerosol deposition is estimated at 0.57 kg ha^{-1} d^{-1} over the delta as a whole. In the delta distributary channels, water supplies 90% of the nitrate and phosphate needs of plants. However, in low productivity areas beyond the edges of the channels (which constitute 90% of the area of the delta) aerosols supply up to 52% of phosphates and 30% of nitrates (but only around 10% of potassium needs). The hitherto ignored atmospheric contribution to nutrient cycling and the nutrient budget of the wetland ecosystem is substantial.

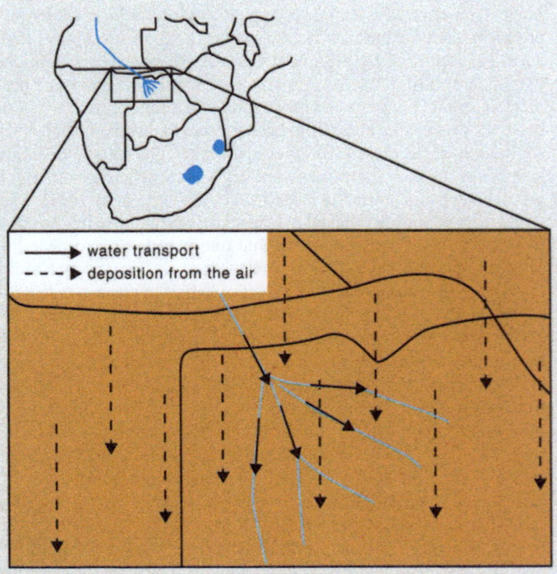

Fig. 4.25. Aerosol transport and atmospheric deposition of nitrogen and phosphorus over the Okavango Delta wetland ecosystem, Botswana (Garstang et al. 1998)

Box 4.10. From North America and Europe to the Sahel: A Cascade of Effects Triggered by Fossil Fuel Combustion

Ulrike Lohmann

How can the use of an air conditioner in a midwestern American home on a hot summer day or a drive down the autobahn in central Germany influence the ability of an African farmer in the Sahel to grow food for his family? This implausible connection demonstrates how an anthropogenic forcing, the combustion of fossil fuels, can cascade through the Earth System to have completely unsuspected impacts elsewhere. Such reverberations are not strict causal chains in which each step is linked by a tight, unique cause-effect relationship. Rather, along the way the original anthropogenically driven changes interact with modes of natural variability in the Earth System as well as with specific local and regional characteristics to produce the observed outcome. However, in many such cases as the one described here there is considerable evidence that the original anthropogenic forcing played a role in the ultimate impact observed in a far distant place on Earth.

The cascade of impacts begins with the production of energy, required both to produce the electricity to run the air conditioner in the North American home and to power the motorcar in Europe. In both cases, fossil fuel combustion was the likely mode of power production (Fig. 4.26). As described earlier in this chapter, fossil fuel combustion influences the Earth System in a number of ways, including the production of sulphate aerosols. Since about 1990, the production of sulphate aerosols has been significantly reduced in many parts of North America and Europe, but in the decades before that time, sulphate aerosols were an important local and regional air quality issue.

In addition to their local and regional effects on air quality, sulphate aerosols change the radiative balance near the Earth's surface in a direction opposite to that of greenhouse gases, leading to a near-surface cooling. The effect is not global, however. The aerosols have a short lifetime in the atmosphere and are deposited back onto the land or sea surface before they have circulated widely around the atmosphere. Thus, the surface cooling is largely a regional effect.

The cooling over Europe and North America in the 1960s, 70s and 80s changed the latitudinal temperature differential between the North Atlantic region and the tropics and subtropics, both in the atmosphere and in the surface layer of the ocean. This in turn changed the atmospheric circulation patterns and led to a shift in the positioning of the African monsoon system over the northern part of the continent. The result was a change in the rainfall patterns over the North African region, and a drying climate in the Sahel during those decades. Land degradation increased during that time and there were dire predictions of the Sahara Desert marching southward. The resulting impacts on the people in the region were severe, with an increase in poverty, malnutrition and starvation (Fig. 4.27). The phenomenon was often cited as a case of grazing and cropping pressure, driven by population increase, overwhelming the resilience of the natural ecosystem, whereas in reality the problem may have been partly triggered by forcings far away from North Africa.

How plausible is this chain of events? Figure 4.28 shows the change in global precipitation patterns simulated when a general circulation model is forced by changes in the observed sulphate aerosol loadings in the Northern Hemisphere compared to the observed precipitation changes around the world during the last century. The agreement for the low latitudes is remarkably good, indicating that sulphate aerosols may indeed be partly responsible for the drought in the Sahel. In addition, when sulphate aerosol loadings were significantly decreased during the 1990s, the rainfall increased again in the Sahel, leading to an easing of the degradation problem and an increase in productivity of the ecosystems of the region. Interestingly, the grazing and cropping patterns of the people in the region did not change significantly during this transition.

Although there is no conclusive proof of the reality of the cascade of events described above, the circumstantial evidence is strong. Given the connectivity of forcings and feedbacks in the natural functioning of the Earth System (cf. Chap. 2), it should not be surprising that anthropogenic forcings of sufficient magnitude will reverberate through the Earth System in ways that are often counterintuitive and difficult to predict.

Fig. 4.26. Combustion of fossil fuels for energy production, leading to the generation of SO_2 and to sulphate aerosol particle production in the atmosphere

Fig. 4.27. A malnourished child is dwarfed by a ruined cereal crop in the Navrongo district of Gambia, where two-thirds of the crops was destroyed by drought (*photo:* FAO photo archives)

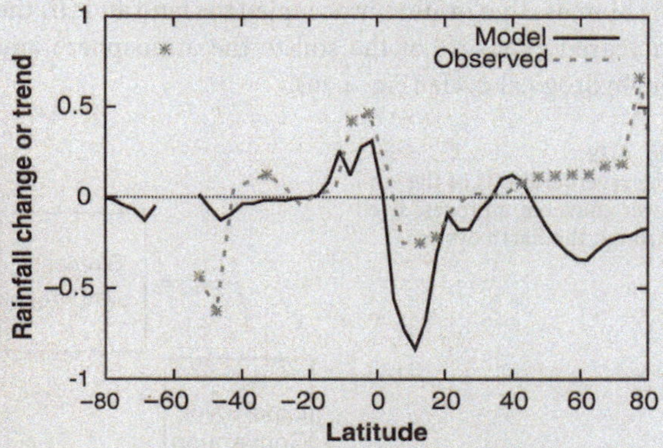

Fig. 4.28. Zonally averaged trend in observed annual-mean precipitation over the period 1901–1998 (mm d^{-1} $century^{-1}$) (*dotted line*) and zonally average difference in annual-mean precipitation (mm d^{-1}) (*solid line*) between present-day and pre-industrial simulations with the CSIRO general circulation model. Points at which the observed trend is significant at the 5% level are shown as *asterisks* (Rotstayn and Lohmann 2002)

through human burning, apparently reducing the area covered by woodlands or thinning the woodlands sufficiently to create more open grasslands. Similarly, the open Australian landscapes that European immigrants found in the 1700s were likely the result of tens of thousands of years of burning by the Aboriginal population (Flannery 1994). The Amazon basin is replete with *terra preta* (black earth) sites that are recognised as an indication of pre-European settlement (Woods and Mann 2000). Such settlement is known to have shaped the species composition of contemporary tropical forests in various regions (Alcorn 1984; Gómez-Pompa et al. 1987).

Contemporary land-use change follows various trajectories after the initial conversion from the natural land cover or from a traditional agricultural system such as slash-and-burn (which when practiced with sufficient fallow period had minimal impact on the natural ecosystem). A common pathway is for less intensive agriculture to be practiced initially following conversion, followed some years later by intensification – the use of fertilisers, irrigation and pest control, for example – to increase yields. This trajectory will become more common as food demand rises and less land is available for conversion. Intensification modifies the cascading effects of land-cover conversion, damping some but intensifying others, such as the addition and mobilisation of nitrogen compounds and their loss to the atmosphere and to waterways. In this section the responses of the Earth System to the land-cover conversion process itself, to the subsequent effects of the changed land cover and to the effects of the intensification of agriculture are discussed in turn.

4.3.1 The Conversion Process

The immediate effects of the land conversion process itself operate primarily in two ways: (*i*) the burning of the biomass that originally occupied the land and (*ii*) the increased exposure of the soil to the atmosphere and the hydrogical cycle (Fig. 4.29).

Biomass Burning

Biomass burning is nearly always used for initial clearing of the original vegetation for agriculture; it is thereafter used for eliminating agricultural waste products on the land. Annual biomass burning has increased by perhaps 30–50% over the past century (Scholes et al. 2003). As was shown in Sect. 2.4.3, the burning is accompanied by the emissions of a wide variety of gases and particles, the exact mix depending on the nature of the fuel, its wetness, and the temperature of the fire. Most prominent of the emissions is the flux to the atmosphere of CO_2.

Carbon cycle. The emission of CO_2 from land-use change is a significant component of the anthropogenic perturbation to the global carbon cycle and is estimated to be about 1.7 Gt C yr^{-1} for the decade of the 1990s (IPCC 2001a). While this figure includes the emissions from the decomposition of the remaining biomass following the fire and the emission from soils due to land-use change, a significant fraction of the total is due to biomass burning. The cascading effects of increasing CO_2 in the atmosphere have been described earlier, and operate irrespective of the source of CO_2, fossil fuel combustion or biomass burning.

Reactive gases. In addition to CO_2, a broad array of trace gases is emitted to the atmosphere from biomass burning and from the soils after the burning. Increased emissions of N_2O and NO have been reported for up to 10 years from converted pastures following forest clearing (Matson et al. 1989). The responses in the atmosphere will be correspondingly long. Given their long lifetimes, these gases not only alter the composition of the atmosphere locally where the burning occurs but are transported long distances and contribute to regional and eventual global modification of the atmosphere. Two examples make the point. First, burning in Amazonia produces O_3 precursor gases that result in a higher tropospheric O_3 maximum over the central South Atlantic Ocean and further afield

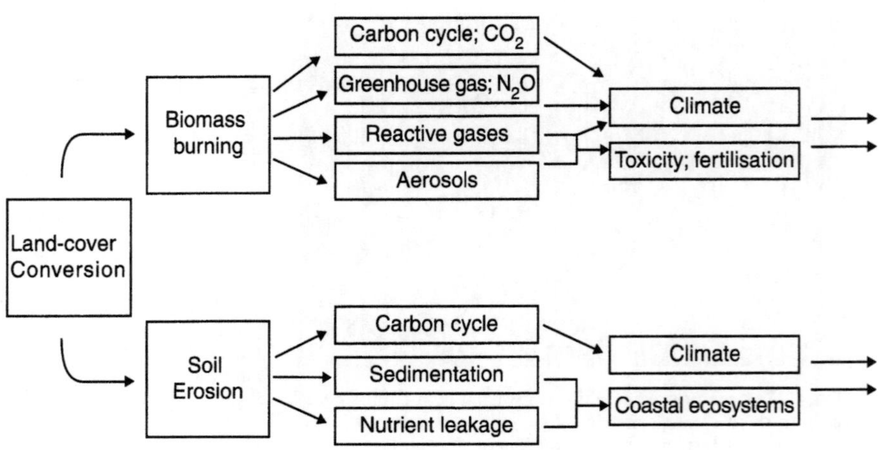

Fig. 4.29. The reverberations of the land-cover conversion process itself through the Earth System

(Penkett et al. 2003). Secondly, among the results of biomass burning in Southeast Asia is the production of CO, an oxidising gas that has a number of implications for Earth System functioning. In early 2001 the burning in Southeast Asia formed plumes of CO that extended all the way across the Pacific Ocean to the west coast of North America (Fig. 4.30).

The emission of trace gases also represents a loss of nutrients from terrestrial ecosystems from which the subsequent agricultural system is derived and thus can become a limiting factor in productivity. It is common for agricultural production to be promoted for a short time, usually several years, from the nutrients contained in the ash from the fires associated with conversion but for productivity to drop sharply thereafter. This can lead to abandonment and further clearing of natural forest or, if sufficient resources are available to the farmer, to an intensification of the agricultural system with the application of artificial fertiliser.

Aerosol particles. Biomass burning generates a range of aerosol particles. The reverberations of aerosol particle emissions through the Earth System were described earlier in this chapter. Their effects on climate and on human health (Sect. 5.3.3) are particularly relevant. By far the most important particles produced in biomass burning are carbon compounds, which can have a warming or cooling effect on climate depending primarily on the size and nature of the particles (black carbon (soot) or organic carbon particles – Sect. 4.2.3). Estimates of the emissions of carbonaceous particles from biomass burning are 42 Tg C yr^{-1} for organic compounds and 4.8 Tg C yr^{-1} for black carbon, but there is much uncertainty associated with these numbers (Scholes et al. 2003). By contrast, most biomass fuels have low sulphur contents and thus the emission of SO_2 from biomass burning is relatively unimportant compared to the emissions from fossil fuel combustion.

Soil Erosion

The direct on-site effects of converting a forest to grazing or cropping include the loss of soil organic carbon via erosion and transport off-site by rivers or via oxidation and loss to the atmosphere. The loss of soil organic matter with conversion of natural ecosystems to agriculture is a prime example of a response to land conversion. Soil organic matter plays an important role in ecosystems, providing the organic substrate for nutrient release, aiding the maintenance of soil structure and water holding capacity and providing protection against erosion. With conversion to permanent agriculture, up to 50% decreases in soil carbon have been reported (Paustian et al. 1997). Figure 4.31 shows a common trajectory for this loss of carbon with time. Initially a sharp decline in soil carbon occurs with, and immediately following, the conversion process. Some time later the level of soil organic matter in the soil normally stabilises and reaches a new equilibrium value.

Carbon cycle. The loss of soil organic matter with conversion of natural ecosystems to agricultural production systems also produces reverberations through the

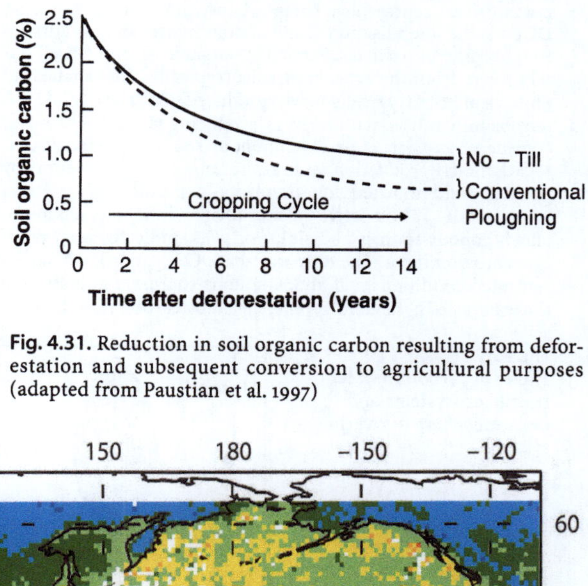

Fig. 4.31. Reduction in soil organic carbon resulting from deforestation and subsequent conversion to agricultural purposes (adapted from Paustian et al. 1997)

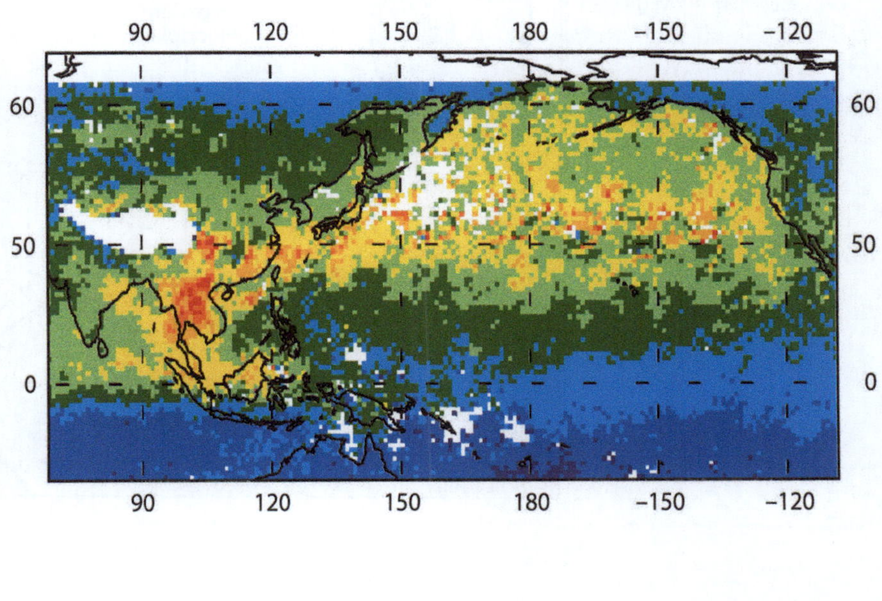

Fig. 4.30. Satellite remote sensing in February 2001 showed strong production of carbon monoxide centered in Thailand, a result of seasonal burning as part of the normal agricultural practices. The carbon monoxide formed a plume that extended all the way across the Pacific Ocean to the west coast of North America (*image:* National Center for Atmospheric Research (NCAR) and National Aeronautics and Space Administration (NASA), USA, 2001)

Box 4.11. Emission of CO_2 from Riverine Systems

Jeffrey E. Richey

The partial pressure of carbon dioxide dissolved in river water (pCO_2) represents a deceptively simple expression of the coupling of the water and carbon cycles between terrestrial and fluvial environments. The distribution of pCO_2 across a river basin is a function of a long sequence of complex biological and weathering processes and interactions, reflecting both internal carbon dynamics and external biogeochemical processes in upstream terrestrial ecosystems. The downstream expression of this coupling is the amount of organic matter and dissolved inorganic carbon mobilised to and through a river system, augmented by in-stream or riparian primary production and respiration. Perhaps the most evocative aspect of pCO_2 is that it is almost always present at concentrations much greater than the atmosphere (that is, it is supersaturated). The question is why, and what are the implications?

What are the sources of pCO_2, both direct and indirect (Fig. 4.32)? Total dissolved inorganic carbon (DIC) is produced via weathering, as the dissolution of primarily carbonate rocks. This process establishes the alkalinity and influences the pH of water, which governs the subsequent partitioning of DIC between pCO_2, bicarbonate, and carbonate ions. The DIC in groundwater is enriched many-fold by the CO_2 produced by the decomposition of organic matter in soils (in productive environments, soil CO_2 may be hundreds of times supersaturated relative to the atmosphere). Hence the DIC entering a stream has both an inorganic weathering component, and an organically-produced respiration component. The land also exports organic matter as dissolved organic carbon (DOC) in groundwater. The DOC available for export to rivers represents a balance between production of *fresh* DOC via the solubilisation of soil organic matter and the adsorption to mineral particles. Particulate organic carbon (POC) enters rivers from the erosion of soils (typically older materials) and as leaf litter (typically newly-produced). Both DOC and POC may be mineralised within rivers, producing pCO_2. These are all considered as external, or allochthonous, sources. The *in situ* (autochthonous) production and respiration of organic matter (by plankton and attached aquatic plants) can both consume and produce pCO_2. The relative balance of autochthonous relative to allochthonous sources and sinks for pCO_2 indicates what processes are dominant. The only way that pCO_2 can exist at supersaturated conditions is if allochthonous sources dominate, and the waters are net heterotrophic, fueled by carbon from land.

In fact, pCO_2 is present at elevated levels in most rivers of the world, from small streams to large rivers. Kempe (1982) called early attention to the elevated levels of pCO_2 in many rivers, and that this was a sensitive indicator of the sources for river respiration. Jones and Mulholland (1998) analysed a time series of elevated pCO_2 in a small temperate stream. Cole and Caraco (2001) computed that the average pCO_2 concentration in 47 rivers averaged 3 230 µatm, or nearly 10 times saturation. Similar conclusions can be drawn from a wide survey of the literature. There are two important consequences of this. The first is that by far the majority of this CO_2 must be derived from the respiration of organic matter of terrestrial origin (allochthonous production). If the pCO_2 were derived from primary production within the water (autochthonous production), the pCO_2 would be near or below equilibrium (which certainly happens in localised environments). The second consequence is that according to the rules of gas exchange, this CO_2 is outgassed (evaded) back to the atmosphere (that is, it becomes a source of CO_2 to the atmosphere). In *toto*, the export of CO_2 and organic matter from land to rivers constitutes a significant sink of terrestrial net ecosystem production.

How large is the return flux (outgassing) of CO_2 to the atmosphere? Telmer and Veizer (1999) computed that outgassing was about 30% of the DIC export in the Ottawa River. Applying that ratio to the global export of DIC to the ocean, they computed that the flux of CO_2 to the atmosphere from rivers would be 0.13 Pg yr^{-1}, or about an order of magnitude higher than early estimates (e.g., Kempe 1982). Cole and Caraco (2001), using a gas exchange coefficient from the Hudson River, 47-river average as representative of flowing waters in general, and assuming that rivers cover ~0.5% of land surface area, computed a global outgassing of ~0.3 Pg C yr^{-1}. More recently, Richey et al. (2002) computed that outgassing of CO_2 from rivers and wetlands of the central Amazonian basin was about 1.2 Mg C ha^{-1} yr^{-1}, an amount comparable to conservative estimates of carbon storage in the Amazon (i.e., an equivalent partitioning of net ecosystem production). Extrapolated across the entire basin, this would produce a flux of about 0.5 Pg yr^{-1} from the Amazon alone. This is an order of magnitude greater than the fluvial export of organic carbon and DIC from the Amazon to the ocean. In contrast to other studies, this calculation emphasised the full drainage network, from first-order streams to the river

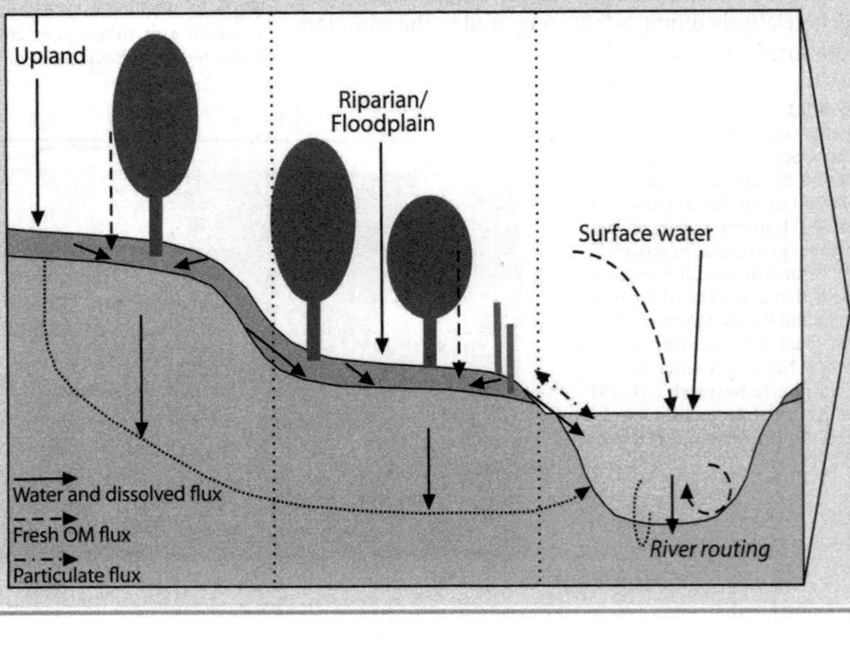

Fig. 4.32. Fluxes of carbon from terrestrial ecosystems and the atmosphere to riverine systems

◀

mainstem and flood-plains, and was done for a specific region of the humid tropics. Assuming that the humid tropics behaves uniformly, then the total outgassing from the tropics would be about 0.9 $Pg\,yr^{-1}$. If the estimates of the tropics are then added to the estimates for more temperate systems, total outgassing likely exceeds 1 $Pg\,yr^{-1}$.

The outgassing of CO_2 is supported by extensive oxidation of organic matter of terrestrial origin within the river systems. This raises a very interesting question. The prevailing wisdom is that riverborne organic matter is already very refractory and not subject to oxidation (after centuries on land). There is evidence based on ^{14}C that the mix of dissolved and particulate organic matter in transport is of variable ages but often quite old, and that the production of CO_2 in rivers is supported by old carbon (Cole and Caraco 2001; Raymond and Bauer 2001). That is, organic C that had resided in soils for centuries to millennia without decomposing is then decomposed in a matter of a few weeks in the riverine environment. Thus, pre-aging and degradation may alter significantly the structure, distributions and quantities of terrestrial organic matter before its delivery to the oceans.

It is necessary to put the outgassing flux into context by relating it back to current views of the role of rivers in the global carbon cycle (Richey et al. 2003). As the main pathway for the ultimate preservation of terrigenous production in modern environments, the transfer of organic matter from the land to the oceans via fluvial systems is a key link in the global carbon cycle, hence the *role* is most typically expressed as the fluvial export of total organic and dissolved inorganic carbon to the ocean. These fluxes are in the range of 0.2–0.4 $Pg\,yr^{-1}$ for DOC, 0.2–0.5 $Pg\,yr^{-1}$ for POC, and 0.4 $Pg\,yr^{-1}$ for DIC. On the order of 0.2–1 $Pg\,yr^{-1}$ may be stored in reservoirs and as part of overall continental sedimentation. As a global aggregate, there would appear to be a net sink (between continental sedimentation and marine sedimentation and dissolution) of ~1 to 1.5 $Pg\,yr^{-1}$. As bulk numbers, these sinks are partially compensated for by the outgassing. But it is likely that these processes are very geographically disperse, with the continental sedimentation occurring in northern temperate regions, and much of the marine sedimentation and outgassing occurring in more tropical regions.

Overall, this sequence of processes suggests that the organic carbon that is being respired is translocated in space and time from its points of origin, such that over long times and large spatial scales, the modern aquatic environment may be connected with the terrestrial conditions of another time (Richey et al. 2003). Linkages between land and water would be stronger than traditionally thought, with river corridors representing a significant downstream translocation of carbon originally fixed on land.

Earth System via its role in the global carbon cycle. It is often assumed that most of this carbon is oxidised and lost to the atmosphere as CO_2. However, some recent analyses (SV Smith et al. 2001) suggests that most of this carbon is lost via erosion rather than oxidation and thus is not transferred to the atmosphere. Rather, it is transported off the land and enters the river networks that transport material to the coastal zone. Much of the eroded carbon may thus not be immediately emitted to the atmosphere, but instead be stored in the large number of large dams and small pondages that modify river flow to the coast. If so, this represents a significant terrestrial sink for carbon as the soil organic matter would be stored in anaerobic conditions away from contact with the atmosphere. This scenario does, on the other hand, raise the potential for increased CH_4 emissions; in addition, recent evidence points to potentially significant CO_2 emissions directly from riverine systems (Box 4.11).

Sedimentation. Another aspect of the mobilisation through the land conversion process of soil organic matter and other materials, in both dissolved and particulate forms, and its riverine transport is the eventual impact on the coastal zone. Although these materials often follow long and complex pathways, undergoing transformations and temporary storage along the way, the impacts on the geomorphology and biogeochemistry of the coastal zone are profound (Sect. 3.3.4). The current global mean annual flux of suspended sediment to the coast is between 16 and 20 × 10^9 tonnes (Ludwig and Probst 1998; Milliman and Syvitski 1992). Stratigraphic data suggests that this flux is probably double the quantity prior to the emergence of significant agriculture. Much larger changes are known from headwater catchments, where conversion to agriculture has increased erosion rates by up to 1 000 times. Much less sediment and particulate nutrients actually reach the coast. Thus, considerable quantities are trapped or immobilised in catchments because of storage on hill slopes, floodplains and in farm dams and reservoirs. About 15% of the contemporary riverine sediment flux is trapped in large reservoirs (Vörösmarty et al. 1997).

Nutrient leakage. Erosion of soil with the conversion process also leads to the leakage of nutrients from the ecosystem, both through the increased emission of nitrogen gases following clearing and the loss via water erosion of dissolved and suspended nitrogen and phosphorus compounds. In many places it is difficult to determine this effect quantitatively as the riverine flux of nitrogen is now dominated by the effects of the application of artificial fertilisers. However, the Amazon Basin may provide one example where the changes in riverine fluxes are still largely dominated by the effects of the clearing process itself. The initial results from the Large-scale Biosphere-Atmosphere Experiment in Amazonia are intriguing (Neill et al. 2001). Using a paired watershed approach, the study suggests that clearing of tropical forest and conversion to pasture results in a decrease of NO_3^- and an increase in PO_4^{3-} in the river systems, precisely the opposite of what would have been predicted from studies in the temperate zone (Fig. 4.33).

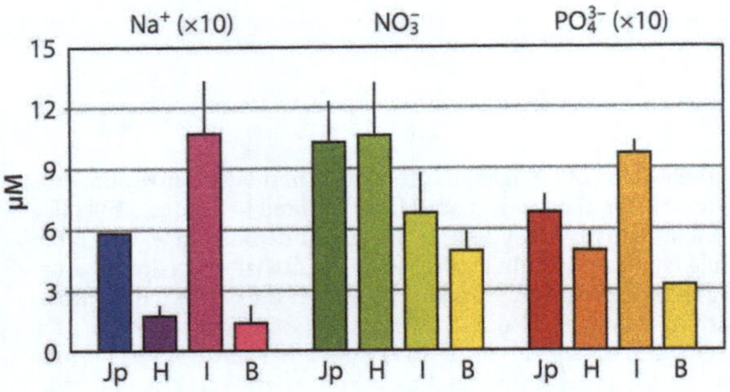

Fig. 4.33. Na^+, NO_3^- and PO_4^{3-} in the Ji-Parana River basin over the discharge hydrographs for the Ji-Parana River *Jp* (upstream to downstream), the headwater rivers *H* (as an average of Comemeracao and Pimento Bueno), impacted rivers *I* (Jaru, Rolim, Machadinho, Urupa), and downstream blackwater river *B* (the Rio Preto) (Richey et al. 2001)

4.3.2 Effects of Changed Land Cover

The reverberations through the Earth System of the conversion of natural ecosystems to agricultural production systems do not end with the conversion process itself. Once the production system is in place and the initial impacts of the conversion process have finished or been stabilised, the main effects on Earth System functioning operate directly through the change in ecosystem structure and indirectly through the associated changes in ecosystem functioning that accompany the change in structure. As discussed in some detail in Chap. 2, the terrestrial biosphere plays an active role in many ways in the functioning of the Earth System and so significant changes in its structure will necessarily have important implications for many aspects of the global environment. The framework for the discussion that follows is shown schematically in Fig. 4.34.

Change in Ecosystem Structure

Many direct effects of changes in the structure of terrestrial ecosystems on the functioning of the Earth System are possible. Two of the most important are the effects on climate of changing albedo and roughness and on biodiversity through changes in the degree of connectivity (lack of fragmentation) within an ecosystem.

Albedo and roughness. The structure of the land surface affects the climate system directly through its reflectance or absorption of incoming solar radiation and through the drag that it exerts on wind flow, that is, its roughness. Land clearance has the immediate physical effect of changing both surface roughness and albedo. The response to these changes is reflected in alterations to the atmospheric processes that lead ultimately to changes in atmospheric circulation and eventually to climatic change. Such feedbacks contributed in part to the existence of savanna vegetation in the Sahel/Sahara region in the early Holocene (Charney 1975; Claussen and Gayler 1997; Claussen et al. 1998).

A contemporary case study concerns the effects of possible reforestation in the high latitudes in response to the Kyoto Protocol. One way of slowing the build-up of CO_2 in the atmosphere is to sequester more of it in terrestrial ecosystems through deliberate changes in ecosystem structure. This is the rationale for the land-use, land-use change and forestry clauses in the Kyoto Protocol. Countries can receive credits against fossil fuel emissions if they deliberately plant forests to sequester carbon. If extensive reforestation were to take place in the high latitudes, however, the overall effect might well be counterproductive. Some exploratory calculations (Betts 2000) have shown that the increase in heat absorbed at the Earth's surface due to the darker surface created by the trees more than offsets the cooling that would result from the uptake of carbon from the atmosphere through the growth of the trees. Although it is unlikely for other reasons that extensive reforestation will take place in the high latitudes, this model experiment highlights the potential importance of changes in ecosystem structure for the functioning of the Earth System.

The other aspect of changing ecosystem structure – changes in surface roughness – can also lead to some unexpected reverberations in the Earth System. A case in point is the impact of phosphate mining on Abbott's booby, an endangered species of sea bird, on Christmas Island in the Indian Ocean. The island was originally covered with moist forest, but decades of mining has converted the landscape into a mosaic of intact forest patches and cleared areas. This pattern has changed the nature of the air flow in the atmospheric boundary layer, in particular, leading to areas of enhanced turbulence in the forest adjacent to the boundaries between forest and cleared area. Abbott's boobies prefer to nest in the tops of the forest trees and many of their favoured nesting areas happen to be in areas of enhanced turbulence. The result has been interference with the bird's breeding patterns and a drop in numbers, leading to their being considered an endangered species (Raupach et al. 1987). This is an example of how land-cover change can create a biodiversity problem by a rather unusual pathway.

Although this case study operates at a local scale and concerns biodiversity, the effects of fragmentation of landscapes may be felt at larger scales with consequences for other aspects of the Earth System. One simulation shows that patterns of landscape fragmentation have impacts on atmospheric circulation at larger scales and may even contribute to changes in regional climate such as the Asian Monsoon (Fu et al. 2002).

Fragmentation. The relationship between fragmentation of landscape structure and biodiversity is complex. Fragmentation can, up to a point, increase overall biodiversity through the creation of a wider range of habitats. This is

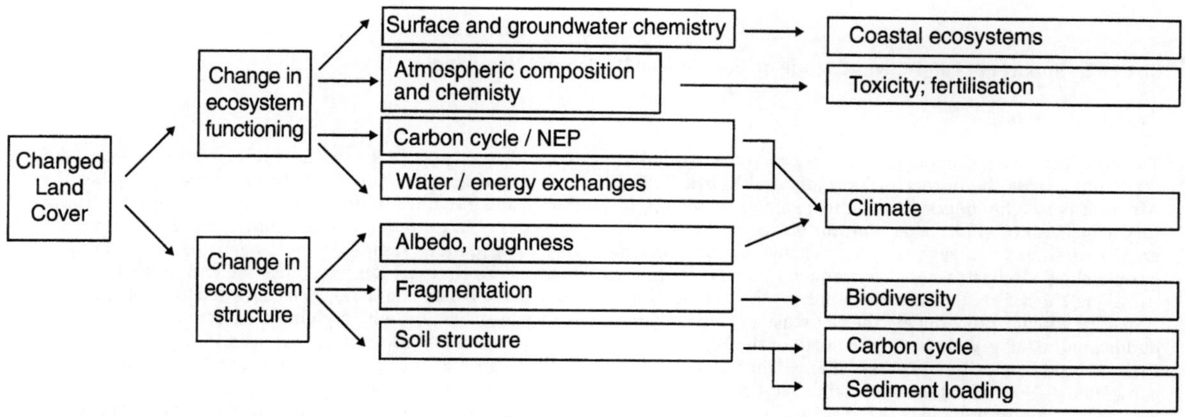

Fig. 4.34. The cascading effects of changed land cover through the Earth System

true primarily for species that do not require large habitat areas or connectivity of habitats to survive. For example, there are some case studies from Europe where the reconstruction of past biodiversity from pollen records has shown that the historical human impact has increased taxonomic diversity at landscape scale through the creation of a greater variety of habitat types (Birks and Line 1992; Berglund 1991; Odgaard 1999; Lotter 1999). In general, in many parts of the world ecosystems that are considered pristine have in fact been affected by human activities in the past and their current biodiversity reflects this influence (Oldfield and Dearing 2003).

Notwithstanding the examples above, the major contemporary response to habitat fragmentation and associated changes in ecosystem connectivity is the loss of biodiversity (Fig. 4.35a). Although the new land cover type following change will support its own assembly of species with its own relationship with connectivity, the overwhelmingly common trend globally is for land cover conversion to lead to significant losses of biodiversity. Conversion of forests to agriculture is considered to be the single most important factor in the current global species extinction rate, which is occurring 100 to 1 000 times more rapidly than before human dominance on Earth (Lawton and May 1995; Sala et al. 1999). Forest clearing fragments natural habitats creating dispersed populations of animal, plant, and microbial species from formerly connected and interacting populations. The extent to which these populations can disperse among patches of habitat to re-supply extinction-prone small local populations is determined by both species characteristics and the size, distribution, and quality of patches and the areas between; the overall relationship between biodiversity and connectivity of landscapes is thus usually highly non-linear (Keitt et al. 1997) (Fig. 4.35b).

An East African example illustrates how land-cover change has disrupted the movement of organisms and led to the loss of populations on a regional scale (Box 4.12). Many of Africa's large herbivores rely on migration pathways to track seasonal changes in climate,

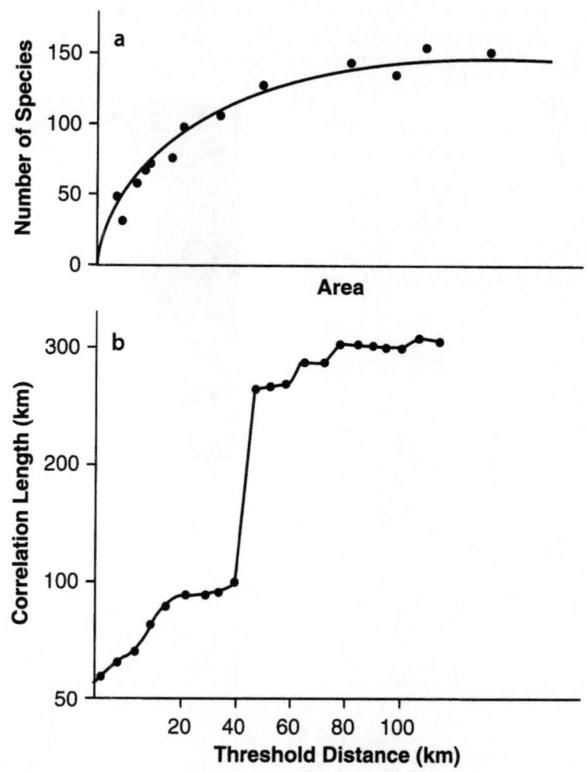

Fig. 4.35. The interaction of habitat and species diversity as shown by the relationships between: **a** habitat fragmentation (patch size) and species loss; **b** habitat connectivity and species dispersal for the landscape patterns of Ponderosa pine and mixed-conifer forests of the southwestern USA (Keitt et al. 1997). Threshold distance is the distance an individual of a species is intrinsically capable of dispersing. Correlation length represents a species' perception of landscape connectivity, that is, the average length that a randomly-placed individual perceives as high-quality, connected habitat (e.g., forest) before a barrier (e.g., agricultural land) is reached

water availability and grasses. In recent decades, with agricultural expansion into wildlife areas, some areas of the Serengeti-Mara ecosystem in Kenya have been converted to intensive wheat farming in response to global market forces. The area where this has occurred is in the traditional migratory pathway of wildebeest and

Box 4.12. Impacts of Land-Cover Change on East African Wildebeest Populations

Suzanne Serneels

The Serengeti-Mara ecosystem is an area of some 25 000 km² stretching across the border of Tanzania and Kenya, in East Africa. It is roughly defined by the movements of the migratory wildebeest (*Connochaetes taurinus mearnsi* Burchell). Migratory wildlife species such as wildebeest, zebra and Thomson's gazelle show similar seasonal movements between habitats, using the short grasslands in the south of the Serengeti National Park in Tanzania during the wet season (January–June) and using the tall grasslands in the north of the Serengeti and in the Masai Mara National Reserve (Kenya) during the dry season (August–November). Another wildebeest population covers a smaller migration range in the Kenyan part of the ecosystem. The Loita Plains make up the wet season range and the main calving area. When the short grasslands are depleted, the herds migrate to the Masai Mara National Reserve, where they meet with the Tanzanian wildebeest population (Fig. 4.36).

Whereas the wildebeest migration in the Tanzanian part of the ecosystem is almost entirely confined to protected areas, the wildebeest population in Kenya resides in unprotected land for most of the year. Land-use changes have mostly been reported and measured in the Kenyan part of the Serengeti-Mara ecosystem (Serneels and Lambin 2001). Most recently, agriculture has expanded rapidly in the rangelands surrounding the Masai Mara National Reserve. Since the end of the 1970s, vast areas in the Loita Plains have been converted to cropland. Large-scale wheat farming, with high input of fertilisers and technology, is now practised on land that used to be the core area of the Kenyan wildebeest wet season range in the Loita Plains. While conversion of land to agriculture started in the early 1970s, the expansion mainly took place from 1984 onward. By 1995, about 50 000 ha had been converted to large-scale cultivation in the Loita Plains, which is about 20% of the wet season range for wildebeest (Serneels et al. 2001) (Fig. 4.37).

The temporal changes in the wildebeest population in the Kenyan part of the Serengeti-Mara ecosystem were analysed, along with their relationship with possible driving forces of

Fig. 4.36.
Wildbeest movements in the Serengeti-Mara ecosystem

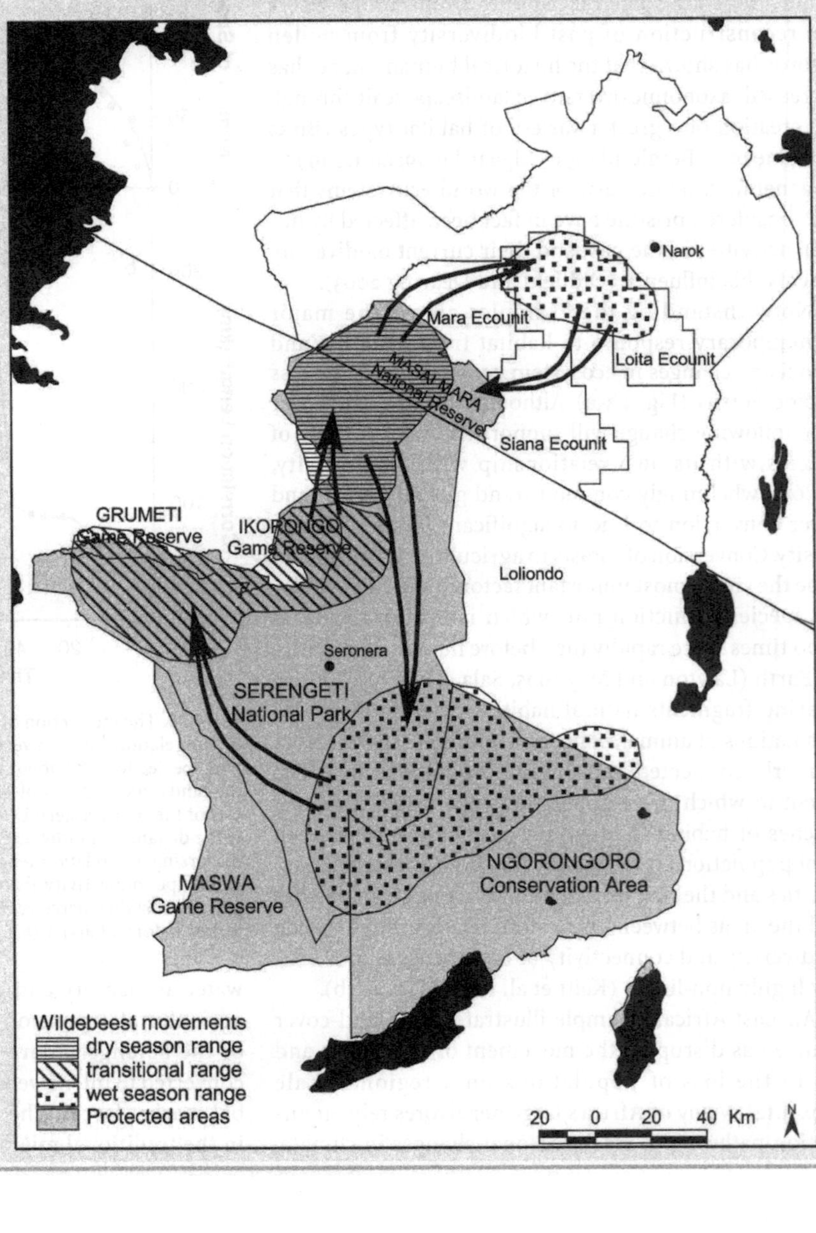

◀

change, such as rainfall, NDVI (Normalised Difference Vegetation Index, a measure of vegetation greenness), livestock numbers and land-cover changes. Changes in the spatial distribution of wildebeest for three periods were compared with spatial changes in livestock distribution and land cover. The analyses were repeated for the Tanzanian part of the ecosystem and results compared. Thus, the relative importance of land-use changes among the different potential driving forces of change in the wildebeest populations was tested.

The wildebeest population in the Kenyan part of the Serengeti-Mara ecosystem has declined drastically over the past 20 years and is currently fluctuating around an estimated population of 31300 animals, which is about 25% of the population size at the end of the 1970s. The Serengeti wildebeest population has not been affected by a downward trend over the last 20 years and is fluctuating around a mean of 1.2 million animals since the late 1970s. The population is regulated by green biomass availability in the dry season (Mduma et al. 1999). There have been only minor changes in land use at the eastern and southern borders of the Serengeti ecosystem. These changes can be related mainly to the development of subsistence or medium-scale agriculture and the changes are scattered in the landscape. They have also taken place away from the main migration routes of wildebeest.

Among all the possible driving forces behind the most recent downward trend, only land-use change showed a clear and concomitant trend over time. Rainfall, NDVI and cattle population showed interannual fluctuations but no trend. The first decline in wildebeest numbers in the Kenyan part of the Serengeti-Mara ecosystem that occurred between 1980 and 1982 was characterised by a uniform decrease in densities in the wet season range, and most probably caused by high wildebeest mortality due to drought conditions. Subsequent declines in the Kenyan wildebeest population are more clearly attributable to changes in land-use, as the decrease in wildebeest densities is limited to those parts of their wet season range that were converted to mechanised agriculture. The expansion of wheat farms forced wildebeest to either use the dryer rangelands or to move to wetter areas where competition with livestock was higher. It is suggested that either option resulted in a decreased per capita availability of food during the wet season. This further suggests that, from the 1980s onward, competition for food in the Kenyan rangelands has put a stress on the wildebeest population throughout the year. Dry season food availability is probably responsible for the smaller interannual fluctuations in wildebeest numbers, while the reduction of the wet season range in the Loita Plains has caused wildebeest numbers to drop considerably since the early 1980s. The shrinking wet season range amplifies the impact of low rainfall during the wet season and prevents the population from recovering from drought impacts, due to limited per capita availability of food.

This study provides compelling evidence that external processes, such as expansion of mechanised agriculture in response to market opportunities and/or policies (Homewood et al. 2001), can have a major impact on the dynamics of ecosystems within protected areas. Over the last decades, the decline in the Kenyan wildebeest population did not seem to affect the much larger Serengeti wildebeest population. However, if more land were to be converted to large-scale farming closer to the Masai Mara National Reserve, the dry season range for both the Kenyan and the Serengeti population would be reduced. This might have serious consequences for both populations and, therefore, for the entire Serengeti-Mara ecosystem.

Fig. 4.37. Conversion of the savanna of the Serengeti in Kenya to wheat farming and the consequent change to the migration pathways of wildebeest (*image:* S. Serneels, U. C. Louvain, pers. comm.)

has disrupted their annual migration to the point where the wildebeest population has declined sharply. On the Tanzanian side of the Serengeti-Mara ecosystem, with different economic forces operating, agricultural expansion has been much slower and fewer land-use changes have occurred. In this region wildebeest migration patterns and numbers have been maintained.

In the longer term the effect of fragmentation will combine strongly with changes in climate to impact biodiversity. As noted in Sect. 4.2.1, many species are already changing their ranges latitudinally and altitudinally in response to climate change. As migration becomes more ubiquitous as an ecological response to changing climate, the nature of the landscape becomes crucial. Much of the human modification and conversion of land cover renders it less easy to traverse for many migrating species, thus creating an insoluble dilemma in terms of the need to migrate and the inability to do so. The inevitable result will be reductions in species abundances as well as extinctions.

Soil structure. One of the most important characteristics of soil structure is the soil organic matter content. This is related to the physical structure of the soil, and hence its infiltration capacity, and also to its nutrient retention qualities. As noted above in Sect. 4.3.1, the land-cover conversion process often leads to extensive soil erosion and hence loss of soil organic matter. Changed land cover may also produce more subtle effects. The level of soil organic matter is to a large degree dependent on the allocation of photosynthate to roots, to the provision of dead plant material from above the soil surface and to microbial activity. All of these are dependent on the nature of the above-ground vegetation. Thus, changed land cover, through the coupling between above- and below-ground processes, can have a significant ongoing effect on the level of soil organic matter (Gregory et al. 1999).

A second effect concerns wind erosion. Insofar as changed land cover exposes more soil more frequently to the atmosphere, wind erosion can result. Model simulations for the midwest of the USA show that wind erosion is highly sensitive to wind speed (Ingram et al. 1996); a 20% increase in wind speed can lead to an eight-fold increase in erosion. This dramatic response is due to a threshold effect that occurs at about 5.5 m s^{-1} (Skidmore 1965), beyond which erosion increases exponentially. Many potential sites of wind erosion in the USA have mean wind speeds now above 4 m s^{-1} but just below the 5.5 m s^{-1} threshold. Thus, the modest increases in wind speed projected for a double-CO_2 climate are likely to lead to increases in erosion.

Change in Ecosystem Functioning

Ecosystem structure and functioning are closely intertwined so that any significant changes in structure carry with them associated changes in functioning. For example, changing a forest to a grassland or vice versa is usually accompanied by significant changes in leaf area, in rooting depth and perhaps in photosynthetic pathway, which can lead to alterations in primary production, nutrient cycling, water transport, disturbance regimes and other aspects of ecosystem functioning. These changes in functioning are known to be important at local and regional scales, but when land-cover change is multiplied around the world and aggregated globally, they may become pervasive enough to impact the functioning of the Earth System as a whole.

Hydrological cycle. Vegetation cover controls water cycling by influencing soil water recharge, horizontal water transfers by sheet flow and river transport and vertical moisture fluxes to the atmosphere through evapotranspiration. Because vegetation cover, vegetation litter and soil organic matter reduce run-off by maintaining soil structure and infiltration rates, clearing usually reduces soil water recharge and increases run-off and erosion. For example, in Brazil it has been shown that forested areas have ten times the infiltration capacity of pastures; deforestation thus leads to a decrease in infiltration and increase in runoff. Such effects have been quantified in numerous paired (one forested, one deforested) catchment studies. In Tanzania annual runoff from a cultivated catchment is 30–36% higher than that from a similar catchment with the original evergreen forest cover (Lorup and Hansen 1997). The effect was more dramatic in a Malaysian study where conversion of tropical forest catchments to cocoa and oil palm plantations resulted in an increase in water yield during the first years after the conversion of 157 and 470% respectively (Abdul Rahim 1988). These and many other such case studies around the world have been undertaken at the small catchment scale. Aggregating these effects to larger (basin) scales is not straightforward, and damping and amplifying effects often lead to surprising results, as will be discussed later in Sect. 5.3.5.

Changes in ecosystem structure almost always change the patterns of water and energy exchange between the land surface and the atmosphere. For example, clearing forests and replacing them with grasslands or crops reduces evapotranspiration by reducing leaf area, thereby decreasing latent heat and increasing sensible heat fluxes to the atmosphere. Changing patterns of surface energy and water fluxes after extensive forest clearing potentially change climate and rainfall patterns, at least at a regional scale. Modelling shows how cumulus convection would be suppressed with forest clearance (Pielke et al. 1997). There is already some observational evidence that deforestation in the Amazon Basin leads to shallow cumulus cloud formation, similar to that observed over natural grasslands in the Basin, compared to the deeper convection normally found over the Amazon forests (Cutrim et al. 1995). Some climate model simulations suggest that complete conversion of forest to grassland in the Amazon Basin could lead to increased surface temperatures, decreased total rainfall, and lengthening of the dry season (Shukla et al. 1990). Such changes might limit subsequent rainforest regeneration in the region.

One of the most comprehensive analyses of the effects of land-cover change on regional climate concerns the implications of historical land-cover change in China on the behaviour of the East Asian Monsoon. The analysis included both the direct (albedo and roughness) and indirect (water and energy exchange) effects on climate together (Fu and Yuan 2001). The results show the surprisingly large effects on atmospheric circulation and climate of changing only land-cover in the model (Box 4.13). Although these are simulated effects from a model experiment, they agree well with observed changes in the East Asian Monsoon in recent decades.

Carbon cycle. Change in ecosystem structure has a long-term effect on the carbon cycle in terms of the altered net ecosystem productivity of the new structure type and perhaps also on altered disturbance regimes of the new vegetation. One of the most common types of structure change is the conversion of forest to agriculture, for which the impact on the carbon cycle can be relatively readily quantified. Most forests are net sinks for atmospheric carbon, the rate of uptake depending on the type and age-class structure of the forest. For example, net ecosystem productivity of moist tropical forests is in the range 1.0 to 5.9 t C ha^{-1} yr^{-1}, that of temperate forests in the range 2.5 to 7.0 t C ha^{-1} yr^{-1} and that of boreal forests from 0 to about 2.5 t C ha^{-1} yr^{-1} (Bolin et al. 2000). Agricultural systems, on the other hand, are likely to be either carbon-neutral or carbon sources when soil emissions are taken into account. The change in ecosystem structure thus has a strong impact on the carbon cycle over a decadal timeframe, or longer. The reverse structural change – from cropland or grassland to forest – has a reverse impact on the carbon cycle. This provides a good scientific rationale for the inclusion of terrestrial carbon sinks in the Kyoto Protocol. When the conversion of cropland or grassland to forest represents a reversal of historic deforestation and the forests are maintained indefinitely on the site, the change in ecosystem structure can have an ongoing impact on the carbon cycle. In a similar way, by comparing the change in net ecosystem productivity associated with a change in ecosystem structure (and by considering the implications of the change in structure for disturbance regimes), the net effect on the carbon cycle can be estimated.

Atmospheric composition and chemistry. At first glance a change in ecosystem structure should have little effect on atmospheric composition and chemistry. However, there are several ways in which the changing functioning of a new assemblage of plants can influence the atmosphere. First, the emission of volatile organic compounds from vegetation is controlled by environmental factors, but is also influenced by the nature of the vegetation itself. Thus, a shift in ecosystem structure, for example from domination by one functional type which is a weak emitter of volatile organic compounds to a functional type which is a strong emitter, can have significant impacts on atmospheric composition. Such is the case when woody plants having higher emissions of isoprene and monoterpene are replaced by annual crops and grasses. Even among tree species, differences may be profound. During the late nineteenth and early twentieth centuries, chestnut blight in the lowland forests of the eastern USA caused widespread change of forest composition resulting in the replacement of chestnut by oak as the dominant species. Oak is a strong emitter of isoprene while chestnut is not. Thus, the result of this change in ecosystem structure has been a strong increase in the emissions of isoprene in the eastern USA (Scholes et al. 2003).

A second effect of changing ecosystem structure is in the emission of trace gases from soils. Nitrogen oxides, for instance, are produced as intermediate compounds or by-products of the processes of nitrification and denitrification (Conrad 1996). Their rate of production is controlled primarily by three environmental factors – nitrogen availability, soil temperature and soil moisture. A change in ecosystem structure, for example, a change from a forest to a grassland, will affect all three of these factors. Significant changes in structure are normally associated with changes in nutrient cycling, and hence in nitrogen availability. Changes in evapotranspiration that occur with a change in leaf area index clearly affect soil moisture, and a change from a closed to an open canopy changes the transfer of heat between the land surface and the atmosphere and thus the temperature of the soil.

Thirdly, as ecosystem composition and structure changes, disturbance regimes may also change. In particular, changes in fire regimes resulting from changing structure affects atmospheric composition through biomass burning emissions. Abandonment of agricultural land and its reversion to scrubland or forest serves as an example. The change in structure often involves an increase in above-ground biomass and hence fuel load. Coupled with the increasing connectivity associated with reversion, fires spread more easily. Such may be happening in the Mediterranean region, where widespread abandonment of agriculture is leading to a re-emergence of the fire-prone scrublands that are typical of the region (Moreno et al. 1998; Moreira et al. 2001).

Box 4.13. Land-Cover Change and Climate in East Asia

Congbin Fu

East Asia is one of the most populous regions on Earth. Owing to a long history of human settlement and agriculture, and more recently urbanisation, industrialisation and very high population densities, natural landscapes have been much changed in China and adjacent counties of East Asia. Human-induced land-use and land-cover changes have been as great in this region as in any other in the world, if not greater. The effects of changing land cover on the general circulation of the atmosphere in the East Asian monsoon region have been significant.

The natural vegetation has been so altered in East Asia over millennia that its reconstruction other than by modelling is not possible. However, it is feasible to specify the equilibrium climax vegetation that may be expected at present based on the prevailing climate with a biome-climate matching approach that is widely used in the international ecological science community (Ojima et al. 2002). Potential vegetation distribution over the region has been simulated in this way using the Century model based on a prescribed vegetation template (Parton et al. 1998). This kind of vegetation cover information is called potential vegetation since human-induced land-cover changes were not included in the simulation. The current vegetation is satellite-derived for 1987–1988 (Meeson et al. 1995); human induced-change is then defined as the difference between potential and current conditions. More than 60% of the region has been affected by conversion of various categories of natural vegetation into farmland, grassland into semi-desert and widespread land degradation (Fig. 4.38a,b).

Changes in surface dynamic parameters, including albedo, surface roughness, leaf area index and fraction of vegetation cover resulting from human-induced land-cover changes have all undergone significant change over large areas of the region (Fig. 4.38c). Simulations using a Regional Climate Model for East Asia (Fu et al. 1999) have shown how, by altering the complex exchanges of water and energy from the surface to the atmosphere, changes in land cover have brought about significant changes to the East Asian monsoon. These include weakening of the summer monsoon low-pressure system over the region and a commensurate increase in anomalous northerly flow which results in the weakening of regional water cycle, i.e. a decrease of atmospheric moisture, precipitation, surface runoff and soil water content (Fig. 4.38d). The consequence of such changes may be a significant factor in explaining the observed trend in aridification that has taken place in many parts of the region during the last century (Fu 1994).

The modification of regional atmospheric circulation described here is the result of land-cover changes only. Land-cover changes in East Asia have produced demonstrable changes to regional ecosystems (Fu and Wen 1999) and significant atmospheric responses that have considerable ramifications for the monsoon circulation of the atmosphere over Asia as a whole, as well as implications for those distant regions of the world where regional climates are linked to changes in the Asian monsoon.

Fig. 4.38a,b. Modelled changes in land cover over East Asia based on the difference between **a** potential vegetation and **b** current vegetation (Fu 2003)

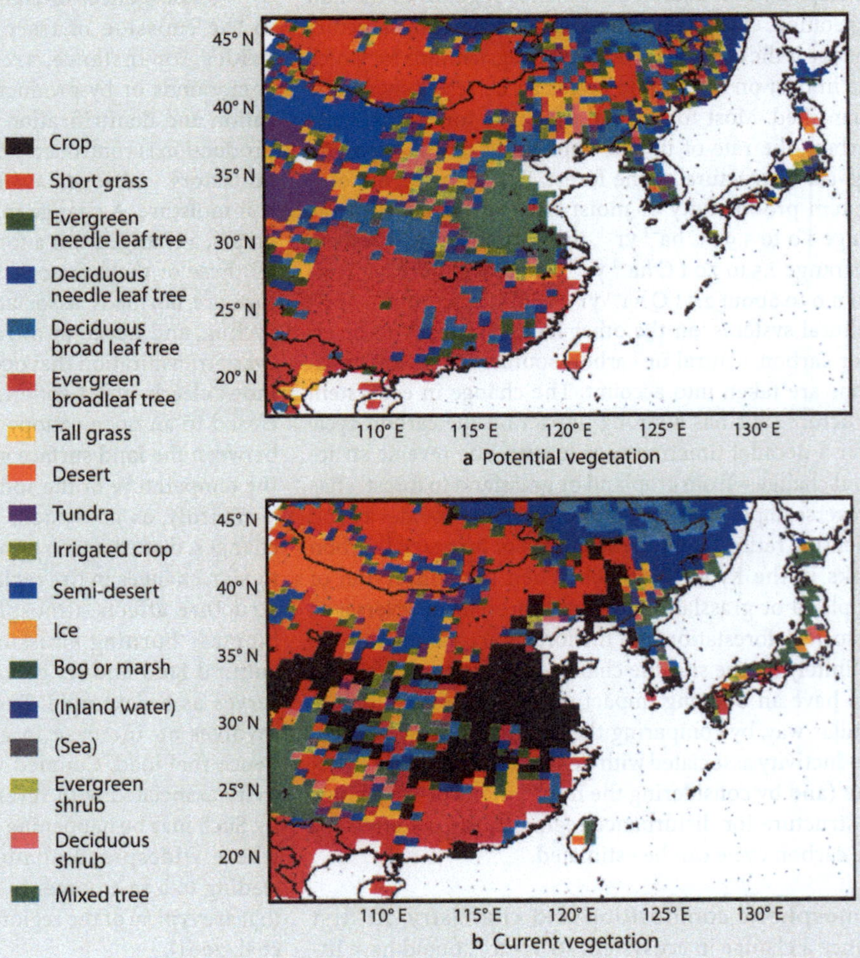

◀

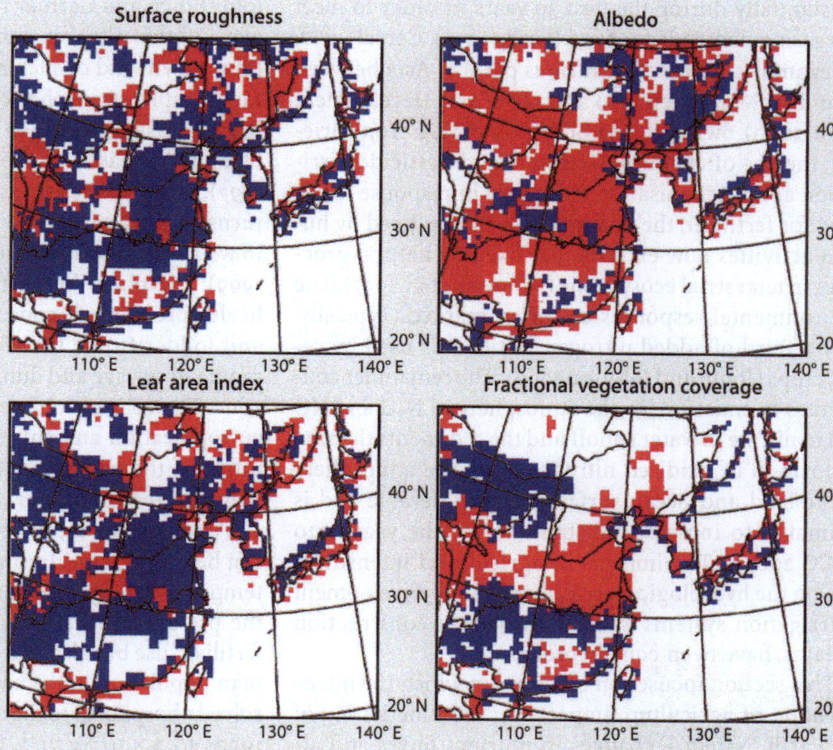

Fig. 4.38c. Changes in surface parameters (*blue:* decrease; *red:* increase) corresponding to the land cover changes over East Asia (Fu 2003)

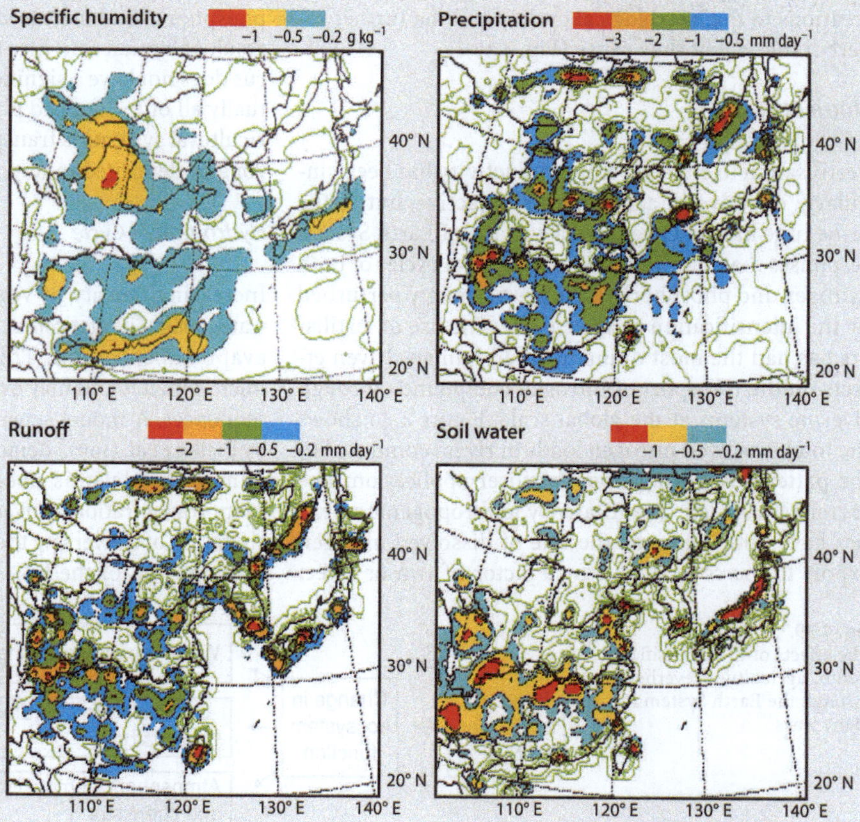

Fig. 4.38d. Summer monsoon climate changes: humidity at 850 hpa, precipitation, surface runoff and soil water content over East Asia (Fu 2003)

4.3.3 Intensification of Agriculture

The rate of intensification of agriculture has increased substantially during the past 30 years in order to meet increasing demands for food (Grigg 1993). Cereal yield, for example, increased in various parts of Asia between 54 and 127% between 1970 and 1995 (Asia Development Bank 2000) owing to use of high-yielding crop varieties, the use of chemical fertilisers and pesticides, irrigation and mechanisation. Primarily in response to the need for fertiliser, the quantity of nitrogen fixed by human activities now exceeds that fixed by natural processes in terrestrial ecosystems (Vitousek et al. 1997). The environmental responses are often marked. Typically, only a third of added nitrogen fertiliser is used by cereal crops (Raun and Johnson 1999). The remainder contributes to emissions to the atmosphere of N_2O and NO and to nitrate in water runoff and the sediments therein. Deposition of oxidised nitrogen from the atmosphere to the land and ocean surfaces is considerable and is estimated to increase substantially by the year 2020 (IPCC 2001a). The impacts of agricultural intensification on the hydrological cycle, through the development of irrigation systems and the associated construction of dams, have been equally profound.

This section focuses on two ways in which the intensification of agriculture impacts on the functioning of the Earth System – changes in nutrient fluxes and alterations to the hydrological cycle – and the further reverberations that they cause (Fig. 4.39).

Nutrient Fluxes

Fertilisation to enhance crop productivity has been singularly effective in achieving its objective, but at the same time has also triggered a cascade of Earth System responses that may be serious. The global cycles of both nitrogen and phosphorus have been strongly perturbed by the intensification of agriculture. The use of fertilisers has had the most dramatic of all human-driven effects on the transport of nitrogen compounds through riverine systems at the global scale. Figure 4.40 shows the total dissolved nitrogen loads in rivers compared to the patterns of nitrogenous fertiliser application. The correlation is high. Quantitatively, anthropogenic activities have now increased the rate of dissolved nitrogen export to the coastal zone by a factor of two or three. Thus, dissolved inorganic nitrogen concentrations in world rivers has increased from 0.13 to 0.33 mg l^{-1} (Meybeck and Ragu 1997). In some highly fertilised catchments, dissolved inorganic nitrogen fluxes have increased 10-fold (Billen and Garnier 1999). In industrialised regions some of this nitrogen comes from atmospheric deposition. Downwind of such regions, in New England, USA, for example, atmospheric deposition of nitrogen oxides has increased more than four-fold since 1900; the NO_3^- flux has increased by the same amount (Jaworski et al. 1997). Some of the excess load of nitrogen in rivers returns to the atmosphere as N_2O during denitrification in wetlands, riparian zones and rivers (Caraco and Cole 1999); percolates into groundwater, where it can cause health problems if consumed; or is removed in crops and fodder (to be transformed subsequently into ammonia in sewage and dung). The rest reaches the coastal zone. The coastal zone response of the system includes eutrophication and shifts in species dominance as the ratios of the key elements N, P and Si change. Anoxia may also result (Box 4.14).

For phosphorus, the perturbation at regional scales can be just as large. In some European river basins contemporary phosphorus fluxes are up to 10-fold those of the pre-agricultural/industrial era, owing primarily to fertiliser use but also as a consequence of increasing human population in the basins. Globally the input of dissolved phosphorus to the oceans had doubled by the early 1990s to 2×10^{12} g yr^{-1}. By the same time particulate phosphorus had increased from 7 to 20×10^{12} g yr^{-1} as a result of human activities (Vollenweider 1989). Phosphorus does not have a significant gaseous phase so that virtually all of the applied phosphorus that is lost from agricultural systems is transported via rivers to the coastal zone or deposited in impoundments along the way.

Hydrological Cycle

Increasing density of vegetation cover (leaf area) and water usage (irrigation) in intensive agriculture enhances evapotranspiration. This impact in turn may alter microclimates and may even have regional climatic consequences. A model sensitivity experiment undertaken by Pielke et al. (1997) demonstrates this effect. A regional climate model was used to simulate the effect on evapotranspiration, convective cloud formation and precipitation of converting the natural landscape in the Great Plains region of the USA from short-grass prairie to a

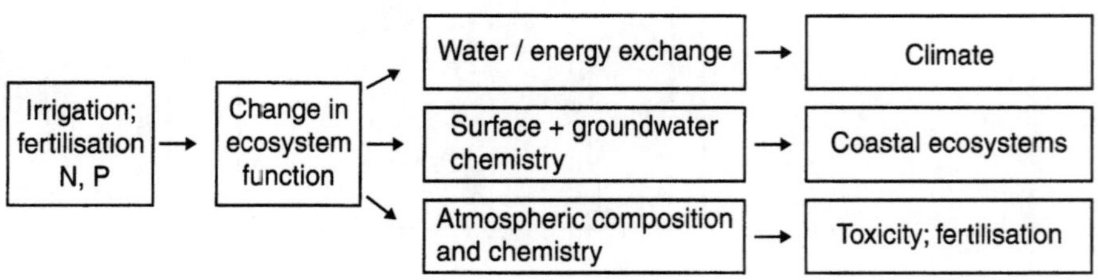

Fig. 4.39.
The effects of the intensification of agriculture reverberate through the Earth System in many ways

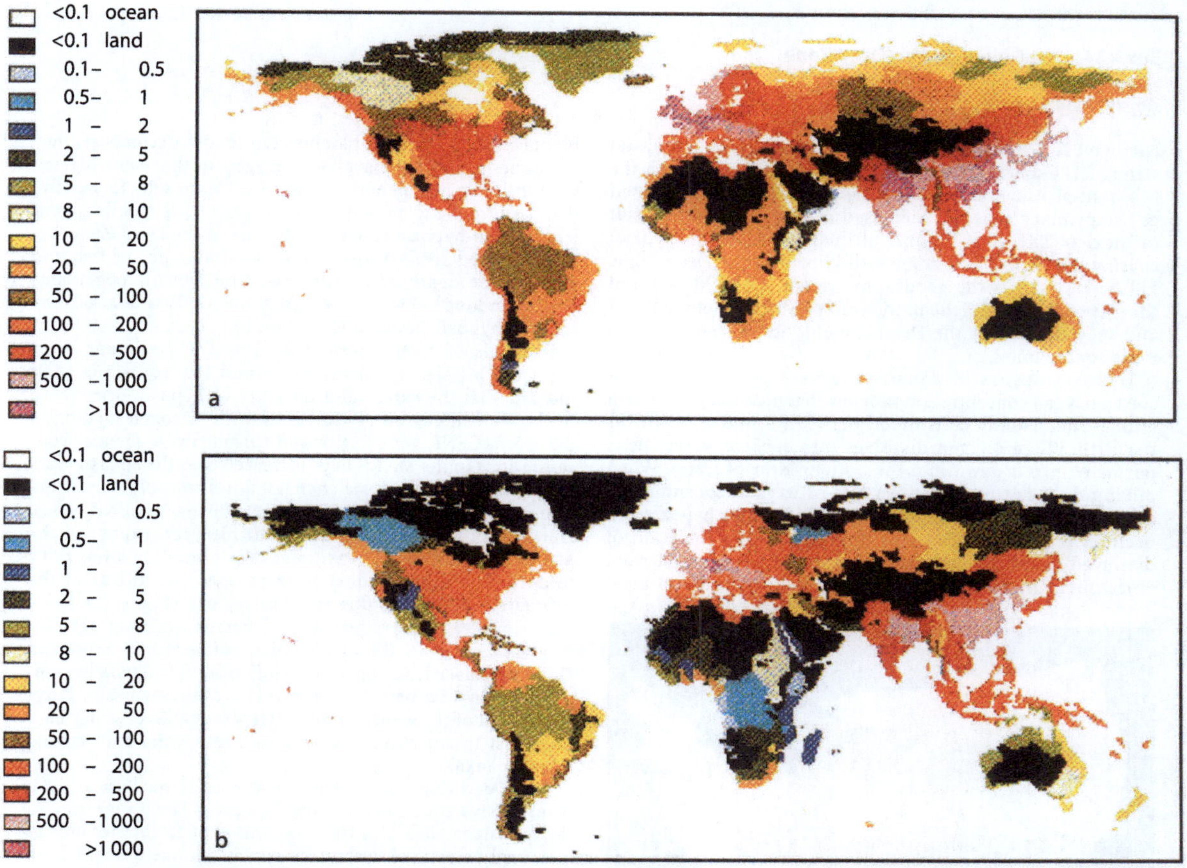

Fig. 4.40. Multiple regression modelling of contemporary dissolved inorganic nitrogen fluxes (kg N $km^{-2} yr^{-1}$) from some rivers (Seitzinger and Kroeze 1998): **a** total dissolved inorganic nitrogen export; and **b** export attributed to fertiliser use

mix of prairie and irrigated crops and shrubs. A particular day was chosen for the simulation and the meteorological conditions of that day were used; only the land cover was changed. As shown in Fig. 4.43, for the natural landscape only, cloud formation was limited to a shallow line of cumulus clouds with no precipitation. In the model simulation for the modified landscape (the current land-cover pattern), the increase in evapotranspiration due to intensive cultivation was enough to produce a thunderstorm system along the dryline. The results of the model simulation agreed with observations; a thunderstorm was indeed observed in the region on the day from which the meteorological parameters were taken (Ziegler et al. 1995; Shaw et al. 1997; Grasso 2000).

The provision of water for irrigation is often achieved through the construction of dams. Impounding of water stabilises natural flow regimes and in so doing, triggers many changes that reverberate through the Earth System. These include disruption of downstream aquatic habitat and ecosystems, replacement of native fisheries, net losses in basin runoff due to evaporative losses, changes in the capacity of the hydrological system to process organic wastes, and changes in nutrient-to-nutrient ratios leading to possible toxic algal blooms in the coastal zone (see Rosenberg et al. 2000). Sedimentation of particulates within reservoirs also leads to a reduction in their transport to the coastal zones of the world, a process known as sediment starving, sometimes with disastrous effects on river deltas and coastlines that are maintained by these sediments (Vörösmarty et al. 2002). In regions with very high erosion rates, the longevity of reservoir use can be considerably reduced by sedimentation in reservoirs. For example, in the Shaanxi Province, China, one-third of reservoir capacity has historically been lost per year relative to new construction and several reservoirs along the Yellow River mainstem have shown 50–80% storage loss within just five to seven years (Zheng and Zhou 1989). The change in flow regime caused by a large impoundment can be seen visually in the hydrograph for the Nile River before and after the construction of the Aswan High Dam (Box 3.5, Fig. 3.44a). This impoundment has had significant effects on river flow characteristics and resulted in the unintended consequence of terminating the millennia-old natural fertilisation associated with annual flooding and silt spreading in the lower reaches of the river.

Even more distant, but nevertheless significant, reverberations are possible in the Earth System as the result of the damming of major rivers. The Three Gorges Dam in China, currently under construction, is one of

Box 4.14. The Gulf of Mexico Dead Zone

Angelina Sanderson

Nitrogen is a limiting nutrient for algal growth in coastal waters (NRC 1993). As a result, heavy nitrate loading via the Mississippi-Atchafalaya River Basin (MARB) is implicated as the primary cause of increased algal growth in the Gulf of Mexico. The algal blooms ultimately result in hypoxia, causing the region to become well known as the Dead Zone. Figure 4.41 shows the geographic range of the MARB and the size of the Dead Zone in the Gulf of Mexico. Between 1993 and 1999 the size of the Dead Zone doubled from 4000 to 8000 square miles.

Hypoxia occurs in aquatic ecosystems when the dissolved oxygen concentration is below that necessary to sustain animal life, usually considered between 0 and 2 millilitres per litre. When oxygen dissolves into surface waters, it is normally mixed down into the bottom layer of water. When mixing does not occur, or the rate of oxygen consumption exceeds the rate of resupply, then bottom water hypoxia can occur. The principle cause of diminished concentration of dissolved oxygen is eutrophication, the process of nutrient enrichment in an ecosystem, which in turn leads to algal blooms. When the algal blooms respire and decompose, they consume much of the available oxygen in the water, leaving very little for marine animals to use. This results in massive die-offs of aquatic animals, or of migration of mobile organisms to non-hypoxic zones. In the case of the Dead Zone, hypoxia-related stress results in a low abundance of fish and shrimp, the death of bottom dwelling benthic organisms and larger long living species, and productivity is reduced during non-hypoxic periods (CENR 2000).

Fig. 4.41. The hypoxic zone (*red*) at the outlet of the Mississippi-Atchafalaya River Basin as measured in a 1999 survey (D. Goolsby and W. Battaglin, US Geological Survey 2000)

Analysis of river nutrient load and of sediment cores from the Mississippi River basin and the Louisiana shelf indicate that the extent and intensity of hypoxia occurrence in the Gulf increased concurrently with nitrogen loading in the river (CENR 2000). Nitrogen (primarily as nitrate) concentrations in the MARB have increased over the last 150 years due primarily to land-use changes, flood control and navigation projects on the river, and climate changes (CENR 2000). Howarth et al. (1996) estimate that nitrogen export to the MARB has increased 2.5–7.5 times over prevailing levels prior to agricultural and industrial practices. Two-thirds of the estimated increase is due to fertiliser use (Fig. 4.42). Principle sources of nitrogen are the basins draining agricultural lands in Iowa, Illinois, Indiana, southern Minnesota and Ohio. Fertiliser leaching and runoff from farmlands is compounded by the loss of mangroves in the river basin. Fifty-six percent of the wetlands in the Mississippi River Valley have been lost to agriculture, navigation, reservoirs and levying (Winger 1986).

The Dead Zone in the Gulf of Mexico is not just a result of excessive nutrient loading, however. The geography of the Louisiana shelf and the large influx of freshwater creates a strongly stratified system, in which relatively warm fresh water gathers above a colder, saltier bottom layer. The bottom layer remains isolated from aeration until the fall and winter months when the surface water cools, and mixing commences. If the Gulf waters were not so strongly stratified, then hypoxia would probably not occur because enough dissolved oxygen would reach the bottom waters (CENR 2000). It is the interaction between human activities and natural processes that drives the Gulf system into a hypoxic state, and creates the Dead Zone.

Fig. 4.42. Nitrogen inputs from fertiliser use in the Mississippi River Basin since the 1960s (based on data from US Geological Survey 2000)

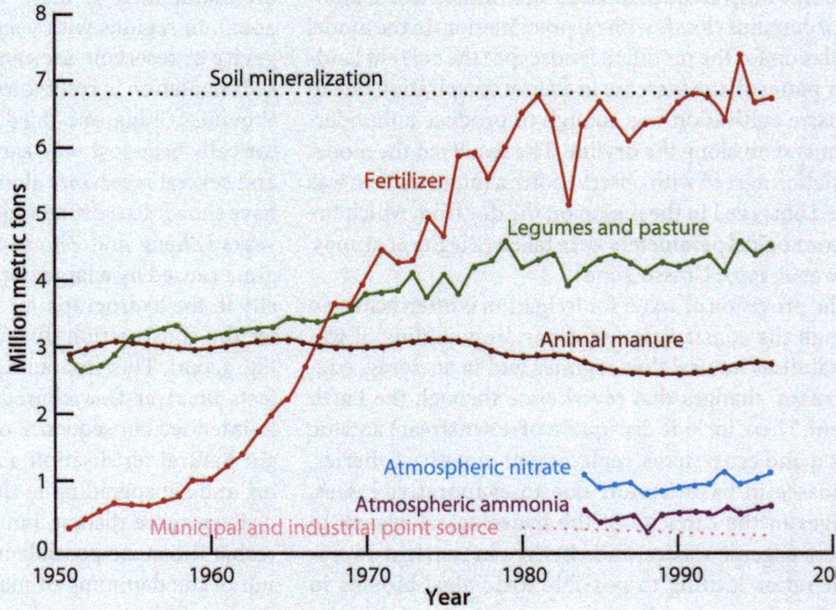

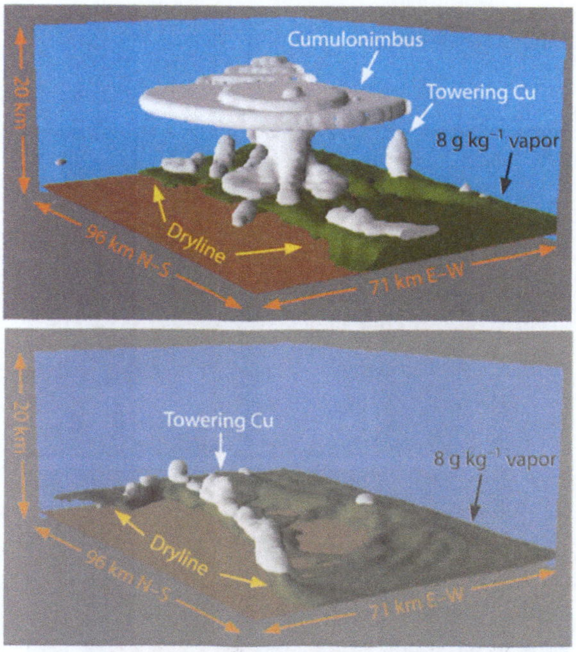

Fig. 4.43. Simulation of the development of a thunderstorm over the current agricultural landscape in the central USA that, given identical large-scale meteorological conditions, may not have developed if the vegetation was that of the pre-agriculture prairies (Pielke et al. 1997). Clouds are shown in *white* and boundary layer humidity in *green*. The actual observations matched the simuation; a thunderstorm was observed over the area for the conditions of the simulation

the largest and most controversial dams ever built. Although much of the controversy focuses on local and regional environmental impacts, the dam's impact on the fisheries and on the biogeochemistry of the South China Sea may be just as important in the longer term (Chen 2002). Continental margins contribute up to half of the total deep ocean particulate organic carbon flux. For the South China Sea, the nutrient-rich intermediate waters are the key to the region's productivity and thus both its role as a major fishery for the Asian region and its contribution to ocean carbon sinks. Nutrients are provided predominantly by upwelling; over four times as much phosphorus and twice as much nitrogen is provided through upwelling than through riverine fluxes. Thus, the direct impact of the Three Gorges Dam and dams on other major rivers draining into the basin on the direct input of nutrients will be minimal. However, the construction of these dams will also reduce the freshwater influx to the sea, thereby reducing the buoyancy effect on the continental shelves and consequently decreasing upwelling, nutrient input, primary productivity and eventually particulate organic carbon flux and fisheries production.

Another distant, but now well-documented, reverberation connects the intensification of agriculture with geohazards. Human activities now have the capability to increase the incidence of earthquakes. Recent advances in understanding of the geological processes associated with earthquakes allow differentiation between natural and human-triggered earthquakes (Gupta 1992, 2002). Most of the human-triggered earthquakes are associated with large impoundments of water; some of these have registered up to 6.0 on the Richter scale. The removal of underground water, often for irrigation, has also been known to trigger earthquakes, as has the removal of oil from underground reservoirs.

A different type of reverberation caused by agricultural intensification is the change in soil chemistry resulting from flooded rice cultivation and the consequent enhancement of CH_4 emissions. Up to 70% of global rice production comes from systems in which the soil is flooded for part of the growing season. This leads to anaerobic conditions for the soil carbon, which in turn leads to the production of CH_4 rather than CO_2 as the organic material decomposes. The overall amount of CH_4 produced this way is estimated to be in the range 30–70 Tg yr^{-1} (Neue 1997). The factors which influence the rate at which CH_4 is produced from a rice paddy include water management, the type and amount of decomposable organic matter incorporated into the soil, the cultivar of rice used, the type and amount of fertiliser applied, and the temperature, soil redox potential and soil pH (Neue 1997). Up to 90% of the CH_4 emitted from rice paddies passes through the rice plant itself, rather than being emitted directly from the soil into the water body and then to the atmosphere (Nouchi et al. 1990).

In summary, the conversion of land cover, the impacts of the changed land cover itself and the subsequent intensification of agriculture reverberate through the Earth System in just as many ways as fossil fuel combustion. However, there is one important difference. Through the mixing power and transport of the atmosphere, many of the effects of fossil fuel combustion are rapidly translated to the global scale; nearly all are at least trans-boundary. The effects of land-use and land-cover change are almost always felt initially at the local scale and thus seem to be limited in scope and are not initially perceived as a global change issue. However, when aggregated over larger space and time scales, and when the impacts further along the chain of reverberations are considered, land-use and land-cover change are a potent driver of changes to Earth System.

Finally, as for fossil fuel emissions, the analysis above has shown how land-cover change, coupled with various types of land use, can impact Earth System functioning along many different pathways. One of the most unexpected pathways concerns the incidence of fire, which is intimately connected to both land cover/use and to weather and climate. The January 2003 bushfires in and around Canberra, Australia, show how decisions on land cover and land use made decades earlier can cascade through time to contribute to a catastrophe (Box 4.15).

Box 4.15. Cascading Impacts of Land Use Through Time: The Canberra Bushfire Disaster

Sandra Lavorel · Will Steffen

The eighteenth of January 2003 has been called Canberra's darkest day (Canberra Times newspaper, 19 January 2003). On that day three large bushfires, which had been burning for about two weeks in the mountain ranges west of Australia's capital city, joined to form a 35-km long fire-front that descended upon the city with extraordinary speed and intensity. The result was a disaster for the city (Fig. 4.44). During a 3–4 hour period on a summer afternoon, the sky turned from apocalyptic orange-red shades to black as night as the fire-front hit the city itself. Over 500 houses were totally destroyed, four people were killed and hundreds of others were taken to hospital with burns, one of the oldest astronomical observatories in Australia was burned beyond repair and the total damage approached 400 million Australian dollars.

The disaster was no simple cause-effect event. Rather, the tragedy was the result of threshold-abrupt change behaviour, building on a complex set of preconditions that depended on land-use and land-cover decisions made over decades coupled with anomalous meterological and climatic conditions of varying scales in space and time, from local to global and from hours to years.

Canberra is a planned city, designed around 1915, built in stages thereafter and growing to a present population of around 300 000. The city is often called Australia's bush capital, reflecting land-cover decisions that were made decades ago. The city is large in area, with substantial areas of park and nature reserves interspersed with lakes and urban and suburban zones. The city literally flows into the landscape around it, a feature that Canberrans have come to value highly.

A prominant feature of southeastern Australian landscapes is that they are fire-prone, being dominated by *Eucalyptus* and *Acacia* species. Upwind of Canberra, to the west, lie extensive areas of mountainous forests, much of it preserved in its natural state as national parkland. The conjunction of the loss of Aboriginal practices of regular burning, decades of park policy to allow natural fire regimes to operate, and possibly effects of changes in climate and atmospheric CO_2 concentrations, has led to a build-up of fuel in the understorey in the form of fire-prone shrubs, woody debris and litter in comparison with what was described by early European colonisers as an open grassy forest having frequent small fires. The present situation sets preconditions for fewer but more intense fires. Today it is unknown whether a policy of frequent control burns, even if it were feasible in this complex landscape, would allow a return to the historical conditions.

Fig. 4.44.
a The fire front approaches the edge of Canberra through parkland; **b** the fire storm enters the suburbs; **c** over 500 houses were burned to the ground; **d** a satellite image of southeastern Australia on 19 January showing active fires (highlighted in *red*) and smoke plumes streaming southeastward across the Tasman Sea

◀

The land-use decision to follow a more *natural* fire regime seemed appropriate throughout the previous century. Fires were usually of moderate intensity and at about 10-year intervals, with the fuel building back up to pre-fire levels in the interim. The city of Canberra was never seriously threatened by the periodic fires in the forests to the west. In the present case, however, the apparently enlightened land-use policy, coupled with extreme climate and weather conditions, caused a threshold to be crossed in terms of fuel load and condition; the conflagration that resulted was without precedent and beyond any expectations.

Southeastern Australia had been experiencing a severe drought for several years, more severe than predicted from the rather moderate El Niño condition that had been prevailing (Fig. 4.45). In the three-four months leading up to the disaster, the relative humidity at the 2 000 m elevation (many of the forests to the west of Canberra lie at 1 200–1 500 m above sea level) was unusually low, about 12% below average. During the same period the forests were subject to extreme heat; the temperature was nearly 3 °C higher than average, and for the preceding year was 1 °C higher than average, even taking the long-term warming trend into account. Coupled with an almost complete lack of rainfall for the summer, these conditions meant that the forests were exceedingly dry and thus set up for a violent fire given the appropriate meteorological conditions.

Late on the seventeenth of January the weather conditions switched from the relatively cool, gentle easterly breezes that were keeping the fires, which had started a week earlier from lightning strikes, up in the mountains and away from the city. By 9 am the next morning the temperature was already 30 °C with northwesterly winds of 30 km hr^{-1}. By early afternoon the temperature had reached 38 °C, the relative humidity was only 2% and the winds were gusting at 70–80 km hr^{-1}. During the course of a few hours, the fires swept down from the mountains about 50 km away and hit the suburbs as a massive fire storm with flames leaping up to 100 m high and with the energy intensity of a large bomb. The fire intensity was estimated to reach 50 000 kW m^{-2} in a pine plantation immediately adjacent to the worst affected suburb. The juxtaposition of pine plantation and suburb and the maintenance of the plantation for recreation instead of harvesting was a land-cover/use choice that contributed to the disaster. Had the plantation been recently harvested or had a grassland been established adjacent to the suburb, the fire intensity would have been much lower as it hit the suburb.

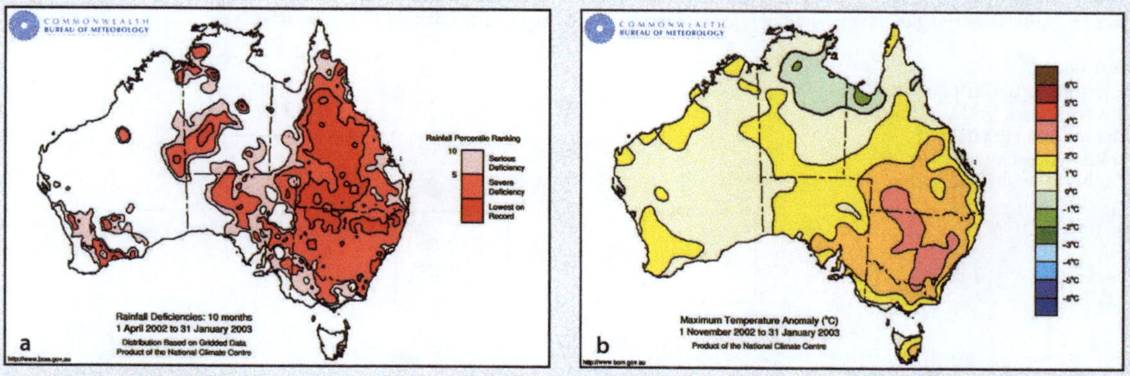

Fig. 4.45a,b. Meteorological data for the 2002/2003 drought in southeastern Australia: **a** rainfall deficit (expressed as percentiles from the mean) for the period April 2002 to January 2003; **b** maximum temperature anomalies for the period November 2002 to January 2003 (Commonwealth Bureau of Meteorology, Australia)

Fig. 4.45c. Meteorological data for the 2002/2003 drought in southeastern Australia: relative humidity anomalies at 700 hPa (corresponding to 2 000 m above sea level) for the period November 2002 to January 2003 (image: reworked from NOAA-CIRES Climate Diagnostic Center)

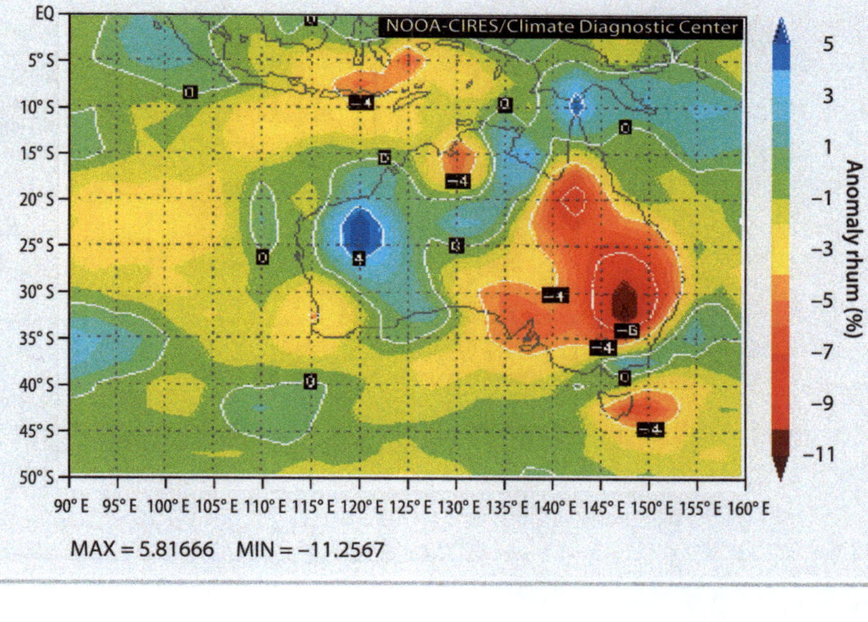

◀

What was the role of global change in this disaster? The extreme climatic conditions that prevailed over southeastern Australia for the months and years leading up to the Canberra disaster appeared to be more closely linked to systemic, global-scale climate change than to the behaviour of the ENSO phenomenon. A recent study (Hoerling and Kumar 2003) presents convincing evidence from both observations and modelling studies that the multi-year droughts (from 1998 through 2002) in the mid-latitudes of both the northern and southern hemispheres were related to a remarkably persistent pattern of sea surface temperature – cold SSTs in the eastern tropical Pacific and warm SSTs in the western tropical Pacific and Indian Oceans. This pattern of strong west-east contrast in SSTs across the Pacific and Indian Oceans resulted in a stable band of high pressure in the atmosphere over the northern and southern mid-latitudes, in turn leading directly to drier conditions. The SST anomalies that forced the drying mid-latitude climate were the largest in the twentieth century, being embedded within a long-term warming trend. Climate attribution studies find that such warm SSTs are beyond what is expected of natural variability and are due in part to increasing greenhouse gas concentrations in the atmosphere. Furthermore, the study suggests that if the west-east contrast in SSTs becomes a more common feature of the climate system, there will be an increased risk for severe and synchronised drying in the mid-latitudes in the future.

The Canberra bushfire disaster was due to a complex cascade of land-cover and land-use decisions made at several times in the last century coupled with climatic teleconnections ultimately forced, at least in part, by the emission of greenhouse gases by human activities around the world (Fig. 4.46). This combination of factors pushed the forests of southeastern Australia across a critical threshold of fuel load and condition, ready for short-term meteorological conditions to trigger violent and uncontrollable wildfires. As an example of how global change impacts will actually be manifest in reality, the Canberra bushfire disaster is typical: a case of system response to multiple, interacting driving forces rather than a simple cause-effect reaction to a changing global mean.

The Canberra bushfire disaster raises questions that challenge global change science in the coming decade: What was the conjunction of biophysical and human conditions that triggered this disaster? Which thresholds were exceeded? How close is the system to these thresholds under its average conditions? Will such catastrophic events become more frequent as global change continues and possibly accelerates? How can policy and management respond to this challenge? What will a future global map of fire regimes look like, and what will be the consequences for the Earth System? Which are the most vulnerable regions to changes in fire regimes?

Fig. 4.46.
A cascading series of events involving both land-use/cover change and climate variability and change led to the Canberra bushfire disaster

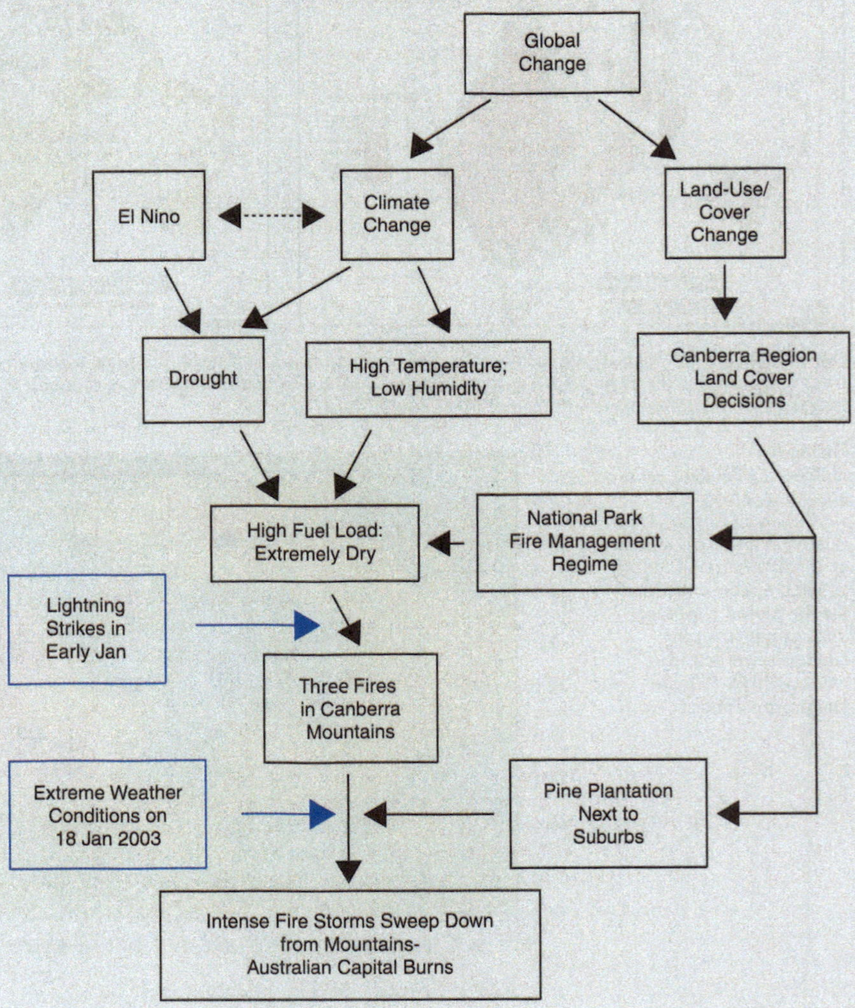

4.4 Multiple and Interacting Changes

Up to now this chapter has focused on a single forcing factor, such as fossil fuel combustion, and dissected in one-by-one fashion the many ways that it can impact the functioning of the Earth System. However, global change occurs for multiple reasons and from interacting forcing mechanisms. The integrated responses of the Earth System are even more complex. Global change interacts with an almost bewildering array of natural variability modes and also with other human-driven effects at a number of scales. Often one component of global change will amplify the effects of another process in the Earth System, and in other cases, components of global change may dampen aspects of Earth System functioning. Especially important are those cases where interacting stresses cause a threshold to be crossed and a step-change in functioning to occur.

In the past decade, the dominant focus of concern in global change research has been on single changes, especially climate change, despite the fact that many environmental changes that are global in extent or effect are underway concomitantly. At the local and regional scale, attention has often been focused on urban air pollution, toxic chemicals, acid rain, and other problems in isolation. In future, the focus will have to be on the interactions among agents and impacts of change at local, regional, and global scales.

In this section the focus is shifted from the drivers of global change to the systems being impacted. Two examples – the massive forest fires that swept parts of Southeast Asia in the late 1990s and the increasing damage to coral reefs – illustrate how multiple, interacting stresses can cause thresholds to be crossed and state changes in systems to occur.

4.4.1 Fires in Southeast Asia

Ecological and biophysical responses (as well as those of social and economic systems) to multiple stresses may be additive, but that is probably the exception rather than the rule. Non-additive, non-linear responses are more common. This complexity of biophysical interactions within ecological and human systems must be fully grasped in order to understand, model and project responses. Similarly apparent is fact that the vulnerability of biophysical systems to global and local environmental changes is controlled as much by human management or inadvertent influences as by the biophysical responses themselves. The extreme fires that raged through parts of Southeast Asia in 1997/98 (Fig. 4.47) provide a classic example of the non-linear response of an ecological system to multiple, interacting stresses.

The fires were largely the result of the interaction of two important Earth System processes – human-driven land-cover change and climate variability. An estimated 45 600 km^2 of forests on the Indonesian islands of Sumatra and Kalimantan (southern Borneo) were burned during the event. The fires were due to land management following conversion of tropical forest to oil palm plantation (deforestation). Normally the slash from the clearing operation is burned during the relatively dry southern monsoon period from June to October. As the wetter northern monsoon phase is established in October/November, the burning activities cease and the fires are extinguished. However, strong ENSO years lead to drought conditions in Southeast Asia during the July-September period, with abnormally low rainfall persisting later in the year. Under these conditions the vegetation is more susceptible to burning and the risk of uncontrolled fires increases sharply.

The 1997/98 episode was undoubtedly the result of the interplay of land-use change with an unusually strong ENSO event. The extensive fires, in addition to causing human health impacts and considerable economic damage to the region, released about 7 Mt of carbon to the atmosphere (Murdiyarso et al. 2002). Only a fraction of this carbon will be replaced through growth of the oil palm plantation, which has a much lower above-ground biomass per unit area than the primary forest which it replaced. In addition, some of the fires spread into areas of peat swamp, which continued to burn for up to a year (Page et al. 2002, Schimel and Baker 2002, Langenfelds et al. 2002). The carbon lost from the peat swamps will likely not be offset by regrowth for many centuries.

4.4.2 Coral Reefs

The current status and projected future of the world's coral reefs (Fig. 4.48) provide an even more complex example of the nature of multiple, interacting stresses on a natural ecosystem and its non-linear responses. Coral reefs are found in shallow water tropical and sub-tropical environments – regions that are characterised by warm temperatures, aragonite supersaturation and high photon flux densities. The reefs are formed by small invertebrates, cnidarians, that are relatives of jellyfish and sea anemones. These animals have algal (dinoflagellate) symbionts that live in their tissues. The symbiosis between these photosynthesising dinoflagellates and their hosts is believed to be critical for the high calcification rates occurring within the corals and which lead to the creation of the reefs themselves (Coates and Jackson 1987). The formation of coral reefs is, in turn, believed to lead to the accumulation of carbonates in shallow-water environments which may lead to feedbacks in the global carbon cycle (Opdyke and Walker 1992).

Fig. 4.47.
The 1997/1998 fires in Southeast Asia: **a** TOMS images showing the aerosol index, a relative measure of the ultraviolet albedo for different wavelengths in the UV spectrum. The *white arrows* indicate the approximate positions of the top and bottom of the mosaic, on the boundary between clear air and the smoke plume (NASA 1997); **b** a wall of fire burning its way through a State forest towards the Batu Putih-Kota Kinabatangan road (*photo:* WWFM, Teoh Teik Hoong, 23 April 1998); and **c** satellite image of the smoke haze spreading over Southeast Asia as a result of the fires (NOAA 14 March 1998)

Despite their appearance as an oasis of diversity in equilibrium with their environment, coral reefs exhibit complex dynamics that are driven by periodic natural disturbances (Nyström et al. 2001) (Table 4.3). Even without any human stresses, reefs are subject to predation and grazing, storms and hurricanes, natural episodes of bleaching, crown-of-thorn outbreaks and natural epidemic diseases. In most cases, the reefs cope with these disturbances and recover to maintain their structure and functioning within the same state. However, as for many ecosystems, coral reefs can exist in several stable states, and changes in disturbance regimes can trigger an abrupt change from one state to another (Sect. 5.3 will contain a related discussion at the scale of the Earth System as a whole).

Over the last four or five decades an increasing number of stresses of human origin have begun to interact with the natural disturbance dynamics of coral reefs. These include increasing nutrient and sediment loadings from on-shore agricultural and industrial activities, changing structure of reef ecosystems due to intensive fishing, and the pressures of rapidly growing tourism, especially the location of tourist facilities on the reefs themselves. One critical difference between many natural disturbances and these recent human pressures is that the latter are generally persistent and slowly accumulate whereas natural disturbances tend to be isolated events at infrequent intervals, allowing the reefs time for recovery in between events.

Global change is adding even more stresses of a quite different nature. Increasing atmospheric CO_2 leads to increased CO_2 concentration in the upper ocean layers, changing carbonate chemistry and thus the ability of reef organisms to create their calcium carbonate shells. A doubling of atmospheric CO_2 is estimated to lead to a 30% reduction in calcification rate of reef organisms (Kleypas et al. 1999) (Fig. 4.12). Warming of the upper layers of the ocean will also affect reefs as coral-forming organisms live in a narrow temperature range, many near their upper limits. Under thermal stress, corals exhibit bleaching, i.e, the corals expel their photosynthesising symbionts and lose their colour. In the worst case, bleaching can lead to the death of the corals in a reef. As this bleaching usually occurs in response to increases in surface water temperature or light availability, coral reefs are particularly sensitive to global warming. These changes are similar to other human pressures in that they are persistent and increase over time. On the other hand, they are significantly different, and potentially more serious, in that they operate at a global scale in a systemic sense, whereas other human pressures are local in origin and vary from place to place in severity.

Fig. 4.48. Coral reefs are a good example of an ecosystem subject to multiple, interacting stresses (*photo:* J. Wellington, World Data Center for Paleoclimatology)

These various pressures act to alter the state of reef ecosystems in two different but interacting ways. First, many of these changes reduce the resilience of reefs to external perturbations, that is, to the magnitude of a disturbance that can be absorbed before the system changes to another state. Secondly, as noted above, human pressures, especially those that operate indirectly through global change, are changing the nature of the disturbance regimes themselves. From the perspective of the long-term natural variability of the Earth System (e.g., CO_2 concentration in the surface waters), the contemporary human-driven changes are faster by at least an order of magnitude than natural variability and are unrelenting in comparison to the more local human pressures. In addition, changing carbonate chemistry and changing surface water temperature act from opposite latitudinal directions and subject coral reefs to an additional pressure.

The fate of coral reefs will not be determined by global change alone (e.g., rising surface water temperature) acting in a simple cause-effect relationship. At present, global change, interacting with the natural disturbance regime of the coral reef ecosystem and with other human-driven stresses, is leading to a higher probability that reef ecosystems will be driven into other states. However, there is growing concern that the systemic global changes are increasing in importance and may well become the dominant determinant of the future of coral reefs within a decade or so. Coral reef ecosystems may already be experiencing the effects of the rise in global temperature over the last few decades. Bleaching of reefs has increased sharply in frequency over this period (Glynn 1993) and it has recently been estimated that 58% of the world's coral reefs are at medium to high risk as a result of changing environmental conditions (Burke et al. 1998). Figure 4.49 shows the extent of coral bleaching and mortality associated with the elevation of sea surface temperatures during the 1997–1998 period, where a strong El Niño event added to the systemic rise in sea surface temperatures that has been occurring the last several decades.

Table 4.3. The disturbance regime of coral reefs (Nyström et al. 2001)

Process	Spatial extent	Frequency	Duration
Predation and grazing	1 – 10 cm	Weeks to months	Minutes to days
Coral collapse (bioerosion)	1 m	Months to years	Days to weeks
Bleaching or disease of individual corals	1 m	Months to years	Days to weeks
Storms	1 – 100 km	Weeks to years	Days
Hurricanes	10 – 1 000 km	Months to decades	Days
Mass bleaching	10 – 1 000 km	Years to decades	Weeks to months
Crown-of-thorns outbreaks	10 – 1 000 km	Years to decades	Months to years
Epidemic disease	10 – 1 000 km	Months to century	Years
Sea-level or temperature change; changing carbonate chemistry	Global	10^4–10^5 years	10^3–10^4 years

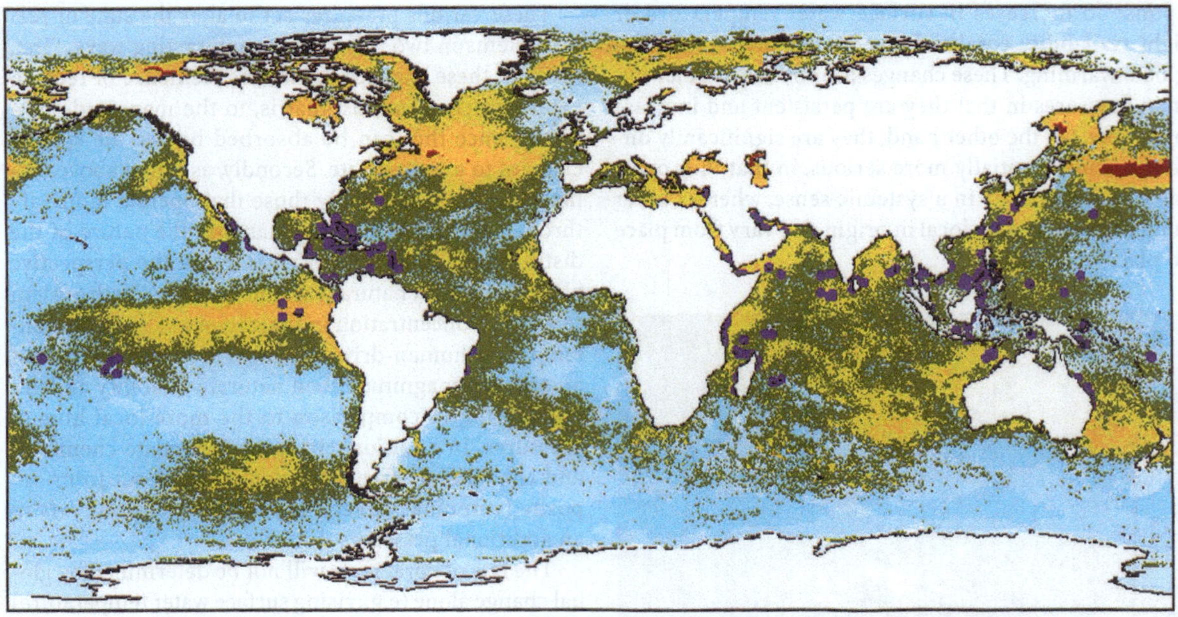

Fig. 4.49. Distribution of sea surface temperature anomalies during the 1997–1998 El Niño period and locations of coral bleaching events (WRI 2001). Exposure for one month at temperatures one or two degrees Celsius higher than the mean averages at the warmest time of year is sufficient to cause the corals to bleach. Actual coral death reaches 95% in some locations

4.5 Integrated Responses of the Earth System to Human Forcing

In Sect. 3.3.6 the direct human alteration of biogeochemical fluxes was considered. This is one component of the changes to Earth's biogeochemical cycles. Another, described in detail earlier in this chapter, is the suite of responses of the Earth System to the human-driven changes in these fluxes. From these two components it may be posited that:

Changed biogeochemical cycling =
human alteration of fluxes
+ Earth System responses

This section deals with the entire equation above by integrating the alteration of fluxes with the Earth System responses in order to examine the perturbed cycle as a whole. The four biogeochemical cycles described in Chap. 3 – carbon, nitrogen, phosphorus and sulphur – are again considered here. In addition, the integrated responses of the two other fundamentally import aspects of the Earth System – the hydrological cycle and the climate system – to the range of human forcings are examined.

4.5.1 The Carbon Cycle

The two most significant human alterations of carbon fluxes are by the combustion of fossil fuels and the removal of carbon from terrestrial ecosystems through land-cover and land-use change. Both of these produce fluxes of CO_2 into the atmosphere. However, measurements show that the annual increase in the amount of CO_2 in the atmosphere is much less than the fluxes, for example, only about 3.2 Pg C yr^{-1} of the total of the nearly 8 Pg C yr^{-1} of estimated emissions from fossil fuel combustion and land-cover and land-use change in the 1990s remained in the atmosphere. The difference is the result of the responses of the Earth System to a wide range of human forcings that resulted in enhanced uptake of CO_2 by both the land and oceans.

Figure 4.50 shows the contemporary carbon cycle for the 1990s with both the direct human alterations of fluxes and the responses of the Earth System included to give a single, integrated budget (Bolin et al. 2000; Prentice et al. 2001). The flux from the land to the atmosphere has been modified from that shown in Fig. 3.53 by merging the direct human-driven flux due to land-use/cover change with the natural fluxes due to respiration processes. The opposing flux, from the atmosphere to the land, has increased by about 3 Pg C yr^{-1} over its pre-industrial value, to give a net flux from the atmosphere to the land of about 1 Pg C yr^{-1}. The ocean has also responded to the increasing concentration of CO_2 in the atmosphere by increasing net uptake by nearly 2 Pg C yr^{-1}.

Apportioning the increased atmosphere-to-land flux of carbon to individual factors is not straightforward as this flux is influenced by many responses of the Earth System to various forcings. Some of the more important factors are:

- the response of photosynthesis of terrestrial biota to increasing atmospheric CO_2 concentration (this effect is well-established both theoretically and through much experimentation, but its magnitude in field conditions is less well known);
- the response of terrestrial biota to increased nitrogen availability (the atmospheric deposition of fixed nitrogen compounds can act as a fertiliser to increase plant growth and thus uptake of carbon);
- the rebound of terrestrial vegetation to past land-use change (in some parts of the northern mid-latitudes, abandonment of agricultural land and the re-establishment of forests leads to a net uptake of carbon as the forests grow and replace vegetation of much less biomass); and
- the response of vegetation to climate variability and change (for example, growing seasons are lengthening in the northern mid and high latitudes, leading to enhanced carbon uptake).

These and other factors are highly interactive, however, and attempting to apportion precisely the terrestrial sink to them individually may prove to be a fruitless exercise.

The net flux of carbon from the atmosphere to the oceans of 1.7 to 2.3 Gt C yr^{-1} is more readily explained. The primary process driving this net flux is a physical one – the increased partial pressure of CO_2 in the lower atmosphere leading to an increased gradient of pCO_2 across the air-sea interface. Although Fig. 4.50 shows that the change in net atmosphere-ocean flux is due to increased dissolution of CO_2 into the ocean, in reality the increased pCO_2 in the atmosphere drives both increased uptake in those areas of the ocean that naturally take up carbon and decreased outgassing in those areas that are naturally carbon sources to the atmosphere.

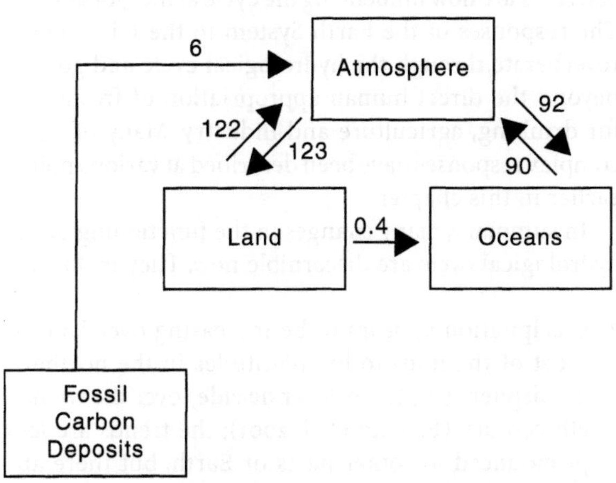

Fig. 4.50. The contemporary carbon cycle based on budgets for the decade of the 1990s (Prentice et al. 2001). All fluxes are given in Pg C yr^{-1} to the nearest whole Pg except for the small land-to-ocean flux. The source term due to land-cover and land-use change has been merged with the respiration term to give a single land-to-atmosphere flux. The carbon cycle is now unbalanced with an accumulation of approximately 3 Pg C yr^{-1} in the atmospheric compartment

The flux of carbon from land to ocean has been maintained at 0.4 Gt C yr^{-1} in Fig. 4.50. This flux has been affected by the responses of the Earth System to several human forcings, but the net effect on the magnitude of the flux is unknown. There is little doubt that additional soil carbon has been mobilised through human activities, but the fate of the carbon is more difficult to determine. Much has been intercepted by dams and impoundments, but some has made its way through the coastal zone and into the oceans.

The responses of the carbon cycle to the human fluxes in terms of increased terrestrial and oceanic uptake may seem small or even trivial compared to the large gross fluxes between the atmosphere and these two compartments. However, it is important to remember that during at least the last 420 000 years the net fluxes between the land and the atmosphere and the ocean and the atmosphere have been very close to zero. Even when the Earth System was moving from a glacial to an interglacial state and hence carbon was being redistributed among the atmospheric, land and ocean pools, the net land-atmosphere and land-ocean fluxes were probably only 0.1 or 0.2 Pg C yr^{-1} on average, an order of magnitude smaller than today's net fluxes.

Much more detailed analyses of the contemporary carbon cycle can be found in Prentice et al. (2001) and Bolin et al. (2000).

4.5.2 Nitrogen, Phosphorus, Sulphur Cycles

Nitrogen

The primary human alteration of nitrogen fluxes is the conversion of non-reactive nitrogen in the atmosphere to reactive forms (both oxidised and reduced forms) and their application to the land surface. However, the human-driven flux (nitrogen fixation) triggers a large number of reverberations through the global nitrogen cycle, leading to significant alterations in other compartments and fluxes in the cycle. The overview by Galloway et al. (1995) provides a useful framework for analysing the changed nitrogen cycle (Box 3.6).

In the preindustrial world the overall amount of nitrogen fixed by natural processes and the denitrification process and consequent return of non-reactive nitrogen to the atmosphere were close to being balanced. Furthermore, the cycles between land and oceans on the one hand and the atmosphere on the other were to a large extent self-contained, i.e., the amount of fixation and denitrification on both land and oceans were approximately balanced. The only major flux between land and oceans was a flux via rivers of about 35 Tg N yr^{-1}, or about 30% or so of the total amount of nitrogen fixed on land. Thus, in general the preindustrial nitrogen cycle could be characterised by the fact that the nitrogen fixed was largely consumed and then dinitrified in the same Earth System compartment – fixation without extensive redistribution.

The human fixation totalling 140 Tg N yr^{-1} represents a doubling, at least, of the natural rate of fixation by terrestrial processes. Much of this additional reactive nitrogen is applied to land (agricultural) systems to increase their productivity. However, substantial amounts of this nitrogen is transported from source via rivers and into the atmosphere. The nitrogen cycle has been altered owing to the facts that:

- the riverine transport of nitrogen from land to sea has increased by about 40 Tg N yr^{-1}, more than double the preindustrial amount with much of this additional nitrogen probably being denitrified in the coastal zone itself;
- the amount of reactive nitrogen emitted to the atmosphere has increased sharply as a result of human activities, from about 29 to 107 Tg N yr^{-1} with the amount of nitrogen transferred from land to ocean via the atmosphere having increased from close to zero to about 18 Tg N yr^{-1} and the amount of reactive nitrogen deposited to the mid-oceans having increased several-fold as a result; and
- some evidence suggests that anthropogenically fixed nitrogen is accumulating in vegetation, soils and groundwater, though quantities are not well known, and it is clear that the overall nitrogen cycle is not in balance in terms of fixation and denitrification processes.

At present, significantly more reactive nitrogen appears to be accumulating in land systems. At the same time significantly more (probably somewhere between 50% and 100%) is moving through the Earth System; extensive redistribution of nitrogen is now occurring around the Earth through atmospheric transport and from land to ocean by river transport.

Phosphorus

The major alteration of fluxes occurs through the mining of phosphate deposits and their conversion to fertilisers or detergents and subsequent application to terrestrial-based systems. The effects of detergents on freshwater ecosystems are well known, but from an Earth System perspective they remain within the land compartment. The major response is thus (*i*) enhanced growth of terrestrial biota due to the application of fertilisers, (*ii*) leakage of some of the applied phosphorus into rivers and eventually to the coastal zone and surface ocean waters, and (*iii*) enhanced growth of marine biota. Given the linkages among the carbon, nitrogen and phosphorus cycles, these responses of the Earth System to increased reactive phosphorus availability interact with changes in the nitrogen and carbon cycles. The magnitude of the increased flux of phosphorus from the land to the coastal zone is significant, but not yet well quantified; a globally consistent effort to quantify C, N and P fluxes in the coastal zone will yield much better quantitative estimates of the phosphorus flux in the near future (Box 6.5).

Sulphur

The primary human-driven flux of sulphur occurs through the combustion of fossil fuels and resultant emission of SO_2 to the atmosphere. The SO_2 undergoes rapid reactions in air leading ultimately to SO_4^{2-} and to sulphuric acid when dissolved in water droplets. The lifetime of sulphur in the atmosphere is short, on the order of days to a week, and so the concentration of SO_2, SO_4^{2-} and intermediates is patchy, clustered around major areas of emissions. The oxidised sulphur compounds are then deposited back to the Earth's surface either through rainfall or dry deposition. Most of the sulphur is deposited back to the land surface, but a significant fraction (ca. 25–30%) is advected to the oceanic atmosphere and deposited on the ocean surface (Charlson et al. 2000). The increased sulphur loading on the land surface is leading to an approximate doubling of the land-ocean flux via rivers to the coastal zone.

4.5.3 The Hydrological Cycle

The hydrological cycle is the life-blood of many of the organisms that inhabit the Earth. At the same time it is, in many ways, the engine of the climate system. Human activities are now influencing the cycle at the global scale. The responses of the Earth System to these influences reverberate through the hydrological cycle and go well beyond the direct human appropriation of freshwater for drinking, agriculture and industry. Many of these complex responses have been described at various points earlier in this chapter.

In summary, many changes in the functioning of the hydrological cycle are discernible now. They include:

- precipitation appears to be increasing over land in most of the mid- to high-latitudes in the northern hemisphere (0.5 to 1.0% per decade) over the twentieth century (Folland et al. 2001); the trends are less pronounced for other parts of Earth, but there appears to be decreasing precipitation in the northern sub-tropics and small increases over tropical lands and tropical oceans;
- in those regions where total precipitation has increased extreme precipitation events are increasing, perhaps by

2 to 4% over the last half of the twentieth century (Folland et al. 2001); by the same token, regions experiencing diminishing total precipitation appear to be experiencing more severe and extended droughts;

- increased aerosol particle loading in the atmosphere is likely affecting the hydrological cycle through changes in precipitation caused by (*i*) the number and size of cloud condensation nuclei in the atmosphere and hence the efficiency of rain droplet formation and (*ii*) diminished evapotranspiration, and hence ultimately diminished precipitation, through decreases in incident solar radiation at the Earth's surface (Box 4.8);
- land-cover change is strongly impacting on the hydrological cycle through changes in the partitioning of incoming solar radiation between evapotranspiration and sensible heat, which in turn affect the amount of water that runs off into riverine systems or infiltrates into soil (Kabat et al. 2004); and
- subtle indirect effects are occurring through the effects of increasing atmospheric CO_2 concentration on the water-use efficiency of terrestrial ecosystems, which ultimately influences the balance between evapotranspiration and soil moisture (Mooney et al. 1999).

4.5.4 The Climate System

It is clear from the discussion throughout this chapter that the climate system is affected by a number of direct human forcings and an even larger range of indirect reverberations through the Earth System arising from other human activities. In addition to the effects of changes in the hydrological cycle, significant effects on the radiative properties of the climate system are clearly discernible. These additional forcings interact with natural forcings such as variations in incoming solar radiation and volcanic eruptions to drive changes to climate. The IPCC Third Assessment Report's summary of the changes in global mean radiative forcing provides a good review (Ramaswamy et al. 2001) (Fig. 4.51):

- carbon dioxide is the most important of the long-lived gases whose concentrations are increasing in the atmosphere as a result of human activities, but CH_4, N_2O and halocarbons (such as CFCs) are also important. All affect the radiation balance of the atmosphere and act to warm the climate near the Earth's surface;

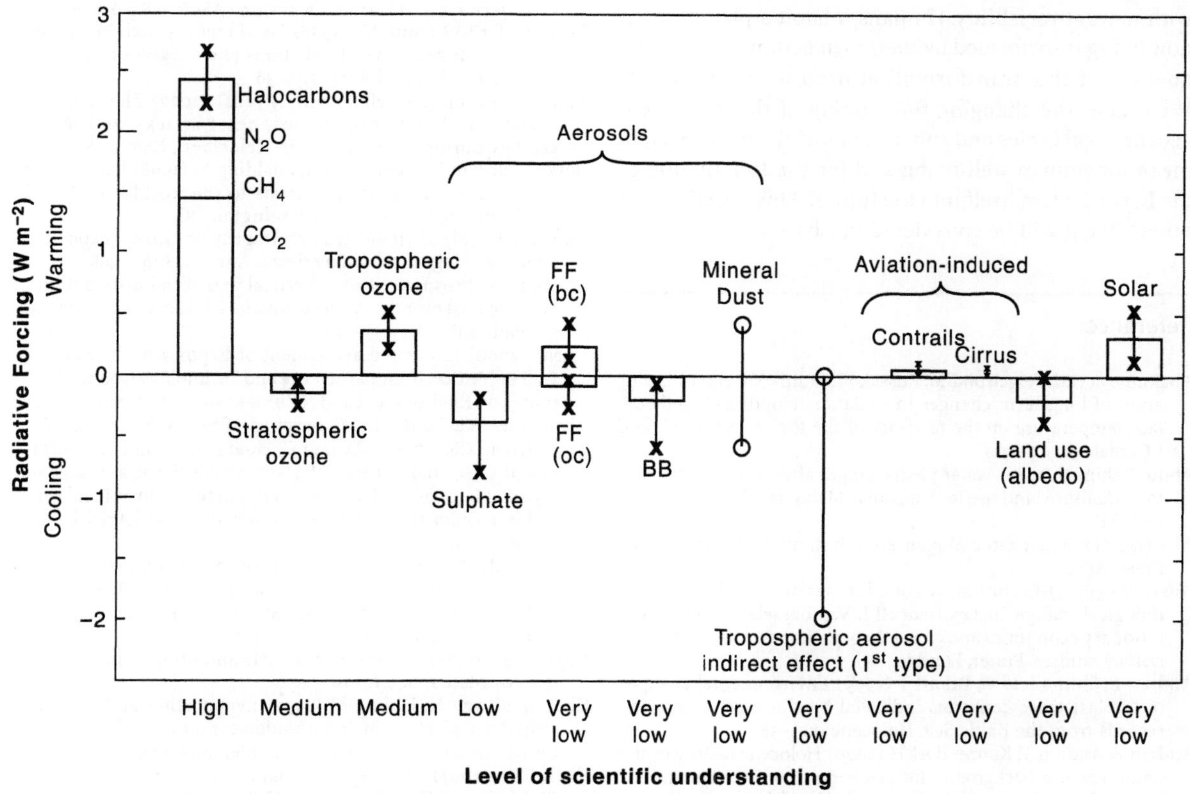

Fig. 4.51. Global mean radiative forcing of the climate system due to a number of agents for the period from pre-industrial (1750) to present (adapted from IPCC 2001a). The height of the *rectangular bar* denotes a central or best estimate value while its absence denotes no best estimate is possible. The *vertical line* about the rectangular bar with *x* delimiters indicates an estimate of the uncertainty range. A *vertical line* without a rectangular bar with no delimiters denotes a forcing for which no central estimate can be given owing to large uncertainties. *FF* means fossil fuel burning while *BB* means biomass burning; *bc* means black carbon and *oc* means organic carbon

- aerosol particles (Box 3.3) influence the climate in complex ways, in addition to their effects via the hydrological cycle. Sulphate and organic carbon particles cool the climate at the surface while black carbon particles cause warming in the troposphere. Aerosol particles indirectly affect the radiative balance by changing the scattering properties of clouds;
- human-driven land-cover change directly affects climate through changes in the reflectance of the Earth's surface, the aerodynamic roughness and the partitioning between latent (evapotranspiration) and sensible heat. Such effects are known to be important for climate locally and regionally and may be significant globally; and
- the magnitude of human-driven radiative forcings of climate are large when compared to changes in natural forcing due to variations in solar radiation.

As analyses of the nature of human activities (Chap. 3) and their effects (this chapter) have shown, the human-environment relationship has changed fundamentally in the last few centuries, and particularly in the last 50 years. These changes do not stop with the initial impact or at the original location of the human activity; they spread through the Earth System in complex ways and interact with natural variability. Humans inhabit a planet that is now being transformed by their own actions. The consequences of this transformation need to be examined. What does the changing functioning of the major biogeochemical cycles and sub-systems of the Earth System mean for human well-being and for the functioning of the Earth System itself into the future? This question and others like it will be considered in Chap. 5.

References

Abakumova GM, Feigelson EM, Russak V, Stadnik VV (1996) Evaluation of long-term changes in radiation, cloudiness, and surface temperature on the territory of the former Soviet Union. J Climate 9:1319–1327

Abdul Rahim N (1988) Water yield changes after forest conversion to agriculture land use in Peninsular Malaysia. J Trop Forest Sci 1:67–84

Alcorn J (1984) Huastec Mayan ethnobotany. University Texas Press, Austin

Allen PM (1994) Evolutionary complex systems: Models of technological change. In: Leydesdorff L, Van Besselaar P (eds) Evolutionary economics and chaos theory: New directions in technology studies. Pinter, London

Andren E, Shimmield G, Brand T (1999) Environmental changes of the last three centuries indicated by siliceous microfossil records from the Baltic Sea. Holocene 9:25–38

Andren E, Andren T, Kunzendorf H (2000) Holocene history of the Baltic Sea as a background for assessing records of human impact in the sediments of the Gotland basin. Holocene 10:687–702

Anon. (1998) Higher temperatures affect krill and other marine organisms. Antarctic Sci 15(4):87

Arnell NW (1999) Climate change and global water resources. Global Environ Chang 9:31–49

Asia Development Bank (2000) Rural Asia: Beyond the green revolution. Manila

Beebee TJC (1995) Amphibian breeding and climate. Nature 374: 219–220

Berendse F, Elberse WT (1990) Competition and nutrient availability in heathland and grassland ecosystems. In: Grace JB, Tilman D (eds) Perspectives on plant competition. Academic Press, San Diego, pp 93–116

Berglund BE (1991) The cultural landscape during 6000 years in Southern Sweden – The Ystad Project. Ecol Bull 41:495

Betts RA (2000) Offset of the potential carbon sink from boreal forestation by decreases in surface albedo. Nature 408:187–190

Billen G, Garnier J (1999) Nitrogen transfers through the Seine drainage network: A budget based on the application of the "Riverstrahler" model. Hydrobiologia 410:139–150

Bird ECF (1993) Submerging coasts: The effects of a rising sea level on coastal environments. John Wiley & Sons Ltd., Chichester

Bird ECF (1996) Coastal erosion and rising sea-level. In: Milliman JD, Haq BU (eds) Sea-level rise and coastal subsidence: Causes, consequences and strategies. Kluwer Academic Publishers, Dordrecht, The Netherlands, pp 87–103

Birks HJB, Line JM (1992) The use of rarefaction analysis for estimating palynological richness from Quarternary pollen-analytical data. Holocene 2:1–10

Bolin B, Sukumar R, Ciais P, Cramer W, Jarvis P, Kheshgi H, Nobre C, Semenov S, Steffen W (2000) Global perspective. In: Watson RT, Noble IR, Bolin B, Ravindranath NH, Verardo DJ, Dokken DJ (eds) Land use, land use change and forestry: A special report of the IPCC. Cambridge University Press, Cambridge, pp 23–51

Bopp L, Monfray P, Aumont O, Dufresne JL, Se Treut H, Madec G, Terray L, Orr JC (2001) Potential impact of climate change on marine export production. Global Biogeochem Cy 15(1):81–99

Brutsaert W, Parlange MB (1998) Hydrologic cycle explains the evaporation paradox. Nature 396:30

Bruun P (1962) Sea level rise as a cause of shore erosion. Journal of Waterways and Harbors Division (ASCE) 88:116–130

Buddemeier RW, Smith SV (1988) Coral reef research in an era of rapidly rising sea level: Predictions and suggestions for long-term research. Coral Reefs 7:51–56

Bunde A, Kropp J, Schellnhuber HJ (eds) (2002) The science of disasters: Climate disruptions, heart attacks, and market crashes. Springer-Verlag, Berlin Heidelberg New York

Burke L, Bryant D, McManus JW, Spalding M (1998) Reefs at risk: A map based indicator of threats to the world's coral reefs. World Resources Institute, Washington DC

Caraco NF, Cole JJ (1999) Human impact on nitrate export: An analysis using world major rivers. Ambio 28:167–170

Carlson TN, Prospero J (1972) Vertical and aerial distribution of Saharan dust over the Western Equatorial North Atlantic Ocean. J Geophys Res 77:5255–5265

CENR (2000) Integrated assessment of hypoxia in the northern Gulf of Mexico. National Science and Technology Council Committee on Environment and Natural Resources, Washington DC

Chameides WL, Yu H, Liu SC, Bergin M, Zhou X, Mearns L, Wang G, Kiang CS, Saylor RD, Luo C, Huang Y, Steiner A, Giorgi F (1999) Case study of the effects of atmospheric aerosols and regional haze on agriculture: An opportunity to enhance crop yields in China through emission controls? P Natl Acad Sci USA 96:13626–13633

Charlson RJ, Anderson TL, McDuff RE (2000) The sulfur cycle. In: Jacobson MC, Charlson RJ, Rodhe H, Orians GH (eds) Earth system science. From biogeochemical cycles to global change. Academic Press, London, pp 343–359

Charney JG (1975) Dynamics of deserts and drought in the Sahel. Q J Roy Meteor Soc 101:193–202

Chattopadhyay N, Hulme M (1997) Evaporation and potential evapotranspiration in India under conditions of recent and future climate change. Agr Forest Meteorol 87:55–73

Chen CTA (2002) The impact of dams on fisheries: Case of the Three Gorges Dam. In: Steffen W, Jäger J, Carson D, Bradshaw C (eds) Challenges of a changing earth: Proceedings of the Global Change Open Science Conference, Amsterdam, The Netherlands, 10–13 July 2001. IGBP Global Change Series. Springer-Verlag, Berlin Heidelberg New York

Cherednik I (1996) Basic methods of soliton theory. World Scientific, Singapore and River Edge New Jersey

Church JA, Gregory JM (2001) Changes in sea level. In: Houghton JT, Ding Y, Griggs DJ, Noguer M, van der Linden PJ, Dai X, Maskell K, Johnson CA (eds) Climate change 2001: The scientific basis. Contribution of Working Group I to the Third Assessment Report of the Intergovernmental Panel on Climate Change. Cambridge University Press, Cambridge, pp 639–696

Claussen M, Gayler V (1997) The greening of the Sahara during the mid-Holocene: Results of an interactive atmosphere-biome model. Global Ecol Biogeogr 6:369–377

Claussen M, Brovkin V, Ganopolski A, Kubatzki C, Petoukhov V (1998) Modelling global terrestrial vegetation-climate interaction. Philos T Roy Soc B 353:53–63

Claussen M, Kubatzki C, Brovkin V, Ganopolski A (1999) Simulation of an abrupt change in Saharan vegetation in the mid-Holocene. Geophys Res Lett 26:2037–2040

Coates AG, Jackson JBC (1987) Clonal growth, algal symbiosis, and reef formation by corals. Paleobiology 13:363–378

Cohen S, Ianetz A, Stanhill G (2002) Evaporative climate changes at Bet Dagan, Israel, 1964–1998. Agr Forest Meteorol 111:83–91

Cole JJ, Caraco NF (2001) Carbon in catchments: Connecting terrestrial carbon losses with aquatic metabolism. Mar Freshwater Res 52(1):101–110

Conover DO, Munch SB (2002) Sustaining fisheries yields over evolutionary time scales. Science 297:94–96

Cooper SR, Brush GS (1991) Long-term history of Chesapeake Bay anoxia. Science 254:992–996

Cooper SR, Brush GS (1993) A 2500 year history of anoxia and eutrophication in Chesapeake Bay. Estuaries 16:617–626

Conrad R (1996) Soil microorganisms as controllers of atmospheric trace gases (H_2, CO, CH, OCS, N_2O, NO). Microbiol Rev 60:609–640

Cowell PJ, Thom BG (1994) Morphodynamics of coastal evolution. In: Carter RWG, Woodroffe CD (eds) Coastal evolution: Late quaternary shoreline morphodynamics. Cambridge University Press, Cambridge, pp 33–86

Cross MC, Hohenberg PC (1993) Pattern formation outside of equilibrium. Rev Mod Phys 65:851–1112

Cutrim E, Martin DW, Rabin R (1995) Enhancement of cumulus clouds over deforested lands in Amazonia. B Am Meteorol Soc 76(10):1801–1805

DeLucia EH, Hamilton JG, Naidu SL, Thomas RB, Andrews JA, Finzi A, Lavine M, Matamala R, Mohan JE, Hendrey GR, Schlesinger WH (1999) Net primary production of a forest ecosystem with experimental CO_2 enrichment. Science 284:1177–1179

Dixon AFG (2000) Insect predator-prey dynamics: ladybird beetles and biological control. Cambridge University Press, New York

Easterling DR, Horton B, Jones PD, Peterson TC, Karl TR, Parker DE, Salinger MJ, Razuvayev V, Plummer N, Jamason P, Folland CK (1997) Maximum and minimum temperature trends for the globe. Science 277:364–367

Edenhofer O (1999) Social conflicts and technological change. PhD Thesis, University of Darmstadt

Edenhofer O, Jaeger C (1996) Power shifts: The dynamics of energy efficiency. Energ Policy 20:513–537

Edwards NT, Tschaplinski TJ, Norby RJ (2002) Stem respiration increases in CO_2-enriched trees. New Phytol 155:239–248

Facchini MC (2002) Clouds, atmospheric chemistry and climate. Institute of Atmospheric and Climate Sciences, National Research Council (ISAC-C.N.R)

FCCC (1999) Article 2, United Nations Framework Convention on Climate Change (UNFCCC). Published for the Climate Change Secretariat by UNEP's Information Unit for Conventions (IUC), France

Fermi E, Pasta J, Ulam S (1974) Studies of nonlinear problems I. In: Newell AC (ed) Nonlinear wave motion. Lectures in applied mathematics (American Mathematical Society) 15, Providence, Rhode Island, pp 143–156

Field RJ, Burgers M (eds) (1985) Oscillations and travelling waves in chemical systems. John Wiley & Sons Ltd., New York

Finzi AC, DeLucia EH, Hamilton JG, Richter DD, Schlesinger WH (2002) The nitrogen budget of a pine forest under free air CO_2 enrichment. Oecologia 132:567–578

Fitter AH, Fitter RSR (2002) Rapid changes in flowering time in British plants. Science 296:1689–1691

Flannery TF (1994) The future eaters. Reed Books, Port Melbourne

Folland CK, Karl TR, Christy JR, Clarke RA, Gruza GV, Jouzel J, Mann ME, Oerlemans J, Salinger MJ, Wang S-W (2001) Observed climate variability and change. In: Houghton JT, Ding Y, Griggs DJ, Noguer M, van der Linden PJ, Dai X, Maskell K, Johnson CA (eds) Climate change 2001: The scientific basis. Contribution of Working Group I to the Third Assessment Report of the Intergovernmental Panel on Climate Change. Cambridge University Press, Cambridge New York, pp 99–181

Forbes DL, Taylor RB (1994) Ice in the shore zone and the geomorphology of cold coasts. Prog Phys Geog 18:59–89

Forbes DL, Orford JD, Taylor RB, Shaw J (1997) Interdecadal variation in shoreline recession on the Atlantic coast of Nova Scotia. In: Proceedings of the Canadian Coastal Conference '97, Guelph, Ontario, Canadian Coastal Science and Engineering Association, Ottawa, Canada, pp 360–374

Formenti P, Andreae MO, Cafmeyer J, Maenhaut W, Holben BN, Lange L, Roberts G, Artaxo P, Lelieveld J (2001) Saharan dust in Brazil and Suriname during the Large-Scale Biosphere-Atmosphere Experiment in Amazonia (LBA) – Cooperative LBA Regional Experiment (CLAIRE) in March 1998. J Geophys Res 106:14919–14934

Fromentin J-M, Planque B (1996) Calanus and environment in the eastern North Atlantic. II. Influence of the North Atlantic Oscillation on *C. finmarchicus* and *C. helgolandicus*. Mar Ecol-Prog Ser 134(1–3):111–118

Fu CB (1994) An aridity trend in China in association with global warming. In: Zepp RG (ed) Climate biosphere interaction: Biogenic emissions and environmental effects of climate change. John Wiley & Sons Ltd., New York, pp 1–17

Fu CB (2003) Potential impacts of human-induced land cover change on East Asia monsoon. Global Planet Change, in press

Fu CB, Wen G (1999) Variation of ecosystems over East Asia in association with seasonal, interannual and decadal monsoon climate variability. Climatic Change 43:477–494

Fu CB, Yuan HL (2001) A virtual numerical experiment to understand the impacts of recovering natural vegetation on summer climate and environmental conditions in East Asia. Chinese Sci Bull 46:1199–1203

Fu CB, Wei HL, Qian Y, Chen M (1999) Documentation on a regional integrated environmental model system (RIEMS). Version 1, TEACOM Report No. 1, START Regional Center for Temperate East Asia, Beijing

Fu F, Harasawa H, Kasyanov V, Kim J-W, Ojima D, Wan Z, Zhao S (2002) Regional-global interactions in East Asia. In: Tyson P, Fuchs R, Fu C, Lebel L, Mitra AP, Odada E, Perry J, Steffen W, Virji H (eds) Global-regional linkages in the Earth System. IGBP Global Change Series. Springer-Verlag, Berlin Heidelberg New York, pp 109–149

Fuehrer J (1996) The critical level for the effect of ozone on crops. In: Karenlampi L, Skarby L (eds) Critical levels for ozone in Europe: Testing and finalising the concepts. UNECE Workshop Report. Kuopio, Finland, pp 27–43

Gaffen DJ, Ross RJ (1999) Climatology and trends of US surface humidity and temperature. J Climate 12:811–828

Galloway J, Levy H, Kasibhatia P (1994) Consequences of population growth and development on deposition of oxidized nitrogen, year 2020. Ambio 23:120–123

Galloway JN, Schlesinger WH, Levy II H, Michaels A, Schnoor JL (1995) Nitrogen fixation: Anthropogenic enhancement-environmental response. Global Biogeochem Cy 9:235–252

Ganopolski A, Rahmstorf S (2002) Abrupt glacial climate changes due to stochastic resonance. Phys Rev Lett 88:038501

Ganopolski A, Petoukhov V, Rahmstorf S, Brovkin V, Claussen M, Eliseev A, Kubatzki C (2001) CLIMBER-2: A climate system model of intermediate complexity. Part II: Validation and sensitivity tests. Clim Dynam 17(10):735–51

Garstang M, Ellery WN, McCarthy TS, Scholes MC, Scholes RJ, Swap RJ, Tyson PD (1998) The contribution of aerosol- and water-borne nutrients to the Okavango Delta ecosystem, Botswana. S Afr J Sci 94:223–229

Glynn P (1993) Coral reef bleaching: Ecological perspectives. Coral Reefs 12:1–17

Golubev VS, Lawrimore JH, Groisman PY, Speranskaya NA, Zhuravin SA, Menne MJ, Peterson TC, Malone RW (2001) Evaporation changes over the contiguous United States and the former USSR: A reassessment. Geophys Res Lett 28:2665–2668

Gómez-Pompa A, Salvador Flores J, Sosa V (1987) The "Pet Kot": A man-made tropical forest of the Maya. Interciencia 12:10–15
Goolsby DA (2000) Mississippi basin nitrogen flux believed to cause gulf hypoxia. EOS Trans Am Geophys Union 29:321–327
Granados J, Körner C (2002) In deep shade elevated CO_2 increases the vigor of tropical climbing plants. Global Change Biol 8:1109–1117
Grasso LD (2000) A numerical simulation of dryline sensitivity to soil moisture. Mon Weather Rev 128:2816–2834
Gregory PJ, Ingram JSI, Campbell B, Goudriaan J, Hunt LA, Landsberg JJ, Linder S, Stafford Smith M, Sutherst RW, Valentin C (1999) Managed production systems. In: Walker B, Steffen W, Canadell J, Ingram J (eds) The terrestrial biosphere and global change. Implications for natural and managed ecosystems. Cambridge University Press, Cambridge, pp 229–270
Grigg OB (1993) The world food problem. Blackwell, Oxford
Grootes PM, Stuiver M, White JWC, Johnson S, Jouzel J (1993) Comparison of oxygen isotope records from the GISP2 and GRIP Greenland ice cores. Nature 366:552–554
Gruber J (1990) Contaminant accumulation during transport through porous media. Water Resour Res 26:99–107
Gunderson CA, Sholtis JD, Wullschleger SD, Tissue DT, Hanson PJ, Norby RJ (2002) Environmental and stomatal control of photosynthetic enhancement in the canopy of a sweetgum (*Liquidambar styraciflua* L.) plantation during three years of CO_2 enrichment. Plant Cell Environ 25:379–393
Gupta HK (1992) Reservoir-induced earthquakes. In: Developments in geotechnical engineering Vol.64. Elsevier Science Publishing Company, New York
Gupta HK (2002) A review of recent studies of triggered earthquakes by artificial water reservoirs with special emphasis on earthquakes in Koyna. Earth-Sci Rev 58(3–4):279–310
Hamilton JG, DeLucia EH, George K, Naidu SL, Finzi AC, Schlesinger WH (2002) Forest carbon balance under elevated CO_2. Oecologia 131:250–260
Harvey N (1986) The Great Barrier Reef: Shallow seismic investigations. Department of Geography, James Cook University of North Queensland, Monograph Series, No. 14 (refereed monograph series)
Hauglustaine DA, Brasseur GP (2001) Evolution of tropospheric ozone under anthropogenic activities and associated radiative forcing of climate. J Geophys Res 106:32337–32360
Haussels R, Klenke T, Kropp J, Ebenhöh W (2001) Observations on the influence of pore space geometry on concentration patterns and transportation properties of dissolved oxygen in a bioactive sandy sediment by a lattice Boltzmann automaton model. Hydrol Process 5(1):81–96
Heath MR, Dunn J, Fraser JG, Hay SJ, Madden H (1999) Field calibration of the Optical Plankton Counter with respect to *Calanus finmarchicus*. Fish Oceanogr 8:13–24
Helbing D, Farkas IJ, Vicsek T (2002) Crowd disasters and simulation of panic situations. In: Bunde A, Kropp J, Schellnhuber HJ (eds) The science of disasters: Climate disruptions, heart attacks, and market crashes. Springer-Verlag, Berlin Heidelberg New York, pp 331–350
Heller EJ (1984) Bound-state eigenfunctions of classically chaotic hamiltonian systems: Scars of periodic orbits. Phys Rev Lett 53:1515–1518
Hendrey GR, Ellsworth DS, Lewin KF, Nagy J (1999) A free-air enrichment system for exposing tall forest vegetation to elevated atmospheric CO_2. Global Change Biol 5:293–310
Heintzenberg J, Raes F, Schwartz SE, Ackermann I, Artaxo P, Bates TS, Benkovitz C, Bigg K, Bond T, Brenguier J-L, Eisele FL, Feichter J, Flossman AI, Fuzzi S, Graf H-F, Hales JM, Herrmann H, Hoffman T, Huebert B, Husar RB, Jaenicke R, Kärcher B, Kaufman Y, Kent GS, Kulmala M, Leck C, Liousse C, Lohmann U, Marticorena B, McMurry P, Noone K, O'Dowd C, Penner JE, Pszenny A, Putaud J-P, Quinn PK, Schurath U, Seinfeld JH, Sievering H, Snider J, Sokolik I, Stratmann F, van Dingenen R, Westphal D, Wexler AS, Wiedensohler A, Winker DM, Wilson J (2003) Tropospheric aerosols. In: Brasseur GP, Prinn RG, Pszenny AAP (eds) The changing atmosphere: An integration and synthesis of a decade of tropospheric chemistry research. The International Global Atmospheric Chemistry Project (IGAC). IGBP Global Change Series. Springer-Verlag, Berlin Heidelberg New York
Herman JR, Bhartia PB, Torres O, Hsu C, Seftor C, Celarier E (1997) Global distribution of UV-absorbing aerosols from Nimbus-7/TOMS data. J Geophys Res 102:16,749–16,759
Hoerling M, Kumar A (2003) The perfect ocean for drought. Science 299:691–694
Hogan J (2002) Will physics crack the market? New Sci 176:16
Holloway GO, Sou T (2002) Has Arctic sea ice rapidly thinned? J Climate 15(13):1691–1701
Holton JR (1992) An introduction to dynamic meteorology, 3rd Ed. Academic Press, San Diego
Homewood K, Lambin EF, Coast E, Kariuki A, Kikula I, Kivelia J, Said M, Serneels S, Thompson M (2001) Long term changes in Serengeti-Mara wildebeest and land cover: Pastoralism, population or policies? P Natl Acad Sci USA 98(22):12544–12549
Howarth RW, Billen G, Swaney D, Townsend A, Jaworski N, Lajtha K, Downing JA, Elmgren R, Caraco N, Jordan T, Berendse F, Freney J, Kudeyarov V, Murdoch P, Zhao-Liang Z (1996) Regional nitrogen budgets and riverine N & P fluxes for the drainages to the North Atlantic Ocean: Natural and human influences. Biogeochemistry 35:75–79
Huang K (1987) Statistical mechanics, 2nd Ed. John Wiley & Sons Ltd., New York
Ingram J, Lee J, Valentin C (1996) The GCTE soil erosion network: A multi-participatory research programme. J Soil Water Conserv 51:377–380
IPCC (2001a) Climate change 2001: The scientific basis. Contribution of Working Group I to the Third Assessment Report of the Intergovernmental Panel on Climate Change. Houghton JT, Ding Y, Griggs DJ, Noguer M, van der Linden PJ, Dai X, Maskell K, Johnson CA (eds) Cambridge University Press, Cambridge New York
IPCC (2001b) Climate Change 2001: Impacts, adaptation and vulnerability. Contribution of Working Group II to the Third Assessment Report of the Intergovernmental Panel on Climate Change. McCarthy JJ, Canziani OF, Leary NA, Dokken DJ, White KS (eds) Cambridge University Press, Cambridge and New York
IPCC (1992) Climate change 1992: The supplementary report to the IPCC Scientific Assessment. Houghton JT, Callander BA, Varney SK (eds) Intergovernmental Panel on Climate Change. Cambridge University Press, Cambridge New York
IPCC (1996) Climate change 1995: The science of climate change. Contribution of Working Group I to the Intergovernmental Panel on Climate Change Second Assessment Report. Houghton JT, Meira Filho LG, Callender BA, Harris N, Kattenberg A, Maskell K (eds) Cambridge University Press, Cambridge New York
Jackson JBC, Kirby MX, Berger WH, Bjorndal KA, Botsford LW, Bourque BJ, Bradbury RH, Cooke R, Erlandson J, Estes JA, Hughes TP, Kidwell S, Lange CB, Lenihan HS, Pandolfi JM, Peterson CH, Steneck RS, Tegner MJ, Warner RR (2001) Historical overfishing and the recent collapse of coastal ecosystems. Science 293:629–637
Jackson JD (1999) Classical electrodynamics, 3rd Ed. John Wiley & Sons Ltd., New York
Jacoby GC, D'Arrigo RD (1997) Tree rings, carbon dioxide, and climatic change. P Natl Acad Sci USA 94:8350–8353
Jaworski NA, Howarth RW, Hetling LJ (1997) Atmospheric deposition of nitrogen oxides onto the landscape contributes to coastal eutrophication in the Northeast United States. Environ Sci Technol 31:1995–2004
Johannessen OM, Shalina EV, Miles MW (1999) Satellite evidence for an arctic sea ice cover in transformation. Science 286: 1937–1939
Jones JB, Mulholland PJ (1998) Carbon dioxide variation in a hardwood forest stream: An integrative measure of whole catchment soil respiration. Ecosystems 1:183–196
Jørgensen BB, Richardson K (1996) Eutrophication in coastal marine ecosystems. Coastal and Estuarine Studies, vol. 52, American Geophysical Union, Washington DC
Kabat P, Claussen M, Dirmeyer PA, Gash JHC, Bravo de Guenni L, Meybeck M, Pielke Sr. RA, Vörösmarty CJ, Hutjes RWA, Lütkemeier S (2004) Vegetation, water, humans and climate: A new perspective on an interactive system. IGBP Global Change Series. Springer-Verlag, Berlin Heidelberg New York

Kaneko K (1989) Pattern dynamics in spatiotemporal chaos. Physica D34:1–7

Kaneko K (1993) Theory and applications of coupled map lattices. John Wiley & Sons Ltd., Chichester New York

Kaneko K, Tsuda I (2001) Complex systems: Chaos and beyond. Springer-Verlag, Berlin Heidelberg New York

Karl TR, Jones PD, Knight RW, Kukla G, Plummer N, Razuvayev V, Gallo KP, Lindseay J, Charlson RJ, Peterson TC (1993) A new perspective on recent global warming – asymmetric trends of daily maximum and minimum temperature. B Am Meteorol Soc 74:1007–1023

Kates WR, Clark WC, Corell R, Hall JM, Jaeger CC, Lowe I, McCarthy JJ, Schellnhuber HJ, Bolin B, Dickson NM, Faucheux S, Gallopin GC, Grübler A, Huntley B, Jäger J, Jodha NS, Kasperson RE, Mabogunje A, Matson P, Mooney H, Moore III B, O'Riordan T, Svedin U (2001) Sustainability science. Science 202:641–642

Keitt TH, Urban DL, Milne BT (1997) Detecting critical scales in fragmented landscapes. Conservation Ecology (online) 1:4. *http://www.consecol.org/vol1/iss1/art4*

Kempe S (1982) Long-term records of CO_2 pressure fluctuations in fresh water. In: Degens ET (ed) Transport of carbon and minerals in major world rivers, Part 1, Vol. 52. Mitt Geol-Palaeont Inst University of Hamburg, pp 91–332

Kleypas JA, Buddemeier RW, Archer D, Gattuso J-P, Langdon C, Opdyke BN (1999) Geochemical consequences of increased atmospheric carbon dioxide on coral reefs. Science 284:118–120

Klyashtorin LB (2001) Climate change and long term fluctuations of commercial catches: The possibility of forecasting. FAO Fisheries Technical Paper 410. Food and Agricultural Organization of the United Nations, Rome

Kolmogorov AN (1941) The local structure of turbulence in an incompressible viscous fluid for very large Reynolds numbers. Dokl Akad Nauk Sssr+ 30(4):301–305

Lachenbruch AH, Marshall BV (1986) Changing climate – geothermal evidence from permafrost in the Alaskan arctic. Science 234:689–696

Langenfelds RL, Francey RJ, Pak BC, Steele LP, Lloyd J, Trudinger CM, Allison CE (2002) Interannual growth rate variations of atmospheric CO_2 and its delta ^{13}C, H_2, CH_4, and CO between 1992 and 1999 linked to biomass burning. Global Biogeochem Cy 16:1–22

Lawrimore JH, Halpert MS, Bell GD, Menne MJ, Lyon B, Schnell RC, Gleason KL, Easterling DR, Thiaw W, Wright WJ, Heim RR, Robinson DA, Alexander L (2001) Climate assessment for 2000. B Am Meteorol Soc 6:1–55

Lawton JH, May RM (eds) (1995) Extinction rates. Oxford University Press, Oxford

Le Fèvre J, Legendre L, Rivkin R (1998) Fluxes of biogenic carbon in the Southern Ocean: Roles of large microphagous zooplankton. J Marine Syst 17:325–345

Leatherman SP (2001) Rating beaches. In: Schwartz M (ed) Encyclopedia of coastal science. Kluwer Academic Publishers, in press

Lelieveld J, Crutzen PJ, Ramanathan V, Andreae MO, Brenninkmeijer CAM, Campos T, Cass GR, Dickerson RR, Fisher H, deGouw JA, Hansel A, Jefferson A, Kley D, de Laat ATJ, Lal S, Lawrence MG, Lobert JM, Mayol-Bracero OL, Mitra AP, Novakov T, Oltmans SJ, Prather KA, Reiner T, Rodhe H, Scheeren H, Sikka D, Williams J (2001) The Indian Ocean experiment: Widespread air pollution from south and southeast Asia. Science 291:1031–1036

Linacre ET (1993a) Data-sparse estimation of lake evaporation, using a simplified Penman equation. Agr Forest Meteorol 64:237–256

Linacre ET (1993b) Estimating US class A pan evaporation from few climate data. Water Int 19:5–14

Lomnitz C, Flores J, Novaro O, Seligman TH, Esquivel R (1999) Seismic coupling of interface modes in sedimentary basins: A recipe for disaster. B Seismol Soc Am 89:14–21

Lorup JK, Hansen E (1997) Effect of land use on the streamflow in the southwestern highlands of Tanzania. In: Rosbjerg D, Boutayeb NE, Gustard A, Kundzewicz ZW, Rasmussen PF (eds) Sustainability of water resources under increasing uncertainty. Proceedings of an International symposium of the fifth scientific assembly of the International Association of Hydrological Sciences (IAHS), Rabat, Morocco, 23 April–3 May 1997. IAHS Press, Wallingford, No. 240, pp 227–236

Lotter AF (1999) Late Glacial and Holocene vegetation history and dynamics as shown by pollen and plant macrofossil analyses in annually laminated sediments from Soppensee, Central Switzerland. Veg Hist Archaeobot 8:165–184

Ludwig W, Probst J-L (1998) River sediment discharge to the oceans: Present day controls of global budgets. Am J Sci 298: 265–295

Luo Y, Medlyn B, Hui D, Ellsworth D, Reynolds J, Katul G (2001) Gross primary productivity in Duke Forest: Modeling synthesis of CO_2 experiment and eddy-flux data. Ecol Appl 11: 239–252

MacIntyre IB, Nurke RB, Stuckenrath R (1977) Thickest recorded Holocene reef section, Isla Periez Core Hole, Alacran Reef, Mexico. Geology 5:749–754

Mann ME, Bradley RS, Hughes MK (1999) Northern Hemisphere temperatures during the past millennium: Inferences, uncertainties, and limitations. Geophys Res Lett 26:759–762

Maslanik JA, Serreze MC, Agnew T (1999) On the record reduction in 1998 western Arctic sea-ice cover. Geophys Res Lett 13: 1905–1908

Mateos JL, Flores J, Novaro O, Seligman TH, Alvarez-Tostado JM (1993) Resonant response models for the Valley of Mexico – II. The trapping of horizontal P waves. Geophys J Int 113:449–462

Mather GK, Terblanche DE, Steffens FE and Fletcher L (1997) Results of the South African cloud seeding experiment using hygroscopic flares. J Appl Meteorol 36:1433–1447

Matson PA, Vitousek PM, Schimel DS (1989) Regional extrapolation of trace gas flux based on soils and ecosystems in exchange of trace gases between terrestrial ecosystems and the atmosphere. In: Andreae MO, Schimel DS (eds) Dahlem Konferenzen. John Wiley & Sons Ltd., Chichester, pp 97–108

Mduma SAR, Sinclair ARE, Hilborn R (1999) Food regulates the Serengeti wildebeest: A 40-year record. J Anim Ecol 68:1101–1122

Meeson BW, Corprew FE, McManus JMP, Myers DM, Closs JW, Sun K-J, Sunday DJ, Sellers PJ (1995) ISLSCP Initiative I – Global data sets for land and atmosphere models, 1987–1988. Vol. 1–5, published on CD by NASA

Meybeck M, Ragu A (1997) Presenting GEMS-GLORI, a compendium for world river discharge to the oceans. Int Ass Hydrol Sci 243:3–14

Milliman JD, Syvitski JPM (1992) Geomorphic/tectonic control of sediment discharge to the ocean: The importance of small mountains and rivers. J Geol 100:525–544

Moonen AC, Ercoli L, Mariotti M, Masoni A (2002) Climate change in Italy indicated by agrometeorological indices over 122 years. Agr Forest Meteorol 111:13–27

Mooney HA, Canadell J, Chapin FS III, Ehleringer JR, Körner Ch, McMurtrie RE, Parton WJ, Pitelka LF, Schulze E-D (1999) Ecosystem physiology responses to global change. In: Walker B, Steffen W, Canadell J, Ingram J (eds) The terrestrial biosphere and global change. Implications for natural and managed ecosystems. Cambridge University Press, Cambridge, pp 141–189

Moreira F, Rego FC, Ferreira PG (2001) Temporal (1958–1995) pattern of change in a cultural landscape of northwestern Portugal: Implications for fire occurence. Landscape Ecol 16:557–567

Moreno JM, Vasquez A, Velez R (1998) Recent history of forest fires in Spain. In: Moreno JM (ed) Large forest fires. Backhuys Publishers, Leiden, pp 159–186

Mund M (1996) Wachstum und oberirdische Biomasse von Fichtenbeständen (*Picea abies* (L.) Karts.) in einer Periode anthropogener Stickstoffeinträge. Diplomarbeit. University of Bayreuth, Germany

Murdiyarso D, Widodo M, Suyamto D (2002) Fire risks in forest carbon projects in Indonesia. Sci China Ser C 45(Supl.):65–74

NASA (1997) Earth probe TOMS: Smoke over Indonesia, September 26, 1997. Earth Science and Image Analysis, NASA-Johnson Space Center

Neill C, Deegan LA, Thomas SM, Cerri C (2001) Deforestation for pasture alters nitrogen and phosphorus in soil solution and stream water of small Amazonian watersheds. Ecol Appl 11(6):1817–1828

Nelson FE (2003) (Un)frozen in time. Science, in press

Nelson R, Winter S (1982) An evolutionary theory of economic change. Belknap, Harvard

Neue HU (1997) Fluxes of methane from rice fields and potential for mitigation. Soil Use Manage 13:258–267
Nicholls RJ, Hoozemans FMJ, Marchand M (1999) Increasing flood risk and wetland losses due to sea-level rise: Regional and global analyses. Global Environ Chang 9:S69–S87
Nixon SW (1995) Coastal marine eutrophication: A definition, social causes, and future concerns. Ophelia 41:199–219
Norby RJ, Wullschleger SD, Gunderson CA, Johnson DW, Ceulemans R (1999) Tree responses to rising CO_2: Implications for the future forest. Plant Cell Environ 22:683–714
Norby RJ, Cotrufo MF, Ineson P, O'Neill EG, Canadell JG (2001) Elevated CO_2, litter chemistry, and decomposition- A synthesis. Oecologia 127:153–165
Norby RJ, Hanson PJ, O'Neill EG, Tschaplinski TJ, Weltzin JF, Hansen RT, Cheng W, Wullschleger SD, Gunderson CA, Edwards NT, Johnson DW (2002) Net primary productivity of a CO_2-enriched deciduous forest and the implications for carbon storage. Ecol Appl 12:1261–1266
Nouchi I, Mariko S, Aoki K (1990) Mechanism of methane transport from the rhizosphere to the atmosphere through rice plants. Plant Physiol 94:59–66
NRC (1993) Managing waste water in coastal urban areas. National Research Council. National Academy Press, Washington DC
Nyström M, Folke C, Moberg F (2001) Coral-reef disturbance and resilience in a human dominated environment. Trends Ecol Evol 15:413–417
Odgaard B (1999) Fossil pollen as a record of past biodiversity. J Biogeogr 26:7–18
Oechel WC, Hastings SJ, Vourlitis G, Jenkins M, Riechers G, Grulke N (1993) Recent change of arctic tundra ecosystems from a net carbon dioxide sink to a source. Nature 361:520–523
Ohmura A, Wild M (2002) Is the hydrological cycle accelerating? Science 298:1345–1346
Ojima D, Lavorel S, Graumlich L, Moran E (2002) Terrestrial human-environment systems: The future of land research in IGBP II. International Geosphere Biosphere Programme. Global Change NewsLetter 50:31–34, Stockholm, Sweden
Oldfield F, Dearing JA (2003) The role of human activities in past environmental change. In: Alverson K, Bradley R, Pedersen T (eds.) Paleoclimate, global change and the future. IGBP Global Change Series. Springer-Verlag, Berlin Heidelberg New York
Opdyke BN, Walker JCG (1992) Return of the coral reef hypothesis: Basin to shelf partitioning of $CaCO_3$ and its effect on atmospheric CO_2. Geology 20:733–736
Page SE, Siegert F, Rieley JO, Boehm H-DV, Jaya A, Limn S (2002) The amount of carbon release from peat and forest fires in Indonesia during 1997. Nature 420:61–65
Panofsky HA (1984) Atmospheric turbulence: Models and methods for engineering applications. John Wiley & Sons Ltd., New York
Parkinson CL, Cavalieri DJ, Gloersen P, Zwally HJ, Comiso JC (1999) Arctic sea ice extents, areas, and trends, 1978–1996. J Geophys Res 9:20837–20856
Parmesan C, Yohe G (2003) A globally coherent fingerprint of climate change impacts across natural systems. Nature 421:37–42
Parton WJ, Hartman M, Ojima D, Schimel D (1998) DAYCENT and its land surface submodel: Description and testing. Global Planet Change 19:35
Pauly D, Christensen V, Dalsgaard J, Froese R, Torres Jr. FC (1998) Fishing down marine food webs. Science 279:860–863
Paustian K, Andrén O, Janzen HH, Lal R, Smith P, Tian G, Tiessen H, Van Noordwijk M, Woomer PL (1997) Agricultural soils as a sink to mitigate CO_2 emissions. Soil Use Manage 13:230–244
Peltier WR (2001) Global glacial isostatic adjusment and modern instrument records of relative sea level history. In: Douglas BC, Kearney MS, Leatherman SP (eds) Sea level rise: History and consequences. Academic Press, International Geophysics Series, Vol. 75
Penkett SA, Law KS, Cox T, Kasibhatla P (2003) Atmospheric photooxidants. In: Brasseur GP, Prinn RG, Pszenny AAP (eds) The changing atmosphere: An integration and synthesis of a decade of tropospheric chemistry research. The International Global Atmospheric Chemistry Project (IGAC). IGBP Global Change Series. Springer-Verlag, Berlin Heidelberg New York
Penman HL (1948) Natural evaporation from open water, bare soil and grass. P Roy Soc Lond A Mat 193:120–145
Percy KE, Awmack CS, Lindroth RL, Kubiske ME, Kopper BJ, Isebrands JG, Pregitzer KS, Hendrey GR, Dickson RE, Zak DR, Oksanen E, Sober J, Harrington R, Karnosky DF (2002) Altered performance of forest pests under atmospheres enriched by CO_2 and O_3. Nature 420:403–407
Peterson TC, Golubev VS, Groisman PY (1995) Evaporation losing its strength. Nature 377:687–688
Petit JR, Jouzel J, Raynaud D, Barkov NI, Barnola J-M, Basile I, Bender M, Chappellaz J, Davis M, Delaygue G, Delmotte M, Kotlyakov VM, Legrand M, Lipenkov VY, Lorius C, Pépin L, Ritz C, Saltzman E, Stievenard M (1999) Climate and atmospheric history of the past 420 000 years from the Vostok ice core, Antarctica. Nature 399:429–436
Petoukhov V, Ganopolski A, Brovkin V, Claussen M, Eliseev A, Kubatzki C, and Rahmstorf S (2000) CLIMBER-2: A climate system model of intermediate complexity. Part I: Model description and performance for present climate. Clim Dynam 16(1):1–17
Petschel-Held G, Lüdeke MKB (2001) Integrating case studies on global change by means of qualitative differential equations. Integrated Assessment 2(3):123–138
Petschel-Held G, Block A, Cassel-Gintz M, Kropp J, Lüdeke MKB, Moldenhauer O, Plöchl M, Schellnhuber HJ (1999) Syndromes of global change – A qualitative modelling approach to assist global environmental management. Environ Model Assess 4(4):295–314
Phillips OL, Vasquez Martinez R, Arroyo L, Baker TR, Killeen T, Lewis SL, Malhi Y, Monteagudo Mendoza A, Neill D, Nunez Vargas P, Alexlades M, Ceron C, Di Flore A, Erwin T, Jardim A, Palaclos W, Saldias M, Vincenti B (2002) Increasing dominance of large lianas in Amazonian forests. Nature 418:770–774
Pielke RA, Lee TJ, Copeland JH, Eastman JL, Ziegler CL, Finley CA (1997) Use of USGS-provided data to improve weather and climate simulations. Ecol Appl 7:3–21
Piketh SJ, Annegarn HJ, Tyson PD (1999a) Lower-tropospheric aerosol loadings over South Africa: The relative impacts of aeolian dust, industrial emissions and biomass burning. J Geophys Res 104:1597–1607
Piketh SJ, Freiman MT, Tyson PD, Annegarn HJ, Helas G (1999b) Transport of aerosol plumes in ribbon-like structures over southern Africa. IUGG 99, Birmingham, England, 19–30 July 1999, A215
Piketh SJ, Tyson PD, Steffen W (2000) Aeolian transport from southern Africa and iron fertilisation of marine biota in the South Indian Ocean. S Afr J Sci 96:244–246
Prentice IC, Farquhar GD, Fasham MJR, Goulden ML, Heimann M, Jaramillo VJ, Khesghi HS, Le Quéré C, Scholes RJ, Wallace DWR (2001) The carbon cycle and atmospheric carbon dioxide. In: Houghton JT, Ding Y, Griggs DJ, Noguer M, van der Linden PJ, Dai X, Maskell K, Johnson CA (eds) Climate change 2001: The scientific basis. Contribution of Working Group I to the Third Assessment Report of the Intergovernmental Panel on Climate Change. Cambridge University Press, Cambridge New York
Prinn RG (1994) Global atmospheric-biospheric chemistry: An overview. In: Prinn RG (ed) Global atmospheric-biospheric chemistry. Plenum Press, New York
Ramanathan V, Crutzen PJ, Kiehl JT, Rosenfeld D (2001a) Aerosols, climate, and the hydrological cycle. Science 294:2119–2124
Ramanathan V, Crutzen PJ, Lelieveld J, Mitra AP, Althausen D, Anderson J, Andreae MO, Cantrell W, Cass GR, Chung CE, Clarke AD, Coakley JA, Collins WD, Conant WC, Dulac F, Heintzenberg J, Heymsfield AJ, Holben B, Howell S, Hudson J, Jayaraman A, Kiehl JT, Krishnamurti TN, Lubin D, McFarquhar G, Novakov T, Ogren JA, Podgorny IA, Prather K, Priestley K, Prospero JM, Quinn PK, Rajeev K, Rasch P, Rupert S, Sadourny R, Satheesh SK, Shaw GE, Sheridan P, Valero FPJ (2001b) The Indian Ocean experiment: An integrated assessment of the climate forcing and effects of the great Indo-Asian haze. J Geophys Res 106:28371–28398
Ramaswamy V, Boucher O, Haigh J, Hauglustaine D, Haywood J, Myhre G, Nakajima T, Shi GY, Solomon S (2001) Radiative forcing of climate change. In: Houghton JT, Ding Y, Griggs DJ, Noguer M, van der Linden PJ, Dai X, Maskell K, Johnson CA (eds) Climate change 2001: The scientific basis. Contribution of Working Group I to the Third Assessment Report of the Intergovernmental Panel on Climate Change. Cambridge University Press, Cambridge New York

Raun WR, Johnson GV (1999) Improving nitrogen use efficiency for cereal production. Agron J 91:357–363

Raupach MR, Bradley EF, Ghadiri H (1987) Wind tunnel investigation into the aerodynamic effect of forest clearing on the nesting of Abbott's Booby on Christmas Island. Report on a study commissioned by the Australian National Parks and Wildlife Service. CSIRO Centre for Environmental Mechanics Technology Rep. 12

Raymond P, Bauer J (2001) Riverine export of aged terrestrial organic matter to the North Atlantic Ocean. Nature 409:497–500

Rice J, Gislason H (1996) Patterns of change in the size spectra of numbers and diversity of the North Sea fish assemblage, as reflected in surveys and models. Ices J Mar Sci 53:1214–1225

Richey JE, Drusche A, Deegan L, Ballester V, Biggs T, Victoria R (2001) Land use changes and the biogoechemistry of river corridors in Amazon. International Geosphere Biosphere Programme, Stockholm, Sweden. IGBP Global Change Newsletter 45:19–22

Richey JE, Melack JM, Aufdenkampe AK, Ballester VM, Hess LL (2002) Outgassing from Amazonian rivers and wetlands as a large tropical source of atmospheric CO_2. Nature 416:617–620

Richey JE, Melack JM, Aufdenkampe AK, Ballester VM, Hess L (2003) Carbon dioxide evasion from central Amazonian wetlands as a significant source of atmospheric CO_2 in the tropics. Nature, in press

Robinson PJ (2000) Temporal trends in United States dew point temperatures. Int J Climatol 20:985–1002

Roderick ML, Farquhar GD (2002) The cause of decreased pan evaporation over the past 50 years. Science 298:1410–1411

Roderick ML, Farquhar GD, Berry SL, Noble IR (2001) On the direct effect of clouds and atmospheric particles on the productivity and structure of vegetation. Oecologia 128:21–30

Rogers A, Ellsworth DS (2002) Photosynthetic acclimation of *Pinus taeda* (loblolly pine) to long-term growth in elevated pCO_2 (FACE). Plant Cell Environ 25:851–858

Root TL, Price JT, Hall, KR, Schneider SH, Rosenzweig C, Pounds JA (2003) Fingerprints of global warming on wild animals and plants. Nature 421:57–60

Rosenberg DM, McCully P, Pringle CM (2000) Global-scale environmental effects of hydrological alterations: Introduction. Bioscience 50:746–751

Rothrock DA, Yu Y, Maykut GA (1999) Thinning of the Arctic sea ice cover. Geophys Res Lett 26:3469–3472

Rotstayn LD, Lohmann U (2002) Tropical rainfall trends and the indirect aerosol effect. J Climate 15:2103–2116

Sabhasri S, Suwarnarat K (1996) Impact of sea level rise on flood control in Bangkok and vicinity. In: Milliman JD, Haq BU (eds) Sea-level rise and coastal subsidence: Causes, consequences and strategies. Kluwer Academic Publishers, Dordrecht, The Netherlands, pp 343–356

Sala OE, Chapin III FS, Gardner RH, Lauenroth WK, Mooney HA, Ramakrishnan PS (1999) Global change, biodiversity and ecological complexity. In: Walker B, Steffen W, Canadell J, Ingram J (eds) The terrestrial biosphere and global change. Implications for natural and managed ecosystems. Cambridge University Press, Cambridge, pp 304–328

Schafer KVR, Oren R, Lai CT, Katul GG (2002) Hydrologic balance in an intact temperate forest ecosystem under ambient and elevated atmospheric CO_2 concentration. Global Change Biol 8:895–911

Scheffer M, Carpenter S, Foley JA, Folke C, Walker B (2001) Catastrophic shifts in ecosystems. Nature 413:591–596

Schellnhuber HJ (1999) Earth system analysis and the second Copernican revolution. Nature 402(Supp):C19–C23

Schimel D, Baker D (2002) Carbon cycle: The wildfire factor. Nature 420:29–30

Schlager W (1999) Scaling for sedimentation rates and drowning of reefs and carbonate platforms. Geology 27:183–186

Scholes MC, Matrai PA, Andreae MO, Smith KA, Manning MR (2003) Biosphere-atmosphere interactions. In: Brasseur GP, Prinn RG, Pszenny AAP (eds) The changing atmosphere: An integration and synthesis of a decade of tropospheric chemistry research. The International Global Atmospheric Chemistry Project (IGAC). IGBP Global Change Series, Springer-Verlag, Berlin Heidelberg New York

Schulze E-D (1995) Herkunft, Wirkung und Verbleib des Stickstoffs in Waldökosystemen. Bayerisches Landesministerium für Ernährung, Landwirtschaft und Forsten. Proceedings of the Symposium on "Waldschäden – Stand der Forschung und Ausblick", Munich May 3, 1995, pp 39–47

Seitzinger SP, Kroeze C (1998) Global distribution of nitrous oxide production and N inputs in freshwater and coastal marine ecosystems. Global Biogeochem Cy 12:93–113

Serneels S, Lambin EF (2001) Impact of land-use changes on the wildebeest migration in the northern part of the Serengeti-Mara ecosystem. J Biogeogr 28(3):391–408

Serneels S, Said MY, Lambin EF (2001) Land cover changes around a major East-African wildlife reserve: The Mara ecosystem (Kenya). Int J Remote Sens 22(17):3397–3420

Shaw BL, Pielke RA, Ziegler CL (1997) A three-dimensional numerical simulation of a Great Plains dryline. Mon Weather Rev 125:1489–1506

Shukla J, Nobre C, Sellers P (1990) Amazon deforestation and climate change. Science 247:1322–1325

Skidmore EL (1965) Assessing wind erosion forces: Directions and relative magnitudes. Soil Sci Soc Am J 29:587–590

Silverberg J, Dosi G, Orsenigo L (1988) Innovation, diversity and diffusion: A self-organisation model. Econ J 98:1032–1054

Smith JB, Schellnhuber HJ, Mirza MMQ (2001) Vulnerability to climate change and reasons for concern: A synthesis. In: McCarthy JJ, Canziani OF, Leary NA, Dokken DJ, White KS (eds) Climate change 2001: Impacts, adaptation and vulnerability. Contribution of Working Group II to the Third Assessment Report of the Intergovernmental Panel on Climate Change. Cambridge University Press, Cambridge

Smith SV, Renwick WH, Buddemeier RW, Crossland CJ (2001) Budgets of soil erosion and deposition for sediments and sedimentary organic carbon across the coterminous United States. Global Biogeochem Cy 15:697–707

Sornette D, Stauffer D, Takayasu H (2002) Market fluctuations II: Multiplicative and percolation models, size effects, and predictions. In: Bunde A, Kropp J, Schellnhuber HJ (eds) The science of disasters: Climate disruptions, heart attacks, and market crashes. Springer-Verlag, Berlin Heidelberg New York, pp 411–435

Stanhill G, Cohen S (2001) Global dimming: A review of the evidence for a widespread and significant reduction in global radiation with discussion of its probable causes and possible agricultural consequences. Agr Forest Meteorol 107:255–278

Steinbock O (1998) Path optimization in chemical and biological systems on the basis of excitation waves. In: Parisi J, Müller SC, Zimmermann W (eds) A perspective look at nonlinear media: From physics to biology and social sciences. Series title: Lecture notes in physics, 503. Springer-Verlag, Berlin Heidelberg New York, pp 179–191

Strain BR, Bazzaz FA (1983) Terrestrial plant communities. In: Lemon ER (ed) CO_2 and plants. Westview Press, Boulder, Colorado, pp 177–222

Swap R, Garstang M, Greco S, Talbot R, Kallbert P (1992) Saharan dust in the Amazon Basin. Tellus 44B:133–149

Szilagyi J, Katul GG, Parlange MB (2001) Evapotranspiration intensifies over the coterminous United States. J Water Res Pl-ASCE 127:354–362

Telmer K, Veizer J (1999) Carbon fluxes, pCO_2 and substrate weathering in a large northern river basin, Canada: Carbon isotope perspectives. Chem Geol 159:61–86

Thom R (1975) Structural stability and morphogenesis: An outline of a general theory of models. Benjamin-Cummings Publishing, Reading

Thomas A (2000) Spatial and temporal characteristics of potential evapotranspiration trends over China. Int J Climatol 20:381–396

Tissue DT, Lewis JD, Wullschleger SD, Amthor JS, Griffin KL, Anderson OR (2002) Leaf respiration at different canopy positions in sweetgum (*Liquidambar styraciflua*) grown in ambient and elevated concentrations of carbon dioxide in the field. Tree Physiol 22:1157–1166

Toth F (2003) Integrated assessment of climate protection strategies. Climatic Change 56(1–2)

Tyson PD, Gatebe CK (2001) The atmosphere, aerosols, trace gases and biogeochemical change in Southern Africa: A regional integration. S Afr J Sci 97:106–118

Tyson PD, Gasse F, Bergonzini L, D'Abreton PC (1997) Aerosols, atmospheric transmissivity and hydrological modelling of climatic change over Africa south of the equator. Int J Climatol 17:1651–1665
Vitousek PM, Aber JD, Howarth RW, Kilens GE, Matson PA, Schindler DW, Schlesinger WH, Tilman DG (1997) Human alteration of the global nitrogen cycle: Sources and consequences. Ecol Appl 7(3):737–750
Vollenweider RA (1989) Global problems of eutrophication and its control. Symposium of Biology Hungary 38:19–41
Vörösmarty CJ, Sharma K, Fekete B, Copeland AH, Holden J, Marble J, Lough JA (1997) The storage and aging of continental runoff in large reservoir systems of the world. Ambio 26:210–219
Vörösmarty CJ, Meybeck M, Fekete B, Sharma K, Green P, Syvitski J (2002) Anthropogenic sediment retention: Major global-scale impact from the population of registered impoundments. Global Planet Change, in press
Walker B, Steffen W, Canadell J, Ingram J (1999) The terrestrial biosphere and global change: Implications for natural and managed ecosystems. Cambridge University Press, Cambridge
Walther G-R, Post E, Convey P, Menzel A, Parmesan C, Beebee T, Frometin J-M, Hoegh-Guldberg O, Bairlein F (2002) Ecological response to recent climate change. Nature 416:389–395
Watson AJ (1997) Surface Ocean-Lower Atmosphere Study (SOLAS). International Geosphere Biosphere Program, Stockholm Sweden, Global Change Newsletter 31
Watson RT, Zinyowera MC, Moss RH, Dokken DJ (1998) The regional impacts of climate change: An assessment of vulnerability. A special report of IPCC Working Group II, Intergovernmental Panel on Climate Change, Cambridge University Press, Cambridge
Winger PV (1986) Forested wetlands of the Southeast: Review of major characteristics and roles in maintaining water quality. US Fish and Wildlife Service Resource Publication No. 13. US Department of the Interior, US Fish and Wildlife Service, Washington DC
Woods W, Mann C (2000) Earthmovers of the Amazon. Science 287:786–789
WRI (2001) Earth trends 2001. World Resources Institute, Washington DC
Wullschleger SD, Gunderson CA, Hanson PJ, Wilson KB, Norby RJ (2002) Sensitivity of stomatal and canopy conductance to elevated CO_2 concentration – interacting variables and perspectives of scale. New Phytol 153:485–496
Yodzis P (1989) Introduction to theoretical ecology. Harper & Row, Cambridge and Philadelphia
Zeeman EC (1976) Catastrophe theory. Sci Am 4:65–83
Zheng Q, Zhou W (1989) Soil erosion in the Yellow River basin and its impacts on reservoir sedimentation and the lower Yellow River. In: Sediment and the environment. IAHS Publ. No. 184. IAHS Press, Wallingford, pp 123–130
Ziegler CL, Martin WJ, Pielke RA, Walko RL (1995) A modeling study of the dryline. J Atmos Sci 52:263–285

Chapter 5

Living with Global Change: Consequences of Changes in the Earth System for Human Well-Being

The changes that are occurring in the functioning of the Earth System, owing at least in part to human activities, have implications for human well-being. Basic goods and services supplied by the planetary life support system, such as sufficiency and quality of food, water resources, air quality, and an environment conducive to human health, are all being affected by global change. These impacts, however, are not the same around the world, but strongly interact with the particular biophysical and socio-economic processes typical of a given region. The concept of vulnerability provides a useful framework within which to study the consequences of global change for human societies. At another level, global change poses potentially serious consequences for the stability of the Earth System itself. The palaeo-record shows that abrupt changes and surprises are common, and that environmental extremes outside the range recorded by instruments in the modern era occur frequently. The stratospheric ozone episode demonstrates that catastrophic failures of the Earth System are not only possible, but that humankind has already narrowly escaped one. Other possible catastrophic failures, such as the slow-down or collapse of the thermohaline circulation in the North Atlantic Ocean, are possible as the Earth System responds to an increasing suite of interacting human forcings.

5.1 Consequences of Global Change

The earliest sections of the book examined ways in which natural processes bring about change in Earth System dynamics at global and regional scales. Later sections considered the manner and degree to which human activities may be viewed as forcing functions driving further change in the System. In this chapter, a different approach is considered, namely that of the way in which humans are becoming increasingly impacted in one way or another by global change and its regional manifestations, and thus how human activities are changing in response to changes in Earth System functioning (Fig. 5.1).

The chapter begins by comparing the standard linear scenario-driven approach to assessing the possible consequences of global change on the Earth System with the need to take into account multiple, complex, nonlinear and interactive forcings of change through the use of vulnerability approaches, including those based on use of palaeo-data and perspectives. The emphasis then moves to issues of vulnerability and sustainability in respect of food, water, air quality and pests and diseases, issues of direct and critical importance for human well-being.

In addition to these well-known impacts, global change presents risks of an entirely different nature. The second part of this chapter examines the connectivity and interaction between different agents of global change and the risks posed by global change to the stability of the Earth System as a whole. The amplification and damping of change are discussed, large systemic changes and their consequences are examined and the possible occurrence of some low-frequency, low probability, high-risk catastrophic events are considered.

5.2 General Approaches for Anticipating the Consequences of Global Change

5.2.1 The Scenario-Driven Approach

The use of scenarios to assess future change has met with considerable success in the past. The approach has most often been used for projecting the consequences of climate change on, for instance, agriculture and food production or water resources. The IPCC Working Group II (IPCC 2001b) has developed and refined the methodology into a powerful tool for assessing the impacts and consequences of climate change over the rest of the twenty-first century. Scenario-driven assessments are equally useful for assessing change based on scenarios other than those of climate, e.g., the impact of land use changes on biodiversity. The approach may be used for investigating many components of global change and in its simplest form follows a linear sequence from glo-

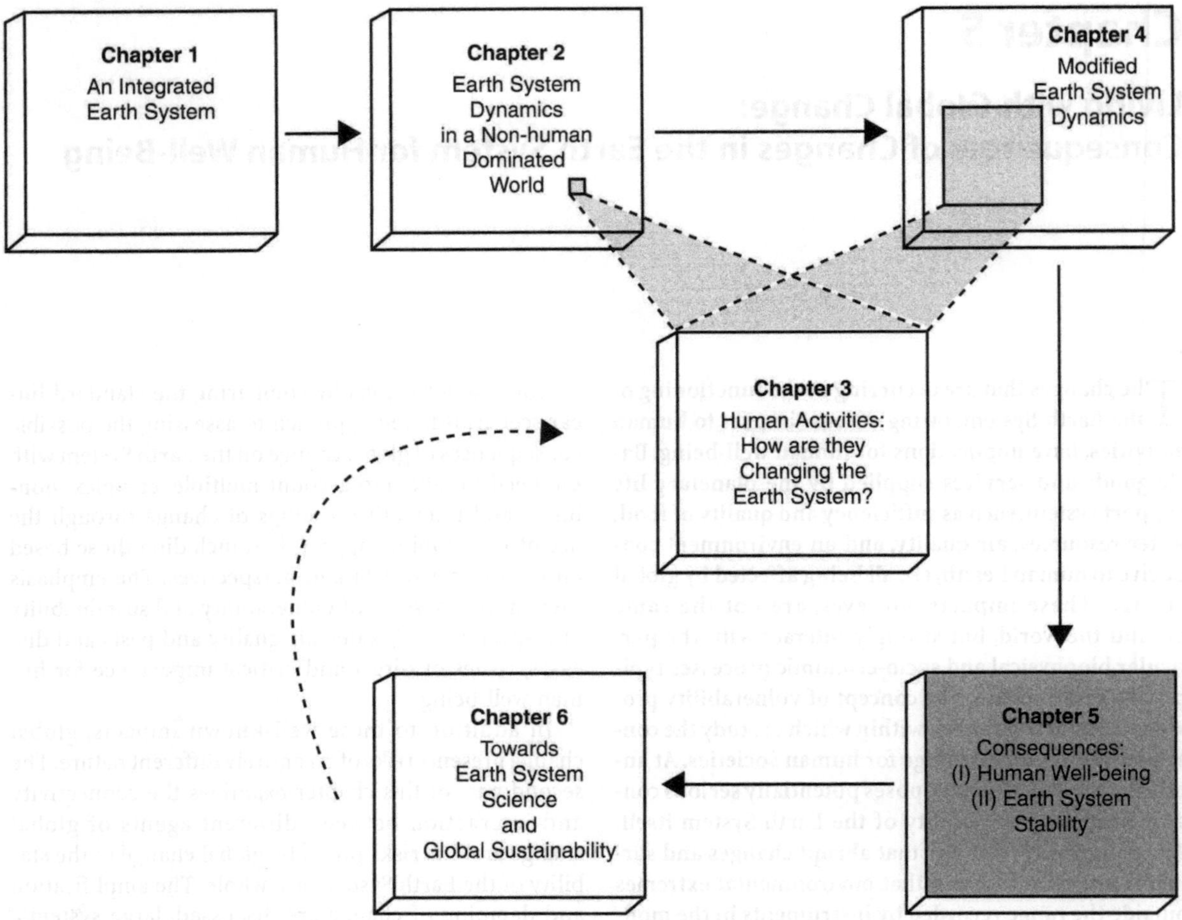

Fig. 5.1. Chapter 5 focus: the consequences of human-driven changes to the Earth System for human well-being and for the functioning of the Earth System itself

Fig. 5.2. A simplified schematic of the scenario-driven approach to global change impacts studies in which a global change scenario is used to drive a model that predicts the impacts on a biophysical system. A second step uses the simulated biophysical changes to estimate the socioeconomic or other human-oriented consequences

bal change scenario through biophysical impact to socio-economic consequence (Fig. 5.2). At each step different models may be used.

Despite the utility of scenario-driven impact studies, they suffer from a number of limitations. One is that each model in the chain carries its own range of uncertainty. When the models' results are concatenated, errors may propagate rapidly to give an overall range of uncertainty so large as to negate the ultimate outcome. Secondly, unless a number of models are used in ensemble mode, with multiple initial scenarios, it is impossible to determine when the system being impacted is near a threshold that would trigger a significant, surprise response, given that non-linearity in the response function is difficult to detect. Most significantly, simple scenario models do not allow the exploration of the effects of multiple interacting stresses. Only when the most sophisticated integrated assessment techniques are employed is this possible. Over the past decade or so, the methodology of integrated assessment modelling has advanced rapidly. In this approach models for both socio-economic and biophysical dynamics are coupled within the same framework. Multiple scenarios of change are used and feedbacks between the human and natural components of the Earth System can be simulated and analysed. An example of an integrated assessment approach, the IMAGE (The Integrated Model to Assess the Global Environment) modelling framework, is described in more detail in Box 5.1.

Scenario-driven approaches to impact assessment, even the most sophisticated of the integrated assessment methods, do not allow the vulnerability or resilience of the impacted systems to be assessed directly. Vulnerability-oriented approaches often achieve greater realism in estimating likely future conditions.

5.2.2 Assessment of Consequences Based on Vulnerability

Vulnerability assessment differs from traditional approaches to impact assessment in a number of important ways (Table 5.1). In essence, impact assessment selects a particular environmental stress of concern (e.g. climate change) and seeks to identify its most important consequences for a variety of social or ecosystem properties. Vulnerability assessment, in contrast, selects a particular group or unit of concern (e.g. landless farmers, boreal forest ecosystems, coastal communities) and seeks to determine the risk of specific adverse outcomes for that unit in the face of a variety of stresses and identifies a range of factors that may reduce response capacity and adaptation to stressors. In principle, the same global change phenomena could be assessed from both perspectives. In practice, impact studies have been most helpful where they have been able to focus on a single stress that dominates system response. However, it is becoming clear that some of the greatest challenges arising from human-environment interactions entail complex system responses to multiple and interacting stresses originating in both the social and environmental realms. Conventional impact assessment practices have been relatively unhelpful in addressing such challenges, primarily because they provide little strategic guidance on which of these multiple stresses a given analysis should consider. Vulnerability assessment offers a maturing strategy to provide such guidance.

Vulnerability to global environmental change has been conceptualised as the risk of adverse outcomes to receptors or exposure units (human groups, ecosystems, and communities) in the face of relevant changes in climate, other environmental variables, and social conditions (Kasperson and Kasperson 2001; Folke et al. 2002; Turner et al. 2003). Vulnerability is emerging as a multidimensional concept involving at least *exposure* – the degree to which a human group or ecosystem comes into contact with particular stresses; *sensitivity* – the degree to which an exposure unit is affected by exposure to any set of stresses; and *resilience* – the ability of the exposure unit to resist or recover from the damage associated with the convergence of multiple stresses. The concepts of preparedness, coping reserve, and adaptive capacity are clearly important. Vulnerability can increase through cumulative events or when multiple stresses weaken the ability of a human group or ecosystem to buffer itself against future adverse events, often through the reduction in coping resources and adaptive capacities. Figure 5.4 shows one conceptual model of vulnerability based on the concepts of exposure, sensitivity and resilience.

Due to its explicit focus on exposure units, vulnerability is an inherently scale-dependent property of systems; it can be considerably different if applied at local, national, regional or global scales. This scale-dependence of vulnerability suggests that much of importance for the efforts of societies to cope with global change will be missed by assessments focused on a narrow range of

Table 5.1. Comparison of vulnerability and scenario approaches to assessing human impacts of global change (Pielke and de Guenni 2004)

Approach	Scenario	Vulnerability
Assumed dominant stress	Climate, recent greenhouse gas emissions to the atmosphere, ocean temperatures, aerosols, etc.	Multiple stresses: climate (historical climate variability), land use and water use, altered disturbance regimes, invasive species, contaminants/pollutants, habitat loss, etc.
Usual timeframe of concern	Long-term, double CO_2, 30–100 years in the future	Short-term (0 to 30 years) and long-term research
Usual scale of concern	Global, sometimes regional, however, there is little evidence to suggest that present models provide realistic, accurate, or precise climate scenarios at local or regional scales	Local, regional, national, and global scales
Major parameters of concern	Spatially averaged changes in mean temperatures and precipitation in fairly large grid cells with some regional scenarios for drought	Potential extreme values in multiple parameters (temperature, precipitation, frost-free days) and additional focus on extreme events (floods, fires, droughts, etc.); measures of uncertainty
Major limitations for developing coping strategies	Focus on single stress limits preparedness for other stresses Results often show gradual ramping of climate change-limiting preparedness for extreme events Results represent only a limited subset of all likely future outcomes – usually unidirectional trends Results are accepted by many scientists, the media, and the public as actual *predictions* Lost in the translation of results is that all models of the distant future have unstated (presently unknowable) levels of certainty or probability	Approach requires detailed data on multiple stresses and their interactions at local, regional, national, and global scales – and many areas lack adequate information Emphasis on short-term issues may limit preparedness for abrupt *threshold* changes in climate some time in the short- or long-term Requires preparedness for a far greater variation of possible futures, including abrupt changes in any direction – this is probably more realistic, yet difficult

Box 5.1. The IMAGE 2 Integrated Assessment Modelling Framework

Rik Leemans

Exploring the long-term dynamics of global environmental change requires an understanding of how the world system could evolve. IMAGE (The Integrated Model to Assess the Global Environment; Alcamo et al. 1998) was developed to quantify the relative importance and long-term impact of major global-change processes and interactions. In doing so, IMAGE explicitly simulates the connections and interactions between the terrestrial biosphere, the anthroposphere and the climate system. Creating a picture of the future depends on estimates of the myriad and complex ways that cultural, economic and socio-political forces interact with the terrestrial biosphere and the climate system. To this end, IMAGE bases its assessments on a number of alternative scenarios of how the future may change. The result is a deeper understanding of how different anthropogenic drivers may influence further global change.

IMAGE is designed to compare reference scenarios with specific mitigation and adaptation scenarios in order to compare the effectiveness of these measures. Scenarios are *what if* representations of how the unknown future might unfold. They form an accepted and valuable tool in analysing how different comprehensive sets of driving forces may influence future global environmental change. One of the major objectives of these assessment models is to improve scientific understanding to provide policy-makers with synthesised information that aids them in developing various policy strategies. The model can also be run in hindcasting mode to identify actions that comply with long-term global change goals.

The socio-economic and energy-use calculations are performed in IMAGE for 17 world regions. The atmospheric and ocean calculations are based on highly aggregated approaches. The land-use and terrestrial carbon calculations are performed on a grid of 0.5 × 0.5 degrees. The innovative aspects of IMAGE consisted of bringing together in a comprehensive framework these highly different components, dimensions and resolutions. The IMAGE simulations are carefully calibrated with observed data for the period 1970–1995. Future trends in socio-economic scenario variables (e.g., demography, wealth, management and technology) have to be defined for each simulation. IMAGE incorporates major global dynamic processes, including several natural interactions and feedbacks, like CO_2 fertilisation and land-use change induced by changed climate.

In 2000 the Intergovernmental Panel on Climate Change (IPCC) published a set of new scenarios in the Special Report on Emissions Scenarios (SRES) (Nakícenovíc et al. 2000) that defined future trajectories of anthropogenic emissions of greenhouse gases in the twenty-first century based on *narrative storylines* for future socio-economic and technological development. The innovative aspect of using narratives lies in the enhanced consistency of future trends in population, wealth, technology, equity and energy use within a scenario. The earlier IPCC emission scenarios were based on expert judgment or literature surveys for every scenario assumption independently. Each SRES narrative now defines characteristic, consistent trends in the input assumptions, recognising explicitly the correlations between the different assumptions.

The quantification of these storylines was based on six different integrated models, including IMAGE. These storylines were constructed along two dimensions, i.e. the degree of globalisation versus regionalisation, and the degree of orientation toward material values versus social and ecological values (Fig. 5.3). The original SRES implementation was defined for four regions (OECD, countries in transition, Asia and rest of the world). This aggregation concealed important regional differences in development as well as in socio-economic and environmental conditions. SRES emphasised emissions from energy use; land-use emissions were poorly represented. Also SRES estimated direct emissions only. Atmospheric, climatic and environmental processes that could alter emissions are neglected.

The SRES narratives are now fully implemented in IMAGE on a much more detailed level considering all energy, industrial and land-use emissions and simultaneously calculating the consequences for atmospheric trace gas concentrations, climate change, land use, ecosystem composition and distribution, sea-level rise and important feedbacks between the compartments (IMAGE team 2001). Basic demographic and economic information is derived from the original SRES data and disaggregated for 17 world regions. The data were checked for consistency by using a population model and world-economy model. By linking the environmental and social aspects of global change, IMAGE is able to achieve new results and couplings that would be impossible to duplicate with simple impact models, which can only illustrate linear cause-and-effect relationships.

In analysing the IMAGE SRES narratives, much attention has been paid to identifying major sources of uncertainties. The different storylines result in largely different sets of energy use and energy carriers, land-use and land-cover patterns, emission trends and levels, trace gas concentrations and climate change. Consequently, sea-level rise and other impacts also differ. Additionally, a sensitivity analysis was done (Leemans et al. 2003) for the narratives with six different climate-change patterns from climate models (GCMs), different climate sensitivities (1.5, 2.5 and 4.5 °C for doubling of greenhouse gas concentrations) and with

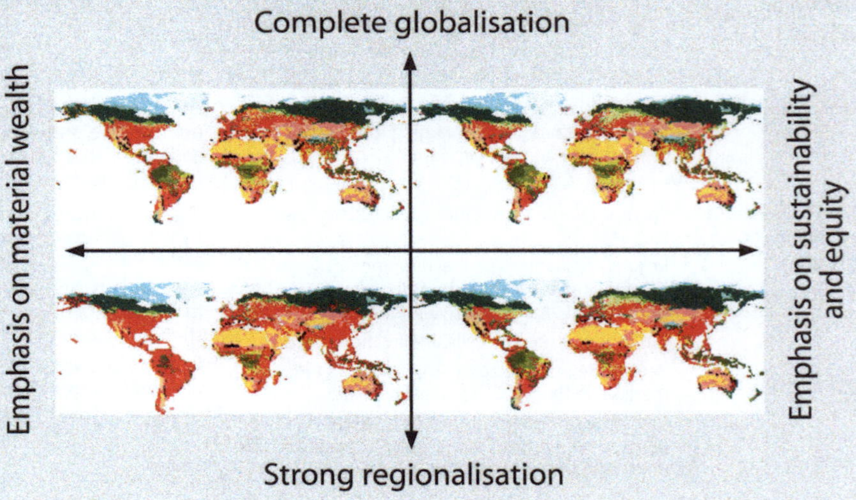

Fig. 5.3.
Land cover in 2100 as projected by the IMAGE model based on the IPCC Special Report on Emission Scenarios (IMAGE team 2001)

◄

or without specific terrestrial biospheric feedback processes (e.g., direct CO_2 effects, soil respiration, plant migration). The results of this sensitivity analysis indicate a possible global mean temperature increase, ranging from 1.5 to 6.0 °C at the end of the twenty-first century. Some of those sensitivity scenarios are certainly unrealistic (e.g. assuming no direct CO_2 effects and immediate vegetation redistribution). The results show that outcomes can easily be manipulated by neglecting important processes and interactions and stress the importance of comprehensive state-of-the-art integrated approaches with an adequate parameterisation of interactions and feedbacks.

The emission pathways are the most widely used aspect of SRES. The IMAGE analysis shows that the SRES narratives dominate greenhouse gas emissions and that these are not strongly influenced by other processes. This finding is encouraging for the way that the IPCC has set up its assessment (developing emission scenarios separately from climate-change and impact assessments) and allows further refinement of local and regional emission scenarios, without accounting for global processes further down the causal chain. The IMAGE analysis shows that this independence falls apart, however, when subsequent components in the causal chain are considered. Atmospheric greenhouse gas concentrations, climate change and vegetation responses strongly interact and cannot be considered in isolation. The added value of these IMAGE results is that the effect of changed land-use patterns, deforestation rates and altered terrestrial carbon fluxes are also considered. The cumulative effects of these changes have a pronounced influence on the final atmospheric concentrations.

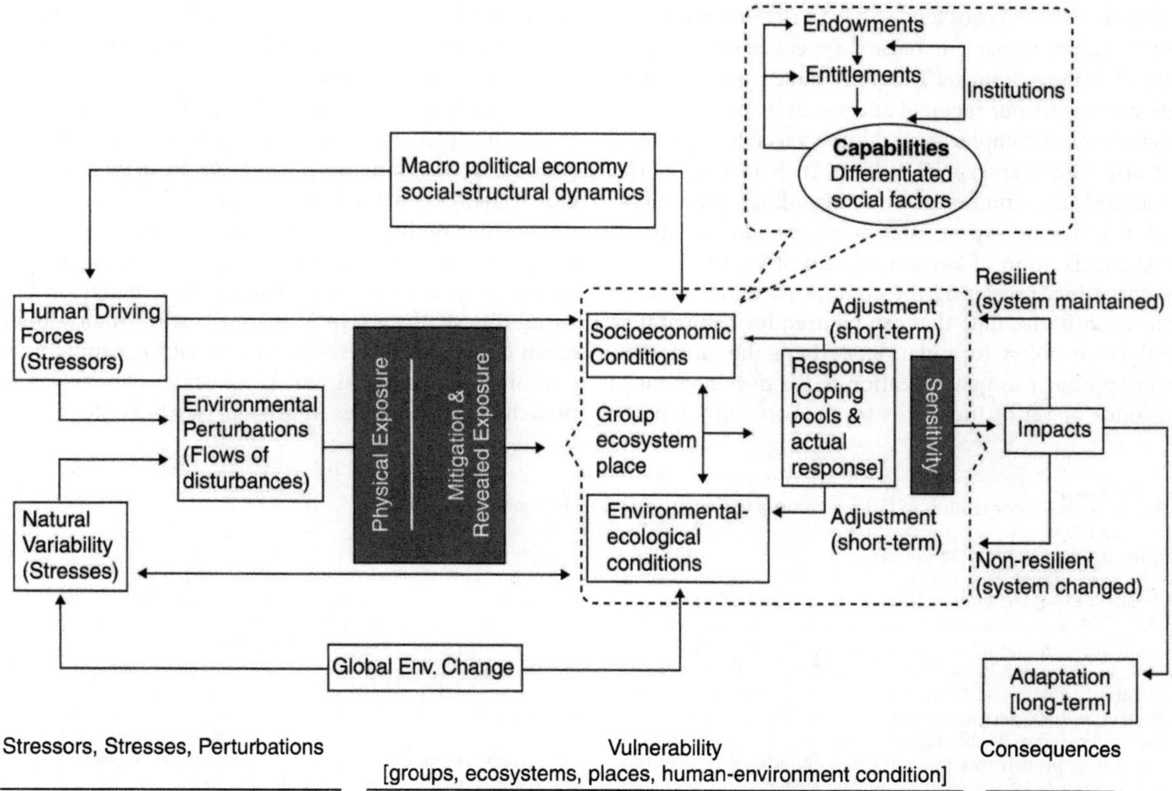

Fig. 5.4. An example of the vulnerability approach to understanding the consequences of global change for human well-being (Turner et al. 2003)

scales, for example only the regional. In particular, it suggests that strategies for reducing vulnerability to global change will require assessments that disaggregate the global or continental analyses adopted for pragmatic reasons in most contemporary work. No inherently superior scale of exposure unit or vulnerability analysis has emerged. It nonetheless appears that many, if not most, useful vulnerability assessments will need to address multiple stresses that interact across a variety of scales.

Within the context of climate change, several approaches based on the vulnerability concept are under development. One such is the *tolerable windows approach* (Bruckner et al. 1999). For the exposure unit considered, potential outcomes are classified either as acceptable or adverse. The assessment then focuses on determining the dynamic combination of environmental and social stresses that could significantly enhance the likelihood of adverse outcomes. It is increasingly recognised that analysis of impacts of environmental change, as well as vulnerability to environmental change, requires development of scenarios of the future that include consideration of not just the environmental stress that may be the primary subject of the investigation, but also encompass other factors that will influence impacts and vulnerability. For example, Carter et al. (1999) describe methods for the development of scenarios of future social, economic and environmental conditions that can be expected to add to or ameliorate stresses that may interact with climate change, or that would influence adaptive capacity.

Conceived in this way, vulnerability analysis can address multiple causes of critical outcomes rather than only the multiple outcomes of a single event. This in turn leads naturally to an evaluation of alternative mitigation and adaptation strategies that could help to avoid such dangerous combinations. Scenarios tying these pieces of the story together become the central output of the vulnerability assessment rather than a peripheral input. Such vulnerability scenarios tend to have a richer texture than conventional impact or hazard type assessments; the vulnerability scenarios generally have more coherent story-lines, greater regional and sectoral specificity, and deeper causal complexity, including variables that characterise human and social systems. Such analyses can be generated in a number of ways, including iterative dialogues involving experts, stakeholders and facilitators.

Quantification of key parameters of vulnerability is essential for translating the conceptual understanding into quantitative data that can be used for comparative analysis or policy formulation. Perhaps the most common approach to quantification is the development of an index or set of indicators to measure vulnerability. One effort (Comfort et al. 1999) is modelled after the United Nations' Human Development Index and aims to develop a standardised all-hazards vulnerability index. In another approach, an Index of Human Insecurity based on a set of 14 indicators has been proposed (Langeweg and Gutierrez-Espeleta 2001). The index is presented on a 1-to-10 scale with each indicator given equal weight. The set of indicators covers a wide range of environmental, economic, social and institutional parameters such as CO_2 emissions, arable land, Gross Domestic Product, literacy, child mortality and an estimation of the level of democracy and human freedoms. In terms of geographical quantification considerable work has also been done on identifying regions of the world that are particularly vulnerable towards the impacts of global change. Such studies include the identification of hot spots of biodiversity loss, of critical regions of threat and of so-called red zones of environmental degradation (Kasperson et al. 1995).

A case study of the vulnerability of farmers in India to climate change and economic globalisation (Box 5.2) shows how the vulnerability approach can be used in a global change context. This example highlights the need to consider multiple stresses and dynamic socio-economic conditions when assessing vulnerability to climate change. Furthermore, it shows that changes to the variability of the Asian Monsoon, rather than global mean climatic changes, will be critical in shaping vulnerability on regional and local scales. Another approach, the syndromes approach, has been described

Box 5.2. The Vulnerability of Indian Farmers to Environmental and Economic Change

Karen O'Brien · Robin Leichenko

A case study of the vulnerability of farmers in India to climate change demonstrates the importance of understanding the socio-economic context, in particular the influence of multiple stresses on vulnerability. The study focused on the interacting effects of climate variability and economic globalisation within the agricultural sector, showing how the two processes intersect and result in double exposure for some districts and farmers. The approach was to accept that climate variability (i.e., the variability of the Asian Monsoon system) is a key factor that affects the viability of farmers but to focus on the changing socio-economic conditions that influence the capacity of farmers to respond to real and potential changes in climate variability, that is, changes in their vulnerability.

India is continuing along a path of economic reform triggered by a balance of payments crisis in 1991. Deregulation of the domestic economy and liberalisation of international trade and investment are parts of the changed policies, and although these may lead to higher incomes and productivity in the agricultural sector in the long run, the shorter term effects may not be positive, especially for the poorer sections of the agricultural sector. For example, decreases in the relative prices of liberalized food grains may improve conditions for consumers in both rural and urban areas, but small-scale producers may experience the negative consequences of import competition and find it difficult to cope with climate variability and potential long-term changes in climate. Ultimately, the removal of trade barriers is likely lead to significant changes in agricultural cropping patterns, thus creating new patterns of vulnerability to climatic impacts.

Some of the most important institutional features of the Indian economy for the rural population have included public investments in rural infrastructure (e.g., irrigation technology) and rural credit programmes. In fact, the current regional disparities in India in agricultural productivity and poverty reduction can be related to the level of public investment in rural infrastructure. To the extent that recent economic reforms may reduce the level of public investment in rural infrastructure, India's less-developed regions will not only have difficulty catching up with more advanced regions, but they will also be more vulnerable to changes in the biophysical environment around them.

It is clear from such detailed analyses that the regional impacts of climate change are a function of more that just the climate changes themselves. For the rural populations in India the multiple, interacting processes involving changes in the regional climate system, liberalisation of the Indian economy, especially trade in agricultural products, and changing levels of public investment in agricultural infrastructure will together have profound effects on the patterns of vulnerability. Adaptation to climate change among India's rural population will have to be addressed within this dynamic socio-economic context.

earlier (Box 3.1) in a broader context as an approach to describe the human-environment relationship more generally, but it is also well placed to study vulnerability to global change at the regional scale.

The current states of vulnerability research and vulnerability assessment exhibit both a potential for substantial synergy in addressing global environmental risks, such as the Indian case study, as well as significant weaknesses which undermine that potential. A substantial base of fundamental knowledge has been created. However, it is highly fragmentary in nature, with competing paradigms, conflicting theory, empirical results often idiosyncratic and tied to particular approaches, and a lack of comparative analyses and findings. However, assessment efforts have increasingly identified vulnerability as a central concern of decision-makers and other interested parties at all levels of governance. Further research in this area will be critically important for Earth System science.

5.2.3 Assessment of Vulnerability Using Palaeo-Data

The bulk of present-day research on the response to future global change of both quasi-natural and highly managed ecosystems, whether it be the scenario-driven or the vulnerability approach, relies largely on a combination of monitoring, experimentation, case studies, large-scale observation campaigns, modelling and analysis set in the contemporary period and recent past. As a result, temporal trajectories are created by developing models that attempt to link spatial patterns and directly observable, short-term changes into a time sequence. Complementary to these studies are the perspectives to be gained from an appraisal of the human-environment relationship as shown in the palaeo-record (Boxes 5.3 and 5.4). Any evaluation of future sustainability or human vulnerability to environmental change requires a deeper understanding of the past interactions between natural variability and human activities in order to

- achieve any realistic concept of pre-human-disturbance environmental baselines,
- identify systems that have been conditioned by or have inherited characteristics from past impacts,
- test hindcasting of predictive models applicable to landscapes impacted by human activity,
- make a robust appraisal of processes operating on longer time-scales than can be documented from direct observations or monitoring,
- understand sensitivities to different combinations of past climate variability and human impacts,
- improve the evaluation of contemporary trends and their future consequences, and
- develop ways of classifying the likely sensitivity of modern human-environment systems to future impacts according to their fundamental dynamical behaviour.

Further development of palaeo-approaches to vulnerability analysis promises to deepen understanding of processes such as land degradation, soil erosion, eutrophication and pollution of both freshwater and marine aquatic systems, surface water acidification, salinisation, non-linear changes in ecosystem structure and functioning and a wide range of multiple stresses arising from the combination of climate variability and human actions.

For most of humanity, the manifestations of global change have meaning only at local or regional scale and this applies in the climate domain as well as to the other processes outlined above. Mitchell and Hulme (1999) critically evaluated the performance of climate models at the regional scale, describing what they term a cascade of uncertainty. Their analysis highlights the intrinsic unpredictability of climate systems subject to the effects of deterministic chaos, the difficulty in simulating the complexity of different forcings on and feedbacks within the Earth System and the inadequacies of the models themselves. With this degree of uncertainty likely to remain inherent in regional climate modelling for the foreseeable future, it is important to consider the potential role of palaeo-records in climate change impact assessment, for palaeo-data become more reliable at the regional than at global level and thus measure reality more accurately.

If global or pan-hemispheric reconstructions of climate variability are examined over the last few centuries, the amplitudes of past decadal or century scale variation fall short of those projected for the next century. At the regional level the variations are often larger and for extreme years/seasons/events (whether they actually occur serially or not) it is often possible to develop credible synoptic situations linked to the phenomena. If future scenarios are then thought of, not as changes in hypothetical means, but as changes in the frequency and persistence of documented extremes, it may be possible to generate scenarios that are both based on reality and make use of the full range of variability that can be reconstructed from past data. This approach also has its flaws and limitations, since future forcings may generate weather patterns and synoptic situations outside the range of past extremes, but these flaws are arguably less severe than the ones that afflict exclusively model-based approaches. Moreover, a strong case can be made for using the full range of evidence available to make some qualitative assessment of system predictability. In the case of well-buffered systems with long histories of no, or rare catastrophic, responses to extreme events, predictability may be relatively high.

Box 5.3. Vulnerability: Past, Present and Future

Frank Oldfield · Bruno Messerli

Archaeological records from around the world provide many examples of flourishing societies that abruptly collapsed. Carefully dated records of environmental change and cultural history point to a strong link between the incidence of drought and the changing fortunes of human societies. In the *upper panel* of the Fig. 5.5 past lake levels at a site in equatorial East Africa have been inferred from stratigraphic changes in a well-dated sediment record. Over the last six centuries, each fall in level reflects a period of drought with severe human consequences as recorded in the oral history of the region. In the *middle panel*, drought in Yucatan, Mexico, is inferred from changes in the stable isotope ratios found in fossil shells of two different species of ostracod (*red* and *blue lines*) preserved in lake sediments. The persistent drought around AD 850 coincides with the collapse of Mayan Civilization. In the *lower panel* the steep fall in carbonate percentage in the marine core represents a major episode of dust deposition that can be directly linked to drought conditions in the nearby region of the Akkadian Empire. The marine and archaeological records can be precisely synchronized by a volcanic ash layer found in both. The episode of drought can thus be shown to coincide with the Akkadian collapse.

These and other palaeo-records link past environmental changes that were within the range of natural variability during the Holocene to responses in human societies, but it is important to realize that these responses reflect socio-economic and cultural factors as well as environmental stress. It is their interactions that determine the nature of the human response. In the case of the Maya for example, overpopulation, excessive urbanization, soil degradation, food security problems and social conflict (Sabloff 1991) are all believed to have increased their vulnerability to the problems created by severe and persistent drought.

In their analysis of human responses to environmental variability, Messerli et al. (2000) trace the changing nature of human-environment interactions in different kinds of cultural and environmental settings ranging from the mobile, prehistoric hunting communities in the Atacama Desert (see also Nunez et al. 2002) to medieval societies in northern Europe and contemporary populations in Bangladesh. From this they develop the concept of trajectories of vulnerability. The question at the beginning of the twenty-first century is: how well prepared are human societies for the environmental changes that lie ahead? In many parts of the world these are likely to be well beyond the range of natural variability experienced by human societies in the past.

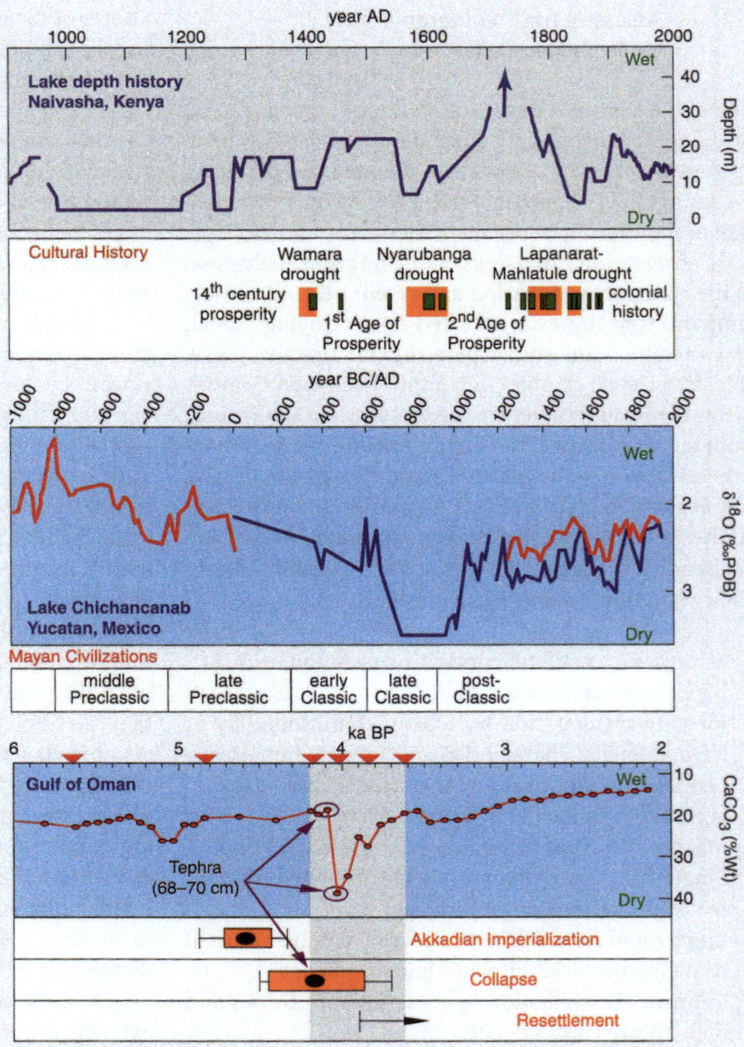

Fig. 5.5.
Past societal collapses: three examples of palaeo-records where the combination of environmental and cultural history, coupled with rigorous chronological constraints, points to a strong link between the incidence of drought and the collapse of human cultures (Verschuren et al. 2000; Hodell et al. 2001; Cullen et al. 2000; adapted from Alverson et al. 2003)

◀

Much of contemporary society's infrastructure – dams, drainage systems, coastal defenses – is designed to withstand extremes defined with respect to recurrence intervals estimated from short term instrumental records. Palaeo-environmental records often document extreme events lying well outside the projected range. Moreover, analyses of future scenarios point to the likelihood of shifts in recurrence intervals that would greatly increase the frequency of the most extreme events (e.g., Milly et al. 2002; Nicholls et al. 1999). The global cost of extreme weather events rose ten-fold over the last 50 years, partly as a result of the expansion of population and infrastructure into more hazard-prone areas, partly as a result of increases in the incidence of extreme events. Although only about a quarter of these losses were insured, the trend has caused severe problems for the global reinsurance industry (IPCC 2001c). These and other observations support the view that the vulnerability of some societies to environmental change is increasing.

In many areas with high population densities and a limited resource base, environmental degradation is already set to increase the vulnerability of societies to the threats posed by future climate change. Continued population growth, the depletion of fossil groundwater at rates greatly exceeding those of recharge, widespread declines in water quality and the loss of soil nutrients through erosion and over-cropping are among the many processes exacerbating these threats. Although projections of future hydrological changes at regional scale are still subject to great uncertainty (Allen and Ingram 2002), they point to the strong likelihood that for at least some of the regions where rainfall is currently seasonal and highly variable, populations will become more vulnerable to recurrent drought (IPCC 2001b). Low lying coastal areas contain some of the most densely settled regions on Earth, and here vulnerability is increasing as a result of rising sea-level, as well as the greater frequency of destructive hurricanes (e.g. Goldenberg et al. 2002).

The example of Bangladesh illustrates how vulnerability to climatic extremes can increase due to increasing population density and over-reliance on limited resources (Hofer and Messerli 1997). The population of the country has doubled between 1961 (55 million) and 1991 (111 million); population density per km^2 has increased from 374 in 1961 to more than 800 in 1994, and the per capita land holding has declined from 0.40 ha in 1950 to 0.16 ha in 1985. Given that more than 80% of the population live in rural areas, every year more people and more small holders are affected adversely by floods and other hazards. Reclamation of former swamps and lakes and the interruption of former feeder channels by buildings, roads, embankments and other infrastructure has led to a loss of storage capacity for the floodwaters of the Brahmaputra, dramatically increasing threats to life and property. All this means that in response to the immediate need to survive, the rapidly growing population has, despite past experience of disasters, increased its own vulnerability to present and future extreme climatic events.

Even now the most extreme floods have, on occasion, affected over 60% of the national territory with loss of more than 2 000 lives. Damaging inundations result from several processes that can overlap and produce cumulative effects: flash floods along the mountain slopes to the north, river floods especially when high rainfall in Assam raises the level of the Brahmaputra, rainwater floods during the monsoon season resulting from high precipitation and saturated soils, and tidal flooding that can affect areas up to 100 km inland. Lateral erosion of the big rivers also makes tens of thousands of people home- and land-less every year. Moreover, disasters due to major cyclones coming from the Bay of Bengal are much more damaging than the floods, with estimated deaths of between 100 000 and 300 000 on at least four occasions since 1876. Finally, Bangladesh is one of the most vulnerable countries to even a slight increase in sea level (cf. Mirza 2002).

By contrast, in those systems which show repeated, non-linear, catastrophic responses to high magnitude events, future behaviour may be more difficult to model or predict.

By linking concern with the past to a research agenda driven by the need to prepare for an uncertain future, the contentious theme of past climate impacts on human societies must be considered (Box 5.3). Oldfield and Dearing (2001) summarised some of the more convincing cases where changing climate may be a decisive factor in shaping the course of cultural and socio-economic development. Each case is interesting in its own right, since for every society, there are quite specific interactions with and responses to climate change that reflect differences in culture and social organisation. The key question here is whether or not there are any generalisations of value for future prediction that can be gleaned from knowledge of past climate-human interactions. One way to approach the question is through the concept of trajectories of vulnerability as human societies change through time.

The concept, developed by Messerli et al. (2000), suggests that vulnerability was high in societies of hunters and gatherers, declined as societies developed more well buffered and productive agrarian-urban mixed systems, but then increases again in regions where population growth and intensifying resource usage begin to diminish the buffering capacity of the natural environment and leaves societies more vulnerable to environmental change and extreme events (Fig. 5.7). The approach builds scenarios for potential implications in the future by drawing insights from detailed case studies of climate-human interactions in the past.

Of particular concern are those processes in the development of human societies that increase vulnerability to extreme events such as droughts and floods. These include population pressure, over-exploitation of finite soil and water resources (for example, groundwater sources) and settlement in areas of high vulnerability, for example, marginal lands or periodically inundated floodplains. During the past century in particular, technological responses to environmental variability have often succeeded in achieving protection against high frequency, low to medium amplitude variations but, by both raising the thresholds of catastrophic impact and encouraging false confidence, they have increased vulnerability to high amplitude, low frequency events (Messerli et al. 2000). If these trends, several of which are accelerating over much of the world, are viewed in the context of past variability, sharp increases in vulnerability for many areas of human settlement appear unavoidable. The devastating impacts of flooding in Bangladesh are a case in point (Fig. 5.8). Messerli et al. (2000) show that flood frequency in Bangladesh has not increased over the last 130 years, but that the area flooded during extreme events has reached much higher levels

Box 5.4. Teleconnections and Human Response in the Past

Peter Tyson

Centennial-scale teleconnections between equatorial east Africa and subtropical southern Africa, similar to those resulting from present-day sub-decadal-scale ENSO variability, have occurred for at least a thousand years and are out of phase between the two regions (Tyson et al. 2002) (Fig. 5.6). In South Africa it would appear that the variability centred at around 100 years has been present for at least 3500 years with little phase or amplitude modulation.

The Kenyan and South African records show a clear link between changing climate and human activities in the pre-industrial world, in contrast to the effect of human activities on global change that comprises such an important focus of this book. The Lake Naivasha proxy data record accords well with the oral history of people living in the region prior to nineteenth century colonisation (Verschuren et al. 2000). Periods associated with drought-induced famine, political unrest and large-scale human migrations matched periods of low levels of Lake Naivasha. The intervening ages of prosperity, agricultural expansion and population growth (Webster 1980) are coeval with high lake levels. Further south, it has been suggested that the medieval warming and wetter conditions around AD 900–1300 may have been responsible for changes in the Iron Age settlement patterns occurring over much of South Africa at the time (Hall 1984; Huffman 1996). The deteriorating conditions associated with the onset of the Little Ice Age may have contributed to the collapse of the nascent Mapungubwe state in the presently semi-arid Shashi-Limpopo basin, the area where the boundaries of Botswana, Zimbabwe and South Africa meet (Vogel 1995; Huffman 1996). In the Kuiseb River delta of Namibia, settlements were likewise abandoned between 1460 and 1640 with the advance of the Little Ice Age (Burgess and Jacobson 1984).

An intriguing possibility is suggested by the probable existence of the out-of-phase rainfall gradient between east and southern Africa in the centuries before the beginning of the Naivasha record. Bantu-speaking agropastoralists first penetrated into the coastal regions of southeastern Africa by about AD 100–200. Thereafter, they expanded rapidly into northeastern interior regions south of the Limpopo River, where numbers of fairly large settlements were in evidence by AD 400–500 (Vogel 1995). Ceramic studies show cultural links between east Africa and the early southeastern coastal settlements and those in the interior (Maggs 1984; Huffman 1989). The Makapansgat climate record indicates almost continuously moister conditions in southern Africa between 1 and AD 500 (Holmgren et al. 1999). Might not the climate gradient between equatorial and subtropical regions, coupled with the need to find additional or new suitable lands for cultivation, have provided an incentive for people to migrate southward to a more supportive environment?

This seems to have been the case at the end of the first millennium AC when further migrations occurred, and when the Naivasha and Makapansgat records show the gradient of change to the south was particularly pronounced. The appearance of a new ceramic style (Hoffman 1989) and linguistic affinities between some of the peoples of southern and east Africa have been argued to reflect the first migration of Sotho-Tswana people into the region (Huffman and Herbert 1994; Iliffe 1995). The climatic gradient towards ameliorating conditions in the south may well have been a significant factor in augmenting the movement of populations in this direction.

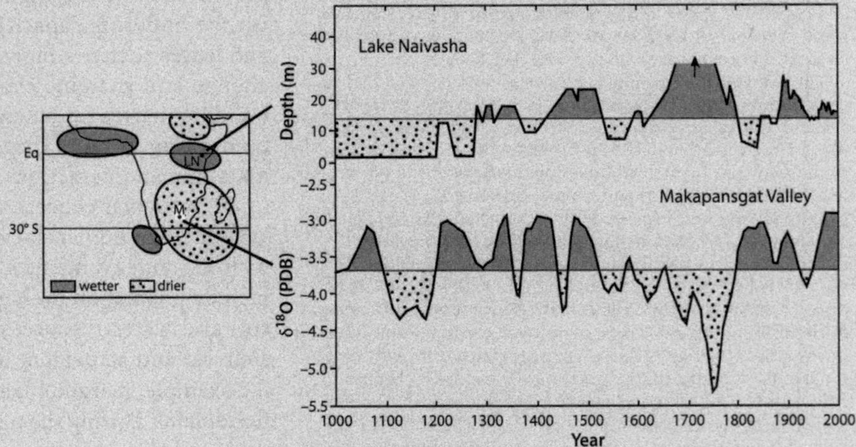

Fig. 5.6. *Left*: Areas in Africa south of the equator showing present-day teleconnection patterns associated with El Niño (*LN* denotes Lake Naivasha, Kenya and *M* the Makapansgat Valley, South Africa); *right*: Makapansgat Valley stalagmite $\delta^{18}O$ variations and the Crescent Island crater, Lake Naivasha lake-level record over the past millennium (Tyson et al. 2002). In the Makapansgat record, high values of $\delta^{18}O$ are associated with warmer, wetter conditions; lower values with cooler, drier climates

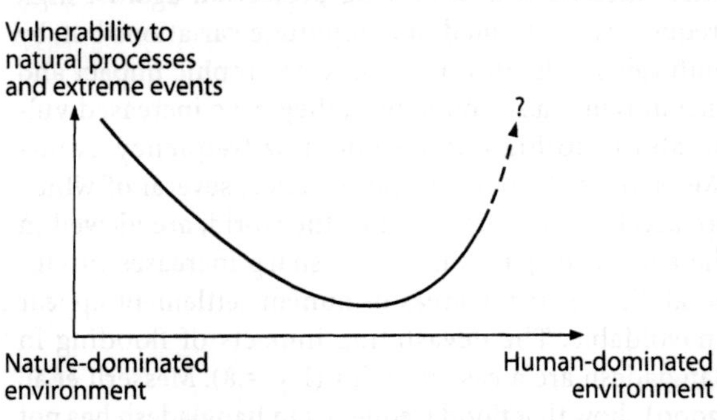

Fig. 5.7. Use of palaeo-data to estimate vulnerability through time as a human society moves from a nature-dominated environment to a human-dominated one (Messerli et al. 2001)

during the last two decades. Over the same period the human population has increased by over 20 million. To consider future vulnerability, three further factors need to be taken into account:

- even the most modest projected rise in mean sea level will have significant impact on the vulnerability of low lying coastal regions;
- in the longer time perspective, the events currently perceived as extreme are relatively modest compared with some of the catastrophic effects of natural river diversions triggered by a combination of major floods and tectonic activity in earlier centuries; and

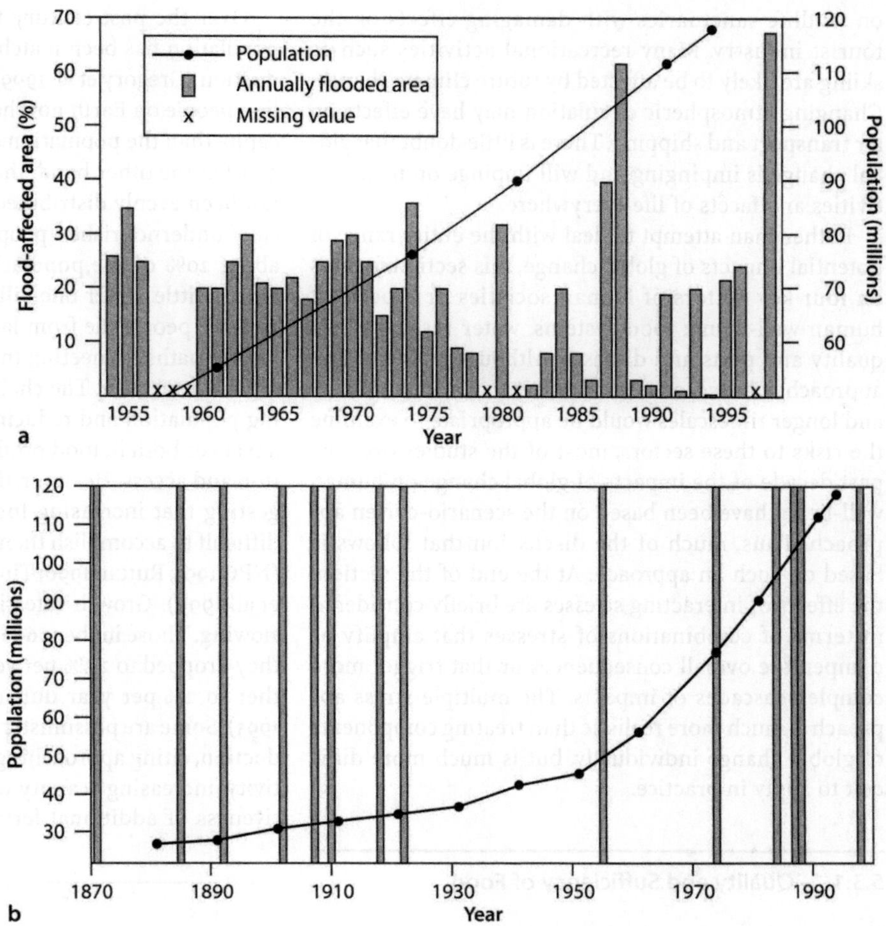

Fig. 5.8. Impacts of floods in Bangladesh: **a** population growth and flood affected areas in Bangladesh; and **b** population growth and major floods in Bangladesh, 1870–1993 (Messerli et al. 2000)

- the possible role played by deforestation and subsequent changes in land cover, soil erosion and hydrology in the upper parts of the Ganges and Brahmaputra catchments is still the subject of much controversy.

The Messerli et al. (2000) review points to importance of extreme weather events rather than upstream human activities as crucial on an event-by-event basis, but evidence from many parts of the world suggests that at the very least, such activities can condition the future behaviour of hydrological systems in such a way that they affect runoff and sediment delivery downstream. There are many cases where the link between human activity and on-site erosion is demonstrable on the small scale but decoupled on the spatial scale of a large river catchment with high levels of sediment storage (Lang et al. 2000; Sidorchuk 2000; Oldfield and Dearing 2003). Nevertheless, by making larger volumes of sediment available for future transport by major floods, human activity changes the downstream balance between sediment deposition and removal, a critical factor in the evolution of the type of deltaic environment typical of much of Bangladesh.

Enhancing one aspect of human life sometimes carries with it the penalty of increased vulnerability to environmental extremes. This is well illustrated in the case of the Middle Yangtze Plain around and to the west of Wuhan. A combination of palaeo-limnological and geomorphological studies and a GIS-based survey in and around Hong Hu, one of the major lakes (352 km^2) that serves to store flood waters from the Yangtze during high flows, shows that as a result of the reclamation of open water and lake-marginal land for some form of cultivation between 1953 and 1976, between 30 and 60 percent of the flood storage capacity of the lake was lost (F. Oldfield, unpublished data; and L. Yu, pers. comm.). There can be little doubt that this loss of flood storage capacity was a major contributor to the magnitude of the flood hazard in the region, which was severely affected by flooding in 1991 and 1998.

5.3 Risks to Key Resources for Human Well-Being

Global change will impact human societies in many ways (IPCC 2001b). Changes in extreme climatic conditions are already having consequences in the insurance and reinsurance industries. Loss of biodiversity has potential impacts on provision of medicines and the pharmaceutical industry, and conversion of savanna and other land types to agricultural activities may encroach on wildlife sanctuaries with damaging effects on the tourist industry. Many recreational activities such as skiing are likely to be affected by future climate change. Changing atmospheric circulation may have effects on air transport and shipping. There is little doubt that global change is impinging and will impinge on many activities and facets of life everywhere.

Rather than attempt to deal with the entire range of potential impacts of global change, this section focuses on four key sectors of human societies or aspects of human well-being: food systems, water resources, air quality and pests and diseases. Although vulnerability approaches based on a combination of contemporary and longer timescales would be appropriate to examine the risks to these sectors, most of the studies over the past decade of the impacts of global change on human well-being have been based on the scenario-driven approach. Thus, much of the discussion that follows is based on such an approach. At the end of the section, the effects of interacting stresses are briefly considered in terms of combinations of stresses that amplify or dampen the overall consequences or that trigger more complex cascades of impacts. The multiple stress approach is much more realistic than treating components of global change individually, but is much more difficult to apply in practice.

5.3.1 Quality and Sufficiency of Food

Terrestrial Resources

Over the next several decades food demand will rise in response to growth of population, growth of per capita income, and attempts to reduce the under-nutrition of the very poor. Further yield increases will be required along with additional growth in foods like meat as consumer preferences shift (Pinstrup-Andersen et al. 1999). Assuming no large change in food distribution, grain production/consumption in the developed countries is estimated to increase by 20% between 1990 and 2030, but in less developed countries, it will need to rise during that same period by about 250% – from about 1.0 billion tonnes per year in the 1990 to 2.5 billion tonnes by 2030 (Gregory et al. 1999) (Table 5.2).

Over the past century the rapid growth in human population has been matched by increases in food production (Gregory et al. 1999) (Fig. 5.9). In fact, the six billion people on Earth now have about 15% more food per capita than the population of four billion about 25 years ago. On the other hand, the increase in production has not been evenly distributed and the number of chronically undernourished people has remained constant at about 20% of the population of less developed countries, a little under one billion people. Every week over 200 000 people die from lack of food (Shah 2002).

The paths to meeting the food demand in the future are far from clear. The challenge of feeding the increasing population and reducing hunger requires dramatic advances both in food production and in food distribution and access. However, there is already evidence suggesting that increasing food production will be more difficult to accomplish than it was in the immediate past (NRC 1991; Ruttan 1996; Tinker 1997; Pinstrup-Andersen et al. 1999). Growth rates in food production are now slowing. Those in the 1960s were at 3% per year, whereas they dropped to 2.3% per year in the 1970s and then further to 2% per year during 1980–1992 (Alexandratos 1995). Some are pessimistic about the future of food production, citing approaching limits to biological productivity, increasing scarcity of water and declining effectiveness of additional fertiliser and pesticide applica-

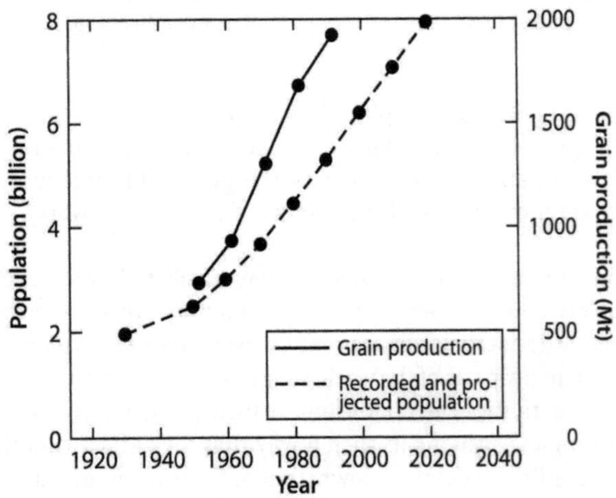

Fig. 5.9. Observed increase in grain production compared to observed and projected population growth (Dyson 1996)

Table 5.2. Predicted changes in grain consumption (Crosson and Anderson 1994)

	1988/1989 (million t)	2030 (million t)	Annual increase (%)
Less developed countries			
Wheat	266	770	2.3
Rice	309	634	1.3
Coarse grains	300	946	3.2
Total	875	2350	2.3
More developed countries			
Total	803	947	0.4

tions as reasons for reduced production levels (Brown and Kane 1994). Even in a world without environmental change, the challenge of continually increasing food production is unprecedented, but in a world of cross-linked global and regional environmental changes, this challenge will become substantially more difficult.

Global change, especially climate change, changing atmospheric composition and chemistry, and negative effects of intensive land use, may all affect the ability to maintain and increase agricultural production. In California, for example, where climate change is likely to reduce winter snow cover and summer water supplies, water for agriculture will come under increasing competition, leading to changes in the crops and acreage under agriculture (Field et al. 1999). In some regions, increased temperatures may have a negative impact on yields and may lead to the need for continuous improvement in crop varieties, since although crops generally grow faster with higher temperature, they may not achieve increased yields owing to differential effects on grain-producing organs (Gregory et al. 1999). The loss of soil fertility and degradation of agricultural lands due to inappropriate management, climate change, and other factors has also become an important issue in some agricultural areas (Matson et al. 1997; NRC 1991). For example, the expansion of irrigated area has contributed to water logging and salinity of some soils. In addition, losses in soil organic matter due to erosion has led to reduced water holding capacity of soils and thus reduced water storage. Reductions in productivity due to changes in air and water quality, some of which emanate from agriculture and may feed back on it, have also raised concerns (Chameides et al. 1994). Any one of these factors alone may impede efforts toward increasing production and yield. Together, these biophysical factors present problems in increasing sustainable agricultural production.

Some factors considered to have a minor impact on crop production in the past may be more important than previously thought. Studies show, for example, that increasing aerosol particle loading and ozone concentrations can have a significant negative impact on crop production. In a case study from East Asia, modelling reveals that the direct effect of regional haze results in a reduction of up to 30% in solar irradiance reaching some of China's most productive agricultural areas (Chameides et al. 1999) (Fig. 5.10). Crop response models suggest an almost 1 : 1 relationship between the percentage increase (decrease) in total surface solar radiation received and the percentage increase (decrease) in yields of rice and wheat. The overall conclusion of the study is that regional haze in China is currently depressing optimal yields of about 70% of the crops grown by up to 30%. The deleterious consequences of increasing tropospheric ozone concentrations on agriculture and health are likewise important for South Asia. At Lahore, Pakistan where 6-hourly mean ozone concentrations sometimes reach to 60 ppb, wheat and rice

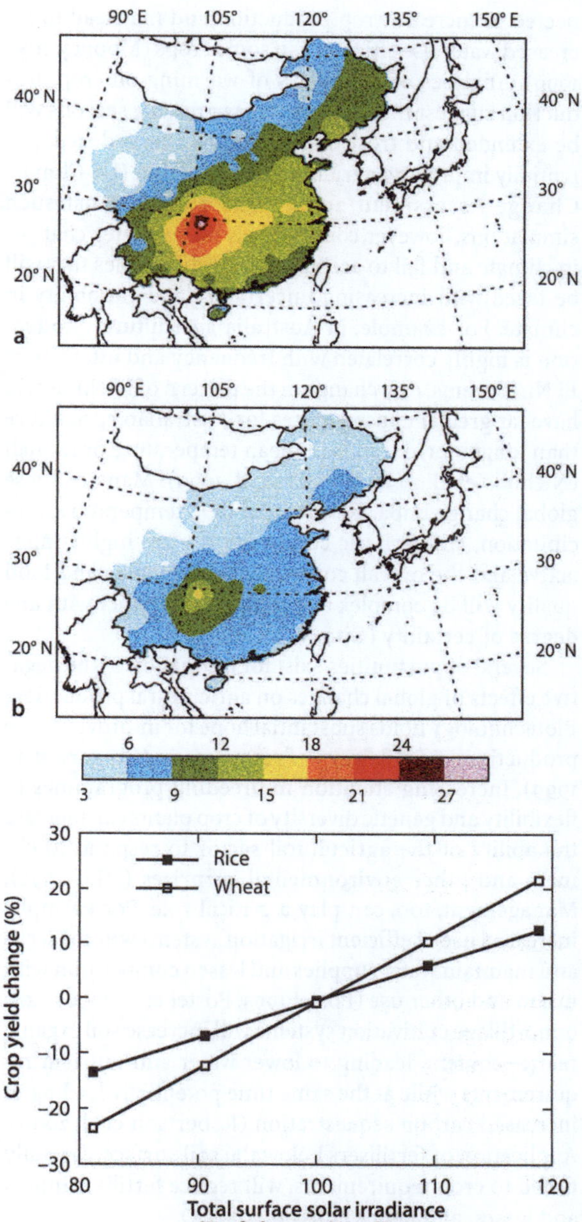

Fig. 5.10. Reduction of agricultural yields in China due to aerosol particles (Chameides et al. 1999). Percent reduction in total surface irradiance over China for summertime conditions for both **a** the cloud-free, high absorption case and **b** the cloud-free, low absorption case. Part **c** shows the model-calculated percentage change in crop yields as a function of the assumed total surface solar irradiance, with 100% representing the observed irradiance. The calculations were carried out by using conditions appropriate for Nanjing

yields decline significantly as a consequence (Maggs et al. 1995). Reductions in two successive seasons ranged from 33% to 46% in wheat and from 37% to 51% in rice. There is also some evidence that increased tropospheric O_3 in the northern mid-latitudes is leading to significant reductions in crop yields (Sect. 4.2.2).

Some global and regional changes, on the other hand, may benefit agriculture. Elevated levels of CO_2 are expected to increase crop production and may lead to increased water use efficiency in some crops (Mooney et al. 1999). Analyses of the effects of warming on crop production suggest that in many areas growing seasons will be extended and frost frequencies will be reduced, potentially improving agricultural production (US Climate Change Assessment 2002; IPCC 2001b). Most such simulations, however, consider only linear or step changes in climate and fail to account for the difficulties that will be faced with increasing uncertainty and variability in climate. For example, in Australia agricultural production is highly correlated with frequency and intensity of El Niño events. Any change in the pattern of El Niños will have far greater consequences for Australian agriculture than longer term trends in mean temperature or rainfall (Nicholls et al. 1996; Pittock et al. 1999). Many of these global change impacts – elevated CO_2, temperature, precipitation, atmospheric composition – are highly interactive and the overall consequences on yields and food quality will be complex and difficult to predict with any degree of certainty (Gitay et al. 2001).

Several opportunities exist for mitigation of the negative effects of global changes on agricultural production. Biotechnology holds substantial hope for improving crop production and efficiency of resource use (Gregory et al. 1999). Increasing attention in breeding programmes to flexibility and genetic diversity of crop plants can increase the ability of the agricultural sector to respond to climate and other environmental surprises (NRC 1992). Management, too, can play a critical role. For example, increased use of efficient irrigation systems will conserve and maintain water supplies and lessen competition with urban and other use (Postel 1993; Postel et al. 1996). Low or no-tillage cultivation systems will increase soil organic matter content, leading to lower water and nutrient requirements while at the same time potentially leading to increased carbon sequestration (Robertson et al. 2000). Application of fertilisers below the soil surface, carefully timed to crop requirements, will reduce fertiliser inputs and losses, and hence emission of N_2O.

Recent trends in precision farming, integrated nutrient management and better crop production management in general (Woomer and Swift 1994; Matson et al. 1997) give hope for increasing efficiency and reducing losses. For example, in the intensive agricultural regions of many parts of the world, the efficiency of food production has risen sharply in the last half-century as shown by the ratio of crops to land (Ausubel 2002) (Fig. 5.11). These gains in productivity have stabilised the amount of land under agricultural production in North America and Western Europe and have led to the abandonment of some marginal lands and their return to forests, thereby incidentally creating a significant sink for atmospheric carbon dioxide (Schimel et al. 2001). Further achievements in efficiency will likely be linked to the introduction of precision farming. While such ap-

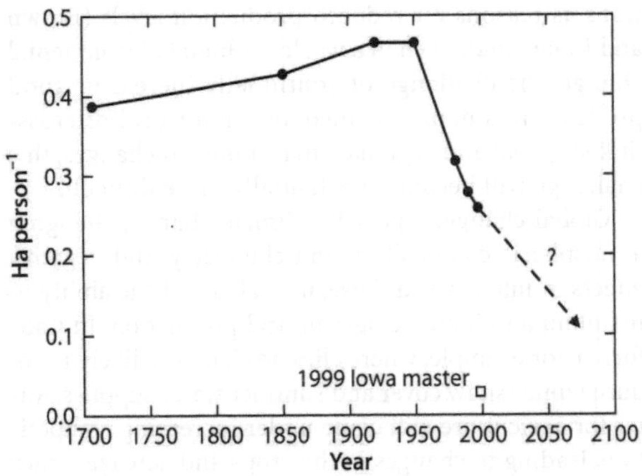

Fig. 5.11. Historical decrease in area of land used to feed a person (Ausubel 2002). The *square* shows the area needed by the Iowa Master Corn Grower of 1999 to supply one person a year's worth of calories. The *dotted line* shows how sustaining the average yield increase by 2 per cent per year extends the decrease in land required to feed a person

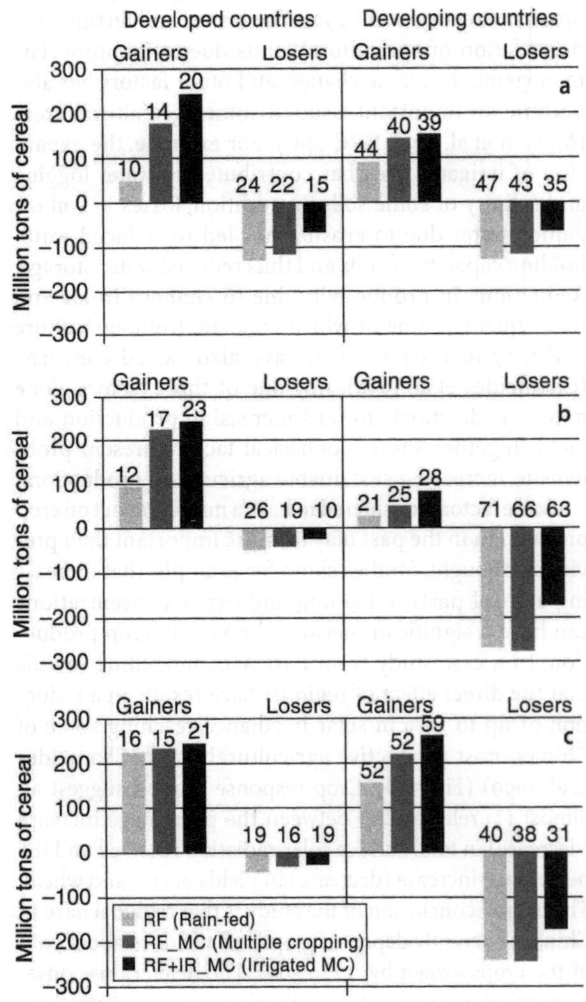

Fig. 5.12. Projected impact of climate change on cereal production in the 2080s: **a** Max-Planck model; **b** Canadian model; **c** Hadley model (modified after Fischer et al. 2001). The number of countries included in analysis is shown above each bar in the chart

proaches are available for many crops and many regions, their present use is minimal, in part because of the requirement for increased information in management.

The recent trends in agricultural production, adoption of improved technologies and world trade suggest that the food challenge will continue to be met in the developed world, but will be severe in many parts of the developing world, especially those countries that currently have average daily per capita calorie deficits of 220 calories or greater. A recent analysis (Shah 2002) of the potential impact of climate change on food security in the twenty-first century highlights the importance of the current disparities in food production capability and of the differential impacts of climate change. Both work in the same direction. Climate change makes the task of improving the condition of most food-insecure populations of the world even more difficult.

The analysis combined the FAO's Agro-Ecological Zone (AEZ) methodology, an analysis of food security by region and country, and three climate change scenarios to estimate changing food security through this century. The projections of climate were obtained from the ECHAM4 model of the Max Planck Institute of Meteorology, Germany; the HadCM2 model of the Hadley Centre for Climate Prediction and Research, UK; and the CGCM1 model of the Canadian Centre for Climate Modelling. All three models predict that the global mean temperature will rise and all suggest that precipitation is likely to come more often in heavy falls and extreme events. The Canadian model projects significantly drier conditions than do the other two, and this has a strong impact on the outcomes of the study.

Globally, climate change is expected to benefit agriculture. The Max Planck and Hadley Centre models suggest that in the 2080s the gain in cereal production will be about 230 million tonnes. However, this aggregate result masks strong regional differences; nearly all of the gains are in the already food-rich countries. All three climate models project increases in grain production of about 200 million tonnes in the developed world. The Max Planck and Hadley Centre models show much smaller net gains – about 30 to 50 million tonnes – for the developing world while the Canadian model shows a large net loss – 170 million tonnes (Fig. 5.12).

In terms of individual countries, the analysis indicates there will be winners and losers (Fig. 5.13). How-

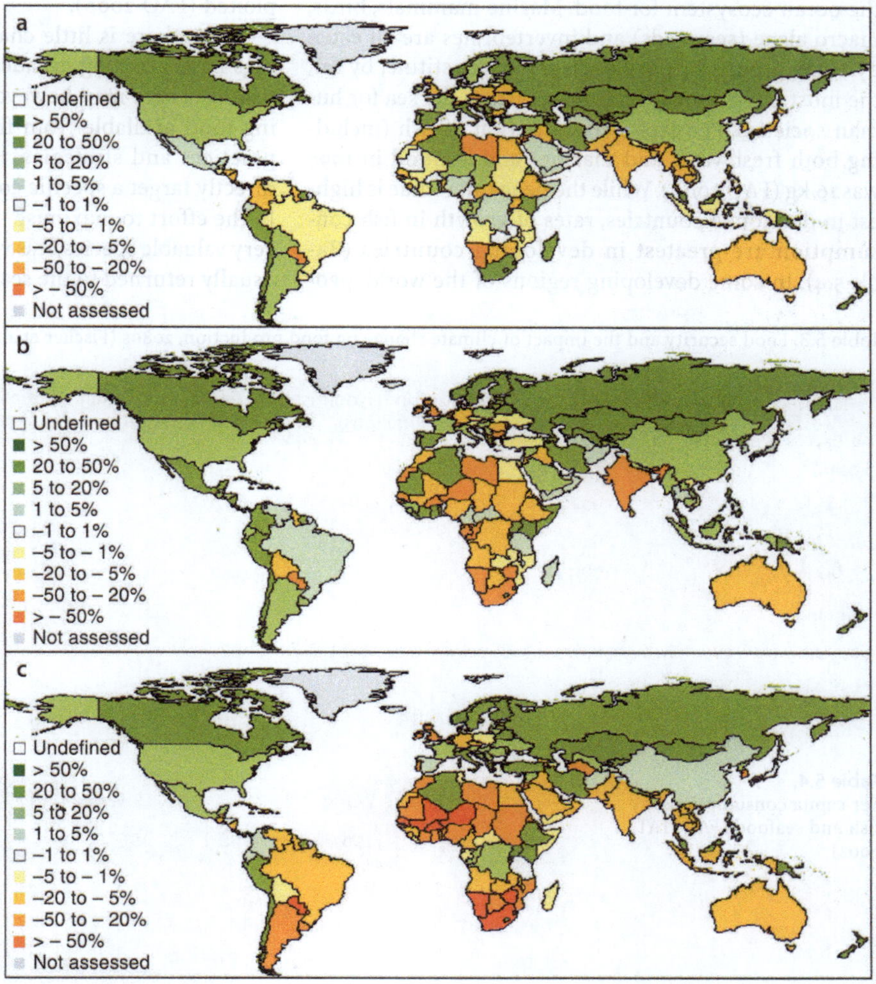

Fig. 5.13. Country-level climate change impacts for the 2080s based on cereal production potential on currently cultivated land: **a** Max Planck model; **b** Hadley model; **c** Canadian model (Fischer et al. 2001)

ever, when the 78 developing countries that have low per capita incomes and account for about 600 of the 800 million undernourished people are considered, the magnitude of the problem becomes apparent (Table 5.3). For all three climate models the number of currently undernourished people in countries whose food production decreases because of climate change (about 400 million in total) is greater than the number in countries which gain. In fact, two of the three models project that the current cereal gap of about 10 million tonnes for those 400 million people will rise markedly to around 130 to 150 million tonnes in the 2080s.

Climate change, coupled with other global changes such as changing atmospheric composition and freshwater availability, has ramifications for issues of fairness and equity. The developing countries of the world, which account for about 80% of the global population, but only about 25% of cumulative global CO_2 emissions over the past 50 years or so, will likely suffer substantially from changing climate in terms of terrestrial food production.

Marine Resources

Different human societies use different components of the ocean ecosystem for food. Marine mammals, birds, macro algae (seaweeds) and invertebrates are all eaten by some groups of people. Fisheries constitute, by far, the most important source of food from the sea for human societies. Per capita consumption of fish (including both freshwater and marine) and seafood in 1998 was 16 kg (FAO 2000). While the per capita value is highest in developed countries, rates of growth in fish consumption are greatest in developing countries (Table 5.4). In some developing regions of the world, people are almost totally dependent upon fish as a protein source (ICLARM 1999). Even at the continental scale, fish are an important source of protein; they contribute 26% of the total animal protein intake in Asia and 21% in Africa (FAO 2000).

Fish and other edible marine resources are not distributed in the same manner as their potential consumers. The resulting trade in fish and seafood products (Table 5.5) is essential to food security. Ignoring possible influences of global change on fish production, human population increases alone will threaten ocean-based food security, especially for the poor. Increasingly fish will become a politically sensitive commodity.

Most commercially exploited fish stocks are today either fully or over-exploited. The total yield from marine fisheries has levelled off at approximately 85 million tonnes per year, after a period of rapid technological development over the last half century (Fig. 5.14). As noted earlier in Chap. 3, of the world's fisheries for which assessments are possible, 47–50% are considered to be fully exploited with no capacity for further development, 15–18% are considered to be overexploited, 9–10% depleted or are recovering from depletion, 21% are moderately exploited and only 4% are considered under-exploited (FAO 2000).

Thus, there is little chance of increasing fish landings in the coming decades by increasing conventional fishing. There may, however, be possibilities of increasing food available from fisheries by changing fishing practices and strategies. For example, many fisheries directly target a specific commercially valuable species. In the effort to maximise the catch of the economically very valuable species, less valuable (*by-catch* species) are usually returned to the sea (*discards*). Most often, these

Table 5.3. Food security and the impact of climate change on food production, 2080s (Fischer et al. 2001)

	Number of countries	Population 1995 (millions)	Under-nourished (millions)	Cereal production 1995 (million t)	Cereal gap 1995 (million t)	Climate impact 2080s (million t)
Losing						
ECHAM4	27	1661	386	362	−12	−60
HadCM2	25	1379	321	277	−10	−156
CGCM1	45	2077	396	467	−12	−135
Winning						
ECHAM4	20	1592	210	481	−6	99
HadCM2	37	2057	275	598	−8	192
CGCM1	17	540	166	100	−6	42

Table 5.4.
Per capita consumption of fish and seafood (FAOSTAT 2002)

	World		Developed		Developing	
	1961	1998	1961	1998	1961	1998
kg yr^{-1}	9.1	16	17.3	23.2	5.3	14
Protein (gm d^{-1})	2.7	4.4	5.2	6.7	1.5	3.8
Fat (gm d^{-1})	0.6	1.0	1.2	1.8	0.3	0.8

Table 5.5. Total fish production, including imports and exports (FAOSTAT 2002)

	Production (mt)	Import		Export	
		Quantity (mt)	Value (1 000$)	Quantity (mt)	Value (1 000$)
1961–1970 (average)					
World	51 872 923	859 500	296 511	841 219	273 147
Developed countries	27 884 429	709 579	247 435	636 860	197 481
Developing countries	23 968 783	149 921	49 076	204 360	75 666
1988–1997 (average)					
World	107 405 232	2 665 621	8 352 905	3 349 022	7 533 558
Developed countries	37 468 576	2 046 829	7 348 566	2 377 718	5 152 394
Developing countries	69 823 060	618 793	1 004 339	971 305	2 381 165

Fig. 5.14.
World capture fisheries and aquaculture production (FAO 2000)

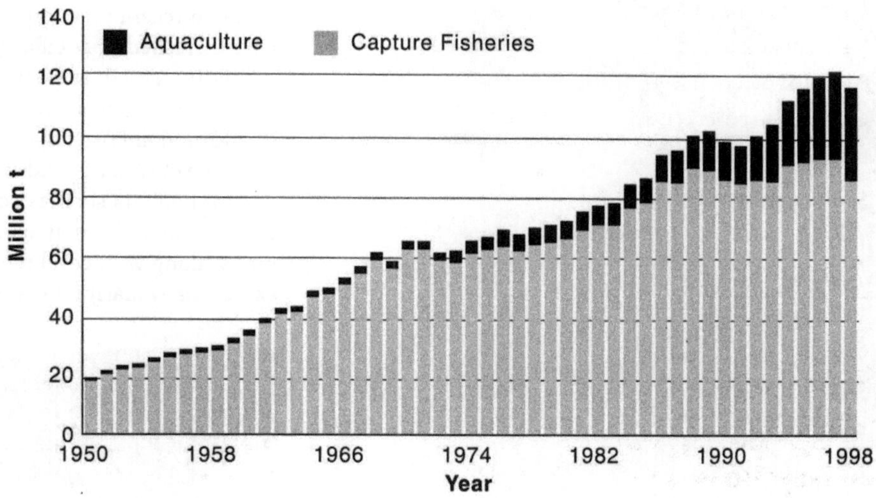

discards are dead or dying when they are returned to the sea. In some fisheries, over 10 kilograms of less valuable and/or undersized by-catch organisms are discarded for every kilogram of landed fish (Table 5.6). Shrimp fisheries generate the greatest by-catch, and pelagic (water column) trawl fisheries the least. On a global scale, it is estimated that approximately 20 million tonnes (equivalent to about a quarter of the world's total annual fish catch) of marine biomass are caught as by-catch and discarded again (Cook 2003). Almost 50% of the by-catch returned to the sea annually takes place in the northwest Pacific or the northeast Atlantic, where the dominant fishers are coming from industrialised countries and fishing primarily for a wealthy and well nourished market. If more of this by-catch was landed and used for consumption, wild fish landings could be increased without further deleterious environmental consequences. As the overall demand for fish increases, and catches increase in value, greater economic value may be placed on the species presently discarded to warrant their landing. The marketplace may force human societies to use the ocean's food resources more frugally.

Another potential mechanism for increasing marine food security is through aquaculture (Fig. 5.15). Since the mid-1980s, when the landings of wild fish began to stagnate, a dramatic increase in fish production through aquaculture has occurred. Most of the global aquaculture production (75%) takes place in low-income, food deficient countries (FAO 2000). Such production may help to feed poor people in developing countries as wild fish prices increase. However, aquaculture most often takes place in coastal regions, where population densities are high and a number of ocean users compete for resources. In addition, intensive fish production through aquaculture places severe demands on the environment through organic and chemical enrichment of local waters and loss of habitats for indigenous species, especially juveniles that rely on mangroves for this important period of their life cycles. Genetic and ecological alteration of natural local species and habitats through interaction of escapees and natural populations also brings about change, often undesirable. Moreover, as fish meal produced from wild fish is most often used for feed, especially for carnivorous fish, aquaculture ironically places additional demands on the wild fish stocks (Naylor et al. 2000).

Despite increasing human pressure on marine resources, the assumption is usually made in assessing future fishing resources that living conditions for fish

Table 5.6. The mean global discard rate (kg discarded/kg landed) listed by gear-type: data are rounded to nearest 0.5 kg (Cook 2003; original data from FAO 1994)

Fishery	kg by-catch per kg landed
Non-pelagic fish trawl	
Bering Sea rock sole	2.61
British Columbia Pacific cod	2.21
Gulf of Alaska flatfish	2.08
Northeast Atlantic dab	2.01
Northeast Atlantic flatfish	1.60
Pelagic fish trawl	
Bering Sea pollock (1988)	0.01
Northeast Atlantic cod	0.00
Bering Sea pollock (1989)	0.00
Gulf of Alaska pollock (1989)	0.00
Bering Sea pollock (1987)	0.00
Shrimp trawl	
Trinidad	14.71
Indonesia	12.01
Australia	11.10
Sri Lanka	10.96
US Gulf of Mexico	10.30
Longline	
Eastern centr. Pacific swordfish (1990)	1.13
Bering Sea Greenland turbot	1.03
Eastern centr. Pacific swordfish (1991)	1.00
Bering Sea sablefish	0.50
Gulf of Alaska cod	0.26
Purse seine	
Northwest Atlantic capelin (1983)	0.81
Northwest Atlantic capelin (1981)	0.37
East central Atlantic sardine	0.03
US Gulf of Mexico menhaden	0.03
West central Pacific tuna	0.00
Danish seine	
Northeast Atlantic haddock	0.50
Northeast Atlantic whiting	0.45
Northeast Atlantic cod	0.36
Pol/trap	
Bering Sea sabelfish	3.51
Bering Sea king crab	3.39
Bering Sea tanner crab	1.78
Northwest Atlantic capelin	0.80
East central Pacific spiny lobster	0.36

will continue to fluctuate within the boundaries experienced during the last 100 years. Living conditions are, however, changing beyond these limits. Habitats for coastal fishing are being changed by a number of human activities (Sect. 3.3.4). Furthermore, systemic changes in the Earth System, such as climate change and the changing chemistry of marine surface waters, are beginning to intersect with the more localised human-driven changes to coastal environments. The impact of these multiple, interacting stresses may first be apparent in coral reefs (cf. Sect. 4.4.2), from which 25% of the fish catch in developing countries is taken. In Asia alone coral reefs provide food for one billion people (Bryant et al. 1998). Bleaching and death of these reefs will most certainly affect these fish landings (Wilkinson et al. 1999). In recent years reports of bleaching of coral reefs have increased, particularly in years with strong El Niño events. Increased sea surface temperatures due to climate change are probably an important contributing factor. In addition, increasing global temperatures are leading to sea level rise. Depending on the rate of sea level rise, coastal habitats such as sea grass beds, mangrove forests and marshes may be adversely affected. These regions, along with coral reefs, are important nursery areas for many marine and estuarine species (Everett 1995).

Fig. 5.15. The increase in aquaculture as a source of marine protein as evidenced by the number of prawn farms along Thailand's coastline in recent years (*photo:* N. Kautsky 2000)

Changes of temperature and circulation patterns in the oceans will influence the geographic distributions of most fish species, their prey and the distribution of their predators. These ecosystem interactions, and the influence of climate change upon them, are still not well understood. In general, the high latitude seas will warm more than those in the tropics, favouring increasing fisheries production in those regions and putting those in the tropics at a relative disadvantage (Everett 1995). However, most marine species exhibit distinct thermal preferences with a well-defined optimal temperature. Thus, as individual species will react differently to a changing environment, the long term impacts of global change will likely be manifest in changing ecosystem structure and trophic level dynamics, with presently unknown consequences for regional and local fishing industries.

Present climate variability plays a strong role in fisheries production. For example, the landings of pink salmon off Alaska are closely correlated with the Pacific Decadal Oscillation Index, an indicator of temperature conditions in the North Pacific (Fig. 5.16a). Even more well known is the interaction between the production of the Peruvian Anchovetta fishery, once the largest single species fishery in the world, and El Niño events (Fig.5.16b). A combination of overexploitation and the 1973 El Niño event led into a sharp decline in the fishery, with a subsequent collapse in the mid-1980s with the intense 1983 El Niño event. The fish population and the fishery recovered in the 1990s but landings dropped sharply again following the recent 1997–1998 El Niño, with a rebound in 2000. Clearly, any increase in the frequency and intensity of El Niño events, as predicted by some climate models, would have a strongly adverse effect on this fishery (Fogarty 2002).

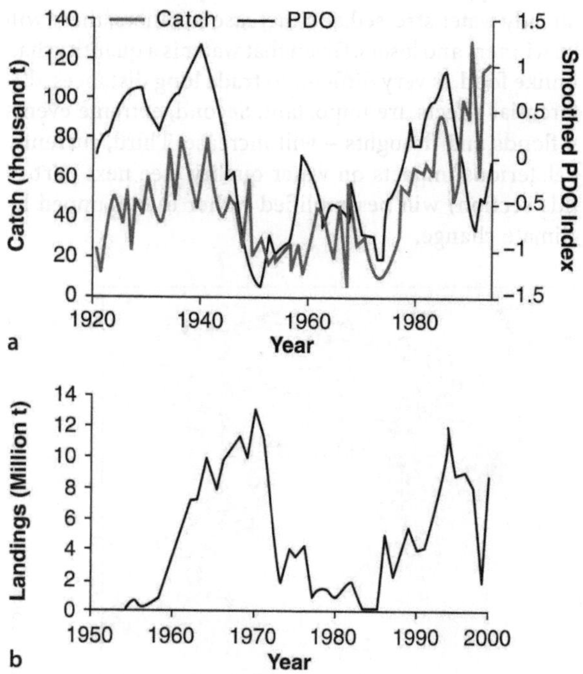

Fig. 5.16. Covariability of fish catches and climate (Fogarty 2002): **a** trends in landings of Alaska pink salmon and the Pacific Decadal Oscillation Index (smoothed using five year running average); and **b** trends in landings in the Peruvian anchovetta fishery with strong El Niño events occurring in 1973, 1983, and 1997–1998

In summary, it is not yet possible to predict with any degree of certainty the impact of global change on future food security from the oceans. However, there can be no doubt that the role of the oceans as a food supplier for human populations will come under increasing pressure in coming decades from both direct human pressures and systemic changes to the functioning of the Earth System.

5.3.2 Water Resources

The consequences of global change on water resources are usually considered among the most severe facing human societies. As described in Sect. 3.3.3, human activities are now having a significant impact on the hydrological cycle at the global scale in many ways. Any analysis of the impact of global change on water resources must take into account the increasing direct influence of human activities on the terrestrial part of the hydrological cycle. In terms of the current human vulnerability to water stress as estimated by the ratio of demand to availability, about 1.75 billion humans, over 25% of the global population, are now considered to be under severe water stress and over 2.0 billion are under moderate or severe water stress (Vörösmarty et al. 2000).

The vulnerability of human societies to strongly variable or declining water resources is closely connected to the need to increase food production. About 70% of the world's current freshwater resource is used for agriculture, but that number approaches 90% in China and India, with extensive irrigation. In other parts of the world, there is growing competition for freshwater between human and ecological needs. Over the past few centuries many natural wetlands have suffered from withdrawals of water upstream for direct human uses. In recognition of the ecosystem goods and services that wetlands and other natural ecosystems provide, there are increasing demands for their restoration through provision of more freshwater. Whatever the socio-economic and cultural setting, the fate of freshwater resources is currently determined by complex interactions among market forces, cultural preferences, institutional dynamics, and regional and international politics. As the twenty-first century progresses, however, the systemic changes occurring to the Earth's environment loom as an ever more important factor in the water resource challenge. Water is not only the most fundamentally important fluid for human existence (and that of other life forms), it is also the life-blood of the Earth System through the operation of the hydrological cycle. This intersection means that the water resource challenge and global change are tightly linked at almost every scale.

An enormous amount of research has already been synthesised on the effects of global change on water resources (IPCC 2001b; Kabat et al. 2004). Here the focus is on three critical aspects: the potential effects of climate change on water *quantity*; the impacts of global change on water *quality*; and a *vulnerability analysis* of global water resources synthesising the effects of systemic changes in the Earth System, primarily climate, with the increasing human demand.

Climate Change Impacts on Water Quantity

The effects of climate change on the hydrological cycle, and on water resources for humans, may already be discernible above natural variability (Fig. 5.17). For the mid and high latitudes in the northern hemisphere, for example, it is very likely that precipitation increased during the twentieth century by 5–10% in many areas. This is not unexpected, however, as increasing temperature leads to increased evaporation from the land surface and a more active hydrological cycle in general. There are some counter-trends, on the other hand. Rainfall has likely decreased by about 3% on average over much of the sub-tropical land areas. There also appears to have been an increase in extreme precipitation events over the past century in the northern hemisphere. In addition, there were some small increases in severe drought or severe wetness in some areas, but these trends were small and appeared to be linked to inter-decadal and multi-decadal climate variability (IPCC 2001a). Potential changes in the frequency of drought events are particularly significant in terms of water resources for humans.

The effect of climate change on water resources in the future is difficult to estimate. Most projections of changes in precipitation and runoff, and hence ultimately the availability of freshwater for human uses, are even less certain than are projections of temperature change. Although it is generally agreed that the hydrological cycle will become more active (but see Sect. 4.2.3 for the potential counteracting effects of aerosol particles), the pattern of projected changes in precipitation and runoff show both increasing and decreasing trends on a regional basis. In most model-based scenarios, precipitation is predicted to increase over the northern mid and high latitudes, especially in winter, but will decrease over parts of Australia, central America and southern Africa. Larger year-to-year variation is expected in most areas. In terms of runoff, which is an important factor for water resources, the estimated trends follow those projected for precipitation fairly well (Fig. 4.11). The scenarios suggest increases in runoff in the northern high latitudes and in Southeast Asia, but decreased runoff in central Asia, the Mediterranean region, southern Africa and Australia. The effects of reduced streamflow and groundwater recharge would impact millions of people. In general, several hundred million to a few billion people are expected to suffer a reduction of water supply by 10% or greater by 2050 (IPCC 2001b).

Table 5.7 provides an overview of potential effects of these projected changes in the hydrological cycle for water resources. Several features are significant. First, there is a high degree of confidence in the direction of many of the trends. That is, it is likely that the impacts on already water stressed countries will not be neutral. Water supplies will decrease in many countries that are already water stressed and increase in others; there will be winners and losers. Given that water is a quantity that, unlike food, is very difficult to trade long distances, differential effects are important. Second, extreme events – floods and droughts – will increase. Third, currently deleterious impacts on water quality (see next part of this section) will be amplified rather than damped by climate change.

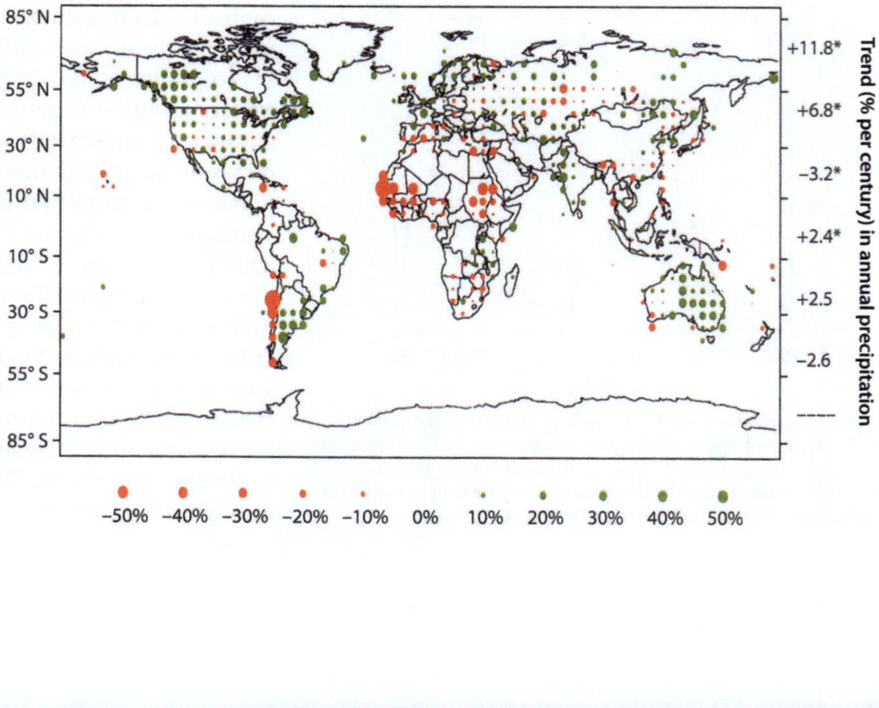

Fig. 5.17. Annual precipitation trends for 1900–1999 (IPCC 2001a). During the 100-year period, calculation of grid cell trends required that at least 66% of the years were without missing data and that there were at least three years of data within each decade except the first and last

Table 5.7. Effects of climate change on water resources (IPCC 2001b)

	2025	2050	2100
CO_2 concentration (ppm)	415–460	460–625	475–1 100
Global mean temperature change from 1990 (°C)	0.4–1.1	0.8–2.6	1.4–5.8
Global mean sea level rise from 1990 (cm)	2–15	5–30	10–90
Water supply	Peak river flow shifts from spring toward winter in basins where snowfall is an important source of water (high confidence)	Water supply decreased in many water-stressed countries, increased in some other water-stressed countries (high confidence)	Water supply effects amplified (high confidence)
Water quality	Water quality degraded by higher temperatures (high confidence) Water quality changes modified by changes in water flow volume (high confidence) Increase in salt-water intrusion into coastal aquifers due to sea level rise (medium confidence)	Water quality effects amplified (high confidence)	Water quality effects amplified (high confidence)
Water demand	Water demand for irrigation will respond to changes in climate: higher temperatures will tend to increase demand (high confidence)	Water demand effects amplified (high confidence)	Water demand effects amplified (high confidence)
Extreme events	Increased flood damage due to more intense precipitation events (high confidence) Increased drought frequency (high confidence)	Further increase in flood damage (high confidence) Further increase in drought events and their impacts	Flood damage several fold higher than "no climate change scenarios"

In terms of extreme events there is some evidence that anthropogenically driven climate change has already changed the number of severe floods around the world. The analysis by Milly et al. (2002) considers 100-year floods (river discharge that has a probability of 0.01 of being exceeded in any given year) for basins larger than 200 000 km^2 with observations that span at least 30 years. The observations demonstrate clearly that the frequency of extreme flood events is increasing. About half of the observational record of 2066 station-years was made after 1953, but 16 of the 21 extreme floods occurred after 1953. The frequency of observed extreme floods was then compared with simulations by climate models of precipitation changes due to the direct radiative effects of greenhouse gases and sulphate aerosols. The sharp increase in floods in the second half of the last century is consistent with the projections of the climate models, which suggest that the increase in frequency of extreme floods will continue into the future (Fig. 5.18).

Changes in the frequency of severe floods have serious implications for water resource infrastructure, such as dams and reservoirs. Such infrastructure has global significance as determinants of river flow and as the major source of water for human consumption, irrigation and industry. Marco-scale hydrological modelling (Arnell 1999) suggests that streamflow in 2050 will increase in high latitudes and in south and east Asia, and decrease in mid-latitudes and the sub-tropics. Despite the fact that different simulations are to some degree model dependent, broad patterns, including regional differences, are clear using different modelling scenarios, as reported by the IPCC (2001b). Hulme et al. (1999) show that by 2050, climate change will have altered streamflow in Europe substantially as the hydrological system amplifies the future precipitation changes suggested. Duration of snow cover in Europe, the midwest of the USA, central Asia and eastern China will be reduced (Arnell 1999), altering the timing and magnitude of peak flows. Peaks in snow cover are likely to move from spring to winter and decrease during summer. Projected changes are much less clear in arid, humid tropical, and sub-tropical regions.

Extreme stream and river flows are also likely to change, though exactly how much is difficult to estimate. Reynard et al. (1998) suggest an increase of 20–21% by 2050 in the magnitude of the 50-year flood in Britain. No estimates are available for changes to extreme flood events (1 000-year return period or greater) of the kind used to estimate dam safety. In certain regions, particularly arid and semi-arid regions, even modest increases in the amount of precipitation in extreme rainfall events can have much greater effects in terms of runoff and streamflow.

In any analysis of the increase or decrease of severe floods, or of another extreme climatic event for that matter, it is important to take care in the use of climate scenarios, especially those based on ensemble averages. A recent study (Palmer and Räisänen 2002) demonstrates how misleading the use of ensemble averages can

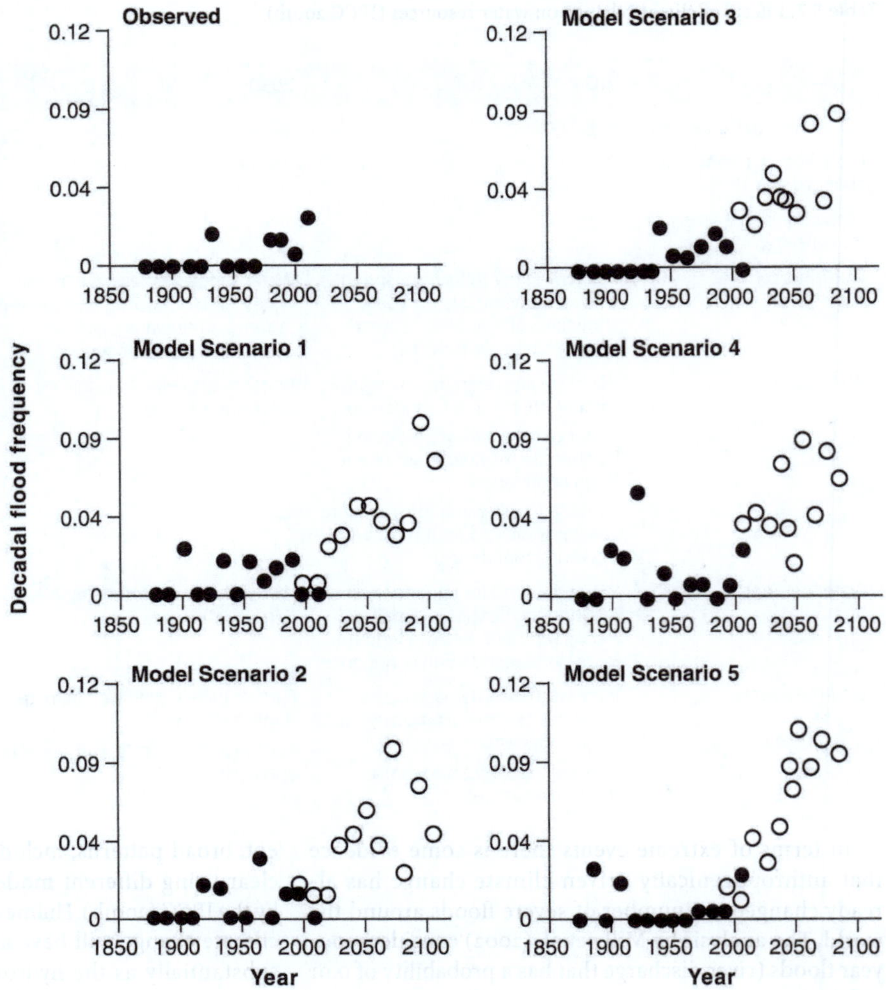

Fig. 5.18. Observed and modelled decadal extratropical severe flood frequencies (Milly et al. 2002). Flood frequency is defined as the ratio of events exceeding the 100-year discharge to the number of station-years of observation. *Filled circles* indicate a station starting and ending as in the historical record; *open circles* denote stations continuing operation once begun. The symbols coincide before the 1980s because no station record ends until 1986. Distinct 100-year flood levels were defined for each of the six data sets with respect to the historical schedule of observations over that data set

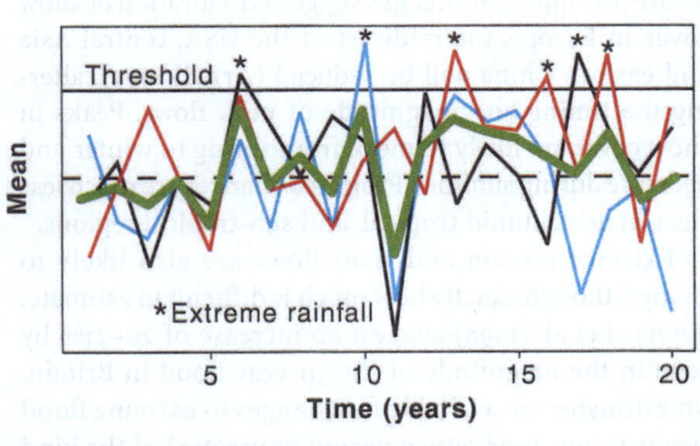

Fig. 5.19. Differences between single-model and ensemble-model forecasts of extreme climate events (Schnur 2002). *Heavy green line* is the ensemble average; *lighter lines* represent individual model simulations

be in considering extreme events. Figure 5.19 shows frequency distributions of extreme events as predicted by three individual scenarios and the ensemble average generated from the same three scenarios. Each of the three individual scenarios shows a finite probability that an extreme event (defined as a rainfall event above a particular threshold) will occur whereas the ensemble average stays beneath the threshold value at all times. Use of a consensus projection or ensemble average masks the actual variability that might be expected and can lead to an underestimation of the risk of an extreme event.

Changes in Water Quality

In addition to quantity, the issue of water quality is of considerable concern. The issue is not new; the earliest human impacts can be traced back 6 000 years or so to the Middle East, where water quality deteriorated due to the effects of land-use change. In Western Europe the earliest changes in river chemistry and associated impacts were due to metals in mining districts as early as the neolithic period; such metal contamination was probably also very high in mining areas during the Middle Age and up to the 1700s. As in many other aspects of global change, however, the most dramatic changes have occurred in the last 50–100 years. In many parts of the Americas, Africa and Australasia, human impacts in terms of river engineering and waste disposal now exceed the influence of natural variability as drivers of change in water quality (Meybeck and Vörösmarty 1999;

Table 5.8.
Major freshwater quality issues at the global scale (Meybeck and Helmer 1992)

Issue	Rivers	Lakes	Reservoirs	Groundwater
Pathogens	xxx	x	x	x
Suspended solids	xx	na	x	na
Decomposable organic matter	xxx	x	xx	x
Eutrophication	x	xx	xxx	na
Nitrate	x	0	0	xxx
Salinisation	x	0	x	xxx
Trace metallic elements	xx	xx	xx	xx
Organic micropollutants	xxx	xx	xx	xxx
Acidification	x	xx	xx	0
Modification of hydrological regime	xx	x	na	x

na: not applicable; 0: rare deterioration; x: occasional or regional deterioration; xx: important deterioration; xxx: severe or global deterioration.

Vörösmarty and Meybeck 2004). The nature of water quality problems has also changed over time. The main problems a century ago were faecal and organic pollutants from untreated human waste-water. Although this is still a problem in some parts of the developing world, it has been largely eliminated in the developed world, to be replaced by agricultural runoff as the major issue of water quality. In developing countries urban and industrial pollution loading on freshwater resources are likely to increase faster than waste-water treatment plants for the next few decades.

Virtually all of the deleterious contemporary effects on water quality are due to direct and indirect human influences (Table 5.8). In terms of global change, the impacts on the current level of water quality and that projected for the near future are dominated by land-cover and land-use change. Systemic changes in the global environment such as climate change, although relatively less important now, will play an increasingly important role in water quality as the century proceeds.

Of the various land-use impacts, agriculture is the largest; agricultural lands are a major contributor of sediments, nutrients, pesticides and coliform bacteria to rivers. For example, the US Environmental Protection Agency (USEPA 1989) estimates that 76% of pollution entering rivers and lakes in the United States is from non-point sources, of which agriculture contributes 64%. The resulting degradation of water quality can be measured in terms of loss of natural aquatic ecosystems and the amenities they provide, a reduction in the supply of useable water, and increases in the costs of treating water for use (Cooper and Lipe 1992; Carpenter et al. 1998).

Non-point agricultural pollution can be either intermittent or continuous and is often linked to the seasonal cycle of agricultural activity. The chemical pollutants, primarily comprised of nitrogen components, phosphorus compounds and pesticides, reach aquatic systems in two ways. Nitrogen-based pollutants, which result from an imbalance between plant requirements and the application of fertiliser, are generally soluble and are thus leached to surface or groundwater. Both phosphorus compounds and pesticides are generally water-insoluble, and reach aquatic systems through adsorption onto fine soil particles and transport via soil erosion. Consequently, any increases in extreme rainfall events and hence erosion will be especially important for loadings of phosphorus and pesticides.

One of the most common ways in which pollutants affect water quality is through eutrophication, which increases the turbidity of the water, induces fish kills, triggers growth of blue-green algae that renders the water unpalatable and increases growth of aquatic weed. All of these effects increase the costs of purifying the water. Pesticides, primarily herbicides and insecticides, are, by contrast, largely man-made substances and thus do not enter natural biogeochemical cycles as nitrogen and phosphorus do. Rather, they exert their influence directly through toxicity to macro-invertebrate fauna (e.g, fish kills) or through gradual bio-accumulation in the body tissues of aquatic organisms such as fish, which may eventually cause toxicity up the food chain to humans (Roux et al. 1994).

Given the importance of human activities for water quality and the reciprocal importance of water quality to human well-being, water quality is inherently of high priority among environmental issues and thus improvements in water quality are more immediately possible than for other components of global change. Indeed, water quality has shown marked improvement in many parts of the developed world (Lomborg 2001). Well-known examples include the cleaning up of the Thames River in the UK and significant improvements in the Baltic Sea due to reductions in effluent discharges from the surrounding countries. Nearly all of these improvements are due to the control of point sources and also to changes in land-cover type and land-use management. For example, replacement of agricultural land with forest almost always leads to improved water quality due to reduced usage of pesticides and fertilisers. Agricultural use of pesticides and fertilisers has decreased in many developed countries since the mid-1980s, with reductions ranging from 17% in pesticide usage in the UK from the mid-1980s to the mid-1990s (Robinson et al. 2000) to 24% in nitrogen fertiliser application in Denmark since 1985 (Olesen 1999).

Major European rivers demonstrate the improvements in water quality achieved in recent decades as well as the changing nature of the remaining contamination. The Rhine River in Europe is a good example of improvements in water quality (Fig. 5.20). Fluxes of phosphate and ammonia have dropped significantly since the mid-1970s; the sources of both are dominated by domestic and industrial sewage sources and have been effectively controlled. It is important to note that these improvements are almost exclusively found in developed countries; much remains to be done to slow and reverse the continuing deterioration of water quality in most parts of the developing world.

However, two water quality issues are still important in industrialised countries. First, the flux of nitrate has remained consistently high, owing to their non-point source originating from the use of fertiliser throughout the large Rhine catchment, which is still largely agricultural. In some areas in the Rhine and other catchments, the nitrate levels are approaching the upper limits set by the World Health Organization for safe drinking water. The second issue is contamination by pesticides, which has grown rapidly since the 1970s. The issue is complex. Although the well-known organic compounds DDT and PCBs have been tightly regulated for a few decades and their concentrations are now decreasing in sediments, there are many other compounds of concern. In the Seine River alone about 100 different, biologically active organic compounds have been identified (Meybeck 2003). The persistence of organic pollutants in continental aquatic systems is almost always high, and their degradation products can sometimes be more toxic than the parent compound. Dealing with such water quality issues can be slow, as it often takes a decade or longer for the effects of a newly introduced compound to be thoroughly analysed and several more decades or longer for any deleterious effects to work their way through food webs from primary producers to predators.

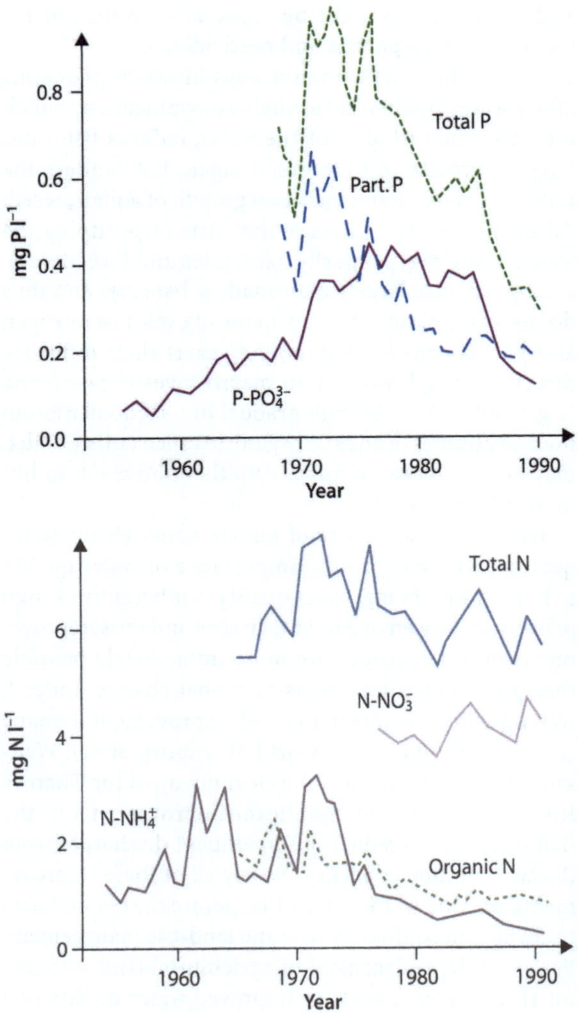

Fig. 5.20. Phosphorus and nitrogen fluxes in the Rhine Basin (based on data from Rhine River Commission)

The nature of land use plays an important role in ameliorating water quality. Wetlands left in their natural state are, in particular, an important land-cover type in terms of maintaining and improving water quality. They help remove sediments through the reduction of velocity of water as it flows through wetlands. In addition, a large variety of aerobic and anaerobic processes that take place in wetlands (e.g., nitrification, denitrification, ammonification, volatilisation) remove, transform and store a large range of chemicals, including excess nutrients, pesticides, herbicides and some heavy metals (McCartney and Acreman 2003). As an example, in one catchment in the UK, wetlands reduced soluble phosphorus concentrations by 50–80% and stored and denitrified 95% of the nitrates discharging into the wetland (Russel and Maltby 1995). The maintenance and restoration of wetlands in developed countries has undoubtedly contributed to the improving water quality there. In developing countries the dominant trend is still towards the conversion of wetlands into other land uses.

The various effects of land-cover type and land-use management will increasingly intersect with systemic changes in the hydrological cycle due to climate change to determine overall water quality. The IPCC synthesis (IPCC 2001d) suggests some of the ways in which changing climate will have an influence on water quality (Table 5.7). One of the most important ways is the change in water flows due to changing climate and the consequent changes in the amount of dilution that occur as a result. In addition, increasing temperature is expected to degrade water quality by changing the rate of reactions upon which water quality is dependent, for exam-

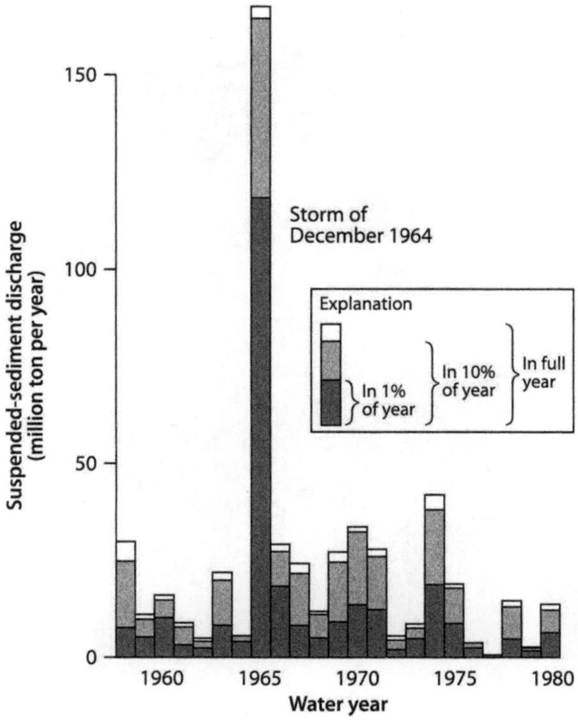

Fig. 5.21. Annual suspended sediment discharge for the Eel River (California, USA) showing frequencies of suspended-sediment discharges within individual years and the importance of infrequent heavy storms in producing large sediment loads (Meade and Parker 1985)

ple, the reactions involved in the eutrophication process. In the coastal zone, salt-water intrusion into aquifers due to sea-level rise and storm surges will also affect quality. All of these effects are expected to be significant by 2025 and to intensify as the century proceeds (IPCC 2001b).

Extreme events will also play a crucial role in future water quality as well as quantity. Rainfall intensity is a key driver of soil erosion and of suspended sediment discharge and thus water quality. If severe rainfall events become more intense and frequent, as projected by most simulations of future climate, huge sediment loadings are possible with severe consequences for water quality. The effects of a three-day storm on sediment loading in the Eel River in California, USA, show the magnitude of increases in sediment loadings that are possible under a changing climate (Fig. 5.21).

Vulnerability of Global Water Resources

Water resource issues are strongly driven by direct human pressures with an increasing overlay of global environmental change. As a result, the vulnerability of humans to water stress over the coming decades will be determined by a combination of interacting factors, most notably owing to increasing demand due to increasing population and human activities and to climate change. An analysis of changing water resources based on these two factors individually and in combination has been carried out recently (Vörösmarty et al. 2000). The study uses the ratio of aggregate water demand (domestic, industrial and agricultural uses) to the water supply as defined by river discharge as an estimate of water stress. Projections of population growth and water demand in the three major usage categories to 2025 coupled with two climate change scenarios for the same period were used to estimate future water stress.

The relative change in demand per discharge is shown in Fig. 5.22. Although climate change will have discernible impacts on freshwater availability by 2025, it is clear that these changes will be overwhelmed by the increase in demand for water resources due to population growth and economic activity alone. In terms of absolute numbers, the current 2.2 billion people living under moderate or severe water stress will rise to 4.0 billion by 2025. The distribution of water stress by continent is even more striking. Africa, Asia and South America all show sharp increases of 73%, 60% and 93% respectively in the ratio of demand and supply. For Africa and South America, climate change is predicted to exacerbate water stress significantly.

That climate change appears relatively unimportant compared to population growth and economic activity in terms of future water stress should be treated with caution for several reasons. First, the period of analysis is from 1985 to 2025, a period where population growth will be at its strongest and climate change will be less rapid than that predicted for the second half of the century. It is possible, and probably even likely, that climate change could become the most important factor affecting water stress from 2050 or so. Secondly, it is still extremely difficult to predict changes in precipitation with any degree of confidence; for the period to 2025, projections of population change are much more certain. Thirdly, water stress is a local and regional scale phenomenon that can be masked if large grid-scale analyses (e.g., country level statistics) are used (Vörösmarty et al. 2001). Projecting changes in precipitation on fine grid-scales is currently beyond the capability of any global climate model; such changes, either positive or negative in terms of water resources, will ultimately combine with local changes in population and economic activity to determine water stress.

5.3.3 Air Quality

While food security and access to reliable water resources are now recognised as global problems, air quality has traditionally been considered a local issue affecting urban areas and their adjacent rural areas only. The issue of acid rain, which became prominent in Western Europe and North America in the 1970s, raised air quality to a regional issue. Increasingly, however, it is

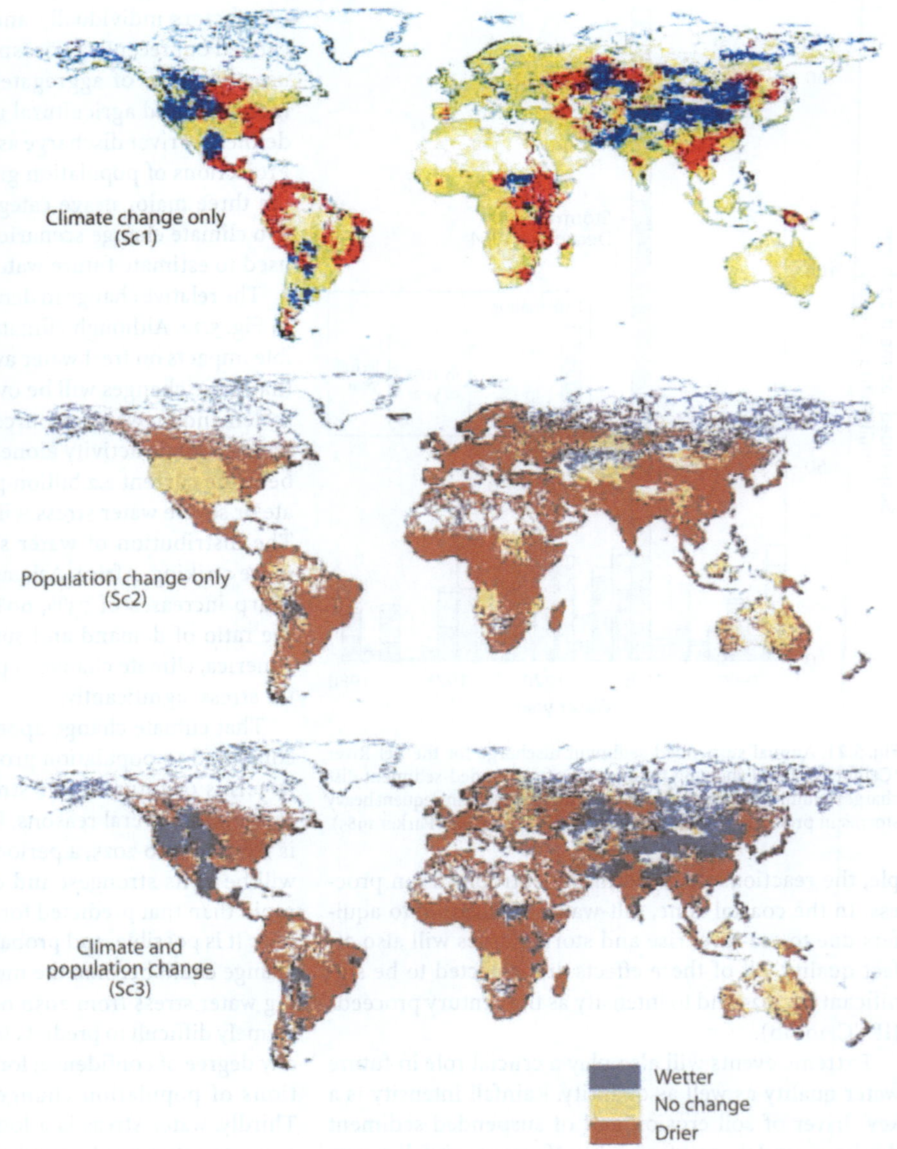

Fig. 5.22. Maps of the change in water reuse index (defined as the quotient of the combination of domestic, industrial and agricultural sectors water demand to the mean annual surface and subsurface runoff accumulated as river discharge) as predicted by the CGCM1/WBM model with climate change alone (*Scenario 1*), population and economic development only (*Scenario 2*), and the effects of all drivers of change (*Scenario 3*) (Vörösmarty et al. 2000). Changes relative to contemporary conditions are shown and a threshold of ±20% is used to highlight areas of substantial change

being realised that intercontinental transport of gases and aerosol particles connects far distant parts of the globe. The emissions from intense point sources, such as megacities, have ramifications well beyond their own urban airsheds. As megacities grow through the first half of the twenty-first century, air quality is set to become a trans-boundary pollution issue of increasing scale. It has become a global change issue in its own right.

The impacts of changing air quality are largely felt through photochemical smog, in which trophospheric ozone plays a key role, and through the increasing concentration of aerosol particles (Box 3.3). Increasing UV-B radiation caused by the loss of stratospheric ozone has also become an issue, although its direct effects have so far been minimal. The impacts of changing air quality on agricultural production have been discussed in Sect. 5.3.1. Here the impacts on human health are briefly summarised. This research is still in its infancy, and much more needs to be learned about the effects, particularly in the long term, of changing atmospheric composition on human health.

Aerosol Particles

Increasing aerosol particle loading can cause adverse health effects related primarily to the respiratory system. Although the specific mechanisms are not known, recent evidence suggests that impacts are more closely related to the mass and number of particles than to their chemical and physical nature. Fine particles (less than 10 microns in diameter) appear to be especially deleterious. Regional scale impacts of aerosol particle loading can be significant. The extensive fires in the Indonesian provinces of Kalimantan (southern part of Borneo) and Sumatra during 1997 (Fig. 4.47) caused a smoke haze episode that had major health impacts. An estimated 20 million people in Indonesia alone suffered from respiratory problems, mainly asthma, upper respiratory tract illness, and skin and eye irritation during the episode. About 210 000 persons were treated clinically, with nearly four times as many acute respiratory illnesses as normal reported in southern Sumatra and a two- to three-fold increase in outpatient visits in Kuching, Sarawak (northwestern Borneo) (Heil and Goldammer 2001).

Tropospheric Ozone

Surface layer levels of O_3 can have adverse health effects, even for short exposure times and relatively low concentrations. Such concentrations occur commonly during sunny periods over large regions in Europe and North America and in other parts of the world where emissions of O_3 precursors are large. In urban areas ozone has long been recognised as an air pollutant damaging to human health. Increasingly, as in the case of aerosols, tropospheric ozone is becoming a global issue as a consequence of long-range transport, primarily through the transport of ozone precursors from one region to another. Examples are the transport of biomass burning products from Brazil and equatorial Africa to the central South Atlantic Ocean (Jacob et al. 1996; Schultz et al. 2000) and the transport of carbon monoxide from its source in Southeast Asia across the North Pacific Ocean to North America (J. Gille, L. Emmons and J. Drummond, pers. comm.). With increasing development in many less-developed parts of the world, interregional transfers of ozone precursors are likely to increase and may be complicated by shifting circulation patterns as the Earth System responds to global change forcings. Controlling O_3 pollution is difficult because of these complex interactions. The effects of long-term exposure to low levels of O_3 are unknown.

UV-B Radiation

Potential health impacts of increased UV-B radiation associated with the stratospheric ozone hole include skin cancer (especially fair-skinned populations), eye cataracts and cancer, and a depression of the immune system towards some infectious diseases and tumours. Fortunately, owing to the large tracts of uninhabited regions beneath the hole (Fig. 5.23), actual health impacts have been slight. In addition, populations at risk in Australia and New Zealand have responded to education campaigns to protect themselves against UV-B radiation, thus decreasing their vulnerability to the impacts of the ozone hole. Nevertheless, the most recent assessment (UNEP 1998) projects significant increases in skin cancer as a result of the ozone hole.

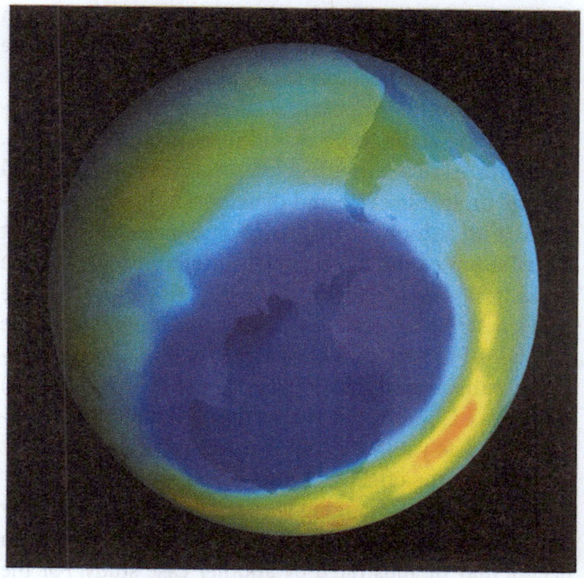

Fig. 5.23. Remote sensing images on 10 September 2000 showing the area of depleted Antarctic ozone (*blue*) over a region of 10.5 million square miles (*image:* NASA, *http://svs.gsfc.nasa.gov*). The Antarctic ozone hole develops each year between late August and early October. Regions with higher levels of ozone are shown in *red*

5.3.4 Pests and Diseases

Sectoral impacts on food systems, water resources and air quality are of direct and obvious importance to human well-being. Human societies depend also on a wide range of ecosystem goods and services (Costanza et al. 1997) that in turn depend on the functioning of terrestrial, marine and coastal ecosystems. Here the impacts of global change on pests and diseases are summarised as an example of the ways in which ecosystem goods and services are already being affected by global change.

Both distribution and abundance of pests and diseases are sensitive to changes in Earth System dynamics. Changes in the temperature regime, in the hydrological cycle and in the ecology of the vectors that transmit diseases may have profound consequences. Some examples demonstrate the point. The case of a forest pest in Alaska highlights the complex nature of interacting drivers that can cause dramatic changes in pest damage, in this case, even changing a forest to a grassland. The spruce bark beetle is a normal component of the long-term successional dynamics of spruce forests in the northern high latitudes. However, the recent strong warming in that region has meant that the beetle can complete its life cycle in one year rather than two, leading to a sharp increase in beetle populations. In addition, around several major towns in Alaska, fire – a natural part of the dynamics of the spruce forests – has been suppressed, leading to stands of very old, weakened trees. The results of these two factors acting together has been outbreaks of spruce bark beetle in these old forests, death of the trees followed by fire, and the conversion of the forests to grasslands (Walker et al. 1999).

Codling moth is a pest that severely limits maize production, the staple food of millions around the world. In South Africa the effect of possible future global change on its life cycle and distribution has been modelled. With a doubling of greenhouse gases, the present distribution of the moth and its annual number of life cycles will change (Schulze 1997a). Not only will the pest's distribution increase, but more importantly, the number of life cycles within a year will increase, sometimes substantially, over large areas (Schulze 1997a) (Fig. 5.24). Quantitatively, how this will affect food productivity has yet to be determined. In the meantime, it is clear that the effect will not be negligible.

The impact of changing land use on water availability and quality, and consequently the implications for human health, may be illustrated by another case study. For the 4 079 km² Mgeni catchment in South Africa, which supplies nearly six million people with water, Kienzle et al. (1997) have modelled the impact of the changes from the undisturbed, baseline vegetation cover to present-day land use on streamflow. They estimated that reductions in streamflow exceeding 60% have occurred, mainly in areas of intensive sugarcane cultivation and exotic afforestation, whereas increases exceeding 100% have been observed in overpopulated areas where livestock overstocking and land degradation have occurred, as well as in urban areas. Future runoff will be affected not only by the present land use, but also by the changes certain to take place in the future as population pressures on the land increase (and as climate continues to change). At present, where human population pressures on the land increase with the development of informal urbanisation and squatter settlements with poor sanitation, coupled with the changes in water supply noted above, water quality deteriorates rapidly and the health hazard increases accordingly. Simulations of the distribution of the pathogen *Escherichia coli* have been made for the Mgeni River basin (Fig. 5.25) using livestock densities and numbers of humans living under poor sanitary conditions in close proximity to streams (Kienzle et al. 1997). The distrubution of the *E. coli* health hazard is seen to be excessively high in places in the Mgeni catchment and is likely to increase in the future unless land use and sanitation can be controlled effectively.

The impacts of global change on human health are not confined to the developing regions of the world. For example, tick-borne diseases are a growing concern in the temperate regions of the northern hemisphere. In Stockholm, Sweden, tick-borne encephalitis is endemic and with an increase in parklands and summer cottages in the region, the number of residents at risk of exposure to the disease has increased considerably over the last two or three decades. From the mid-1980s a vaccine against tick-borne encephalitis has become available, and the Stockholm Country authorities subsequently embarked on an extensive and successful campaign to immunise large numbers of the local residents. Much to their surprise, however, the incidence of the disease did not decrease through the 1990s (Fig. 5.26).

The solution to the mystery lay in the recent increases of temperature that have occurred across much of the northern high latitudes. This increase has led to the northward expansion of the area inhabited by tick (Fig. 5.26). In addition, owing to the warming in the

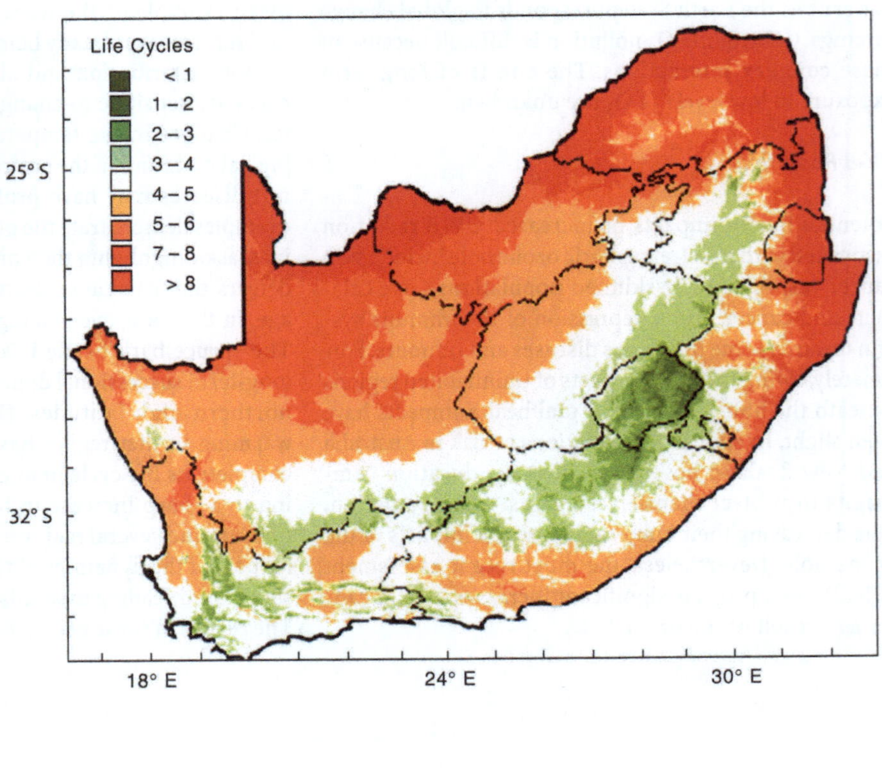

Fig. 5.24. Agricultural Catchments Research Unit (ACRU) model simulations, forced by transient-run HadCM2 model doubling of greenhouse gases, showing changes in the possible number of life cycles of Codling Moth per annum with future climate changes in South Africa (Schulze 1997a)

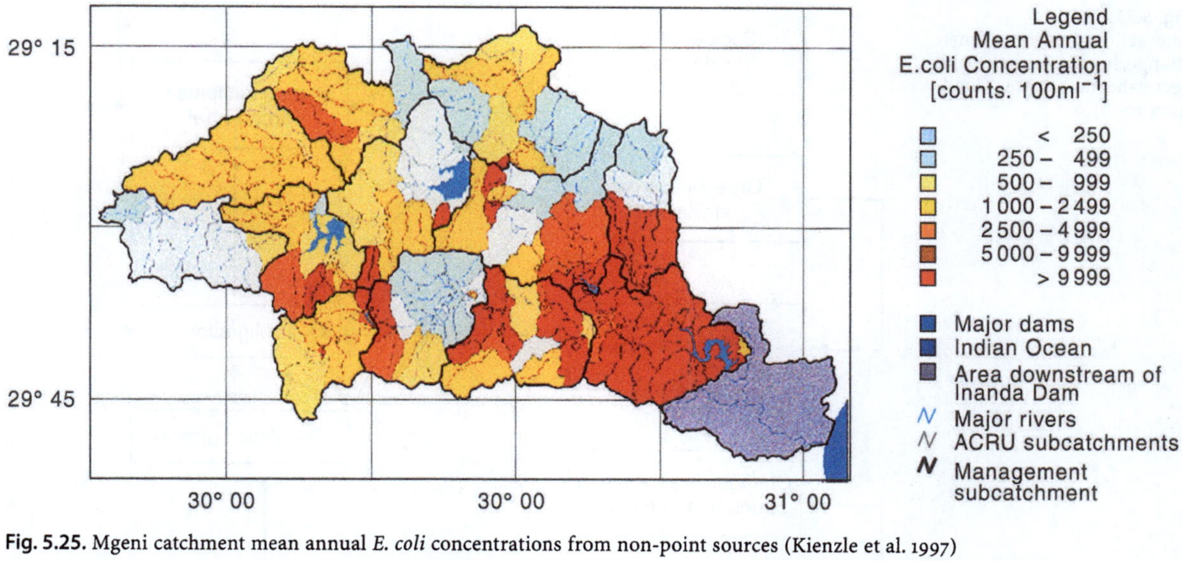

Fig. 5.25. Mgeni catchment mean annual *E. coli* concentrations from non-point sources (Kienzle et al. 1997)

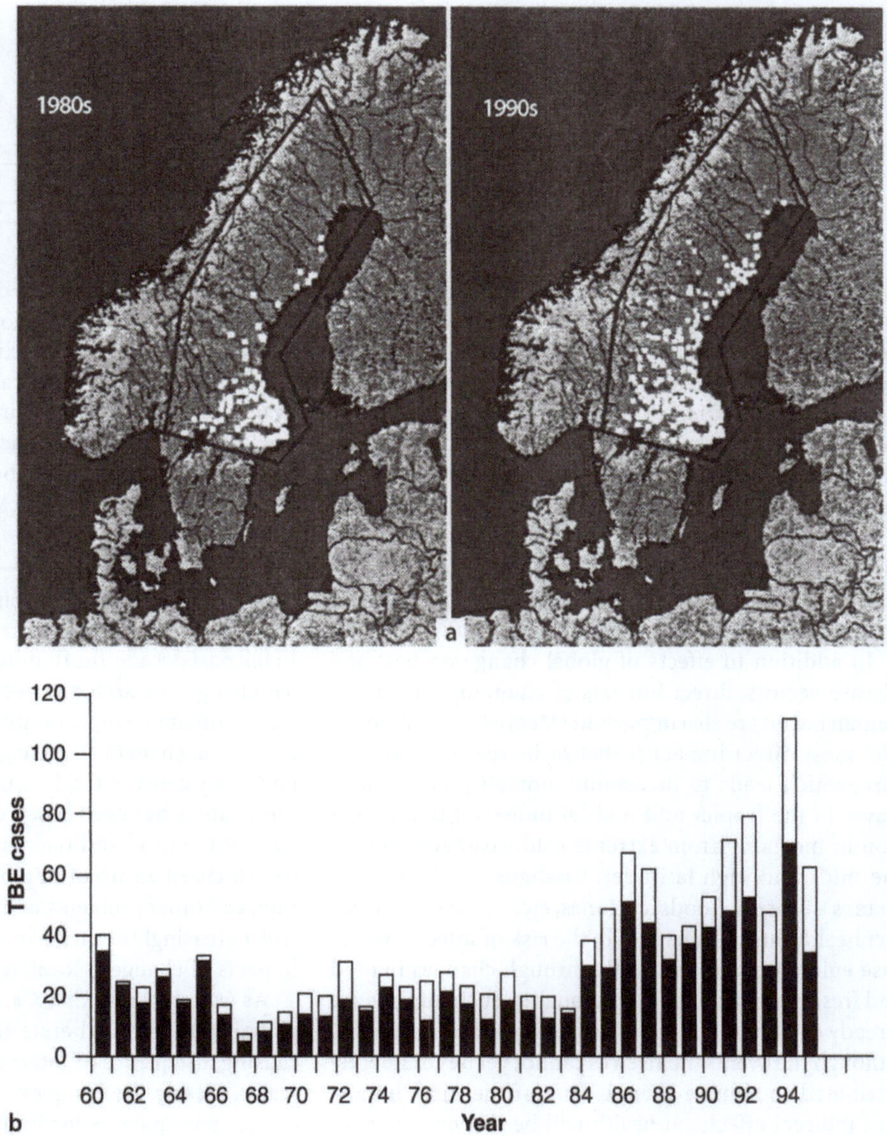

Fig. 5.26. Increasing range and population size of ticks in Sweden: **a** northward shift of tick distribution in Sweden over the last decade; and **b** increase in the incidence of tick-borne encephalitis in Stockholm County during the study period 1960–1995 (*lower half of column*) compared with the rest of the country (*upper part of the column*) (Lindgren 2000)

Fig. 5.27.
Interactions between climate change, human societies and vector-borne diseases (Lindgren 2000)

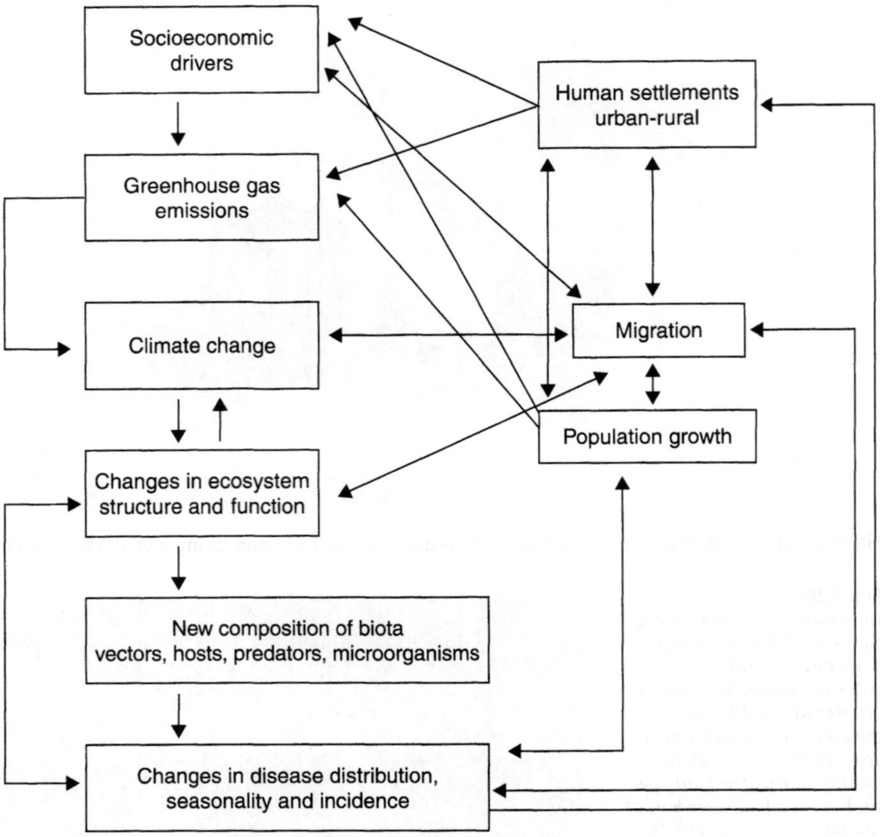

Stockholm region, the tick is now able to complete its life cycle in a single season, whereas formerly it required two to do so. The population of the ticks has risen sharply in mid-central Sweden. Thus, the effects of an active and successful vaccination campaign in Stockholm County have been offset by greater exposure of inhabitants to an increasing number of ticks in parks and outdoor areas due to the warmer conditions. This case study (Lindgren 2000) is an excellent example of the need, in some parts of the world, to factor in even now global change and its consequences for pest distribution and dynamics in developing public health campaigns (Fig. 5.27).

In addition to effects of global change on pest and disease vectors, direct impacts of changing climate on human health are also important (McMichael et al. 2001). The most direct impact is that of increasing temperature, which leads to increasing mortality due to heat waves in the tropics and mid-latitudes but to a reduction in mortality from extreme cold weather events in the mid- and high latitudes. Changes in climatic extremes – storms, floods, cyclones, etc. – also directly affect health through changes in the risk of infectious disease epidemics and indirectly through changes in food and freshwater availability and quality. Such impacts are already evident in the recent increase in extreme events, although the events themselves cannot yet be confidently attributed to climate change. One of the most important indirect effects on health will be through changes in the quality and quantity of food and thus to the nutritional intake of populations. Projections of changes in food provision due to climate change suggest that populations already vulnerable to this indirect risk to health will become increasingly so as the century proceeds. When all of these factors are considered, negative health impacts of global change are expected to outweigh positive ones (McMichael et al. 2001).

5.3.5 Amplifying, Damping and Multiple Effects

In the past decade, the dominant focus of concern in global change research has been on single changes, especially climate change, despite the fact that many environmental changes that are global in extent or effect are underway concomitantly. Much of the analysis earlier in this chapter has been based on such an approach. Similarly, at the local and regional scale, attention has often been focused on urban air pollution, toxic chemicals, acid rain, and other problems in isolation. In future, the focus will increasingly be on the interactions among agents and impacts of change at local, regional, and global scales.

As was shown in Chap. 4, the various components of global change reverberate through the Earth System, causing a sequence of interacting effects at a number of scales. Clearly, the complex nature of global change has direct consequences for impacts on human well-being.

In some cases the various components of global change may dampen each other, thus diminishing the impact on humans of each change if it had acted independently. In other cases the changes amplify one another, with potentially devastating consequences. In reality, the overall consequences for humans will be a mix of amplifying and damping effects that may cascade into the future. These features are illustrated in several examples below. The hydrological cycle features in each example, as the impact of global change on it are likely to have profound consequences for humans in many parts of the world.

The first two examples are from southern Africa, where water resources are already an important issue and are likely to increase in importance with continuing increases in population coupled with projected changes in regional climate. In fact, one of the more robust aspects of projections of future changes in the hydrological cycle is that precipitation and runoff are both expected to decrease in southern Africa. Given the strong population growth in this region, it is particularly appropriate that changes in the hydrological cycle there be examined in more detail.

The first example demonstrates how one component of the Earth System may dampen the effect of another. Such is the case when the effect of land use changes is to increase runoff to compensate for diminished rainfall. Modelling of a South African river catchment (using the HadCM2 model) suggest that with a doubling of greenhouse gases, in the Mgeni River basin of southeast South Africa, catchment runoff is likely to diminish by 30% in the upper basin, while remaining relatively unaffected in the lower catchment (Schulze and Perks 2000). In the lower reaches of the basin, increased runoff induced by intensified land use will offset the diminution by the lower rainfalls predicted. Ironically, by clearing forests and woodlands, societies may actually buffer themselves against a projected decrease in rainfall.

On the other hand, the intersection of land-use change and rainfall may have an amplifying effect when climate variability or extreme events are considered. An example from southernmost Africa illustrates the point. Mean annual temperatures are likely to increase by 1–2 °C over much of southern Africa with continuing global change (Murphy and Mitchell 1995; Joubert and Kohler 1996). These increases are amplified into mean annual potential evaporation increases of 5–20% south of 10° S (Hulme 1996). With a doubling of greenhouse gases, subtropical southern Africa is likely to experience diminished rainfall and negative hydrological consequences (Meigh et al. 1998). Changes in extremes rather than in mean conditions will have the greatest impact. One simulation suggests an increase of up to 40% of rainfall events with >12.5 mm per rain day in parts of the summer rainfall region of South Africa (Joubert 1995). Such changes will cascade through the hydrological system exacerbating stormflows, changing sediment yields and shortening return periods of extreme events.

One of the clearest manifestations of amplification of change by the hydrological system is in the enhancement of rainfall variability into runoff variability on time scales ranging from the individual event to inter-annual and longer. Inter-annual runoff variability varies from less than 50% in the wetter eastern to more than 300% in the more arid western parts of South Africa (Schulze 1997a). The behaviour of drainage basins in South Africa during El Niño events clearly demonstrates this amplifying effect. During El Niño events some regions of the world experience much smaller rainfall than average. Southern Africa experienced this during the 1982/1983 El Niño event when in two-thirds of the region rainfall was reduced by about 50–80% of the long term median (Fig. 5.28a). However, the response of river runoff (Fig. 5.28b) was highly non-linear with the reductions in rainfall highly amplified. The simulated yearly runoff was only 20–50% of the long-term median over much of the region with considerable areas generating less than 20% of normal runoff (Schulze 1997b). The patchiness of the response illustrates also the high dependence of runoff on antecedent wet or dry conditions. Climatologically, a unit change in rainfall results in tripling of runoff over much of southernmost Africa; in certain areas it may be quintupled or more (Schulze 2000; Schulze and Perks 2000). Such highly non-linear responses of river catchments to rainfall variability are typical of arid and semi-arid regions characterised by low river runoff, such as the Sahel and Australia (similar findings have been reported in 28 catchments in Australia (Chiew et al. 1995)). This amplification effect can lead to seasonal dryness, intermittent flow, potential drought, and in the most extreme cases, even to famine.

A case study from medieval Germany shows how the amplifying effect of land use-climate variability interactions can have consequences for a century into the future (Bork et al. 1998) (Fig. 5.29). For hundreds of years from about AD 600 the increasing population of Germany had steadily cleared forests and woodlands for agriculture; by the early 1300s only 15% of the land remained forested. As shown in the examples above, changed land cover can have profound effects on runoff, in this case increasing runoff by up to 60%. When a rare (1 000 year) rainfall event hit central Europe in 1342, the flow rates in rivers through Germany were enormous, exceeding maximum twentieth century flow rates by 50 to 200 times. The cascading impacts through the human/environment system were profound and continued for over a century. Floods wiped out crops, villages and infrastructure such as bridges. Soil loss was extreme. Agricultural production subsequently dropped and famines ensued. A few years later the bubonic plague arrived in Germany and more than a third of the already weakened population died. For over 100 years, agricultural land was abandoned and the amount of forest and woodland increased by three times, again changing the rates of evapotranspiration, runoff and soil erosion. The situation finally stabilised again in the late sixteenth century.

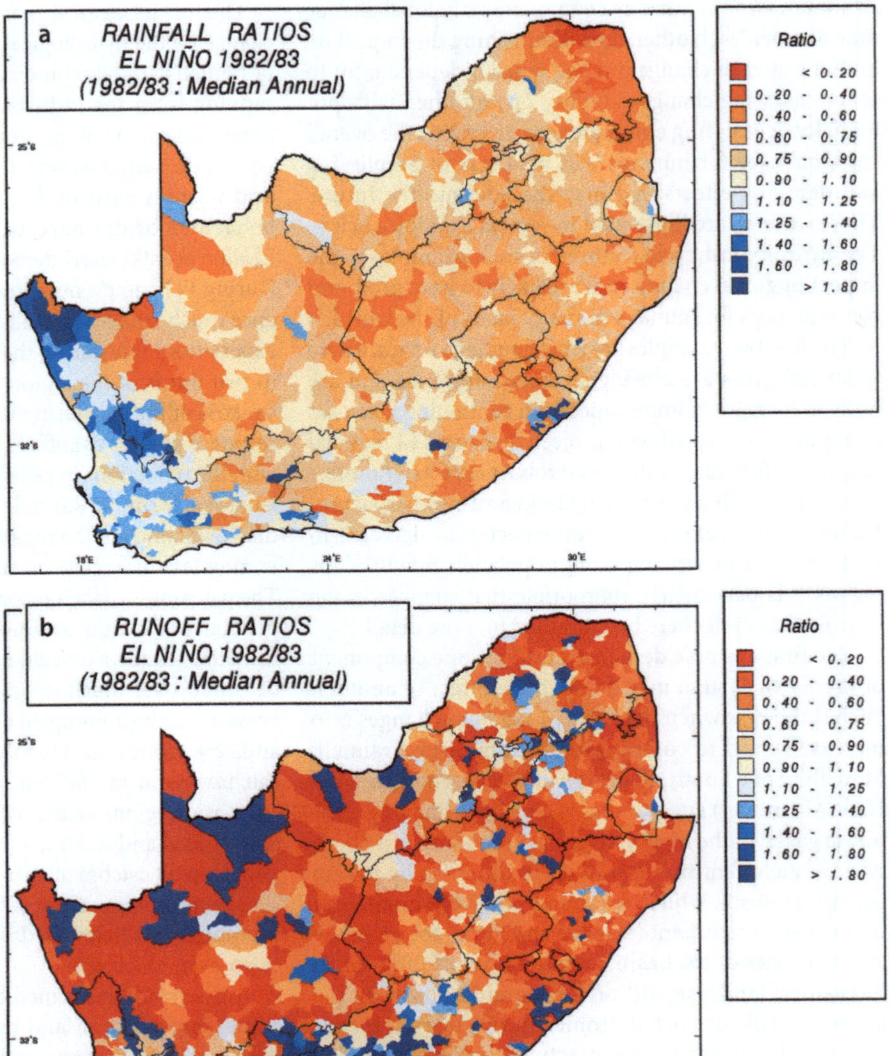

Fig. 5.28.
a Rainfall ratios during the 1982–1983 El Niño event; and **b** runoff ratios for the same El Niño event (Schulze 2000)

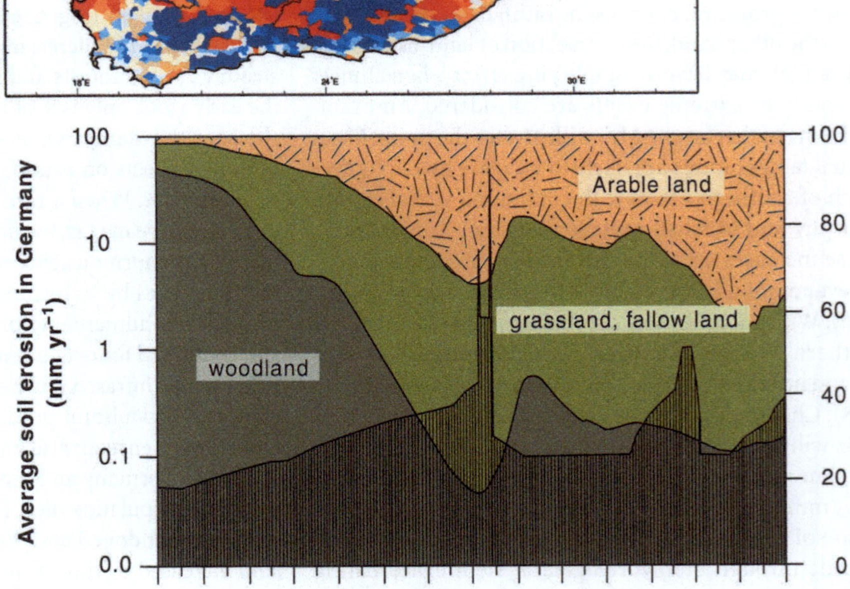

Fig. 5.29.
Climate variability, land-use change and soil erosion in Medieval Germany (without Alps) (Bork et al. 1998)

5.4 Risks Facing the Earth System as a Whole

Nearly all of the consequences of global change for human well-being discussed above deal with particular sectors of society. A sectoral approach to impacts and consequences lends itself to economic analysis and interpretation. Much of the debate surrounding the severity of such impacts then revolves around cost-benefit analyses of mitigation of global change compared to adapting to its consequences. Some argue that the costs of mitigation far outweigh the benefits, and therefore the appropriate societal response is to adapt to the coming changes. However, all such analyses implicitly assume that global change will consist of only moderate changes to the natural variability of the global environment within an overall envelope of well-buffered stability.

As shown in Chap. 2, the palaeo-record shows that abrupt changes and surprises are a common feature of the Earth System and that environmental extremes beyond those recorded during the period of instrumental record occur frequently. In fact, human societies have developed over a period of time in Earth System history (the Holocene) that is relatively short and relatively stable over many parts of the world when compared to the total range of variability characteristic of the last 400 000 years (Fig. 1.3).

The concept of thresholds and abrupt changes (Fig. 5.30) is crucial to understanding the nature of the risks to the Earth System as a whole (Box 4.1). This type of change arises when a well-buffered system is forced beyond a certain limit. Until the time that the threshold is approached, it appears that the system is unresponsive to the forcing function. However, when the threshold is passed, the response can be sudden and severe. In fact, the system can move to another state very quickly when the threshold is passed, a state which may prove to be difficult to reverse or may even be irreversible. Changes of this nature are especially dangerous in the context of global change. Societies can have little or no warning that a forcing factor is approaching such a threshold, and by the time that the change in Earth System functioning is observed, it will likely be too late to avert the major change. Many of the examples of catastrophic failures discussed below show features of threshold-abrupt change behaviour.

5.4.1 Catastrophic Failures

The stratospheric ozone episode demonstrates that catastrophic failures of the Earth System are not only possible, but that recently humankind narrowly escaped one. Other catastrophic failures are possible as the Earth System as a whole adjusts to an ever-increasing suite of interacting human forcings.

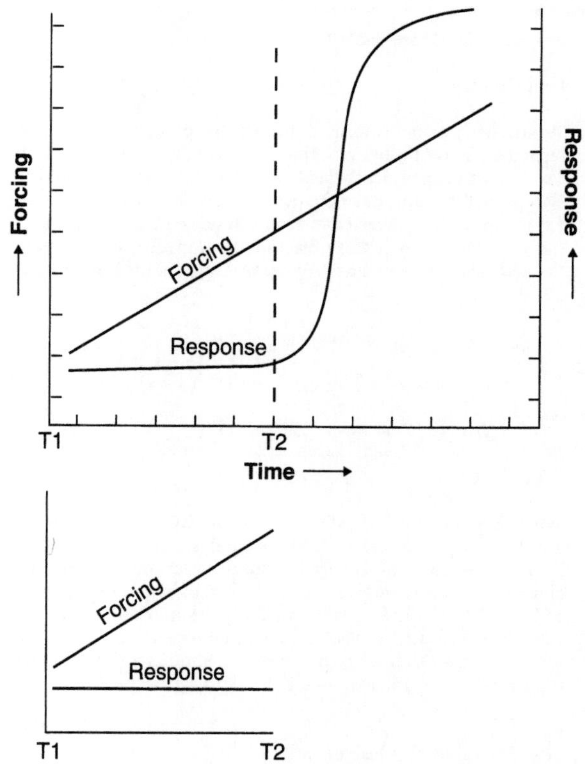

Fig. 5.30. Thresholds and abrupt changes. Many processes within the Earth System are well-buffered and appear to be unresponsive to a forcing factor (e.g., between *T1* and *T2* in *lower figure*) until a threshold is crossed and then a major change occurs abruptly (see also Box 4.1.)

Ozone Hole

The development of the ozone hole was an unforeseen and unintended consequence of widespread use of chlorofluorohydrocarbons (CFCs) as refrigerants (Box 5.5). Had bromofluorocarbons been used instead, the result could have been catastrophic. In terms of function as a refrigerant or insulator, bromofluorocarbons are as effective as chlorofluorocarbons. However, on an atom for atom basis, bromine is about 100 times more effective at destroying ozone than is chlorine. As the Nobel Laureate Paul Crutzen (1995) has written:

> This brings up the nightmarish thought that if the chemical industry had developed organobromine compounds instead of the CFCs – or, alternatively, if chlorine chemistry would have run more like that of bromine – then without any preparedness, we would have been faced with a catastrophic ozone hole everywhere and at all seasons during the 1970s, probably before the atmospheric chemists had developed the necessary knowledge to identify the problem and the appropriate techniques for the necessary critical measurements. Noting that nobody had given any thought to the atmospheric consequences of the release of Cl or Br before 1974, I can only conclude that mankind has been extremely lucky.

Humankind was also lucky in a number of other ways. First, much of the stratospheric chemistry needed to respond to the ozone hole crisis had been developed in

Box 5.5. The Ozone Hole

Paul Crutzen

Stratospheric ozone is formed through the photolysis of O_2 and recombination of the two resulting O atoms with O_2 ($3O_2 \longrightarrow 2O_3$). These reactions are clearly beyond human control. Reactions are also needed to reproduce O_2, otherwise within 10 000 years all oxygen would be converted to ozone. Besides the *Chapman reactions* $O + O_3 \longrightarrow 2O_2$, these reactions also involve several reactive radicals. The ozone destroying reaction chains can be written as

$$\begin{array}{lllll} & X + O_3 & \longrightarrow & XO + O_2 & \\ & O_3 + h\nu & \longrightarrow & O + O_2 & (\lambda < 1140\ \text{nm}) \\ & O + XO & \longrightarrow & X + O_2 & \\ \hline \text{net:} & 2O_3 & \longrightarrow & 3O_2 & \end{array}$$

where X stands for OH, NO, Cl or Br, and XO correspondingly for HO_2, NO_2, ClO and BrO. These catalysts are influenced by human activities, especially by the production of industrial chlorine, which is transferred to the stratosphere in the form of CCl_4, CH_3CCl_3 and most importantly the chlorofluorocarbon ($CFCl_3$ and CF_2Cl_2) gases. The current content of chlorine in the stratosphere, about 3 nmol mol^{-1}, is about six times higher than what is naturally supplied by CH_3Cl.

For a long time it was believed that chemical loss of ozone by reactive chlorine would mostly take place in the 25–50 km height region and that at lower altitudes in the stratosphere, which contains most ozone, only relatively little loss would take place. The reason is that the NO_x and the ClO_x radicals react to form $ClONO_2$ and HCl:

$$\begin{array}{lll} ClO + NO_2 + M & \longrightarrow & ClONO_2 + M \\ ClO + NO & \longrightarrow & Cl + NO_2 \\ Cl + CH_4 & \longrightarrow & HCl + CH_3 \end{array}$$

Most inorganic chlorine is normally present as HCl and $ClONO_2$, which do not react with each other and with ozone in the gas phase, thus protecting ozone from otherwise much larger destruction.

This favourite situation does not always exist. In 1985 scientists from the British Antarctic Survey presented their observations showing total ozone depletions over the Antarctic by more then 50% during the late winter/springtime months September to November, with ozone depletions taking place in the 14–22 km height region where normally maximum ozone concentrations are found (Fig. 5.31). Within a few weeks after polar sunrise almost all ozone is destroyed, creating the *ozone hole.*

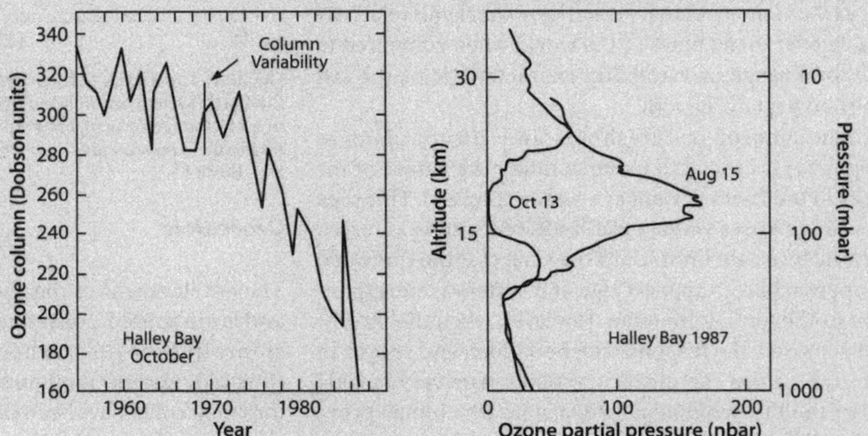

Fig. 5.31. *Left*: The rapid decrease in total ozone column over the Antarctic from 1956 to 1988 (100 Dobson units correspond to a layer of ozone 1 mm thick if it could be compressed to standard temperature and pressure at the Earth's surface); *right:* altitudinal dependence of ozone loss between August and October 1987 (the drastic ozone losses constitute the so-called ozone hole) (Farman et al. 1985; Hofmann et al. 1989)

earlier research undertaken to evaluate the impact of supersonic aircraft on the atmosphere. Second, scientists of the British Antarctic Survey had done careful, systematic measurements of ozone concentrations over Antarctica for many years so that a clear trend could be detected (Fig. 5.33). Third, the areas of the world most affected by stratospheric ozone are very lightly populated.

Stratospheric ozone depletion is an example of a powerful, non-linear feedback system, as well as a clear case of an anthropogenically driven chemical instability in the Earth System. The rapidity with which the ozone hole developed is characteristic of threshold-abrupt change behaviour. Luckily, in this case the damage is reversible, albeit over a considerable period of time.

Cleansing Efficiency of the Atmosphere

A wide range of substances, of both natural and anthropogenic origin, is released from the Earth's surface into the atmosphere. However, in the long term, these do not accumulate in the atmosphere but rather are converted into other forms, in the case of pollutants into less harmful substances, and rained out of the atmosphere. Thus, the atmosphere can be thought of as a giant washing machine that has the ability to clean itself of harmful substances that are introduced. This capability is sometimes referred to as the cleansing efficiency of the atmosphere.

In more technical terms, the atmosphere is an oxidising medium and its ability to oxidise a wide range of

◄

How was this possible? Nobody had expected this; in fact, it was believed that at high latitudes ozone in the lower stratosphere was largely chemically inert.

It only took some two years of research to identify the main processes that lead to these large ozone depletions and to show that the CFCs were the culprits. The explanation involves each of the following necessary conditions:

First, low temperatures, below about -80 °C, are needed to produce ice particles consisting of nitric acid and water vapour. In this process also the NO_x catalysts are removed from the stratosphere through the reactions

$$NO + O_3 \longrightarrow NO_2 + O_2$$

$$NO_2 + NO_3 + M \longrightarrow N_2O_5 + M$$

$$N_2O_5 + H_2O \longrightarrow 2HNO_3$$

thereby producing HNO_3, which is incorporated in the particles. Secondly, on the surface of the ice particles HCl and $ClONO_2$ react with each other to produce Cl_2 and HNO_3; the latter is immediately incorporated in the particles. Thirdly, after the return of daylight after the polar night, Cl_2 is photolysed to produce 2 Cl atoms. Fourthly, the chlorine atoms start a catalytic chain of reactions, leading to the destruction of ozone:

$$Cl + O_3 \longrightarrow ClO + O_2$$

$$Cl + O_3 \longrightarrow ClO + O_2$$

$$ClO + ClO + M \longrightarrow Cl_2O_2 + M$$

$$Cl_2O_2 + h\nu \longrightarrow Cl + ClO_2$$

$$Cl + ClO_2 \longrightarrow 2Cl + O_2$$

$$\text{net: } 2O_3 \longrightarrow 3O_2$$

The breakdown of ozone is proportional to the square of the ClO concentrations. As these grew for a long time by more than 4% per year, ozone loss increased by 8% from one year to the next. Also, because there is now about six times more chlorine, about 3 nmol mol^{-1}, in the stratosphere compared to natural conditions when chlorine was solely provided by CH_3Cl, the ozone depletion is now 36 times greater than prior to the 1930s when CFC production started. Earlier chlorine-catalysed ozone destruction was unimportant. Finally, enhanced ClO concentrations are advected to the lower stratosphere by downwind transport from the middle and upper stratosphere within a meteorologically stable vortex with the pole more or less at the centre. This is important because at the higher altitudes more organic chlorine is converted to much more reactive inorganic chlorine gases, including the ozone-destroying catalysts Cl, ClO, and Cl_2O_2.

All five factors have to come together to cause the ozone hole (Fig. 5.32). It is not surprising that the ozone hole was not predicted. This experience shows the critical importance of measurements. What other surprises may lie ahead involving instabilities in other parts of the complex Earth System?

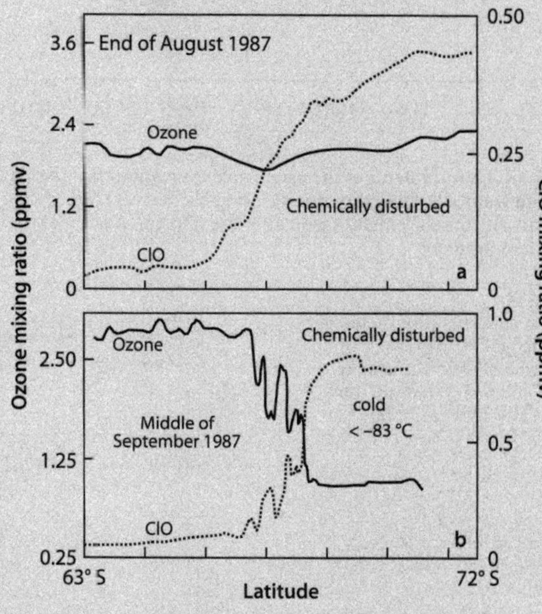

Fig. 5.32. High concentrations of ClO radicals and simultaneous rapid ozone destruction in winter with very low temperatures (Anderson et al. 1989)

chemical species and facilitate their removal from the atmosphere is also referred to as its oxidising efficiency. Removal of substances from the atmosphere is thus effected by the increased solubility of the oxidised products, which are then removed by precipitation. The oxidation reaction, and not the subsequent dissolution and removal via precipitation, is usually the rate-limiting step. The OH (hydroxyl) radical, a highly reactive chemical species, is the primary oxidising agent, and is often called the detergent of the atmosphere. As the OH radical is involved in a wide range of reactions, changes in the composition of reactive gases in the atmosphere, such as NO_x, could affect the concentration of OH and thus the atmosphere's cleansing efficiency, an important Earth System process.

In general, the concentration of the OH radical and hence the oxidising efficiency of the atmosphere is controlled by O_3, water vapour, UV radiation and levels of trace gases with which it reacts, such as CH_4, CO and volatile organic compounds. The concentrations of OH at any one time are very low and highly variable due in part to its rapid cycling with other HO_x species, such as HO_2 and H. Given that the concentration of OH depends on many other species that are influenced by human activities, it is important to determine if the OH concentration is changing in recent decades. Direct measurement of OH concentration is extremely difficult so most estimates of its concentration are based on measured concentrations of some species with which it reacts, atmospheric transport models and simulations of the reaction chemistry. Such models are sensitive to the assumed changes in emissions of compounds such as NO_x, CO, CH_4 and hydrocarbons, all affected by human activities.

Despite the difficulties, estimates of OH concentration have improved in recent years (Fig. 5.34). The consensus of the several modelling studies that have been undertaken is that globally averaged OH concentration has remained relatively constant with a maximum estimated decrease of around 20% (Wang and Jacob 1998). Two observationally-based estimates from an indirect method using measurements of the concentration of methyl chloroform (CH_3CCl_3) also show relatively small changes but in opposite directions. Estimates by Krol et al. (1998) show a small increase in OH concentration of about 0.46% per year for the period 1978–1993. An estimate by Prinn et al. (2001), on the other hand, suggests that the concentration of OH has decreased globally, but especially in the northern hemisphere, over the last 23 years. These latter estimates suggest that the OH radical concentration increased from 1978, when measurements began, peaked in the 1988–1990 period, and has decreased since then at an accelerating rate (Fig. 5.35). The significance of this recent observed decrease is difficult to determine at present. The trend is not long enough or pronounced enough to determine if it is part of natural variability or whether anthropogenic influences play a role. In addition, the methodology requires that emissions of CH_3CCl_3 must be very well known, which is a matter for debate.

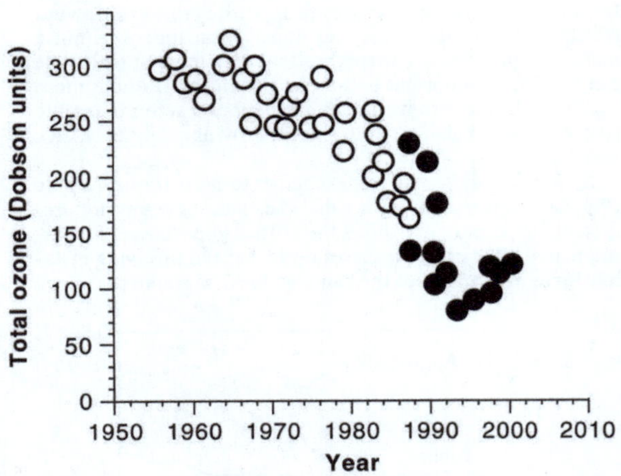

Fig. 5.33. Measurements of the ozone hole over Antarctica (based on data from J. D. Shanklin, British Antarctic Survey). *Filled circles* indicate measurements taken after the implementation of the Montreal Protocol

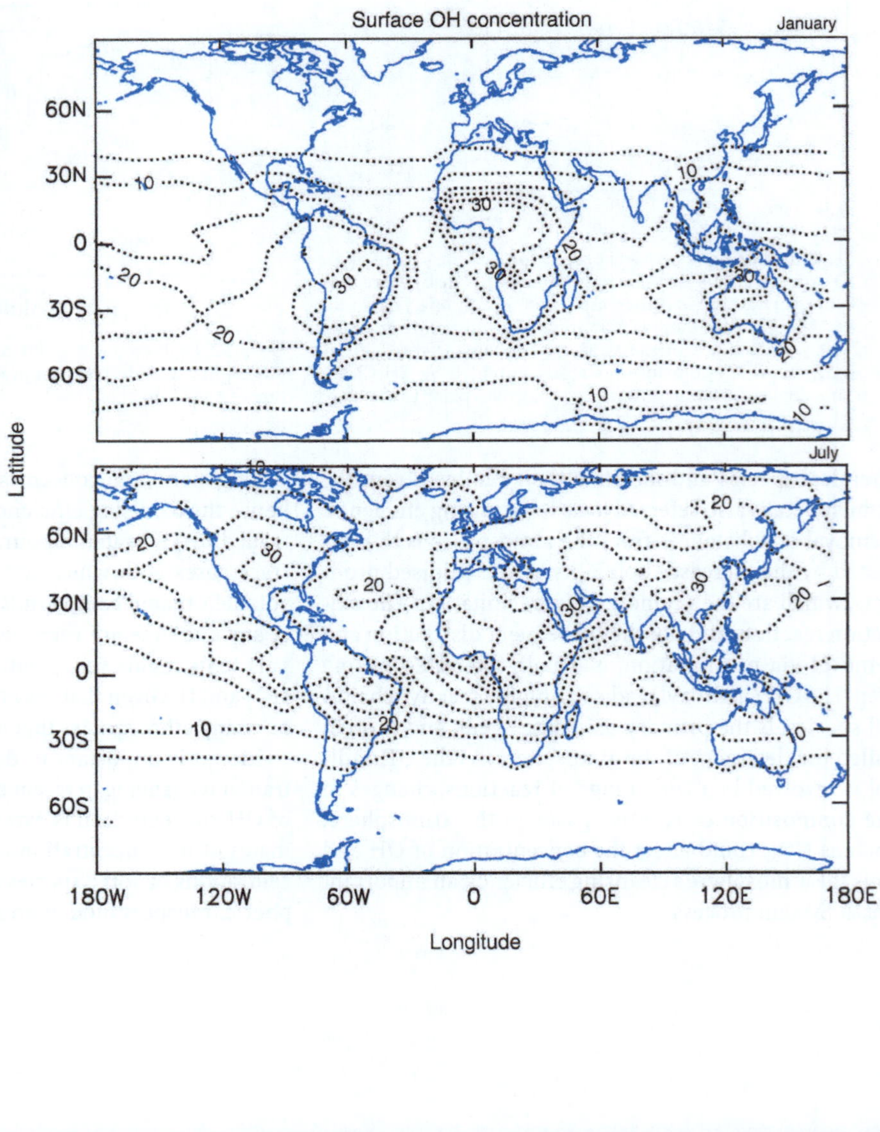

Fig. 5.34. Global concentration fields of OH (10^6 molecules cm^{-3}) in the atmospheric surface layers: *top* January, *bottom* July (Spivakovsky et al. 2000)

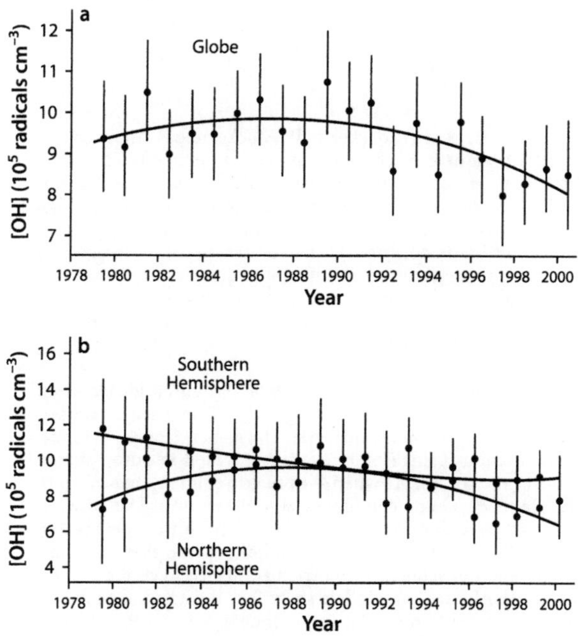

Fig. 5.35. **a** Annual global and **b** hemispheric OH concentrations estimated from the annualised content method with 1σ error bars (*thick* for estimated error (ε_i) *thin* for total) estimated from the Kalman filter and the Monte Carlo method (Prinn et al. 2001; Prinn et al. 1995). For 2000 only six monthly running-mean observations are available, so ε_i is larger. Also shown are polynomial fits to these annual concentrations (*solid lines*). Trend accelerations represented by these polynomials are significantly non-zero

The relatively small changes in the OH concentration observed up to now may be due to the intrinsic robustness of the tropospheric chemical system or, alternatively, may be due to offsetting anthropogenic impacts. For example, the increase of CO in the atmosphere leads to a decrease in the OH concentration while emission of NO_x leads to an increase. The two processes do not directly offset one another, however, as CO is more evenly dispersed globally while NO_x production and destruction in the atmosphere is a more local phenomenon. In addition, emission of other traces gases such as CH_4 and changes in stratospheric ozone concentration also affect OH abundance. In general, surface emissions tend to be balanced by changes in stratospheric ozone. However, a recent analysis has shown that meteorological variability, which affects the concentration of water vapour in the atmosphere, is also important, and is perhaps the dominant influence when compared to the effects of anthropogenic emissions of precursors (Dentener et al. 2003).

In summary, there is yet no clear consensus on the reason for the relatively small changes in OH concentration observed up to now despite the significant anthropogenic emissions of trace gases that affect OH chemistry. If they are due to offsetting anthropogenic impacts, future changes in the cleansing efficiency of the atmosphere will depend significantly on future trends in trace gas emission and ultimately on population growth, economic development and technological changes (Brasseur et al. 2003). If the relatively small changes are dominated by meteorological variability instead of chemistry, however, OH abundance and hence the cleansing efficiency of the atmosphere may be very well buffered in the face of anthropogenic perturbations and thus unlikely to experience any significant or abrupt changes (Dentener et al. 2003).

Thermohaline Circulation in the North Atlantic Ocean

Another example of a potential catastrophic perturbation in the Earth System concerns the North Atlantic thermohaline circulation and its effect on the climate of northern Europe. A great deal of heat is transported globally with the movement of ocean water. The northern North Atlantic Ocean, for example, is a recipient of heat by this process via the Gulf Stream. As these warm surface waters reach the high latitudes north of Iceland and cool, they release their heat to the atmosphere and the surface waters cool and sink (the so-called North Atlantic Deep Water Formation). Driven by predominantly westerly winds, it is this delivery of heat to northern latitudes that makes human existence at 60° N a much more pleasant experience in Scandinavia than in northern Canada or Siberia.

In the past this circulation pattern has not been constant. At least some of the abrupt climate changes seen in palaeo-records are probably associated with a slowdown or reorganisation of this pattern of oceanic circulation (Rühlemann et al. 1999). A change in this circulation pattern would entail a decrease in heat transport to and subsequent cooling of northern latitudes and a concomitant warming of the tropical Atlantic Ocean. Such responses have been seen at various times in the geological record (Bond et al. 1999), suggesting that the Earth System may have had equilibrium states, or preferred modes of operation, other than that of the present. When the Earth System is switching between states, for example when coming out of a glacial state into an interglacial, the North Atlantic oceanic circulation pattern appears to flicker or flip-flop between the two states and in the process triggers abrupt climate changes in the region and perhaps around the northern hemisphere. Some of these changes are extremely rapid, with temperature rises of around 10 °C in a decade.

Could abrupt changes in the North Atlantic occur with present and projected forcings? Box 5.6 analyses this possibility on the basis of an understanding of thermohaline circulation derived from both palaeo-studies and modelling results. The conclusion is that a weakening of the circulation is possible and that the crossing of a threshold leading to irreversible changes in ocean circulation cannot be excluded within the range of projected climate change over the next century. Until more is known about the inherent stability of North Atlantic thermohaline cir-

Box 5.6. Thermohaline Circulation: Past Changes and Future Surprises?

Stefan Rahmstorf · Thomas F. Stocker

Ocean currents driven by surface fluxes of heat and freshwater and subsequent interior mixing of heat and salt are referred to as thermohaline circulation (Webb and Suginohara 2001; Bryden and Imawaki 2001; Rahmstorf 2003). Thermohaline circulation is complex, but can be summarised as a global-scale deep overturning of water masses, with sinking motions occurring in the northern Atlantic and around Antarctica and a transport of relatively warm surface waters towards these sinking regions (Schmitz 1995). A heat transport of up to 10^{15} W is associated with this circulation in the Atlantic (Ganachaud and Wunsch 2000), rivalling that of the atmosphere.

The opposing effects of cooling and freshening on density and associated feedbacks, together with the threshold behaviour of oceanic convection, make the thermohaline circulation a fragile system that can respond in a highly non-linear fashion to changes in surface climate. There is strong evidence that this has repeatedly occurred in the past, and reason for concern that it might happen again in the future (Broecker 1987, 1997; Alley et al. 2003).

The best evidence for major past thermohaline circulation changes comes from the last glacial (120 000–10 000 years BP (see Clark et al. 2002; Rahmstorf 2002 for reviews). Two main types of abrupt and large climate shifts have occurred during that period, Dansgaard-Oeschger events and Heinrich events. Sediment records show that both of these were associated with major circulation changes in the North Atlantic.

Dansgaard-Oeschger events typically start with an abrupt warming (by up to 10 °C in Greenland) within a few decades or less, followed by gradual cooling over several hundred or thousand years (Fig. 5.36). The cooling phase often ends with an abrupt final temperature drop back to cold (*stadial*) conditions. Twenty-four such events occurred during the last glacial, and they have been documented at many sites around the world. The amplitude is largest around the northern Atlantic, while the South Atlantic responds with the opposite sign and a different temporal characteristic. Slow cooling starts when the north is abruptly warming, a behaviour that has been likened to a *bipolar seesaw* (Broecker 1998; Stocker 1998). Many of the observed features of these events can be explained by considering them as switches between different modes of the Atlantic thermohaline circulation, characterised by different locations of deep water formation (Broecker et al. 1985; Ganopolski and Rahmstorf 2001). The sudden warming is, according to this theory, caused by a northward push of warm Atlantic waters into the Nordic Seas. What triggers these mode switches is still unclear, but one hypothesis that is being debated proposes a mechanism associated with stochastic resonance (Alley et al. 2001).

Heinrich events are massive surges of the Laurentide Ice Sheet through Hudson Strait (Hemming 2003). They have a variable spacing of several thousand years. The icebergs released to the North Atlantic during Heinrich events leave tell-tale drop-stones in the ocean sediments when they melt. Sediment data suggest that Heinrich events completely shut down or at least drastically reduced the formation of North Atlantic Deep Water (NADW), and thus a major component of the thermohaline circulation. Consequently, Heinrich events led to cooling in the North Atlantic region. Models show that the amount of freshwater input due to the melting continental ice – on the order of 0.1 Sv, as suggested by the palaeo-climatic records (Clark et al. 2001) – is indeed sufficient to stop NADW formation. Palaeoclimatic data again suggest an Atlantic see-saw effect, consistent with the idea that the temperature changes are mainly caused by a reduction in interhemispheric heat transport in the Atlantic Ocean.

Palaeo-climate records provide no indication that the Atlantic thermohaline circulation (THC) has undergone large and rapid changes during the last 8 000 years, most likely because the continental ice sheets have become too small to produce instabilities and catastrophic freshwater discharges into the Atlantic Ocean, and because the circulation is more stable in a warm climate (Ganopolski and Rahmstorf 2001). However, global warming involves a number of new possibilities to alter substantially the freshwater balance of the Atlantic

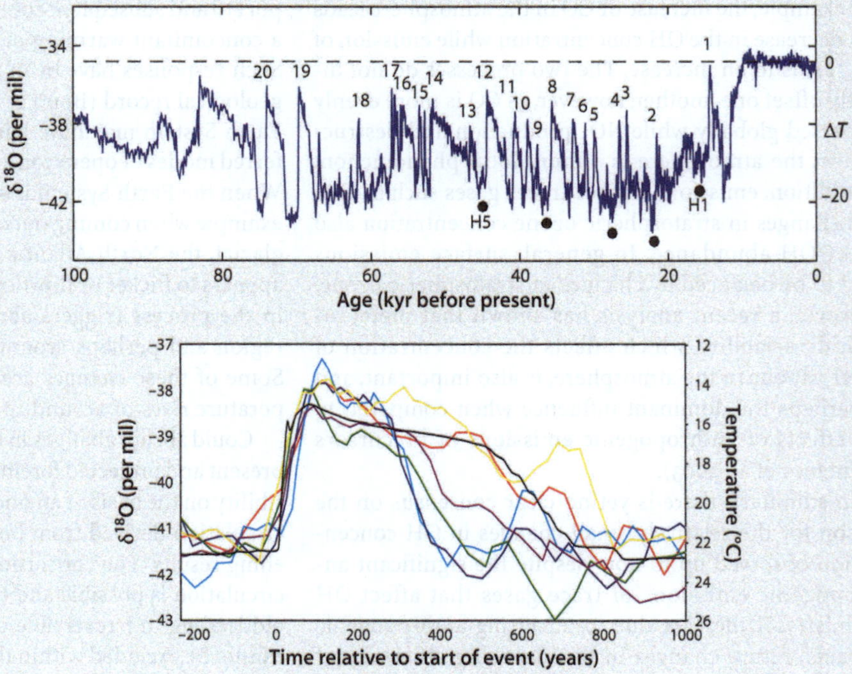

Fig. 5.36. Record of δ18O (per mil, *scale on left*) from the Greenland Ice Sheet Project (GRIP) ice core, a proxy for atmospheric temperature over Greenland (approximate temperature range, in °C relative to Holocene average, is given on the *right*), showing the relatively stable Holocene climate in Greenland during the past 10 000 years and Dansgaard-Oeschger (D/O) warm events (numbered) during the preceding colder glacial climate. The *lower panel* shows a close-up of several of the more recent events superimposed (*coloured lines*). The *black line* shows a model-simulated D/O event (Ganopolski and Rahmstorf 2001)

◀

in the future and thereby trigger THC variability or even collapses.

When air temperature rises, surface waters also tend to warm up, an effect which is enhanced particularly in the high latitudes via the snow/ice albedo feedback. In addition, the hydrological cycle may be accelerated in a warmer atmosphere because of increased evaporation, larger atmospheric moisture capacity and stronger meridional moisture transport (Dixon et al. 1999). The observed increase in river runoff in the high latitudes may be due to this effect (Peterson et al. 2002).

These effects tend to reduce the THC because heating and freshening both decrease surface water density. Indeed, the majority of coupled climate models indicates a reduction of the THC from 10% to 80% in response to increasing CO_2 concentrations in the atmosphere for the next 100 years (IPCC 2001a) (see figure). Knowledge, however, is still incomplete and significant uncertainties persist. This is illustrated by the large spread of simulated THC changes. Some models exhibit strong enough feedback mechanisms which stabilise the THC, at least for some decades. These are associated with changing modes of climate variability in a warmer world (Latif et al. 2000; Delworth and Dixon 2000).

Long-term simulations with different climate models suggest that the maximum projected CO_2 concentration may constitute a threshold for the Atlantic THC beyond which the circulation stops (Manabe and Stouffer 1993; Stocker and Schmittner 1997) (Fig. 5.37). In these early simulations a threshold was found between 2× and 4× pre-industrial CO_2 concentration. Although the existence of the threshold and its location critically depend on the climate sensitivity of the coupled models, the details of the hydrological cycle and other parameterisations (Rahmstorf and Ganopolski 1999), such a potentially irreversible change in circulation patterns cannot be excluded to occur in the future (Alley et al. 2003). A reduction or complete collapse would decrease the meridional heat transport in the Atlantic, and hence a regional cooling is superimposed on the global warming. Because of the uncertainty where this threshold lies, it is not known whether the combined effect leads to a net warming or net cooling in the regions most affected by the meridional heat transport of the Atlantic THC.

Model simulations indicate that the threshold may be crossed if the forcing is strong enough and applied for long enough. The threshold may well lie within the range of warming that is expected under business-as-usual in the next 100 years or less. However, prediction of the location of this threshold is impossible at the moment because of fundamental limitations of the predictability of non-linear events such as a THC collapse (Knutti and Stocker 2002). Nevertheless, current models suggest that the risk of major ocean circulation changes becomes significant for the more pessimistic warming scenarios, but can be greatly reduced if global warming is limited to the lower end of the IPCC range. Because of the interplay between the forcing and the limited rate of heat uptake in the ocean, also the rate of increase in CO_2 matters: the ocean-atmosphere system appears less stable under faster perturbations (Stocker and Schmittner 1997). This indicates that humankind has a choice: early action significantly widens policy options regarding future fossil fuel emissions scenarios.

Based on present knowledge of the climate system, the following results appear not to depend on model type, resolution or parameterisations:

- the Atlantic THC can have multiple equilibria which implies thresholds;
- reorganisations of the THC can be triggered by changes in the surface heat and freshwater fluxes;
- most models indicate a weakening of the THC in the next 100 years. This implies an approach towards possible thresholds;
- crossing of thresholds and associated irreversible changes of ocean circulation *cannot be excluded* within the range of projected climate changes of the next century.

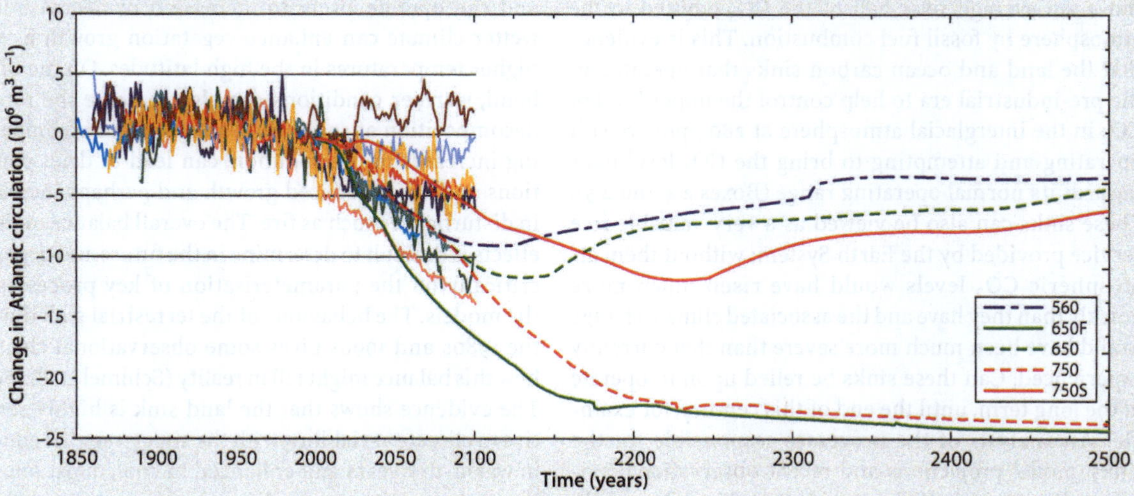

Fig. 5.37. Composite of changes in meridional overturning in the Atlantic Ocean simulated to 2100 by a set of comprehensive coupled climate models (*fine lines*) (Cubasch et al. 2001; IPCC 2001a). To illustrate the possible long-term behaviour of the thermohaline circulation, simulations using a coupled model of reduced complexity are overlaid (Stocker and Schmittner 1997). They use artificial CO_2 emissions scenarios that are identified in the inset. Carbon dioxide increases by rates of 0.5, 1 and 2% per year up to maximum concentrations of 560, 650 and 750 ppm, and constant thereafter. Depending on the rate of CO_2-increase and the maximum CO_2 concentration, and hence the warming, the THC crosses a threshold beyond which the circulation stops and remains collapsed

culation and about the nature of global change forcings, however, this remains a surprise of low probability. Nevertheless, the palaeo-record and model experiments show that the possibility cannot be dismissed on the grounds of lack of precedent or implausibility.

On the contrary, recent observational evidence provides some support to the model projections that the North Atlantic thermohaline circulation could weaken or shut down later this century. About one-third of the cold water formed in the North Atlantic, after it has sunk to the lower levels of the Arctic and Greenland Seas, begins its journey southward by spilling through the channel between Iceland and the Faroe Islands. This through-flow has decreased by 20% over the last 50 years (Hansen et al. 2001). More research is needed to determine whether this observation is matched by a slowdown in other branches of the circulation around Iceland or whether it might be part of long-term natural variation. However, given its consistency with many model projections, the connection to anthropogenically-driven changes to the Earth System cannot be discounted. Should this indeed be an early warning sign of an impending abrupt change in oceanic circulation, the human consequences of such a change would be incalculable in their severity.

Collapse of Carbon Sinks

A change in the capacity of the terrestrial and marine sinks to slow the build-up of atmospheric CO_2 is a further example of a potentially catastrophic perturbation in the Earth System. Currently land and ocean sinks remove, on average, over half of the CO_2 emitted to the atmosphere by fossil fuel combustion. This is evidence that the land and ocean carbon sinks that operated in the pre-industrial era to help control the upper level of CO_2 in the interglacial atmosphere at 280 ppm are still operating and attempting to bring the CO_2 level back towards its normal operating range (Boxes 2.4 and 2.5). These sinks can also be viewed as a very valuable free service provided by the Earth System; without them atmospheric CO_2 levels would have risen much more quickly than they have and the associated climate change would have been much more severe than that currently experienced. Can these sinks be relied upon to operate in the long term, until the end of this century, for example? An analysis of the processes responsible for the sinks, model projections and recent observational evidence all suggest that it is possible that this ability might weaken or fail later this century. Figure 5.38 shows the simulated behaviour of both the land and ocean carbon sinks for the rest of the twenty-first century. The simulations are based on the responses of the processes responsible for the sinks to projected changes in atmospheric CO_2 concentration and the associated changes in climate (Cramer et al. 2001; Orr et al. 2001).

The land sink (Fig. 5.38a) shows increasing strength throughout the next 50 years but then will likely either stabilise at a new equilibrium level or perhaps even weaken during the second half of the century. In these simulations two processes are primarily responsible for the land sink: increasing uptake of CO_2 by vegetation as atmospheric CO_2 concentration increases (the CO_2 fertilisation effect) and the response of vegetation and soil carbon to changes in temperature and precipitation. The *grey shaded area* in Figure 5.38a shows the effect of CO_2 fertilisation alone while the *solid lines* show the projections based on both atmospheric CO_2 and climatic change.

These model projections are clearly highly sensitive to the parameterisation of the CO_2 fertilisation effect. A large number of experiments shows conclusively that terrestrial vegetation takes up significantly more CO_2 as the concentration of the gas in the atmosphere increases (Mooney et al. 1999; Hamilton et al. 2002) (Box 4.5). However, the magnitude of this effect in *in situ* ecosystems, which are subject to a variety of other growth-limiting factors, is still difficult to determine. Recent syntheses of the location and causes of the terrestrial carbon sink (Schimel et al. 2001; Goodale et al. 2002) suggest that land-use change effects (abandonment of agricultural land and regrowth of forests coupled with fire suppression) are likely to be just as important as CO_2 fertilisation. This conclusion has obvious implications for projections of the behaviour of the land sink into the future; in particular, it suggests that sink strengths based on very large CO_2 fertilisation effects may be significant overestimates.

The effects of climate on the land sink are multiple and can operate either to increase it or decrease it. A wetter climate can enhance vegetation growth as will higher temperatures in the high latitudes. On the other hand, warmer conditions can also increase the rate of decomposition of soil carbon and, without accompanying increases in precipitation, can lead to drier conditions and hence reduced growth and perhaps increases in disturbances such as fire. The overall balance of these effects is difficult to determine in the future and depends critically on the parameterisation of key processes in the models. The behaviour of the terrestrial sink during the 1980s and 1990s gives some observational clues to how this balance might fall in reality (Schimel et al. 2001). The evidence shows that the land sink is highly sensitive to climate variability, with the sink strength reduced in warm, dry years and enhanced in cool, moist ones. If these observations are indicative of the general behaviour of the land sink, it will likely weaken due to climate change during the rest of the century.

One of the simulations (Fig. 5.38a) shows a sharp decrease in the strength of the land sink towards the end of the century, with the land possibly becoming a source of carbon by 2100. This is largely due to a strong response function of soil carbon decomposition to in-

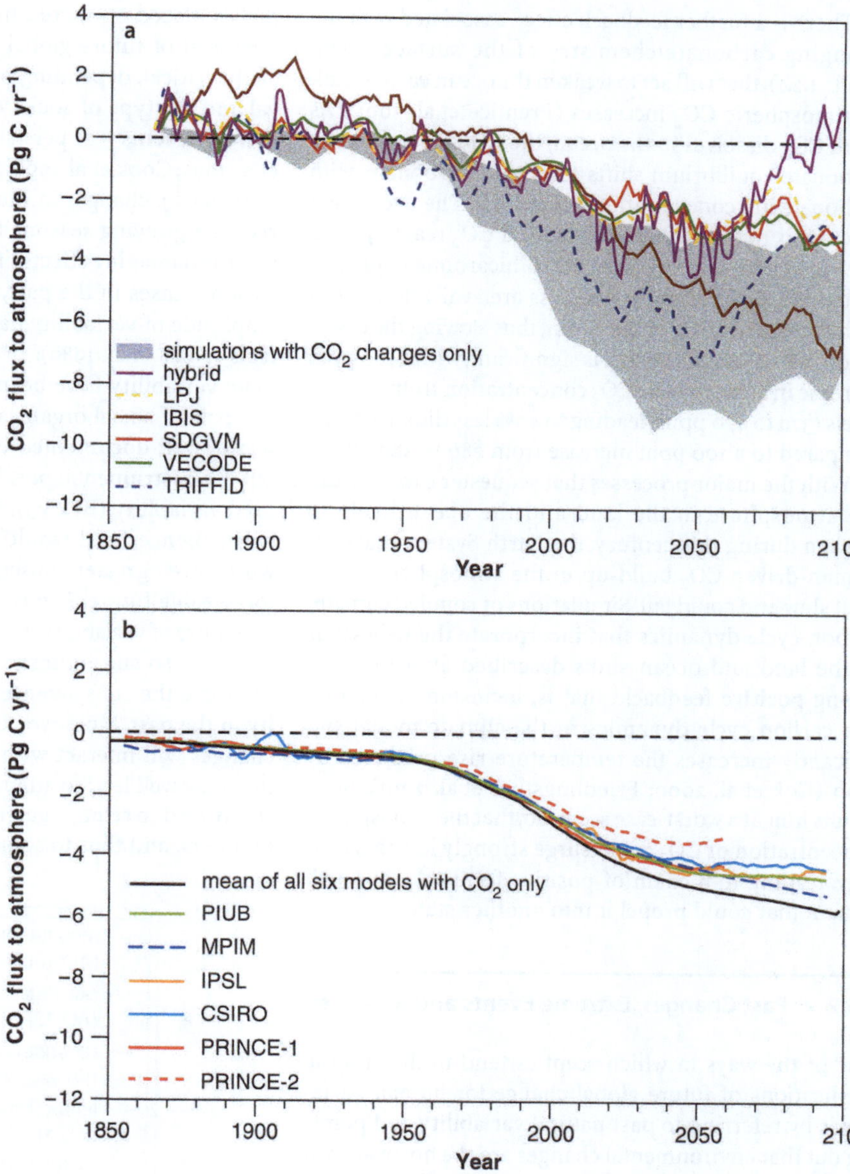

Fig. 5.38.
Projections of **a** terrestrial and **b** marine CO_2 uptake through the twenty-first century (Cramer et al. 2001; IPCC 2001a). In panel **a** six dynamic global vegetation models were driven by atmospheric CO_2 concentration projections of the IPCC SAR IS92a scenario and by simulated climate changes obtained from the Hadley Centre climate model with CO_2 and sulphate aerosol forcing from IS92a. The *grey shaded area* represents the response of the vegetation models to increasing atmospheric CO_2 alone, without any associated climate change. In panel **b** six ocean carbon models were used to estimate the combined impact of increasing atmospheric CO_2 and changing climate on marine CO_2 uptake

creasing temperature. In addition, the combination of increasing temperature and drying conditions triggers dieback in the Amazon Basin and converts the rainforest into a grassland. This conversion releases significant amounts of carbon to the atmosphere.

Overall the message is clear. Many lines of evidence based on both process-based models and observations suggest that the land sink strength will level off around the middle of the century and could drop thereafter.

Ocean uptake of CO_2 is also sensitive to both increasing CO_2 and climate, as shown by the results of model simulations (Prentice et al. 2001) (Fig. 5.38b). The pattern is similar as for the terrestrial sink; the ocean sink increases strongly during the first half of the century but then increases at a smaller rate during the second half. Most models show a continuing but small increase in sink strength at 2100.

Much of the behaviour of the ocean sink is driven by the increasing solubility of CO_2 in seawater with the increasing partial pressure of CO_2 in the atmosphere. The effect of including climate change as well as atmospheric CO_2 is to modify the CO_2-only results towards a somewhat weaker sink (6 to 25% weaker by 2050 compared to 1990). The processes responsible for the weakening sink are (*i*) decreasing solubility of CO_2 with increasing temperature of surface waters; (*ii*) increased stratification of the upper ocean; and (*iii*) perhaps an effect of changing ocean biology although this is very uncertain. At present ocean carbon models or coupled ocean-atmosphere models treat the biological pump in a fairly simply way. Critical processes, such as interaction with nutrient cycles (N, P, Fe, Si), are not yet included. Early sensitivity studies suggest that inclusion of more complex biology may weaken the ocean sink somewhat further (Prentice et al. 2001).

There is a further feedback effect, associated with the changing carbonate chemistry of the surface ocean (Sect. 4.2.1), that will act to weaken the ocean carbon sink as atmospheric CO_2 increases (Prentice et al. 2001). As more CO_2 dissolves in the ocean, the CO_2-biocarbonate-carbonate equilibrium shifts towards bicarbonate, with carbonate ion concentration decreasing. The overall net reaction is one molecule of dissolved CO_2 reacting with one carbonate ion to produce two bicarbonate ions. As carbonate ions are consumed, less are available to react with any further CO_2 in the ocean, thus slowing the dissolution rate of CO_2. The effect is significant, with a 100 ppm increase in atmospheric CO_2 concentration from current levels (370 to 470 ppm) leading to 40% less dissolved CO_2 compared to a 100 ppm increase from 280 to 380 ppm.

With the major processes that sequester carbon from the atmosphere in the land and the ocean likely to weaken during this century, the Earth System brake on human-driven CO_2 build-up in the atmosphere will at least slow and could fail. Simulations of coupled climate-carbon cycle dynamics that incorporate the behaviour of the land and ocean sinks described above show a strong positive feedback; that is, inclusion of interactive carbon cycle dynamics in the climate model significantly increases the temperature rise predicted by 2100 (Cox et al. 2000; Friedlingstein et al. 2001). Such results hint at a worst-case scenario, that the atmospheric concentration of CO_2 could surge strongly late this century, leading to a chain of positive feedbacks in Earth System that could propel it into another state.

5.4.2 Past Changes, Extreme Events and Surprises

One of the ways in which sceptics tend to dismiss the implications of future global change for human societies is by referring to past natural variability and pointing out that environmental changes are the norm and do not depend on anthropogenic forcing. In addressing this point, the reality of past environmental changes is acknowledged. However, it is useful to explore the likely future human consequences of change within the range of natural variability that has been characteristic of the period preceding major human impacts, but subject to the same range of natural forces and feedbacks that operate today. In addition, it is necessary to use the range of recent variability as a context for evaluating the human implications of likely future changes. Finally, it is useful to identify, in the palaeo-record, some of the potential changes that may be unlikely, but are plausible, not without precedent, and would have truly devastating consequences were they to occur again in the future.

One of the clearest messages from the past is that human societies are most strongly impacted by extreme events at regional scale, rather than by smooth changes in temperature at global scale. To that extent, the emphasis placed on increasing global temperature in the portrayal of future global change is misleading. Especially critical, depending on the region, the time interval and the type of society, has been the occurrence of extreme events – of persistent, severe droughts (Hodell et al. 1995; Cook et al. 1999; Stahle et al. 2000; Verschuren et al. 2000), changes in the intensity and recurrence of reduced growing seasons for crops (Landsteiner 1999) and the damaging effects of floods (Messerli et al. 2000). In some cases in the past, an increase in the pace and amplitude of variability has had severe and decisive effects (Nials et al. 1989). In all cases, the impacts of climate variability have been mediated by the prevailing patterns of social organisation. In many regions of the world, well-documented extreme events in the recent but pre-instrumental past have exceeded directly measured variability (Box 5.3). They have had major human consequences and would have, in many parts of the world, even greater consequences at present and in the foreseeable future. Pointing to the ubiquity and intrinsic nature of variability is thus no cause for complacency; the more so since human actions have often served to reinforce the consequences of natural climate variability in the past. Moreover, in the future, ongoing natural changes will interact with anthropogenic ones in ways that may well lead to attenuation in some cases, but will surely lead to reinforcement through positive feedbacks in others, and thus to some unpleasant surprises.

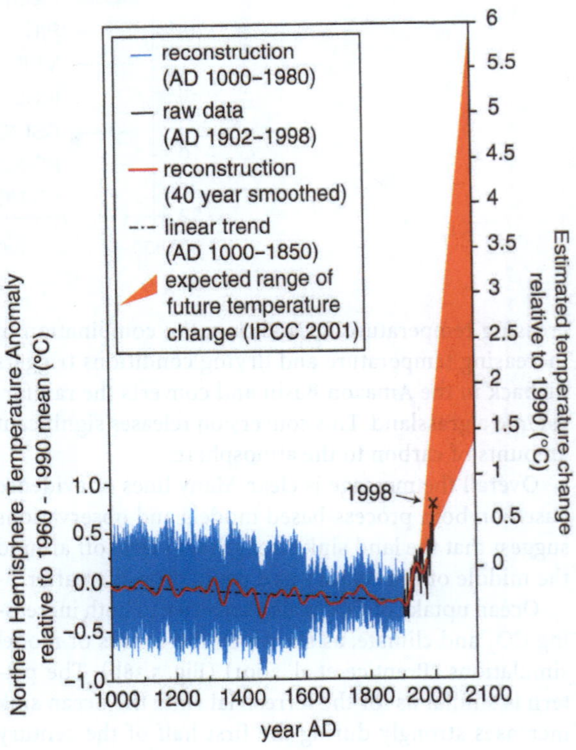

Fig. 5.39. The range of future projections of temperature change lie far outside the range of historic global average temperature for at least the last millennium (Mann et al. 1999; IPCC 2001a)

If past changes in climate are portrayed in the way that projected future changes are most often expressed, some comparisons between past and likely future scales of change can be made. The range of variability in mean annual temperature over the northern hemisphere as a whole during the last 1 000 years is around 1 °C (Mann et al. 1999), yet it is within this narrow range of variability that severe impacts on human populations at the regional scale have taken place. Even the most conservative estimates of future temperature increase exceed this range of variability and the current bandwidth of consensus projections for the end of the present century exceeds it by several degrees (Fig. 5.39). Knowledge of past variability thus serves not as a recipe for dismissing the consequences of the future changes, but as a context for evaluating their likely magnitude and human implications.

A frequently considered possibility, given some credibility by changes in mode during recent decades, is that El Niño events may change significantly in frequency, intensity or duration. There is no consensus on the probability of such changes occurring, but the palaeo-record does show major changes in its periodicity (Fig. 5.40) and even indicates that the ENSO phenomenon as currently characterised probably began some 5 000 years ago. At the very least, it is necessary to recognise that ENSO is not a stable phenomenon and that an examination of past behaviour leads one to expect rather than preclude a dramatic change in its mode in the future (Box 5.7).

An Example of a Surprise

The themes of connectivity in the Earth System and the reverberation of human-driven changes through the System have been highlighted earlier in this volume. At the scale of the Earth System as a whole, an examination of changes in the past can give hints of the types of surprises that might be possible in the next century or two, surprises that would have profound consequences for human societies.

A shutdown of the thermohaline circulation (Box 5.6) would not only affect societies in Scandinavia, but would

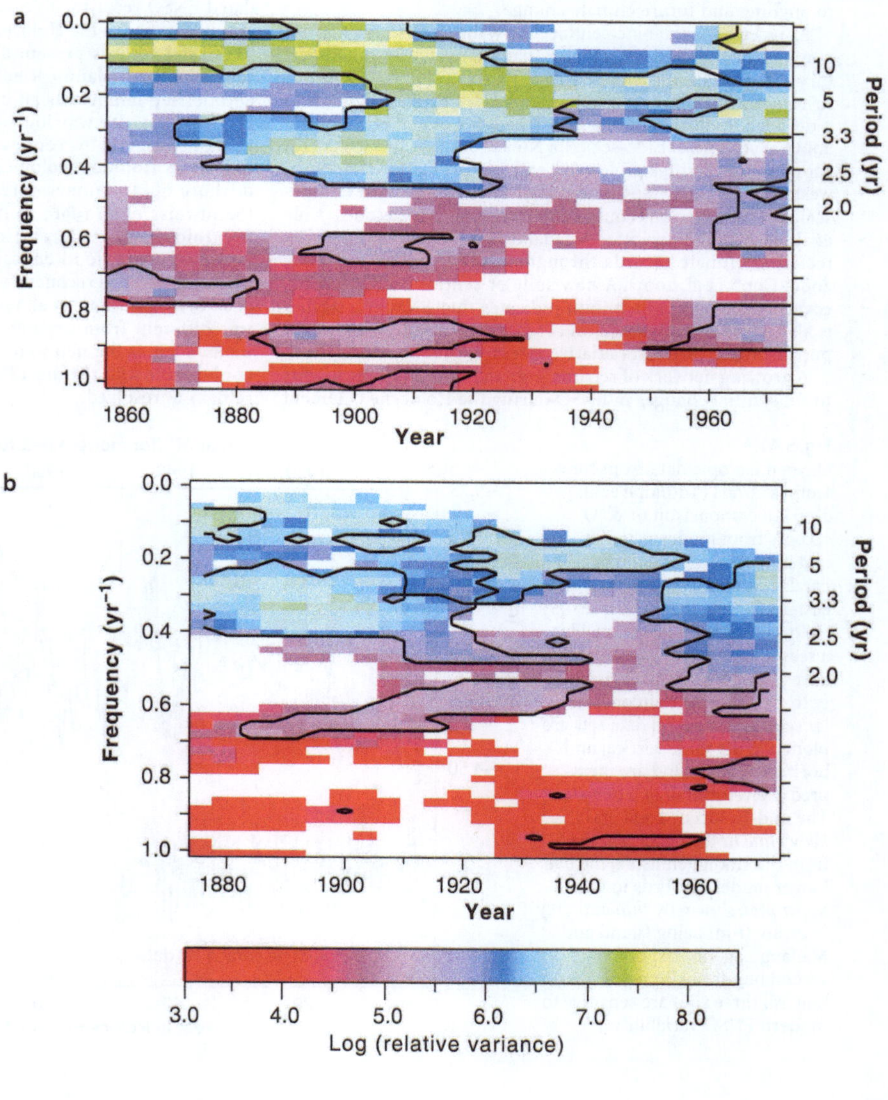

Fig. 5.40. Periodicity of El Niño events as given by evolutionary spectral analysis results: **a** Maiana and **b** Niño 3.4 SST records (Urban et al. 2000). Analysis was performed on 40-year segments of data overlapped by four years, using multi-taper methods with red noise background assumptions. *Coloured regions* show variance significant above the median background (>50%) and *black contours enclose regions* with variance significant at >90%. These spectra share several general features: significant nineteenth-century decadal variance, strong variance at 2.9-year periods in the early twentieth-century, attenuation of 2–4-year variance between 1920–1955, and the general progression from decadal towards higher-frequency interannual variance over the length of the record. The low-frequency variance concentration in the final years of the coral record reflects the step-function change in 1976

Box 5.7. Variation of ENSO through Time: How Well Is El Niño Understood?

Julia Cole

Several decades of intensive research on ENSO have generated detailed knowledge of the modern characteristics of the phenomenon. Over the last half of the twentieth century, a canonical picture of ENSO has emerged, with a characteristic fingerprint of climate anomalies within and beyond the tropical Pacific and a typical interannual (2–7 yr) time scale of variability (Trenberth et al. 1998; Wallace et al. 1998). Intensive observation of the tropical Pacific ocean and atmosphere has led to the identification of physical mechanisms that explain the major temporal and spatial features of ENSO. These mechanisms provide the underpinning for predictive models that are capable of moderate skill in forecasting the state of ENSO several months in advance (Zebiak and Cane 1987; Barnston et al. 1999).

Despite such notable progress, ENSO is far from being perfectly understood. Mounting evidence from palaeo-climatic studies indicates that ENSO (and its global impacts) have behaved very differently in the past. Numerical simulations support the argument that ENSO responds sensitively to changes in climate forcing and background state. Moreover, palaeo-climate and modelling studies support the likelihood of substantial internal (unforced) variability in ENSO. Both internal and external sources of ENSO variability are likely to complicate efforts to predict how ENSO and its teleconnections will respond to ongoing and future climate change.

As recently as the nineteenth century, the time scale of ENSO may have been significantly longer than during the twentieth century. A central Pacific coral record (Urban et al. 2000) (Fig. 2.13a) reveals that the familiar interannual beat of ENSO lengthens to a decadal rhythm between 1840 (the start of the record) and about 1890. The decade-scale La Niña from 1855 to 1863 corresponds closely with a prolonged, extreme drought in the southwestern USA (1855–1865), implying that the impacts of decade-scale ENSO variations could similarly persist for decades (Cole et al. 2002). The imprint of decadal ENSO variations can be detected in climate records throughout the tropics (Cole et al. 2000; Cobb et al. 2001). A new suite of central Pacific records covering the past millennium indicates that ENSO variability is also strongly modulated on century time scales, with no apparent link to external radiative forcing (Cobb et al. 2003).

A growing network of records from the tropical Pacific points to substantial changes in ENSO during the Holocene (Clement et al. 2000; Cole 2001). Many of these are discontinuous and/or not located ideally for ENSO studies, but together they offer strong evidence that ENSO behaved very differently before about 5 000–7 000 years ago. In the western Pacific, corals from New Guinea dating to ~6 500 years BP show little interannual variability and a regular (but weak) seasonal cycle. This situation contrasts sharply with modern coral records, in which weak seasonality is overwhelmed by substantial year-to-year variance (Tudhope et al. 2001) (Fig. 5.41). Coral records from the Great Barrier Reef, Australia, dating to ~5 300 yr BP confirm that the interannual hydrologic variance associated with today's ENSO extremes is absent, and temperatures are higher (Gagan et al. 1998).

Eastern Pacific records add both detail and complexity to the question of mid-Holocene ENSO variability. In southern Ecuador a series of cores from Lago Pallcacocha (Rodbell et al. 1999; Moy et al. 2002) (Fig. 2.13b) records sediment inputs associated with intense rainfall; recent layers correlate well with the twentieth-century record of moderate to strong El Niño events. Interannual ENSO variations were absent before about 7 000 years ago, and the frequency of these events is modulated on millennial time scales, with maximum intensity around 1 200 years ago. This site provides the only continuous, well-dated ENSO record with a strong calibration to modern variability that spans the Holocene.

Coastal records present a more heterogeneous picture. In the Galapagos Islands, Riedinger et al. (2002) use carbonate and detrital laminae as an index of El Niño activity and suggest that activity was low before 4 000 years ago, increasing to modern levels by 3 000 years ago. In northern Peru, biotic indicators from archeological sites indicate warm and stable offshore ocean temperatures between 8 000–5 000 years ago (Sandweiss et al. 1996; Andrus et al. 2002). However, sedimentological interpretations at nearby sites suggest an unusual lack of flood deposits between 8 900–5 700 years ago that is difficult to reconcile with warmer ocean waters (Keefer et al. 1998; Fontugne et al. 1999). Clearly, early Holocene ENSO was different from present – certainly weaker, although key issues such as the nature of the background state (warmer or cooler) and the timing of the shift to modern conditions remain unresolved.

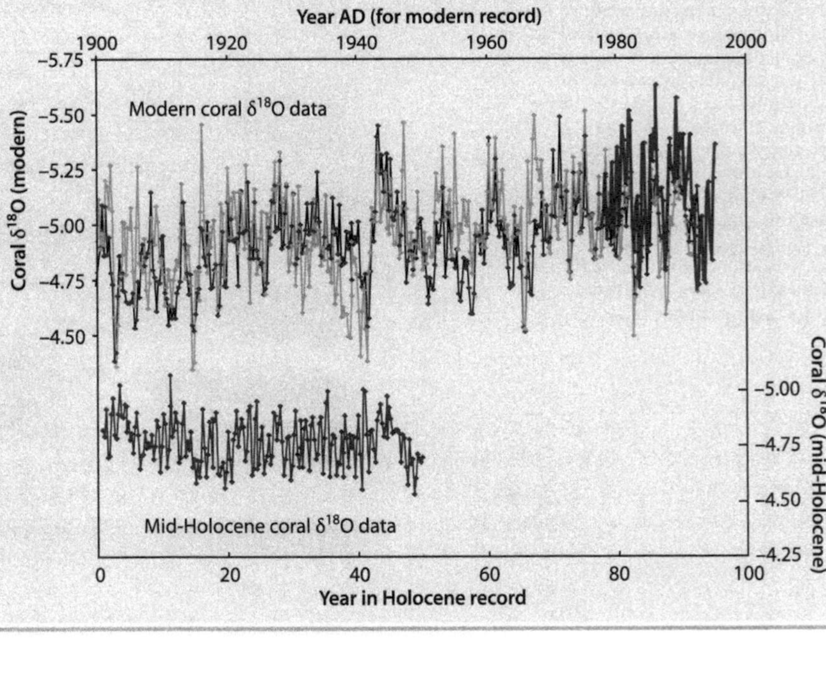

Fig. 5.41. Oxygen isotopic data from New Guinea corals (Tudhope et al. 2001): a comparison of $\delta^{18}O$ records from modern (*top lines*) and mid-Holocene (6500 yr BP) corals (*bottom line*) from northeastern New Guinea. The mid-Holocene data are dominated by a seasonal cycle; the modern sites have strong interannual variability from ENSO overlain on a similar seasonal cycle. All data sets are plotted at the same vertical and horizontal scales and are measured at a resolution of 4 per year. The mid-Holocene coral and the *heavy line* in the *upper plot* are from the Huon Peninsula; the two longer modern records in the *upper plot*, shown by *thin* and *grey lines*, are from Laing Island and Madang Lagoon, respectively, several hundred kilometres distant. All three sites are sensitive to modern ENSO variability

◀

Numerical simulations support the sensitivity of ENSO to changing background climate and global forcings in the past. Clement et al. (1999, 2000) show that changes in orbital forcing of seasonal solar irradiance can alter the frequency of ENSO warm and cold events. In particular, when perihelion (the point on the Earth's orbit closest to the Sun) occurs during boreal summer/autumn, the trade winds in that season strengthen, which inhibits the development of warm El Niño anomalies. At the same time, the Asian monsoon intensifies; in a global coupled ocean-atmosphere model, this response also enhances Pacific summer trade winds (Liu et al. 2000). Both of these responses to orbital forcing should weaken ENSO variability in the early-mid Holocene and strengthen it as perihelion moves towards boreal winter in the latest Holocene. Stronger trade winds also generate cooler background conditions in global coupled models, yielding a more La Niña-like background state consistent with many palaeo-climatic records (Otto-Bliesner 1999; Clement et al. 2000; Cole 2001).

The last glacial interval provides an obvious target for examining the relationship between mean climate state and ENSO variability. Unfortunately, few ENSO-sensitive palaeoclimate records contain the necessary resolution to address this question. One exception is the suite of records from New Guinea corals (Tudhope et al. 2001), which include glacial intervals of intermediate intensity dating to 40 000, 85 000, 112 000, and 130 000 years before present. These results suggest that glacial boundary conditions suppress interannual ENSO variability; potential mechanisms include weaker ocean-atmosphere interactions and stronger trade winds. Numerical simulations of last glacial maximum conditions, by contrast, indicate strengthened ENSO variability (B. Otto-Bliesner, pers. comm.). These results are not directly comparable as both orbital forcings and background climate would have been different between the last glacial maximum and the less extreme conditions during the intervals covered by corals. Palaeo-oceanographic data have attempted to address this question using sediment samples averaging several centuries each. Inferred gradients among sites and shifts in convective systems suggest, paradoxically, that glacial stages are characterised by more El Niño-like (warm Pacific) conditions, and interglacials are cooler and more La Niña-like (Stott et al. 2002; Koutavas et al. 2002).

Much of the interest in ENSO stems from its somewhat predictable pattern of extratropical impacts, but recent palaeo-climate records show clearly that the global footprint of ENSO has not remained constant (Fig. 2.12) (see also Cole and Cook 1998). Climate model studies support this observation; changes in the background state influence the propagation of anomalies through the atmosphere and ocean (Meehl and Branstator 1992; Otto-Bliesner 1999). Unstable teleconnection patterns complicate both palaeo-climate reconstruction of ENSO (from regions outside the tropical Pacific) and the prediction of future ENSO impacts.

What does the history of ENSO suggest about its behaviour in the future? ENSO has responded to past changes in external forcings and background conditions; in addition, ENSO has substantial internal variability. Furthermore, ENSO teleconnections vary through time, certainly in response to changes in background conditions and perhaps independently of such variability. Current climate models can capture basic elements of modern ENSO, but the diversity of behaviour seen in the palaeo-climate record provides a fresh and compelling set of challenges to climate prediction. Models need to simulate with skill both internally and externally generated ENSO variability, and the necessary global background conditions, to predict ENSO behaviour in the coming century. Amidst the uncertainties, one conclusion is unavoidable: ENSO will change.

also have reverberations around the planet. There is good evidence, based on the Grip and Byrd ice core data correlated using methane (Blunier et al. 1998), that a bipolar see-saw ocean circulation system operates in the Atlantic. The shutdown of the deep-water formation in the north would lead to the delivery of additional heat to the high latitudes of the southern hemisphere. Sufficient additional heat could result in the surging, melting and then eventual collapse of the West Antarctic Ice Sheet. The resulting sea level rise would be catastrophic, inundating low-lying islands in the Indian and Pacific Oceans and flooding many vulnerable coastal regions, such as much of Bangladesh.

Such a chain of events is not implausible. Each of the individual events has happened in the past. The point is that it is highly unlikely that abrupt changes will happen in isolation. The connectivity of the Earth System means that reverberations will occur not only at local and regional levels, but at the global scale as well, with little chance at present of being able to predict the paths of these reverberations and their ultimate consequences for human societies.

5.4.3 Human Perceptions of Global Change

Virtually all of the discussions on the consequences of global change for human well-being focus on the material and physical aspects of such change – provision of food and water, security of infrastructure, implications for human health, and so on. Virtually no analyses consider the psychological impacts or consequences of global change on individual humans and on their societies. Many in the scientific community may consider these aspects to be irrational and inconsequential. Yet, in the final analysis, it will be the human *perceptions* of global change and the risks associated with it that will determine societal responses.

The perception of global change itself is strongly rooted in fundamental, almost religious, perceptions of the place of humanity in the natural world. Ironically, both of two major perspectives are based on the understanding that humans are an *integral part* of the natural world, of the Earth System. One perspective holds that the Earth System is so vast and robust that human activities cannot possibly influence it in any significant or long-term way (e.g., anthropogenic CO_2 increase will cause only an inconsequential change to the more pronounced natural rhythms of climate), or that human ingenuity will solve any serious problems before they arise. The other perspective holds that there are certain boundaries or thresholds of the Earth's natural dynamics – the Achilles heels of the Earth System – that humanity cannot approach or cross without causing significant and perhaps irreversible change to Earth System functioning with potentially catastrophic consequences for human societies. Similarly, perceptions of the *consequences* of global change, some of which are unfolding even now, are causing growing concerns in many people. At the heart of these concerns is the fundamental place of humanity in the natural world. The following three examples illustrate the point.

Snow at the equator is one of the geographical features of Earth that has fascinated humans for centuries. In the western world the snows of Kilimanjaro have been popularised by the writing of Ernest Hemingway. Yet in just a few generations the summit glaciers and snowfields may disappear, probably due to anthropogenic climate change (Fig. 5.42). At present rates, Kilimanjaro will be snowfree by 2015 or 2020, within the lifetime of many people already on the planet. It may be possible to calculate the economic consequences in terms of loss of tourist income to East African countries, but by far the most significant consequence will be the psychological one, the fact that through a myriad of daily activities far removed from tropical Africa humanity can so influence the global environment that such striking changes can occur so rapidly.

By almost any analysis, the economy of Switzerland will be very little affected by global change, or could even benefit from it. However, the Swiss Alps will likely be strongly affected as most of the glaciers will either retreat significantly or disappear altogether. Indeed, many glaciers are already experiencing rapid rates of retreat. The importance of Switzerland's mountainous landscapes for the psyche of its people cannot be overestimated, and thus the effects of global change on the perception of the Swiss population of their homeland may be profound. It will benefit the Swiss people little if global change improves the Swiss economy but destroys the soul of the nation (C. Jaeger, pers. comm.).

The loss of biological diversity is an aspect of global change that may prove to be just as important as climate change. The consequences of the homogenisation of Earth's biological fabric on the functioning of ecosystems, and ultimately on the functioning of the Earth System as a whole, is a critical scientific question (Loreau et al. 2001). Yet the human perception that significant disruption of Earth's complex webs of life is an inherently dangerous pathway may be more important than scientific knowledge in dealing with the issue. In southern China bees have disappeared from a region where the production of fruit is an important economic activity. The local population has responded by pollinating the trees themselves (Fig. 5.43). The economic losses of having to pollinate fruit trees manually can be calculated, but the image of humans climbing trees to pollinate flowers penetrates to the fundamental nature of the problem far more effectively.

Fig. 5.42. Mount Kilimanjaro with *inset* showing rate of glacier loss over the last century with the projected disappearance of all ice around 2020 (Alverson et al. 2001)

Fig. 5.43. In Maoxian county, near the border between China and Nepal, people pollinate apple trees by hand because the bees which were pollinating these trees have become extinct. It takes 20–25 people to perform the work of two bee colonies. (*photos:* ICIMOD)

These and other examples demonstrate that any complete analysis of the consequences of global change must go well beyond scientific and economic considerations to fundamental moral and ethical issues (Bradley 2002). It must cut to the core of human-environment relationships, and to the perceptions of the place of humanity in the natural world. Human civilisations around the world and all that they entail – accumulated wisdom, literature, art and music handed down over generations – have developed in the context of a naturally varying global environment, a planetary life support system with a pace and patterns of change within which civilisations could develop, grow and prosper. Now, however, human activities are changing the natural rhythms of Earth in ways that are only partly understood and onto trajectories that may lead to unknown states of the planetary environment, some possibly much less hospitable for human and other forms of life. After all of the scientific syntheses and assessments are done, after all of the economic analyses of global change consequences are laid out in cost-benefit terms, after all of the debates between sceptics and believers have generated even more heat, one basic ethical question lies at the heart of the global change debate. How should the generations of humans now inhabiting Earth respond to the fact that their activities are creating risks, possibly very big risks, for the future of Earth's life support system?

References

Alcamo J, Leemans R, Kreileman E (1998) Global change scenarios of the 21st Century: Results from the IMAGE 2.1 model. Elsevier Science Ltd., Oxford, UK

Alexandratos N (ed) (1995) World agriculture: Towards 2010, an FAO study. Food and Agriculture Organization of the United Nations, Rome and John Wiley & Sons Ltd., Chichester, UK

Allen MR, Ingram WJ (2002) Constraints on future changes in climate and the hydrological cycle. Nature 419:224–232

Alley RB, Anandakrishnan S, Jung P (2001) Stochastic resonance in the North Atlantic. Paleoceanography 16:190–198

Alley RB, Marotzke J, Nordhaus WD, Overpeck JT, Peteet DM, Pielke RA Jr, Pierrehumbert RT, Rhines PB, Stocker TF, Talley LD, Wallace JM (2003) Abrupt climate change. Science 299:2005–2010

Alverson K, Bradley R, Briffa K, Cole J, Hughes M, Larocque I, Pedersen T, Thompson L, Tudhope S (2001) Disappearing evidence: The need for a global paleoclimate observing system. Global Change Newsletter No. 46, International Geosphere Biosphere Programme, Stockholm, Sweden, pp 2–6

Alverson K, Bradley R, Pedersen T (eds) (2003) Paleoclimate, global change and the future. IGBP Global Change Series. Springer-Verlag Berlin Heidelberg New York

Anderson JG, Brune WH, Proffitt MH (1989) Ozone destruction by chlorine radicals within the Antarctic vortex: The spatial and temporal evolution of $ClO\text{-}O_3$ anticorrelation based on in situ ER-2 data. J Geophys Res 94:11465–11479

Andrus CF, Crowe DE, Sandweiss DH, Reitz EJ, Romanek CS (2002) Otolith $\delta^{18}O$ record of mid-Holocene sea surface temperatures in Peru. Science 295:1508–1511

Arnell NW (1999) Climate change and global water resources. Global Environ Chang 9:S31–S49

Ausubel J (2002) Maglevs and the vision of St. Hubert. In: Steffen W, Jäger J, Carson D, Bradshaw C (eds) Challenges of a changing earth: Proceedings of the Global Change Open Science Conference. Amsterdam, The Netherlands, 10–13 July 2001. Springer-Verlag Berlin Heidelberg New York, pp 175–182

Barnston AG, Glantz MH, He Y (1999) Predictive skill of statistical and dynamical climate models in SST forecasts during the 1997–98 El Niño episode and the 1998 La Niña onset. B Am Meteorol Soc 80:217–243

Blunier TJ, Chappellaz J, Schwander J, Dälenbach A, Stauffer B, Stocker TF, Raynaud D, Jouzel J, Clausen HB, Hammer CU, Johnsen SJ (1998) Asynchrony of Antarctic and Greenland climate change during the last glacial period. Nature 394:739–743

Bond G, Showers W, Elliot M, Evans M, Lotti R, Hajdas I, Bonani G, Johnsen SJ (1999) The North Atlantic's 1–2 kyr climate rhythm: Relation to Heinrich events, Dansgaard/Oeschger cycles and the Little Ice Age. In: Clark PU, Webb RS, Keigwin LD (eds) Mechanisms of global climate change at millenial time scales. American Geophysical Union, Washington, USA, pp 35–58

Bork H-R, Bork H, Dalchow C, Faust B, Piorr H-P, Schatz Th (1998) Landschaftsentwicklung in Mitteleuropa. Wirkungen des Menschen auf Landschaften. Klett-Perthes, Gotha

Bradley RS (2002) Climate change-past, present and future: A personal perspective. In: Steffen W, Jäger J, Carson D, Bradshaw C (eds) Challenges of a changing earth: Proceedings of the Global Change Open Science Conference. Amsterdam, The Netherlands, 10–13 July 2001. IGBP Global Change Series. Springer-Verlag Berlin Heidelberg New York, pp 109–112

Brasseur GP, Artaxo P, Barrie LA, Delmas RJ, Galbally I, Hao WM, Harriss RC, Isaksen ISA, Jacob DJ, Kolb CE, Prather M, Rodhe H, Schwela D, Steffen W, Wuebbles DJ (2003) An integrated view of the causes and impacts of atmospheric changes. In: Brasseur GP, Prinn RG, Pszenny AAP (eds) The changing atmosphere: An integration and synthesis of a decade of tropospheric chemistry research. IGBP Global Change Series. Springer-Verlag Berlin Heidelberg New York, pp 207–230

Broecker WS (1987) Unpleasant surprises in the greenhouse? Nature 328:123–126

Broecker WS (1997) Thermohaline circulation, the Achilles heel of our climate system: Will man-made CO_2 upset the current balance? Science 278:1582–1588
Broecker WS (1998) Paleocean circulation during the last deglaciation: A bipolar seesaw? Paleoceanography 13:119–121
Broecker WS, Peteet DM, Rind D (1985) Does the ocean-atmosphere system have more than one stable mode of operation? Nature 315:21–25
Brown L, Kane H (1994) Full house: Reassessing the Earth's population carrying capacity. WW Norton, New York
Bruckner Th, Petschel-Held G, Tóth FL, Füssel H-M, Helm C, Leimbach M (1999) Climate change decision-support and the tolerable windows approach. Environ Model Assess 4:217–234
Bryant D, Burke L, McManus J, Spalding M (1998) Reefs at risk: a map-based indicator of threats to the world's coral reefs. World Resources Institute, Washington DC
Bryden HL, Imawaki S (2001) Ocean heat transport. In: Siedler G, Church J, Gould J (eds) Ocean circulation and climate: Observing and modelling the global ocean. Academic Press, New York, pp 455–474
Burgess RL, Jacobson L (1984) Archaeological sediments from a shell midden near Wortel Dam, Walvis Bay. Palaeoecol Afr 16:429–435
Carpenter SR, Bolgrien D, Lathrop RC, Stow CA, Reed T, Wilson MA (1998) Ecological and economic analysis of lake eutrophication by nonpoint pollution. Aust J Ecol 23:68–79
Carter TR, Hulme M, Viner D (eds) (1999) Representing uncertainty in climate change scenarios and impact studies. Proceedings of ECLAT-2 Helsinki Workshop. Climatic Research Unit, Norwich, UK, pp 38–53
Chameides WL, Kasibhatla PS, Yienger J, Levy II H (1994) Growth of continental-scale metro-agro-plexes, regional ozone pollution, and world food production. Science 264:74–77
Chameides WL, Yu H, Liu SC, Bergin M, Zhou X, Mearns L, Wang, G, Kiang CS, Saylor RD, Luo C, Steiner A, Giorgi F (1999) Case study of the effects of atmospheric aerosols and regional haze on agriculture: An opportunity to enhance crop yields in China through emission controls. P Natl Acad Sci USA 96: 13626–13633
Chiew FHS, Whetton PH, McMahon TA, Pittock AB (1995) Simulation of the impacts of climate change on runoff and soil moisture in Australian catchments. J Hydrol 167:121–147
Clark PU, Marshall SJ, Clarke GKC, Hostetler SW, Licciardi JM, Teller JT (2001) Freshwater forcing of abrupt climate change during the last glaciation. Science 293:283–287
Clark PU, Pisias NG, Stocker TF, Weaver AJ (2002) The role of the thermohaline circulation in abrupt climate change. Nature 415:863–869
Clement AC, Seager R, Cane MA (1999) Orbital controls on the El Niño/Southern Oscillation and tropical climate. Paleoceanography 14:441–456
Clement AC, Seager R, Cane MA (2000) Suppression of El Niño during the mid-Holocene by changes in the Earth's orbit. Paleoceanography 15:731–737
Cobb KM, Charles CD, Hunter DE (2001) A central tropical Pacific coral demonstrates Pacific, Indian, and Atlantic decadal climate connections. Geophys Res Lett 28:2209–2212
Cobb KM, Charles CD, Cheng H, Edwards RL (2003) El Niño/Southern Oscillation and tropical Pacific climate during the last millennium. Nature 424:271–276
Cole JE (2001) A slow dance for El Niño. Science 291:1496–1497
Cole JE, Cook ER (1998) The changing relationship between ENSO variability and moisture balance in the continental United States. Geophys Res Lett 25:4529–4532
Cole JE, Dunbar RB, McClanahan TR, Muthiga NA (2000) Tropical Pacific forcing of decadal SST variability in the western Indian Ocean over the past two centuries. Science 287:617–619
Cole JE, Overpeck JT, Cook ER (2002) Multiyear La Niñas and prolonged US drought. Geophys Res Lett 29:10.1029/2001GL013561, 25-1-25-4
Comfort L, Wisner B, Cutter S, Pulwarty R, Hewitt K, Oliver-Smith A, Wiener J, Fordham M, Peacock W, Krimgold F (1999) Reframing disaster policy: The global evolution of vulnerable communities. Env Hazards 1:39–44
Cook R (2003) The magnitude and impact of by-catch mortality by fishing gear. In: Sinclair M, Valimarsson G (eds) Responsible fisheries in the marine ecosystem. FAO and CABI Publishing, Rome and Oxford, pp 219–233
Cook ER, Meko DM, Stahle DW, Cleaveland MK (1999) Drought reconstructions for the continental United States. J Climate 12:1145–1162
Cooper CM, Lipe WM (1992) Water quality and agriculture: Mississippi experiences. J Soil Water Conserv 47:220–223
Costanza R, d'Arge R, de Groot R, Farber S, Grasso M, Hannon B, Limburg K, Naeem S, O'Neill R, Paruelo J, Raskin R, Sutton P, van den Belt M (1997) The value of the world's ecosystem services and natural capital. Nature 387:253–260
Cox PM, Betts RA, Jones CD, Spall SA, Totterdell IJ (2000) Acceleration of global warming due to carbon-cycle feedbacks in a coupled model. Nature 408:184–187
Cramer W, Bondeau A, Woodward FI, Prentice IC, Betts RA, Brovkin V, Cox PM, Fisher V, Foley JA, Friend AD, Kucharik C, Lomas MR, Ramankutty N, Sitch S, Smith B, White A, Young-Molling C (2001) Global response of terrestrial ecosystem structure and function to CO_2 and climate change: Results from six dynamic global vegetation models. Global Change Biol 7:357–373
Crosson P, Anderson JR (1994) Demand and supply: Trends in global agriculture. Food Policy 19:105–119
Crutzen P (1995) My life with O_3, NO_x and other YZO_xs. Les Prix Nobel (The Nobel Prizes) 1995. Stockholm: Almqvist & Wiksell International, pp 123–157
Cubasch U, Meehl GA, Boer GJ, Stouffer RJ, Dix M, Noda A, Raper S, Senior CA, Yap KS (2001) Projections of future climate change. In: Houghton JT, Ding Y, Griggs DJ, Noguer M, van der Linden PJ, Dai X, Maskell K, Johnson CA (eds) Climate change 2001: The scientific basis. Contribution of Working Group I to the Third Assessment Report of the Intergovernmental Panel on Climate Change. Cambridge University Press, Cambridge New York
Cullen HM, deMenocal PB, Hemming S, Hemming G, Brown FH, Guilderson T, Sirocko F (2000) Climate change and the collapse of the Akkadian empire: Evidence from the deep sea. Geology 28:379–382
Delworth TL, Dixon KW (2000) Implications of the recent trend in the Arctic/north Atlantic Oscillation for the North Atlantic thermohaline circulation. J Climate 13:3721–3727
Dentener F, Peters W, Krol M, van Weele M, Bergamaschi P, Lelieveld J (2003) Inter-annual variability and trend of CH_4 lifetime as a measure for OH changes in the 1979–1993 time period. J Geophys Res, in press
Dixon KW, Delworth TL, Spelman MJ, Stouffer RJ (1999) The influence of transient surface fluxes on North Atlantic overturning in a coupled GCM climate change experiment. Geophys Res Lett 26:2749–2752
Dyson T (1996) Population and food: Global trends and future prospects. New York, Routledge
Everett J (1995) Fisheries. In: Watson RT, Zinyowera MC, Moss RH, Dokken DJ (eds) Climate change 1995. Impacts, adaptations and mitigation of climate change: Scientific- technical analyses. Contribution of Working Group II to the Second Assessment Report of the Intergovernmental Panel on Climate Change, pp 511–538
FAO (1994) A global assessment of fisheries by-catch and discards. Prepared by Alverson DI, Freeberg MH, Murawski SA, Pope JG. Food and Agriculture Organization of the United Nations, Rome. FAO Fisheries Technical Paper No. 339
FAO (2000) The state of world fisheries and aquaculture. Food and Agricultural Organization of the United Nations, Rome. *http://www.fao.org/sof/sofia/index_en.html*, 30 April 2002
FAOSTAT (2002) Statistical Database. Food and Agriculture Organization of the United Nations, Rome. *http://www.apps.fao.org*, 12 August 2002
Farman JC, Gardiner BG, Shanklin JD (1985) Large losses of total ozone in Antarctica reveal seasonal interaction. Nature 315: 207–210
Field CB, Daily GC, Davis FW, Gaines S, Matson PA, Melack J, Miller NL (1999) Confronting climate change in California: Ecological impacts on the Golden State. Union of Concerned Scientists, Cambridge, MA and Ecological Society of America, Washington, DC, USA

Fischer G, Shah M, van Velthuizen H, Nachtergaele F (eds) (2001) Global agro-ecological assessment for agriculture in the 21st Century. Summary Report of the IIASA Land Use Project, IIASA, Laxenburg, Austria

Fogarty MJ (2002) Climate variability and ocean ecosystem dynamics: Implications for sustainability. In: Steffen W, Jäger J, Carson D, Bradshaw C (eds) Challenges of a changing earth: Proceedings of the Global Change Open Science Conference. Amsterdam, The Netherlands, 10–13 July 2001. IGBP Global Change Series. Springer-Verlag Berlin Heidelberg New York, pp 27–29

Folke C, Carpenter S, Elmqvist T, Gunderson L, Holling CS, Walker B, Bengtsson J, Berkes F, Colding J, Danell K, Falkenmark M, Gordon L, Kasperson R, Kautsky N, Kinzig A, Levin S, Mäler K-G, Moberg F, Ohlsson L, Olsson P, Ostrom E, Reid W, Rockström J, Savenije H, Svedin U (2002) Resilience and sustainable development: Building adaptive capacity in a world of transformations. ICSU Series on Science for Sustainable Development, No. 3. Scientific background paper commissioned by the Enviromental Advisory Council of the Swedish Government in preparation for WSSD. Paris, International Council for Science (ICSU). *http://www.icsu.org/Library/WSSD-Rep/Vol3.pdf*

Fontugne M, Usselmann P, Lavallee D, Julien M, Hatte C (1999) El Niño variability in the coastal desert of southern Peru during the mid-Holocene. Quaternary Res 52:171–179

Friedlingstein P, Bopp L, Ciais P, Dufresne J-L, Fairhead L, LeTreut H, Monfray P, Orr J (2001) Positive feedback between future climate change and the carbon cycle. Note du Pole de Modelisation, Geophys Res Lett 28:1543–1546

Gagan MK, Ayliffe LK, Hopley D, Cali JA, Mortimer GE, Chappell J, McCulloch MT, Head MJ (1998) Temperature and surface-ocean water balance of the mid-Holocene tropical western Pacific. Science 279:1014–1018

Ganopolski A, Rahmstorf S (2001) Rapid changes of glacial climate simulated in a coupled climate model. Nature 409: 153–158

Ganachaud A, Wunsch C (2000) Improved estimates of global ocean circulation, heat transport and mixing from hydrographic data. Nature 408:453–457

Gitay H, Brown S, Easterling W, Jallow B (2001) Ecosystems and their goods and services. In: McCarthy JJ, Canziani OF, Leary NA, Dokken DJ, White KS (eds) Climate change 2001: Impacts, adaptation and vulnerability. Contribution of Working Group II to the Third Assessment Report of the Intergovernmental Panel on Climate Change. Cambridge University Press, Cambridge and New York

Goldenberg SB, Landsea CW, Mestas-Nunez AM, Gray WM (2002) The recent increase in Atlantic hurricane activity: Causes and implications. Science 293:471–474

Goodale CL, Apps MJ, Birdsey RA, Field CB, Heath LS, Houghton RA, Jenkins JC, Kohlmaier GH, Kurz W, Liu S, Nabuurs G-J, Nilsson S, Shvidenko AZ (2002) Forest carbon sinks in the northern hemisphere. Ecol Appl 12:891–899

Gregory PJ, Ingram JSI, Campbell B, Goudriaan J, Hunt LA, Landsberg JJ, Linder S, Stafford Smith M, Sutherst RW, Valentin C (1999) Managed production systems. In: Walker B, Steffen W, Canadell J, Ingram J (eds) Global change and the terrestrial biosphere: Implications for natural and managed ecosystems. A synthesis of GCTE and related research. IGBP Book Series No. 4, Cambridge University Press, Cambridge, pp 229–270

Hall M (1984) Prehistoric farming in the Mfolozi and Hluhluwe valleys of southeast Africa: An archaeobotanical survey. J Archaeol Sci 11:223–235

Hamilton JG, DeLucia EH, George K, Naidu SL, Finzi AC, Schlesinger WH (2002) Forest carbon balance under elevated CO_2. Oecologia 131:250–260

Hansen B, Turrell WR, Osterhus S (2001) Decreasing overflow from the Nordic seas into the Atlantic Ocean through the Faroe Bank channel since 1950. Nature 411:927–930

Heil A, Goldammer JG (2001) Smoke-haze pollution: A review of the 1997 episode in Southeast Asia. Regional Environmental Change 2:24–37

Hemming SR (2003) Heinrich events: Massive Late Pleistocene detritus layers of the North Atlantic and their global impact. Rev Geophys, in press

Hodell DA, Curtis JH, Brenner M (1995) Possible role of climate in the collapse of Classic Maya civilization. Nature 375:391–394

Hodell DA, Brenner M, Curtis JH, Guilderson T (2001) Solar forcing of drought frequency in the Mayan lowlands. Science 292:1367–1370

Hofer T, Messerli B (1997) Floods in Bangladesh. Process understanding and development strategies. A synthesis paper for SDC and UNU. Institute of Geography, University of Bern, Switzerland

Hofmann DJ, Harder JW, Rosen JM, Hereford JV, Carpenter JR (1989) Ozone profile measurements at McMurdo Station, Antarctica, during the spring of 1987. J Geophys Res 94:16527–16536

Holmgren K, Karlen W, Lauritzen SE, Lee-Thorp JA, Partridge TC, Piketh S, Repinski P, Stevenson C, Svanered O, Tyson PD (1999) A 3000-year high-resolution stalagmite-based record of palaeoclimate for north-eastern South Africa. Holocene 9:295–309

Huffman TN (1989) Iron age migrations. Johannesburg: Witwatersrand University Press

Huffman TN (1996) Archaeological evidence for climatic change during the last 2 000 years in southern Africa. Quatern Int 33: 55–60

Huffman TN, Herbert RK (1994) New perspectives on Eastern Bantu. Azania 29–30:1–10

Hulme M (ed) (1996) Climate change and southern Africa. Climatic Research Unit, University of East Anglia, Norwich, UK

Hulme M, Barrow EM, Arnell NW, Harrison PA, Johns TC, Downing TE (1999) Relative impacts of human-induced climate change and natural climate variability. Nature 397:688–691

ICLARM (The World Fish Center) (1999) Aquatic resources research in developing countries: Data and evaluation by region and resource system. Supplement to the ICLARM Strategic Plan 2000–2020, ICLARM working document No. 4 on CD

Iliffe J (1995) Africans: The history of a continent. Cambridge University Press, Cambridge

IMAGE Team (2001) The IMAGE 2.2 implementation of the SRES scenarios: A comprehensive analysis of emissions, climate change and impacts in the 21st century. RIVM CD-ROM Publication 481508018, National Institute of Public Health and the Environment, Bilthoven, Netherlands

IPCC (2001a) Climate change 2001:The scientific basis. Contribution of Working Group I to the Third Assessment Report of the Intergovernmental Panel on Climate Change. Houghton JT, Ding Y, Griggs DJ, Noguer M, van der Linden PJ, Dai X, Maskell K,Johnson CA (eds) Cambridge University Press, Cambridge and New York

IPCC (2001b) Climate change 2001: Impacts, adaptation and vulnerability. Contribution of Working Group II to the Third Assessment Report of the Intergovernmental Panel on Climate Change. McCarthy JJ, Canziani OF, Leary NA, Dokken DJ, White KS (eds) Cambridge University Press, Cambridge and New York

IPCC (2001c) Climate change 2001: Mitigation. Contribution of Working Group III to the Third Assessment Report of the Intergovernmental Panel on Climate Change. Metz B, Davidson O, Swart R, Pan J (eds) Cambridge University Press, Cambridge and New York

IPCC (2001d) Climate change 2001: Synthesis report. Based on draft prepared by Watson RT, Albritton DL, Barker T, Bashmakov IA, Canziani O, Christ R, Cubasch U, Davidson O, Gitay H, Griggs D, Houghton J, House J, Kundzewicz Z, Lal M, Leary N, Magadza C, McCarthy JJ, Mitchell FB, Moreira JR, Munasinghe M, Noble I, Pachauri R, Pittock B, Prather M, Richels RG, Robinson JB, Sathaye J, Schneider S, Scholes R, Stocker T, Sundararaman N, Swart R, Taniguchi T, Zhou D and many IPCC authors and reviewers. Cambridge University Press, Cambridge and New York

Jacob DJ, Heikes BG, Fan S-M, Logan JA, Mauzerall DL, Bradshaw JD, Singh HB, Gregory GL, Talbot RW, Blake DR, Sachse GW (1996) Origin of ozone and NO_x in the tropical troposphere: A photochemical analysis of aircraft observations over the South Atlantic basin. J Geophys Res 101:24235–24250

Joubert A (1995) Simulations of Southern African climate by early-generation general circulation models. S Afr J Sci 91:85–91

Joubert AM, Kohler MO (1996) Projected temperature increases over Southern Africa due to increasing levels of greenhouse gases and sulphate aerosols. S Afr J Sci 92:524–526

Kabat P, Claussen M, Dirmeyer PA, Gash JHC, de Guenni LB, Meybeck M, Pielke Sr. RA, Vörösmarty C, Hutjes RWA, Luetkemeier S (eds) (2004) Vegetation, water, humans and the climate: A new perspective on an interactive system. IGBP Global Change Series. Springer-Verlag Berlin Heidelberg New York

Kasperson JX, Kasperson RE (2001) Summary of international workshop on Vulnerability and Global Environmental Change, 17–19 May 2001, Stockholm Environment Institute (SEI), Stockholm, Sweden. SEI Risk and Vulnerability Programme Report 2001-01. Stockholm: Stockholm Environment Institute. *http://www.sei.se/risk/workshop.html*

Kasperson JX, Kasperson RE, Turner BL II (eds) (1995) Regions at risk: Comparisons of threatened environments. United Nations University, Tokyo

Keefer DK, deFrance SD, Moseley ME, Richardson III JB, Satterlee DR, Day-Lewis A (1998) Early maritime economy and El Niño events at Quebrada Tacahuay, Peru. Science 281:1833–1835

Kienzle SW, Lorentz SA, Schulze RE (1997) Hydrology and water quality of the Mgeni Catchment. Water Research Commission, Pretoria, South Africa

Knutti R, Stocker TF (2002) Limited predictability of the future thermohaline circulation close to an instability threshold. J Climate 15:179–186

Koutavas A, Lynch-Stieglitz J, Marchitto TM, Sachs JP (2002) El Niño-like pattern in ice age tropical Pacific sea surface temperature. Science 297:226–230

Krol M, van Leeuwen P-J, Lelieveld J (1998) Global OH trend inferred from methylchloroform measurements. J Geophys Res 103:10697–10711

Lang A, Preston N, Dickau R, Bork H-R, Maeckel R (2000) Land use and climate impacts on fluvial systems during the period of agriculture: Examples from the Rhine catchment. PAGES (Past Global Changes) Newsletter 8(3):11–13

Landsteiner E (1999) The crisis of wine production in the late sixteenth century in central Europe. Climatic Change 43:323–334

Langeweg F, Gutierrez-Espeleta EE (2001) Human security and vulnerability in a scenario context: Challenges for UNEP's global environmental outlook. IHDP Update: Newsletter of the International Human Dimensions Programme on Global Environmental Change, No. 2/2001, pp 11–12

Latif M, Roeckner E, Mikolajewicz U, Voss R (2000) Tropical stabilization of the thermohaline circulation in a greenhouse warming simulation. J Climate 13:1809–1813

Leemans R, Eickhout BJ, Strengers B, Bouwman AF, Schaeffer M (2003) The consequences of uncertainties in land use, climate and vegetation responses on terrestrial carbon. Sci China Ser C 45:126–141

Lindgren E (2000) The new environmental context for disease transmission, with case studies on climate change and tick-borne encephalitis. PhD Thesis. Natural Resources Management, Department of Systems Ecology, Stockholm University, Sweden

Liu Z, Kutzbach JE, Wu L (2000) Modeling climate shift of El Niño variability in the Holocene. Geophys Res Lett 27:2265–2268

Lomborg B (2001) The skeptical environmentalist: Measuring the real state of the world. Cambridge University Press, Cambridge

Loreau M, Naeem S, Inchausti P, Bengtsson J, Grime JP, Hector A, Hooper DU, Huston MA, Raffaelli D, Schmid B, Tilman D, Wardle DA (2001) Biodiversity and ecosystem functioning: Current knowledge and future challenges. Science 294:804–808

Maggs R, Wahid A, Shamshi SRA, Ashmore MR (1995) Effects of ambient air pollution on wheat and rice yield in Pakistan. Water Air Soil Poll 85:1311–1316

Maggs T (1984) The Iron Age south of the Zambezi. In: Klein RG (ed) Southern African prehistory and palaeoenvironments. AA Balkema, Rotterdam, pp 329–360

Manabe S, Stouffer RJ (1993) Century-scale effects of increased atmospheric CO_2 on the ocean-atmosphere system. Nature 364:215–218

Mann ME, Bradley RS, Hughes MK (1999) Northern hemisphere temperatures during the past millennium: inferences, uncertainties, and limitations. Geophys Res Lett 26:759–762

Matson PA, Parton WJ, Power AG, Swift MJ (1997) Agricultural intensification and ecosystem properties. Science 277:504–509

McCartney MP, Acreman MC (2003) Wetlands and water resources. In: Maltby E (ed) The wetlands handbook. Blackwells, Oxford, UK, in press

McMichael A, Githeko A, Akhtar R, Carcavallo R, Gubler D, Haines, A, Kovats RS, Martens P, Patz J, Sasaki A, Ebi KL, Focks D, Kalkstein L, Lindgren E, Lindsay S, Sturrock R, Confalonieri U, Woodward A (2001) Human health. In: Climate change 2001: Impacts, adaptation, and vulnerability. Contribution of Working Group II to the Third Assessment Report of the Intergovernmental Panel on Climate Change. Cambridge University Press, Cambridge, pp 451–485

Meade RH, Parker RS (1985) Sediments in rivers of the United States. In: Natural water summary, 1984. US Geological Survey Water, supply paper 2275, USGS, Reston, USA, pp 40–60

Meehl GA, Branstator GW (1992) Coupled climate model simulation of El Niño/Southern Oscillation: Implications for paleoclimate. In: Diaz H, Markgraf V (eds) El Niño: Historical and paleoclimatic aspects of the Southern Oscillation. Cambridge University Press, Cambridge, pp 69–91

Meigh JR, McKenzie AA, Austin BN, Bradford RB, Reynard NS (1998) Assessment of global water resources – phase II. Estimates for present and future water availability for eastern and southern Africa. DFID Report 98/4. Institute of Hydrology, Wallingford, UK

Messerli B, Grosjean M, Hofer T, Nunez L, Pfister Ch (2000) From nature-dominated to human-dominated environmental changes. Quaternary Sci Rev 19:459–479

Messerli B, Grosjean M, Hofer T, Nunez L, Pfister C (2001) From nature-dominated to human-dominated environmental changes. In: Ehlers E, Krafft T (eds) Understanding the Earth System, compartments, processes and interactions. Springer-Verlag Berlin Heidelberg New York, pp 195–208

Meybeck M (2003) Global analysis of river systems: From Earth System controls to Anthropocene controls. Philos T Roy Soc B, in press

Meybeck M, Helmer R (1992) An introduction to water quality. In: Chapman D (ed) Assessment of the quality of the aquatic environment through water, biota and sediment. D. Chapman & Hall, London, pp 1–17

Meybeck M, Vörösmarty C (1999) Global transfer of carbon by rivers. Global Change Newsletter No. 37, International Geosphere Biosphere Programme, Stockholm, Sweden, pp 12–14

Milly PCD, Wetherald RT, Dunne KA, Delworth TL (2002) Increasing risk of great floods in a changing climate. Nature 415:514–517

Mirza MMQ (2002) Global warming and changes in the probability of occurrence of floods in Bangladesh and implications. Global Environ Chang 12:127–138

Mitchell T, Hulme M (1999) Predicting regional climate change: Living with uncertainty. Prog Phys Geog 23:57–78

Mooney H, Canadell J, Chapin FS, Ehleringer J, Körner Ch, McMurtrie R, Parton W, Pitelka L, Schulze E-D (1999) Ecosystem physiology responses to global change. In: Walker BH, Steffen W, Canadell J, Ingram J (eds) The terrestrial biosphere and global change: Implications for natural and managed ecosystems. IGBP Book Series No. 4, Cambridge University Press, Cambridge, pp 141–189

Moy CM, Seltzer GO, Rodbell DT, Anderson DM (2002)Variability of El Niño/Southern Oscillation activity at millennial timescales during the Holocene epoch. Nature 420:162–165

Murphy JM, Mitchell JBF (1995) Transient response of the Hadley centre coupled ocean-atmosphere model to increasing carbon dioxide. Part 2. Spatial and temporal structure of the response. J Climate 8:57–80

Nakícenovíc N, Alcamo J, Davis G, de Vries B, Fenhann J, Gaffin S, Gregory K, Grübler A, Jung TY, Kram T, la Rovere EE, Michaelis L, Mori S, Morita T, Pepper W, Pitcher H, Price L, Riahi K, Roehrl A, Rogner H-H, Sankovski A, Schlesinger ME, Shukla PR, Smith S, Swart RJ, van Rooyen S, Victor N, Dadi Z (2000) Special report on emissions scenarios. Intergovernmental Panel on Climate Change, Cambridge University Press, Cambridge

Naylor RL, Goldburg RJ, Mooney H, Beveridge M, Clay J, Folke N, Kautsky N, Lubcheno J, Primavera J, Williams M (2000) Nature's subsidies to shrimp and salmon farming. Science 282: 883–884

Nials FL, Gregory DA, Graybill DA (1989) Salt river stream flow and Hohokam irrigation systems. In: Graybill DA, Gregory DA, Nials FL, Gasser R, Miksicek C and Szuter C (eds) The 1982–1992 excavations at Las Colinas: Environment and subsistence 5. Arizona State Museum Archaeological Series, pp 59–78

Nicholls N, Lavery B, Fredericksen C, Drosdowsky W, Torok S (1996) Recent apparent changes in relationships between the El Niño-Southern Oscillation and Australian rainfall and temperature. Geophys Res Lett 23:3357–3360

Nicholls RJ, Hoozemans FMJ, Marchand M (1999) Increasing flood risk and wetland losses due to global sea-level rise: Regional and global analysis. Global Environ Chang 9:69–87

NRC (1991) Toward sustainability: A plan for collaborative research on agriculture and natural resources management. Panel for Collaborative Research Support for AID's Sustainable Agriculture and Natural Resource Management. National Research Council. National Academy Press, Washington, USA

NRC (1992) Pesticides in the diets of infants and children. National Research Council. National Academy Press, Washington, USA

Nunez L, Grosjean M, Cartajena I (2002) Human occupations and climate change in the Puna de Atacama, Chile. Science 298: 821–824

Oldfield F, Dearing JA (2001) Biome 300 and HITE: A PAGES perspective. Land Use and Land Cover Change (LUCC) Newsletter 7:7–8

Oldfield F, Dearing JA (2003) The role of human activities in past environmental change. In: Alverson K, Bradley R, Pedersen T (eds) Paleoclimate, global change and the future. IGBP Global Change Series. Springer-Verlag Berlin Heidelberg New York, pp 143–162

Olesen US (1999) Agriculture in Denmark, a source of nutrient pollution of the water environment. In: van de Kraats JA (ed) Farming without harming: The impact of agricultural pollution on water systems. Drukkerij Belser, Lelystad, Netherlands, pp 33–44

Orr J, Maier-Reimer E, Mikolajewicz U, Monfray P, Sarmiento JL, Toggweiler JR, Taylor NK, Palmer J, Gruber N, Sabine CL, Le Quéré C, Key RM, Boutin J (2001) Estimates of anthropogenic carbon uptake from four 3-D global ocean models. Global Biogeochem Cy 15:43–60

Otto-Bliesner B (1999) El Niño/La Niña and Sahel precipitation during the middle Holocene. Geophys Res Lett 26:87–90

Palmer T, Räisänen J (2002) Quantifying the risk of extreme seasonal precipitation events in a changing climate. Nature 415:512–514

Peterson BJ, Holmes RM, McClelland JW, Vörösmarty CJ, Lammers RB, Shiklomanov AI, Shiklomanov IA, Rahmstorf S (2002) Increasing river discharge to the Arctic Ocean. Science 298:2171–2173

Pielke RA Sr, de Guenni LB (eds) (2004) How to evaluate vulnerability in changing environmental conditions? In: Kabat P, Claussen M, Dirmeyer PA, Gash JHC, de Guenni LB, Meybeck M, Pielke Sr. RA, Vörösmarty C, Hutjes RWA, Luetkemeier S (eds) Vegetation, water, humans and the climate: A new perspective on an interactive system. IGBP Global Change Series. Springer-Verlag Berlin Heidelberg New York

Pinstrup-Andersen P, Pandaya-Lorch R, Rosengrant MW (1999) World food prospects: Critical issues for a twenty-first century. International Food Policy Research Institute, Washington, USA

Pittock AB, Allen RJ, Hennessy KJ, McInnes KL, Suppiah R, Walsh KJ, Whetton PH (1999) Climate change, climate hazards and policy responses in Australia. In: Downing TE, Oltshoorn AA, Tol RSL (eds) Climate, change and risk. Routledge, London, pp 19–59

Postel S (1993) Facing water scarcity. In: State of the world, 1993. A Worldwatch Institute Report, W. W. Norton & Co., New York and London

Postel S, Daily GC, Ehrlich P (1996) Human appropriation of renewable fresh water. Science 271:785–788

Prentice IC, Farquhar GD, Fasham MJR, Goulden ML, Heimann M, Jaramillo VJ, Khesghi HS, Le Quéré C, Scholes RJ, Wallace DWR (2001) The carbon cycle and atmospheric carbon dioxide. In: Houghton JT, Ding Y, Griggs DJ, Noguer M, van der Linden PJ, Dai X, Maskell K, Johnson CA (eds) Climate change 2001: The scientific basis. Contribution of Working Group I to the Third Assessment Report of the Intergovernmental Panel on Climate Change. Cambridge University Press, Cambridge and New York

Prinn RG, Weiss RF, Miller BR, Huang J, Alyea FN, Cunnold DM, Fraser PJ, Hartley DE, Simmonds PG (1995) Atmospheric trends and lifetimes of CH_3CCl_3 and global OH concentrations. Science 269:187–192

Prinn RG, Huang J, Weiss RF, Cunnold DM, Fraser PJ, Simmonds PG, McCulloch A, Harth C, Salameh P, O'Doherty S, Wang RHJ, Porter L, Miller BR (2001) Evidence for substantial variations of atmospheric hydroxyl radicals in the past two decades. Science 292:1882–1887

Rahmstorf S (2002) Ocean circulation and climate during the past 120,000 years. Nature 419:207–214

Rahmstorf S (2003) Thermohaline circulation: The current climate. Nature 421:699

Rahmstorf S, Ganopolski A (1999) Long-term global warming scenarios computed with an efficient coupled climate model. Climatic Change 43:353–367

Reynard NS, Prudhomme C, Crooks SM (1998) The potential impacts of climate change on the flood characteristics of a large catchment in the UK. In: Proceedings of the Second International Conference on Climate and Water, Espoo, Finland, August 1998. Helsinki University of Technology, Helsinki, Finland, pp 320–332

Riedinger MA, Steinitz-Kannan M, Last WM, Brenner M (2002) A ~6100 yr record of El Niño activity from the Galapagos Islands. J Paleolimnol 27:1–7

Robertson GP, Paul EA, Harwood RR (2000) Greenhouse gases and intensive agriculture: Contributions of individual gases to radiative warming of the atmosphere. Science 289:1922

Robinson M, Boardman J, Evans R, Heppell K, Packman JC, Leeks GJL (2000) Land use change. In: Acreman MC (ed) The hydrology of the UK. Routledge, London, pp 30–54

Rodbell DT, Seltzer GO, Anderson DM, Abbott MB, Enfield DB, Newman JH (1999) An ~15 000-year record of El Niño-driven alluviation in southwestern Ecuador. Science 283:516–520

Roux DJ, Thirion C, Smidt M, Everett MJ (1994) A procedure for assessing biotic integrity in rivers – application to three river systems flowing through the Kruger National Park, South Africa. Institute for Water Quality Studies Report No. N 0000/00REQ/0894. Pretoria, South Africa

Rühlemann C, Mulitza S, Müller PJ, Wefer G, Zahn R (1999) Warming of the tropical Atlantic Ocean and slow down of thermocline circulation during the last deglaciation. Nature 402:511–514

Russel M, Maltby E (1995) The role of hydrologic regime on phosphorous dynamics in a seasonally waterlogged soil. In: Hughes JMR, Heathwaite AL (eds) Hydrology and hydrochemistry of British Wetlands. John Wiley & Sons Ltd., Chichester, UK, pp 245–260

Ruttan VW (1996) Population growth, environmental change and technical innovation: Implications for sustainable growth in agricultural production. In: Ahlburg DA, Kelly AC, Oppenheim Mason K (eds) The impact of population growth on well-being in developing countries. Springer-Verlag Berlin Heidelberg New York

Sabloff JA (1991) Die Maya. Archaeologie einer Hochkultur. Spectrum der Wissenschaft, Heidelberg, 205, 1st edition. Scientific American Library

Sandweiss DH, Richardson JB, Reitz EJ, Rollins HB, Maasch KA (1996) Geoarcheological evidence from Peru for a 5 000 years B.P. onset of El Niño. Science 273:1531–1533

Schimel DS, House JI, Hubbarde KA, Bousquet P, Ciais P, Peylin P, Braswell BH, Apps MJ, Baker D, Bondeau A, Canadell J, Churkina G, Cramer W, Denning AS, Field CB, Friedlingstein P, Goodale C, Heimann M, Houghton RA, Melillo JM, Moore B III, Murdiyarso D, Noble I, Pacala SW, Prentice IC, Raupach MR, Rayner PJ, Scholes RJ, Steffen WL, Wirth C (2001) Recent patterns and mechanisms of carbon exchange by terrestrial ecosystems. Nature 414:169–172

Schmitz W (1995) On the interbasin-scale thermohaline circulation. Rev Geophys 33:151–173

Schnur R (2002) The investment forecast. Nature 415:483–484

Schultz M, Jacob DJ, Bradshaw JD, Sandholm ST, Dibb JE, Talbot RW, Singh HB (2000) Chemical NO_X budget in the upper troposphere over the tropical South Pacific. J Geophys Res 105: 6669–6679

Schulze RE (1997a) Impacts of global climate change in a hydrologically vulnerable region: Challenges to South African hydrologists. Prog Phys Geog 21:113–136
Schulze RE (1997b) South African atlas of agrohydrology and climatology. WRC Report TT82/96. Water Research Commission, Pretoria, South Africa
Schulze RE (2000) Modelling hydrological responses to landuse and climatic change: A southern African perspective. Ambio 29:12–22
Schulze RE, Perks LA (2000) Assessment of the impact of climate change on hydrology and water resources in South Africa. Report to the South African Country Studies for Climate Change Programme, ACRUcons Report 33, School of Bioresources Engineering and Environmental Hydrology, University of Natal, Pietermaritzburg
Shah M (2002) Food in the 21st century: Global climate of disparities. In: Steffen W, Jäger J, Carson D, Bradshaw C (eds) Challenges of a changing earth: Proceedings of the Global Change Open Science Conference. Amsterdam, Netherlands, 10–13 July 2001. IGBP Global Change Series. Springer-Verlag Berlin Heidelberg New York, pp 31–38
Sidorchuk A (2000) Past erosion and sedimentation within drainage basins on the Russian plain. PAGES (Past Global Changes) Newsletter 8(3):19
Spivakovsky C, Logan JA, Montzka SA, Balkanski YJ, Foreman-Fowler M, Jones DBA, Horowitz LW, Fusco AC, Brenninkmeijer CAM, Prather MJ, Wofsy SC, McElroy MB (2000) Three-dimensional climatological distribution of tropospheric OH: Update and evaluation. J Geophys Res 105:8931–8980
Stahle DW, Cook ER, Cleaveland MK, Therrell MD, Meko DM, Grissino-Mayer HD, Watson E (2000) Tree-ring data document 16th century megadrought over North America. EOS Trans Am Geophys Union 81:121–125
Stocker TF (1998) The seesaw effect. Science 282:61–62
Stocker TF, Schmitter A (1997) Rate of global warming determines the stability of the ocean-atmosphere system. Nature 388:862–865
Stott L, Poulsen C, Lund S, Thunell R (2002) Super ENSO and global climate oscillations at millennial time scales. Science 297:222–226
Tinker I (1997) Street foods: Urban food and employment in developing countries. Oxford University Press, New York
Trenberth KE, Branstator GW, Karoly D, Kumar A, Lau N-C, Ropelewski CF (1998) Progress during TOGA in understanding and modeling global teleconnections associated with tropical sea surface temperatures. J Geophys Res 103:14291–14324
Tudhope AW, Chilcott CP, McCulloch MT, Cook ER, Chappell J, Ellam RM, Lea DW, Lough JM, Shimmield GB (2001) Variability in the El Niño Southern Oscillation through a glacial-interglacial cycle. Science 291:1511–1517
Turner BL II, Kasperson RE, Matson PA, McCarthy JJ, Corell RW, Christensen L, Eckley N, Kasperson JX, Luers A, Martello ML, Polsky C, Pulsipher A, Schiller A (2003) A framework for vulnerability analysis in sustainability science. P Natl Acad Sci USA, in press
Tyson PD, Lee-Thorp J, Holmgren K, Thackeray JF (2002) Changing gradients of climate change in southern Africa during the past millennium: Their implications for population movements. Climatic Change 52:129–135
UNEP (1998) Environmental effects of ozone depletion: 1998 assessment. United Nations Environment Programme, Nairobi, Kenya
Urban FE, Cole JE, Overpeck JT (2000) Influence of mean climate change on climate variability from a 155-year tropical Pacific coral record. Nature 407:989–993
US Climate Change Assessment (2002) United States national assessment of the potential consequences of climate variability and change: A detailed overview of the consequences of climate change and mechanisms for adaptation. US Global Change Research Program Office. *www.usgcrp.gov/usgcrp/nacc/default.htm*, 30 April 2002
USEPA (1989) Report to Congress: Water quality of the nation's lakes. EPA 440/5-89-003. US Environmental Protection Agency, Washington, USA
Verschuren D, Laird KR, Cumming BF (2000) Rainfall and drought in equatorial east Africa during the past 1100 years. Nature 403:410–414
Vogel CH (1995) People and drought in South Africa: Reaction and mitigation. In: Binns T (ed) People and the environment in Africa. John Wiley & Sons Ltd., Chichester, pp 249–256
Vörösmarty CJ, Meybeck M (2004) Responses of continental aquatic systems at the global scale: New paradigms, new methods. In: Kabat P, Claussen M, Dirmeyer PA, Gash JHC, de Guenni LB, Meybeck M, Pielke RA Sr., Vörösmarty C, Hutjes RWA, Luetkemeier S (eds) Vegetation, water, humans and the climate: A new perspective on an interactive system. IGBP Global Change Series. Springer-Verlag Berlin Heidelberg New York
Vörösmarty CJ, Green P, Salisbury J, Lammers RB (2000) Global water resources: Vulnerability from climate change and population growth. Science 289:284–288
Vörösmarty CJ, Hinzman L, Peterson BJ, Bromwich DL, Hamilton L, Morison J, Romanovsky V, Sturm M, Webb R (2001) The hydrologic cycle and its role in arctic and global environmental change: A rationale and strategy for synthesis study. ARCUS, Fairbanks, USA
Walker BH, Steffen WL, Langridge J (1999) Interactive and integrated effects of global change on terrestrial ecosystems. In: Walker BH, Steffen W, Canadell J, Ingram J (eds) The terrestrial biosphere and global change: Implications for natural and managed ecosystems. IGBP Book Series No. 4, Cambridge University Press, Cambridge, pp 329–375
Wallace JM, Rasmusson EM, Mitchell TP, Kousky VE, Sarachik ES, von Storch H (1998) The structure and evolution of ENSO-related climate variability in the tropical Pacific: Lessons from TOGA. J Geophys Res-Oceans 103(C7):14241–14259
Wang Y, Jacob DJ (1998) Anthropogenic forcing on tropospheric ozone and OH since pre-industrial times. J Geophys Res 103: 31123–31135
Webb DJ, Suginohara N (2001) The interior circulation of the ocean. In: Siedler G, Church J, Gould J (eds) Ocean circulation and climate: Observing and modelling the global ocean. Academic Press, New York, USA, pp 205–214
Webster JB (1980) Drought, migration and chronology in the Lake Malawi littoral. Transafr J Hist 9:70–90
Wilkinson C, Linden O, Cesar H, Hodgson G, Rubens J, Strong AE (1999) Ecological and socio-economic impacts of 1998 coral mortality in the Indian Ocean: An ENSO impact and a warning of future change? Ambio 28:188–196
Woomer PL, Swift MJ (1994) Biological management of tropical soil fertility. John Wiley & Sons Ltd., Chichester, pp 47–80
Zebiak SE, Cane MA (1987) A model El Niño/Southern Oscillation. Mon Weather Rev 115:2262–2278

Chapter 6

Towards Earth System Science and Global Sustainability

Based on the latest scientific understanding from IGBP and related research, an attempt has been made in this book to show how the Earth System functions, how human activities have now become a global-scale force in their own right, and what the implications of global change are for human well-being and for the Earth's life support system. This last chapter looks forward by asking a series of questions that science must help to answer over the coming decades if global sustainability is to be achieved. Two over-arching questions, however, still dominate: first, what the nature of changes in the Earth System will be over the next decades and secondly, what the implications of these changes for humankind will be. Other critical questions follow. What type and scale of management responses – from prevention and adaptation to more proactive geo-engineering approaches – are consistent with the scientific knowledge base? How must science itself change to tackle the challenges that lie ahead; how can an innovative and integrative Earth System science be built? Global change is a reality. Human-induced changes over the last two centuries have transformed the Earth in both systemic and cumulative ways, and have pushed it beyond its natural operating domain. The continuing transformation of the Earth System may lead to extremes and rates of change that are much more significant than smooth alterations in global means. The interactions between the likely accelerating changes to the Earth System over the coming decades and the growing needs of a rapidly expanding human population give a sense of urgency to realising the goals of Earth System science and global sustainability.

6.1 From Climate Change to Earth System Science

The dominant themes of this book have been the nature and human implications of changes in the Earth System in the past and over the next decades. Nested within this are issues concerned with, for example, the extent to which anthropogenic forcing of global climate is already detectable and the ways in which such forcing may interact with natural variability. A growing consensus among global change scientists favours the view that anthropogenic influence on climate is already detectable and that likely future changes will have major human consequences that must be viewed within the context of a growing world population and an accelerating sequence of environmental transformations that involve much more than climate change alone.

The range of scientific methodologies available for addressing these issues is powerful and varied, including experiments, coordinated observation and monitoring programmes, modelling and the reconstruction of environmental changes in the past. These have been the tools used by the several projects within IGBP and the many other research groups working in parallel. The preceding chapters have presented a synthesis of the results obtained and their likely human consequences. This final chapter (Fig. 6.1) addresses the following questions:

- what are the main scientific conclusions and the implications for humanity that have emerged from this research?
- what constitutes good stewardship of the Earth System? Which environmental management strategies are most consistent with the current state of scientific knowledge? To what extent can future dangers be identified and avoided? What key preventative measures and responses should be taken?
- what are the essential characteristics of the Earth System science required both to improve understanding of global change and to contribute most effectively to future policy and management?
- finally, how, through research and communication strategies, can Earth System science best enhance the sustainability of the planet's environmental systems upon which the future well-being of human popu-lations depend?

6.2 The Knowledge Base

Somewhat more than a decade ago it was recognised that the Earth behaves as a system in which the oceans, atmosphere and land, as well as the living and non-living parts therein were all connected. While accepted by many, this working hypothesis seldom formed the basis for global change research. Little understanding existed of how the Earth worked as a system, how the parts were connected, or even about the importance of the various component parts

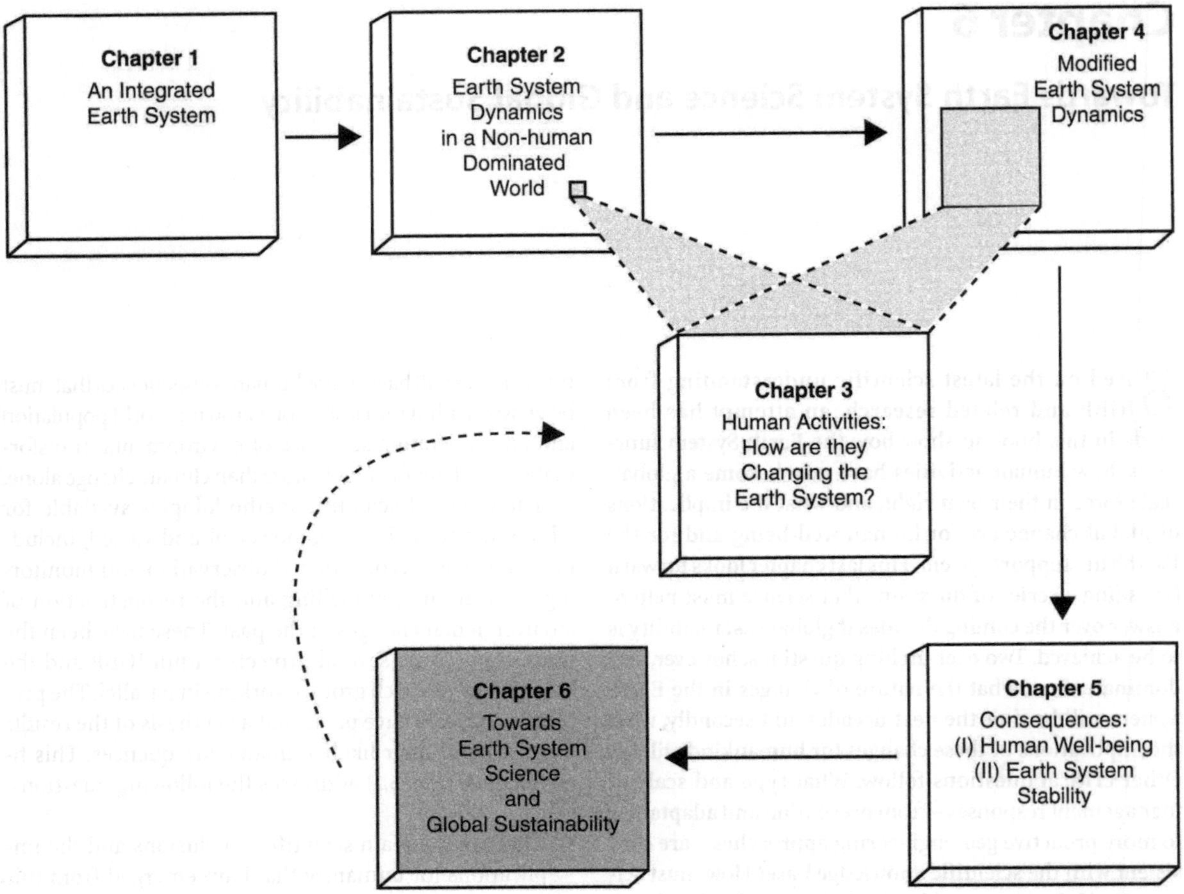

Fig. 6.1. Chapter 6 focus: the scientific challenge to create a truly integrated Earth System science and the societal challenge to deal effectively with global change

of the system. Feedback mechanisms were not always clearly understood, nor were the dynamics controlling the system.

Over the intervening years much has been learned. In many respects former uncertainties about the nature and future course of global change have been reduced. In others, the realisation that uncertainty is an inherent part of the system has gained credence. Over the last 10 years, the understanding of how humans are bringing about global change has undergone a quantum leap. Attempts to separate natural and anthropogenically induced variability in the Earth System have proved to be successful in many respects. The decade has been one of scientific challenge, achievement and excitement.

The scientific landscape is very different now from that of the late 1980s. In general, global change research has confirmed many of the hypotheses and much of the sketchy understanding of a decade ago, adding a wealth of quantitative detail and process-level understanding at all scales. Largely through a significant increase in the ability to unravel the past, the understanding of the natural dynamics of the Earth System has advanced greatly. It is now clear that global change is one of the paramount environmental issues facing humankind at the beginning of the new millennium.

The task of synthesising a decade of global change research has been daunting, but the rewards have been great. Detailed results have been presented earlier in this volume; even more detailed advances in global change research are available in the individual core project syntheses and that of START (Global Change System for Analysis, Research and Training) published in the IGBP book series. In this concluding chapter only generalised highlights are presented, the so-called big-picture findings. They are based on detailed, quantitative science that has been published by a multitude of scientists working worldwide over the past 10 years and longer. Each of the five major research findings presented here is supported by one or two examples, drawn from the earlier chapters of this volume where they are described in more detail and where additional references are given.

6.2.1 Biology in the Earth System

The Earth is a system that life itself helps to modulate. Biological processes interact with physical and chemical processes to create the planetary environment, but biology plays a much stronger role than previously thought in the functioning of the Earth System.

An example from North Africa during the early and mid-Holocene, described in Sect. 2.3.1, illustrates the important role of biological processes in modulating the Earth's environment. At that time the Sahel and Sahara regions were significantly wetter than they are today. An early hypothesis suggested that the climate shift was caused by small changes in tropical insolation due to variations in the Earth's orbit during that period (Kutzbach and Guetter 1986). Simulations by general circulation models with modified orbital forcing showed that these orbital changes led to an amplification of the African monsoon but the change in rainfall was not large enough to drive the change in vegetation observed.

The key to simulating correctly the wetter climate and the occurrence of vegetation much further north than today was to include both changes in sea-surface temperature and land-surface characteristics in an interactive way. The results of a coupled atmosphere-ocean-vegetation model strongly suggested that the success of the simulation was due to the inclusion of the interaction between the albedo of the Sahara region and the atmospheric circulation, along with factors connected to moisture convergence (and hence sea-surface temperature), convective precipitation and soil moisture (Claussen and Gayler 1997). These simulations indicated that biogeo-physical feedbacks – primarily those associated with albedo – tend to amplify the climate change due to variations in orbital forcing.

Many such specific examples of biological influence on the Earth's environment can be found. A greater overall challenge is to establish the role played by biological processes in the long-term functioning of the Earth System as a whole, for example, as defined by the pattern of glacial-interglacial cycling revealed in the Vostok ice core. To what extent can the succession of bounded and tightly coupled oscillations in the Earth's climate and atmospheric composition through the glacial-interglacial cycles be simulated and understood? How important are biological processes as integral, active components in Earth System functioning, rather than just as passive elements reacting to physico-chemical changes? How critical are connections between the terrestrial biosphere and ocean processes?

Recent studies (Bopp et al. 2003) take a step towards answering these questions. They show that an explicitly simulated dust field for the LGM (Last Glacial Maximum) can provide enough iron to the Southern Ocean to stimulate a diatom bloom, increasing export production and drawing down CO_2 substantially. In the simulation several factors are found to contribute to higher atmospheric dust content at LGM – increased atmospheric transport as a result of stronger winds; a reduced hydrological cycle, especially lowered precipitation linked to cold ocean conditions; and reduced C_3 plant growth linked to both lower atmospheric CO_2 and drier climatic conditions (Mahowald et al. 1999). The simulated dust field generates loadings consistent with evidence for dust deposition in an extensive data set of marine and ice-core records. Equally, the simulated spatial field of change in export production to the deep ocean is compatible with the pattern of changes shown by a data set based on a range of marine sediment proxies. In this example, the synergy between the state of the terrestrial biosphere and changed atmospheric processes is crucial for generating greater dust entrainment, while the enhanced primary productivity of marine diatoms provides the fuel for the biological pump sequestering carbon in the deep ocean. Both terrestrial and marine biota are vital components of the system of interactions that has been simulated and checked against palaeo-data.

Claquin et al. (2002) show that enhanced atmospheric dust loading at LGM has also had a major effect on net radiative forcing. As Overpeck et al. (1996) suggested, the effect would have been positive over bright areas such as ice sheets, but in the simulation of Claquin et al., the largest loadings and radiative forcing are over the tropics, where the net effect (taking relatively bright land and relatively dark oceans together) is large and negative – comparable to that of the lowering of CO_2. Alternative simulations designed to allow for the main uncertainties involved indicate that this is a robust conclusion, but it cannot be readily translated into surface cooling. Taking this further step will involve quantifying additional feedbacks within the Earth System.

In the example given above two aspects of atmospheric dust loading are explored: the first leads to an enhanced biological pump (hence lower atmospheric CO_2 through iron fertilisation), the second to reduced radiative forcing. They show some of the complex ways in which the physical, chemical and biological properties of land surfaces, atmosphere and oceans interact with the mutually reinforcing feedbacks that created the minimum atmospheric trace gas concentrations and climate of the LGM (Fig. 6.2). They also illustrate the growing synergy between the development of complex interactive models and their testing against global palaeo-data sets.

6.2.2 The Nature of Global Change

Global change is much more than climate change. It is real, it is happening now *and it is accelerating. Human activities are significantly influencing the functioning of the Earth System in many ways; anthropogenically driven changes are clearly identifiable beyond natural variability and are equal to some of the great forces of nature in their extent and impact.*

The changes in the human-nature relationship that have occurred over the last several centuries and that are giving rise to global change are complex and profound. As discussed in Chap. 3, they are almost certainly unprecedented in the history of Earth. The expansion of humankind, both in numbers and per capita exploitation of Earth's resources,

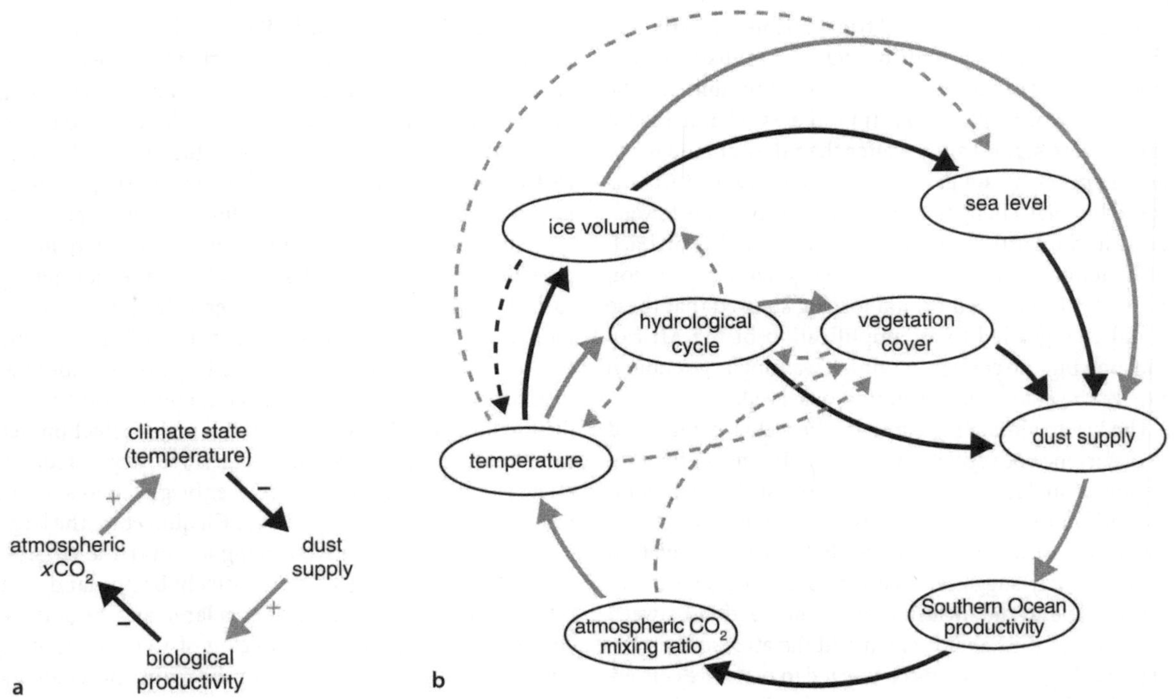

Fig. 6.2. Feedbacks in the climate system (Ridgwell and Watson 2002). Different components of the Earth system can directly interact in three possible ways; a positive influence (i.e., an increase in one component directly results in an increase in a second), a negative influence (i.e., an increase in one component directly results in a decrease in a second), or no influence at all. An even number (including zero) of negative influences occurring within any given closed loop gives rise to a positive feedback, the operation of which will act to amplify an initial perturbation. Conversely, an odd number of negative influences gives rise to a negative feedback, which will tend to dampen any perturbation. In the two schematics, positive influences are shown in *grey*, and negative in *black*. **a** Schematic diagram of a simplified feedback system, involving dust, the strength of the biological pump, CO_2, and climatic state (represented by mean global surface temperature). Because there is an even number of negative influences (= 2), this represents a positive feedback, with the potential to amplify an initial perturbation (in either direction). **b** Schematic diagram of the hypothetical glacial dust-CO_2-climate feedback system, with explicit representation of the various dust mechanisms that we have identified. Primary interactions in the dust-CO_2-climate subcycle are indicated by *thick solid lines*, while additional interactions (peripheral to the argument) are shown *dotted* for clarity. For instance, the two-way interaction between temperature and ice volume is the *ice-albedo* feedback. Four main (positive) feedback loops exist in the system: (1) dust supply ⟶ productivity ⟶ xCO_2 ⟶ temperature ⟶ ice volume ⟶ sea level ⟶ dust supply (four negative interactions), (2) dust supply ⟶ productivity ⟶ xCO_2 ⟶ temperature ⟶ hydrological cycle ⟶ vegetation ⟶ dust supply (two negative interactions), (3) dust supply ⟶ productivity ⟶ xCO_2 ⟶ temperature ⟶ hydrological cycle ⟶ dust supply (two negative interactions), (4) dust supply ⟶ productivity ⟶ xCO_2 ⟶ temperature ⟶ ice volume ⟶ dust supply (two negative interactions)

has been remarkable. During the past three centuries human population increased tenfold to six billion. Concomitant with this population increase, the rate of consumption has risen even more sharply. Just as rapid and profound are other changes sweeping across human societies, many through the process often termed *globalisation*.

The top half of Fig. 6.3 shows examples of some of these changes in human activities over the past few hundred years; it captures the changing nature of human societies at this pivotal time in the development of the human-environment relationship. One feature stands out as absolutely remarkable. The second half of the twentieth century is unique in the entire history of human existence on Earth. Many human activities reached take-off points sometime in the twentieth century and have accelerated sharply towards the end of the century. The last 50 years have without doubt seen the most rapid transformation of the human relationship with the natural world in the history of the species.

The pressure on the Earth System from this increasing human enterprise is intensifying sharply (Crutzen 2002):

- Whereas petroleum was only discovered in the last 150 years, already humankind has exhausted almost 40% of known oil reserves that took over several hundred million years to generate.
- It is estimated that nearly 50% of the land surface has been transformed by direct human action, with significant consequences for biodiversity, nutrient cycling, soil structure and biology, as well as climate.
- More nitrogen is now fixed into available forms through the production of fertilisers and burning of fossil fuels than is fixed naturally in all terrestrial systems.
- More than half of all accessible freshwater is appropriated for human purposes.
- The concentrations of several climatically important greenhouse gases, such as CO_2, N_2O and CH_4, have substantially increased in the atmosphere.
- Coastal wetlands have also been significantly affected by human activities, with the loss of 50% of the world's mangrove ecosystems.

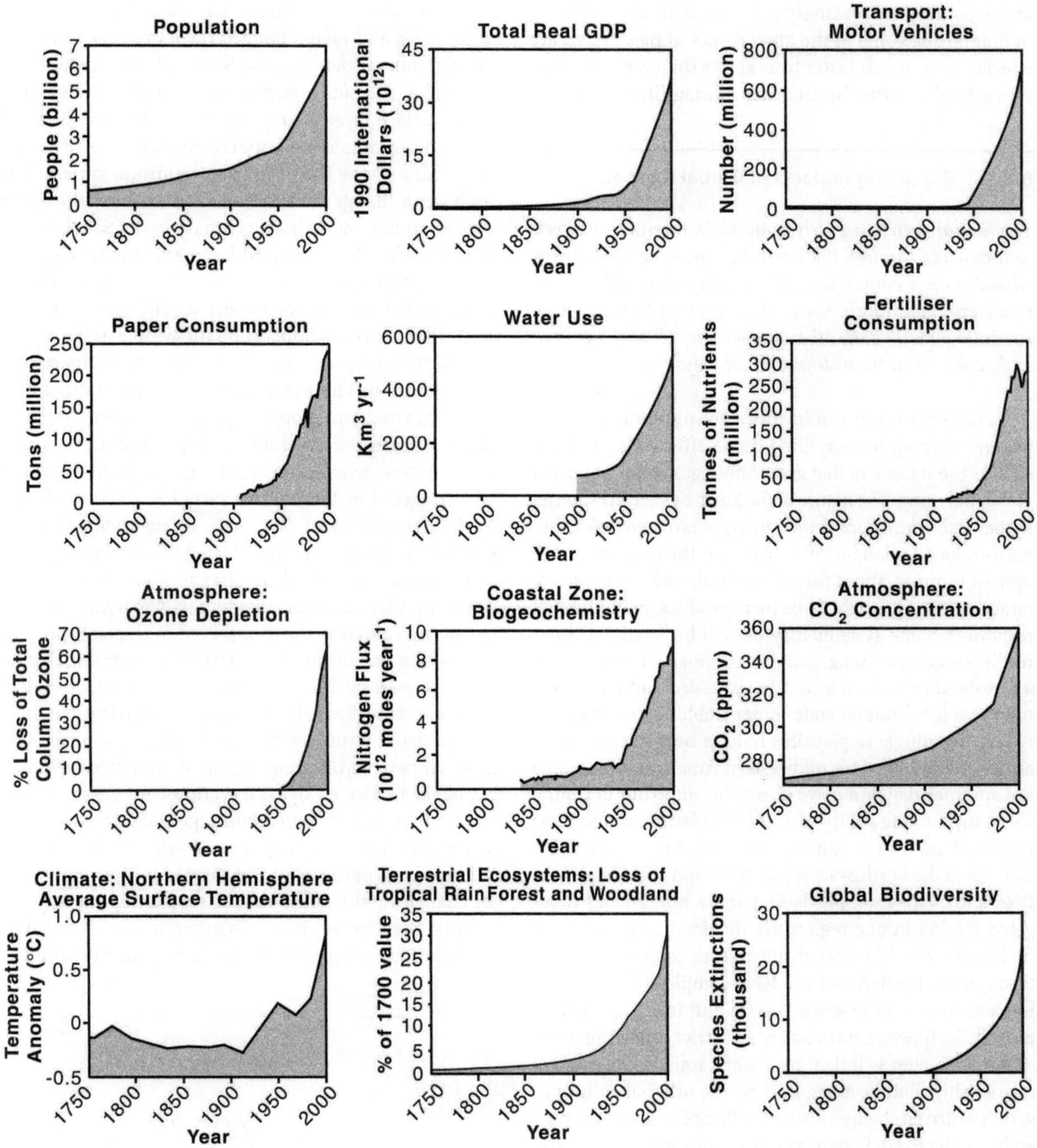

Fig. 6.3. The increasing rate of change in human activity since the beginning of the Industrial Revolution (*top six panels*) (US Bureau of the Census 2000; Nordhaus 1997; UNEP 2000; Pulp and Paper International 1993; Shiklomanov 1990; IFIA 2002). The *bottom six panels* show some of the global-scale changes in the Earth System as a result of the dramatic increase in human activity (based on data from J. D. Shanklin, British Antarctic Survey; Mackenzie et al. 2002; Etheridge et al. 1996; Mann et al. 1999; Richards 1990; WRI 1990; Wilson 1992)

- Up to 50% of marine fish stocks for which information is available are fully exploited, 15–18% are overexploited and 9–10% have been depleted or are recovering from depletion.
- Extinction rates are increasing sharply in marine and terrestrial ecosystems around the world; the Earth is now in the midst of the sixth great extinction event in its history, but the first one caused by the activities of a biological species (*Homo sapiens*).

The bottom half of Fig. 6.3 shows that the impacts of these accelerating human changes are now clearly discernible at the level of the Earth System as a whole. Many key indicators of the functioning of the Earth System are now showing responses that are, at least in part, clearly driven by the changing human imprint on the planet. All components of the global environment, oceans, coastal zone, atmosphere and land, are being influenced. In summary, human impacts on the Earth are becoming increasingly equivalent to or exceeding in magnitude some of the great forces of nature and are operating on much faster time scales than rates of natural variability, often by an order of magnitude or more.

6.2.3 Cascading Impacts of Global Change

The human enterprise drives multiple, interacting effects that cascade through the Earth System in complex ways. Global change cannot be understood in terms of a simple cause-effect paradigm. Cascading effects of human activities interact with each other and with local- and regional-scale changes in multidimensional ways.

Earth System functioning does not respond to anthropogenic forcing in simplified cause-effect relationships, such as the statement that greenhouse gas emissions cause global warming. The nature of the Earth System's responses to increasing anthropogenic forcing is more complex. Human-induced changes often cascade through the Earth System to bring about further multiple collateral and interacting changes, which in turn produce many further responses. Some systems may be well buffered and show resilience against change; others may not be. In some cases a threshold of resilience may be exceeded and the system may pass into another state, either stable or unstable.

The seemingly implausible linkage between the use of an air conditioner in a midwestern American home on a hot summer day or a drive down the autobahn in central Germany and the ability of an African farmer in the Sahel to grow food for his family is one such case of cascading changes in the Earth System (Rotstayn and Lohmann 2002) (Box 4.10). This example demonstrates how an anthropogenic forcing in one region, via the combustion of fossil fuels and the production of sulphate aerosols, can cascade through the Earth System to have completely unforeseen impacts in distant places on Earth and in regions far removed. Such reverberations are not strict causal chains in which each step is linked by a tight, unique cause-effect relationship. Rather, along the way the original anthropogenically driven changes interact with modes of natural variability in the Earth System as well as with specific local and regional characteristics to produce the observed outcome.

The cascade of impacts leading to the Sahelian drought conditions was most probably initiated with the production of energy and hence sulphate aerosol particles in North America and Europe. In addition to their local and regional environmental effects, sulphate aerosol particles change the radiative balance near the Earth's surface in a direction opposite to that of greenhouse gases, leading to a near-surface cooling. The effect is not global, however. The particles have a short lifetime in the atmosphere and are deposited back onto the land or sea surface before they have circulated widely around the atmosphere. Thus, the surface cooling is largely a regional effect.

The cooling over Europe and North America in the 1960s, 1970s and 1980s changed the latitudinal temperature differential between the North Atlantic region and the tropics and subtropics, both in the atmosphere and in the surface layer of the ocean. This in turn changed atmospheric circulation patterns and led to a shift in the positioning of the North African monsoon system. The result was a change in the rainfall patterns over the North African region and a drying climate in the Sahel during those decades (Rotstayn and Lohmann 2002) (Fig. 6.4). Land degradation increased during that time, leading to concern that the Sahara Desert would expand southwards. The resulting impacts on the people in the region were severe, with an increase in poverty, malnutrition and starvation. The phenomenon was often cited as a case of grazing and cropping pressure, driven by population increase, overwhelming the resilience of the natural ecosystem, whereas in reality the problem may have been triggered by forcings far away from North Africa.

The plausability of this chain of events needs to be questioned. Trends in rainfall during the 1901–1998 period as simulated by a general circulation model driven by the change in sulphate aerosol loading in North America and Europe are similar to those observed in tropical and subtropical regions in northern Africa. In addition, during the 1990s with the advent of emission controls in source regions and the dimunition of sulphate particle loadings, the simulated return of the North African monsoon to its earlier pattern, bringing increased rainfall to the Sahel, was again replicated by observations. An easing of the degradation and an increase in regional ecosystem productivity followed. During the transition period grazing and cropping patterns of the people in the region did not change significantly, further supporting the suggestion that the aerosol-driven changes in rainfall variability might have been the primary factor causing drought.

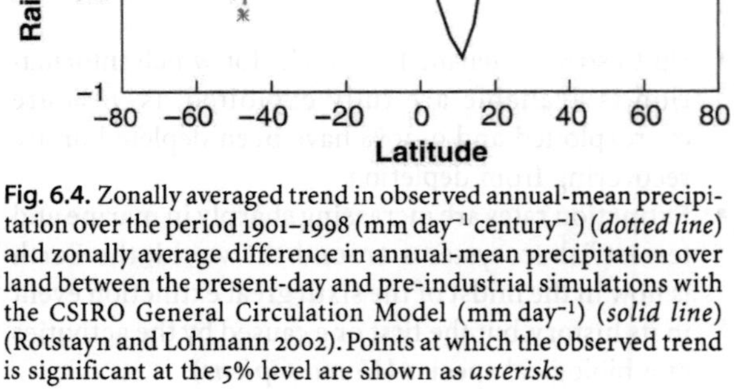

Fig. 6.4. Zonally averaged trend in observed annual-mean precipitation over the period 1901–1998 (mm day^{-1} $century^{-1}$) (*dotted line*) and zonally average difference in annual-mean precipitation over land between the present-day and pre-industrial simulations with the CSIRO General Circulation Model (mm day^{-1}) (*solid line*) (Rotstayn and Lohmann 2002). Points at which the observed trend is significant at the 5% level are shown as *asterisks*

6.2.4 Thresholds and Abrupt Change

The Earth's dynamics are characterised by critical thresholds and abrupt changes. The Earth System has operated in different quasi-stable states, with abrupt changes occurring between them over the last half million years. Human activities clearly have the potential to switch the Earth System to alternative modes of operation that may prove irreversible and could inadvertently trigger changes with catastrophic consequences.

The palaeo-record shows that abrupt changes are a common feature of the Earth System and that extremes outside those observed during the period of instrumental records are not only possible but should be expected under global change. As discussed in more detail in Sect. 5.4.1, catastrophic failures are possible as the Earth System as a whole adjusts to an accelerating suite of interacting human and natural forcings.

A potential catastrophic abrupt change in the Earth System is the slowing or collapse of the North Atlantic thermohaline circulation (Box 5.6), which could shift the Gulf Stream southwards and significantly diminish the amount of heat transported across Scandinavia and northern Europe. At present these regions are habitable largely because of the warmth transmitted from ocean to atmosphere as the Gulf Stream surface waters cool north of Iceland and then sink to depth. The predominantly westerly winds then transport the warm air across northern Europe.

This feature of the global thermohaline circulation is known to have been unstable in the past. During the last 75 000 years the North Atlantic region has experienced strikingly abrupt changes in climate, with swings of up to 10 °C in only a decade. These abrupt changes have occurred when the Earth is in a glacial state or moving from the glacial to the interglacial state. They appear as repeated flickers consisting of rapid increases in temperature followed by a relaxation of the system to glacial conditions. Although the cause of this flickering is not known precisely, it is probably related to a slowing or reorganisation of the thermohaline circulation in the North Atlantic (Clark et al. 2002).

The question of whether the current human perturbation of the Earth System can trigger instabilities in the North Atlantic thermohaline circulation in the coming decades must be posed. The palaeo-evidence cannot be used as a direct analogue. The abrupt changes during the last 75 000 years occurred when the Earth was in a glacial state and were manifest as abrupt rises in temperature (Fig. 6.5). The Earth is currently in an interglacial state and the concern is not with an abrupt rise in temperature but rather with a cooling associated with a switch of the deep water formation to south of Iceland. However, there are plausible scenarios. For example, an increase in rainfall and a melting of Arctic sea ice associated with rising temperatures could lead to a layer of fresher, less dense surface water in the seas north of Iceland, slowing the formation of deep water and returning the regional climate to the colder, glacial state.

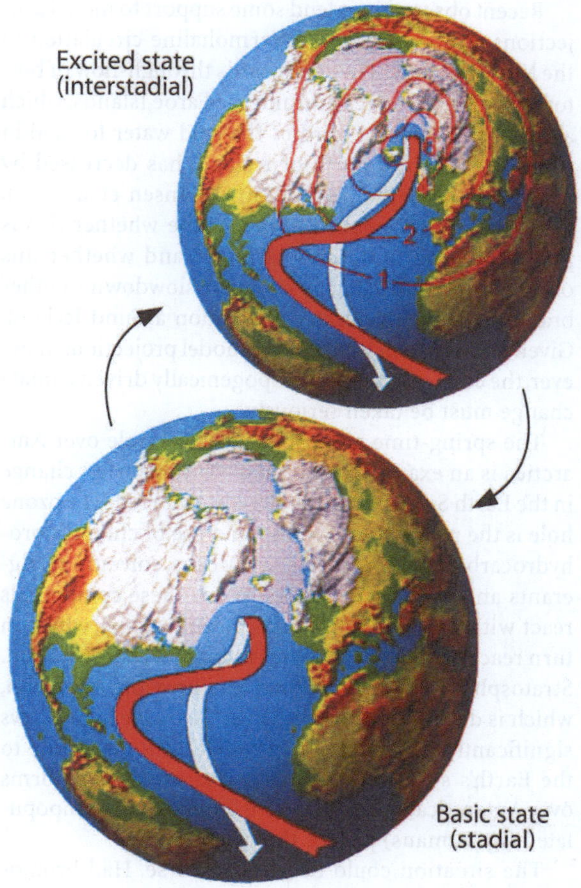

Fig. 6.5. Schematic diagram of two glacial climate states (Ganopolski and Rahmstorf 2002). *Top:* the metastable *warm* or *interstadial* mode; *bottom:* the stable *cold* or *stadial* mode. *Contours* show the surface air temperature difference relative to the stable state. Ocean circulation is shown schematically, surface currents in *red* and deep currents in *light blue*. Ice sheets shown are prescribed from Peltier (1994)

Model studies suggest that climatic changes associated with the current anthropogenic forcings indeed have the potential to switch ocean circulation in the North Atlantic, although the models differ widely in the simulated timing and amount of cooling (Knutti and Stocker 2002). Intriguingly, the rate at which anthropogenic perturbations occur, especially the emission of CO_2 to the atmosphere, is important in determining the state of the thermohaline circulation (Stocker and Schmittner 1997). Until more is known about the nature and rate of changes to the hydrological cycle in the North Atlantic region, this projection remains a surprise of low probability. Nevertheless, it certainly cannot be dismissed on the grounds of implausibility or lack of precedent. Should abrupt cooling actually occur this century, it could have catastrophic consequences for northern Europe.

Recent observations lend some support to model projections of a slowdown of thermohaline circulation in the North Atlantic. The southwards through-flow of bottom water between Iceland and the Faroe Islands, which accounts for about a third of the cold water formed in the North Atlantic north of Iceland, has decreased by about 20% over the last 50 years (Hansen et al. 2001). More research is needed to determine whether this is part of long-term natural variation and whether this observation is matched by a similar slowdown in other branches of the deepwater circulation around Iceland. Given its consistency with many model projections, however, the connection to anthropogenically driven climate change must be taken seriously.

The spring-time stratospheric ozone hole over Antarctica is an example of a human-driven abrupt change in the Earth System that has already occurred. The ozone hole is the result of the wipespread use of chlorofluorohydrocarbons as aerosols in spray cans, solvents, refrigerants and as foaming agents. When these compounds react with sunlight, they form chlorine atoms, which in turn react with ozone and thus reduce its concentration. Stratospheric ozone acts a filter for ultraviolet radiation, which is deleterious for life. Thus, the ozone hole allows significantly more ultraviolet radiation to penetrate to the Earth's surface. Fortunately, the ozone hole forms over Antarctica, so its effects are limited to that unpopulated (by humans) part of the planet.

The situation could have been worse. Had bromofluorocarbons been used instead of chlorofluorocarbons, there would have been an ozone hole around the entire Earth for all seasons. Bromine is about 100 times more effective than chlorine at destroying ozone. Such a scenario would have been a truly catastrophic abrupt change in the Earth System as humankind would have been totally unprepared for it, given the lack of atmospheric chemical understanding at that time. There is no *a priori* reason why bromine was not used instead of chlorine in such compounds, so humankind was simply lucky that it did not already trigger a disastrous change to the global environment (Crutzen 1995).

6.2.5 A No-Analogue State

The Earth is currently operating in a no-analogue state. In terms of key environmental parameters, the Earth System has recently moved well outside the range of the natural variability exhibited over at least the last half million years. The nature *of changes now occurring* simultaneously *in the Earth System, their* magnitudes *and* rates of change *are unprecedented.*

The Earth System operates in cycles that have well-defined time scales and set points that limit the magnitudes of its rhythmic changes (Chap. 1 and Fig. 1.3). From the perspective of the Earth System, the question of how these recent human-driven changes compare with the natural variability of the system in terms of magnitudes and rates must be asked. The increase in atmospheric CO_2 concentration, which has become somewhat of an icon for global change, provides a useful indicator with which to evaluate the rate and magnitude of human-driven change.

Analysis of the Vostok ice core data from Antarctica suggests that over the past 420 000 years the atmospheric CO_2 concentration has oscillated in a regular pattern over approximately 100 000 year cycles by about 100 ppmV, between about 180 and 280 ppmV. Viewed against that pattern, the human imprint on atmospheric CO_2 concentration is unmistakable. Atmospheric CO_2 concentration now stands at 370 ppmV, almost 100 ppmV above the previous maximum level. In effect, the human perturbation to atmospheric CO_2 (ca. 100 ppmV) now equals in magnitude the entire operating range of CO_2 between its glacial and interglacial extremes. In addition, the new concentration has been reached at a rate at least 10 and possibly 100 times faster than at any other time during the previous 420 000 years (Falkowski et al. 2000). In this case, human-driven changes are clearly well outside the range of natural variability exhibited by the Earth System for the last half-million years at least.

There is increasing evidence that the Earth's climate is now also changing beyond the patterns of natural variability, at least partly in response to anthropogenically driven changes to atmospheric composition. The last half of the twentieth century experienced warming at a hemispheric, and almost certainly a global, scale that was unprecedented for the last millennium at least. Even taking into account the large uncertainties associated with proxy estimates of temperature for the early part of the millennium, 1998 was the warmest year for the past 1 000 years (Mann et al. 1999).

The changes in atmospheric CO_2 concentration and temperature, however, are only two of the many globally significant human-driven changes that are occurring simultaneously (Fig. 6.6). Before the era of significant human influence on the global environment, perturbations in atmospheric composition and climate may have been somewhat buffered by a relatively untransformed terrestrial biosphere with its own dynamic rhythms. For the past few hundred years, however, the terrestrial biosphere has been under increasing pressure from human activities. About half of the ice-free land surface of the planet has been converted to simpler ecosystems or substantially modified by direct human activities (Sect. 3.3.1). Most of the remaining half is managed to some extent by humans. A significant fraction of the Earth's terrestrial productivity, somewhere between 10 and 50%, is co-opted by humans.

At the same time the terrestrial freshwater cycle has been largely remade by human activities. About 40% of the total global runoff to the oceans is now intercepted by dams and pondages (Sect. 3.3.3). Sediment delivery to the coastal seas has been increased in some catchments

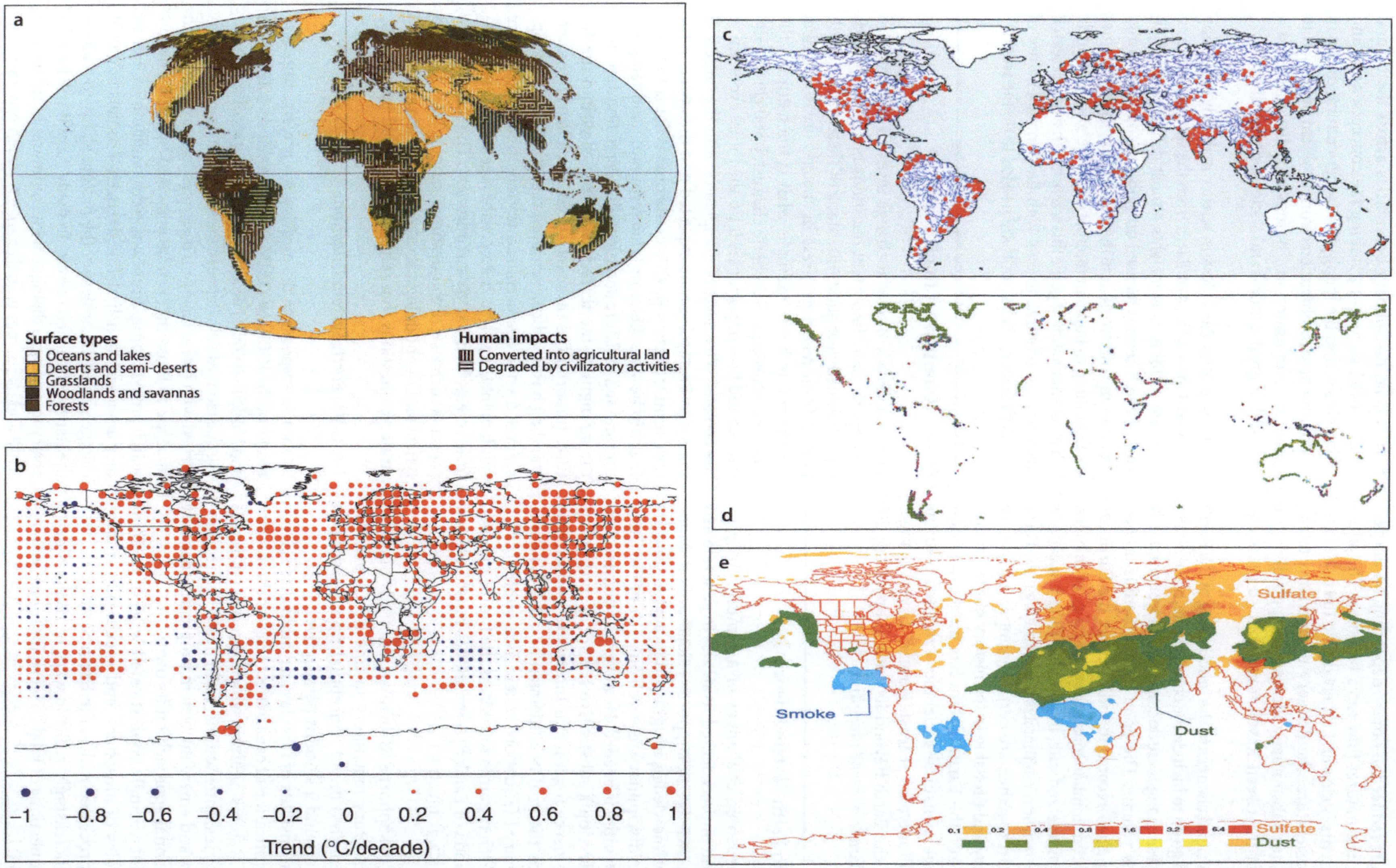

Fig. 6.6. Human-driven changes at the global scale that are affecting many aspects of the Earth System: **a** human transformation of Earth's terrestrial surface (Schellnhuber 1998); **b** global pattern of temperature change (IPCC 2001); **c** global distribution of 622 large reservoirs, classified as those with a maximum capacity greater than 0.5 km^3 (Vörösmarty et al. 1997) (see also Fig. 3.43); **d** coastal zone modification as represented by global estimates of relatively pristine or undisturbed coastal regions (LOICZ 2002) (see also Fig. 3.50); and **e** calculated global fields of mineral dust, sulphate and smoke as determined with a chemical transport model using actual meteorology (*http://www.nrlmty.navy.mil/aerosol/* and Heintzenberg et al. 2003) (see also Fig. 3.38c)

and drastically reduced in others, depending on the perturbation to the natural flow rates by dams. The situation for nutrients is clearer. Largely due to the use of fertilisers in agriculture, the delivery of nutrients to the coastal seas has substantially increased, especially since the advent of the Green Revolution about 50 years ago (Sect. 3.3.4).

The list of rapid changes to the functioning of the Earth System could go on to include the coastal zone, biodiversity and so on. These changes are documented more completely earlier in this volume. The point is that there is no evidence from palaeo-records to indicate that the Earth has ever experienced simultaneously the wide range of changes, in both magnitude and rate, in so many different aspects of the global environment that it is experiencing now. These changes, taken together, are certainly unprecedented in human history and probably in the history of the Earth.

In summary, the Earth System has moved well outside the range of natural variability exhibited over the last half million years at least. Human-driven changes are pushing the Earth beyond its natural operating domain into planetary *terra incognita*.

6.3 Making Earth System Science

The unprecedented challenges of a rapidly changing Earth demand unprecedented strategies to generate new scientific knowledge to support societal action. Much exciting new knowledge, insight and understanding are flowing from disciplinary research on parts of the Earth System. Even more is needed. However, the biggest challenge is to develop a substantive *science of integration*, putting the pieces together in innovative and incisive ways towards the goal of understanding the dynamics of the planetary life support system as a whole.

The research required to address the future-oriented research agenda of Earth System science must have new characteristics. It must:

- continue to support and facilitate the study of pieces of the planetary machinery in fine detail, but also place them into broad conceptual frameworks that are built around a systems perspective;
- embed the insights of this classical analytical science – the identification of cause-effect relationships – into complex systems analysis which directly addresses the synergies, interactions, switches/triggers, nonlinearities and emergent system properties that defy the traditional approach on its own;
- develop increasingly sophisticated ways of examining the roles of biospheric and specifically anthropogenic processes within the Earth System;
- develop increasingly realistic ways of responding to 'what if?' questions of vital concern for future sustainability; and
- transcend disciplinary boundaries across the natural and social sciences by linking the concepts, skills and insights across the biophysical/socio-cultural divide in exciting new combinations from problem definition to the communication of findings to policymakers in a world of great cultural and socioeconomic diversity.

An integrative Earth System science is already beginning to unfold. Observations of Earth from the surface and from space are yielding new insights almost daily, interdisciplinary research centres focused on global change are springing up around the world, and the global change programmes are beginning to build an international science framework. What questions should guide this science? What new research strategies are required to deal with the complex nature of the Earth System? What tools are needed to do the work?

6.3.1 Questions at the Frontier

The past decade of global change research has provided answers to many important questions but in the process has generated new questions. Many of these naturally follow on from recent progress in specific areas of global change research. For example, what is the nature of the coupling between atmospheric chemistry and climate dynamics and how important is it for the future evolution of the climate? What level of compliance with the targets of the Kyoto Protocol can be expected and thus what is the near-term trajectory of CO_2 emissions? What is the relationship between heterotrophic respiration and temperature and soil moisture and how will this relationship affect the strength of the terrestrial carbon sink into the future? Important as these questions are, a level of understanding has now been achieved that allows the development of a set of overarching questions, pitched at a more fundamental level, to help set the agenda for systems-level research on Earth's environment. The following list presents such a set of 23 questions, ranging from the analytical and methodological in research to normative and strategic questions that rely on value-based judgments.

The GAIM Earth System Questions

In 2001, in response to the evolving science, structure and results of the IGBP, the GAIM (Global Analysis, Integration and Modelling) Task Force developed a new set of over-arching questions as a challenge to the scientific community concerned with global change. These questions are not limited in scope to those that can be answered by individual research projects, programmes, or even communities. Rather, they are meant to help define the overall context of global change science regardless of the present ability to address the issues articulated therein. In many cases it will be necessary to develop a dialogue with communities far beyond IGBP and its partner global change research programmes.

Analytic questions:

1. What are the vital organs of the ecosphere in view of operation and evolution?
2. What are the major dynamical patterns, teleconnections and feedback loops in the planetary machinery?
3. What are the critical elements (thresholds, bottlenecks, switches) in the Earth System?
4. What are the characteristic regimes and time-scales of natural planetary variability?
5. What are the anthropogenic disturbance regimes and teleperturbations that matter at the Earth System level?
6. Which are the vital ecosphere organs and critical planetary elements that can actually be transformed by human action?
7. Which are the most vulnerable regions under global change?
8. How are abrupt and extreme events processed through nature-society interactions?

Operational questions:

9. What are the principles for constructing "macroscopes", i.e., representations of the Earth System that aggregate away the details while retaining all systems-order items?
10. What levels of complexity and resolution have to be achieved in Earth System modelling?
11. Is it possible to describe the Earth System as a composition of weakly coupled organs and regions, and to reconstruct the planetary machinery from these parts?
12. What might be the most effective global strategy for generating, processing and integrating relevant Earth System data sets?
13. What are the best techniques for analysing and possibly predicting irregular events?
14. What are the most appropriate methodologies for integrating natural-science and social-science knowledge?

Normative questions:

15. What are the general criteria and principles for distinguishing non-sustainable and sustainable futures?
16. What is the carrying capacity of the Earth?
17. What are the accessible but intolerable domains in the co-evolution space of nature and humanity?
18. What kind of nature do modern societies want?
19. What are the equity principles that should govern global environmental management?

Strategic questions:

20. What is the optimal mix of adaptation and mitigation measures to respond to global change?
21. What is the optimal decomposition of the planetary surface into nature reserves and managed areas?
22. What are the options and caveats for technological fixes like geoengineering and genetic modification?
23. What is the structure of an effective and efficient system of global environment and development institutions?

For some questions, GAIM is already in a position to begin to address the issues involved. In other cases, close collaboration will be necessary with IGBP projects, and with WCRP (World Climate Research Programme), IHDP (International Human Dimensions Programme on Global Environmental Change), DIVERSITAS (an international programme of biodiversity science) and others. There are additional questions whose answers will depend not on scientific research, but rather on social or philosophical considerations.

The 23 GAIM Earth System questions are not limited to those that can be answered directly by the scientific community itself. Rather, the questions help to define the overall context of global change science from an Earth System perspective, regardless of the present ability to address the issues embodied in the questions. Several of the questions challenge perceptions and values of society in general and demand dialogues among a wide range of communities if they are to be addressed. The 23 questions, in fact, anticipate the advent of a unified Earth System science and encompass the natural and socioeconomic dimensions in an integrated fashion.

6.3.2 Coping with Complexity and Irregularity

Most environmental systems are characterised by a multitude of non-linear internal interactions and external forcings. As a consequence, they do not often behave regularly and predictably, but rather exhibit chaotic dynamics or abrupt transitions to new modes of operation under appropriate forcing. These systems usually become even more complicated when human activities are involved, which introduces a further element of indeterminacy. All these generic difficulties are reflected and severely amplified at the Earth System level, where the connections and feedbacks of millions of entangled local, regional and global processes have to be considered. Typical examples of Earth System complexity and irregularity are the intricate patterns in space and time of carbon fluxes between the Earth's surface and the atmosphere, and the abrupt and seemingly erratic temperature oscillations documented in the Greenland ice-core records. Nature-society interactions also tend to come in complicated functional patterns that defy the power of standard scientific analysis.

In order to cope with these challenges, it is first useful to develop a typology of Earth System irregularities (Schellnhuber 2002). This facilitates the recognition of the types of complexity or irregularity that exist and helps to identify the appropriate tool(s) needed to study, understand and perhaps project the behaviour of the System into the future.

Non-Linearity

Non-linear behaviour is ubiquitous throughout the Earth System; indeed it is much more difficult to find examples of linear than non-linear processes. In its simplest form non-linearity means that a response to a forcing function does not behave in a strictly proportional manner. In terms of the Earth System three types of non-linear behaviour that are particularly important were outlined in Sect. 2.6. Classic examples are those in which the amplitude of a response in the Earth System is massively out of proportion to the original forcing function, such as the shift from the glacial to the interglacial state originally triggered by small changes in solar radiation due to Milankovitch forcing. Similar effects happen at sub-global scales. The browning of the Sahara, described in Sect. 2.6.4, was ultimately driven by a small, subtle change in the Earth's orbit and the resulting small change in the distribution of solar radiation in the West African region.

Thresholds, Hysteresis and Irreversible Changes

Abrupt changes that occur when some aspect of system functioning crosses a threshold are a common feature of Earth System dynamics. For example, forest fires that affect millions of hectares occur when the combination of the amount of fuel (woody material) and its dryness reach a critical value. This abrupt change is reversible, albeit over decades or centuries. Another example of a system reaching criticality is the melting of ice sheets towards the end of glacial periods. Such changes can occur relatively rapidly, in the context of Earth System dynamics, but can be reversed only over many millennia. An example of an irreversible change is the extinction of a biological species. This can also result from a threshold effect. When landscapes become fragmented due to human-driven land-cover change, the viability of some species can become threatened. The process is highly non-linear, with little or no decrease in populations until a critical level of fragmentation is achieved, at which point extinction can rapidly follow (Lavorel et al. 1995) (Fig. 6.7).

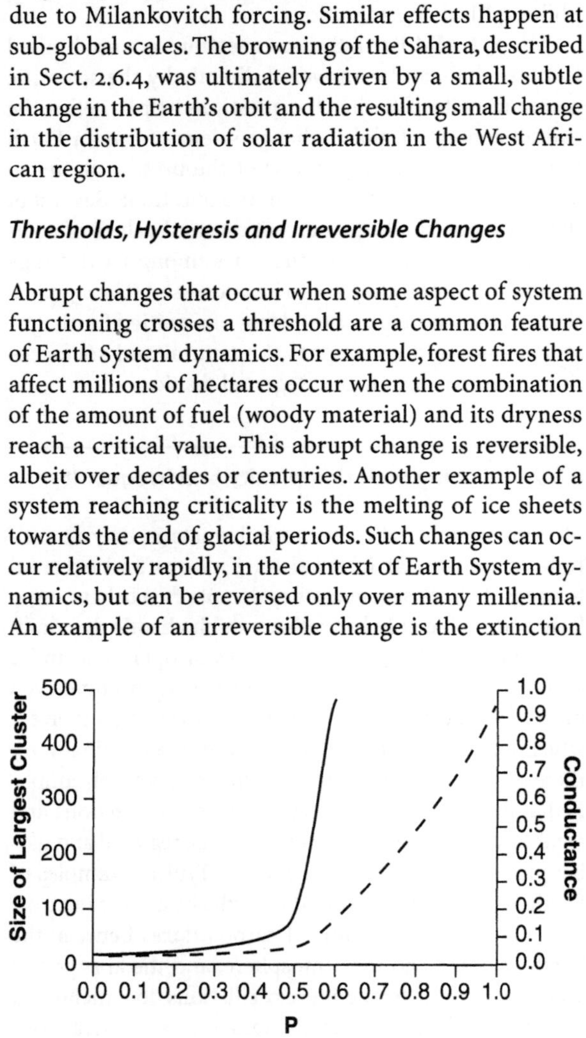

Fig. 6.7. Change in conductivity (*dashed line*) and average cluster size (*solid line*) with the fraction of the landscape containing suitable habitat (*P*) (adapted from Plotnick and Gardner 1993). *Curves* are estimated from gridded landscapes with connectivity between habitat sites established by contact along one or more edges of the four adjacent sites. A connectivity threshold exists at $P = 0.5928$, producing a sudden increase in average cluster size. Assuming that movement is restricted to adjacent habitat sites, conductance below the critical threshold is nearly zero. Above the critical threshold conductance rapidly increases as resistance to movement declines

Indeterminacy

The situation where the behaviour of a system cannot be precisely predicted is called *indeterminacy*, or sometimes *intrinsic uncertainty*, as opposed to uncertainty that can be reduced or eliminated through further study and analysis. It is important to make this distinction. One of the common positions taken by sceptics of global change and by some political leaders is that the future course of global change needs to be predictable with a high degree of certainty before any action towards mitigation can be taken. This position is based on a false understanding of the nature of Earth System. Rather, the intrinsic inability to make precise predictions about the future course of the global environment needs to be recognised, as does the consequent need to view system behaviour in terms of probabilities and risks of various trajectories and outcomes.

The best known example of indeterminacy in Earth System science is the role of human behaviour in the functioning of the System. For example, the precise timing and magnitude of the economic collapse of the former Soviet Union, and the consequent reduction in greenhouse gas emissions from that part of the world, were impossible to predict. All projections of the future dynamics of the Earth System are, to a significant degree, dependent on assumptions about the behaviour of humans at a number of scales. Human behaviour aside, there continue to be debates about the ultimate determinacy of the physical climate system itself and the degree to which the future course of the global climate can be accurately predicted even if perfect understanding of the underlying processes and feedbacks is achieved.

Complexity and Emergent Properties

Complexity is a very general term that can be used in connection with all of the other irregularities described in this section. However, it also has a narrower interpretation. The term complexity can be applied to the feature of systems by which cause-effect relationships between individual components at the sub-system level are not additive or aggregate in simple ways when all of the components are linked to form the system. Emergent properties of the system as a whole appear.

The most striking example of emergent behaviour of the Earth System as a whole is the regular pattern of glacial-interglacial dynamics revealed in the Vostok ice core record (Petit et al. 1999). There are many physical, chemical and biological processes and properties that combine to control the Vostok dynamics – for example, the effect of temperature on the solubility of CO_2 in seawater; the radiative properties of simple atmospheric gases like H_2O, CO_2 and CH_4; the response of terrestrial heterotrophic respiration to moisture and heat; and the relationship between heat, precipitation and pressure in controlling ice sheet dynamics. Much is known about these and many other such processes on their own. However, nobody could have predicted that their combination to form the Earth System could have led to the observed elegant cycling of materials and energy between the glacial and interglacial states.

Scaling Effects

Finally, scaling effects can provide their own type of complexity. In many cases processes and systems are studied and well-understood at fine or local scales and, when aggregated to the global scale, can also be understood in terms of overall relationships. The intermediate scale, called the mesoscale by atmospheric and ocean scientists and the landscape or regional scales by terrestrial ecologists, has proven to be much more difficult to understand and simulate. The reasons for this are well known. Local-scale phenomena are often deterministic enough that they can be at least partially understood using the more standard tools of the environmental sciences while global-scale processes at decadal or longer time scales can be understood in the aggregate as many of the underlying complexities have been averaged away.

However, to make progress Earth System science must be able to deal with the hitherto difficult, sub-global scale. Examples of these difficulties abound. A well-known example concerns carbon fluxes between the land surface and the atmosphere. These can be measured directly at the local scale using eddy correlation techniques and then scaled up to regional and global scales. Alternatively, land-atmosphere fluxes can be inferred, at least to regional scale, using inverse techniques and measured CO_2 concentrations in the atmosphere. These two approaches do not agree quantitatively. The problem lies in the fact that there is no reliable methodology yet to determine carbon fluxes directly at the regional scale and thus validate one or the other technique, or perhaps neither.

How can this daunting complexity and irregularity in the global environment be tackled scientifically? Earth System science can take advantage of the considerable progress recently made in fields outside the usual arena of global change research, such as non-linear dynamics, complex systems analysis, statistical physics, scientific computing or artificial intelligence research. By combining, for example, arguments from bifurcation theory and stochastic analysis, the wild temperature fluctuations during the last ice age can be explained to a large degree (Ganopolski and Claussen 2000). There are now methods developed in biophysics that try to anticipate when critical system thresholds will be crossed by detecting warning signs of the imminent phase transition. This approach is particularly relevant in Earth System analysis for identifying the switch and choke points in the planetary machinery that might be activated inadvertently by human activities. Science can even benefit from the existence of strong non-linearities in the Earth System by devising an inverse sustainability strategy that calculates the critical anthropogenic perturbations to be avoided at all costs.

In addition to these more visionary tools, Earth System science will still need to build on many of those developed and used successfully during the previous decade of global change research. These tools will undoubtedly need to be developed further, as well as integrated into more powerful combinations of multi-technique approaches that tackle system complexities directly.

6.4 The Earth System Science Toolkit

Making Earth System science happen demands an extensive toolkit comprised of existing and developing techniques, such as fully coupled three-dimensional atmosphere-ocean general circulation models, and novel methods and approaches such as those described above. The challenge is to deploy the toolkit not as a series of independent instruments that examine parts of the planetary machinery in isolation, but as an interlinked suite of probes and processors that sense and interpret Earth System behaviour in a holistic way.

6.4.1 Palaeo-Science

The only way to investigate Earth System processes operating on timescales longer than the period of contemporary instrumental records is through palaeo-environmental research. The Earth System leaves traces of and clues to its behaviour in a wide variety of archives – marine and lake sediments, ice caps, tree rings, long-lived corals and archaeological remains and other historical records. Although palaeo-science has achieved remarkable success in shaping the present understanding of Earth System dynamics, significant challenges remain in building a more integrative global system of palaeo-observation and in recovering key archives of past change before they disappear in the current era of accelerating change (Box 6.1). Several new directions are emerging which will make palaeo-science an even more powerful tool in the coming decades.

Box 6.1. Disappearing Evidence: The Need to Protect and Study Key Palaeo-Archives

Frank Oldfield · Keith Alverson

A major obstacle to producing reliable predictions of environmental change and its ecological impacts is a lack of data on timescales longer than the short period of instrumental measurements in the case of climate, or the often even shorter period of quantitative observations in the case of present day ecosystems. Recently established global observation systems will need to be continuously operated for many decades before they begin to provide information on long term processes and interactions. By contrast, carefully calibrated proxy records from natural archives of past environmental variability and ecosystem change can provide a great deal of relevant information now. Unfortunately, some of the most valuable palaeo-environmental archives are disappearing as a consequence of climate change and others are being rapidly degraded, largely due to human activities.

All *mountain glaciers* in tropical and temperate latitudes, except those in Scandinavia, are now rapidly retreating. This implies degradation of the palaeo-record from summit areas of consistent snow accumulation, as well as the exposure of new geomorphological and stratigraphic evidence for past environmental changes, as the glacier retreats. Many *coral reefs* are under increasing pressure, both from local factors such as sedimentation, tourism and over-fishing, as well as from global factors such as warming ocean surface waters and changing ocean chemistry due to increasing concentrations of CO_2 in sea water. Coral bleaching, arising mainly from elevated sea surface temperatures and, to a lesser extent, high light levels, has caused widespread coral mortality in recent years (van Woesik 2001). The loss of corals as palaeo-environmental archives could have serious implications for our ability to reconstruct past changes in sea surface temperatures and salinity, as well as the frequency, amplitude and teleconnection patterns of ENSO (e.g., Gagan et al. 2000). The possibility of using *tree ring records* from the lowland tropics is, ironically, disappearing due to deforestation, just as methods are being developed for extracting palaeo-climatic information from several tropical trees species (Briffa 2000). The transfer functions used to relate the composition of *biological species assemblages in lakes* to climate are now being altered as direct human impacts on lake ecology are beginning to override climate as the dominant influence (e.g., Lotter 1998). This makes it more difficult to develop the inferential basis for reconstructing past climate change from lake sediment sequences. *Coastal environments* that preserve detailed, continuous and sensitive records of past variations in sea-level, bridging the gap between the culmination of the post-glacial eustatic rise and the beginning of tide gauge records, are rare and increasingly threatened, in part because of the extent to which low lying coastal areas are modified by human activities. Such archives are essential for disentangling the relative importance of late Holocene climate variability on sea-level (Goodwin 2003). *Boreal peat lands*, especially those that have become isolated from ground- and surface-water inputs and are therefore dependent on direct precipitation, are among the most sensitive palaeo-environmental archives for reconstructing past changes in moisture status (e.g., Barber and Charman 2003). Threats to these sites include drainage for cultivation and extensive peat cutting for fuel and moss litter. They are in urgent need of protection, especially where they are in close proximity to centres of population. At the same time, many *coastal tropical wetlands* capable of providing long term records of past environmental variability in regions where all too little is known at present are under threat from both drainage and fire (Immirzi and Maltby 1992).

In some of the examples given above, the urgent need is for protection of those archives of greatest value against further losses and degradation. In other cases, preservation is not possible and the urgent need is to extract the information they contain before the opportunity to do so is gone. The case of tropical glaciers is especially important. Ice core data from these glaciers has been crucial in determining the past variability of the hydrological cycle in the low latitudes. This aspect of climate variability has been particularly important for the viability of human societies in the past and is likely to remain so in the future, since many of the most vulnerable groups live in the low latitudes, where water resources are becoming an increasingly serious issue. Yet tropical glaciers are disappearing at an alarming rate. The summit ice area on Mt Kilimanjaro, for example, has decreased by 81% from 1912 to 2000 (Fig. 6.8). If this trend continues, there will be no ice, and hence no opportunity to obtain further palaeo-climatic information, by 2015.

Alverson et al. (2001, 2003) use some of the examples described here to argue for the establishment of a Global Palaeo Observing System (GPOS), designed, wherever possible, to conserve those endangered natural archives that contain records of past environmental variability on socially-relevant time scales and to expedite the urgent study of the palaeo-environmental record held in those archives, the disappearance of which is imminent and inevitable.

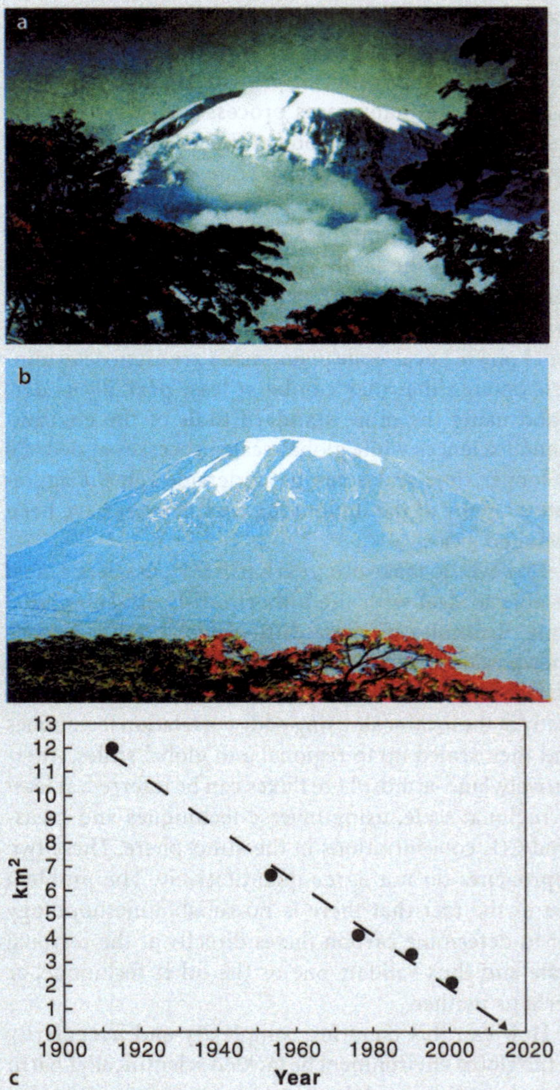

Fig. 6.8. Photographs of ice on the summit of Kilimanjaro, **a** in 1970; **b** in 2000; **c** projected disappearance of the glacier on top of Kilimanjaro by 2020 (photos and data from Thompson et al. 2002, adapted by Alverson et al. 2001)

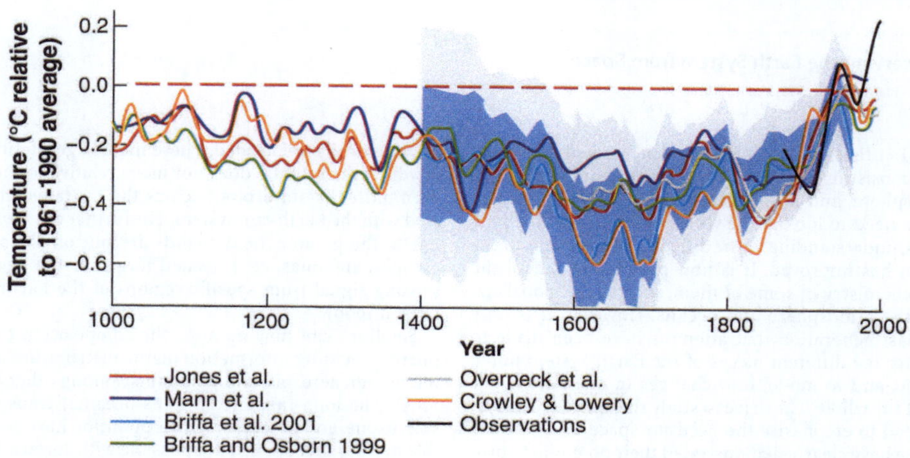

Fig. 6.9. Comparison of six large-scale reconstructions of northern hemisphere mean annual surface temperature, all calibrated with linear regression against the 1881–1960 mean April-September observed temperature averaged over land areas north of 20° N. All series have been smoothed with a 50-year Gaussian-weighted filter and are anomalies from 1961–1990 mean. Observed temperature for 1871–1997 (*black*) are from from Jones et al. (1999); circum-Arctic temperature proxies for 1600–1990 (*grey*) from Overpeck et al. (1997); northern hemisphere temperature proxies for 1000–1980 (*red*) from Jones et al. (1998); global temperature and non-temperature proxies for 1000–1980 (*dark blue*) from Mann et al. (1998, 1999); three northern Eurasian tree ring width chronologies for 1000–1987 (*green*) from Briffa and Osborn (1999); 13 northern hemisphere temperature proxies for 1000–1987 (*orange*) from Crowley and Lowery (2000) (but excluding the two low-resolution records used by them); and the northern hemisphere reconstruction from the age-banded analysis for 1402–1960 (*light blue*), bounded by the one and two sigma standard errors, from Briffa et al. (2001)

First, there is a strong move towards both multi-proxy and interdisciplinary approaches. A good example of the former is the reconstruction of northern hemisphere surface temperature from a number of proxies (Fig. 6.9). The fact that there is general agreement among a range of proxies gives confidence that the record faithfully captures the environmental changes of the past. In addition, differences between records provides insights into the individual processes on which the proxy relationships are based and their behaviour through time. Where the lines of evidence available for a given period span a range of environmental archives and proxies, both physico-chemical and biological, it becomes possible to explore the interactions between them, including ecosystem responses and feedbacks.

Secondly, an increasingly important aspect of the palaeo-record is its capacity to provide, through a wide range of well-calibrated data, spatially explicit and carefully dated reconstructions against which model simulations can be tested. As already illustrated in several sections of this book, much of the new knowledge about the complex interactions and non-linear behaviour of the Earth System, including biophysical and biogeochemical feedbacks, has come through the iterative synergy between model development and the comparison of hindcast simulations with palaeo-data, either for critical time slices or during periods of Earth System transformation. Building on this new knowledge in future will make a major contribution to the insight needed for improving the scientific contribution to modelling future global change.

Thirdly, the increasing linkages between palaeo-scientists and those studying contemporary and future changes are already providing an invaluable historical context against which modern change can be evaluated. The challenge for the future in palaeo-science is to adopt a hypothesis-driven approach, synthesise records from a range of proxies, construct quantitatively calibrated and well-constrained chronologies, and harmonise these records with contemporary observations. The ultimate goal is to provide a consistent and continuous data stream from the distant past, through the recent past and contemporary period to the future, using prognostic models. An early example of this approach – the record of atmospheric CO_2 concentration from 420 000 BP (the Vostok ice core record), through the period of contemporary measurement (the Mauna Loa record) and into the future via the IPCC projections (Fig. 1.1) gives a striking insight into the potential for human activities to affect Earth System functioning. Such a perspective is not possible without the long-time record.

Finally, palaeo-science is placing increasing emphasis on the more recent past, where the interplay between environmental variability and human civilisations provides intriguing insights into the factors that influence societal vulnerability to environmental change. There are many examples from the past several thousand years of flourishing societies that abruptly collapsed (Box 5.3). In many cases these collapses are tightly correlated with sudden and persistent environmental changes, often involving the hydrological cycle. Even more collaborative research between palaeo-environmental scientists, anthropologists and archaeologists is required to determine if modern societies, with their high population densities, over-reliance on limited resources and inter-linked economies and communication systems, are more vulnerable to environmental change than were societies in the past.

Box 6.2. Observing the Earth System from Space

Jack A. Kaye

The recognition that the Earth needs to be studied as a complete system consisting of an interactive atmosphere, ocean, biosphere, cryosphere, and lithosphere has led scientists to also recognise the need to look at the whole system. Over the past few centuries, understanding of the different components of the Earth System has improved. It is now possible to model the physics and chemistry of some of them, such as the global circulation of the atmosphere and oceans. However, it is only within the past generation that attempts have been made to weave together the different pieces of the Earth System into a unified whole, and to model how changes in one part of the system affect the whole. The drive to study the entire Earth System has helped to emphasise the need for space-based observations. These have clearly demonstrated their potential to provide critically needed observations over the entire Earth and have had a significant impact on the discipline of Earth System science. The inclusion of a large number of satellite images and satellite-derived data in this book attests to the revolution in Earth System observation that satellite observations have helped enable over the last two decades.

Over the past 40 years, satellite observations have dramatically increased knowledge of the Earth System. From the first images in the early 1960s that showed the distribution of clouds over the Earth, satellites now track nearly all aspects of the Earth System over all its regions (Stevens and Kelley 1992; Gurney et al. 1993). The last several years have seen an enormous enhancement in space-based remote sensing of the Earth through launches of research satellites such as the European Space Agency's ENVISAT, the National Space Development Agency of Japan's ADEOS-II (now Midori-II) satellite, as well as several platforms of the US National Aeronautics and Space Administration (NASA) Earth Observing System (EOS). Improvements in data transmission and processing now enable significant amounts of data to become available in near real-time (Earth Observatory 2003), allowing satellite data to play important roles in forecasting and disaster management (CEOS 2003a), as well as for studies of climate and global change. The importance of satellite-based remote sensing of the Earth is reflected by the many nations that carry out such programmes, both individually and in partnership (CEOS 2003b; WMO 2003).

Space-based remote sensing observations have a number of advantages over more conventional means of observation. With a single platform in low Earth orbit, one or more instruments can view the entire Earth in a single day – and even get multiple views of many regions of the Earth. While surface-based research observations typically require the presence of multiple locations in stable and developed geopolitical regimes with good access to power, data links and trained personnel for operation and maintenance, a single satellite can provide observations over all areas of the surface (developed and undeveloped countries, stable and unstable political situations, land and ocean, difficult and easy in which to work) with a single instrument. Further, a single, well characterised instrument can be used to study the entire planet, reducing the inter-instrument calibration issues associated with surface-based networks.

However, tradeoffs exist in using satellites to observe the Earth. Satellites in low Earth orbit move rapidly, typically at 7 km s^{-1}, which precludes the possibility of sampling the short-term (e.g., more frequent than diurnal) evolution of environmental parameters with a single satellite. Satellite observations of the surface in most wavelength regions are limited by the presence of clouds and aerosols, which may contaminate a field of view. Depending on the spatial resolution of the satellite, uncertainty may arise from the presence of variability within the satellite's field of view. Once satellite instruments are launched, they are subject to degradation in orbit, which needs to be tracked over the lifetime of the mission. In order to obtain more frequent measurements with a single satellite, higher altitude orbits are needed. These include geostationary orbits, in which the satellite does not move relative to the ground, or even more distant orbits, such as the Lagrange points associated with the Earth-sun system. The further out the orbit is from Earth, the greater the demands are put on the systems (telescopes, antennas, etc.) needed to gather the desired remote sensing signal from specific regions of the Earth (Wertz and Larson 1999).

Satellites can now monitor the composition of the atmosphere, providing information on the distribution of ozone, water vapour, aerosols, and pollutants, among other things (Kaye 1995). The long-range (e.g., trans-oceanic) transport of aerosols, ozone, and precursors of air pollution have been made visible in ways that are just not possible with surface-based observations. Satellites can determine the presence of clouds, and provide quantitative information about land cover and land use and also biological productivity – on land and in the ocean. Ocean surface height, winds, temperature, and chlorophyll amounts over the ocean can all be measured, allowing scientists unprecedented opportunities to study not only the physical components of phenomena such as El Niño, but to demonstrate how the distribution of biological activity responds to changes in the physical state of the ocean.

More recently, satellite observations have increased in their ability to measure in the vertical. Through use of active remote sensing, in which the satellite brings its own energy source to transmit a signal to Earth and then measures the reflected and scattered signal which returns to the spacecraft, satellites are now able to better determine the vertical variation in parameters of interest. Radar altimetry and lidar are the primary such techniques and have been used to study sea surface height and land surface deformation, and will increasingly be used to study ice surface height (through the NASA ICESat launched in January 2003) and cloud and aerosol distributions (to be studied through the Cloudsat and Cloud Aerosol Lidar and Infrared Pathfinder Satellite (CALIPSO) missions to be launched in 2004 by NASA together with the Canadian Space Agency and French Centre National d'Etudes Spatiales (CNES), respectively, as well as with ICESat). Space-based observations of vegetation structure and biomass constitute a new and significant challenge for radar and lidar systems that could yield valuable insight on these critical environmental parameters. More recently, variations in the Earth's gravitational field are being sampled by the US-German Gravity Recovery and Climate Experiment (GRACE) to provide information about the spatial and temporal variations in the Earth's gravitational field, and this information can be used to study variations in the distribution of water.

Satellite observations, originally developed for research, have become critical components of the global environmental monitoring system. They constitute a major fraction of the information used to initialise the models used for numerical weather prediction. Until recently, primarily observations of the vertical profiles of water vapour and temperature were used in this way, but an increasing set of parameters are now being used, such as measurements of precipitation, ocean surface winds, and ozone distributions. Satellite observations also have become the basis for multi-decadal studies of climate change, atmospheric chemical change, and land cover change. Long-term data sets of ozone and aerosols have been critical in documenting both global and polar ozone depletion, as well as the effects of large volcanic eruptions on the distribution of stratospheric aerosols. Satellites have also provided a quantitative data set of the variation of both total and spectrally-resolved solar irradiance over more than two 11-year solar cycles – measurements that would be impossible from the ground because of the need to make them above the Earth's interfering atmosphere. According to passive microwave satellite observations, the distribution of sea ice has been shown to be changing in very different ways in the Arctic and Antarctic over the past two decades

◀

(Fig. 6.10). This shows a statistically significant decrease in sea ice extent in the former, with evidence of an increase in the latter. High resolution land cover and land use observations from the Landsat series of satellites going back to 1972 have shown not only long-term changes in land use associated with deforestation and in some cases reforestation, but have provided regional information about the growth of urban and suburban regions at the expense of surrounding agricultural lands.

Future advances in satellite observations of the Earth over the next decade are expected because of advances in technology, platform technology, data transmission and algorithms. Technology holds the promise of reducing the size and power needs of instruments, allowing for significant reductions in cost of future instruments. The Advanced Land Imager instrument flown by NASA on its EO-1 spacecraft meets many of the goals of the imaging instruments aboard the Landsat satellite at dramatically reduced size and cost. Satellites are increasingly being flown in multi-national constellations of several spacecraft, allowing for the benefits of multi-instrument synergy without the cost, complexity, and risk associated with the use of large multi-instrument platforms. Developments such as large synthetic apertures for microwave observations and long-duration multi-frequency lidars will enable the benefits of active remote sensing to be applied to an increasing suite of environmental parameters. Furthermore, combinations of such sensors with the large telescopes needed for use in higher altitude orbits will facilitate active measurements from these orbits, allowing scientists of the future to reap the benefits of both more frequent temporal sampling and higher vertical resolution from a single instrument than is possible today.

Fig. 6.10. Deviation in monthly sea ice extent from long-term monthly means for Arctic (*top panel*) and Antarctic (*bottom panel*) over more than 20 years using data from satellite microwave radiometers (C. Parkinson, NASA/GSFC, pers. comm.)

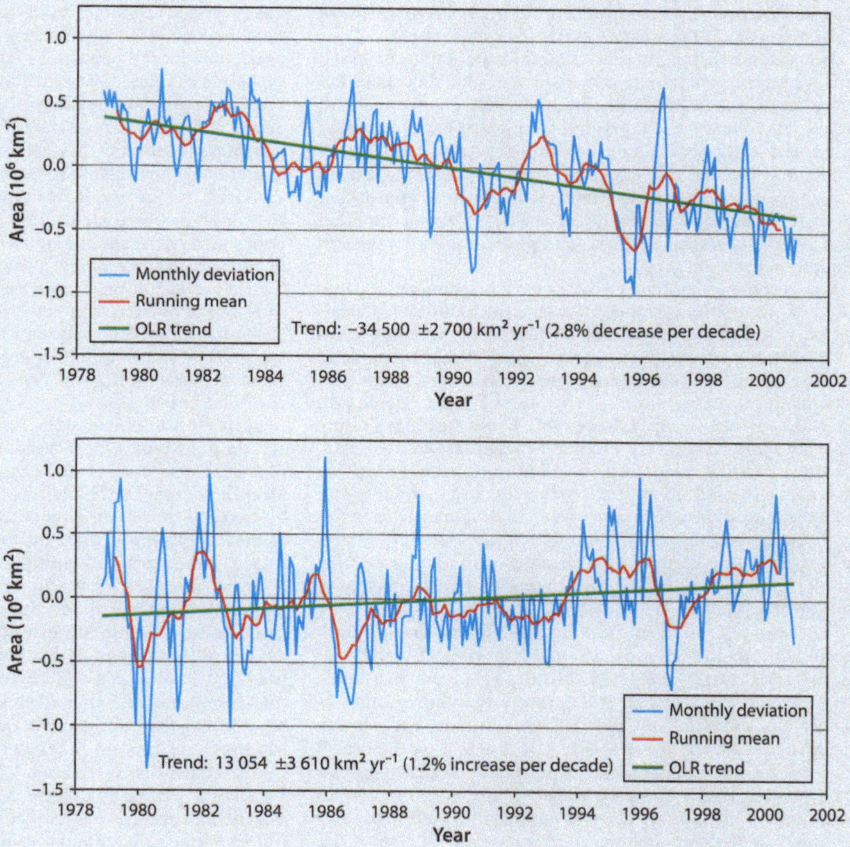

Realising the full potential of palaeo-science will depend on the continued development and maintenance of high quality data archives. Ensuring high data quality, as well as harmonising and improving access to palaeo-data archives, are vital tasks that require sustained and dedicated resourcing. Given the acceleration of international networking among scientists and scientific institutions as a central strategy for undertaking Earth System science, the provision of standardised, well-documented and easily accessible databases will become even more important in the future.

6.4.2 Contemporary Observation and Monitoring

Observation of the Earth from space has revolutionised human perspectives and understanding of the planet, and the expanding array of sophisticated remote sensors promises a flood of valuable data, complementing observations made on the Earth's surface (Box 6.2). Creative approaches for documenting the role of people in the Earth System can illuminate the human dimensions of planetary dynamics (Box 6.3). Well-conceived and

Box 6.3. Integrated Land-Change Science: Remote Sensing and the Human Dimensions

B. L. Turner II

The human impress on the terrestrial biosphere, from degradation and depletion of resources to the functioning of ecosystems, is registered significantly through land-use and land-cover change. To model and project this change in ways useful for comprehensive understanding of the Earth System and sustainability requires land-use and land-cover dynamics to be treated as a coupled human-environment system. A significant data source for this coupling is remote sensing, especially satellite imagery, which enhances analysis and assessment longitudinally and in terms of spatial specificity. The linkage of the remote sensing, biophysical and human sciences to tackle the coupled system has generated an emergent *integrated land-change science* (Turner 2002). This inter- or multi-disciplinary approach to land change analysis has proven successful in regard to land cover (biophysical attributes registered in remotely sensed imagery) and problematic for land-use change (human purpose or intent of land cover), hindering the assessment of the human part of the coupled system in question.

Can remote sensing provide useful information to social scientists studying the human dimensions of global change? Recent innovations in land-use and land-cover change research suggest that there are opportunities to adapt remote sensing techniques to that portion of the coupled system to which the human sciences provide expertise (Fox et al. 2003; Liverman et al. 1998; Walsh and Crews-Meyer 2002). Various approaches are under active investigation, two of which warrant attention here: geomantics-social analysis and modelling to the pixel in remotely sensed imagery.

Socio-economic data can be combined with geomantics (Geographic Information Systems, geoscience and automated information systems) to explore various means of detecting signals not readily apparent in the imagery to gain insights about land change related, for example, to market, subsistence, policy, and urbanisation processes and the vulnerability of people and places to various hazards. Recent advances include: linking household lifecycles to the kind and magnitude of deforestation in Amazonia; detecting vegetation impacts of stocking strategies in the Karoo of South Africa linked to ethnicity and socio-economic circumstances of the land manager; connecting foreign direct investment and agricultural land productivity to explain the pattern of urban expansion in southern China; and determining the location of potential disease outbreaks linked to ENSO.

Two examples illustrate the approaches in more detail. Recent work from Mali indicates that strategies of land burning not only can be detected, but the resulting fire-landscape pattern can be used to determine the vulnerability of villages and crops to fires (Laris 2002). Fires set early in the cropping season are low intensity and create a patchwork of low-level biomass that serves to buffer the landscape from large fires late in the cropping season. Mali policy directed to reduce or stop burning changes the amount and pattern of early fires, increasing the probability of village and crop losses from late fires. In another example, principle component analysis has been used with high-temporal resolution imagery to detect areas of relative loss and gain in the Miombo vegetation in Neno, Malawi (Haan et al. 2000). Part of this deforestation is attributed to breakdown in social institutions that once controlled the collection of an edible and highly valued insect that migrates seasonally to the local forests. With this breakdown, the insect has been over-harvested, and with its reduced number, trees are felled to access them rather than shaken as in the past. Changes in resource governance and control have been instrumental in leading to a specific type of deforestation, a cause-consequence first detected by special analysis of satellite imagery.

A large number of efforts are underway to model the coupled human-environment system linked directly to the pixel in remotely sensed imagery, expanding the social dimension of the problem (Fox et al. 2003; Parker et al. 2002; Walsh and Crews-Meyer 2002). Much of this work has been directed to modelling tropical deforestation and demonstrating the important role of spatial inertia; deforestation is amplified on lands (pixels) adjacent to recent deforestation. Much more depth of understanding has been gained, however, as illustrated in the following examples.

Several efforts are underway to use biophysical, socioeconomic, and remote sensing data to model the deforestation and agricultural change in the southern part of the Yucatán Peninsula pixel-by-pixel. Government census and various spatial data (e.g., roads, soil type) are used in a logit model of the study region to estimate the probability of deforestation of each pixel in the landscape as a function of the explanatory variables (Geoghegan et al. 2003). The results are consistent with previous aggregate econometric studies. For instance, the higher the elevation of a pixel, the further a pixel is from a road or market, or the smaller the population level of the area, the smaller the probability of deforestation. Reducing the scale of analysis reveals more. A random sample of household survey data, including surveys of distant cultivated fields linked to remotely sensed imagery, is used in a survival analysis to identify the effect of household-level explanatory variables on the probability of both the location and timing of deforestation. The results indicate that in addition to identifying several variables relevant for policy analysis, including household demographics, proximity to roads, and government provision of agricultural support, the deforestation process is characterised by non-linear duration dependence over time, with the probability of forest clearance first decreasing and then increasing with the passage of time (Vance and Geoghegan 2002). The two models help to explain the location of deforestation, cultivation, and successional growth.

Scale dependent models, multi-level models, and spatial simulation models are also being used to examine land dynamics. Walsh et al. (2001) characterised plant biomass variation in northeast Thailand at nine different spatial scales that ranged from 30–1050 m. Using spatial agglomeration techniques, population distribution models, a remote sensing image time-series, and multiple regression and canonical correlation analyses, variation in plant biomass was assessed relative to biophysical and demographic variables and found to be scale dependent. Results indicated the importance of community-level, population variables (e.g., number of households) at finer scales and environmental gradients (e.g., soil moisture potential) at coarser scales. Models were generated to examine the intra-annual variation in plant biomass and the nature of biomass change within a region characterised by the cultivation of rainfed, lowland paddy rice and upland field crops, primarily cassava and sugar cane. Similar to scale dependent analyses, multi-level models (or Generalised Linear Mixed Models) have been used to study land dynamics in Ecuador by combining community- and household-level independent variables in the same equation to consider household decisions within a larger context. Finally, spatial simulations have been generated to assess land dynamics in Ecuador through cellular automata and agent-based models. Messina and Walsh (2001) developed spatially-explicit simulations of deforestation in the Ecuadorian Amazon using a cellular automata approach. A satellite image time-series was used to develop land transition or growth rules for the model, neighbourhoods were used to consider contextual effects, and initial conditions were set by satellite classification. The effects of geographic accessibility, resource endowments, and urbanisation were integrated into the model to estimate land composition and spatial pattern for historical and future time periods and relative to development scenarios.

-deployed global monitoring networks can provide powerful early warning systems for global change. Some innovative concepts have already been developed and first concrete steps taken towards developing a global macroscope (Schellnhuber 1999), a truly integrated system for observing the most critical features of Earth System dynamics.

Meeting the challenge of observing the critical states of and trends in the Earth System will first require new initiatives to bring partial and fragmented but nevertheless useful current observing programmes together into a common global framework. Such work is essential for bridging the large gap between the present state of observational systems and the urgent requirement to build a unified global system to continuously monitor the state of the Earth System. One of the most ambitious and successful of such efforts is the project on Global Observation of Forest Cover (GOFC) and Global Observation of Land-cover Dynamics (GOLD) (GOFC-GOLD 2002). The vision of GOFC-GOLD is to build a system for the continuous, internally consistent, global-scale observation of land cover. An initial step towards this vision was the construction of a global map of percentage tree cover (Fig. 6.11). Several important aspects of Earth System functioning, such as the role of terrestrial ecosystems in the global carbon cycle and the nature of biogeophysical feedbacks of the land surface to climate, are related to the degree of tree cover, emphasising its value as a key indicator of the state of the Earth System.

To meet the challenges of observing societal aspects of global change, a more comprehensive global information base must be developed to guide informed economic, social and environmental decisions and action in transitions to sustainability. In order to undertake science for sustainable development, more integrated data sets and modelling tools will be needed to provide systematic, structured analyses of global change issues. In particular, there will be a demand for data of a new quality in economics and social sciences. This has led to the idea of a Sustainability Geoscope (*www.sustainability-geoscope.net*). Built on a combination of remote sensing and on-the ground-observations, the Geoscope provides a framework for an observation and monitoring scheme on a global scale, including economic, social, environmental and institutional components.

In addition to new initiatives of various types like GOFC-GOLD and the Geoscope, there is a need for an overall framework into which the various components of an Earth System Observing System can contribute. The first steps toward such a system have been taken with the formation of the IGOS (Integrated Global Observing Strategy) Partnership (*www.igospartners.org*). The partnership brings together the efforts of a number of international bodies concerned with the observational component of global environmental issues, both from a research and a long-term operational programme perspective. Membership includes the Earth observation satellite community, the *in situ* observation systems and the international global change programmes IGBP and WCRP. The primary goal of the IGOS Partnership is to undertake a strategic planning process towards the harmonisation and integration of the large number of important but fragmented observations of parts of the Earth System. Although the task of developing a global monitoring system that is effective enough to provide early warning of abrupt changes in the Earth System is truly daunting, IGOS-P represents the first attempt at building the institutional structure needed to achieve this ambitious goal.

In contrast to the rapid development of the remote sensing approach for Earth observation and the institu-

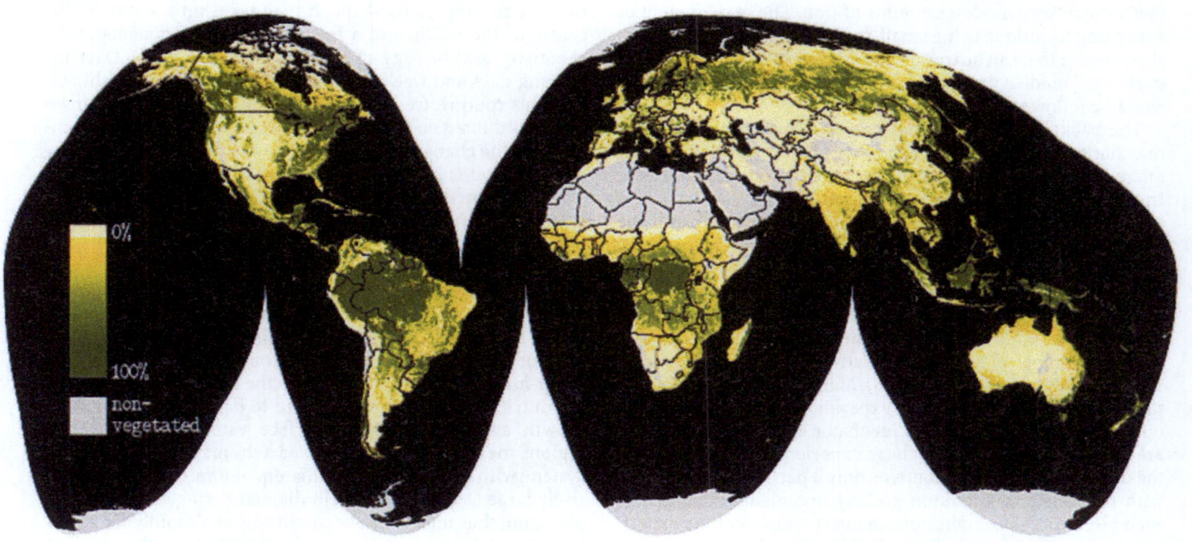

Fig. 6.11. A prototype data set estimating percentage tree cover from 10 to 80% based on satellite data acquired by the Advanced Very High Resolution Radiometer in 1992–1993 (DeFries et al. 2000). Percentage tree cover is likely to be underestimated in areas with significant cloud cover throughout the year. The spatial resolution of 1 km precludes the detection of finer forest fragments

tional advances towards building truly integrated observing systems, there remain significant challenges in building the surface-based component of Earth observing systems. The deterioration of the global hydrological observing system is a striking example (Meybeck and Vörösmarty 2004). Ironically, just as water resource issues are being recognised as arguably the most important facing humanity in the first part of the twenty-first century, hydrological observations are in a rapid and ongoing decline. For example, the fraction of Earth's drainage basins for which hydrological data is available has dropped to 45% in the 1990s; more is known about land surface runoff 20 years ago than today. There is only sporadic and non-systematic information about inland water quality (e.g., Vörösmarty et al. 1997); there is presently no global database or observation system for the fluxes of sediment to the coastal zone. Observation of changes in aquatic biodiversity is virtually non-existent. Building, rebuilding and maintaining such ground-based observation components are critically important for the Earth System science toolkit, yet severe legal, economic and political constraints appear to be accelerating their demise.

6.4.3 Earth System Experimentation

Simulation of future environmental conditions on Earth provides the means to study the structure and functioning of ecosystems under new combinations of atmosphere and climate. Addition and removal of species provide insights into the responses of biological systems as their complexity changes. Manipulation of element flows, such as the addition of iron to nutrient-poor areas of the oceans, mimics system responses to changes in biogeochemical cycling.

Terrestrial ecosystems are already experiencing significantly altered environmental conditions. How will they respond to accelerating changes in atmospheric composition and climate? Recent advances in experimental techniques are achieving an unprecedented level of realism in subjecting terrestrial ecosystems to changes in their abiotic environment. Whole intact ecosystems in the field can now be fumigated with elevated atmospheric CO_2 in the absence of any enclosures that would alter the micrometeorology (the Free-Air Carbon Dioxide Enrichment (FACE) experiments) (Hendrey et al. 1999) (Box 4.5). In addition,

Box 6.4. Iron Fertilisation of Ocean Ecosystems

Andrew J. Watson

One of the most exciting developments in marine sciences in the last decade has been the introduction of open-ocean ecosystem manipulation experiments, enabling the effects of a single change to the natural marine ecosystem to be studied over a period of weeks. The experiment-and-control technique, fundamental to scientific enquiry, is comparatively easy to apply to plants and ecosystems on the land, but in the ocean the constant movement and mixing of the water makes such experiments technically challenging. However, in the special case of iron enrichment, the very low concentrations of Fe (of order 0.1 nM) found in some open ocean regions means that it is possible to enrich an experimental area of hundreds of square kilometres with very modest amounts of iron. The development of a labelling technique using small amounts of sulphur hexafluoride, a tracer that can be tracked in real time from a ship (Watson et al. 1991), made it practical to perform *patch* experiments that could be followed for several weeks.

The background to the iron release experiments has been the continuing debate about the role of iron as an important micro-nutrient, in short supply in large regions of the ocean. Iron limitation of the Southern Ocean marine biota was first suggested as long ago as the 1930s (Gran 1931), but until the advent of *ultra-clean* measurement techniques in the 1980s, the true, sub-nanomolar, concentrations of iron in the ocean were unknown. In the late 1980s, Martin and colleagues published papers showing that in laboratory incubations of phytoplankton in bottles, addition of iron apparently caused substantially increased biological production relative to controls (Martin and Fitzwater 1988; Martin et al. 1990). Many marine biologists remained unconvinced by these experiments however, pointing out that conditions inside an incubator on the deck of a ship are radically different from those experienced by plankton in the ocean. The incubator contains only a part of the ecosystem, with the larger zooplankton grazing organisms excluded for example, and this complicates the interpretation of the experiments. In particular, the large increases in the incubators of plankton biomass and uptake of nutrients and CO_2, could not be reliably extrapolated to the open ocean.

To begin an iron enrichment experiment, large tanks of seawater in which iron salts are dissolved are prepared on the deck of the research vessel. Iron sulphate has been used so far; by lowering the pH of the sea water in the tank it dissolves readily. A solution of the tracer that will be used to track the patch is also prepared, and to begin the experiment, the two solutions are pumped over the stern of the ship at constant rates while the vessel steams a pattern around a drifting buoy. Typically a few tonnes of iron sulphate are used, released over the course of a day or so into a patch a of order ten kilometres across. Once the release is over, the analytical instruments in the ship's laboratories are turned on and the ship steams a survey over the patch, mapping its shape. It then takes up a station in the centre of the patch, and a full suite of measurements of the chemistry and biology of the ecosystem are made. Over the ensuing days and weeks, the team of scientists on the ship repeat this routine, frequently re-mapping the patch as it grows bigger, is diluted and carried along by the currents, while documenting the changes occuring inside the patch and comparing them to what is happening to the undisturbed ecosystem outside. Concentrations of dissolved iron are continually monitored, and if these fall too low, a re-infusion with inorganic Fe may be performed.

At the time of writing, such experiments have been performed in three major regions of the world ocean – the Equatorial Pacific (Martin et al. 1994; Coale et al. 1996), Southern Ocean (Boyd et al. 2000; Watson et al. 2000) and most recently the Northeast Pacific. All these regions are so-called *high-nutrient-low-chlorophyll* areas, where the nitrate and phosphate, the nutrients classically considered to be limiting to plankton growth, are abundant in the surface waters. In each of these regions the addition of iron caused a dramatic shift in the ecosystem, with the initiation of a diatom bloom. Diatoms are relatively large (tens of microns in diameter) single-celled phytoplankton, that under bloom conditions can rapidly fix a large biomass and remove carbon and nutrients from the water, resulting in drawdown of the classical nutrients and, potentially, uptake of carbon dioxide from the atmosphere.

whole ecosystems can be heated via cables in the soils and infrared heaters above the vegetation to study the impacts of a warmer climate on ecosystems (Rustad et al. 2001).

Experimentation does not stop with the manipulation of conditions in the abiotic environment. The structures of ecosystems themselves are now being experimentally altered to gain insights into the roles that they play in Earth System functioning. Experimental research on biodiversity has shifted from model ecosystems in controlled environments to the manipulation of biodiversity in the field. Species-removal experiments (Wardle et al. 1999) are now conducted to understand better the role that so-called keystone species play in ecosystem functioning. Entirely artificial grassland and forest ecosystems have been built up in the field through the selection of a given number of species from a large pool (Tilman et al. 1996; Hector et al. 1999). By manipulating both the number and type of species, the importance for ecosystem functioning of the number of species alone compared with the role of particular species types can be elucidated.

Ecosystem structure has been experimentally manipulated in the ocean also, but by very different techniques. By altering the biogeochemistry of large areas through the addition of iron to the surface waters, changes in both the productivity and structure of marine ecosystems can be studied (Box 6.4).

Nature itself and human activities have both provided inadvertent experiments that can give important new insights into Earth System functioning. Natural disturbances, such as volcanic eruptions and the subsequent lava flows, can eliminate existing terrestrial ecosystems, thereby allowing the study of the change in functioning as ecosystems recolonise destroyed areas and slowly build up system complexity. Humans have inadvertently conducted many experiments by simplifying ecosystems through land-cover conversion, through the introduction of alien species into new ecosystems and by the fertilisation of large areas of land and coastal seas through the alteration of biogeochemical cycles. The Earth System science research community can potentially learn an enormous amount from these perturbation experiments. In fact, the Earth as a whole is now a global test bed as humanity accelerates its unintended planetary-scale experiment with its own life support system (Deevey 1967; Oldfield and Dearing 2003).

◄

The fate of the diatom blooms produced by iron enrichment varies from experiment to experiment. In the warm waters of the equatorial Pacific the blooms are apparently quite short-lived and fade in the space of two weeks, but in the Southern Ocean they last much longer. Figure 6.12 shows the Southern Ocean Iron Enrichment Experiment (SOIREE) bloom created in the Southern Ocean, as seen by the SeaWiFs satellite six weeks after the experiment began (Abraham et al. 2000), with high chlorophyll concentrations still apparent.

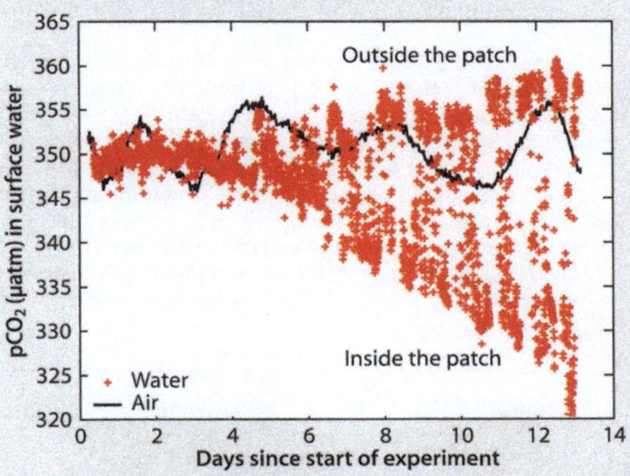

Fig. 6.12. Measurements of partial pressure of carbon dioxide (pCO_2) in the surface of the ocean in the two weeks following the release of iron during the Southern Ocean Iron Enrichment Experiment (SOIREE). The *red crosses* indicate the pCO_2 variability at the water surface as it was continuously recorded by the vessel as it steamed into and out of the fertilised patch of water. The *wavy black line* corresponds to the actual CO_2 partial pressure in the overlying air (which follows changes in the atmospheric pressure). The figure shows steadily decreasing pCO_2 within the patch, as it was utilised by phytoplankton stimulated by the iron additions

These experiments have brought about a *paradigm shift* in marine science. To an extent that almost no one would have believed in the late eighties, it has become clear that iron is a critical limiting nutrient for the marine biota. Its supply has a fundamental influence on the state of the marine ecosystem, switching it from a *recycling mode* in which the plankton community is relatively self-contained, to *export mode*, in which carbon and nutrients are stripped from the surface waters and exported to depth (Landry et al. 1997). This has important consequences for understanding of the Earth System, for the biological export in the Southern Ocean is a key factor in setting natural concentrations of CO_2 in the atmosphere, and the supply of iron in atmospheric dust to the Southern Ocean has an important role in explaining why CO_2 varied between glacial and modern pre-industrial times (Martin 1990; Watson et al. 2000).

Most interest, particularly among the popular press, has however been generated by the suggestion first made by Gribbin (1988) that deliberate fertilisation of the oceans by iron could be used to sequester carbon dioxide from the atmosphere and help ameliorate global warming. The potential of this technique to sequester atmospheric CO_2 has been greatly exaggerated in some quarters. It is true that fertilisation of the Southern Ocean (but not the equatorial Pacific) could sequester significant quantities of atmospheric carbon dioxide, so it is possible that deliberate fertilisation could have some limited role as part of a future strategy to deal with climate change and carbon sequestration. However, it has been known for more than a decade that there are dynamic limitations to how quickly the ocean can take up CO_2 (Peng and Broecker 1991). In practice, even if massive geo-engineering (cf. Sect. 6.5.2) were to be undertaken, it could not solve the climate change problem. The proposal to use the oceans in this way raises deep ethical issues (who has the right to use our common heritage in this way?) as well as environmental concerns (little is known about possible unintended consequences of large scale fertilisation). These questions need further research and public debate before any large scale programme should be undertaken.

6.4.4 Global Networks

Networking has been a tool for scientific research long before the advent of global change research, although it is now becoming one of the most important research approaches in Earth System science (Ingram et al. 1999). In its most basic form, a network brings together groups of researchers to exchange ideas and to undertake joint research. This brings three major benefits. First, networks facilitate research that could not be accomplished by independent researchers working on their own, for example, experiments where many individual sites or studies are required, enabling the generation of synthesised results. Secondly, participation in a network facilitates the lateral transfer across research groups of methodology, technology and experience. Thirdly, added value of a resource or financial nature can be obtained through the sharing of models, data and research facilities, all of which can be expensive for individual groups to develop and maintain or obtain from elsewhere.

Box 6.5. Studying the Biogeochemistry of Coastal Seas

Chris Crossland

The movement of important elements, such as carbon, nitrogen and phosphorus, from the land to the ocean is a critical process in the functioning of the Earth System, both in its natural and its human-dominated states. A strong argument can be made that the high load of materials from land and the human influences along the coastal interface cause much of the net reaction of the materials to occur within the landward margin of the coastal sea rather than on the shelf. Material reactions and transformations will differ at local scales and changing trends in the net biogeochemical processes within the coastal system require relatively fine scale measurements to understand variability and the effects of changing climatic and human forcing. Given the strong heterogeneity of the coastal zone, studying fluxes at the global scale is difficult and can only be effectively undertaken using a networking approach. A program of assessment has been established involving coastal scientists from all global regions (*http://www.nioz.nl/loicz/*) who developed and applied a globally applicable methodology for estimating these fluxes, especially within the bays and estuaries of the inner coastal zone.

The strategy was to develop a methodology that (*i*) depended on existing or easily obtainable data; (*ii*) did not require large amounts of data, and (*iii*) was widely applicable and uniform so as to allow effective comparison among sites. This last criterion was difficult to meet as the budget sites exhibit an exceptional range of heterogeneity. The sites vary dramatically in their characteristics, from lagoons and estuaries less than 1 km² in area to the 106 km² East China Sea; from sites that are decimetres in depth to sites that are hundreds of metres deep; from sites that are virtually devoid of loadings from land to sites that receive heavy loads of inorganic nutrients derived from human wastes, agriculture and other sources; from sites that are river-dominated estuaries to hypersaline embayments; from tropical to arctic climatic zones. In addition, the sites vary widely in terms of the quantity and quality of data available.

Despite these enormous variations, the methodology has proven to be robust and effective. The approach aims to estimate the carbon, nitrogen and phosphorus fluxes into and out of the coastal zone, including the emission of gases to the atmosphere. The approach is based on the fundamental concept of the conservation of mass. This is straightforward for water volume and salt, but becomes complicated for nutrients as they undergo reactions in the coastal waters. Here the approach has been to measure fluxes for one of the nutrients (usually phosphorus) and then use scaling ratios to relate these fluxes to those of carbon and nitrogen compounds. Figure 6.13 shows some of the results.

An important part of the networking effort has been the communication of both the methodology and the results. A series of more than 20 workshops has been held to introduce the budget methodology to scientific communities in various parts of the world. A web site has been established to summarise and update the budgeting procedures, provide tools for implementing the procedures, disseminate various forms of teaching materials and post existing budgets as they are developing (*http://data.ecology.su.se/MNODE*). More than 200 site budgets have been developed by nearly 180 scientists around the world (Fig. 6.14).

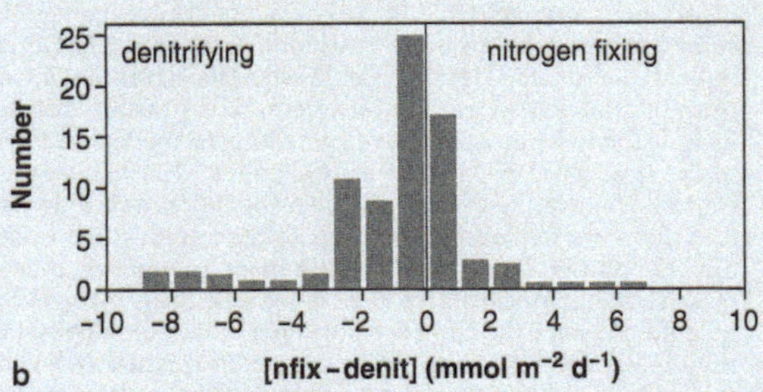

Fig. 6.13. Frequency distributions of $[p-r]$ and $[nfix-denit]$ at the budget sites. $[p-r]$ is primary production minus respiration and $[nfix-denit]$ is nitrogen fixation minus denitrification (adapted from Smith et al. 2002)

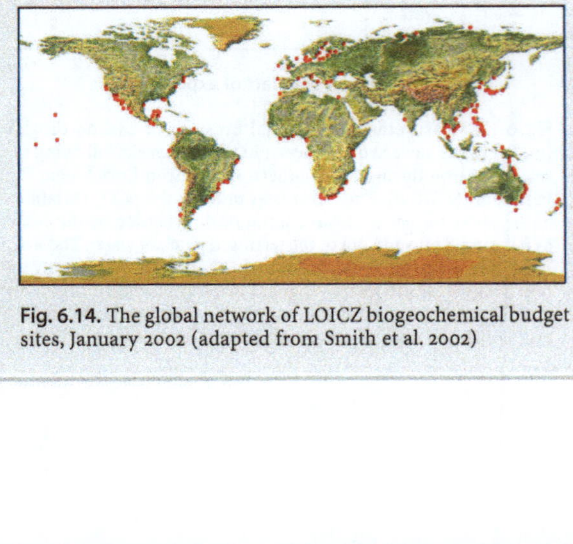

Fig. 6.14. The global network of LOICZ biogeochemical budget sites, January 2002 (adapted from Smith et al. 2002)

Scientific networks studying aspects of the global environment will continue to form the backbone of the emerging new science required to study Earth as a system. Planetary patterns emerge more clearly when small-scale or site-specific measurements and process studies are carried out in a consistent and comparative way across the globe. Major challenges of site representivity and stratification, scaling up and aggregation, and interpolation and interpretation stand between the individual studies and their application to Earth System questions. Nowhere is this more important than in research on the coastal zone, which can be visualised as a strip of land and coastal sea about 50 km in width and about 500 000 km long with considerable heterogeneity in both dimensions (Box 6.5).

The rapid development of new communication and organisational tools such as the internet and institutional networking promise astounding advances in the capability to link the best sceintific minds together to work on common problems. Virtual institutions or centres, such as the Tyndall Centre for Climate Change Research in the United Kingdom, bring together scientists from an impressively wide range of disciplines to work together without having to leave their home institutions. Such institutionally light and flexible networks of scientists can bring to bear a powerful team in a quick and resource efficient way. The international global change programmes (DIVERSITAS, IGBP, IHDP, WCRP), through their Earth System Science Partnership, aim to build similar networks of scientists and institutions on a vast international scale, in effect to build a system of science to tackle the most urgent problems of global change.

6.4.5 Integrated Regional Studies

Global change is evolving in very different ways in different regions of the world. Increasing and variable rates of population growth, rapid rates of land-cover conversion, dense human settlements, evolution of megacities, large-scale industrialisation and economic development are all leading to rapid changes in regional socioeconomic conditions and in terrestrial, marine and atmospheric systems, with implications for the regional environment and resource development as well as for the global environment. Because regions are normally defined by common biophysical characteristics (e.g., the Asian monsoon) and broadly common socioeconomic characteristics, regional-scale studies of global change and sustainable development lend themselves to the undertaking of vulnerability analyses, identification of hotspots of risk and identification and simulation of syndromes of environmental degradation.

From an Earth System perspective, regions manifest significantly different dynamics and thus changes in regional biophysical, biogeochemical and anthropogenic components may produce considerably different consequences for the Earth System at the global scale. Regions are not closed systems and thus the linkages between regional change and the Earth System are critical. Regions may function as switch or choke points and small changes in regional systems may lead to profound changes in the ways that the Earth System operates (Fig. 6.15). Regional studies can contribute substantially to the reconstruction of global dynamics from regional patterns; in effect, integrated regional studies represent a unique way to reconstruct the Earth System from its components and are thus an essential part of the Earth System science toolkit.

To be effective as a technique for Earth System aggregation, the integrated regional studies must have a number of characteristics. They should: (*i*) be built around regions that may function as switch or choke points, in which small changes in regional systems may lead to profound changes in the ways in which the Earth System operates; (*ii*) show how the region functions as an entity and how that functioning might change; (*iii*) transcend disciplinary boundaries across natural and social sciences and address all relevant aspects of marine, terrestrial, atmospheric, social, economic and cultural components and processes within and across the region; and (*iv*) be designed to facilitate the two-way linkage between the region and the global system.

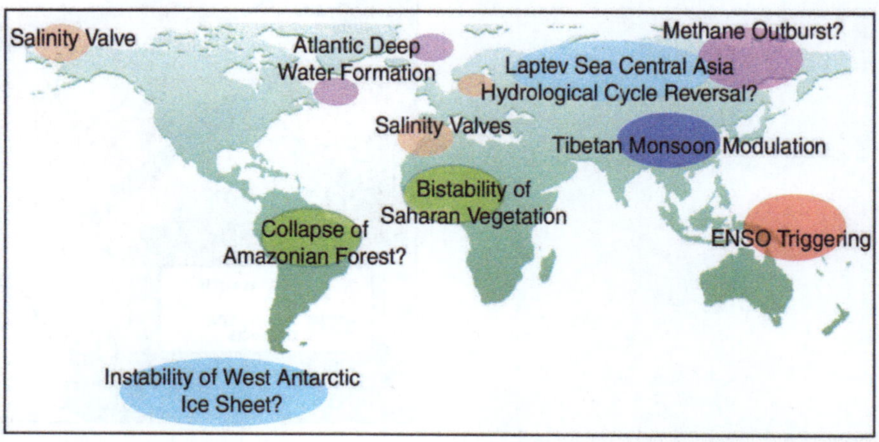

Fig. 6.15. Critical region analysis of hot spots or switch and choke points; an early attempt at identifying parts of the Earth where changes at the regional scale can cause significant changes in the functioning of the Earth System as a whole (Schellnhuber 2002)

Box 6.6. LBA: The Large-scale Biosphere-Atmosphere Experiment in Amazonia

Carlos A. Nobre

The Amazon Basin contains the world's largest extent of tropical rainforest, harbouring 20 to 25% of the planet's biodiversity. The region receives abundant rainfall and freshwater, has over 100 Gt of carbon stored in vegetation and soil, a multitude of ecosystems, and considerable biological and ethnic diversity. It is both a major resource for the countries of the region, and an important feature in the functioning of the Earth System. However, the natural system is under unprecedented pressure. Increasing rates of forest conversion since the 1970s has given rise to concerns about the environmental consequences of deforestation. At the same time demographic, economic and social pressures within the Amazonian countries continue to push for further development of the basin. Regional development policies (colonisation projects, subsidised cattle ranching, etc.) and, more recently, demand for tropical timber are the main causes of deforestation. Today in Brazilian Amazonia alone about 15% of the tropical forest has been cut down and the annual deforestation rates (15 000 to 20 000 km^2) are unsustainably high. Until recently, scientific knowledge to underpin the sustainable development of the region was lacking.

The Large-scale Biosphere-Atmosphere Experiment in Amazonia (LBA), an international research initiative led by Brazil (Nobre et al. 2002) to study the region holistically, was initiated in the early 1990s. The aim was to understand the climatological, ecological, biogeochemical, and hydrological functioning of Amazonia, the impact of land use and future climate change on those functions, and the interactions between Amazonia and the Earth System. The programme is large in scale, comprised of over 100 closely-linked and integrated studies involving about 600 scientists and students from South and North America, Europe and Japan.

Two key questions, central to the work of the LBA, integrate the studies in the physical, chemical, biological, and human sciences. They are:

- How does Amazonia currently function as a regional entity with respect to the cycles of water, energy, carbon, trace gases and nutrients?
- How will changes in land use and climate affect the biological, chemical and physical functions of Amazonia, including sustainable development in the region and the influence of Amazonia on global climate?

In many ways the LBA is a model for integrated regional studies. The reasons for the programme's success to date include the facts that:

- The project has built from its inception a framework for putting together the individual component studies to address long-standing issues and controversies raised at the outset. Various techniques are used to scale up the individual studies to the basin level: remote sensing, modelling, and the use of transects based on ecoclimatic (precipitation) and land-use intensity gradients (Fig. 6.16).
- A modular approach has been used and has facilitated the inclusion of many research groups that can bring their own specific expertise, and often resources, to bear on individual aspects of the regional problems. Given the desired level of integration of the individual, disciplinary-oriented studies, LBA is also an experiment on the building of inter-disciplinarity in Earth System science.
- National and regional development concerns have been effectively integrated with global environmental questions. LBA research has contributed strongly to the efforts of the international global change programmes while at the same time supporting the Brazilian government to develop an appropriate scientific basis for policies on sustainable use of Amazonian natural resources.
- From the early stages of the planning phase, LBA has attracted scientists from within Brazil, from the broader Amazon Basin region and from the international scientific community. This international collaboration has been fundamental in the rapid improvement of scientific capacity within Brazil and in the growing appreciation in the international scientific community of the nexus between regional development and global environment.
- The enhancement of research capacities and networks within and between the Amazonian countries is helping advance education and applied research into sustainable development, and helping in the process of formulating policies for sustainable development of the region. The development of a scientific community capable of *local to global to local thinking* is essential for the region and will help meet the challenge of avoiding one-way migration of talented young scientists.

Given its success, the LBA serves as an example to be emulated elsewhere for building institutions dedicated to enhancing the undertaking of science for sustainable development, and for strengthening institutions that bridge the widening scientific and technological gaps between developing and developed countries.

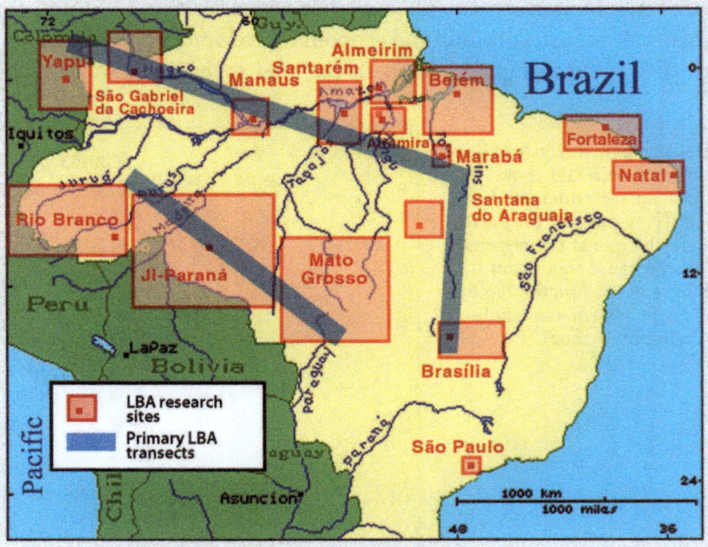

Fig. 6.16.
A central component of the LBA research strategy is based on two ecological transects with ecoclimatic and land-use intensity gradients across the basin (Nobre et al. 2001)

Currently the only example of an integrated regional study is the Large-scale Biosphere-Atmosphere Experiment in Amazonia (LBA) (Box 6.6). However, other efforts are underway to build similar projects in other critical regions of the planet. One such is an integrated regional study of Monsoon Asia. This will be a complex, multi-layered study of change in Asia and its implications for the Earth System, involving the most populous countries on Earth. Another will be a major, long-term effort in Antarctica and the surrounding Southern Ocean focusing on the role of Antarctica in the climate system, in biogeochemical cycles and in the sustainable provision of marine protein.

6.4.6 Simulating Earth System Dynamics

Observations, experimental results and large regional studies must eventually be brought together in coherent ways to understand and simulate the dynamics of the Earth System as a whole. A vast array of mathematical models has already been designed to simulate the Earth System or parts of it. Simple, stylised Earth System models focus on description and understanding of the major features of the planetary machinery while comprehensive Earth System simulators are being assembled from the most sophisticated component modules. In between, Earth System models of intermediate complexity have already proven to be effective tools for both hindcasting and forecasting Earth System behaviour. Even more innovative approaches will be required, however, to include the complexity of human and societal behaviour as an integral part of Earth System dynamics.

The range of Earth System modelling approaches can be classified in terms of their level of integration (Claussen et al. 2002) (Fig. 6.17). Simple conceptual models, sometimes called tutorial models, are based on a small number of mechanistic processes, but nevertheless can exhibit complex behaviour when the processes are coupled. Their main purpose is to demonstrate the plausibility of processes and

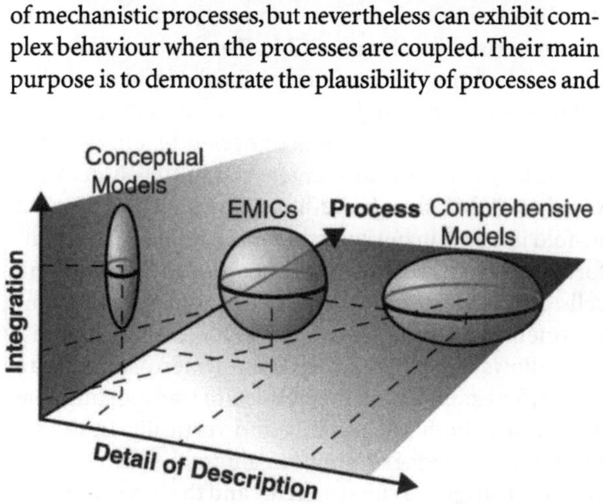

Fig. 6.17. A pictorial classification of Earth System models, from simple conceptual models to comprehensive models, or General Circulation Models (GCMs). The models are classified according to their level of detail, treatment of process and degree of integration (Claussen et al. 2002)

feedbacks; they are generally not used as predictive tools. Another type of simple model is the box model (usually one-dimensional), which is used to extend and interpolate the results of comprehensive models over long time periods and with variable boundary conditions.

At the other end of the Earth System modelling spectrum are the well-known comprehensive general circulation models (GCMs) or, more recently, coupled general circulation models (CGCMs). They originated from short-term weather forecast models, but have been extensively developed towards more complete Earth System models. GCMs divide the Earth System into a spatially explicit set of grid cells (both horizontally and vertically) and use a mass and energy balance approach to simulate mathematically the physical dynamics of the atmosphere and the oceans and their coupling. Interactions with the terrestrial biosphere are usually treated as a lower boundary condition although more recent versions of GCMs are including more interactive versions of biosphere-atmosphere coupling. Figure 6.18 shows how additional processes are being systematically incorporated into the framework of a GCM. Two of the more immediate challenges are (*i*) to incorporate the dynamics of major biogeochemical cycles, such as the carbon cycle, in a fully interactive way and (*ii*) to integrate chemistry into the atmospheric dynamics of the models.

GCMs have proven to be a very powerful tool in simulating and understanding climate variability and change. As more and more processes are included in the models, their ability to simulate correctly the contemporary variability in climate has improved significantly (Sect. 3.4.2). GCMs are the basis on which the IPCC projections of future climate change over the next century are constructed. Given their complexity and demand on computer time, however, current GCMs cannot be used for long-time simulations of Earth System dynamics, for example, over thousands or hundreds of thousands of years (glacial-interglacial cycles). This is actually a major constraint, as the palaeo-records have shown ranges of variability in major aspects of Earth System functioning that lie outside the bounds evident in the instrumental record (Sect. 2.2 and Sect. 2.6) and provide warnings of abrupt changes, extreme events and potential surprises that present significant risks for the Earth System in the Anthropocene Era (Sect. 5.4).

In the spectrum of Earth System modelling approaches, a newer genre of model has been developed in the last decade or so that lies between the simple conceptual models and GCMs (Claussen et al. 2002). Complementary to both of these other approaches, Earth System Models of Intermediate Complexity (EMICs) aim to simulate the dynamics of the natural Earth System (i.e., humans and their activities are prescribed as external driving forces) albeit at a reduced level of resolution and at a higher level of parameterisation compared to GCMs (Box 6.7). One of the strengths of EMICs is that most of them emphasise the interactions among proc-

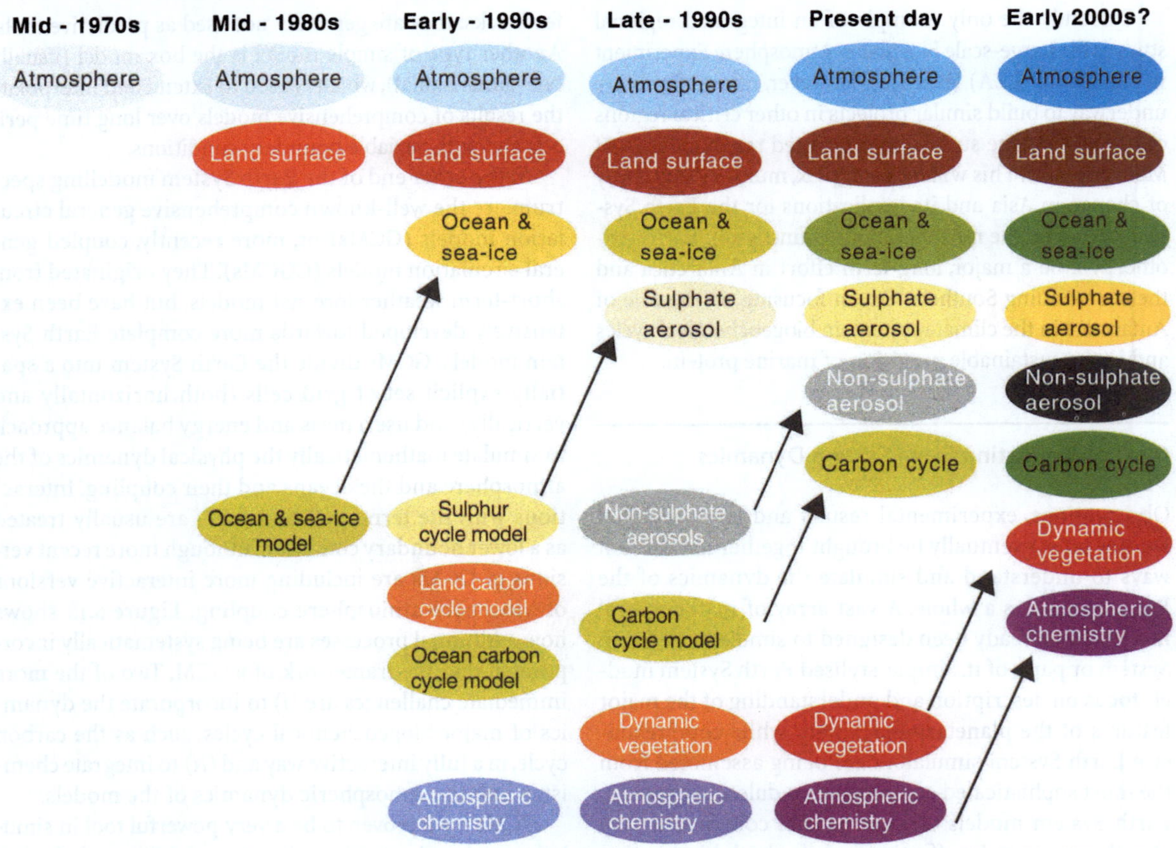

Fig. 6.18. The development of climate models over the last 25 years showing how the different components are first developed separately and later coupled into comprehensive climate models (IPCC 2001). Not shown are box or 1- and 2-dimensional simplified models, which played an important role in early climate research

esses and thus capture many of the forcings and feedbacks between biogeochemical cycles and physical dynamics. In addition, their relative simplicity compared to fully comprehensive models allows simulations up to hundreds and thousands of years, so allowing EMICs to be tested against many of the long-term palaeo-records.

The complementarity of GCMs and EMICs is evident in their uses. GCMs are the most appropriate Earth System model to simulate changes in the climate system over a few centuries, from the beginning of the Industrial Revolution, for example, to the end of the twenty-first century. They are commonly used to generate scenarios of relevance for the time scales of policymakers and resource managers. Their higher spatial resolution allows application to many global change impact studies. On the other hand, EMICs are useful for simulating some aspects of the complexity of the Earth System that GCMs cannot yet do, such as thresholds and abrupt change, bifurcation points and flickering. In addition, they can be used to explore the parameter or phase space of the Earth System at some level of completeness and thus provide insights on where critical thresholds might lie and, through multiple ensemble runs, generate information on probabilities of events or sensitivities of changes in drivers. This can provide guidance for more detailed investigations to be undertaken by GCMs.

Perhaps the most ambitious of all the Earth System models is that of Japan's Frontiers Program on Global Change (*www.jamstec.go.jp/frsgc/*). The model is based on the GCM approach but the simulations are carried out at significantly higher spatial resolution, allowing the direct simulation of dynamics that are normally parameterised in current GCMs. The higher resolution is possible through use of the the Earth Simulator, a distributed memory parallel computing system that consists of 640 processor nodes connected in a high speed network. It is the world's largest computing system devoted to global change studies. The Simulator allows a 10-fold increase in horizontal spatial resolution over full-form GCMs, from grid cells of about 100–200 km to grid cells of 10 km for both atmospheric and oceanic compartments and to about 50 vertical layers in the oceans. This allows the direct simulation of meso-scale ocean eddies, for example, and their effects of transporting heat and salinity both horizontally and vertically. In the atmosphere the higher resolution allows the direct simulation of convective cloud systems and the heavy rainfalls associated with them. Projections of future climate states carried out on the Earth Simulator may be especially useful for impact studies given their ability to project climate change on a regional scale.

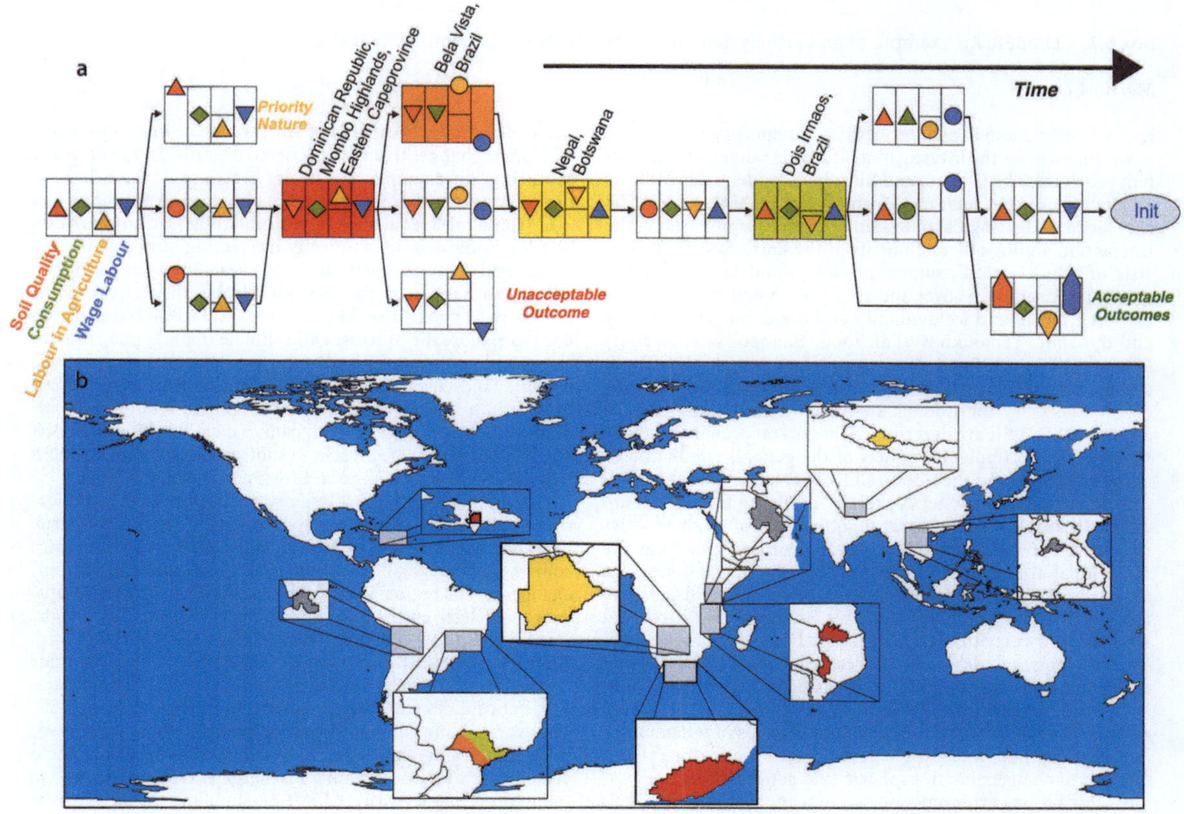

Fig. 6.19. Symbolic dynamics of the Sahel Syndrome (Schellnhuber 2002): **a** the temporal dynamics of **b** various regions of the Earth where the syndromes approach (Box 3.1) is applied using qualitative differential equations. The symbols in the *first box* (*left side* **a**) represent four important parameters associated with the Sahel Syndrome. Unacceptable outcomes are shown by the symbols lying outside of their boxes. The approach uses rudimentary and fragmentary knowledge to generate qualitative diagnostics indicating whether a particular region will suffer significant land degradation through time

Comparison of EMICs, GCMs and the higher resolution models run on the Earth Simulator raises a more fundamental issue of model development. One approach to the development of Earth System models is to couple complex or full-form models of parts of the Earth System. An example is the coupling of dynamic carbon cycle models to GCMs to capture the interactions between changing climate and the processes that control the carbon cycle (Cox et al. 2000; Friedlingstein et al. 2001). Such coupling is almost always a complex affair, involving technical issues of computer code as well as fundamental scientific issues of the conceptualisation of processes and the compatibility of time and space scales in the simulations. There are concerns that such complexities may bias the outcomes of the coupled model runs and make their interpretation very difficult.

An alternative approach is to use a single modelling framework, such as an existing GCM, and then build more complex subcomponents into the model (Fig. 6.18). This is, in effect, how many GCMs have evolved to incorporate many more of the processes known to be important for the climate system. This avoids technical problems of coupling pre-existing but separate models, but questions of compatible levels of complexity of model components and interpretation of results still remain. Development of completely new Earth System simulators, such as EMICs and Japan's Earth Simulator, in bottom-up fashion *de novo* represents an attempt to overcome these problems.

Even more severe challenges face the developers of Earth System models when the human dimensions are included in a realistic and comprehensive way. Much important information is in the form of incomplete, disjointed and qualitative data rather than the more precise quantitative data and well-defined relationships on which biophyscial scientists build mathematical expressions into their models. New tools, including pattern recognition algorithms, neural networks and fuzzy logic, are being employed to deal with these challenges in terms of Earth System simulation (Schellnhuber 2002). Certain techniques can be used, in particular, to construct weakly predictive models of complex nature-society patterns like degradation syndromes (Box 3.1). The syndromes approach uses some of these techniques to overcome problems caused by uncertainties in data sets, process descriptions and simulation techniques instrumental to Earth System research. The development of an evolution tree (Fig. 6.19) is a way of assimilating a

Box 6.7. CLIMBER: An Example of an Earth System Model of Intermediate Complexity (EMIC)

Martin Claussen

Earth System Models of Intermediate Complexity (EMIC) are powerful tools for the investigation of a wide range of Earth System problems. The Climate and Biosphere model (CLIMBER) is an example of one such model and is designed as a scientific tool for investigating Earth System dynamics, emphasising the interaction among all components of the Earth System. It consists of fully coupled components that simulate atmospheric dynamics, vegetation cover and structure dynamics, terrestrial carbon cycle, inland ice dynamics and ocean biogeochemistry and dynamics (Petoukhov et al. 2000; Ganopolski et al. 2001) (Fig. 6.20). CLIMBER is a comprehensive model regarding Earth system interactions. However, because of computational efficiency necessary for systems analysis, the individual components in CLIMBER are described in a less comprehensive manner than, for example, in models of the general circulation of the atmosphere and the ocean. CLIMBER has a spatial resolution finer than hemispheric, thereby resolving the continent-ocean distribution, but coarser than the scale of weather activity. Hence synoptic variability is parameterised as well as the meridional atmospheric circulation and the relation between long-term large-scale azonal temperature fields and pressure fields in the long planetary-scale waves. Vegetation cover and structure are described by the fractional coverage of a grid cell by trees, grass and desert and by leaf area index (LAI). Carbon in vegetation is aggregated into pools of fast, green and slow, structural biomass and woody residues and humus. The vegetation cover and structure respond to changes in temperature and precipitation, including a sensitivity of net primary production to changes in atmospheric CO_2 concentration.

CLIMBER has achieved some significant successes in simulating changes in Earth System functioning in the past. For example, the model is capable of simulating the long-term natural climate changes that have occurred over the past 9 000 years, capturing, for example, the strong greening of the Sahara in the early and mid-Holocene (from about 9 000 to 6 000 years ago) (Ganopolski et al. 1998) and simulating the changes in carbon fluxes between dynamically interacting atmosphere, vegetation and ocean during the last 9 000 years (Brovkin et al. 2002). Other foci have been the last interglacial, the Eemian, some 125 000 years ago (Kubatzki et al. 2000). CLIMBER has also been used to explore the processes and climate impacts that are associated with natural solar and volcanic forcing as well as anthropogenic land-cover change and CO_2 emissions during the last millennium (Bauer et al. 2003). Theoretical studies have addressed the role of vegetation dynamics including biogeophysical and biogeochemical feedbacks at subtropical and high northern latitudes (Claussen et al. 2001; Brovkin et al. 2003).

One of CLIMBER's most interesting uses has been to explore the processes that trigger abrupt changes in the Earth System. Examples are the abrupt desertification of the northern Africa some 5 500 years ago (Claussen et al. 1999), possible (abrupt) changes in this region in the future (Claussen et al. 2003), Dansgaard/Oeschger events (Ganopolski and Rahmstorf 2001, 2002) (see also Box 5.6) and Heinrich events (Calov et al. 2002) (Fig. 6.21). Heinrich events, related to large-scale surges of the Laurentide ice sheet over northern America, represent some of the most dramatic abrupt climate changes occurring during the last glacial. The average time between simulated events is about 7 000 years, while the surging phase of each event lasts only several hundred years, with a total ice volume discharge corresponding to 5 to 10 metres of sea level rise. CLIMBER simulates the ice surges as internal oscillations of the ice sheet.

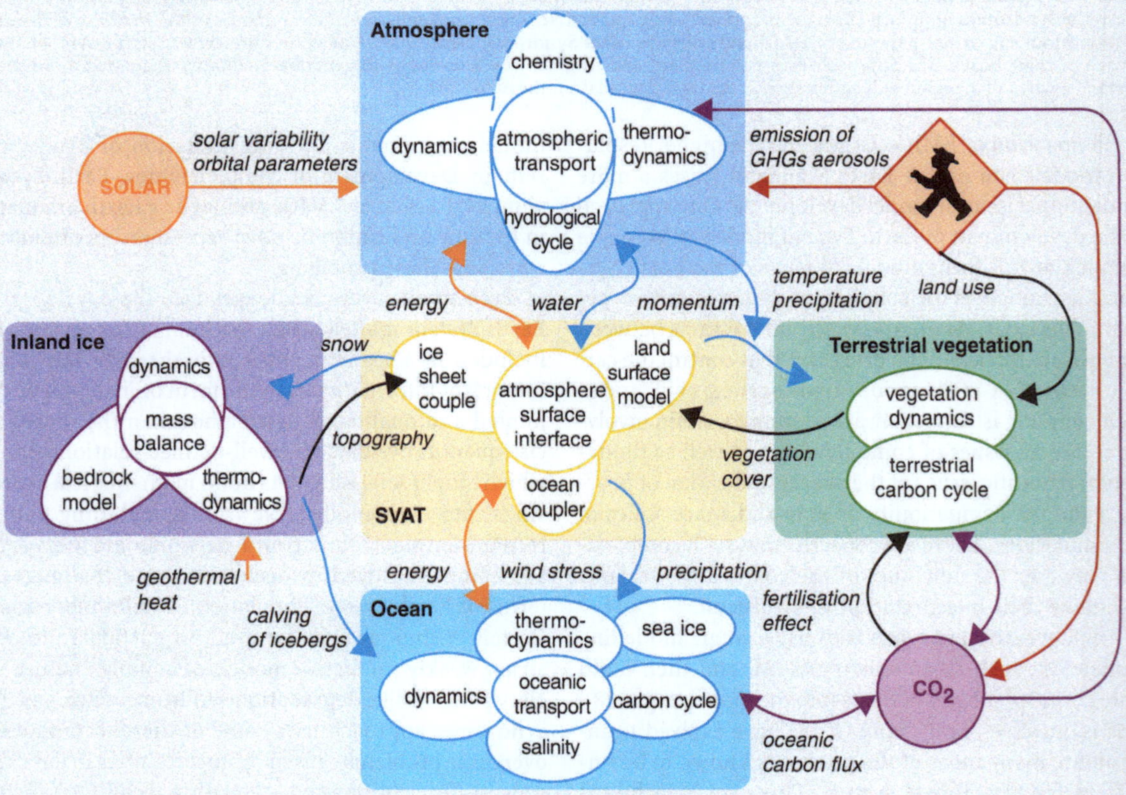

Fig. 6.20. Structural diagram of CLIMBER, an example of an Earth System Model of Intermediate Complexity

◄

An interesting riddle posed by palaeo-climatic evidence concerns the almost synchronous discharges of icebergs from the American and European ice sheets during Heinrich events. These findings appear to contradict the idea that Heinrich events and other ice surges are just internal oscillations of ice sheets, since in this case, surges from different ice sheets should be independent of each other. However, in strongly non-linear systems, such as the Earth System, processes can readily be synchronised by a very weak external forcing via so-called stochastic resonance. This idea has been demonstrated to work in the case of Dansgaard/Oeschger events (Ganopolski and Rahmstorf 2002), and it seems to work for Heinrich events too. Some weak signal with a periodicity of about 1500 years, which seems to be a pertinent signal during the last glacial, induces only small-scale surges, but on average, each second micro event provokes a more extended instability of the ice sheet in the eastern part of Hudson Strait. Every second or third event of these surges triggers large-scale instabilities which are known as Heinrich events. In this way more than one half of the Heinrich events are synchronised. The origin of this tiny synchronising external signal is not known. It could be an external solar signal, or it could also be an internal signal, for example, a sea level rise due to an iceberg discharge from other ice sheets. It is indeed not the signal that is important, but the possibility of a synchronisation between ice surges from different ice sheets.

These examples show the power of EMICs in simulating some highly non-linear features of the Earth System. Such capability is an important piece of the Earth System science toolkit in terms of exploring Earth System phase space for abrupt and potentially catastrophic changes in Earth System functioning.

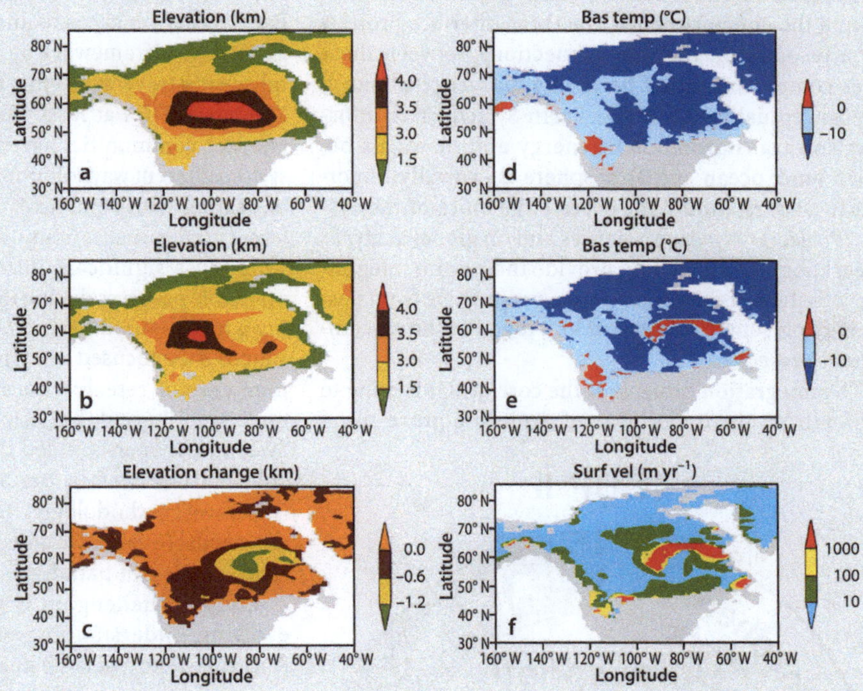

Fig. 6.21. Dynamics of a simulated Heinrich event: **a** elevation of the Laurentian Ice sheet before and **b** after a Heinrich event; **c** differences between elevations before and after this event; **d** temperature (corrected for pressure melting) at the bottom of the ice sheet before and **e** during the same event; **f** ice-surface velocity (Calov et al. 2002)

wide range of imprecise, fragmentary, qualitative yet useful data. An entirely new methodology for integrated assessment of global change problems is the incorporation of large groups of real and proxy decision-makers into the simulation process itself.

6.4.7 Global Integration, Synthesis and Communication

The Earth System toolkit described in the preceding sections represents an exceptionally powerful approach to understanding the workings of the planet's life support system. Many new insights and novel conceptualisations of the Earth System will be proposed, debated and eventually accepted, refined or disproved. Knowledge of Earth System dynamics will expand at an accelerating rate. However, society expects that the scientific community will eventually address directly the simple yet profound questions that it asks about the global environment. Thus, no toolkit would be complete without explicit methodologies to synthesise, integrate and communicate the expanding knowledge base on the Earth System to underpin the evolution towards global sustainability.

The IGBP synthesis project, of which this volume is a part, is an attempt to bring together what has been learned over the past decade about global change and the Earth System and to communicate it to broad audiences within and outside the scientific community. Similar volumes are being produced for components of the Earth System, including atmospheric chemistry, ocean biogeochemistry, terrestrial ecology, the hydrological cycle, the coastal zone and regional-global linkages. In addition, this work contributes directly to international assessments of the current state of scientific knowledge on key aspects of the global environment, the best known of which is the Intergovernmental Panel on Climate Change (IPCC).

Synthesis still falls short, however, of reaching the level of integration required to understand the dynamics of the Earth System in a holistic way. Research must be designed for integration from the start of the work. The new structure of IGBP (Fig. 6.22) is designed with global integration as the ultimate goal. With an overall theme of the Earth's biogeochemical cycling as its central focus, IGBP has built its programme around research in the three major components of the Earth System – land, ocean and atmosphere – and the three interfaces linking the compartments. The three interface projects are a recognition that the connections between these three compartments are becoming the weakest links in the conceptualisation of the Earth System. The transport and transformation of energy and materials between land, ocean and atmosphere is crucially important in understanding the overall dynamics of the System. Projects on palaeo-sciences and on global analysis, integration and modelling provide the central integrating activities, organised in a time perspective from past through present to future, that knit together the six component projects.

The integration projects at the core of IGBP bring to the forefront the importance of exploring processes at the system level that cannot be identified or described without treating the Earth System as a single unit. In this way, focus is placed on the processes that regulate the interactions between land, atmosphere, ocean, and society. In addition, only by considering the whole System within the same, internally consistent framework can one explore and assess sensitivities to perturbations (natural and anthropogenic), feedbacks within the System, and interactions between components that may have been previously considered independent but that control the behaviour of the System as a whole.

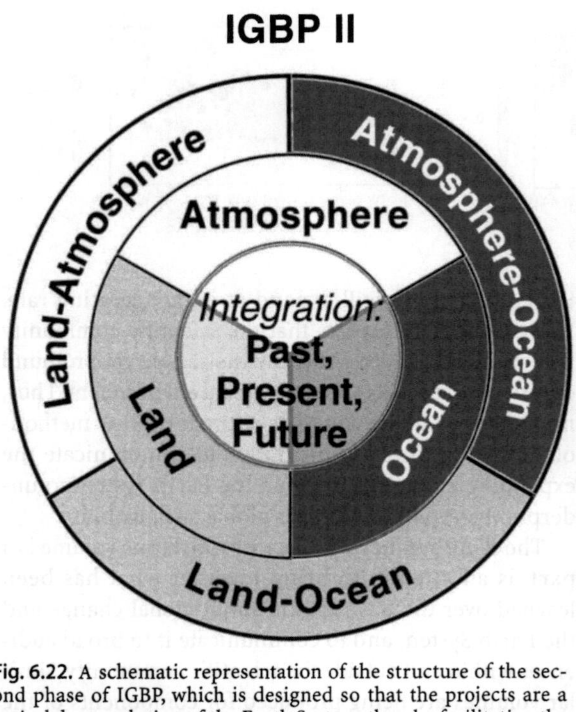

Fig. 6.22. A schematic representation of the structure of the second phase of IGBP, which is designed so that the projects are a logical deconvolution of the Earth System, thereby facilitating the integration of their work into a overall understanding of Earth System dynamics

One of the most pressing and challenging aspects of global integration is the need to understand the Earth System in the Anthropocene, that is, as a fully coupled and interactive system built of tightly interwoven biophysical and human components. Putting people into Earth System analysis requires the construction of a single, coherent framework built jointly by social and natural scientists. Why has this been so difficult to achieve? The answer lies at least partly in the fact that the two research communities have developed different research styles, different ways of framing questions, different ways of dealing with complexity and heterogeneity and different conceptualisations of the Earth System itself. Despite these significant differences, there has been remarkable progress during the last decade towards building a truly unified Earth System science. The recent advances have focused on: (*i*) the development of a common core of research questions, shared concepts and understandings of the basic characteristics of the Earth System; (*ii*) open-minded discussion and debate to resolve differences in perceptions of problems; and (*iii*) novel methodologies to combine techniques, data and results (qualitative and quantitative) across assumptions, scales and paradigms.

Equally challenging is the communication of the evolving understanding of the Earth System in the Anthropocene to a large number of audiences, from scientists themselves to a curious public sector eager to learn more about changes to their own environment, changes that they can sense but don't fully understand. One of the most innovative ways of communicating this science is through the development of an Earth System Atlas, a platform for the dissemination of peer-reviewed information on the evolving nature of the Earth System. The overall aim of the Atlas is to provide a quantitative and geographically referenced tool that can help the user to visualise the extent and rate of global change, in effect to bring to life the wide-spread and unprecedented changes occurring in the global environment.

Beyond the Atlas project, communication must become an integral part of Earth System science. Global change lies at the nexus between science and society, so efforts to transfer the expanding knowledge base to a wide variety of audiences – policymakers, resource managers, the general public – must be intensified. Equally important is the need to equip the next generation of scientists with the tools required to undertake Earth System science. Courses focusing on global change and Earth's environment are already expanding rapidly in universities around the world, spawning younger scientists who are challenging their elders with their breadth of understanding of Earth System dynamics. Finally, the divide between the developed and the developing worlds, noted earlier in this volume, also afflicts the scientific world in all aspects – research, education, facilities and communication. Special, dedicated efforts are required in all of these areas to ensure that Earth System science becomes truly global.

6.5 Towards Global Sustainability

This volume has made a strong case that human-driven changes are pushing the Earth System beyond its natural operating domain into planetary *terra incognita*. Management strategies for global sustainability are urgently required. Earth System science is the key to implementing any approach towards good planetary management, as it can provide critical insights into the feasibility, risks, trade-offs and timeliness of any proposed strategy.

6.5.1 Advancing Sectoral Wisdom

In 1987 the Bruntland Commission's *Our Common Future* detailed the challenges to the environment and sustainability arising from activities within particular sectors: energy, industry, agriculture, urban systems, living resources and human population (WCED 1987). In the years since the Bruntland report, substantial progress has been made, yet much, much more is necessary (NRC 1999). In industry, for example, there have been considerable improvements in reducing and reusing materials and reducing wastes. This trend towards dematerialisation must be accomplished universally and at much greater rates. In the energy sector, gradual progress has been made in increasing efficiency and in development of alternatives to fossil fuel sources, but critical air pollution and global greenhouse gas problems resulting from fossil fuel combustion continue to grow around the world. Dramatic increases in energy efficiency, decarbonisation and the development and utilisation of new sustainable energy technologies, such as a hydrogen-based energy system, are needed.

In agriculture, continued increases in food production and improvements in distribution and access are needed urgently as the world's human population continues to grow and dietary preferences change. The challenge of increasing agricultural production is substantial and much evidence suggests that the gains of the past decades will be difficult to repeat in the next decades. These and other sectoral challenges are as significant today as they were in 1987.

While the importance of individual sectors has not diminished, it has become clear that sectoral interaction and interdependency is increasing. The consequences are cumulative, sometimes non-linear, and may be associated with abrupt changes and the crossing of critical thresholds. Some of the world's most critical resources, in particular water, air, and ecosystems, are at risk because of the connectivity of sectoral activities. Water resources, for example, reflect the demands and activities of all sectors. Strategies for management of water are increasingly attempting to take integrative, regional approaches. Likewise, management of air quality recognises increasingly the multi-sectoral and regional, rather than local, determinants of air quality. In terms of sustaining ecosystem goods and services, the importance of overlapping sectoral activities has led to the development of more integrated landscape planning that reconciles ecosystem processes with human social and economic activities.

Yet there is an even more profound management challenge that goes beyond the sectoral approach, individually or integrated. For the first time since the evolutionary appearance of humans on the planet, human activities themselves are creating risks of planetary dimensions, risks that could trigger abrupt, catastrophic and possibly irreversible changes. No longer can sectoral or regional changes be viewed as occurring within an overall envelope of well-buffered global stability. Stewardship of the Earth System itself has now become an issue of great import.

6.5.2 Stewardship of the Earth System

Humanity is already managing the planet, but in an unconnected and haphazard way driven ultimately by individual and group needs and desires. As a result of the innumerable human activities that perturb and transform the global environment, the Earth System is being pushed beyond its natural operating domain. Many of these global changes are accelerating as the consumption-based Western way of life becomes more widely adopted by a rapidly growing world population. The management challenges to achieve a sustainable future are unprecedented.

The word management or the term planetary management often raise concern. It is important to differentiate between attempts to manage the functioning of the Earth System (e.g., geo-engineering proposals) and attempts to manage human activities at the global scale (e.g., the Kyoto Protocol) so as to lessen their impact on the Earth System and thus to allow it to function in a more natural mode. Attempts to manage the functioning of the Earth System itself are fraught with difficulties, perhaps insurmountable. Systems theory suggests that complex systems can never be managed; they can only be perturbed and the outcomes observed. Futhermore, many of these outcomes will likely be unpredictable, even with a vast amount of information on the system, leading to unintended and potentially severe consequences. This property of complex systems is manifest in the Earth System, for instance, as the instrinsic level of unpredictability in climate. There is a further problem. Humans and their societies and institutions are embedded in and are an integral part of the Earth System. Humans cannot fully stand outside the human-environment system as they attempt to analyse it, and thus cannot be in position to manage the Earth System in any objective fashion. Humans cannot be dispassionate observers and objective managers. In the text describing various potential management strategies that follows, attempts are made to differentiate the two meanings of the word management.

Business-as-Usual

This strategy is built around the belief that no explicit measures need to be taken with respect to global change, either to attempt to slow or prevent it or to adapt to it. The argument that a business-as-usual approach will be adequate to deal with global change (e.g., Lomborg 2001) is based on several assumptions: (*i*) threats to the global environment are overstated or misrepresented by green organisations and sometimes even by the scientific community; (*ii*) the Earth System is more robust towards perturbations than most people, including many scientists, believe; (*iii*) the existing market-oriented economic system will be able to solve any environmental problems, including global-scale ones; and (*iv*) resources that might be used to mitigate global change would be better spent on more pressing problems, such as provision of clean water, improvement of public sanitation systems in the developing world, etc. Although the business-as-usual strategy appears to be a safe, conservative approach to the problem, it involves taking great risks.

In considering management strategies, the Earth System features of most relevance are those that provide ecosystem goods and services for humans and, more generally, provide a stable and reliable life support system for human societies. The major issue to be faced is whether human activities can transform the Earth System to such an extent that the processes, feedbacks and interactions that currently operate will be altered sufficiently to threaten the planet's life support system.

The results presented in this volume suggest that the Earth System is exceedingly complex and has many ways in which it interacts with human activities. Rather than being either simply robust or fragile, the Earth System is better viewed as having Achilles heels, or hot spots that are modes of operation or critical processes that are particularly susceptible to abrupt changes triggered by human activities and which, in turn, may lead to changes that, at worst are catastrophic, or at best will still render the planet less hospitable for human life. The stratospheric ozone episode is an example of such an Achilles heel. The more that is learned about the nature of the Earth System and the human pressures on it, the lower is the probability that the Earth is robust in every important respect. Thus, to advocate a business-as-usual strategy towards global change is a high-risk approach, not a safely conservative one. The science presented here argues strongly that a business-as-usual approach is not appropriate to deal with global change and that explicit management strategies are needed. The most well-known of these is adaptation.

Adaptation

Adaptation to global change is based on two fundamental assumptions: (*i*) global change will be moderate and slow enough that human societies have both the time and economic resources to adapt to these changes rather than to proactively slow or prevent them (mitigation); and (*ii*) enough can be learned about the nature of global change that adaptive measures can be taken in a timely and effective fashion.

A common approach to adaptation is to first use a global change scenario of some sort to estimate the impacts of global change on the system in question (Sect. 5.2.1) and then to undertake management strategies that anticipate these impacts and adapt the system to them. The interplay between climate and agricultural production illustrates this approach. Seasonal forecasts of climate, particularly of El Niño events, are now becoming available. These enable farmers to adapt to the projected conditions by varying their planting date or switching to different crop types. For longer-term climate change, management strategies to adapt agricultural production might be to switch to different crop types depending on the scenario, to establish irrigation systems if a drying climate is projected, to move elements of the agricultural system to a location projected to be more favourable, or to develop new varieties of crops that are better adapted to the projected climatic conditions.

A contrasting approach to adaptation focuses on the resilience of the system under study (Gunderson and Holling 2002; Folke et al. 2002a,b) (Box 6.8). One of the assumptions behind this approach is that there is still considerable uncertainty associated with global change projections and, moreover, systems will be subject to multiple, interacting stresses. It will be almost impossible to project realistically the behaviour of these stresses into the future. Thus, a better approach is to focus on the system itself and to understand and improve, if possible, its inherent resilience towards perturbations or impacts of various types. When the system becomes less resilient for some reason (e.g., removal of primary pro-

Box 6.8. Enhancing Resilience for Adapting to Global Change

Carl Folke

Two fundamental errors underpin past global change thinking and practices dealing with such change. The first is an implicit assumption that responses of ecosystems to human use are linear and predictable and can be controlled. The second is the assumption, largely subconscious, that the human and natural systems are separate and can be treated independently. The evidence accumulated in diverse regions all over the world suggests that natural and social systems behave in non-linear ways and exhibit marked thresholds in their dynamics. Furthermore it shows that social-ecological systems act as strongly coupled, integrated complex systems and that their management is faced with uncertainty and surprise.

In recent years several research groups have concentrated on trying to understand the dynamics of ecological systems and their linkages to social systems using the concept of resilience as a unifying theme (Gunderson and Holling 2002). *Resilience* is the magnitude of disturbance that can be tolerated before an ecosystem moves into a different state with a different set of controls, i.e., the major processes and functions of the system are changed to the degree that a different set of ecosystem services, or even disservices, are generated by the system. Resilience of complex systems has three properties: (*i*) the amount of change the system can undergo and still remain within the same state; (*ii*) the degree to which the system is capable of self-organisation (compared with lack of organisation or organisation forced by external factors) and (*iii*) the degree to which the system can build the capacity to learn and adapt (*www.resalliance.org*) (Carpenter et al. 2001).

Vulnerability is the antonym of resilience. In a vulnerable system even a small change may shift the ecosystem into an undesirable state, a shift that may cause severe social and economic consequences.

Although stochastic events like storms or fire can trigger shifts between ecosystem states, recent studies of lakes, rangelands, oceans, coral reefs and forests show that it is the erosion of resilience by human action that usually paves the way for a shift to an alternate and often less desirable state with a reduced capacity to support social and economic development (Scheffer et al. 2001) (Fig. 6.23). For example, in lakes, water clarity often seems hardly affected by increased human-induced nutrient concentrations until a critical threshold is passed, at which point the lake shifts abruptly from clear to turbid waters. With this increase in turbidity, submerged plants disappear. The associated loss of animal diversity and reduction of the high algal biomass makes this state undesirable as activities like recreation or fisheries are affected by the eutrophied state.

Similar human-induced shifts take place between states in rangelands, usually between dominance of grasses and dominance of woody plants (small trees and shrubs). The shifts are driven by fire and grazing pressure under highly variable rainfall conditions. Persistent high grazing pressure for sheep or cattle production precludes fire. Above a certain density of woody plants, there is insufficient grass fuel to permit a fire and the rangeland shifts to the less productive (from a human use perspective) woody plant state. This pattern persists even when grazing animals are removed. It can take decades for the woody plant community to re-structure and open up sufficiently to allow fire back into the system.

Coral reefs in the Caribbean region have undergone dramatic changes, from a state dominated by hard coral to one dominated by fleshy algae. The changes have been driven by a combination of nutrient increase and overfishing of algae grazing organisms with hurricanes and disease. The loss of grazer diversity through overfishing resulted in eroded resilience and increased vulnerability (Jackson et al. 2001). A disturbance event that would have been absorbed by a resilient coral reef became the trigger that caused the ecosystem to shift from a coral-dominated state to one dominated by algae (Nyström et al. 2000) (Fig. 6.23).

Such shifts, from one ecosystem state to another, may be virtually irreversible as in desertification, or periodic, as in the recurrent outbreaks of forest pests or diseases such as influenza. The new state may not generate the same level, or even the same type, of life-support functions as before, and thereby cause social and economic disruption. Furthermore, loss of resilience in landscapes and seascapes may impact the climate system through a reduced capacity to retain stocks of carbon and other greenhouse gas precursors in terrestrial and aquatic systems when they are subject to disturbance events and compounded perturbations.

Managing for resilience enhances the likelihood of sustaining development in the face of global change where the future is unpredictable and surprise is likely. The resilience perspective shifts policies from those that aspire to control change in systems assumed to be stable, to managing the capacity of social-ecological systems to cope with, adapt to, and shape change (Folke et al. 2002a,b). Adaptive capacity concerns the learning aspect of system behaviour in response to disturbance and reflects the ability of a social-ecological system to cope with novel situations without losing options for the future (Berkes et al. 2003). Adaptive capacity in ecological systems is related to genetic diversity, biological diversity, and the heterogeneity of landscape mosaics. In social systems, the existence of institutions and social networks that learn and store knowledge and experience, create flexibility in problem solving and balance power among interest groups play an important role in adaptive capacity. Adaptive co-management systems (collaboration of a diverse set of stakeholders operating at different levels, often in networks, from local users, to municipalities, to regional and national organisations, and also international bodies, involved in ecosystem management) are emerging in many places to deal with the challenges of managing uncertainty and surprise in the face of global change (Olsson, Folke and Berkes, pers. comm.).

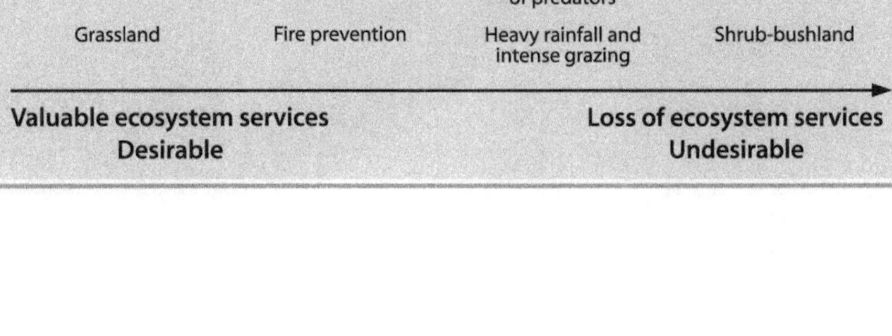

Fig. 6.23. Shifts in ecosystem state due to human-induced loss of resilience for three systems and their implications for ecosystem services

ductivity at a high rate), it becomes vulnerable to changes that it could have otherwise absorbed without significant changes in functioning. This approach to adaptation aims to determine in detail what features or characteristics confer resilience to systems and to support and enhance these features as an insurance against perturbations, such as those arising from global change, that are difficult to predict and thus adapt to specifically.

Although the adaptation and resilience approaches have many attractive features, they have one major weakness in connection with global change. They cannot prevent undesirable changes in the human-environment system when global change drives that system across a critical threshold. For example, the state of coral reefs has often been connected with local and regional scale driving forces such as fishing and tourism pressure, eutrophication from agricultural effluents and sedimentation from land-use changes in the hinterland. All of these are amenable to management changes that increase the resilience of the coral reef ecosystem towards uncontrollable perturbations like storms. However, there is growing evidence (R. Buddemeier, pers. comm.) that these factors are now being overshadowed by the long-term, chronic pressures of changing carbonate chemistry in seawater due to dissolution of CO_2 from the atmosphere and the increased temperature of surface waters due to global warming. No type of local or regional management can build resilience towards this type of forcing factor should a chemical or physical threshold be crossed. The only solution is mitigation of the global-scale driving force.

A third, and highly controversial, type of adaptation is to employ technological solutions to global change at the planetary scale via geo-engineering (Schneider 2001). Examples of geo-engineering (Fig. 6.24) range from the fanciful to more feasible actions already being seriously debated. They include the orbiting of giant mirrors around the Earth to deflect incoming solar radiation; the injection of aerosols into the atmosphere to counteract greenhouse gases; injection of propane into the stratosphere for the neutralisation of chlorine atoms that endanger the ozone hole; stimulation of the marine carbon pump through iron fertilisation of plankton; massive increase in the storage of water on the terrestrial surface to slow or eliminate sea level rise; injection of CO_2 sequestered from industrial effluents into the deep ocean or into geological formations; and massive reforestation to reduce the amount of CO_2 in the atmosphere. The geo-engineering approach represents an attempt to manage Earth System functioning itself and is thus fraught with the severe shortcomings of attempting to manage any complex system, as was emphasised earlier.

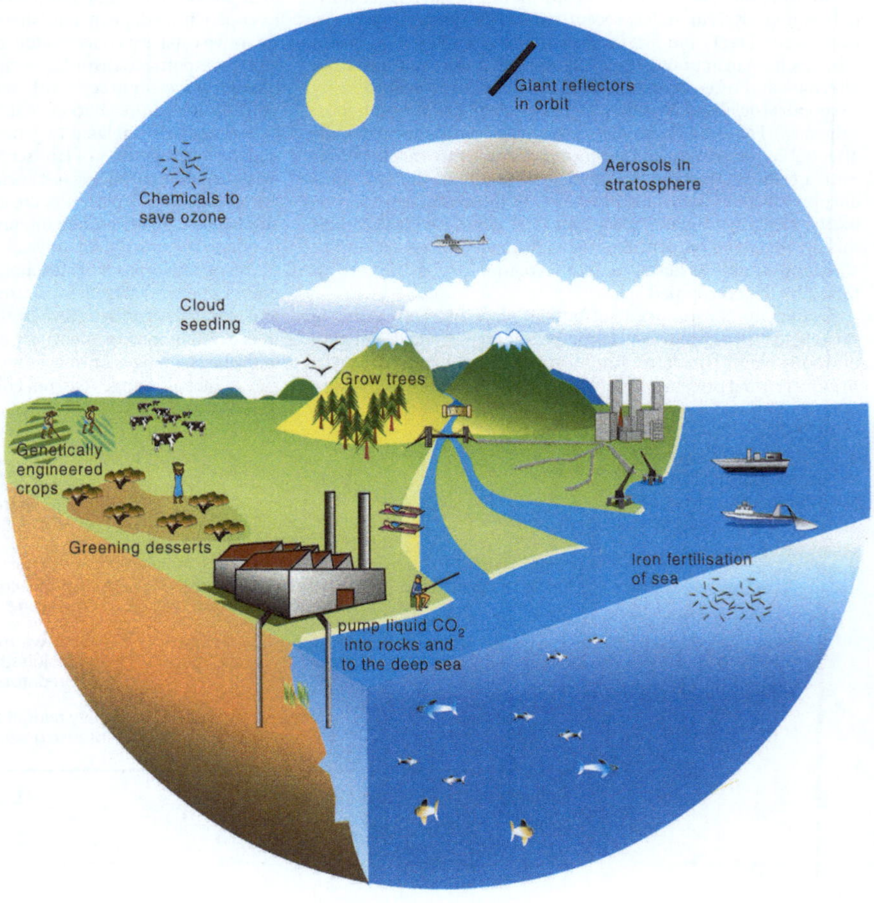

Fig. 6.24. Schematic representation of various geo-engineering proposals to manage parts of the Earth System (adapted from Keith 2001)

It is imperative that the use of technology for global-scale adaptation should not be confused with the use of technology for the mitigation of global change (Steffen 2002). As an analogy, the problems of oil spills on marine and coastal ecosystems can be addressed in two contrasting ways. The spills can be largely avoided in the first place by improving ship safety technology or by re-routing cruise tracks away from potentially dangerous obstacles. Alternatively, the impacts of an oil spill once it has occurred can be ameliorated by improved technologies for treating injured wildlife, cleaning beaches, etc. Clearly, it is far preferable to prevent the oil spill in the first place than to try to clean up after it.

The same principle is even more applicable to the case of global change and the Earth System. It is exceedingly dangerous to apply simplistic technological solutions based on cause-effect logic within the context of complex, non-linear systems that are, as yet, poorly understood. In addition, there are serious ethical and moral considerations that will need to be debated and overcome across a very wide range of cultures and belief systems before any sort of global geo-engineering fix can be attempted. Two examples illustrate some of these points.

It is sometimes proposed that aerosols can be released high in the atmosphere to counteract the effects of greenhouse gases. This proposal, however, is based on a simplistic and incomplete understanding of the role of aerosols in Earth System functioning; it accounts only for the direct radiative effects of aerosols. Aerosols also have indirect effects in terms of their property as cloud condensation nuclei and hence on the hydrological cycle and climate system. This indirect effect, at present poorly understood, may be at least as important on climate as the direct effect. In addition, aerosol-inducing cooling will trigger teleconnections in the climate system that may lead to unpleasant surprises around the planet, for example, the shifting of monsoons and rainfall patterns (Rotstayn and Lohmann 2002). In addition to their climatic effects aerosols are important for the chemistry of the atmosphere; their role in heterogeneous reaction chemistry is poorly understood at present, but is undoubtedly important for the maintenance of the oxidising efficiency of the atmosphere and hence its ability to cleanse itself of pollutants. Thus, the scientific knowledge base of the many complex roles that aerosols play in the Earth System is still much too weak to even contemplate widespread release of aerosols to cool climate.

A second example is the use of terrestrial carbon sinks to remove CO_2 from the atmosphere. This methodology, in fact, is already being used to some extent and is an important component of the Kyoto Protocol. It may be viewed as one of the most benign and non-controversial types of (bio)geo-engineering. In addition to its climate role, enhanced terrestrial carbon sinks through reforestation, for example, can have other collateral benefits such as reduction of soil erosion and provision of biofuels if systems are established on a sustainable basis.

Nevertheless, there are dangers with even this type of geo-engineering. Often the enhancement of terrestrial sinks is considered as a substitute for the reduction of fossil fuel emissions; in fact, this is precisely what is done in the accounting procedures for compliance with the Kyoto Protocol. From an Earth System perspective, fundamentally flawed logic stands behind the approach of equivalence or substitutability of terrestrial sinks and fossil fuel emissions. The latter represent the injection of a previously isolated source of carbon into the active atmosphere-terrestrial-ocean carbon cycle, a net addition of carbon to the active system. On the other hand, use of terrestrial sinks is simply a reallocation of carbon within the active atmosphere-terrestrial-ocean carbon cycle, and not a net removal of carbon from the active system. Thus, care must taken to manage terrestrial sinks carefully to prevent the case whereby enhanced carbon uptake during the first half of the twenty-first century, for example, becomes the source for enhanced emissions during the second half of the century, exacerbating rather than helping to solve the atmospheric CO_2 problem. Most scientists and the IPCC are well aware of these characteristics of terrestrial carbon sinks, but it is not clear that policymakers or resource managers are. The geo-engineering approach of using terrestrial carbon sinks to remove CO_2 from the atmosphere, so often touted as a win-win solution, is not without complexities that could lead to unpleasant surprises in the future.

Serious research questions, both biophysical and socio-cultural, remain to be considered before informed and responsible interventions in the functioning of the Earth System may be contemplated, if ever. Even if a sufficient understanding is achieved, serious management challenges will remain.

Mitigation

All three of the adaptation approaches to planetary management discussed so far are end-of-pipe pollution approaches. That is, they do not tackle the problem of global change at its source, but rather react to the problem in various ways after it has occurred. A contrasting approach is mitigation, where the objective is to try to diminish or eliminate the problem itself rather than react to it. Put simply, the approach is to take some of the pressure off the Earth System by lessening the intensity of many of the drivers of change described in Chap. 3. In this sense, mitigation represents an attempt to manage human activities and not the functioning of the Earth System itself, with the aim to allow the System to function in a more natural way.

Technological advances comprise a powerful tool for mitigation and the wisdom of their use stands in strong contrast to the dangers of post-hoc technological fixes via geo-engineering described above. Many examples

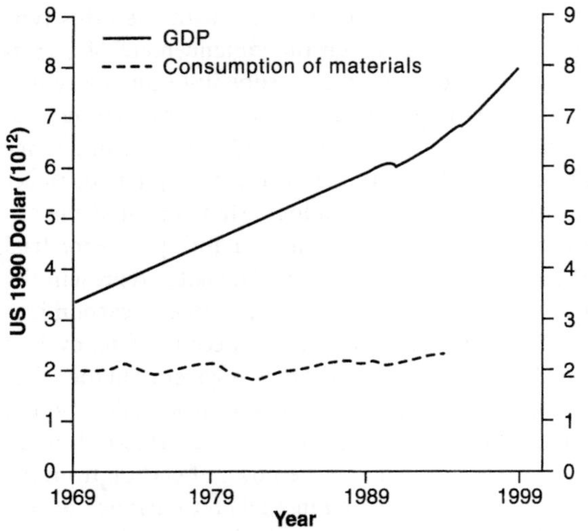

Fig. 6.25. Stabilisation of the amount of material flowing through the USA economy despite continued economic growth (Matos and Wagner 1998; Bureau of Economic Analysis 2000)

exist of technology-driven improvements for mitigating global change. In many parts of the world motorcars have become more fuel-efficient and the production of effluents per kilometre driven has dropped significantly; aeroplanes have also become substantially more fuel-efficient. More revolutionary modes of transport promise even greater gains in energy efficiency (Ausubel 2002). Many household appliances consume substantially less energy than their predecessors. In terms of the production of energy, renewable sources like wind power are now becoming reliable and well-accepted contributors to national grids in many countries. In general, a trend of dematerialisation can be seen in many advanced economies (Fig. 6.25), in which the value and amount of economic activity continue to grow but the amount of physical material flowing through the economy does not. This leads to a decoupling of economic growth and pollution, an evolving form of mitigation against global change. Mitigation measures ex-

Box 6.9. Globalisation and Environmental Change

William C. Clark · Jill Jäger · Robert W. Kates

The questions of how globalisation affects the global environment and how the environment is being globalised require answering. Globalism has been defined as "networks of interdependence at intercontinental distances… linked through flows and influences of capital and goods, information and ideas, people and force, as well as environmentally and biologically relevant substances…" (Keohane and Nye 2000). The related process of globalisation "refers to the widening, deepening and speeding up of [such] global interconnectedness" (Held et al. 1999). In this view, neither globalism nor globalisation is new. Kates (2003) suggests six major periods of globalisation: the earliest prehistoric period in which humans spread out of Africa around the world; the pre-modern period of early empires and world religions; the early modern period of Western expansion; the modern industrial era; the contemporary period from 1945 to the present; and the future, in particular the first half of the twenty-first century.

Focusing on the contemporary and near future period, Clark (2000) points to three particular linkages at intercontinental scales between society and environment that define globalisation:

- the flow of energy, materials and organisms through the environment couple the actions of people in one place with threats and opportunities faced by people elsewhere;
- the flow of ideas regarding environment from one place to another;
- the changing configuration of actors, norms and expectations that have emerged as a result of attempts by societies to deal with the flows of energy, materials and organisms, as well as the flows of ideas.

As demonstrated in all chapters of this book, the planet's environment is shaped by complex linkages between the atmosphere, ocean, soil and biota. It has also been shown that humans now play a major role in influencing flows of energy and materials. Chapter 4 has demonstrated that changes in one part of the Earth System can cascade through the System. The impacts of long distance transport of materials have been shown to be more complex and pernicious than originally supposed. Materials entrained in the global circulation of the atmosphere, in particular carbon monoxide, carbon dioxide, methane, nitrous oxide, chlorofluorocarbons, persistent organic pollutants, and fine particles, emitted locally may be distributed globally. Emissions of materials capable of being transported long distances have increased dramatically since the mid-twentieth century.

The most dramatic globalisation in biotic linkages has involved movements of pests or disease organisms from their native habitats (usually by human transportation processes) to places where they have never been and where the resident people or other biota they value have evolved no protection from the invader. The impacts of such long-distance invasions are extensive and represent a major cause of land-use transformation and species extinction. The rates of invasion are not well documented, but it appears that there has been a rapid acceleration of incidence and extents due to increases in commerce, tourism, and travel in general (Baskin 2002). As a result of increasing population, consumption, and economic connectivity of the world, an increasing number of human activities undertaken in more and more parts of the world are having impacts at transcontinental scales. Added to this, the impacts are also increasingly interactive (e.g., Gorham 1996).

The modern idea of a mutual *global* interdependence between nature and society traces back at least to the work of Vernadsky (1945), and has increasingly taken hold in political circles over the last generation (Turner et al. 1990). One might expect that the emerging idea of a globally connected environment would lead to the parallel emergence of a demand for comparably global environmental policy, management and research. To a significant degree this is actually happening, as evidenced by the growth in international environmental treaties and agreements (Mitchell 2003), as well as the significantly expanded global environmental research and monitoring system reflected throughout the contributions to this volume. Supported by modern trends enhancing the rate and extent of information transfer, the idea of environmental globalisation has helped promote the emergence of a surprisingly rational system of global environmental management (The Social Learning Group 2001). That system, however, remains highly incomplete and often ineffective (Speth 2003). This suggests that, in globalisation parlance, contemporary ideas about global environmental interdependence almost certainly need to be more widely, deeply, and rapidly adopted before they are adequate to the challenges that confront humanity.

One of the most promising such ideas to have emerged over the last two decades is that of *sustainable development*. This plicitly designed to combat global change can thus build on current trends in technology and can accelerate them by stipulating performance guidelines and setting increasingly stringent emission targets for the future.

A closely related aspect of mitigation is the use of improved management techniques for activities that are important drivers of global change. Again, there are many positive trends currently under way so that explicit mitigation measures for global change have a strong base upon which to build. Transport and agriculture provide good examples. In addition to gains in efficiency due to technological advances themselves, other gains come from management of, for example, traffic patterns through road design and the control of traffic flow during busy periods. The aviation industry, through codeshare flights and increased passenger loadings, is moving more people for less effluent per person-kilometre flown. The agricultural industry is also continually improving its efficiency. Based on increased knowledge of the nutrient dynamics of crops, less and less fertiliser is used to grow more crops. Better irrigation techniques are also increasing the water use efficiency of agricultural systems. Under the term *precision farming*, these advances promise to increase agricultural production without the need for more land and with relatively less fertiliser and irrigation required (Ausubel 2002).

Taken together, more efficient technology and more effective resource management coupled with trends towards increased social learning and global interconnectivity (Box 6.9) offer hope that society may be on the threshold of a great transformation that could meet the challenge of global change (Fig. 6.26). Some projections of the future depict a worldwide transition to much more sustainable global society (Raskin et al. 2002), aimed at achieving the four great human aspirations: peace, freedom, material well-being and environmental health. Such a scenario for the future includes, in addition to vastly improved technology and resource management, signifi-

◀

idea encompasses global environmental concerns but explicitly links them to locally situated concerns about meeting human needs for food security, energy, industrial production, habitation and resource conservation in ways that conserve key life support services over the long run (as discussed at the UN Conference on Environment and Development in Rio de Janeiro, 1992, and the World Summit on Sustainable Development in Johannesburg, 2002). Central to the sustainability idea – and a new contribution to environmental thinking – is the growing consensus that effective efforts to rationalise human aspirations and environmental limits over the long run require not only an understanding of globalisation but also a commitment to democratisation. The future globalisation of environmental ideas may thus be tightly bound to the symbiotic globalisation of ideas about civil society and human rights in efforts to promote sustainability (Khagram et al. 2003).

Global governance relationships include formal treaties among states governing intercontinental environmental issues; they also include a wide range of relationships governing the interactions within and between governments, businesses and environmental advocates around the world. At the nation state level, the single greatest contribution of globalisation may well be the copying of environmental regulations from one country to another (Brickman et al. 1985). Internationally, over 700 multilateral environmental agreements exist, covering perhaps 250 distinct problems. Roughly two-thirds of these have been created since 1970 (Mitchell 2003). In addition, the networks of connections running through intergovernmental organisations dealing with environmental issues have *thickened* considerably over the last several decades. Here the term thicken refers to the increases in variety, strength and density of long distance relationships among actors, the number of actors involved and the speed of change in society that those changed relationships helped to induce.

Significant intercontinental governance initiatives dealing with the environment have also been launched in the private sector (e.g., World Business Council on Sustainable Development and the International Chamber of Commerce). Many individual firms with multi-national operations have also brought environmental norms and considerations into their standard operating procedures, which often has the effect of transplanting higher standards from one part of the world into others where government actions on the environment may lag behind. Furthermore, in recent decades, non-governmental organisations have played an increasing role in the globalisation of environmental governance. They have shown a remarkable ability to utilise emerging information technologies and to be fast in organising *ad hoc* coalitions to address particular issues. Finally, the recent changes of environmental governance have seen the emergence of networks of actors that cut across environment/development concerns to lie with coalitions of actors, united by their advocacy of particular programs but not constrained within particular sectors or even levels of governance. While not all of the governance linkages noted here involve intercontinental scales, many do. An accelerating thickening of globalism in environmental governance is certainly underway.

In his discussion of how globalisation affects the prospects for sustainability, Kates (2003) concludes that there are five causal elements: population, consumption, technology, ideas and governance. He argues that globalisation will encourage a decline in the population growth rate, accelerate consumption through increases in consumption and production, increase trade and transport, and spread ideas and images of Western material standards of living. Globalisation facilitates the creation and diffusion of technologies that lessen the needs for energy and materials per unit of production or consumption, create fewer toxic substances or pollutants, or substitute information for energy and material use. However, technologies can also have negative effects. Similarly the spread of ideas and environmental governance can also have dual impacts.

Kates concludes that in preserving life-support systems the crucial issue will be globalisation's impact on current and projected production and consumption of energy and materials that are environmentally degrading and resource depleting. The absolute growth in consumption of energy and materials seems likely, in his view, to overwhelm the steady global progress in technical efficiency. He argues that the globalisation of environmental ideas and of environmental governance have been facilitated by increasing interconnectedness but may well be insufficient to counter the major threats to the planetary life-support systems associated with the economic dimensions of globalisation. For globalisation to encourage a transition toward sustainability, it will be necessary to strengthen efforts to assist the transition and dampen those movements that hinder.

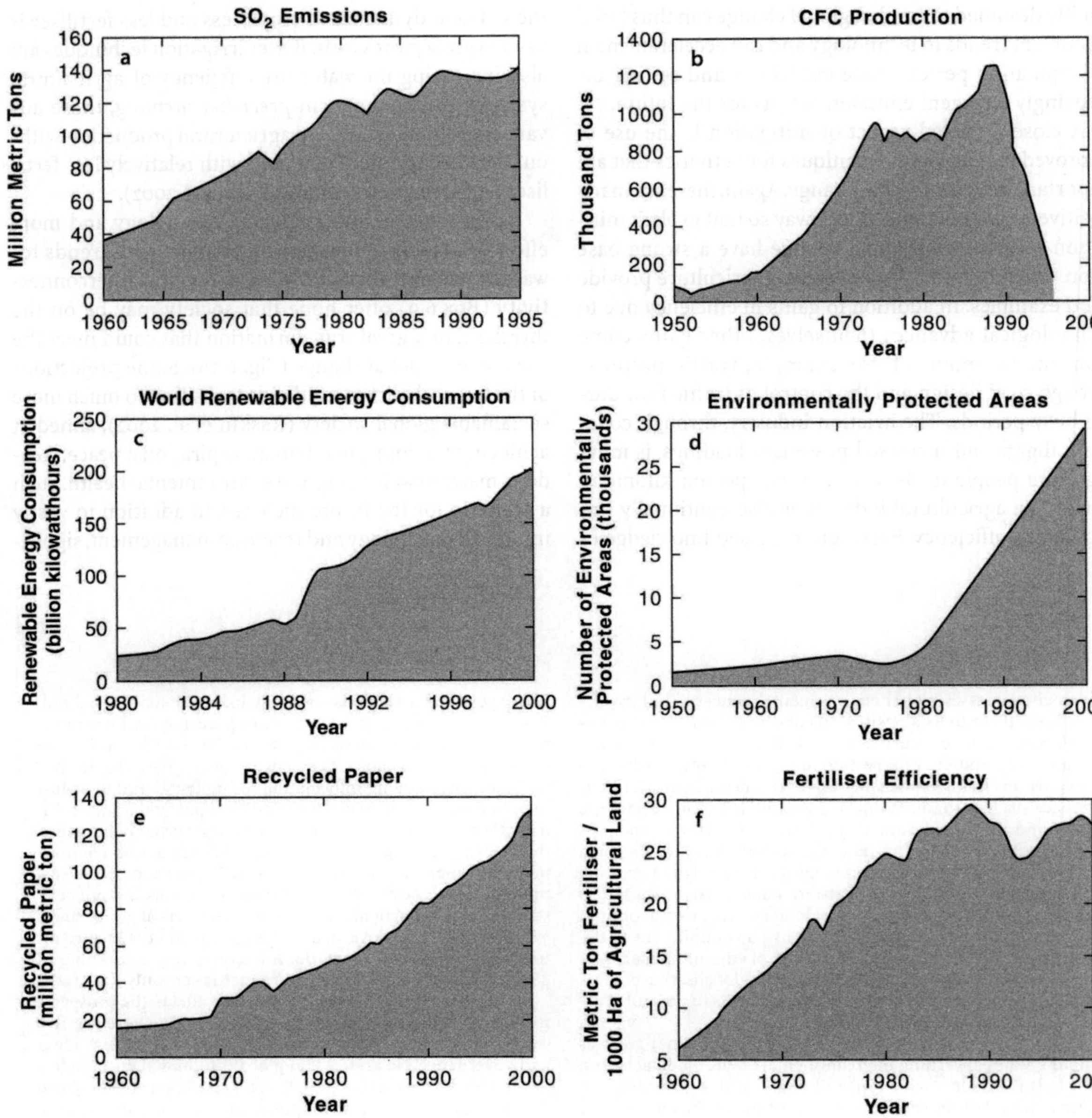

Fig. 6.26. Improvement in several environmental indicators, suggesting that societies may be meeting the challenge of global change: **a** SO_2 emissions, 1960–1995 (WRI 1999); **b** CFC production, 1950–1992 (based on data from Philippe Rekacewicz, UNEP GRID-Arendal); **c** world renewable energy consumption, 1980–2000 (EIA, *http://www.eia.doe.gov/*); **d** increase in number of environmentally protected areas, 1950–2000 (WRI 1992; WCMC 1999); **e** amount of paper recycled, 1960–2000 (FAOSTAT 2002); and **f** increase in fertiliser efficiency, 1960–2000 (FAOSTAT 2002)

cant changes in governance, institutions and value systems. In effect, such a transition – to the so-called planetary phase of human existence – would be the fourth major historical era, following the stone age, early civilisation and the modern era.

To achieve the type of great transition described above, global change will need to be limited to tolerable rates and magnitudes. Advances in technology and management are essential but on their own will not be enough. Given the sheer (and growing) volume of individual actions that ultimately drive global change, alterations in societal values and individual behaviour, particularly with regard to consumption, may also be required to slow or reverse current trends in the global environment. Analyses of the impacts of the entire global population of six billion (not to mention the projected eight-nine billion expected later this century) suggest that humanity is already overshooting the capacity of Earth's biological systems to provide resources and to absorb effluents indefinitely (Wackernagel et al. 2002). Furthermore, if all humans succeed in living in future at the same material standards as those currently prevailing in the United States, humanity will in effect need several planet Earths to be able to exist in an ecologically sustainable fashion. These projections strongly suggest that new approaches – strategies which are more integrated, more long-term in outlook, more attuned to the natural dynamics of the Earth System and more visionary – are required if humanity is to deal with global change in an effective way.

New Paradigms for Global Sustainability

All of the management approaches described above are extensions of known approaches that have been used at sub-global scales. However, what differentiates global change from all other environmental problems is that it operates at the level of the Earth System itself, of the planetary life support system. In addition to specific management approaches, an overall, comprehensive, internally consistent strategy for management of the Earth System is required. Much has been written on the issue of management of the global environment or aspects of it (Clark and Munn 1986; Turner et al. 1990; NRC 1999; The Social Learning Group 2001). One of the more conceptually thorough and comprehensive treatments is that by Schellnhuber (1998), which provides a useful framework for the following discussion.

The first step needed to address the issue of global-scale management is to ask some fundamental questions about the nature of the human response to global change. What kind of world do we live in now? What is the nature of the planetary life support system? What kind of world do we want to live in? What must we do to get there? Variants of these questions are recognisable as some of the 23 questions described in Sect. 6.3.1 as the central challenges for Earth System science over the coming decades. They are also directly relevant to the use of this expanding knowledge base for global management.

Perhaps the paradigm that is closest to present-day management methods for environmental problems is the one based on a small number of environmental qualities or aggregated functions defined as sustainability indicators. The management objective is to maintain the evolution of the Earth System within a safe range of these indicators. This approach is what Schellnhuber calls *standardisation*. An example of the use of this approach today is the definition and maintenance of minimum air quality standards for urban airsheds. In terms of the Earth System, an application of this approach might be to maintain the global load of air pollutants to within a certain range that ensures that the cleansing efficiency of the atmosphere is not overwhelmed. As is often the case in setting standards, environmental goals are traded off against economic realities.

Sustainability indicators/standards may seem an attractive and feasible approach given their present use for controlling local and regional emissions. In fact, they are the only tool that current legal systems recognise so that offenders can be identified and punished. However, an indicators/standards approach to global sustainability raises serious challenges from an Earth System perspective. Standards are almost always established with a certain time horizon in mind. Cases may well arise where desired standards over a short term can nevertheless lead to larger, long-term accumulated effects that can trigger irreversible changes in the Earth System. Even more serious is the fact that environmental standards are usually set with a single objective in mind. The standards approach raises the danger that unexpected side effects or interactions in a highly non-linear system are not taken into account and acceptable standards in one area, set in a simple cause-effect way, can lead to surprises elsewhere.

In an ideal world the most attractive of all of the management approaches would aim to maximise generalised utility over a prescribed time period. Here utility is defined in a broad, normative way through millions of acts and interactions by individuals around the world. This approach is referred to as *optimisation* (Schellnhuber 1998). It will, of course, be tempting for economists to define utility in only an economic sense, but for global management purposes it is much broader. The specification of the time period raises an interesting point. This management approach can be defined as maximising utility at every point as the Earth System evolves into the future, or it could be defined as the maximisation of utility at the end of a given time period. The latter would allow, for example, the heavy exploitation of resources, including fossil fuels, by the developing world during the first part of the twenty-first century in order to obtain an overall larger utility function for the Earth System at the end of the century.

Attractive though it sounds, there are weaknesses in the maximised utility approach. Clearly some trade-offs will have to be made in order to define an aggregated utility for the Earth as a whole. These problems have already arisen in the climate change issue during IPCC deliberations, for example, when the question has been raised of whether it is possible to trade off consumer goods in the developed world against human well-being in the developing world. A further difficulty with the approach arises with the scale of interest. Maximising the utility of human-environment systems at the local scale everywhere may still lead to problems, as there is no guarantee that such solutions, when aggregated, are also globally optimal. It is conceivable that, through teleconnections and emergent properties, undesirable changes, especially abrupt changes, could be triggered even when local utility is maximised. In fact, the production of chlorofluorocarbons can be viewed as an increase in local utility by the replacement of various chemicals by compounds less harmful, at least on a local scale. The effects of chlorofluorocarbons on stratospheric chemistry on the global scale, and the resulting abrupt change in the Earth System, could not have been imagined by those who introduced them.

Consideration of abrupt changes with potentially catastrophic effects leads to the most intriguing of all approaches to Earth management, one in which the primary emphasis is placed on the precautionary principle of preventing the worst from happening (an approach called *pessimisation* by Schellnhuber (1998)). Over the past decade the realisation that Earth's natural variability is characterised by abrupt changes with catastrophic consequences for human societies has grown from the fringes of global change science to a mainstream concern. The precautionary principle is directly relevant to this characteristic of the Earth System, as it aims at managing human activities so as to steer the global environment away from any such disastrous pathways. The ozone hole episode (Sect. 5.4.1) was a narrow escape from just such a pathway. A related question is whether other management approaches, for example the one aimed at maximising general utility, should be followed if they push the global environment close to such thresholds.

The temporal aspect of the precautionary principle again raises important questions. Figure 6.27, adapted from Schellnhuber (1998), shows a hypothetical analogy of a boat approaching a fatal drop over a waterfall. The point is that the boat's captain must make his decision on the course well before he is fully aware of the dangers that lie ahead. Human societies are already in precisely this situation with their unintended experiment with the Earth System. Even more important is whether a global management strategy aimed at avoiding catastrophic changes can insure itself against future unsound management. Can the increasingly global human society ensure itself against the future leadership of a major country, for example, attempting to push the Earth System too close to a dangerous threshold for its own national gain?

A more practical application of the precautionary principle is the notion of guardrails (WGBU 1996), that the highest priority in any form of Earth System management is to avoid abrupt and irreversible changes from which human societies would find it difficult or impossible to recover. The guardrails concept calls into question the widely quoted three pillars structure to sustainability, in which economic, social, and environmental aspects can be traded off against one another. If a critical environmental threshold is crossed, economic and social considerations count for little. Sound environmental management requires that the crucial guardrails of Earth System dynamics are respected and avoided, regardless of their short-term economic and social implications. However, as abrupt, potentially catastrophic changes are low probability, high impact events, they present significant challenges to decision-making systems in contemporary human societies.

In a world in which the disparity between the wealthy and the poor, both within and between countries, is growing, equity issues are important in any consideration of global environmental management. Management approaches focussing on equity aim to achieve a relative balance amongst the various participants who have a stake in global management (i.e, virtually the entire human population), but taking into the account the nature of Earth System dynamics. The inequities amongst groups of people who currently reside on the planet are becoming well documented, are apparently increasing, are serious and need to be addressed (Sect. 3.4.1). From an Earth System perspective, equally important is equity in a temporal dimension, between current and future generations, a central feature of the Brundtland Commission's report on sustainability (WCED 1987).

Fig. 6.27. Decision-making amid uncertainty (Schellnhuber 1998). A boat is situated upstream of a bifurcation in a river – one branch going over a harmless rapids and the other going over a big waterfall. The captain must make a decision as to which side of the river he should follow before he is fully aware of the consequences. In this analogy, the role of Earth System science is to provide the captain with an accurate map of the river so that he is aware of the dangers ahead and can chart a safe course

Inter-generational equity raises difficult questions for Earth System management. By its very nature, management of the Earth System requires choices to be made at critical points, for example, on the mix of source reduction and sink enhancement that will be employed to tackle the carbon cycle issue. Such choices (e.g., reductions in fossil fuel emissions) necessarily rule out options for future generations, while at the same time new opportunities will emerge. The point is that humanity is now entering a new era in its relationship with its own life support system. All humans inherit, in terms of culture and history, the legacy of those that have gone before them. However, the very environment around children and grandchildren will increasingly be a product, at least in part, of the parent generation. Being able to weigh up the options being lost against those being gained, in both number and quality, is the central question in the inter-generational equity debate.

A final potential management approach is built around a desire for a stable equilibrium in the global environment. The term and concept have in the past produced debates between those who seek a stable or ideal natural environment and those who view such goals as a threat to an economic system that depends on ever-expanding consumption, mobility, recreation and other activity. The debate takes three common forms. First, there are those who seek preservation of a perceived idealised natural/human global environment – a stabilisation of the *status quo*. They often perceive as natural those environments that have been modified by humans thousands of years ago. Secondly, a growing number of people are advocating restoration of previously altered environments and restoration ecology is thus a rapidly growing field of science. Thirdly, many propose that the Earth System be deconvoluted into areas that are largely natural and protected and those that are exploited for human purposes. The appropriate mix of these, aimed at some sort of longer-term stabilisation, is an intriguing scientific question (and one of the 23 Earth System questions considered in Sect. 6.3.1).

A stabilisation approach has problems on longer time scales. For example, the Vostok ice core data show that the Earth System has not operated over the last 420 000 years in an equilibrium fashion, but rather has regularly oscillated between two primary states. Finer-detailed palaeo-records show that the Earth is never static and it is almost impossible to define an equilibrium state; variability abounds at nearly all spatial and temporal scales. Thus, it might be more appropriate in the stabilisation paradigm to consider the rates and magnitudes of perturbations within regular patterns of variability rather than stabilisation around perceived equilibrium points.

It is clear that the current state of the global environment is moving far beyond known modes of variability. Thus, the question arises as to how management might move the Earth System from its present no-analogue state to something that might be viewed as stabilisation. A practical example might be the stabilisation of the mean surface temperature of Earth, and thus the stabilisation of the concentrations and patterns of greenhouse gases and aerosols in the atmosphere. The proposed stabilisation state can be approached rapidly in a crash-landing style or approached more gently in the so-called soft landing mode. The rate of management change that might be initiated is important. In respect of greenhouse gases, if the approach to stabilisation of atmospheric CO_2 is too slow, positive feedbacks from increasing soil respiration rates and decreasing solubility in the oceans may lead, for example, to a surge in concentration that is impossible to stabilise through decarbonisation of the energy sector. Finally, the costs of stabilisation must also be considered. These include not only the economic costs, but also the social, ethical and ecological costs of the management process (or lack thereof).

In reality human societies will not be able to adopt any one of these approaches in their pure forms. Even if they desired to do so, lack of complete scientific understanding and limits to predictability may prevent the full implementation of the paradigm. Thus, management of the Earth System will almost surely be a mix of current and more visionary approaches implemented by an imperfect and only partially efficient decision-making system based on an evolving, but never fully complete knowledge base. In addition, the state and dynamics of the Earth System itself will be changing due to natural variability and managed and unmanaged human perturbations. In these conditions, adaptive management is the only feasible way forward.

Adaptive Management

In a broad sense adaptive management is an interactive process of learning by doing and of doing based on learning. In the context of the above discussion on more theoretical approaches to management, it implies that combinations of approaches will be required. For example, the overall strategy must be to avoid catastrophic state changes, but within that overall strategy a number of the other approaches may also be applied. The climate system could be stabilised at some agreed level of change while also taking equity and economic and social cost considerations into account.

Described in terms more commonly used at smaller scales, adaptive management means that environment and development decisions are tentatively made on the basis of imperfect information, but constantly revised according to the influx of relevant new scientific findings. The technique is already being applied to resource management issues at local scales, where the feedbacks between outcomes of management decisions and generation of new knowledge based on the outcomes are tight enough in both space and time scales to make the system work. However, contemplation of the use of adaptive management for the Earth System as a whole raises unprecedented challenges. Three of these are especially demanding.

First, the global environment itself is a system of intricate complexity built over billions of years of evolution. The notion that humanity can somehow understand it well enough to be able to manage it in a sensitive and intelligent way seems well beyond the present state of scientific understanding. How can the long and complex chains of reverberations through the Earth System that inevitably will be triggered by management actions be understood well enough to lead to improved management?

Box 6.10. Sustainability Science[1]

Jill Jäger

Sustainability concerns have occupied a place on the global agenda since at least the 1980s, with publication of the International Union for the Conservation of Nature's (IUCN) *World Conservation Strategy* and the Brundtland Commission's report *Our Common Future* (IUCN 1980; WCED 1987). The prominence of that place has been rising, however. UN Secretary-General Kofi Annan reflected a growing consensus when he wrote in his Millennium Report to the General Assembly that "Freedom from want, freedom from fear, and the freedom of future generations to sustain their lives on this planet" are the three grand challenges facing the international community at the dawn of the twenty-first century (Annan 2000). Science and technology are increasingly recognised to be central to both the origins of Secretary-General Annan's three challenges, and to the prospects for successfully dealing with them.

Though visions of sustainability vary across regions and circumstances, a broad international consensus has emerged that its goals should be to foster a transition toward development paths that meet human needs while preserving the Earth's life support systems and alleviating hunger and poverty – i.e. that integrate the three pillars of environmental, social and economic sustainability.

Calls for strengthening science and technology programmes targeted on sustainable development built slowly during the 1990s following the UN Conference on Environment and Development (UNCED) in Rio. Many of the earliest and most thoughtful contributions to this discourse came from the developing world through the work of individual scholars and of institutions. A number of national academies of science or other advisory bodies also addressed this topic (e.g. Rocha-Miranda 2000; WBGU 1997; NRC 1999; Science Council of Japan 2000; UK Royal Society 2000). The global change programmes also contributed with a set of core questions (Sahagian and Schellnhuber 2002; Sect. 6.3.1).

In October 2000 two dozen scientists, drawn from the natural and social sciences and from across the world, met in Friibergh, Sweden to explore how science and technology could better facilitate a transition to sustainability (Kates et al. 2001). The participants in the Friibergh meeting agreed that sustainability science focuses on the dynamic interactions between nature and society. With a view toward promoting the research necessary to achieve understanding of these dynamic interactions, the following core questions were proposed:

- How can the dynamic interactions between nature and society – including lags and inertia – be better incorporated in emerging models and conceptualisations that integrate the Earth System, human development, and sustainability?
- How are long-term trends in environment and development, including consumption and population, reshaping nature-society interactions in ways relevant to sustainability?
- What determines the vulnerability or resilience of the nature-society system in particular kinds of places and for particular types of ecosystems and human livelihoods?
- Can scientifically meaningful *limits* or *boundaries* be defined that would provide effective warning of conditions beyond which the nature-society systems incur a significantly increased risk of serious degradation?
- What systems of incentive structures – including markets, rules, norms and scientific information – can most effectively improve social capacity to guide interactions between nature and society toward more sustainable trajectories?
- How can today's operational systems for monitoring and reporting on environmental and social conditions be integrated or extended to provide more useful guidance for efforts to navigate a transition toward sustainability?
- How can today's relatively independent activities of research planning, monitoring, assessment, and decision support be better integrated into systems for adaptive management and societal learning?

Following the Friibergh meeting, the Initiative on Science and Technology for Sustainability, ISTS, (*http://sustainabilityscience.org*) held a series of meetings to discuss regional perspectives on these core questions, as well as the research strategies and infrastructure requirements that they imply. In parallel, as preparations accelerated for the World Summit on Sustainable Development, numerous other activities intensified the discussions on science, technology and sustainability. These have been documented in ICSU/ISTS/TWAS (2002).

The substantive focus of much of the science needed to promote sustainable development will have to be on the complex, dynamic interactions between nature and society (*socio-ecological* systems), rather than on either the social or environmental sides of this interaction. Moreover, some of the most important interactions will occur in particular places, or particular enterprises and times. Science and technology for sustainable development therefore needs to be *place-based* or *enterprise-based*, embedded in the particular characteristics of distinct locations or contexts. This means that science and technology will have to broaden where it looks for knowledge, reaching beyond the essential bodies of specialised scholarship to include endogenously generated knowledge, innovations and practices. Devising approaches for evaluating which lessons can usefully be transferred from one setting to another is a major challenge facing the field.

Despite the need for a place-based or highly contextualised character of much of the science and technology needed to promote sustainability, the need to deepen and strengthen work on certain core concepts has been raised repeatedly during the discussions on sustainability science with emphasis on the following themes:

- Adaptiveness, vulnerability and resilience in complex socio-ecological systems
- Sustainability in complex production-consumption systems
- Institutions for sustainable development

For knowledge to be effective in advancing sustainable development goals, it must be accountable to more than peer review (Cash et al. 2003). In particular, it must be sufficiently reliable (or *credible*) to justify people risking action upon it, relevant to decision makers' needs (i.e. *salient*) and democratic in its choice of issues to address, expertise to consider and participants to engage (i.e. socially and politically *legitimate*). These three properties are tightly interdependent and efforts to enhance one may often undermine the others. In particular, a simple focus on maximising one of these attributes (e.g., is the science *credible*?) is an insufficient and counterproductive strategy for contributing to real world problem solving where a mix of all three attributes is essential. The interdependence of saliency, credibility and legitimacy poses substantial challenges to the design of institutions for mobilising research and development, assessment and decision-support for sustainable development.

The prospects for successfully navigating transitions toward sustainability will depend in large part on an improved dialogue between the science and technology community and problem-solvers pursuing sustainability goals. Significantly, this needs to be done in ways that enhance the ability of problem-solvers at all levels to harness science and technology from anywhere in the world in meeting their goals. It will be essential to understand

[1] Based on the Consensus Report and Background Document, Mexico City Synthesis Conference, May 2002: Science and Technology for Sustainable Development, ICSU Series on Science for Sustainable Development No. 9, International Council for Science, Paris.

◀

what sorts of institutions can best perform these complex bridging roles (i.e. act as *boundary organisations*) – between science and policy, and across scales and across the social and natural science disciplines – under a wide range of social circumstances. In addition, in a rapidly changing world of interdependence, such institutions need to be agile. There is a clear demand for systematic efforts to analyse comparatively the performance of experiments in the design of institutions for linking knowledge and action to identify how and under what conditions some boundary organisations work better than others, and above all to help the groups running the existing institutions to learn from one another.

In the spirit of the discussions before and during the World Summit on Sustainable Development, it is clear that sustainability science will need strong partnerships with a wide range of social actors in order to address the critical challenges of sustainable development. Partnerships are needed in order to: strengthen the ability of locally-based initiatives to harness science and technology from around the world in support of their efforts to solve their most urgent sustainable development problems; facilitate engagement of young scientists and technologists in efforts to support environmentally sustainable human development around the world and build a global community of scientists and engineers for sustainable development.

Furthermore, science and technology cannot effectively contribute to sustainable development without basic scientific and technological capacity. It is necessary to build capacity in interdisciplinary research, understand complex systems, deal with irreducible uncertainty, and integrate across fields of knowledge, as well as harness and build capacity for technological innovation and diffusion in both the private and public sectors. With a few important but relatively small and under-funded exceptions, efforts to sustain the lives of future generations on this planet still lack dedicated, problem-driven and solution-oriented research and development systems with attendant funding mechanisms for research and technology innovation.

A great deal of the help that science and technology can provide to sustainable development must emerge from solution-focused research and development conducted in close collaboration with local stakeholders and decision-makers. The transcendent challenge is to help promote the relatively local (place- or enterprise-based) dialogues from which meaningful priorities can emerge, and to put in place the local support systems that will allow those priorities to be implemented. Locally- or regionally-focused universities and NGOs will almost certainly occupy a central place in those systems, and need to reach out to link with the larger science and technology community.

Science and technology will be severely hampered in promoting sustainability until it has developed a much firmer empirical foundation for its efforts than is available today. A determined effort to move from case studies and pilot projects toward a body of comparative, critically evaluated knowledge is therefore urgently needed. In addition, progress toward sustainability will require a constant feedback from observations. Such observation provides a reference for theoretical debates and models on strategies for vulnerability reduction, and metrics for measuring success. In order to ensure a data stream needed to form the empirical basis of sustainability science, the observations of the natural sciences and of economic reporting should be augmented in the fields of socio-economic indicators, world views and society-biosphere interactions. An observation system for sustainability science will need to be based on a large sample of comparative regional studies, emphasising meaningful, relevant and practical indicators. Standards for documentation and access to data will also have to be developed.

Secondly, the time scales of Earth System processes present severe constraints on the knowledge feedback processes required for adaptive management. Many processes have such long lag times, or are so well buffered, that a management-induced perturbation may show no known effects for decades or longer but still have consequences later. The management action may have seemed sensible, benign and entirely appropriate at the time undertaken, but during the future, unbeknown, may be pushing the Earth System towards a threshold and abrupt change that would be impossible to prevent once the problem is recognised. The use of, at the time thought to be benign, chloroflurocarbons is a case in point. Has the contemporary large and rapid increase, from an Earth System perspective, of CO_2 in the atmosphere already pushed the Earth System dangerously close or perhaps even beyond a critical threshold? How can this be known given our present state of understanding? What are the implications for management of such questions?

Finally, the spatial scales of teleconnections in the Earth System mean that management decisions in one part of the world affect environment and development in others. For example, attempts at adaptive management to deal with the problems of land degradation in the Sahel region of Africa may be negated by sulphate aerosol emissions in the industrialised countries of the North (Box 4.10). The challenges to environment, development and legal institutions to deal with the complexities of Earth System management are formidable. Issues such as whether poor nations in one region could obtain legal redress from wealthy nations in other regions are likely to arise.

Earth System science is essential for any attempts at adaptive management at the planetary scale to succeed. There are already powerful tools to support management approaches. For example, integrated Earth System models allow many different scenarios of interacting natural and human-driven changes to be developed and evaluated. Such models, of course, need to be rigorously tested against empirical reality, both past and present, to increase confidence in their projections. In addition, the models and the scenario development that follows from them must evolve further through integrated interdisciplinary research, and in continuing dialogues between the scientific community and policymakers at a variety of levels. However, much more science is needed to support sustainable development at the global scale (Box 6.10). The most challenging scientific task of all is to build a common international framework for Earth System science that can harness the potential synergies that will arise from the interactions of tens of thousands of investigators, research groups and institutions around the world. The ultimate challenge, however, is directed towards the governance and management communities as they must deal with the implications of Earth System

Box 6.11. The Amsterdam Declaration

The year 2001 marked a critical turning point in the development of Earth System science. On 10–13 July of that year over 1400 members of the scientific, policy, resource management and media communities gathered in Amsterdam to hear many of the world's leading global change scientists present the latest scientific understanding of the many facets of global environmental change and examine their implications for humanity's aspirations for improved well-being (Global Change Open Science Conference, Amsterdam). The Conference also pointed towards the new approaches needed to study Earth as a system, and explored the role of science in the quest for global sustainability.

The Conference directly addressed some important questions related to the nature of the Earth System and the implications of human activities for the global environment. How much of global change is natural and how much is due to human activities? Is the Earth well-buffered against human-driven change or can human activities drive it into rapid and irreversible changes? What role does biodiversity have in creating and maintaining the Earth System and what are the implications of biodiversity loss? What is happening to the world's climate? Can humanity feed, water, clothe and house its expanding population while controlling or reducing global environmental change? Can technology spare the planet? What will it take to achieve global sustainability?

Participation in the Conference was an indicator of the first steps towards building a global system of Earth System science. Researchers from 105 countries around the world, representing all of the inhabited continents, attended. Of these more than 400, just under a third of the total number of participants, came from 62 developing countries. About 150 students participated in the event. Efforts to communicate the exciting science presented in Amsterdam were equally successful. Fifty-five journalists attended the Conference and many more reported from a distance. About 250 stories from the meeting were reported in both the electronic and print media, including reports in the major international news agencies Reuters and Associated Press. Most importantly, the scientific community itself signalled to the rest of the world that it is now ready to meet the challenge of knowledge production for global sustainability through the release of the Amsterdam Declaration on Global Change.

The Amsterdam Declaration on Global Change

The scientific communities of four international global change research programmes – the International Geosphere-Biosphere Programme (IGBP), the International Human Dimensions Programme on Global Environmental Change (IHDP), the World Climate Research Programme (WCRP) and the international biodiversity programme DIVERSITAS – recognise that, in addition to the threat of significant climate change, there is growing concern over the ever-increasing human modification of other aspects of the global environment and the consequent implications for human well-being. Basic goods and services supplied by the planetary life support system, such as food, water, clean air and an environment conducive to human health, are being affected increasingly by global change.

Research carried out over the past decade under the auspices of the four programmes to address these concerns has shown that:

- The Earth System behaves as a single, self-regulating system comprised of physical, chemical, biological and human components. The interactions and feedbacks between the component parts are complex and exhibit multi-scale temporal and spatial variability. The understanding of the natural dynamics of the Earth System has advanced greatly in recent years and provides a sound basis for evaluating the effects and consequences of human-driven change.
- Human activities are significantly influencing Earth's environment in many ways in addition to greenhouse gas emissions and climate change. Anthropogenic changes to Earth's land surface, oceans, coasts and atmosphere and to biological diversity, the water cycle and biogeochemical cycles are clearly identifiable beyond natural variability. They are equal to some of the great forces of nature in their extent and impact. Many are accelerating. Global change is real and is happening *now*.
- Global change cannot be understood in terms of a simple cause-effect paradigm. Human-driven changes cause multiple effects that cascade through the Earth System in complex ways. These effects interact with each other and with local- and regional-scale changes in multidimensional patterns that are difficult to understand and even more difficult to predict. Surprises abound.
- Earth System dynamics are characterised by critical thresholds and abrupt changes. Human activities could inadvertently trigger such changes with severe consequences for Earth's environment and inhabitants. The Earth System has operated in different states over the last half million years, with abrupt transitions (a decade or less) sometimes occurring between them. Human activities have the potential to switch the Earth System to alternative modes of operation that may prove irreversible and less hospitable to humans and other life. The probability of a human-driven abrupt change in Earth's environment has yet to be quantified but is not negligible.
- In terms of some key environmental parameters, the Earth System has moved well outside the range of the natural variability exhibited over the last half million years at least. The *nature* of changes now occurring *simultaneously* in the Earth System, their *magnitudes* and *rates of change* are unprecedented. The Earth is currently operating in a *no-analogue state.*

On this basis the international global change programmes urge governments, public and private institutions and people of the world to agree that:

- An ethical framework for global stewardship and strategies for Earth System management are urgently needed. The accelerating human transformation of the Earth's environment is not sustainable. Therefore, the *business-as-usual* way of dealing with the Earth System is *not* an option. It has to be replaced – as soon as possible – by deliberate strategies of good management that sustain the Earth's environment while meeting social and economic development objectives.
- A new system of global environmental science is required. This is beginning to evolve from complementary approaches of the international global change research programmes and needs strengthening and further development. It will draw strongly on the existing and expanding disciplinary base of global change science; integrate across disciplines, environment and development issues and the natural and social sciences; collaborate across national boundaries on the basis of shared and secure infrastructure; intensify efforts to enable the full involvement of developing country scientists; and employ the complementary strengths of nations and regions to build an efficient international system of global environmental science.

The global change programmes are committed to working closely with other sectors of society and across all nations and cultures to meet the challenge of a changing Earth. New partnerships are forming among university, industrial and governmental research institutions. Dialogues are increasing between the scientific community and policymakers at a number of levels. Action is required to formalise, consolidate and strengthen the initiatives being developed. The common goal must be to develop the essential knowledge base needed to respond effectively and quickly to the great challenge of global change.

Berrien Moore III	Chair, IGBP
Arild Underdal	Chair, IHDP
Peter Lemke	Chair, WCRP
Michel Loreau	Co-Chair, DIVERSITAS

Challenges of a Changing Earth: Global Change Open Science Conference, Amsterdam, The Netherlands 13 July 2001 science. How can a large group of independent nations with differing cultures, values, wealth, social organisation and world views come together to manage their own single, connected life support system in a coherent and effective way?

6.5.3 Challenges of a Changing Earth

Much exciting science has been carried out and much has been achieved in several decades of global change science. Yet much remains to be done. The past decade of research, summarised in this volume and presented at the Global Change Open Science Conference in Amsterdam in 2001 (Steffen et al. 2002) (Box 6.11), has unveiled more and more about the complex and interrelated nature of the Earth System, and about the ways in which human activities are impacting the System. Many of the results presented in this volume, exciting and provocative though they are, are still fragmentary and raise further questions that demand to be answered. Much of the current understanding remains at the hypothesis stage, triggering tantalising ideas about how the Earth System operates, but lacking confirmation. Nevertheless, the global change research of the past decade has demonstrated beyond doubt one exceptionally important fact: the Earth System has moved well outside the range of natural variability exhibited over the last half million years at least. The nature of changes now occurring simultaneously in the global environment, their magnitudes and rates, are unprecedented in human history and probably in the history of the planet. As has been said repeatedly, the Earth is now operating in a no-analogue state.

Although science has vastly improved the understanding of the nature of global change, it is much more difficult to discern the implications of the changes. They are cascading through the Earth's environment in ways that are difficult to understand and often impossible to predict. Surprises abound. The human-driven changes to the global environment will, at least, require societies to develop a multitude of creative responses and adaptation strategies. At worst, they may drive the Earth itself into a different state that may be much less hospitable to humans and other forms of life.

As global change assumes a more central place in human affairs, science must accept the responsibility of developing and communicating the essential knowledge base societies can use to debate, consider and ultimately decide on how to respond to global change. Ultimately, upon this science the preservation of the global support system for life on Earth depends. The challenge of ensuring a sustainable future is daunting. It can be met, but only with a new and even more vigorous approach to studying and managing an integrated Earth System.

References

Abraham ER, Law CS, Boyd PW, Lavender SJ, Maldonado MT, Bowie AR (2000) Importance of stirring in the development of an iron-fertilized phytoplankton bloom. Nature 407:727–730

Alverson K, Bradley R, Briffa K, Cole J, Hughes M, Larocque I, Pedersen T, Thompson L, Tudhope S (2001) Disappearing evidence: The need for a global paleoclimate observing system. Global Change Newsletter No. 46, International Geosphere Biosphere Programme, Stockholm, Sweden, pp 2–6

Alverson K, Bradley R, Pedersen T (eds) (2003) Paleoclimate, global change and the future. IGBP Global Change Series. Springer-Verlag Berlin Heidelberg New York

Annan K (2000) We, the peoples: The role of the United Nations in the 21st century. United Nations, New York, USA, *http://www.un.org/millennium/sg/report/full.htm*

Ausubel J (2002) Maglevs and the vision of St. Hubert. In: Steffen W, Jäger J, Carson D, Bradshaw C (eds) Challenges of a changing earth: Proceedings of the Global Change Open Science Conference. Amsterdam, The Netherlands, 10–13 July 2001. Springer-Verlag Berlin Heidelberg New York, pp 175–182

Barber KE, Charman D (2003) Holocene palaeoclimate records from peatlands. In: Mackay AW, Battarbee RW, Birks HJB, Oldfield F (eds) Global change in the Holocene. Arnold, London, in press

Baskin Y (2002) A plague of rats and rubber-vines: The growing threat of species invasions. Island Press, Washington

Bauer E, Claussen M, Brovkin V, Huenerbein A (2003) Assessing climate forcings of the Earth system for the past millennium. Geophys Res Lett, in press

Berkes F, Colding J, Folke C (eds) (2003) Navigating social-ecological systems: Building resilience for complexity and change. Cambridge University Press, Cambridge

Bopp L, Kohfeld KE, Le Quéré C, Aumont O (2003) Dust impact on marine biota and atmospheric CO_2 during glacial periods. Paleoceanography, in press

Boyd PW, Watson AJ, Law CS, Abraham E, Trull T, Murdoch R, Bakker DCE, Bowie AR, Buesseler K, Chang H, Charette M, Croot P, Downing K, Frew R, Gall M, Hadfield M, Hall J, Harvey M, Jameson G, La Roche J, Liddicoat M, Ling R, Maldonado M, McKay RM, Nodder S, Pickmere S, Pridmore R, Rintoul S, Safi K, Sutton P, Strzepek R, Tanneberger K, Turner S, Waite A, Zeldis J (2000) A mesoscale phytoplankton bloom in the polar Southern Ocean stimulated by iron fertilization. Nature 407:695–702

Brickman R, Jasanoff S, Ilgen T (1985) Controlling chemicals: The politics of regulation in Europe and the United States. Cornell University Press, Ithaca

Briffa KR (2000) Annual climate variability in the Holocene: Interpreting the message of ancient trees. Quaternary Sci Rev 19:87–105

Briffa KR, Osborn TJ (1999) Seeing the wood from the trees. Science 284:926–927

Briffa KR, Osborn TJ, Schweingruber FH, Harris IC, Jones PD, Shiyatov SG, Vaganov EA (2001) Low-frequency temperature variations from a northern tree ring density network. J Geophys Res 106D:2929–2941

Brovkin V, Bendtsen J, Claussen M, Ganopolski A, Kubatzki C, Petoukhov V, Andreev A (2002) Carbon cycle, vegetation and climate dynamics in the Holocene: Experiments with the CLIMBER-2 model. Global Biogeochem Cy 16:1139, doi:10.1029/2001GB001662

Brovkin V, Levis S, Loutre MF, Crucifix M, Claussen M, Ganopolski A, Kubatzki C, Petoukhov V (2003) Stability analysis of the climate-vegetation system in the northern high latitudes. Climatic Change 57:119–138

Bureau of Economic Analysis (2000) The world economy: A millennial perspective. *http://www.bea.doc.gov/bea/dn/0898nip3/table1.htm*, 30 Nov 2002

Calov R, Ganopolski A, Petoukhov V, Claussen M, Greve R (2002) Large-scale instabilities of the Laurentide ice sheet simulated in a fully coupled climate-system model. Geophys Res Lett 29: 2216, doi:10.1029/2002GL016078

Carpenter SR, Walker B, Anderies JM, Abel N (2001) From metaphor to measurement: Resilience of what to what? Ecosystems 4:765–781

Cash D, Clark WC, Alcock F, Dickson N, Eckley N, Guston D, Jäger J, Mitchell R (2003) Knowledge systems for sustainable development. P Natl Acad Sci USA, in press

CEOS (2003a) Committee for Earth Observations Satellites. *http://disaster.ceos.org*

CEOS (2003b) Committee for Earth Observations Satellites. *http://www.ceos.org*

Claquin T, Roelandt C, Kohfeld KE, Harrison SP, Tegen I, Prentice IC, Balkanski Y, Bergametti G, Hansson M, Mahowald N, Rodhe H, Schulz M (2002) Radiative forcing of climate by ice-age atmospheric dust. Clim Dynam, in press

Clark WC (2000) Environmental globalization. In: Nye JS, Donahue J (eds) Governance in a globalizing world. Brookings Institution Press, Washington, pp 86–108

Clark WC, Munn RE (eds) (1986) Sustainable development of the biosphere. Cambridge University Press, Cambridge

Clark PU, Pisias NG, Stocker TF, Weaver AJ (2002) The role of the thermohaline circulation in abrupt climate change. Nature 415: 863–869

Claussen M, Gayler V (1997) The greening of the Sahara during the mid-Holocene: Results of an interactive atmosphere-biome model. Global Ecol Biogeogr 6:369–377

Claussen M, Kubatzki C, Brovkin V, Ganopolski A, Hoelzmann P, Pachur HJ (1999) Simulation of an abrupt change in Saharan vegetation at the end of the mid-Holocene. Geophys Res Lett 24:2037–2040

Claussen M, Brovkin V, Petoukhov V, Ganopolski A (2001) Biogeophysical versus biogeochemical feedbacks of large-scale land-cover change. Geophys Res Lett 26:1011–1014

Claussen M, Mysak LA, Weaver AJ, Crucifix M, Fichefet T, Loutre M-F, Weber SL, Alcamo J, Alexeev VA, Berger A, Calov R, Ganopolski A, Goosse G, Lohmann G, Lunkeit F, Mokhov II, Petoukhov V, Stone P, Wang Z (2002) Earth system models of intermediate complexity: Closing the gap in the spectrum of climate system models. Clim Dynam 18:579–586

Claussen M, Brovkin V, Ganopolski A, Kubatzki C, Petoukhov V (2003) Climate change in northern Africa: The past is not the future. Climatic Change 57:99–118

Coale KH, Johnson KS, Fitzwater SE, Gordon RM, Tanner S, Chavez FP, Ferioli L, Sakamoto C, Rodgers P, Millero F, Steinberg P, Nightengale P, Cooper D, Cachlan WP, Landry MR, Constantinou J, Rollwagen G, Trasvina A, Kudela R (1996) A massive phytoplankton bloom induced by an ecosystem-scale iron fertilization experiment in the equatorial Pacific Ocean. Nature 383:495–501

Cox PM, Betts RA, Jones CD, Spall SA, Totterdell IJ (2000) Acceleration of global warming due to carbon cycle feedbacks in a coupled climate model. Nature 408:184–197

Crowley TJ, Lowery TS (2000) How warm was the Medieval warm period? Ambio 29:51–54

Crutzen PJ (1995) My life with O_3, NO_x and other YZO_xs. Les Prix Nobel (The Nobel Prizes) 1995. Stockholm: Almqvist & Wiksell International, pp 123–157

Crutzen PJ (2002) Geology of Mankind. Nature 415:23

Deevey ES (1967) Coaxing history to conduct experiments. Bioscience 19:40–43

DeFries R, Hansen M, Townshend JRG, Janetos AC, Loveland TR (2000) A new global 1 km data set of percent tree cover derived from remote sensing. Global Change Biol 6:247–254

Earth Observatory (2003) Available at: *http://earthobservatory.nasa.gov*

Etheridge DM, Steele LP, Langenfelds RL, Francey RJ, Barnola J-M, Morgan VI (1996) Natural and anthropogenic changes in atmospheric CO_2 over the last 1000 years from air in Antarctic ice and firn. J Geophys Res 101:4115–4128

Falkowski P, Scholes RJ, Boyle E, Canadell J, Canfield D, Elser J, Gruber N, Hibbard K, Högberg P, Linder S, Mackenzie FT, Moore B III, Pedersen T, Rosenthal Y, Seitzinger S, Smetacek V, Steffen W (2000) The global carbon cycle: A test of knowledge of Earth as a system. Science 290:291–296

FAOSTAT (2002) Statistical Database. Food and Agriculture Organization of the United Nations, Rome. *http://www.apps.fao.org*, 12 Aug 2002

Folke C, Carpenter S, Elmqvist T, Gunderson L, Holling CS, Walker B (2002a) Resilience and sustainable development: Building adaptive capacity in a world of transformations. Ambio 31: 437–440

Folke C, Colding J, Berkes F (2002b) Building resilience for adaptive capacity in social-ecological systems. In: Berkes F, Colding J, Folke C (eds) Navigating social-ecological systems: Building resilience for complexity and change. Cambridge University Press, Cambridge

Fox J, Mishar V, Rindfuss R, Walsh S (eds) (2003) People and the environment: Approaches for linking household and community survey to remote sensing and GIS. Kluwer, Amsterdam, Netherlands, in press

Friedlingstein P, Bopp L, Ciais P, Dufrene J-L, Fairhead L, LeTreut H, Monfray P, Orr J (2001) Positive feedback between future climate change and the carbon cycle. Geophys Res Lett 28: 1543–1546

Gagan MK, Ayliffe LK, Beck JW, Cole JE, Druffel ERM, Dunbar RB, Schrag DP (2000) New views of tropical paleoclimates from corals. Quaternary Sci Rev 19:45–64

Ganopolski A, Claussen M (2000) Simulation of Mid-Holocene and LGM climates with a climate system model of intermediate complexity. Proceedings of the Third PMIP Workshop, WCRP-111, pp 201–204

Ganopolski A, Rahmstorf S (2001) Simulation of rapid glacial climate changes in a coupled climate model. Nature 409: 153–158

Ganopolski A, Rahmstorf S (2002) Abrupt glacial climate changes due to stochastic resonance. Phys Rev Lett 88: 038501-1–038501-4

Ganopolski A, Rahmstorf S, Petoukhov V, Claussen M (1998) Simulation of modern and glacial climates with a coupled global climate model. Nature 391:351–356

Ganopolski A, Petoukhov V, Rahmstorf S, Brovkin V, Claussen M, Eliseev A, Kubatzki C (2001) CLIMBER-2: A climate system model of intermediate complexity. Part II: model sensitivity. Clim Dynam 17:735–751

Geoghegan J, Schneider L, Vance C (2003) Temporal dynamics and spatial scales: Modeling deforestation in the southern Yucatan peninsular region. Geojournal, in press

GOFC-GOLD (2002) Global Observation for forest and land cover dynamics. *http://www.fao.org/gtos/gofc-gold/*, 21 Oct 2002

Goodwin ID (2003) Unravelling climatic influences on late Holocene sea level variability. In: Mackay AW, Battarbee RW, Birks HJB, Oldfield F (eds) Global change in the Holocene. Arnold, London, UK, in press

Gorham E (1996) Lakes under a three-pronged attack. Nature 381:109–110

Gran HH (1931) On the conditions for the production of plankton in the sea. Rapports et Proces verbaux des Réunions, Conseil International pour l'exploration de la Mer 75:37–46

Gribbin J (1988) Any old iron. Nature 331:570

Gunderson LH, Holling CS (eds) (2002) Panarchy: Understanding transformations in human and natural systems. Island Press, Washington, USA

Gurney RJ, Foster JL, Parkinson CL (eds) (1993) Atlas of satellite observations related to global change. Cambridge University Press, Cambridge

Haan N, Gumbo D, Eastman JR, Toledano J, Snel M (2000) Linking geomantics and participatory social analysis for environmental monitoring: Case studies from Malawi. Cartographica 37:21–32

Hansen B, Turrell WR, Osterhus S (2001) Decreasing overflow from the Nordic seas into the Atlantic Ocean through the Faroe Bank channel since 1950. Nature 411:927–930

Hector A, Schmid B, Beierkuhnlein C, Caldiera MC, Diemer M, Dimitrakopoulos PG, Finn JA, Freitas H, Giller PS, Good J, Harris P, Högberg P, Huss-Danell K, Joshi J, Jumpponen A, Körner C, Leadley PW, Loreau M, Minns A, Mulder CPH, O'Donovan G, Otway SJ, Pereira JS, Prinz A, Read DJ, Scherer-Lorenzen M, Schulze E-D, Siamantziouras A-SD, Spehn EM, Terry AC, Troumbis AY, Woodward FI, Yachi S, Lawton JH (1999) Plant diversity and productivity experiments in European grasslands. Science 286:1123–1127

Heintzenberg, J, Raes F, Schwartz SE, Ackermann I, Artaxo P, Bates TS, Benkovitz C, Bigg K, Bond T, Brenguier J-L, Eisele FL, Feichter J, Flossman AI, Fuzzi S, Graf H-F, Hales JM, Herrmann H, Hoffman T, Huebert B, Husar RB, Jaenicke R, Kärcher B, Kaufman Y, Kent GS, Kulmala M, Leck C, Liousse C, Lohmann U, Marticorena B, McMurry P, Noone K, O'Dowd C, Penner JE, Pszenny A, Putaud J-P, Quinn PK, Schurath U, Seinfeld JH, Sievering H, Snider J, Sokolik I, Stratmann F, van Dingenen R, Westphal D, Wexler AS, Wiedensohler A, Winker DM, Wilson J (2003) Tropospheric aerosols. In: Brasseur GP, Prinn RG, Pszenny AAP (eds) The changing atmosphere: An integration and synthesis of a decade of tropospheric chemistry research. IGBP Global Change Series. Springer-Verlag Berlin Heidelberg New York, pp 125–156

Held DA, McGrew DG, Perraton J (1999) Global transformations: Politics, economics, and culture. Stanford University Press, Stanford

Hendrey GR, Ellsworth DS, Lewin KR, Nagy J (1999) A free-air enrichment system for exposing tall forest vegetation to elevated atmospheric CO_2. Global Change Biol 5:293–310

ICSU/ISTS/TWAS (International Council for Science, Initiative on Science and Technology for Sustainability, Third World Academy of Sciences) (2002) Science and technology for sustainable development. ICSU Series on Science for Sustainable Development, No. 9. ICSU, Paris. Available at : *http://www.icsu.org/Library/WSSD-Rep/Vol9.pdf*

IFIA (2002) Fertilizer indicators. International Fertilizer Industry Association. *http://www.fertilizer.org/ifa/statistics/STATSIND/cnworld.asp*, 25 Oct 2002

Immirzi CP, Maltby E (1992) The global status of peatlands and their role in carbon cycling. Friends of the Earth Trust Ltd., London

Ingram JSI, Canadell J, Elliott T, Hunt LA, Linder S, Murdiyarso D, Stafford Smith M, Valentin C (1999) Networks and consortia. In: Walker BH, Steffen W, Canadell J, Ingram J (eds) The terrestrial biosphere and global change: Implications for natural and managed ecosystems. IGBP Book Series No. 4, Cambridge University Press, Cambridge, pp 45–65

IPCC (2001) Climate change 2001:The scientific basis. Contribution of Working Group I to the Third Assessment Report of the Intergovernmental Panel on Climate Change. Houghton JT, Ding Y, Griggs DJ, Noguer M, van der Linden PJ, Dai X, Maskell K, Johnson CA (eds) Cambridge University Press, Cambridge and New York

IUCN (1980) World conservation strategy: Living resource conservation for sustainable development. International Union for the Conservation of Nature, Gland, Switzerland

Jackson JBC, Kirb MX, Berher WH, Bjorndal KA, Botsford LW, Bourque BJ, Bradbury RH, Cooke R, Erlandsson J, Estes JA, Hughes TP, Kidwell S, Lange CB, Lenihan HS, Pandolfi JM, Peterson CH, Steneck RS, Tegner MJ, Warner RR (2001) Historical overfishing and the recent collapse of coastal ecosystems. Science 293:629–638

Jones PD, Briffa KR, Barnett TP, Tett SFB (1998) High-resolution palaeoclimatic records for the last millennium: Interpretation, integration and comparison with General Circulation Model control-run temperatures. Holocene 8:455–471

Jones PD, New M, Parker DE, Martin S, Rigor IG (1999) Surface air temperature and its changes over the past 150 years. Rev Geophys 37:173–199

Kates RW (2003) The nexus and the neem tree: Globalization and a transition toward sustainability. In: Speth JG (ed) Worlds apart: Globalization and the environment. Island Press, Washington, in press

Kates RW, Clark WC, Corell R, Hall JM, Jaeger CC, Lowe I, McCarthy JJ, Schellnhuber HJ, Bolin B, Dickson NM, Faucheux S, Gallopin G, Grübler A, Huntley B, Jäger J, Jodha NS, Kasperson RE, Mabogunje A, Matson P, Mooney H, Moore B III, O'Riordan T, Svedin U (2001) Sustainability science. Science 292:641–642

Kaye JA (1995) Space-based data in atmospheric chemistry. In: Barker JR (ed) Progress and problems in atmospheric chemistry. World Scientific, Singapore, pp 569–615

Keith DW (2001) Geoengineering. Nature 409:420

Keohane RO, Nye JS (2000) Introduction. In: Nye JS, Donahue J (eds) Governance in a globalizing world. Brookings Institution Press, Washington, pp 1–41

Khagram S, Clark WC, Raad DF (2003) From the environment and human security to sustainable security and development. J Hum Dev, in press

Knutti R, Stocker TF (2002) Limited predictability of the future thermohaline circulation close to an instability threshold. J Climate 15:179–186

Kubatzki C, Montoya M, Rahmstorf S, Ganopolski A, Claussen M (2000) Comparison of a coupled global model of intermediate complexity and an AOGCM for the last interglacial. Clim Dynam 16:799–814

Kutzbach JE, Guetter PJ (1986) The influence of changing orbital parameters and surface boundary conditions on climate simulations for the past 18 000 years. J Atmos Sci 43:1726–1759

Landry MR, Barber RT, Bidigare RR, Chai F, Coale KH, Dam HG, Lewis MR, Lindley ST, McCarthy JJ, Roman MR, Stoecker DK, Verity PG, White JR (1997) Iron and grazing constraints on primary production in the central equatorial Pacific: An EqPac synthesis. Limnol Oceanogr 42:405–418

Laris P (2002) Burning the seasonal mosaic: Preventing burning strategies in the wooded savanna of southern Mali. Hum Ecol 30:155–186

Lavorel S, Gardner RH, O'Neil RV (1995) Dispersal of annual plants in hierarchically structured landscapes. Landscape Ecol 10: 277–289

Liverman D, Moran EF, Rindfuss RR, Stern PC (eds) (1998) People and pixels: Linking remote sensing and social science. National Academy Press, Washington

LOICZ (2002) Coastal typology project of the IGBP/LOICZ project. Land-Ocean Interactions in the Coastal Zone International Project Office, Texel, Netherlands, *http://www.nioz.nl/loicz/res.htm*

Lomborg B (2001) The skeptical environmentalist: Measuring the real state of the world. Cambridge University Press, Cambridge

Lotter AF (1998) The recent eutrophication of Baldeggersee (Switzerland) as assessed by fossil diatom assemblages. Holocene 8:395–405

Mackenzie FT, Ver LM, Lerman A (2002) Century-scale nitrogen and phosphorus controls of the carbon cycle. Chem Geol 190: 13–32

Mahowald N, Kohfeld K, Mansson M (1999) Dust sources and deposition during the last glacial maximum and current climate: A comparison of model results with paleodata from ice cores and marine sediments. J Geophys Res 104:15895–15916

Mann ME, Bradley RS, Hughes MK (1998) Global-scale temperature patterns and climate forcing over the past six centuries. Nature 392:779–787

Mann ME, Bradley RS, Hughes MK (1999) Northern hemisphere temperatures during the past millennium: Inferences, uncertainties, and limitations. Geophys Res Lett 26:759–762

Martin JH (1990) Glacial-interglacial CO_2 change: The iron hypothesis. Paleoceanography 5:1–13

Martin JH, Fitzwater SE (1988) Iron-deficiency limits phytoplankton growth in the Northeast Pacific subarctic. Nature 331:341–343

Martin JH, Gordon RM, Fitzwater SE (1990) Iron in Antarctic waters. Nature 345:156–158

Martin JH, Coale KH, Johnson KS, Fitzwater SE, Gordon RM, Tanner SJ, Hunter CN, Elrod VA, Nowicki JL, Coley TL, Barber RT, Lindley S, Watson AJ, Vanscoy K, Law CS, Liddicoat MI, Ling R, Stanton T, Stockel J, Collins C, Anderson A, Bidigare R, Ondrusek M, Latasa M, Millero FJ, Lee K, Yao W, Zhang JZ, Friederich G, Sakamoto C, Chavez F, Buck K, Kolber Z, Greene R, Falkowski P, Chisholm SW, Hoge F, Swift R, Yungel J, Turner S, Nightingale P, Hatton A, Liss P, Tindale NW (1994) Testing the iron hypothesis in ecosystems of the equatorial Pacific Ocean. Nature 371:123–129

Matos G, Wagner L (1998) Consumption of materials in the United States, 1900–1995. Annu Rev Energ Env 23:107–122

Messina JP, Walsh SJ (2001) 2.5D morphogenesis: Modeling landuse and landcover dynamics in the Ecuadorian Amazon. Plant Ecol 156:75–88

Meybeck M, Vörösmarty CJ (2004) The integrity of river and drainage basin systems: Challenges from environmental change. In: Kabat P, Claussen M, Dirmeyer PA, Gash JHC, de Guenni LB, Meybeck M, Pielke Sr. RA, Vörösmarty C, Hutjes RWA, Luetkemeier S (eds) Vegetation, water, humans and the climate: A new perspective on an interactive system. IGBP Global Change Series. Springer-Verlag, Berlin Heidelberg New York, in press

Mitchell R (2003) International environmental agreements. Annual Review of Environment and Natural Resources, in press

Nobre CA, Wickland D, Kabat PI (2001) The Large Scale Biosphere-Atmosphere Experiment in Amazonia (LBA). Global Change Newsletter No. 45, International Geosphere Biosphere Programme, Stockholm, Sweden, pp 2–4

Nobre CA, Artaxo P, Silva Dias MAF, Victoria RL, Nobre AD, Krug T (2002) The Amazon Basin and land-cover change: A future in the balance? In: Steffen W, Jäger J, Carson D, Bradshaw C (eds) Challenges of a changing earth: Proceedings of the Global Change Open Science Conference. Amsterdam, The Netherlands, 10–13 July 2001. Springer-Verlag, Berlin Heidelberg New York, pp 137–141

Nordhaus WD (1997) Do real wage and output series capture reality? The history of lighting suggests not. In: Bresnahan T, Gordon R (eds) The economics of new goods. University of Chicago Press, Chicago

NRC (1999) Our common journey: A transition toward sustainability. Board on Sustainable Development, National Research Council, National Academy Press, Washington

Nyström M, Folke C, Moberg F (2000) Coral-reef disturbance and resilience in a human-dominated environment. Trends Ecol Evol 15:413–417

Oldfield F, Dearing JA (2003) The role of human activities in past environmental change. In: Alverson K, Bradley R, Pedersen T (eds) Paleoclimate, global change and the future. IGBP Global Change Series. Springer-Verlag, Berlin Heidelberg New York, pp 143–162

Overpeck J, Rind D, Lacis A, Healy R (1996) Possible role of dust-induced regional warming in abrupt climate change during the last glacial period. Nature 384:447–449

Overpeck J, Hughen K, Hardy D, Bradley R, Case R, Douglas M, Finney B, Gajewski K, Jacoby G, Jennings A, Lamoureux S, Lasca A, MacDonald G, Moore J, Retelle M, Smith S, Wolfe A, Zielinski G (1997) Arctic environmental change of the last four centuries. Science 278:1251–1256

Parker D, Berger T, Manson S, McConnell WJ (eds) (2002) Agent-based models of land-use and land-cover change. LUCC Report Series No. 6, LUCC (Land Use and Cover Change) International Project Office, Louvain-la-Neuve, Belgium

Peltier WR (1994) Ice age paleotopography. Science 265:195–201

Peng T-H, Broecker WS (1991) Dynamic limitations on the Antarctic iron fertilization strategy. Nature 349:227–229

Petit JR, Jouzel J, Raynaud D, Barkov NI, Barnola J-M, Basile I, Bender M, Chappellaz J, Davis M, Delaygue G, Delmotte M, Kotlyakov VM, Legrand M, Lipenkov VY, Lorius C, Pépin L, Ritz C, Saltzman E, Stievenard M (1999) Climate and atmospheric history of the past 420 000 years from the Vostok ice core, Antarctica. Nature 399:429–436

Petoukhov V, Ganopolski A, Brovkin V, Claussen M, Eliseev A, Kubatzki C, Rahmstorf S (2000) CLIMBER-2: A climate system model of intermediate complexity. Part I: Model description and performance for present climate. Clim Dynam 16:1–17

Plotnick RE, Gardner R (1993) Lattices and landscapes. In: Gardner RH (ed) Lectures on mathematics in the life sciences: Predicting spatial effects in ecological systems. American Mathematical Society, Providence, USA, pp 129–157

Pulp and Paper International (1993) PPI's international fact and price book. In: FAO forest product yearbook 1960–1991. Food and Agriculture Organization of the United Nations, Rome

Raskin P, Banuri T, Gallopín G, Gutman P, Hammond A, Kates R, Swart R (2002) Great transition. The promise and lure of the times ahead. Report of the Global Scenario Group, Stockholm Environment Institute, Stockholm, Sweden

Richards J (1990) Land transformation. In: Turner BL II, Clark WC, Kates RW, Richards JF, Mathews JT, Meyer WB (eds) The Earth as transformed by human action: Global and regional changes in the biosphere over the past 300 years. Cambridge University Press, Cambridge, pp 163–178

Ridgwell AJ, Watson AJ (2002) Feedback between aeolian dust, climate, and atmospheric CO_2 in glacial time. Paleoceanography 17:10.1029/2001PA000729

Rocha-Miranda CE (ed) (2000) Transition to global sustainability: The contributions of Brazilian science. Academia Brasiliera de Ciências, Rio de Janeiro. *http://sustainabilityscience.org/keydocs/brazilsci.htm*

Rotstayn LD, Lohmann U (2002) Tropical rainfall trends and the indirect aerosol effect. J Climate 15:2103–2116

Rustad LE, Campbell JL, Marion GM, Norby RL, Mitchell MJ, Hartley AE, Cornelissen JHC, Gurevitch J, GCTE-NEWS (2001) A meta-analysis of the response of soil respiration, net N mineralization and aboveground plant growth to experimental ecosystem warming. Oecologia 126:543–562

Sahagian D, Schellnhuber J (2002) GAIM in 2002 and beyond: A benchmark in the continuing evolution of global change research. Global Change Newsletter No. 50, International Geosphere-Biosphere Programme, Stockholm, Sweden, pp 7–10. *http://www.igbp.kva.se/uploads/3_GAIM.pdf*

Scheffer M, Carpenter SR, Foley J, Folke C, Walker B (2001) Catastrophic shifts in ecosystems. Nature 413:591–596

Schellnhuber HJ (1998) Discourse: Earth System analysis: The scope of the challenge. In: Schellnhuber HJ, Wetzel V (eds) Earth System analysis. Springer-Verlag Berlin Heidelberg New York, pp 3–195

Schellnhuber HJ (1999) 'Earth System' analysis and the second Copernican revolution. Nature 402:C19–C23

Schellnhuber HJ (2002) Coping with Earth System complexity and irregularity. In: Steffen W, Jäger J, Carson D, Bradshaw C (eds) Challenges of a changing earth: Proceedings of the Global Change Open Science Conference. Amsterdam, The Netherlands, 10–13 July 2001. Springer-Verlag Berlin Heidelberg New York, pp 151–156

Schneider SH (2001) Earth systems engineering and management. Nature 409:417–421

Science Council of Japan (2000) Towards a comprehensive solution to problems in education and the environment based on a recognition of human dignity and self-worth. Science Council of Japan, Tokyo

Shiklomanov IA (1990) Global water resources. Nat Resour 26(3)

Smith SV, LOICZ Modelling Team (2002) Carbon-nitrogen-phosphorus fluxes in the coastal zone: The global approach. Global Change Newsletter No. 49, International Geosphere-Biosphere Programme, Stockholm, Sweden, pp 7–11

Speth JG (2003) Perspectives on the Johannesburg Summit. Environment 45:24–29

Steffen W (2002) Will technology spare the planet? In: Steffen W, Jäger J, Carson D, Bradshaw C (eds) Challenges of a changing earth: Proceedings of the Global Change Open Science Conference. Amsterdam, The Netherlands, 10–13 July 2001. Springer-Verlag Berlin Heidelberg New York, pp 189–191

Steffen W, Jäger J, Carson D, Bradshaw C (eds) (2002) Challenges of a changing earth: Proceedings of the Global Change Open Science Conference. Amsterdam, The Netherlands, 10–13 July 2001. Springer-Verlag Berlin Heidelberg New York

Stevens PR, Kelley KW (1992) Embracing Earth: New views of our changing planet. Chronicle Books, San Francisco

Stocker TF, Schmittner A (1997) Influence of CO_2 emission rates on the stability of the thermohaline circulation. Nature 388:862–865

The Social Learning Group (2001) Leaning to manage global environmental risks. MIT Press, Cambridge

Thompson LG, Mosley-Thompson E, Davis ME, Henderson KA, Breche HH, Zagorodnov VS, Mashiotta TA, Li P-N, Mikhalenko VN, Hardy DR, Beer J (2002) Kilimanjaro ice core record: Evidence of Holocene climate change in tropical Africa. Science 298:589–593

Tilman D, Wedin D, Knops J (1996) Productivity and sustainability influenced by biodiversity in grassland ecosystems. Nature 379:718–720

Turner BL II (2002) Toward integrated land-change science: Advances in 1.5 decades of sustained international research on land-use and land-cover change. In: Steffen W, Jäger J, Carson D, Bradshaw C (eds) Challenges of a changing earth: Proceedings of the Global Change Open Science Conference. Amsterdam, The Netherlands, 10–13 July 2001. Springer-Verlag Berlin Heidelberg New York, pp 21–26

Turner BL II, Clark WC, Kates RW, Richards JF, Mathews JT, Meyer WB (1990) The Earth as transformed by human action: Global and regional changes in the biosphere over the past 300 years. Cambridge University Press, Cambridge

UK Royal Society (2000) Towards sustainable consumption: A European perspective. London

UNEP (2000) Global environmental outlook 2000. United Nations Environment Programme, Nairobi, Kenya

US Bureau of the Census (2000) International database. *http://www.census.gov/ipc/www/worldpop.htm*, data updated 10 May 2000

Vance C, Geoghegan J (2002) Temporal and spatial modelling of tropical deforestation: A survival analysis linking satellite and household survey data. Agr Econ 27:317–332

van Woesik R (2001) Coral bleaching: Transcending spatial and temporal scales. Trends Ecol Evol 16:119–121

Vernadsky W (1945) The biosphere and the noosphere. Am Sci 33:1–12

Vörösmarty CJ, Wasson R, Richey JE (eds) (1997) Modeling the transport and transformation of terrestrial materials to freshwater and coastal ecosystems. Workshop report and recommendations for IGBP inter-core project collaboration. IGBP Report No. 39, IGBP Secretariat, Stockholm

Wackernagel M, Schulz NB, Deumling D, Linares AC, Jenkins M, Kapos V, Monfreda C, Loh J, Myers N, Norgaard R, Randers J (2002) Tracking the ecological overshoot of the human economy. P Natl Acad Sci USA 99:9266–9271

Walsh SJ, Crawford TW, Welsh WF, Crews-Meyer KA (2001) A multiscale analysis of LULC and NDVI variation in Nang Rong District, Northeast Thailand. Agr Ecosyst Environ 85:47–64

Walsh SJ, Crews-Meyer KA (eds) (2002) Linking people, place, and policy: A GIScience approach. Kluwer Academic Publisher, Boston

Wardle DA, Bonner KI, Barker GM, Yeates GW, Nicholson KS, Bardgett RD, Watson RN, Ghani A (1999) Plant removals in perennial grassland: Vegetation dynamics, decomposers, soil biodiversity and ecosystem properties. Ecol Monogr 69:535–568

Watson AJ, Liss PS, Duce RA (1991) Design of a small-scale iron fertilisation experiment. Limnol Oceanogr 36:1960–1965

Watson AJ, Bakker DCE, Ridgwell AJ, Boyd PW, Law CS (2000) Effect of iron supply on Southern Ocean CO_2 uptake and implications for glacial atmospheric CO_2. Nature 407:730–733

WCED (1987) Our common future (The Brundtland Report). World Commission on Environment and Development. Oxford University Press, New York

WCMC (1999) Protected areas database. World Conservation Monitoring Centre, unpublished data of the WCMC, Cambridge

Wertz JR, Larson WJ (eds) (1999) Space mission analysis and design, 3rd Ed. Space Technology Library, Microcosm Press, El Segundo

WGBU (1996) World in transition: Ways towards global environmental solutions. Annual Report 1995. German Advisory Council on Global Change. Springer-Verlag, Berlin Heidelberg New York

WGBU (1997) World in transition: The research challenge. Annual Report 1996. German Advisory Council on Global Change. Springer-Verlag, Berlin Heidelberg New York, *http://www.wbgu.de/wbgu_publications.html*

Wilson EO (1992) The diversity of life. Allen Lane, the Penguin Press

WMO (2003) World Meteorological Organization. *http://alto-stratus.wmo.ch/sat/stations/asp_htx_idc/Missionsearch.asp*

WRI (1990) Forest and rangelands. In: A guide to the global environment. World Resources Institute, Washington, pp 101–120

WRI (1992) Environmental almanac. World Resources Institute. Houghton Mifflin Company, Boston

WRI (1999) World resources 1998–1999. World Resources Institute, Washington DC, *http://www.wri.org/wri/wr-98-99/acidrain.htm*

Appendix

The International Geosphere-Biosphere Programme (IGBP) is an international scientific research programme that brings together researchers from biology, physics, chemistry, geology and other disciplines to undertake a study of global change in the context of the Earth System. IGBP's goal is to undertake a systems analysis of planetary composition and dynamics, focusing on the interactive physical, chemical and biological processes that define Earth System dynamics, the changes that are occurring in these dynamics and the role of human activities in these changes.

IGBP was established in 1986 by the International Council for Science (ICSU) with over 500 scientists involved in the planning phase. A growing number of scientists, estimated at 10 000 currently participates in IGBP research, contributing their work to the programme on a voluntary basis. IGBP adds value to these individual projects through the development of international agreed research agendas, scientific networks, common experimental protocols, and frameworks for strategic resource allocation, as well as by model intercomparisons and synthesis and integration activities.

The IGBP research effort is built around eight projects: three oriented towards the three major Earth System compartments – land, ocean and atmosphere; three concentrating on the interfaces that transport and transform matter and energy between the three compartments; and two focusing on the changing environment of the planet as a whole, from past through present to the future. The projects are:

IGAC (International Global Atmospheric Chemistry).
Objective: To understand the role of atmospheric chemistry in the climate system and to determine the effects of changing regional emissions and depositions, long-range transport and chemical transformations on air quality.

SOLAS (Surface Ocean – Lower Atmosphere Study).
Objective: To achieve a quantitative understanding of the key biogeochemical-physical interactions and feedbacks between the ocean and the atmosphere, and how this coupled system affects and is affected by climate and environmental change.

The ***Ocean Project***, implemented through a partnership of the ***GLOBEC*** (Global Ocean Ecosystem) and ***IMBER*** (Integrated Marine Biogeochemistry and Ecosystem Research) projects.
GLOBEC objective: To advance understanding of the structure and functioning of the global ocean ecosystem, its major subsystems, and its response to physical forcing so that a capability can be developed to forecast the responses of the marine ecosystem to global change. IMBER objective: To determine how oceanic biogeochemical cycles, marine ecosystems and their interactions respond to global change and, in turn, feed back to the Earth System.

LOICZ (Land-Ocean Interactions in the Coastal Zone).
Objective: To assess, model and predict the changes in and resilience of the global coastal zone under multiple forcing and as an integral part of the Earth System, including the contribution of, and consequences for, human use of the coastal zone.

The ***Land Project***.
Objective: To identify the nature and magnitude of changes in land systems; to determine the consequences of these changes for the provision of ecosystem services and for the functioning of the Earth System; and to determine the vulnerability or sustainability of changing land systems. Note: The existing ***LUCC*** (Land-Use/Cover Change) project will complete its work in October 2005. LUCC objective: To improve understanding of land-use and land-cover change dynamics and their relationships with global environmental change.

ILEAPS (Integrated Land Ecosystem-Atmosphere Processes Study).
Objective: To understand how interacting physical, chemical and biological processes transport and transform energy and matter through the land-atmosphere interface; to determine the implications of these processes for the Earth System; and to determine how human activities are influencing this coupled system.

PAGES (Past Global Changes).
Objective: To provide a quantitative understanding of the Earth's environment in the geologically recent past and to define the envelope of natural environmental variability against which anthropogenic impacts on the Earth System may be assessed.

GAIM (Global Analysis, Integration and Modelling).
Objective: To advance the study of the coupled dynamics of the Earth System using as tools both data and models.

Co-sponsors of various IGBP projects include the Committee on Atmospheric Chemistry and Global Pollution (CACGP), the Intergovernmental Oceangraphic Commission (IOC), the International Human Dimensions Programme on Global Environmental Change (IHDP), the Scientific Committee on Oceanographic Research (SCOR) and the World Climate Research Programme (WCRP).

IGBP contributes to broader global change and Earth System questions through its participation in the Earth System Science Partnership (ESSP), formed by four international global change research programmes: DIVERSITAS (a global programme of biodiversity science), IGBP, IHDP and WCRP. The goal of the ESSP is to undertake the integrated study of the Earth System, the changes that are occurring to the System and the implications of these changes for global sustainability. The ESSP undertakes four types of activity:

- Earth System analysis and modelling, via collaboration among existing projects/activities of the four constituent programmes.
- Joint projects on issues of global sustainability, designed to address the global change aspects of a small number of critical issues for human well-being: carbon cycle/energy systems, food systems, water resources and human health.
- Regional activities, including (*i*) research, capacity building and networking carried out by the Global Change System for Analysis, Research and Training (START) and (*ii*) integrated regional studies.
- Global change open science conferences, the first of which was Challenges of a Changing Earth, held in Amsterdam in July 2001. The second open science conference is scheduled for 2006.

IGBP works closely with the International Group of Funding Agencies (IGFA) on common issues concerning resources for global change science. Observational aspects of IGBP's work are facilitated by participation in the Integrated Global Observing Strategy Partnership (IGOS-P).

The scientific programme of IGBP is coordinated by a central secretariat, which is hosted by the Royal Swedish Academy of Sciences in Stockholm. International project offices located around the world coordinate more focused aspects of the programme's scientific activities undertaken at the project level.

IGBP Secretariat
Royal Swedish Academy of Sciences
PO Box 50005
Lilla Frescativägen 4
S-104 05 Stockholm
Sweden
Tel: +46-8-16-64-48
Fax: +46-8-16-64-05
E-mail: sec@igbp.kva.se

Further information on IGBP can be obtained from the programme's website, which also contains links to the IGBP projects and to the ESSP and the other global change research programmes:

www.igbp.kva.se

Acknowledgements

The authors are grateful to a large number of people for contributions of various types. The Chair of IGBP for the period 1998 through 2001, Berrien Moore III, provided the leadership and overall guidance of the IGBP synthesis project, to which this volume contributes. Guy Brasseur, the Chair of IGBP from 2002, suggested the inclusion of expert boxes, which have significantly strengthened the book. Three of the authors are acknowledged individually for particular contributions. Pamela Matson developed the overall structure of the volume. Angelina Sanderson carried out most of the literature searches and much of the editorial work, particularly with regard to many of the composite figures. Peter Tyson's careful proofreading enhanced the style and presentation of the volume.

This book would not have been possible without strong support from Sweden. The Royal Swedish Academy of Sciences in Stockholm, which hosts the IGBP Secretariat, provided a large amount of in-kind support throughout the synthesis project. In addition, Professor Erling Norrby, the Standing Secretary of the Academy, helped to raise significant financial resources in support of the project. The Millennium Committee of Sweden and the Swedish Foundation for Strategic Environmental Research (MISTRA) are gratefully acknowledged for providing this financial support. The International Group of Funding Agencies (IGFA) is acknowledged with thanks for its strong, ongoing support of the global change research on which this volume is based.

The authors are grateful for the many helpful reviews of the volume, or parts thereof, carried out by the Scientific Committee of the IGBP and by the staff of the International Project Offices of the IGBP core projects. The entire volume was reviewed by Paul Crutzen, Ian Lowe and Colin Prentice. These careful, detailed and thoughtful reviews led to major improvements to the scientific quality of the volume.

The staff of the IGBP Secretariat in Stockholm are thanked for their many contributions to the synthesis project, above and beyond their normal duties in support of the programme. Particular thanks for their work on the figures in this volume are given to A. Bastås, J. Bellamy and P. Nilsson. Other Secretariat staff who supported the synthesis project are W. Broadgate, S. Eliott, J. Morais, S. Roger, N. Swanberg, C. Widlund, C. Wilson-Boss and E. Wännman.

Many people have written pieces of text, advised on text, or contributed material on which the authors have directly drawn or have helped the authors locate material (figures, text, references) needed for the volume. In this regard John Gould is particularly thanked for his excellent text on ocean circulation (Sect. 2.5.1). Others acknowledged for their assistance are:
Y. P. Abrol, I. J. Ackermann, H. Akimoto, J. Allen, S. J. Allen, K. Alverson, R. Anderson, T. L. Anderson, M. O. Andreae, E. C. Apel, R. Avissar, P. Artaxo, C. S. Atherton, E. L. Atlas, E. Balbon, D. Baldocchi, B. Balino, M. Barange, R. T. Barber, J.-M. Barnola, L. A. Barrie, P. J. Bartlein, B. Bass, N. R. Bates, T. S. Bates, A. Becker, A. C. M. Beljaars, C. Benkovitz, A. K. Betts, R. A. Betts, K. Bigg, T. Blunier, H.-J. Bolle, T. C. Bond, A. Bondeau, M. Bonell, P. W. Boyd, R. S. Bradley, G. P. Brasseur, L. Bravo de Guenni, J.-L. Brenguier, C. A. M. Brenninkmeijer, P. G. Brewer, K. R. Briffa, W. Broadgate, R. Brown, H. Bugmann, J. P. Burrows, J. H. Butler, J. G. Calvert, B. Campbell, J. Canadell, N. Carslaw, D. J. Carson, F. S. Chapin III, T. Chase, C.-T. A. Chen, J. M. Chen, W. Chen, P. Ciccioli, S. A. Cieslik, M. Claussen, C. Clerbaux, J. Cole, Y. C. Collingham, E. Cook, E. Cortijo, D. S. Covert, P. M. Cox, R. A. Cox, W. Cramer, T. L. Crawford, D. R. Crosley, P. J. Crutzen, A. D. Culf, P. Czepiel, R. D'Arrigo, C. I. Davidson, J. Dearing, R. J. Delmas, F. J. Dentener, R. G. Derwent, M. Diepenbroek, P. A. Dirmeyer, N. Dittert, R. J. Dobosy, H. Dolman, S. C. Doney, J. R. Drummond, R. A. Duce, H. W. Ducklow, C. M. Eakin, J. R. Ehleringer, F. L. Eisele, T. Elliott, S. Emerson, D. J. Erickson III, P. G. Falkowski, M. J. R. Fasham, R. A. Feddes, F. C. Fehsenfeld, J. Feichter, J. Fishman, D. R. Fitzjarrald, A. I. Flossman, T. Foken, J. A. Foley, M. J. Follows, R. Foster, L. Francois, R. Francois, B. Frenzel, S. Frolking, C. Fu, R. Fuchs, S. Fuzzo, S. Gadgil, I. E. Galbally, L. Gallardo-Klenner, R. H. Gardner, J. H. C. Gash, Z. Gedalov, H. Geist, L. Gordon, J. Goudriaan, J. Gould, T. Graedel, H.-F. Graf, C. Granier, W. B. Grant, P. J. Gregory, E. C. Grimm, A. E. Guenther, H. Gupta, V. Gupta, F. Habets, J. M. Hales, D. O. Hall, F. G. Hall, S. Halldin, W. M. Hao, N. Hanan, R. Hanson, H. Harasawa, R. J. Harding, P. J. Harrison, R. C. Harriss, D. Heard, J. Heintzenberg, H. Herrmann, M. Hoepffner, H. Hoff, T. Hoffman, S. Houweling, O. Hov, G. H. Huang, B. J. Huebert, M. K. Hughes, L. A. Hunt, B. Huntley,

R. B. Husar, M. F. Hutchinson, R. W. A. Hutjes, J. S. I. Ingram, I. S. A. Isaksen, D. J. Jacob, R. Jaenicke, R. A. Jahnke, B. Jähne, C. Jeandel, F. Joos, P. Jöckel, P. Kabat, N. Kalra, M. Kanakidou, D. M. Karl, V. Kasyanov, B. Kärcher, Y. Kaufman, G. S. Kent, Y. H. Kerr, A. Kettle, V. Khattatov, R. P. Kiene, S. Kienzle, J.-W. Kim, M. Koike, C. E. Kolb, Y. Kondo, R. Koster, Ch. Körner, H. Kremer, S. Krishnaswami, B. Kruijt, V. Krysanova, C. Kull, M. Kulmala, M. D. Kumar, K. R. Kumar, L. Labeyrie, P. Lacarrére, J.-P. Lacaux, J.-F. Lamarque, K. Lambeck, E. F. Lambin, J. J. Landsberg, J. Langridge, A. Larigauderie, I. Larocque, W. K. Lauenroth, K. S. Law, E. M. Laws, L. Lebel, C. Leck, R. Leemans, L. Legendre, J. Lelieveld, E. Lepers, D. P. Lettenmaier, S. Linder, E. Lindgren, K. Lindsay, C. Lioussem, P. S. Liss, C. Liu, K. K. Liu, L. Liu, O. Llinás, K. Lochte, U. Lohmann, S. Lorentz, D. C. Lowe, S. Lütkemeier, R. Macdonald, S. Madronich, N. Mahowald, G. Malin, M. R. Manning, A. Marenco, J. A. Marengo, B. Marticorena, L. A. Martinelli, J.-C. Marty, P. A. Matrai, E. Mayorga, W. McConnell, J. J. McDonnell, W. R. McGillis, J. McManus, P. H. McMurry, R. E. McMurtrie, R. H. Meade, J.-C. Menaut, M. Meybeck, A. F. Michaels, J. M. Miller, J. C. Miquel, A. P. Mitra, J. Moncrieff, P. S. Monks, S. A. Montzka, J. L. Moody, H. A. Mooney, J. K. Moore, E. Moran, A. R. Mosier, J. Morais, D. Murdiyarso, J. W. Murray, S. W. A. Naqvi, H.-U. Neue, S. Neuer, B. Nijssen, I. R. Noble, C. A. Nobre, J. Noilhan, Y. Nojiri, K. Noone, F. O'Connor, E. Odada, C. O'Dowd, D. Ojima, Y. Ono, J. J. Orlando, J. C. Orr, T. J. Osborn, J. Overpeck, H. W. Paerl, D. Paillard, P. I. Palmer, D. D. Parrish, W. J. Parton, T. F. Pedersen, S. A. Penkett, J. E. Penner, L. Perks, J. Perry, G. Petschel-Held, J-R. Petit, K. Pickering, R. Pielke Sr., R. A. Pielke Jr., L. F. Pitelka, A. J. Pitman, J. Plane, U. F. Platt, J. Polcher, M. J. Prather, A-H. Prieur-Richard, S. D. Prince, R. G. Prinn, A. A. P. Pszenny, J. P. Putaud, P. K. Quinn, R. Rabin, F. Raes, O. Ragueneau, P. S. Ramakrishnan, N. Ramankutty, D. Raynaud, M. R. Raupach, J. E. Richey, R. T. Rivkin, H. Rodhe, G.-J. Roelofs, D. Sahagian, O. E. Sala, J. Schafer, H. Schlager, M. C. Scholes, R. J. Scholes, E.-D. Schulze, R. E. Schulze, U. Schurath, S. E. Schwartz, D. Schwela, P. Seakins, W. Seiler, J. H. Seinfeld, G. Shimmield, H. H. Shugart, H. Sievering, M. A. Silva Dias, H. B. Singh, K. A. Smith, J. Snider, I. Sokolik, R. Spahni, M. Stafford Smith, C. Steffen, T. Stocker, A. Stohl, T. J. Stohlgren, F. Stratmann, R. W. Sutherst, E. Swietlicki, G. Szejwach, A. M. Thompson, P. Tréguer, N. B. A. Trivett, J. L. Turon, G. S. Tyndall, R. Valentini, R. Van Dingenen, C. Valentin, M. Velayutham, A. Vetrov, R. L. Victoria, H. Virji, P. Viterbo, C. Vogel, C. J. Vörösmarty, M. Wackernagel, C. Waelbroeck, B. H. Walker, S. Walters, Z. Wan, C. Wang, A. J. Watson, C. P. Weaver, T. Webb III, F. Wechsung, R. F. Weiss, D. Werth, D. L. Westphal, A. S. Wexler, C. Whitlock, A. Wiedensohler, J. W. Williams, R. G. Williams, S. G. Willis, J. Wilson, D. M. Winker, C. S. Wong, F. I. Woodward, D. J. Wuebbles, F. Wulff, Y. Xue, T. Yasunari, Y. Yokohama, L. Yu, X. Zeng, X. Zhang, S. Zhao.

The authors and publishers would like to thank those who provided illustrative material and are grateful to the following for permission to reproduce copyright material (for which acknowledgement and citation are made in figure captions and lists of references):

(1.7a) Vitousek PM (1994) Beyond global warming – ecology and global change. Ecology 75(7):1861–1876, reprinted courtesy of the Ecological Society of America

(2.2) Reprinted from Jacobson M, Charlson RJ, Rodhe H, Gordon OH. Earth System science: From biogeochemical cycles to glacial change. 527 pp, © 2000, with permission from Elsevier Science

(2.8) Reprinted from Quaternary Science Review, Alley RB. The Younger Dryas cold interval as viewed from Central Greenland. pp 213–226, © 2000, with permission from Elsevier Science

(2.9) Climate Dynamics. The cold event 8 200 years ago documented in oxygen isotope records of precipitation in Europe and Greenland. Von Grafenstein U, Erlenkeuser H, Muller J, Jouzel J, Johnsen S, 14:73–81, Fig. 5, 1998, © Springer-Verlag Berlin Heidelberg

(2.10, 2.14) Reprinted from Quaternary Science Reviews, Gasse F. Hydrological changes in the African tropics since the Last Glacial Maximum. pp 189–211, © 2000, with permission from Elsevier Science

(2.15) Wilson EO (1992) The diversity of life. Allen Lane, the Penguin Press, p 191, reproduced by permission of Penguin Books Ltd.

(2.16) Reprinted with permission from Johnson TC, Scholz CA, Talbot MR, Kelts K, Ssemanda I, McGill JW. Late Pleistocene desiccation of Lake Victoria and rapid evolution of cichlid fishes. Science 273:1091–1093, Copyright 1996 American Association for the Advancement of Science

(2.18) Reprinted with permission from Ganopolski A, Kubatzki C, Claussen M, Brovkin V, Petoukhov V. The influence of vegetation-atmosphere-ocean interaction on climate during the mid-Holocene. Science 280:1916–1919, Copyright 1998 American Association for the Advancement of Science

(2.21) Reprinted with permission from Cramer W, Bondeau A, Woodward FI, Prentice IC, Betts RA, Brovkin V, Cox PM, Fisher V, Foley JA, Friend AD, Kucharik C, Lomas MR, Ramankutty N, Sitch S, Smith B, White A, Young-Molling C (2001) Global response of terrestrial ecosystem structure and function to CO_2 and climate change: Results from six dynamic global vegetation models. Global Change Biology 7:357–373, © Blackwell Publishing Ltd.

(2.24, 2.25, 2.26) Reprinted with permission from Loreau M, Naeem S, Inchausti P, Bengtsson J, Grime JP, Hector A, Hooper DU, Huston MA, Raffaelli D, Schmid B, Tilman D, Wardle DA. Biodiversity and ecosystem functioning: Current knowledge and future challenges. Science 294:804–808, Copyright 2001 American Association for the Advancement of Science

(2.28b) Oecologia, A global analysis of root distributions for terrestrial biomes 1996. Jackson RB, Canadell J, Ehleringer JR, Mooney HA, Sala OE, Schulze ED, Oecologia 108:389–411; Figure 2, p 392, 1996, © Springer-Verlag Berlin Heidelberg

(2.30) Takahashi T, et al. (1997): Global air-sea flux of CO_2: An estimate based on measurements of sea-air pCO_2 difference. *Proceedings of the National Academy of Science* 94:8292–8299, © 1997, National Academy of Sciences, U.S.A.

(2.35) Reprinted from Earth System Science: From biogeochemical cycles to global change. Jacobson M, Charlson RJ, Rodhe H, Gordon OH. London, UK: Academic Press, 527 pp, © 2000, with permission from Elsevier Science

(2.36) Reprinted with permission from Cramer W, Kicklighter DW, Bondeau A, Moore B III, Churkina G, Nemry B, Ruimy A, Schloss AL, The participants of the Potsdam NPP Model Intercomparison (1999) Comparing global models of terrestrial net primary productivity (NPP): Overview and key results. Global Change Biology 5(Suppl. 1):1–15, © Blackwell Publishing Ltd.

(2.37) Vegetation, water, humans and climate: A new perspective on an interactive system. Does land surface matter in weather and climate? Claussen M, © Springer-Verlag Berlin Heidelberg
(2.39) Reprinted with permission from Mooney HA, Vitousek PM, Matson PA (1987) Exchange of materials between the biosphere and atmosphere. Science 238:926–932, © 1987, American Association for the Advancement of Science
(2.41) Reprinted with permission from Milich L (1999) The role of methane in global warming: Where might mitigation strategies be focused? Global Environmental Change 9:179–201, © 1999 with permission from Elsevier Science
(2.44) Reprinted with permission from Apel JR (1987) Principles of ocean physics. Academic Press, © 1987 with permission from Elsevier Science
(2.51) Reprinted with permission from Milliman JD, Meade RH (1983) World-wide delivery of river sediment to the oceans. Journal of Geology 91:1–21, © University of Chicago Press
(2.56) Reprinted with permission from Watson RT, Noble IR, Bolin B, Ravindranath NH, Verardo DJ, Dokken DJ (eds) (2000) Land use, land-use change, and forestry. A Special Report of the IPCC. Cambridge University Press, Cambridge, UK, © International Panel on Climate Change
(2.58) Reprinted with permission Black TA, den Hartog G, Neumann HH, Blanken PD, Yang PC, Russell C, Nesic Z, Lee X, Chen SG, Staebler R, Novak MD (1996) Annual cycles of water vapour and carbon dioxide fluxes in and above a boreal aspen forest. Global Change Biology 2:219–229, © 1996 American Association for the Advancement of Science
(2.60) Reprinted with permission from DeMenocal PB, Ortiz J, Guilderson T, Adkins J, Sarnthein M, Baker L, Yarusinki M (2000) Abrupt onset and termination of the African Humid Period: Rapid climate response to gradual insolation forcing. Quaternary Science Review 19:347–361, © 2000 with permission from Elsevier Science

(3.6) Organisation for Economic Co-operation and Development (OECD) (2001) The Environmental outlook for the chemicals industry. Figure 4, p 29
(3.8) Total final consumption by fuel – the World, © OECD/IEA, 2002
(3.10, 3.11) Reprinted with permission Food and Agricultural Organization of the United Nations (FAO). The state of food and agriculture 2000
(3.15b, 3.27) Reprinted with permission from Food and Agriculture Organization of the United Nations (FAO). State of the world's forest 2001, © 2001
(3.16) World Tourism Organization (2001) Tourism highlights 2001
(3.17) Reprinted with permission from Geist HJ, Lambin EF (2002) Proximate causes and underlying forces of tropical deforestation. BioScience 52(2):143–150, © American Institute of Biological Sciences
(3.21) FAO photo, R. Faidutti 1990
(3.25) Klein Goldewijk K, Battjes R (1997) One Hundred year database for integrated environmental assessments. Reprinted with permission from National Institute for Public Health and the Environment (RIVM), Bilthoven, Netherlands
(3.29c) Reprinted with permission from Martínez-Cortizas A, Pontevedra-Pombal X, García-Rodeja E, Nóvoa-Muñoz JD, Shotyk W. Mercury in a Spanish peat bog: Archive of climate change and atmospheric metal deposition. Science 284:939–942, © 1999 American Association for the Advancement of Science
(3.29e) Reprinted with permission from Weiss D, Shotyk W, Appleby PG, Kramers JG, Cherbukin AK (1999) Atmospheric Pb deposition since the industrial revolution recorded by five Swiss peat profiles: enrichment factors, fluxes, isotopic composition and sources. Environmental Science and Technology 33:1340–1352, © American Chemical Society
(3.49) Reprinted with permission from Mackenzie FT, Ver LM, Lerman A (2002) Century-scale nitrogen and phosphorus controls of the carbon cycle. Chemical Geology, Vol. 190, © 2002 with permission from Elsevier Science
(3.51) Reid WVC, Miller KR (1989) The scientific basis for the conservation of biodiversity. World Resources Institute, Washington DC
(3.60c) Stallings B (ed) (1995) Global change, regional response: The new international context of development. Cambridge, © Cambridge University Press
(3.63) Reproduced with permission from Jacobson AD, DeOliveira JAA, Barange M, Cisneros-Mata MA, Felix-Uraga R, Hunter JR, Kim JY, Matsuura Y, Niquen M, Porteiro C, Rothschild B, Sanchez RP, Serra R, Uriarte A and Wada T (2001) Surplus production, variability, and climate change in the great sardine and anchovy fisheries. Can J Fish Aquat Sci 58:1891–1903
(3.65) Reproduced with permission from Food and Agricultural Organization of the United Nations (FAO). The state of world fisheries and aquaculture 2000
(3.66h) Worldwatch Institute. State of the World 1994. Copyright 1994, *www.worldwatch.org*

(4.7) Reproduced with permission from Fromentin J-M, Planque B (1996) *Calanus* and environment in the eastern North Atlantic. II. Influence of the North Atlantic Oscillation on *C. finmarchicus* and *C. helgolandicus*. Marine Ecology Progress Series 134, No. 1–3 (1996): 111–118
(4.13, 4.16, 4.21) Mooney et al. In: The terrestrial biosphere and global change. Implications for natural and managed ecosystems. 1999, Cambridge University Press
(4.22) Reprinted with permission Ramanathan V, Crutzen PJ, Kiehl JT, Rosenfeld D. Aerosols, climate, and the hydrological cycle. Science 294:2119–2124, Copyright 2001, American Association for the Advancement of Science
(4.31) Paustian K, Andrén O, et al. (1997) Agricultural soils as a sink to mitigate CO_2 emissions. Soil Use and Management (13):230–244, © British Society of Soil Science
(4.35) Keitt TH, Urban DL and Milne BT (1997) Detecting critical scales in fragmented landscapes. Conservation Ecology (online), courtesy of Ecological Society of America
(4.43) Pielke RA, Lee TJ, Copeland JH, Eastman JL, Ziegler CL, and Finley CA (1997) Use of USGS-provided data to improve weather and climate simulations. Ecological Applications 7:3–21, courtesy of Ecological Society of America
(4.47) Photo from World Wide Fund for Nature, Malaysia

(5.9) Dyson T. Population and food: Global trends and future prospects. New York, Routledge, Thompson Publishing Services, 1996
(5.10) Chameides WL, Yu H, Liu SC, Bergin M, Zhou X, Mearns L, Wang G, Kiang CS, Saylor RD, Luo C, Steiner A, Giorgi F (1999) Case study of the effects of atmospheric aerosols and regional haze on agriculture: An opportunity to enhance crop yields in China trough emission controls. Proceedings of the National Academy of Sciences of the United States of America, 96 Issue 24, 13626–12633, © (1999) National Academy of Sciences, U.S.A.
(5.11) Reproduced with permission from Ausubel J (2002) Maglevs and the vision of St. Hubert. In: Steffen W, Jäger J, Carson D, Bradshaw C (eds) Challenges of a changing earth: proceedings of the Global Change Open Science Conference. Amsterdam: The Netherlands, 10–13 July 2001, Fig. 33.4, © Springer-Verlag Berlin Heidelberg
(5.14) Reproduced with permission from Food and Agricultural Organization of the United Nations (FAO). The state of world fisheries and aquaculture 2000
(5.16) Reproduced with permission from Fogarty MJ (2002) Climate variability and ocean ecosystem dynamics: implications for sustainability. In: Steffen W, Carson DC, Jäger J, Bradshaw C (eds) Challenges of a changing Earth. Proceedings of the Global Change Open Science Conference, Amsterdam, NL, 10–13 July 2001, pp 27–29. Figure 4.1 and 4.2, © Springer-Verlag Berlin Heidelberg
(5.24) Schulze RE (1997) Impacts of global climate change in a hydrologically vulnerable region: Challenges to South African hydrologists. Progress in Physical Geography 21:113–136, with permission of Arnold Publishers
(5.38) Reproduced with permission from Cramer W, Bondeau A, Woodward FI, Prentice IC, Betts RA, Brovkin V, Cox PM, Fisher V, Foley JA, Friend AD, Kucharik C, Lomas MR, Ramankutty N, Sitch S, Smith B, White A, Young-Molling C (2001) Global response of terrestrial ecosystem structure and function to CO_2 and climate change: Results from six dynamic global vegetation models. Global Change Biology 7: 357–373, © Blackwell Publishing Ltd.

(6.6a, 6.27) Reproduced with permission from Schellnhuber HJ (1998) Discourse: Earth System analysis: The scope of the challenge. In: Earth System Analysis. Schellnhuber HJ, Wetzel V (eds) pp 3–195, Fig. 20a, © Springer-Verlag Berlin Heidelberg

(6.7) Plotnick RE, Gardner RH. Lattices and landscapes. Predicting spatial effects in ecological systems. (Gardner RH, ed), Lectures on Mathematics in the Life Sciences, Vol. 23. with permission from American Mathematical Society, Providence, RI (1993)

(6.11) Reproduced with permission from DeFries R, Hansen M, Townshend JRG, Janetos AC, Loveland TR (2000) A new global 1 km data set of percent tree cover derived from remote sensing. Global Change Biology 6: 247–254, © Blackwell Publishing Ltd.

(6.15, 6.19) Reproduced with permission from Schellnhuber HJ (2002) Coping with Earth System complexity and irregularity. In: Steffen W, Jäger J, Carson D, Bradshaw C (eds) Challenges of a changing Earth: Proceedings of the Global Change Open Science Conference, Amsterdam, The Netherlands, 10–13 July 2001, Fig. 28.1 and 28.3, © Springer-Verlag Berlin Heidelberg

(6.26) Downing R, Ramankutty R, and Shah J (1997) RAINS-ASIA: An assessment model for acid deposition in Asia. The World Bank, Washington, D.C., p 11

Every effort has been made to trace and acknowledge copyright holders. Should any infringements have occurred, apologies are tended and omissions will be rectified in the event of a reprint of the book.

Index

A

B

C

D

E

G

H

I

J

K

L

M

N

O

P

Q

R

S

T

Y

Z

MIX
Papier aus verantwortungsvollen Quellen
Paper from responsible sources
FSC® C105338

If you have any concerns about our products,
you can contact us on
ProductSafety@springernature.com

In case Publisher is established outside the EU,
the EU authorized representative is:
Springer Nature Customer Service Center GmbH
Europaplatz 3, 69115 Heidelberg, Germany

Printed by Libri Plureos GmbH
in Hamburg, Germany